Computer Science Foundations and Applied Logic

Editorial Board

Erika Abraham, Department of Computer Science, RWTH Aachen University, Aachen, Nordrhein-Westfalen, Germany

Olaf Beyersdorff, Friedrich Schiller University Jena, Jena, Thüringen, Germany

Jasmin Blanchette, Garching, Germany

Armin Biere, Informatik, ETH Zentrum, RZH, Computer Systems Institute, Zürich, Switzerland

Sam Buss, Department of Mathematics, University of California, San Diego, CA, USA

Matthew England [iD], Engineering and Computing, Coventry University, Coventry, UK

Jacques Fleuriot, The University of Edinburgh, Scotland, Selkirkshire, UK

Pascal Fontaine, University of Lorraine, Villers Les Nancy Cedex, France

Arie Gurfinkel, Pittsburgh, PA, USA

Marijn Heule, Algorithms, University of Texas, Austin, TX, USA

Reinhard Kahle, Departamento de Matematica, Universidade Nova de Lisboa, Caparica, Portugal

Phokion Kolaitis, University of California, Santa Cruz, CA, USA

Antonina Kolokolova, Department of Computer Science, Memorial University of Newfoundland, St. John's, NL, Canada

Ralph Matthes, Universität München, München, Germany

Assia Mahboubi, Institut National de Recherche en Informatique et en Automatique, Nantes Cedex 3, France

Jakob Nordström, Stockholm, Sweden

Prakash Panangaden, School of Computer Science, McGill University, Montreal, QC, Canada

Kristin Yvonne Rozier, Champaign, IL, USA

Thomas Studer, Institute of Computer Science and Applied Mathematics, University of Berne, Berne, Switzerland

Cesare Tinelli [iD], The University of Iowa, Iowa City, IA, USA

Computer Science Foundations and Applied Logic is a growing series that focuses on the foundations of computing and their interaction with applied logic, including how science overall is driven by this. Thus, applications of computer science to mathematical logic, as well as applications of mathematical logic to computer science, will yield many topics of interest. Among other areas, it will cover combinations of logical reasoning and machine learning as applied to AI, mathematics, physics, as well as other areas of science and engineering. The series (previously known as *Progress in Computer Science and Applied Logic*, *https://www.springer.com/series/4814*) welcomes proposals for research monographs, textbooks and polished lectures, and professional text/references. The scientific content will be of strong interest to both computer scientists and logicians.

Nikolai Kosmatov · Virgile Prevosto ·
Julien Signoles
Editors

Guide to Software Verification with Frama-C

Core Components, Usages, and Applications

Editors
Nikolai Kosmatov
Thales Research and Technology
Palaiseau, France

Virgile Prevosto
Université Paris-Saclay, CEA, List
Palaiseau, France

Julien Signoles
Université Paris-Saclay, CEA, List
Palaiseau, France

ISSN 2731-5754 ISSN 2731-5762 (electronic)
Computer Science Foundations and Applied Logic
ISBN 978-3-031-55607-4 ISBN 978-3-031-55608-1 (eBook)
https://doi.org/10.1007/978-3-031-55608-1

© The Editor(s) (if applicable) and The Author(s), under exclusive license to Springer Nature Switzerland AG 2024

This work is subject to copyright. All rights are solely and exclusively licensed by the Publisher, whether the whole or part of the material is concerned, specifically the rights of translation, reprinting, reuse of illustrations, recitation, broadcasting, reproduction on microfilms or in any other physical way, and transmission or information storage and retrieval, electronic adaptation, computer software, or by similar or dissimilar methodology now known or hereafter developed.
The use of general descriptive names, registered names, trademarks, service marks, etc. in this publication does not imply, even in the absence of a specific statement, that such names are exempt from the relevant protective laws and regulations and therefore free for general use.
The publisher, the authors and the editors are safe to assume that the advice and information in this book are believed to be true and accurate at the date of publication. Neither the publisher nor the authors or the editors give a warranty, expressed or implied, with respect to the material contained herein or for any errors or omissions that may have been made. The publisher remains neutral with regard to jurisdictional claims in published maps and institutional affiliations.

This Springer imprint is published by the registered company Springer Nature Switzerland AG
The registered company address is: Gewerbestrasse 11, 6330 Cham, Switzerland

Paper in this product is recyclable.

Ce qui se conçoit bien s'énonce clairement.
Nicolas Boileau

In English:

What is well conceived is clearly stated.
Nicolas Boileau

Foreword

The C programming language continues to be used in foundational software such as device drivers, networking software, operating systems, and controllers, and such software is of growing importance as computers are used in more and more products and cyber-physical systems. C is an attractive programming language for such systems because it provides low-level access to machine capabilities, explicit storage allocation, and pointer arithmetic. However, these same features make it difficult to use C safely and reliably. Programming safely and reliably in language such as C benefits greatly from tool support, which can record a programmer's decisions and check that they are carried out consistently and without running into any semantic problems.

A central goal of my work on the Java Modeling Language (JML) has been to explore the extent to which such tools could benefit programmers. While many consider formal method tools to be too expensive to use, a basic premise of my work has been that, by sharing a common specification language, many different tools could be used without much added effort, and that the sum of the benefits of these tools would outweigh the costs of specification.

The editors and authors of this book have been working on many of these questions but in the context of the C programming language. These excellent computer scientists have made significant progress toward answering these fundamental questions, in part by building the Frama-C platform. This platform not only takes advantage of a common specification language, it also is innovative in the way in which it combines tools.

The editors are all well-published in the field of formal methods and have considerable experience in building such tools. Nikolai Kosmatov is a research engineer at Thales Research and Technology in Palaiseau, France. He is well-known for his work both on Frama-C and PathCrawler, a structural test generation tool. Virgile Prevosto and Julien Signoles are research engineers at CEA List, France. CEA is the French Alternative Energies and Atomic Energy Commission, which is concerned with both safety and security. Virgile Prevosto is well-known for his work on the ANSI C Specification Language (ACSL) and automated theorem proving. Julien Signoles is well-known for his work both on Frama-C and in applications of Frama-C in cybersecurity.

I am sure that readers interested in either formal methods or tool support for low-level languages will learn a lot from all of the research these editors have gathered into this book.

October 2023

Gary T. Leavens
Professor of Computer Science
University of Central Florida
Orlando, FL, USA

Preface

The C programming language is a cornerstone of computer science. Designed by Ritchie and Thompson at Bell Labs as a key element of Unix engineering, it was rapidly adopted by system-level programmers for its portability, efficiency, and relative ease of use compared to assembly languages. It remains today, nearly 50 years after its creation, still widely used in software engineering. But C is a hard language to wield. Its native design choices giving a large freedom to the developer, the same that gave its popularity, clash with the requirements of modern development practices such as strong typing, encapsulation, or genericity. Given its ubiquity in software engineering, this had noticeable safety and, more recently, cybersecurity impacts. The use of verification techniques—and, in the case of systems with high-confidence requirements, formal methods—can address these shortcomings.

Formal methods are a set of techniques based on logic, mathematics, and theoretical computer science that are used for specifying, developing, and verifying software and hardware systems. By relying on solid theoretical foundations, formal methods are able to provide strong guarantees about those systems.

Frama-C is an open-source tool dedicated to the analysis of source code written in C, by means of formal methods. Frama-C is provided[1] by the List institute of CEA, the French Alternative Energies and Atomic Energy Commission, at Université Paris-Saclay in France. Frama-C was born in 2004 and its first public release was published in 2008. It did not come out of nowhere: it was created to succeed Caveat,[2] provided by CEA List, and Caduceus,[3] provided by Inria, the French National Institute for Research in Digital Science and Technology. These were two tools developed by two different teams in the same neighborhood, the *plateau de*

---

[1] See https://frama-c.com.

[2] P. Baudin, A. Pacalet, J. Raguideau, D. Schoen, and N. Williams. *Caveat: a tool for software validation.* In International Conference on Dependable Systems and Networks (DSN), 2002.

[3] J.-C. Filliâtre, and C. Marché. *The Why/Krakatoa/Caduceus Platform for Deductive Program Verification.* In International Conference on Computer Aided Verification (CAV), 2007.

Saclay,[4] situated 20 km south of Paris in France. They shared the same goal: proving properties of C code.

Frama-C was aimed at taking the best of both tools, while overcoming their limitations. From a technical side, it means being extensible, and not limited to one verification technique only. From a non-technical side, it means adopting an open-source model, and providing a tool usable both by academia for teaching formal methods and conducting advanced research, and by engineers for concrete industrial applications. In particular, Caveat was used since the late 1990s to prove functional correctness of the control-command code of the A380 plane by Airbus.[5] A mandatory goal was to replace Caveat by Frama-C without impacting Airbus' methodology and results.

Frama-C does not come alone: it implements the specification language ACSL[6] that allows users to attach formal specifications to C code: Frama-C can then verify whether the code is correct with respect to its specifications. ACSL is designed jointly by CEA List and Inria since 2006. It is inspired by the specification languages of Caveat and Caduceus, but also by JML[7] for Java. All of them belong to the family of behavioral interface specification languages,[8] the first of which was Eiffel[9] in the 1980s.

For almost 20 years, Frama-C and ACSL have been developed and extended. Even if they are still evolving and a lot remains to be done for improving them, the initial goal is reached. Indeed, on the one hand, they are now widely used in academia for teaching, prototyping, and research in the field of practical formal methods. On the other hand, they are also used in various industries, such as transport, energy, defense, smart cards, and space. In particular, while their initial focus was the functional safety of code in the most critical industries, they now also focus on security properties of non-critical code. Nowadays, concrete applications range from discovering hardly-catchable defects in legacy, open-source, or recently-developed software components to helping security audits and certifying safety- or security-critical systems at the highest level of standards in the avionic, nuclear, or smart card industries.

---

[4] Sometimes referred to as *Silicon Plateau*.

[5] F. Randimbivololona, J. Souyris, P. Baudin, A. Pacalet, J. Raguideau, and D. Schoen. *Applying Formal Proof Techniques to Avionics Software: A Pragmatic Approach*. In International Symposium on Formal Methods in the Development of Computing Systems (FM), 1999.

[6] See https://github.com/acsl-language/acsl.

[7] G. T. Leavens, A. L. Baker, and C. Ruby. *JML: A Notation for Detailed Design*. In Behavioral Specifications of Businesses and Systems. Springer, 1999.

[8] J. Hatcliff, G. T. Leavens, K. R. M. Leino, P. Müller, and M. Parkinson. *Behavioral Interface Specification Languages*. In Computing Surveys, 2012.

[9] B. Meyer. *Eiffel: A language and environment for software engineering*. In Journal of Systems and Software, 1988.

Purpose

Frama-C is today one of the most popular toolsets for program analysis and verification. We believe that it is a good time to take a step back and to provide a large panorama of what Frama-C and ACSL have achieved until now and how they can be used. This is the purpose of this book. Indeed, it covers a wide range of usages of Frama-C and ACSL, from the most standard ones, which are typically learned by master students, to a few advanced ones, currently at the state of the art of research in formal methods, through experience reports on applications in the industry that could benefit verification engineers. Unless otherwise specified, this book is based on Frama-C 28.0-Nickel, which can be installed through the Opam package manager[10] if it is not easily provided in your favorite operating system. This book also comes with supplementary materials, e.g., examples of code and solutions to exercises, which are publicly available online in the companion repository https://git.frama-c.com/pub/frama-c-book-companion/.

This book aims at providing (1) a tutorial for the main Frama-C analyzers and ACSL, (2) the basics of the underlying techniques, (3) examples of more advanced techniques and methodological insights, and (4) a few hints about how it is, or could be, used in practice for verifying properties on medium to large-size C code.

Audience

Therefore, this book has multiple audiences, not necessarily mutually exclusive: (1) students following courses in formal methods, who would like to learn program analysis and verification methods in practice by means of Frama-C; (2) teachers of such courses, who need to demonstrate various practical usages and techniques, or seek a source of examples and exercises; (2) verification engineers, not necessarily familiar with formal methods, who would like to understand how Frama-C could be integrated into their daily practice; (3) Frama-C practitioners, already familiar with some part(s) of the tool, who need new verification approaches or want to increase their level of expertise; and (4) researchers and experts in formal methods, who would like to understand how Frama-C works internally. Since the audiences are quite broad, the next paragraph gives some hints about the target audiences for each chapter. Some chapters also give more precise target audience indications for some of their sections.

---

[10] See https://www.frama-c.com/html/get-frama-c.html for detailed instructions on how to get Frama-C.

Outline

This book contains 16 chapters, organized in three parts. Even if the reader may read any chapter independently from the rest of the book, in particular if they are already used to formal methods or Frama-C, some chapters are certainly easier to understand in depth if some of the preceding chapters have been read earlier. Figure 1 shows the dependency graph between the different chapters. Additionally, some chapters are easier to read than some others. In particular, a few sections and chapters contain technical content that may be difficult to grasp for readers who are not yet familiar with formal methods and notation: the same figure also indicates a difficulty level for each chapter. Level 1 means that the whole content of the chapter is accessible to every target audience of the book. Level 1+3 means that most sections of the chapter are also accessible to everyone, even if some particular sections are more technical but may be easily skipped if you are not interested in advanced features and usages of Frama-C. Level 2 means that reading the chapter requires to have understood some previous parts of the book (of level 1 or 2) to follow it in depth. Level 3 means that most sections of the chapter are quite technical: they target students, experts, and researchers who need a deep understanding of advanced features and usages and are not afraid of technical content.

Part I of the book, *Core Components and Analyzers*, contains six chapters that cover the main components of Frama-C. The first two chapters cover the basics.

- Chapter 1, "Formally Expressing What a Program Should Do: The ACSL Language", introduces in depth the ACSL specification language.
- Chapter 2, "The Heart of Frama-C: The Frama-C Kernel", covers the part of Frama-C, namely its kernel, that provides a common basis to all Frama-C analyzers.

The four other chapters cover the core Frama-C analyzers.

- Chapter 3, "Abstract Interpretation with the Eva Plug-in", presents the Eva plug-in dedicated to value analysis by means of abstract interpretation.
- Chapter 4, "Formally Verifying that a Program Does What It Should: The Wp Plug-in", presents the Wp plug-in dedicated to proving properties by means of deductive verification.
- Chapter 5, "Runtime Annotation Checking with Frama-C: The E-ACSL Plug-in", presents the E-ACSL plug-in dedicated to checking properties at runtime by means of runtime annotation checking.
- Chapter 6, "Test Generation with PathCrawler", presents PathCrawler, the test generation tool of Frama-C.

Part II of the book, *Advanced Usages and Analyses*, also contains six chapters.

- Chapter 7, "The Art of Developing Frama-C Plug-ins", presents how to extend Frama-C with new analyzers developed in the OCaml programming language.
- Chapter 8, "Tools for Program Understanding", introduces a set of Frama-C tools that help users explore and better understand their code and the results of Frama-C, notably with Eva.
- Chapter 9, "Combining Analyses Within Frama-C", presents how to exploit one of the strengths of Frama-C: integrating several analyzers together when using one analyzer only is not enough.
- Chapter 10, "Specification and Verification of High-Level Properties", explains how to specify and verify properties that are not easily expressible within the ACSL specification language.
- Chapter 11, "Advanced Memory and Shape Analysis", explains how it is possible to verify memory properties over complex data-structures, such as linked lists.
- Chapter 12, "Analysis of Embedded Numerical Programs in the Presence of Numerical Filters", explains how it is possible to verify embedded programs that strongly rely on floating-point numbers.

Part III of the book, *Case Studies and Industrial Applications*, contains four chapters that show usages of Frama-C in practice.

- Chapter 13, "An Exercise in Mind Reading: Automatic Contract Inference for Frama-C", shows how to extend Frama-C for inferring contracts automatically in order to reduce the verification cost when using Wp.
- Chapter 14, "Exploring Frama-C Resources by Verifying Space Software", presents how Frama-C has been used to verify software of satellite launcher projects at IAE (Institute of Aeronautics and Space) in Brazil.
- Chapter 15, "Ten Years of Industrial Experiments with Frama-C at Mitsubishi Electric R&D Centre Europe", presents how Mitsubishi Electric R&D Centre Europe (MERCE) has used Frama-C for safety and security applications in different industrial domains.
- Chapter 16, "Proof of Security Properties: Application to JavaCard Virtual Machine", shows how Thales has successfully used Frama-C for verifying the security properties of a JavaCard virtual machine in order to be compliant with the most rigorous level of the Common Criteria certification (EAL7).

Afterthoughts

Writing and editing this book has required a long and significant effort. Also, Frama-C is rapidly evolving, with two releases a year. Therefore, even if we

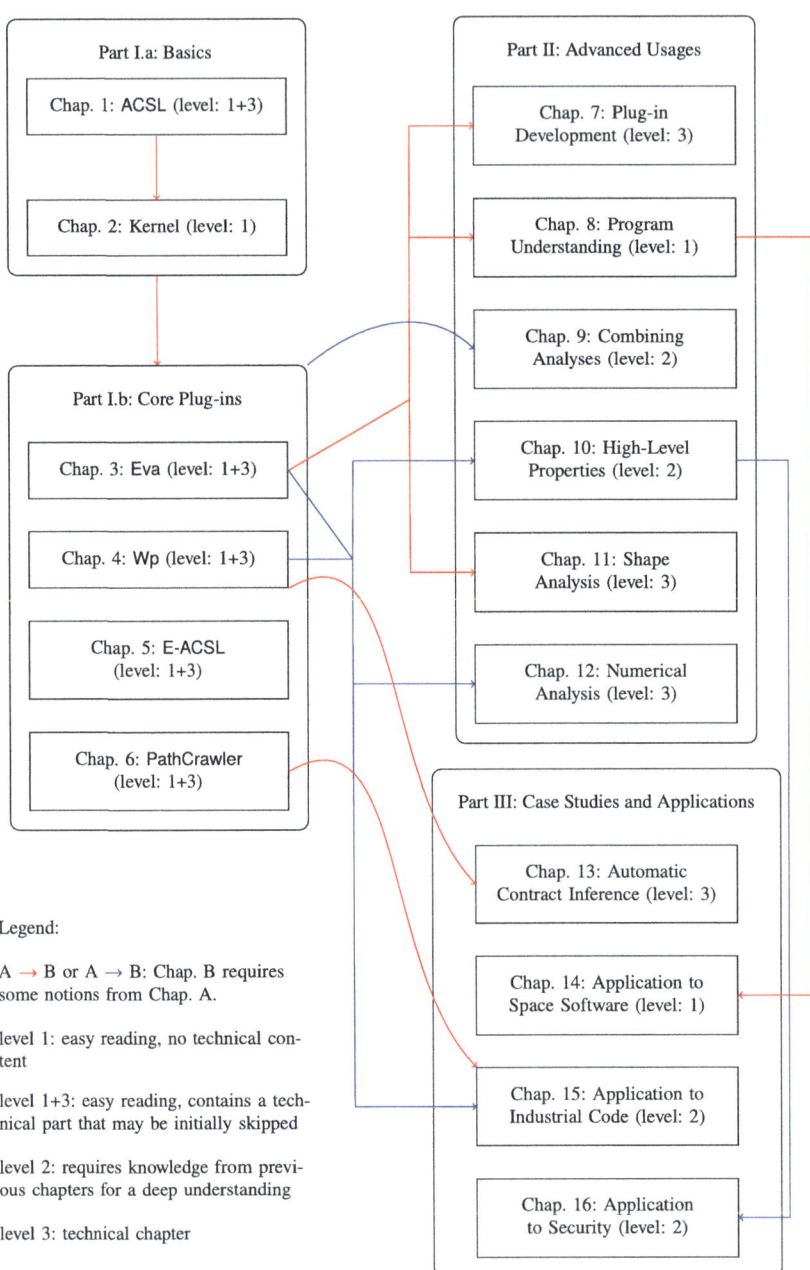

Fig. 1 Chapter dependency graph

Preface

did our best to produce a high-quality book, some outdated content or undetected mistakes might be sparsely present. If you discover any of them, or wish to suggest an improvement, feel free to contact the editors: we will maintain a list of updates on the companion repository at https://git.frama-c.com/pub/frama-c-book-companion/.

Palaiseau, France
December 2023

Nikolai Kosmatov
Virgile Prevosto
Julien Signoles

Acknowledgements

First and foremost, we would like to thank the people who have initiated this journey. Indeed, Frama-C would very probably not have come to light if the then head of *Laboratoire de Sûreté des Logiciels* at CEA List, Jacques Raguideau, and his successor, Fabrice Derepas, had not encouraged Benjamin Monate and Pascal Cuoq, back in 2004, to start the development of Caveat's successor from scratch.

Since that time, many computer scientists, at CEA List and elsewhere, have contributed code to the platform[11] or participated in the design of its components, including interns, Ph.D. students, post-docs, permanent researchers, and engineers. Even more have used Frama-C, and, through their feedback, helped shape it to its actual form.

Work on Frama-C has been funded in great part through collaborative research projects, notably by ANR, the French National Research Agency, by the European Commission through its successive funding programs for research and innovation, from FP6 to Horizon Europe, by BPIFrance, and by DARPA, the Defense Advanced Research Projects Agency. This includes notably the ANR projects *CAT* (C Analysis Toolbox) and *U3CAT* (Unification of Critical C Code Analysis Techniques), which have seen the first releases of Frama-C, and the current *SecOPERA* and *MedSecurance* projects funded through Horizon Europe.

We warmly thank the authors who have contributed to this book and accepted to share their expertise and experience with the readers. We are also very grateful to the reviewers, whose comments have been very helpful for bringing this book up to a high level of quality. Finally, we would also like to thank Springer for accepting to publish this book, as well as for their help and support during its preparation.

---

[11] See https://opam.ocaml.org/packages/frama-c/ for the list of contributors to the components of the main release. (It does not include contributors to plug-ins distributed separately.)

List of Reviewers

Wolfgang Ahrendt, Chalmers University of Technology, Sweden
Nicolas Berthier, OCamlPro, France
Cristian Cadar, Imperial College London, UK
David Cok, Safer Software, LLC, USA
Mickaël Delahaye, DGA MI, France
Claire Dross, AdaCore, France
Jean-Christophe Filliâtre, LMF, CNRS, Université Paris-Saclay, France
Laurent Fuchs, XLIM, Université de Poitiers, CNRS, France
Arnaud Gotlieb, Simula Research Laboratory, Norway
Guillaume Hiet, CentraleSupélec, Inria, CNRS, University of Rennes, IRISA, France
Marieke Huisman, University of Twente, The Netherlands
Éric Jenn, IRT Saint Exupéry, France
Nikolai Kosmatov, Thales Research and Technology, Palaiseau, France
Pascale Le Gall, CentraleSupélec, Université Paris-Saclay, France
Frédéric Loulergue, Université d'Orléans, France
Claude Marché, LMF, Inria, Université Paris-Saclay, France
Raphaël Monat, CRIStAL, Inria, Université Lille, France
David Monniaux, Verimag, CNRS, Université Grenoble-Alpes, France
Laurent Mounier, Verimag, Université Grenoble-Alpes, France
Patricia Mouy, ANSSI and Université Paris-Saclay, CEA, List, France
Guillaume Melquiond, LMF, Inria, Université Paris-Saclay, France
David Mentré, Mitsubishi Electric R&D Centre Europe (MERCE), France
Yannick Moy, AdaCore, France
Marie-Laure Potet, Verimag, Grenoble INP/Université Grenoble-Alpes, France
Virgile Prevosto, Université Paris-Saclay, CEA, List, France
Virgile Robles, Tarides, France

Mihaela Sighireanu, LMF, ENS Paris-Saclay, Université Paris-Saclay, France
Julien Signoles, Université Paris-Saclay, CEA, List, France
Laura Titolo, National Institute of Aerospace, USA
Virginie Wiels, ONERA, France
Boris Yakobowski, AdaCore, France

Contents

Part I Core Components and Analyzers

1 **Formally Expressing What a Program Should Do: The ACSL Language** 3
  Allan Blanchard, Claude Marché, and Virgile Prevosto

2 **The Heart of Frama-C: The Frama-C Kernel** 81
  André Maroneze, Virgile Prevosto, and Julien Signoles

3 **Abstract Interpretation with the Eva Plug-in** 131
  David Bühler, André Maroneze, and Valentin Perrelle

4 **Formally Verifying that a Program Does What It Should: The Wp Plug-in** 187
  Allan Blanchard, François Bobot, Patrick Baudin, and Loïc Correnson

5 **Runtime Annotation Checking with Frama-C: The E-ACSL Plug-in** 263
  Thibaut Benjamin and Julien Signoles

6 **Test Generation with PathCrawler** 305
  Nicky Williams and Nikolai Kosmatov

Part II Advanced Usages and Analyses

7 **The Art of Developing Frama-C Plug-ins** 341
  François Bobot, André Maroneze, Virgile Prevosto, and Julien Signoles

8 **Tools for Program Understanding** 403
  André Maroneze and Valentin Perrelle

| 9 | Combining Analyses Within Frama-C | 423 |

Nikolai Kosmatov, Artjom Plaunov, Subash Shankar, and Julien Signoles

| 10 | Specification and Verification of High-Level Properties | 457 |

Lionel Blatter, Nikolai Kosmatov, Virgile Prevosto, and Virgile Robles

| 11 | Advanced Memory and Shape Analyses | 487 |

Matthieu Lemerre, Xavier Rival, Olivier Nicole, and Hugo Illous

| 12 | Analysis of Embedded Numerical Programs in the Presence of Numerical Filters | 521 |

Franck Védrine, Pierre-Yves Piriou, and Vincent David

Part III Case Studies and Industrial Applications

| 13 | An Exercise in Mind Reading: Automatic Contract Inference for Frama-C | 553 |

Jesper Amilon, Zafer Esen, Dilian Gurov, Christian Lidström, and Philipp Rümmer

| 14 | Exploring Frama-C Resources by Verifying Space Software | 583 |

Rovedy Aparecida Busquim e Silva, Nanci Naomi Arai, Luciana Akemi Burgareli, Jose Maria Parente de Oliveira, and Jorge Sousa Pinto

| 15 | Ten Years of Industrial Experiments with Frama-C at Mitsubishi Electric R&D Centre Europe | 617 |

Éric Lavillonnière, David Mentré, and Benoît Boyer

| 16 | Proof of Security Properties: Application to JavaCard Virtual Machine | 659 |

Adel Djoudi, Martin Hána, and Nikolai Kosmatov

Index 685

Contributors

Jesper Amilon KTH Royal Institute of Technology, Stockholm, Sweden

Nanci Naomi Arai Division of Aerodynamics, Control and Structures (ACE), Institute of Aeronautics and Space (IAE), SP, Brazil

Patrick Baudin Université Paris-Saclay, CEA, List, Palaiseau, France

Thibaut Benjamin Université Paris-Saclay, CEA, List, Palaiseau, France; University of Cambridge, Cambridge, UK

Allan Blanchard Université Paris-Saclay, CEA, List, Palaiseau, France

Lionel Blatter Max Planck Institute for Security and Privacy, Bochum, Germany

François Bobot Université Paris-Saclay, CEA, List, Palaiseau, France

Benoît Boyer Mitsubishi Electric R&D Centre Europe (MERCE), Rennes, France

David Bühler Université Paris-Saclay, CEA, List, Palaiseau, France

Luciana Akemi Burgareli Division of Aerodynamics, Control and Structures (ACE), Institute of Aeronautics and Space (IAE), SP, Brazil

Rovedy Aparecida Busquim e Silva Division of Aerodynamics, Control and Structures (ACE), Institute of Aeronautics and Space (IAE), SP, Brazil

Loïc Correnson Université Paris-Saclay, CEA, List, Palaiseau, France

Vincent David IRSN, Fontenay-aux-Roses, France

Adel Djoudi Thales Digital Identity & Security, Meudon, France

Zafer Esen Uppsala University, Uppsala, Sweden

Dilian Gurov KTH Royal Institute of Technology, Stockholm, Sweden

Martin Hána Thales Digital Identity & Security, Prague, Czech Republic

Hugo Illous Université Paris-Saclay, CEA, List, Palaiseau, France;
Inria and Département d'informatique de l'ENS, Paris, France

Nikolai Kosmatov Thales Research & Technology, Palaiseau, France

Éric Lavillonnière Mitsubishi Electric R&D Centre Europe (MERCE), Rennes, France

Matthieu Lemerre Université Paris-Saclay, CEA, Palaiseau, France

Christian Lidström KTH Royal Institute of Technology, Stockholm, Sweden

Claude Marché LMF, CNRS, Inria, Université Paris-Saclay, Gif-sur-Yvette, France

André Maroneze Université Paris-Saclay, CEA, List, Palaiseau, France

David Mentré Mitsubishi Electric R&D Centre Europe (MERCE), Rennes, France

Olivier Nicole Université Paris-Saclay, CEA, List, Palaiseau, France;
Inria and Département d'informatique de l'ENS, Paris, France

Jose Maria Parente de Oliveira Division of Computer Science, Aeronautics Institute of Technology (ITA), SP, Brazil

Valentin Perrelle Université Paris-Saclay, CEA, List, Palaiseau, France

Pierre-Yves Piriou EDF R&D PRISME Performance, Risque Industriel et Surveillance pour la Maintenance et l'Exploitation, Chatou, France

Artjom Plaunov City University of New York, New York, USA

Virgile Prevosto Université Paris-Saclay, CEA, List, Palaiseau, France

Xavier Rival Inria and Département d'informatique de l'ENS, CNRS, PSL University, Paris, France

Virgile Robles Tarides, Paris, France

Philipp Rümmer University of Regensburg, Regensburg, Germany

Subash Shankar City University of New York, New York, USA

Julien Signoles Université Paris-Saclay, CEA, List, Palaiseau, France

Jorge Sousa Pinto High-Assurance Software Laboratory (HASLab), Institute for Systems and Computer Engineering, Technology and Science (INESC TEC) and University of Minho, Braga, Portugal

Franck Védrine Université Paris-Saclay, CEA, List, Palaiseau, France

Nicky Williams Université Paris-Saclay, CEA, List, Palaiseau, France

Part I
Core Components and Analyzers

Chapter 1
Formally Expressing What a Program Should Do: The ACSL Language

Allan Blanchard, Claude Marché, and Virgile Prevosto

Abstract This chapter presents ACSL, the ANSI/ISO C Specification Language, focusing on its current implementation within the Frama-C framework. As its name suggests, ACSL is meant to express precisely and unambiguously the expected behavior of a piece of C code. It plays a central role in Frama-C, as nearly all plug-ins eventually manipulate ACSL specifications, either to generate properties that are to be verified, or to assess that the code is conforming to these specifications. It is thus very important to have a clear view of ACSL's semantics in order to be sure that what you check with Frama-C is really what you mean. This chapter describes the language in an agnostic way, independently of the various verification plug-ins that are implemented in the platform, which are described in more details in other chapters. It contains many examples and exercises that introduce the main features of the language and insists on the most common pitfalls that users, even experienced ones, may encounter.

Keywords Formal specification · Function contracts · Semantics · First-order logic · ANSI/ISO C specification language

This chapter focuses on the *ANSI-ISO C Specification Language*, abbreviated as ACSL. It is the formal language supported by Frama-C for writing specifications in the form of code annotations. As the term *formal* suggests, the meaning of the ACSL annotations is intended to be precise and unambiguous in a mathematical sense. This aspect is essential to guarantee that all Frama-C plug-ins that make use of such annotations, whether they are generating them or checking them for validity, agree on their meaning in the utmost precise sense.

A. Blanchard (✉) · V. Prevosto
Université Paris-Saclay, CEA, List, Palaiseau, France
e-mail: allan.blanchard@cea.fr

V. Prevosto
e-mail: virgile.prevosto@cea.fr

C. Marché
LMF, CNRS, Inria, Université Paris-Saclay, Gif-sur-Yvette, France
e-mail: claude.marche@inria.fr

```
1  /*@ ensures \result == x;
2    @ ensures x == \old(x)+1;
3    @ ensures \result == x+1;
4    @ ensures \exists int z; z > x ==> z > \result;
5    @ ensures \forall int y; y > x ==> y > \result;
6    @*/
7  int f(int x) {
8    x = x+1;
9    return x;
10 }
```

Fig. 1.1 A toy example illustrating some common pitfalls of ACSL. Which of the five post-conditions are valid? Check your answers with the correct solutions given in Sect. 1.9

This chapter is not meant as a replacement of the ACSL reference manual [1], which gives an exhaustive presentation of ACSL. It is also not meant to replace any tutorial on verification using formal proofs, like one does when using the Wp plugin, even if formal verification heavily relies on ACSL annotations. Instead, this chapter aims at giving an overview of the important features of ACSL, in particular regarding its semantics, independently of any plug-in. It is not meant to cover all aspects of ACSL either, and in particular it does not consider any feature that is not yet implemented in the Frama-C framework version 28.0-Nickel. It has partly the form of a tutorial, with a series of exercises, so as to allow the readers to check their understanding of the semantics. Special care is taken to clarify the common pitfalls that users may encounter. To give a quick illustration of such pitfalls, where even an experienced user may fall into, consider the simple C code of Fig. 1.1 annotated with an ACSL contract made of five different post-conditions. Identify which post-conditions among them are valid ones according to your understanding of ACSL. Please take time to think about this short exercise: if your answers are not the correct ones, as given in Sect. 1.9, then this chapter is certainly worth reading!

The structure of this chapter is as follows. Section 1.1 starts by general considerations about formal specification languages, including historical facts and some positioning with respect to similar existing specification languages for programming languages other than C. Section 1.2 starts the technical study with the central notion of function contracts. Section 1.3 goes into more details on the underlying core logic of ACSL, with which the formulas are written. Section 1.4 considers the annotations that can be placed directly inside the code forming the body of functions, which are mainly assertions and loop invariants. When verifying C code, talking about memory is crucial, hence Sect. 1.5 presents built-in predicates and functions related to this aspect. Section 1.6 presents ghost code and how it is handled by Frama-C. Section 1.7 describes how to introduce user-defined logic functions and predicates through global ACSL annotations. Finally, Sect. 1.8 summarizes the kinds of specifications that can readily be written in ACSL and gives an overview of existing Frama-C plug-ins that target other kinds of properties that would be difficult to express in pure ACSL. Notions and pitfalls are sometimes illustrated by exercises.

The answers for these exercises, including the one in Fig. 1.1, are grouped together in Sect. 1.9.

Some paragraphs of this chapter attempt to provide, at least partly, a formal semantics of ACSL annotations. These parts, namely Sects. 1.2.3, 1.3.3, 1.4.3 and 1.4.6, can be skipped, they are not essential to understand the rest.

1.1 On Formal Specification Languages

The design of formal specification languages for specifying the intended behavior of programs has a long and rich history. It builds upon the respective pioneer working of Floyd [15] and Hoare [20], at the end of the 1960s, on axiomatic semantics of programs. Another landmark of this history is the concept of *design-by-contract* introduced by Meyer in 1985 [33]. The underlying concepts were brought to concrete realizations in the 1990s, first with the design of the Eiffel language [32], quickly followed by the Larch specification language, designed and used for specifying contracts of SmallTalk programs [9] and C++ programs [24]. Static verification of contracts initially started with a checker for Modula-3 programs, giving birth to the notion of *extended static checkers* [29]. The reader interested in details should read the survey of Hatcliff, Leavens, Leino, Müller and Parkinson [18] about what they call *Behavioral Interface Specification Languages* (BISL) in general. At the beginning of the 2000s, the JML language [6, 10, 25, 27] was designed as a BISL for the Java programming language, largely inspired by Eiffel and Larch. The ESC-Java static checker was designed as a variant of ESC-Modula-3 for Java [11].

The ACSL language is a BISL for the C programming language. Although it was designed independently of Frama-C, its main implementation is as the specification language accepted by the Frama-C kernel. Historically, ACSL was inspired by the specification language of Caduceus, a prototype tool for the deductive verification of C source code, designed by Filliâtre and Marché [14], on top of the Why framework [13], nowadays superseded by Why3 [4]. The language of Caduceus was itself inspired by JML. A visible historical heritage from this long lineage is the idea of putting ACSL annotations inside special form of C comments, starting by /*@ or //@.

As an indirect descendant of the specification language of Why, ACSL inherited a feature that is widely shared with other Why descendants: the fact that the underlying logic is basically first-order classical logic, with total functions. This aspect must be opposed to another possible choice that would have been using only *executable expressions* in the annotations, which is for example the choice of Eiffel and JML.[1] The impact of this design choice for the user is discussed in Sect. 1.3.4. The latter alternative has the advantage of making it easier to design a *run-time assertion checker*, which is intended to check the validity of specifications at run-time. Among the descendants of Why, Krakatoa [30] is a tool for verification of Java

---

[1] It can be argued that not all JML assertions are executable, see [8, 26].

code, annotated with a non-executable variant of JML. Although built on top of Why3, the Spark2014 [31] environment for the verification of Ada code made the choice of an executable semantics for the annotations. The fact that ACSL follows a non-executable semantics has advantages for deductive verification of annotations, but drawbacks for dynamic verification. This is the reason of the existence of the E-ACSL variant described in Chap. 5. The choice of having an executable semantics or not, and also the choice of a logic of total functions, have also an important impact on the adoption of the annotation language by the users. For more details on this aspect of user adoption we recommend reading papers by Chalin [7]. More details on the implications of the choice of an executable versus a non-executable semantics are found in the survey by Hatcliff et al. [18], and also in a paper of Kosmatov, Marché, Moy and Signoles [22] for the specific case of Why, Frama-C and Spark2014.

1.2 Function Contracts

As for any BISL, the central concept in ACSL is the notion of *function contract*. In this first technical section, we start by introducing the basic form of contracts in Sect. 1.2.1. Section 1.2.2 presents useful syntactic sugar for writing contracts. Section 1.2.3 discusses a formal setting in which we describe the semantics of contracts. As already mentioned above at the end of introduction, this last section may be skipped on a first reading of this chapter. It is intended to provide a formal justification to the later discussions on pitfalls to avoid. We focus on *side effects* in Sect. 1.2.4, that is how one can specify what a function modifies in the memory. Section 1.2.5 introduces the notion of *behavior*, which allows one to properly separate, in ACSL contracts, different cases for the expected behavior of the underlying function.

To present this section in a more friendly manner, we restrict ourselves to a subset of C where all variables and parameters are of type **int**. Similarly, the return type of functions is assumed to be **int**, or just **void** when they return nothing. This restriction will be lifted in Sect. 1.3.

1.2.1 Basic Function Contracts

The basic, yet general form of a contract for an arbitrary function f with parameters $x_1, \ldots, x_k$ is

```
1 /*@ requires P;
2   @ ensures Q;
3   @*/
4 int f(int x_1,..., int x_k)
```

The symbol P after the keyword **requires** denotes an expression called a *precondition*, whereas the symbol Q after the keyword **ensures** denotes a *post-*

condition. Roughly speaking, P and Q are Boolean expressions similar to expressions that can be given as conditions of an **if** or a **while** statement. Yet, they are not the same as C conditions, and we call them instead *propositions*, or equivalently *formulas*. A first important difference between propositions and C conditions is that propositions *cannot have side effects*, i.e., they cannot modify any variable or memory location. As a consequence, they cannot contain assignment operators. Another difference is that they cannot contain any C function calls.[2] In some sense, the language of propositions is a subset of C expressions, but there are also additional constructs, like quantification. We will present the language of propositions in detail in Sect. 1.3.

A contract on a function like f above is an agreement between the *caller* (the code that calls f) and the *callee* (f itself). In the C program under consideration, let us assume we have some function call to f, say of the form

```
int x = f(e₁,...,eₖ);
```

then the contract specifies that, after evaluating each expression e_i to some value v_i, the pre-condition P of f must hold, when the value of each parameter x_i is set to v_i. This is a constraint that the caller must guarantee. On the other hand, when the callee returns, its part of the contract is to guarantee that the post-condition Q holds. For the post-condition Q to bring interesting information to the caller, it should be able to talk about the returned value of the call, that is stored into the variable x above. For that purpose, the special ACSL keyword **\result** can be used in the formula Q. Naturally, the **\result** keyword is meaningless for a function that returns **void**.

Integer square root

Consider the following simple program

```
1  /*@ requires x >= 0;
2    @ ensures \result * \result <= x &&
3    @         x < (\result+1)*(\result+1);
4    @*/
5  int integer_sqrt(int x) { ... }
6
7  void test(void) {
8    int y = integer_sqrt(42);
9  }
```

At the location of the call `integer_sqrt(42)`, the contract given to `integer_sqrt` requires the argument 42 to satisfy the pre-condition, which is the case. The returned value n is then guaranteed to satisfy the post-condition, that is its square is smaller or equal to 42, whereas the square of $n + 1$ is larger than 42. The returned value 6 would be a convenient answer.

Generally speaking, the post-condition formally expresses that the function computes the square root of its argument, rounded towards 0.

---

[2] It is possible to define (and use) purely logical functions, as described later in Sect. 1.7.

The post-condition of a contract specifies a property on its returned value (and, as we will see later on, its effect on global variables), but there is no obligation to specify the result uniquely.

Middle values
Consider the following simple program
```
1 /*@ requires x < y;
2   @ ensures x <= \result && \result < y;
3   @*/
4 int in_middle(int x, int y) { ... }
5
6 void test(void) {
7   int m = in_middle(2009,2023);
8 }
```

According to the contract, since the arguments 2009 and 2023 satisfy the pre-condition, the function `in_middle` must return a number n in between. The returned value $n = 2016$, exactly the middle of 2009 and 2023 would be a convenient answer, yet it is not the only possible answer.

1.2.2 Additional Constructs and Syntactic Sugar for Contracts

On the example of integer square root above, one may wonder whether 6 was the only possible answer for the test call, according to the contract given? On a first glance, since the contract does not say that the result is non-negative, one may think that the function may return either the square root or the negation of the square root of its argument. Yet, the way the post-condition is formulated in fact implies that, for example neither -7 nor -6 can satisfy the post-condition when $x = 42$. However, even if the given post-condition logically implies that the result is non-negative, it is a good practice to state such property explicitly, for example as

```
1 /*@ requires x >= 0;
2   @ ensures \result >= 0 &&
3   @         \result * \result <= x &&
4   @         x < (\result+1)*(\result+1);
5   @*/
6 int integer_sqrt(int x) { ... }
```

Stating several post-conditions, separated by conjunctions && inside a single large formula, becomes quickly poorly readable. For this reason, ACSL allows one to state several pre-conditions and several post-conditions. Moreover each of these can be given a name, useful for further reference. For example, the same contract as above could be written

1 Formally Expressing What a Program Should Do: The ACSL Language

```
/*@ requires ArgumentIsNonNegative: x >= 0;
  @ ensures ResultIsNonNegative: \result >= 0;
  @ ensures ResultIsSqrt:
  @             \result * \result <= x < (\result+1)*(\result+1);
  @*/
int integer_sqrt(int x) { ... }
```

Notice that the second post-condition ResultIsSqrt uses another handy syntactic sugar of ACSL, namely the *chaining* of comparison operators: for example a <= x < b is a shortcut for a <= x && x < b.

With respect to the language of C expressions, the language of propositions of ACSL is augmented with several specific constructs. We will review most of them in Sect. 1.3, but for now let us introduce one of the most common of them: the *logical implication* symbol. It is a Boolean binary operator written ==>. An expression of the form e_1 ==> e_2 is logically equivalent to !e_1 || e_2, meaning that it is true whenever either e_1 is false or e_2 is true. In others words, e_2 must hold as soon as e_1 holds. Implication is the recommended way to express *cases* in contracts.

> **Integer Square Root continued**
> Imagine we want to provide an integer square root function that does not require its argument to be non-negative, but simply would return 0 when given a negative argument. Its contract could be stated as follows:
>
> ```
> /*@ ensures ResultIsNonNegative: \result >= 0;
>   @ ensures NegativeArgument: x < 0 ==> \result == 0;
>   @ ensures ResultIsSqrt: x >= 0 ==>
>   @             \result * \result <= x < (\result+1)*(\result+1);
>   @*/
> int integer_sqrt(int x) { ... }
> ```

> **Exercise 1 Middle value**
> Consider again the "middle values" example above. Propose a contract which would guarantee that there is always only one possible return value. Try to find a solution without using the division operator.

1.2.3 Towards a Formal Semantics of Contracts

As said in the introduction of Sect. 1.2, the present section may be skipped in a first reading. Its purpose is to formalize, at least partially, the semantics of contracts, so as to rely on a solid basis for later discussions on potential pitfalls in understanding the meaning of contracts. In particular, these discussions may rely on notions introduced in the current section.

Indeed, the informal meaning of contracts, as described in previous sections, may sometimes reveal itself too imprecise for understanding the subtleties that may show up on specific features or corner cases. In this section, we thus propose a more formal presentation that should help to disambiguate subtle situations and also offer a reference for a general agreement on the meaning of ACSL specifications. Yet, giving a fully formal semantics to ACSL would be a huge task, far beyond the objective of this chapter. An earlier attempt for providing a formal semantics to ACSL was done by Herms [19], using the Coq proof assistant. We propose here only a partial formal presentation of such a semantics, the interested reader should refer to the thesis manuscript above for more details.

Generally speaking, giving a precise semantics to ACSL constructs should be built upon an already given precise semantics of the C language itself. This is already a great challenge in itself, given the numerous traps, ambiguities and under-specified behaviors of C code. There exists several projects proposing such a formal semantics for C: the CompCert project [3] and the CH2O project [23] propose semantics using the Coq proof assistant, AutoCorres [17] formalizes C within Isabelle/HOL, and KCC [12] is built upon the Maude environment. It is important to be aware of the ambiguities in C when one wants to formalize the expected behavior of a program. These ambiguities can be classified in three groups:

- *Implementation-defined behavior* concerns cases where the execution is non-ambiguously defined, not from a standard (say the C11 standard [21]), but from implementation choices. In other words, these are cases where a standard is ambiguous and leaves free choices for say compiler implementations. Typical examples include the number of bits in the integer types, and their endianness. As a concrete example, the execution of the code

```
short s = 0x1234;
char  c = *((char*)&s);
```

on a big-endian architecture gives 0x12 as the result value in c, but 0x34 instead on a little-endian architecture. In Frama-C, these architecture-dependent parameters are described using the so-called *machdep* mechanism (see Chap. 2). Somehow, the semantics is thus parameterized by the *machdep* configuration.

- *Unspecified behavior* concerns cases where the standard is not precise, yet the behavior of the code in question will do something reasonable (and will never *crash*, that is: stopping execution abruptly) among the different possibilities. A typical example is the order of evaluation of operands. As a concrete example, if f is defined as:

```
int x;

//@ requires x < INT_MAX;
int f() {
    x++;
    return x;
}
```

then (in the absence of overflow) the expression f() - f() would give -1 if evaluation proceeds from left to right, and 1 otherwise.[3] It means that a formal semantics might make a choice, or be *non-deterministic*, providing several possible outcomes to a given program.

- *Undefined behavior* is the case where what will happen is not predictable, and the program could even crash. Typical cases are: accessing an array out of its bounds, performing a signed integer overflow, performing multiple updates to a variable in one statement, or writing to a non-properly allocated memory location. When formalizing the semantics, one could choose to block execution (simulating a crash) for all such cases, but also, instead, introduce a notion of undefined value to allow execution to continue.

The lack of a unique and non-ambiguous semantics for C is a challenge to define a semantics for ACSL, which we will discuss below after a quick introduction about how one can formalize semantics via *judgments* and *inductive rules*.

1.2.3.1 Formalizing Rules for Operational Semantics

Without the ambition of giving a fully formal semantics of C here, we assume we already have such a semantics, given in terms of a *small-step execution judgment* [28]. It is a mathematical binary relation of the form

$$(C, i) \rightarrow (C', i')$$

meaning that in the *execution context* C a single atomic step of execution of the instruction i leads to the new execution context C' and the next instruction to execute is i'. Without giving details yet, the context C is assumed to record all the information about the current state of the program or system, that is the global data (global variables, data allocated on the memory heap) and the data local to a function call (function parameters, local variables, typically allocated on the stack). A judgment of similar shape can be given for expressions as

$$(C, e) \rightarrow (C', e')$$

meaning that evaluation of expression e in context C reduces, in one step, to evaluation of expression e' in context C'. Given such judgments, a formal semantics can be given by *inductive rules*, which tell how one can derive a valid judgment from a possibly empty set of already known valid judgments (called *premises*). For example, a rule without premise for a simple assignment of a value v to a variable x is typically written as

$$\overline{(C, \mathtt{x} = v) \rightarrow (C[\mathtt{x} \leftarrow v], v)}$$

---

[3] On the other hand, C semantics guarantees that there's no interleaving between the calls, so that the second increment cannot occur before the first call returns.

where $C[x \leftarrow v]$ means that the value recorded for x in context C is to be replaced by v. The rule thus states that the value v becomes the new value of x in the context, and the resulting value of the instruction is v itself (indeed, we remind the reader here that in C code, the assignment instruction returns the value assigned). Other examples are rules for the execution of a sequence $i_1; i_2$, saying first that i_1 should be executed:

$$\frac{(C, i_1) \to (C', i'_1)}{(C, i_1; i_2) \to (C', i'_1; i_2)}$$

and then, as soon as execution of i_1 is finished (giving a value v):

$$\overline{(C, v; i_2) \to (C, i_2)}$$

where the value v is thrown away and execution continues with i_2.

Definition 1.1 Given a context C and a statement i, an *execution* of i in C is a finite or infinite sequence of pairs (C_k, i_k) such that $C_0 = C$, $i_0 = i$ and for any k, $(C_k, i_k) \to (C_{k+1}, i_{k+1})$. Such a sequence is *complete* if either it is infinite, or it is finite, ending at (C_n, i_n), and there is no reduction rule that applies to (C_n, i_n). If i_n is a value v, we say that execution ends normally, otherwise we say that the execution *blocks*.

Exercise 2 Semantics of if and while statements
Propose rules for defining the semantics of the `if` and the `while` statements.

Notice that a formal semantics could be *non-deterministic*, that is there could be possibly several sequences of execution that start from the same state (C, i). The classification of ambiguities given above correspond to different situations with respect to the semantics of executions:

- The semantic rules should be parameterized with implementation-defined behaviors, for example parameterized by sizes of integer types. But, when those parameters are fixed, the semantics should be deterministic.
- Regarding unspecified behaviors, the semantic rules could be designed to restrict executions to be deterministic (e.g., by choosing an evaluation order for function arguments) or to remain non-deterministic.
- Regarding undefined behaviors, the semantics could propose evaluation rules, for example by introducing an explicit representation of undefined values. Yet, an ultimately defensive semantics should simply not provide any possible execution rules for undefined cases, which means that execution will *block* in the sense of the definition above.

Definition 1.2 (*Safety of executions*) A complete (in the sense of Definition 1.1) execution is called *safe* if it is either infinite or it ends normally.

1 Formally Expressing What a Program Should Do: The ACSL Language

The latter definition justifies that blocking on any undefined behavior is ultimately defensive: the *safety* of a given execution would mean that no undefined behavior was met at all during that execution. Yet, for defining the semantics of ACSL, it would be overly restrictive to consider only safe executions.

1.2.3.2 Semantics of Function Calls and Contracts

Giving rules for the execution of a function call is a bit more involved than for other statements, because of the handling of local contexts. Let's now say that the execution context C is a pair (Σ, Π) where Σ stores the global data, whereas Π is a stack of *call frames* $\pi_1, \ldots, \pi_k$, each π_i denoting the local data of one pending function call. We assume that the innermost call frame, say the top of the stack, is the π_1 on the left. A rule for calling a function f having parameters $x_1, \ldots, x_k$ and a body b can then be stated as

$$\frac{\pi = \{x_1 \leftarrow v_1, \ldots, x_k \leftarrow v_k\}}{(\Sigma, \Pi, f(v_1, \ldots, v_k)) \rightarrow (\Sigma, \pi \cdot \Pi, b)}$$

which means that the execution should proceed by executing the body b of f, in a context where the global part is the same but the stack of call frames is augmented with a new frame, made of the parameters of f, initialized with the values given as arguments of the call. Another rule can now be given to handle the end of the execution of the body, that is when a `return` statement is reached:

$$\frac{}{(\Sigma, \pi \cdot \Pi, \texttt{return } v) \rightarrow (\Sigma, \Pi, v)}$$

Notice in particular that the local frame for execution of the body, π, is thrown away.

Exercise 3 Semantics of local variables
Give rules for the declaration of a local variable of type **int** with an initializer.

The last element we need to formally define the semantics of ACSL contracts is a formal semantics for the propositions given as arguments to the **requires** and **ensures** clauses. We will go into more details about that in Sect. 1.3, but for now we only assume that such a semantics is already defined, using a judgment denoted as

$$\Sigma, \Pi \models \varphi$$

which means that in the global context Σ and stack Π the proposition φ holds. A natural question that arises is what would be the meaning of a formula containing unspecified or undefined expressions, such as an access to an uninitialized variable or a division by zero. This is an important question indeed, that will be addressed in Sect. 1.3.4. From now on, please assume that the semantics is *total*, that is in a given context Σ, Π, a formula φ must be either true or false, but cannot be anything like "unspecified" or "undefined".

The semantics of a contract on a function f with parameters $x_1 \ldots, x_k$, body b, pre-condition P and post-condition Q is then given by extending the data in each local frame: in addition to the local data in a frame π, we store the post-condition and also a copy of the initial frame. The rules above for function calls are extended accordingly as follows. First the rule for executing the call becomes

$$\frac{\pi = \{x_1 \leftarrow v_1, \ldots, x_k \leftarrow v_k\} \quad \Sigma, \pi \cdot \Pi \models P}{(\Sigma, \Pi, f(v_1, \ldots, v_k)) \to (\Sigma, (\pi, Q, \pi) \cdot \Pi, b)}$$

where the important addition is the extra premise that requires to check that the pre-condition holds in the context of the global data and the new local frame containing the parameters of f. The rule for the return of the call becomes

$$\frac{\Sigma, \pi_0 \cup \{\texttt{\textbackslash result} \leftarrow v\} \cdot \Pi \models Q}{(\Sigma, (\pi, Q, \pi_0) \cdot \Pi, \texttt{return}\ v) \to (\Sigma, \Pi, v)}$$

where the important addition is the premise that requires to check that the post-condition holds in the context of the global data and the local frame containing the parameters of f.

It is worth noting that the semantics of **ACSL** contracts given here is an execution semantics, in contrast with an *axiomatic* semantics that will be discussed in Sect. 1.4.6.1.

A sample execution
Consider the annotated program

```
1  int x;
2
3  /*@ requires x > 0 && y > 0;
4   @ ensures \result >= x;
5   @ ensures \result <= x+y;
6   @*/
7  int f(int y) {
8    x = x+y;
9    y = y+10;
10   return x+3;
11 }
```

Let Q denote the conjunction of the post-conditions. The execution of statement x = f(1) in the context where x is 2 proceeds as follows. First, the expression f(1) itself must be evaluated:

$(\{x = 2\}, []), \texttt{f(1)} \to$
$(\{x = 2\}, [\{y = 1\}, Q, \{y = 1\}]), \texttt{x = x+y; y = y+10; return x+3}$

because the pre-condition holds: $\{x = 2\}, [\{y = 1\}] \models x > 0 \;\&\&\; y > 0$. Then the execution continues as

$(\{x = 2\}, [\{y = 1\}, Q, \{y = 1\}]),$ x = x+y; y = y+10; return x+3 $\rightarrow$
$(\{x = 2\}, [\{y = 1\}, Q, \{y = 1\}]),$ x = 3; y = y+10; return x+3 $\rightarrow$
$(\{x = 3\}, [\{y = 1\}, Q, \{y = 1\}]),$ 3; y = y+10; return x+3 $\rightarrow$
$(\{x = 3\}, [\{y = 1\}, Q, \{y = 1\}]),$ y = y+10; return x+3 $\rightarrow$
$(\{x = 3\}, [\{y = 1\}, Q, \{y = 1\}]),$ y = 11; return x+3 $\rightarrow$
$(\{x = 3\}, [\{y = 11\}, Q, \{y = 1\}]),$ 11; return x+3 $\rightarrow$
$(\{x = 3\}, [\{y = 11\}, Q, \{y = 1\}]),$ return x+3 $\rightarrow$
$(\{x = 3\}, [\{y = 11\}, Q, \{y = 1\}]),$ return 6

and then at this point, the first post-condition holds because 6 >= 3, but not the second because 6 <= 3+1 is false (notice that we took the value of y in the copy of the initial call state). The execution thus blocks. If only the first post-condition was given, then execution would proceed to

$$\rightarrow (\{x = 3\}, []), 6$$

and finally the execution of x = f(1) would give the normal final state

$$\rightarrow (\{x = 6\}, []), 6$$

As illustrated by the previous example, it is important to notice that **when evaluating a post-condition, the values of the parameters are taken in the initial call context**. In other words: **the local modifications of the parameters are not seen by the callee**. As another example, you could go back to the introductory example of Fig. 1.1, it should now be clear that the first post-condition is wrong, whereas the third one is valid: the fact that x is modified inside the function body is invisible in the contract.

1.2.3.3 Validity and Conformity of ACSL Annotations

As can be seen in the rules given before, examples of blocking executions are given by invalid pre-conditions at some function call, or invalid post-conditions at a return, because no rule can proceed with execution. The formal definition of the validity of an ACSL annotation is thus exactly chosen as the fact that the evaluation of this annotation during execution does not block.

Definition 1.3 (*Validity of an ACSL annotation*) An ACSL annotation in a given program is said to be *universally valid* (or *valid* for short) if no execution of that program ever blocks on this annotation.

In other words, by definition, an invalid pre- or post-condition is one that leads to a blocking execution.

Definition 1.4 (*Conformity of a function contract*) An annotated function of the form

```
/*@ requires P;
  @ ensures Q;
  @*/
int f(int x₁,...,int xₖ) { b }
```

conforms to its contract when for any context (Σ, Π), and for any values $v_1, \ldots, v_k$, if $\pi = \{x_1 \leftarrow v_1, \ldots, x_k \leftarrow v_k\}$ and $\Sigma, \pi \cdot \Pi \models P$, any *safe* execution of the body b in context $(\Sigma, \pi \cdot \Pi)$ is non-blocking.

Please note carefully the following implications of this definition: for a function that conforms to its contract

- execution of the body can be unsafe, and in such a case the execution may reach the end without satisfying the post-condition
- execution of the body can be infinite, that is it will never return, and nothing is imposed on the post-condition.

Definition 1.5 (*Partial correctness*) An annotated function of the form

```
/*@ requires P;
  @ ensures Q;
  @*/
int f(int x₁,...,int xₖ) { body }
```

is *partially correct* (or *correct* for simplicity) when for any context (Σ, Π) where the pre-condition holds, the execution of the body is safe and non-blocking.

In other words, a function equipped with a contract is partially correct if it conforms to its contract and the pre-condition is sufficient to guarantee the safety of its execution. Yet, it does not mean that its executions must always terminate (that's why it is called *partial* correctness). We will discuss termination in Sect. 1.4.7.

Conformity and division
The annotated function

```
/*@ requires 0 <= x <= 1000;
  @ ensures 1 <= \result <= 1000;
  @*/
int f(int x) {
  return 1000 / x;
}
```

conforms to its contract, but it is not correct, because if x is zero its execution is unsafe (undefined behavior). On the other hand

1   Formally Expressing What a Program Should Do: The ACSL Language   17

```
/*@ requires 1 <= x <= 1000;
  @ ensures 1 <= \result <= 1000;
  @*/
int f(int x) {
   return 1000 / x;
}
```

is correct.

Conformity and integer overflow
The function
```
/*@ ensures \result < x;
  @*/
int f(int x) {
   return x-1;
}
```
conforms to its contract, but it is not correct because of a potential signed overflow on x-1 (undefined behavior). Correctness can be ensured by adding a pre-condition stating that x should not be INT_MIN.

Conformity and non-initialization
The function
```
/*@ requires 0 <= x <= 1000;
  @ ensures \result < x;
  @*/
int f(int x) {
   int y;
   if (x > 0) { y = x-1; }
   return y;
}
```
conforms to its contract, but it is not correct because of potential use of an uninitialized value for y (undefined behavior). That is, if the conditional is false, the execution is unsafe, there is no way to ensures that the post-condition holds. Yet this is not required for conformity. In contrast,

```
/*@ requires 0 <= x <= 1000;
  @ ensures 0 <= \result <= 999;
  @*/
int f(int x) {
   int y = 0;
   if (x > 0) { y = x-1; }
   return y;
}
```

is correct, that is both safe, because y is initialized, and conforming to its contract because in each branch of the conditional the post-condition holds.

Conformity and termination
The function

```
int x;

/*@ ensures x == 0;
  @*/
void f(void) {
  while (x) {
    printf("x=%d\n", x);
  }
}
```

is correct, because the only case where it terminates is when x is zero,[4] and in that case the post-condition holds.

Exercise 4 Threshold
The threshold function below aims at reducing a value n to a given interval min..max

```
/*@ ensures ResultBound: min <= \result <= max;
  @ ensures NoThresholdCase: min <= n <= max ==> \result == n;
  @*/
int threshold(int min, int max, int n) {
  if (n < min) { return min; }
  if (n > max) { return max; }
  return n;
}
```

1. For each post-condition, explain if it is valid. If it is not, propose a suitable pre-condition that would make the function conform to its contract.
2. Is the contract enforcing a unique result for any input? If no, propose additional post-conditions to enforce the uniqueness of the result.

---

[4] Note that this example uses a printf statement to avoid any unspecified behavior related to the absence of side effect in the loop, allowing a compiler to completely remove the loop [21, Section 6.8.5 Iteration statements]

Exercise 5 Semantics of ensures clauses
Which ensures clauses are valid?

```
1  int x;
2
3  /*@ requires Pre1: x >= 0 && y >= 0;
4   @ ensures Post1: x >= 3;
5   @ ensures Post2: y >= 1;
6   @ ensures Post3: x >= y+3;
7   @*/
8  void f(int y) {
9    y = x + y + 1;
10   x = y + 2;
11 }
```

1.2.4 Specifying Side Effects: Framing and Referring to Former Values

We say that a function has *side effects* when it is modifying some global data, such as a global variable. In this case, it is expected that the contract explicitly states what these side effects are. Two ACSL constructs are dedicated to this aspect: the **assigns** clause and the **\old** modifier.

1.2.4.1 The **assigns** Clause

An **assigns** clause in a contract comes in addition to the **requires** and **ensures** clauses. Yet, instead of introducing a proposition, it is associated with a collection of *left-values*, that are the expressions that can be used on the left of an assignment operator. Roughly speaking, these are thus variable names, or structure, array or pointer accesses. We restrict ourselves to variables in this section. The general form of a contract is thus extended as

```
1  /*@ requires P;
2   @ assigns y₁,...,yₘ;
3   @ ensures Q;
4   @*/
5  int f(int x₁,...,int xₖ)
```

The meaning of the **assigns** clause is that the variables y_j given in the list are potentially modified by a call to f. We say potentially here because the clause does not require the value of y_j to change: a call to f may or may not assign to y_j, or assign to it but in fact leave it with the same value at exit as before the call.

Basic usage of assigns clause
Consider the toy example of annotated code below

```
1 int w,x,y,z;
2
3 /*@ assigns x,y,z;
4   @*/
5 void f(int b) {
6   x += w;
7   if (b) { y += 1; } else { z += 1; }
8 }
```

The code of f conforms to the **assigns** clause, because:

- The variable w is not modified by any call to f, so it does not need to be listed;
- The variable x is assigned by any call, and even if it remains the same when w is 0, it must be listed in the **assigns** clause;
- The variables y and z are modified or not, depending on the argument b, they both have to be listed since they are potentially modified.

As seen in the toy example above, the list of variables in an **assigns** clause is in general an over-approximation of the set of side effects: for example, no call to f can modify both y and z. This over-approximation property (which is already an idea present in JML [25]) is an important thing to remember about **assigns** clauses. In essence, the purpose of the **assigns** clause is not mainly to tell that the variables given are potentially modified, but **that any variable that is not mentioned in the clause is unchanged**. From the callee point of view of the contract, it promises that the called function is not changing the values of the variables that are not listed.

A special case to mention is the case where a function does not modify anything. A special construct **\nothing** can be used in an **assigns** clause in such case. It is important to notice that providing no **assigns** clause is **not** equivalent to state

```
1   @ assigns \nothing;
```

On the contrary, when no **assigns** clause is given then any global variable in the scope of the considered function might be modified. In practice, the absence of **assigns** clause is a large source of imprecision in a contract, and the Frama-C kernel emits a warning in such a case.[5]

---

[5] Notice that ACSL does not have any equivalent to the assign \everything of JML, which is somehow the default, and anyway is too imprecise in most of the cases.

Exercise 6 assigns clause
Consider the code

```
1  int x,y,z;
2
3  void f() {
4    x += 1; y += 1; x -= 1;
5  }
```

Which of the following **assigns** clauses are valid, in the sense that f would conform to the contract?

```
1  //@ assigns \nothing;
2  //@ assigns x;
3  //@ assigns y;
4  //@ assigns z;
5  //@ assigns x,y,z;
```

1.2.4.2 The \old Construct

A complement of the **assigns** clause is the **\old** construct, to denote the value of a variable at start of a function call. This construct is usable in the formulas given as **ensures** clauses, but not elsewhere. For example, a suitable contract for the toy example above could be:

```
1  int w,x,y,z;
2
3  /*@ assigns x,y,z;
4    @ ensures ChangeOfX: x == \old(x) + w;
5    @ ensures ChangeOfY: b ==> y == \old(y) + 1 && z == \old(z);
6    @ ensures ChangeOfZ: ! b ==> y == \old(y) && z == \old(z) + 1;
7    @*/
8  void f(int b) {
9    x += w;
10   if (b) { y += 1; } else { z += 1; }
11 }
```

The argument of **\old** is not limited to variables only, but can be any logic term or any logic formula φ. In such a case, it is as if the **\old** was applied to each occurrence of a variable inside φ. A toy example is

```
1  int x,y,z;
2
3  /*@ assigns x,y;
4    @ ensures \old(y+z >= 0) ==> x >= \old(x);
5    @*/
6  void f(int n) {
7    y += z;
8    x += y;
9  }
```

where \old(y+z >= 0) is equivalent to \old(y)+\old(z) >= 0. In fact from the code, we know that y == \old(y)+\old(z) and x == \old(x)+y, implying that the **ensures** clause holds.

An important pitfall to avoid related to \old values of variables comes from the important remark made at end of Sect. 1.2.3.2: the effects potentially applied to parameters are never visible in a contract. Consider the trivial code

```
1 /*@ ensures \result > n;
2   @ ensures n > \old(n);
3   @ ensures \result == n;
4   @    @*/
5 int f(int n) {
6   n += 1;
7   return n;
8 }
```

Here \result is indeed the initial value of n plus 1, but since the modification of n is not visible in the contract, all occurrences of n in the **ensures** clauses are indeed equal to \old(n). As a consequence, the first post-condition is valid but the others are not. As another example, you could go back to the introductory example of Fig. 1.1, it should now be clear that the second post-condition is wrong: the values denoted by x and \old(x) are indeed identical.

The \old construct is a particular case of the \at construct that is discussed in more details in Sect. 1.4.2. See that section in particular for other pitfalls to avoid, and exercises.

1.2.5 Contracts with Multiple Cases: Behaviors

For the sake of readability and traceability, it is possible to structure contracts as a finite set of different *contract cases*, i.e., where the functional behavior is decomposed into complementary situations. This was already implicit in the previous example of integer square root where two different **ensures** clauses were given depending on whether the argument was negative or not. To specify such contract cases in a clearer manner, ACSL provides the notion of *named behaviors*, or just *behaviors* for short, in contracts. A general form of a contract with behaviors is

```
1  /*@ requires P;
2   @ behavior b₁:
3   @    assumes A₁;
4   @    ensures Q₁;
5       ⋮
6   @ behavior bₘ:
7   @    assumes Aₘ;
8   @    ensures Qₘ;
9   @*/
10 int f(int x₁,...,int xₖ)
```

1 Formally Expressing What a Program Should Do: The ACSL Language

A simple way to understand the meaning of such a contract is as syntactic sugar for

```
/*@ requires P;
  @ ensures \old(A₁) ==> Q₁;
    ⋮
  @ ensures \old(Aₘ) ==> Qₘ;
  @*/
int f(int x₁,...,int xₖ)
```

An additional interesting feature that comes with named behaviors is the ability to specify that the given set of behaviors is *complete*, and that they are *mutually disjoint*. If ones adds, to the contract above, the line

```
    ⋮
  @ complete behaviors;
  @*/
```

then it specifies that the formula $P ==> (A_1 || \cdots || A_m)$ holds: in any allowed call state, at least one of the case applies. Similarly, adding the line

```
    ⋮
  @ disjoint behaviors;
  @*/
```

specifies that for any distinct indices i and j, the formula $P ==>\ !(A_i\ \&\&\ A_j)$ holds: the cases i and j do not apply at the same time. Consequently, specifying both complete and disjoint behaviors expresses that exactly one case applies at a time.

Integer Square Root continued
Consider the following contract with behaviors for our integer square root function.

```
/*@ behavior NegativeArgument:
  @    assumes x < 0;
  @    ensures \result == 0;
  @ behavior NonNegativeArgument:
  @    assumes x >= 0;
  @    ensures ResultIsNonNegative: \result >= 0;
  @    ensures ResultIsSquare:
  @       \result * \result <= x < (\result+1)*(\result+1);
  @ complete behaviors;
  @ disjoint behaviors;
  @*/
int integer_sqrt(int x) { ... }
```

The extra clauses for completeness and disjointness are valid, because the two assumptions on lines 2 and 5 are mutually exclusive.

Exercise 7 Difference between requires and assumes clauses
Consider the following piece of code.

```
/*@ requires 5 <= x <= 100;
  @ behavior B:
  @    assumes x >= 20;
  @    ensures \result >= 30;
  @*/
int f(int x) { return x+10 }

//@ ensures \result >= 15;
int g(void) {
  return f(7);
}
```

Is the function f correct (in the sense of Definition 1.5)? Does it conform to its contract? And the function g? Same questions when one exchanges the formulas in the **requires** and the **assumes** clauses of f.

The usage of behaviors allows the addition of **requires** clauses to each case, and also **assigns** clauses. We refer to the ACSL reference manual [1, Sect. 2.3.3] for details. It is worth insisting on the meaning of **assigns** clauses in behaviors. Let us assume a contract of the general form below.

```
/*@ behavior b₁:
  @    assumes A₁;
  @    assigns L₁;
  @    ensures Q₁;
       ⋮
  @ behavior bₘ:
  @    assumes Aₘ;
  @    assigns Lₘ;
  @    ensures Qₘ;
  @*/
int f(int x₁,...,int xₖ)
```

One has to remember that an **assigns** clause for a list L of left-values means that anything outside L is unchanged by a function call. In the context of named behaviors, this must be understood as: for any function call for which the pre-condition holds and the assumed A_j holds, any left-value outside of L_j is unmodified. It is not trivial to write an equivalent version of such a contract without behaviors: in the case that the set of behavior is complete, one could state a global **assigns** clause as

```
@ assigns L₁,...,Lₘ;
```

and then add additional **ensures** clauses in behaviors stating what is unchanged, say for any variable x which belongs to one of the sets $L_2, \ldots, L_m$ but not to L_1:

1 Formally Expressing What a Program Should Do: The ACSL Language

```
/*@ behavior b₁:
  @    assumes A₁;
  @    ...
  @    ensures x == \old(x);
```

Notice that if a contract with behaviors contains also a global **assigns** clause on a set L, then each **assigns** clause in behaviors should contain a subset of L. For example, the toy example of Sect. 1.2.4.2 could be either written as

```
int w,x,y,z;

/*@ assigns x,y,z;
  @ ensures ChangeOfX: x == \old(x) + w;
  @ behavior ChangeOfY:
  @    assumes b;
  @    assigns x,y;
  @    ensures y == \old(y) + 1;
  @ behavior ChangeOfZ:
  @    assumes ! b;
  @    assigns x,z;
  @    ensures z == \old(z) + 1;
  @*/
void f(int b) {
  x += w;
  if (b) { y += 1; } else { z += 1; }
}
```

or as

```
int w,x,y,z;

/*@ assigns x,y,z;
  @ ensures ChangeOfX: x == \old(x) + w;
  @ behavior ChangeOfY:
  @    assumes b;
  @    ensures y == \old(y) + 1;
  @    ensures z == \old(z);
  @ behavior ChangeOfZ:
  @    assumes ! b;
  @    ensures z == \old(z) + 1;
  @    ensures y == \old(y);
  @*/
void f(int b) { ...
```

1.3 The Core Logic of ACSL

This section presents in more details the underlying logic of ACSL, that is the language of propositions used in contract clauses. From now on, unlike in the previous section, we do not restrict ourselves anymore to programs using only the type **int**. We start by describing the syntax of the proposition language in Sect. 1.3.1, and the built-in theories in Sect. 1.3.2. We then discuss the semantics of ACSL formulas by presenting a partial formalization in Sect. 1.3.3. Finally, in Sect. 1.3.4, we present in more details the issues related to under-specified expressions, which are a major sources of misunderstanding and pitfalls.

1.3.1 Terms and Formulas in ACSL

The underlying logic setting of ACSL is first-order classical logic, also known as predicate logic. In this setting, *formulas* (also called *propositions*) are constructed by connectives for conjunction, disjunction and negation, with symbols identical to those in C (respectively &&, || and !), but also the *implication* (==>) and the logical *equivalence* (<==>). These connectives relate atomic propositions which are made of *predicate symbols* applied to *terms*. A fundamental predicate symbol is the equality, denoted as in C as ==. The language of terms is made of constants, variables, and *function symbols* applied to terms. The function symbols include few access functions inherited from C, namely pointer dereferencing (*t), structure or union field access (t.fieldname) and the array access (t[t]).

This logic is *typed*, meaning for example that equality can be stated only between two terms of the same type. The set of types contains the C types defined in the program under consideration, and *logic types*. Logic types include built-in logic types as described in Sect. 1.3.2 and user-defined types as described in Sect. 1.7.

The language of propositions is augmented with *quantifiers*. The notation

$$\backslash\text{forall } \tau\ x\,;\ P$$

denotes the *universal quantification* of proposition P over the variable x of type τ. It is valid whenever P is valid for all the possible values of type τ for x. The notation

$$\backslash\text{exists } \tau\ x\,;\ P$$

denotes the *existential quantification* of proposition P over the variable x of type τ. It is valid whenever P is valid for at least one value of type τ for x.

Examples: Quantifiers
The formula

```
\forall unsigned short x; \exists int y; y > 2*x
```

reads as "for any value x from the domain of type **unsigned short**, there exists a value y in the domain of type **int** such that y is larger than two times x." Assuming typical sizes for **short** and **int**, this is indeed a tautology.

The formula

```
\forall size_t i; 0 <= i < 100 ==> 'A' <= t[i] <= 'Z'
```

reads as "for any index i whose value lies in the range 0..99 (bounds included), the value stored in array t at index i is a character between A and Z." As such, it is not a tautology: it holds or not depending on the context where t occurs.

1 Formally Expressing What a Program Should Do: The ACSL Language

The language of ACSL formulas also offers additional constructs to ease the writing of specifications, including local bindings, written as

$$\texttt{\textbackslash let}\ x = t\ ;\ P$$

and conditionals written as in C as

$$(\ t\ )\ ?\ P_1\ :\ P_2$$

These are detailed in ACSL reference manual [1, Sect. 2.2]. When a formula is written as a post-condition of a contract, then the special keyword **\result** is available as well (as presented in Sect. 1.2.2) together with the special construct **\old(t)** as presented in Sect. 1.2.4.

Exercise 8 Sortedness of a section of an array
Propose an ACSL formula that, for a given array t of **int**s and two indices a and b, expresses that the section of t between a (included) and b (not included), is sorted in strictly increasing order.

Combining quantifiers and logic connectives is not as simple as it may seem, and there are classical traps in which many may fall one day or another. A classical one is the combination of the quantifiers and the implication. Imagine a common pattern where one wants to specify a property about all values of an array of size l, for example that all values are non-negative. This should be written using a universal quantification and an implication as

```
\forall size_t i; 0 <= i < l ==> a[i] >= 0
```

This is the proper way to specify the expected property only on the valid indices. A less common but possible pattern is to specify a property valid for at least one valid index of an array, instead of all of them. Without care one may write something like

```
\exists size_t i; 0 <= i < l ==> a[i] >= 0
```

which unfortunately does not mean what is intended. Instead, this formula is *always valid*, because it suffices to choose -1 as the value of i to make the implication valid (since its premise would be false). The proper way to write the intended property is with a conjunction instead of an implication, as

```
\exists size_t i; 0 <= i < l && a[i] >= 0
```

In fact, stating a formula of the form

$$\texttt{\textbackslash exists}\ \tau\ x\ ;\ P \texttt{==>} Q$$

is almost certainly an error. Generally speaking, writing formulas in ACSL should be always done with special care, because no tool is able to tell whether what you wrote really means what you intended it to mean. Reviewing written formulas by a third party is good practice. As another example, you could go back to the introductory example of Fig. 1.1, it should now be clear that the fourth post-condition is valid, because the premise of the implication can be made false, say by taking z equal to x. On the other hand, the fifth post-condition is not valid, because it does not hold when z is equal to x+1.

1.3.2 Built-In Logic Theories of ACSL

There are fundamental built-in theories in ACSL that we review below.

1.3.2.1 Equality

The main fundamental theory is the one of equality, involving a binary predicate written ==. It is defined on any type, more precisely t_1==t_2 is a well-typed formula whenever t_1 and t_2 have the same type. This binary predicate is reflexive, symmetric and transitive, that is the following properties hold on any type τ.

```
\forall τ x;     x == x
\forall τ x,y;   x == y ==> y == x
\forall τ x,y,z; x == y ==> y == z ==> x == z
```

It is also a *congruence*, meaning that for any function symbol f and predicate symbol p of some arity n the following properties hold.

```
\forall τ₁ x₁,y₁,...,τₙ xₙ,yₙ;
   x₁ == y₁ && ··· && xₙ == yₙ ==> f(x₁,...,xₙ) == f(y₁,...,yₙ)
\forall τ₁ x₁,y₁,...,τₙ xₙ,yₙ;
   x₁ == y₁ && ··· && xₙ == yₙ ==> p(x₁,...,xₙ) <==> p(y₁,...,yₙ)
```

1.3.2.2 Integer Arithmetic

The second fundamental built-in theory is the theory of mathematical integers, or *Integer Arithmetic*, whose type is written **integer**. Is it essential to remember that this arithmetic theory is significantly different from the arithmetic of machine integers which is subject to undesirable behavior like overflow. So one has to remember that in ACSL formulas, operators like + or * denote the ideal mathematical operations that never overflow. This is indeed a major pitfall when writing specifications, illustrated in the following exercise.

Exercise 9 Bounded versus unbounded integers
Among the propositions below, which of them hold?

- `\forall integer x; x*x >= 0`
- `\forall integer x; \exists integer y; x+y == 0`
- `\forall integer x; x < 0 ==> 0 < -x`

Same question when replacing **integer** by **int**.

The subtleties in the meaning of unbounded versus bounded integers in the annotations must be kept in mind, since they commonly have an impact on the conformity of contracts. This is again illustrated via an exercise as follows.

Exercise 10 ACSL integers, conformity and safety
Consider the following annotated C function.

```
/*@ ensures \result == x+y;
  @*/
int f(int x, int y) {
    return x+y;
}
```

Is it safe? Does it conform to its contract? Are the answers the same if one replaces **int** by **unsigned int**?

1.3.2.3 Real Arithmetic

Another built-in theory in ACSL is the theory of mathematical real numbers, or *Real Arithmetic*. The type of real numbers is written as **real**. This theory is significantly different from the floating-point numbers that are typically used to represent real quantities in C programs. In particular, as for integer arithmetic, it is essential to remember that the operations in this theory can never "overflow".

Indeed, the primary usage of real numbers in ACSL annotations is for specifying the behavior of programs involving computation on floating-point numbers. In ACSL formulas, any C variable of type **float** or **double** is interpreted as the real number it represents. This not only means that the same care as for integers must be taken regarding overflows, but one should also be very cautious regarding the rounding errors occurring in the code, which do not exist on real numbers in the specification. For example, the function

```
/*@ ensures \result * 10.0 == 1.0;
  @*/
float f(void) {
    return 0.1;
}
```

does not conform to its contract. Indeed, since the floating-point representation of 0.1 is not exact, it is a number which, when multiplied by 10, is not equal to 1 in the mathematical real arithmetic. See the ACSL manual [1, Sect. 2.2.5] for extra operations provided to state properties on floating-point computations.

Exercise 11 Real versus Floating-Point Arithmetic
Consider the following annotated C function.

```
void f(float f, double d) {
  float ff = f * f;
  double dd = d * d;
  //@ check R: \forall real r; r * r != 3.0;
  //@ check D: d * d != 3.0;
  //@ check DD: dd != 3.0;
  //@ check F: f * f != 3.0;
  //@ check FF: ff != 3.0;
}
```

Which of the **check** assertions[6] are valid?

There are indeed many specific pitfalls to avoid when formally specifying floating-point computations. For more information and advice on specifying floating-point programs using ACSL, we recommend reading a paper written in 2011 by Boldo and Marché [5].

1.3.3 Semantics of Propositions

As Sect. 1.2.3, this section may be skipped on a first read. We have already seen that the semantics of formulas is described using a judgment of the form

$$\Sigma, \Pi \models \varphi$$

meaning that formula φ holds in the global context Σ and local context Π. A similar judgment is considered for terms, under the form

$$\Sigma, \Pi \models t \Rightarrow v$$

meaning that term t evaluates to the value v.

Notice that the semantics is formalized only on well-typed terms and formulas, the well-typedness could be defined using typing judgments.

Typical rules for connectives are

---

[6] Clauses **check** are described in Sect. 1.4.1. Here we just ask whether the corresponding formulas are valid after execution of the function's body.

1 Formally Expressing What a Program Should Do: The ACSL Language

$$\frac{\Sigma, \Pi \models \varphi_1 \quad \Sigma, \Pi \models \varphi_2}{\Sigma, \Pi \models \varphi_1 \ \&\&\ \varphi_2}$$

$$\frac{\Sigma, \Pi \models \varphi_1}{\Sigma, \Pi \models \varphi_1\ ||\ \varphi_2}$$

$$\frac{\Sigma, \Pi \models \varphi_2}{\Sigma, \Pi \models \varphi_1\ ||\ \varphi_2}$$

respectively for conjunction and disjunction. A rule for universal quantification is

$$\frac{\text{for any value } v \text{ of type } \tau,\ \Sigma, \{x = v\} \cdot \Pi \models \varphi}{\Sigma, \Pi \models \backslash\texttt{forall}\ \tau\ x;\ \varphi}$$

with somehow an infinite set of premises.

Typical rules for terms include rules for variables and for built-in operators, such as

$$\frac{\Sigma(x) = v}{\Sigma, \Pi \models x \Rightarrow v}$$

$$\frac{\Sigma, \Pi \models t_1 \Rightarrow v_1 \quad \Sigma, \Pi \models t_2 \Rightarrow v_2 \quad v_1 +_\mathbb{Z} v_2 = v}{\Sigma, \Pi \models t_1 + t_2 \Rightarrow v}$$

where $+_\mathbb{Z}$ is the mathematical addition in integer arithmetic. Indeed, as seen for the addition above, interpretation of predicate symbols and function symbols is done according to some underlying model, and as such the interpretation of integer operators are the operations in the classical mathematical model $\mathbb{Z}$, whereas the real operations are interpreted in the classical mathematical set of real numbers $\mathbb{R}$. It is important to have in mind that these are different from the arithmetic of (bounded) machine integers and floating-point numbers.

Exercise 12 Semantics of \let and conditional expressions
Propose rules for the semantics of the ACSL local binding "$\backslash\texttt{let}\ x = t;\ \varphi$" and the ACSL conditional expression "$(t)?\varphi_1 : \varphi_2$".

1.3.4 Total Functions and Underspecified Terms

A major difficulty for users when writing specification is caused by the *totality* aspect of the logic: first-order logic is a logic of *total functions*. It means that, unlike C functions, all function symbols and all predicate symbols denote mathematical functions that are *total*, that is: they are defined for all possible values of their parameters (according to their respective types). A first classical example is the division operator: even if in C, making a division by zero is an error that would (usually) lead to a crash of the program, dividing by zero in the logic is not an error: division is a total function. In terms of formal semantics, we have

$$\frac{\Sigma, \Pi \models t_1 \Rightarrow v_1 \quad \Sigma, \Pi \models t_2 \Rightarrow v_2 \quad div(v_1, v_2) = v}{\Sigma, \Pi \models t_1/t_2 \Rightarrow v}$$

where div is some total binary function, for which the only thing known is that it coincides with mathematical integer division when the second argument is not zero. This is an *under-specified* total function. Notice the difference between under-specification of logic functions and some unspecified behavior of C code (such as the order of evaluation of function arguments): the result of say $div(1, 0)$ is not specified but is yet some integer value. Consider for example the proposition x == y/0 for some variables x and y. It is a well-formed, well-typed proposition. As the result of a total function returning an integer, the term y/0 denotes some integer, although it is not determined which integer it is (it is under-specified). Some traps may show up when working with under-specified terms. Consider for example the formula 1/0 == 1/0: even if 1/0 is under-specified, this formula is necessarily a valid statement in ACSL, as an instance of the reflexivity axiom for equality. Similarly, an array access term a[i] in the logic always denotes some value, even if the index i is not within the valid array bounds, so that a[i] == a[i] always holds. Also, if p is some pointer variable, the statement *p == *p necessarily holds, even if p does not point to a memory location that is valid for reading, or even if it is valid for reading but not initialized.

Exercise 13 Under-specified terms
Among the propositions below, which of them are valid (in any context)?

- 1/0 == 2/0
- 1/(1-1) == 1/(2-2)
- x == y ==> x/z == y/z

Exercise 14 Undefined behavior versus under-specification
Consider the following C program
```
1  /*@ ensures E1: \result == \result;
2   @ ensures E2: x > 0 ==> \result == 1;
3   @ ensures E3: x <= 0 ==> \result != \result;
4   @ ensures E4: x > 0;
5   @*/
6  int f(int x) {
7     int y;
8     //@ check A1: y == y;
```

```
9   if (x > 0) {
10      y = 1;
11      //@ check A2: x != y;
12  }
13  else {
14      //@ check A3: x != y;
15      x = y;
16      //@ check A4: x != y;
17  }
18  return y;
19 }
```

Is it safe? Which of the **check** and **ensures** clauses are valid?

Exercise 15 Floating-point numbers in logic formulas
Consider the following program fragment

```
1  /*@ ensures E1: \result == \result;
2   @ ensures E2: \eq_float(\result,\result);
3   @*/
4  float f(float x, float y) {
5    return (x / y);
6  }
```

Is it safe? Notice that **\eq_float** is a built-in ACSL predicate implementing floating-point equality. Are the post-conditions valid?

1.4 Code Annotations

This section is dedicated to ACSL annotations that are put inside the code as part of any C function's body. Those annotations can be used to state properties that are not expressible at the level of contracts, e.g., when concerning local variables. An important other use of code annotations is when trying to statically prove a contract valid, e.g., using the Wp plugin (see Chap. 4). In that case they may be required to complete the proof, the typical case being the use of *loop invariants*.

Section 1.4.1 presents in-code *assertions* and Sect. 1.4.2 shows how to refer to past values using *labels*. Section 1.4.3 discusses their semantics and possible traps. Section 1.4.4 presents *loop invariants* and Sect. 1.4.6 discusses the important differences between operational semantics of annotations and the so-called *axiomatic semantics*. Section 1.4.7 focuses on how to specify *termination* of loops and recursive calls.

1.4.1 Code Assertions

The simplest form of code annotations are *assertions*. They can be placed at any position allowed for a C statement. The general form is one of the following three clauses

```
1 //@ check A;
2 //@ assert A;
3 //@ admit A;
```

Informally, their meaning is the same: whenever execution reaches this control point, the proposition A holds. The formula A can be any ACSL proposition, as those used in contracts. A minor difference though is that the **\old** and **\result** constructions are not allowed. Referring to past values of variables in assertions is done as described in the next section. As can be seen in the example below, the **check** clauses can naturally talk about local variables. Also, unlike what happens in contracts, the effects on function parameters are observable in code annotations.

Simple usage of check clause
The following piece of code conforms to its annotations.

```
1  /*@ requires x >= 0;
2    @ ensures x == \old(x);
3    @*/
4  void f(int x) {
5    int y = x;
6    x = x+1;
7    //@ check x >= 1 && x > y;
8    y = y+2;
9    //@ check x < y && y >= 2;
10 }
```

The three forms of assertions indeed have the same semantics: the proposition coming afterwards must hold whenever execution reaches the control point. The difference is visible only when considering axiomatic semantics (introduced later in Sect. 1.4.6.1): the proposition introduced by an **assert** clause is visible in latter statements, whereas the proposition introduced by a **check** clause remains invisible in the remainder, which can be used for debugging or for controlling the size of the proof contexts for deductive verification for example. In the context of formal proof, the **admit** clause is significantly different from the two other clauses: it states that the proposition associated to it can be assumed valid in the remainder of the code. Formally, it states that executions that do not satisfy this proposition are considered unsafe, as if it was an undefined behavior. It somehow has the effect of stating an axiom inside the code, and as such it is a construction that should be used with particular care, similarly as the axiomatic blocks that are discussed later in Sect. 1.7.4, but it can be useful to state hypotheses about the environment for example. The clauses **assert** and **admit** are in practice only needed when performing formal verification of contracts, for example with the Wp plug-in.

1 Formally Expressing What a Program Should Do: The ACSL Language   35

1.4.2 Referring to the Past Using Labels

Assertions in the code can refer to past values of variables thanks to the construct \at, which is similar to the \old construct already introduced in Sect. 1.2.4. The \at construct takes two arguments: a term (or a formula) and a *label* of the code. A more or less general form is thus

```
int f(...) {
   ...
   Lab:
   ...
   //@ assert ... \at(x,Lab) ...;
   ...
}
```

The meaning of the \at construct is to denote the value of its first argument as it was when execution traversed the corresponding label. For this to make sense, some scoping rules must be respected, as described in the ACSL reference manual [1, Sect. 2.4.3].

In addition to referring to labels in the C code, ACSL also provides default labels, as described in the ACSL reference manual [1, Sect. 2.4.3]. These include the label **Here** which always denotes the program point where the related proposition is placed,[7] and the label **Pre** which denotes, inside any code annotation, the entry point of the enclosing function. The label **Old** is also one of such pre-defined labels, so that indeed \old(φ) is just a syntactic sugar for \at(φ,Old).

As already said for \old, applying an \at to a term or a formula means applying it to each of the variables it contains. Indeed, unlike in Sect. 1.2, we now consider arbitrary data types. This needs a more precise definition: \at(φ,L) is a shortcut for φ where each *left-value l* of φ is replaced by \at(l,L). Left-values are the objects that can occur on the left part of an assignment. These include variables but also the other form of referencing memory locations: pointer access *p, structure access s.f and s->f, array access a[i]. That notation should be used carefully then, since \at(*p,L) is usually different from *\at(p,L), \at(s->f) is usually different from \at(s,L)->f, and if a has a pointer type, \at(a[i],L) is usually different from \at(a,L)[i], a[\at(i,L)] and \at(a,L)[\at(i,L)]. An additional difficulty arises here: the semantics of ACSL specifies [1, Sect. 2.2.7 Structures, Unions and Arrays in logic] that structures and arrays in logic formulas denote respectively purely functional records and purely functional maps. It means that, if s is a structure (and not a pointer to a structure), then in fact \at(s.f,L) is the same as \at(s,L).f, and if s is an array (and not a pointer) then \at(a[i],L) is the same as \at(a,L)[\at(i,L)].

Semantics of \at
Consider the following code.

---

[7] It seems useless at first, but we will see two uses of it later on.

```
1  int a[2], *p;
2  struct S {int f;};
3  struct S s, *q;
4
5  /*@ requires a[0] == 0 && a[1] == 1 && p == (int*)a;
6   @ requires s.f == 5 && q == &s;
7   @*/
8  void f() {
9    int i = 0;
10   L:
11   i = 1;
12   a[0] += 10; a[1] += 10; s.f += 10;
13   //@ check C1: \at(a[i],L) == 0;
14   //@ check C2: \at(a,L)[i] == 1;
15   //@ check C3: a[\at(i,L)] == 10;
16   //@ check C4: \at(p[i],L) == 0;
17   //@ check C5: \at(p,L)[i] == 11;
18   //@ check C6: p[\at(i,L)] == 10;
19   //@ check C7: \at(s.f,L) == 5;
20   //@ check C8: \at(s,L).f == 5;
21   //@ check C9: \at(q->f,L) == 5;
22   //@ check C10: \at(q,L)->f == 15;
23  }
```

All the given checks are valid. Notice the difference between C2 and C5, emphasizing why \at(p,L) and \at(a,L) are different. Similarly, notice the difference between C8 and C10.

Exercise 16 Semantics of \old
Consider the following code:

```
1   int x, a[2], *p;
2
3   /*@ requires x == 0 && i == 0;
4    @ requires a[0] == 2 && a[1] == 3 && p == (int*)&a;
5    @ ensures E1: \old(a)[x] == ?;
6    @ ensures E2: \old(a)[i] == ?;
7    @ ensures E3: \old(p)[x] == ?;
8    @ ensures E4: \old(p)[i] == ?;
9    @ ensures E5: \old(x + i + a[x] + a[i] + p[x] + p[i]) == ?;
10   @*/
11  void f(int i) {
12    x += 1; i += 1; a[0] += 2; a[1] += 3;
13  }
```

give appropriate values for each question mark to make all the post-conditions valid.

1 Formally Expressing What a Program Should Do: The ACSL Language

A common mistake is the capture of a variable inside an \at term relating to a label where this variable does not have the intended value, or even does not exist at all. An example is as follows.

Some pitfalls in the usage of \at
Consider the code
```
void f(int *p, int s) {
  L: {
    for (int i = 0; i < s; i++) {
      p[i] = p[i]+1;
      //@ assert p[i] == \at(p[i],L)+1;
    }
  }
}
```

the use of `\at(p[i],L)` is ill-formed here since variable `i` is not visible at label L. Consider now the variant

```
void f(int *p, int s) {
  int i=0;
  L: {
    for (i = 0; i < s; i++) {
      p[i] = p[i]+1;
      //@ assert p[i] == \at(p[i],L)+1;
    }
  }
}
```

The term `\at(p[i],L)` is now well-formed, but it does not mean the value of the array p at index i because i is 0 at label L, so it denotes the initial value of `p[0]` instead. Here are other suitable ways to specify the intended meaning, that is that the new value of `p[i]` is its former value plus 1. The first one makes use of another `\at` inside, referring to the label **Here** (this is indeed one example of use of that predefined label). It is as follows.

```
//@ assert p[i] == \spsat(p[\at(i,Here)],L)+1;
```

Now the internal `\at` prevent considering the value of i at label L. A second solution is as follows, making use of a local binding in the logic to escape the capture of i in the `\at`.

```
//@ assert \let j = i; p[i] == \at(p[j],L)+1;
```

Notice finally that if p was not a pointer but a global array, then another solution would simply to write

```
//@ assert p[i] == \at(p,L)[i]+1;
```

because, as said above, `\at(p,L)` would denote the logical map obtained from p at label L.

Notice that in some specific cases, it is possible that an execution reaches an annotation mentioning a label under \at, although the execution never passed through the corresponding label. This can happen for example if a program used a goto statement, as in the following program fragment.

```
int f() {
  int x = 0;
  goto L2;
  ++x;
  L1:
  ++x;
  L2:
  //@ assert x > \at(x,L1);
}
```

In such a case, the term \at(x,L1) has an under-specified value, as explained in Sect. 1.3.4.

1.4.3 Formalizing Semantics of Assertions and Labels

This section can be skipped on a first read. As a continuation of Sect. 1.2.3, let us give formal rules for the semantics of assertions and \at. Giving a rule for assertions alone is quite straightforward:

$$\frac{C \models \varphi}{(C, \mathsf{assert}\ \varphi) \to (C, \mathsf{void})}$$

For \at, we need to expand the context of judgments to record the past program state. One relatively simple way is to say that C is not only representing one configuration but a family of configurations indexed by labels. Let us denote such a family as $\mathcal{C}$ and write $\mathcal{C}_L$ to denote the configuration at label L. The rule for getting the value of a variable, as a term in the logic, can be expressed as

$$\frac{\mathcal{C}_{Here}(x) = v}{\mathcal{C} \vdash x \Rightarrow v}$$

and the similar rule for a variable under \at being

$$\frac{\mathcal{C}_L(x) = v}{\mathcal{C} \vdash \mathsf{\backslash at}(x, L) \Rightarrow v}$$

We complete the semantics with a rule for traversing labels during executions:

$$\overline{(\mathcal{C}, L:i) \to (\mathcal{C}[\mathcal{C}_L \leftarrow \mathcal{C}_{Here}], i)}$$

which acts as making a "copy" of the current configuration $\mathcal{C}_{Here}$ into the "name" $\mathcal{C}_L$.

1.4.4 Loop Annotations

The other important form of code annotations are the annotations one can attach to **while** loops, **for** loops or **do...while** loops. These loop annotations are essential when one wants to formally prove properties on the behavior of loops, notably when using the Wp plug-in of Frama-C. These annotations are inserted just before the loop they are attached to.

The main kind of clauses that can be given as a loop annotation is a *loop invariant*. For a **while** loop it has the following general shape.

```
1   //@ loop invariant I;
2   while (c) { body }
```

The informal meaning is somehow similar to assertions, except that the invariant must hold at each loop iteration, more precisely at each execution point just before executing the condition c. Notice the importance of saying *before* here, since the expression c may have side effects.

The general shape for a **do...while** loop is the following.

```
1   //@ loop invariant I;
2   do { body } while (c);
```

In this case, the invariant must hold each time we reach **do**, i.e., before entering the loop and *after* each execution of condition c that does not evaluate to 0. Notice that this means the invariant does not need to hold when exiting the loop.

The general shape for a **for** loop is the following.

```
1   //@ loop invariant I;
2   for (init; c; increment) { body }
```

In this case, the invariant must hold *after* the initialization statement, and then at each iteration *before* executing the condition c.

Loop invariants
Consider the following code.

```
void f(void) {
  int x=0;
  /*@ loop invariant OK1: 0 <= x <= 10;
    @ loop invariant Wrong1: x < 10;
    @*/
  while (x < 10) { ++x; }
  /*@ loop invariant OK2: 10 <= x < 20;
    @ loop invariant Wrong21: x < 19;
    @ loop invariant Wrong22: 11 <= x;
    @*/
  do { ++x; } while (x < 20);
  /*@ loop invariant OK3: 0 <= x <= 10;
    @*/
  for (x=0; x<10; ++x);
}
```

Each invariant with name starting with OK is valid, but the others are not. The invariant Wrong1 is invalid because when exiting the first loop, that is when x is 10 and the condition is checked for the last time, it does not hold. In contrast, the invariant OK2 is right to state that x is smaller than 20, because, for a do ... while loop, the invariant is not required to hold when the condition is false (and in fact OK2 does not hold when exiting the loop). The invariant Wrong21 is invalid, because x can reach 19. The invariant Wrong22 is invalid because x is 10 when entering the second loop, and even if it is incremented before the condition, the semantics given above requires the invariant to hold before entering the loop, and not only when checking the condition.

Exercise 17 Loop invariants
In the following, which optimal values would you give to constants A_i and B_i to make the loop invariants valid?

```
void f(void) {
  int x = 0;
  //@ loop invariant A1 <= x <= B1;
  while (++x < 10);

  x = 0;
  //@ loop invariant A2 <= x <= B2;
  while (x++ < 10);

  x = 0;
  //@ loop invariant A3 <= x <= B3;
  do {} while (++x < 10);

```

```
14    x = 0;
15    //@ loop invariant A4 <= x <= B4;
16    do {} while (x++ < 10);
17
18    //@ loop invariant A5 <= x <= B5;
19    for(x=0; ++x < 10; );
20
21    //@ loop invariant A6 <= x <= B6;
22    for(x=0; x++ < 10; );
23  }
```

1.4.5 Loop Invariants in Presence of **break** and **continue** Statements

Special attention should be given to loop annotations when **break** or **continue** statements are present. When the loop body executes a **break** statement, it is a design choice of ACSL that the loop is exited **without any additional check for the invariant**. When the loop body executes a **continue** statement, then, except for **do...while** loops, the invariant is going to be checked immediately, because it is just before the condition that will be executed. For **do..while** loops, the condition will be executed and then the invariant checked.

Exercise 18 Loop invariants with break statements
In the following, which optimal values to give to constants A_i and B_i to make the loop invariants valid?

```
1   void f(void) {
2     int x = 0;
3     //@ loop invariant A1 <= x <= B1;
4     while (1) { if (++x >= 10) break; }
5
6     x = 0;
7     //@ loop invariant A2 <= x <= B2;
8     while (1) { if (x++ >= 10) break; }
9
10    x = 0;
11    //@ loop invariant A3 <= x <= B3;
12    do { if (++x >= 10) break;} while (1);
13
14    x = 0;
15    //@ loop invariant A4 <= x <= B4;
16    do { if (x++ >= 10) break;} while (1);
17
18    //@ loop invariant A5 <= x <= B5;
```

```
19    for(x=0; ; ){ if (++x >= 10) break;}
20
21    //@ loop invariant A6 <= x <= B6;
22    for(x=0; ; ){ if (x++ >= 10) break;
23  }
```

As emphasized by the previous exercise, it is indeed always possible to transform a loop in a loop without condition but a **break** statement as follows. A loop

```
1   //@ loop invariant I;
2   while (c) { body }
```

is equivalent to

```
1   //@ loop invariant I;
2   while (1) { if (!c) break; body }
```

a loop

```
1   //@ loop invariant I;
2   do { body } while (c);
```

is equivalent to

```
1   //@ loop invariant I;
2   do { body ; if (!c) break; } while (1);
```

and

```
1   //@ loop invariant I;
2   for (init; c; increment) { body }
```

is equivalent to

```
1   //@ loop invariant I;
2   for (init; ; increment) { if (!c) break; body}
```

Another kind of clauses in loop annotations is a framing clause similar to the **assigns** clause of contracts (see Sect. 1.2.4). It is stated after the **loop assigns** keyword. It is intended to provide (an over-approximation of) the side effects of the loop iteration. It somehow behaves like a loop invariant in the sense that it should hold at each iteration. More precisely, at each loop iteration, all memory locations that are disjoint from the locations declared in the clause are unmodified since the loop entry. This semantics permits writing annotations such as

```
1   int a[10];
2
3   void f(void) {
4      /*@ loop assigns i,a[0..i-1];
5         @*/
6      for (int i = 0; i < 10; ++i) { a[i] = 0; };
7   }
```

1 Formally Expressing What a Program Should Do: The ACSL Language

1.4.6 Formal Semantics of Loop Invariants

This section can be skipped on a first read. We can extend our formal rules for the semantics of ACSL with the following rules for invariants in loops.

The rule for **while** loops is as follows.

$$\frac{C \models \varphi}{\begin{array}{l}(C, \texttt{loop invariant } \varphi \texttt{ while } c \texttt{ \{ } b \texttt{ \}}) \\ \rightarrow (C, \texttt{if } (c)\{b\texttt{;loop invariant } \varphi \texttt{ while } c \texttt{ \{ } b \texttt{ \}}\})\end{array}}$$

Notice how this rule expresses that the invariant must hold immediately at loop entry, even before checking the condition c. In particular, the loop invariant must hold even if no iterations at all are performed. Moreover, it is worth noting that in the case c contains side effects, the invariant φ must hold before these side effects take place.

The rule for **for** loops is expressed by referring to **while** loops, as follows.

$$\begin{array}{l}(C, \texttt{loop invariant } \varphi \texttt{ for } (i, c, s) \texttt{ \{ } b \texttt{ \}}) \\ \rightarrow (C, i\texttt{;loop invariant } \varphi \texttt{ while } c \texttt{ \{ } b \texttt{ ; } s\}\})\end{array}$$

Notice how this implies that the invariant φ is checked before each execution of c but after the initialization statement i.

The rule for **do...while** loops is as follows.

$$\frac{C \models \varphi}{\begin{array}{l}(C, \texttt{loop invariant } \varphi \texttt{ do \{ } b \texttt{ \} while } (c)) \\ \rightarrow (C, b; \texttt{ if } (c) \texttt{ \{ loop invariant } \varphi \texttt{ do \{ } b \texttt{ \} while } (c)\})\end{array}}$$

Notice that, in contrast to other kinds of loops, the invariant φ is checked after the possible side effects of the loop condition c.

1.4.6.1 Axiomatic Semantics of ACSL

So far, our presentation of a formal semantics for ACSL annotations relies on extending an operational semantics into a blocking semantics, making the execution block whenever an invalid annotation is encountered.

A different kind of formal semantics would be an *axiomatic semantics* which would express the validity of annotations in terms of validity of *Hoare triples* [28]. Roughly speaking, a Hoare triple is a judgment denoted as $\{P\} s \{Q\}$, whose intended meaning is that in any program configuration where the pre-condition P holds, if execution of statement s terminates, then the post-condition Q holds in the resulting configuration. A typical rule for such judgments is the rule for the while loops as follows (assuming the condition c has no side effect for simplicity).

$$\frac{\{I \wedge c\}\, b\, \{I\}}{\{I\}\, (\texttt{loop invariant}\ I\ \texttt{while}\ (c)\ \{\, b\, \})\, \{I \wedge \neg c\}}$$

Such a rule makes explicit that assuming I holds before entering the loop, and provided it is shown preserved by a loop iteration, one deduces it holds at loop exit. A Hoare-style rule for **do..while** loops is quite similar (again assuming the condition c has no side effect):

$$\frac{\{I\}\, b\, \{(c \Rightarrow I) \wedge (\neg c \Rightarrow Q)\}}{\{I\}\, (\texttt{loop invariant}\ I\ \texttt{do}\ \{\, b\, \}\ \texttt{while}\ (c)\,)\, \{Q \wedge \neg c\}}$$

It emphasizes that, assuming the invariant I holds when entering the loop, assuming I is preserved by an iteration when the side effect free condition c holds, and assuming it entails some post-condition Q when c does not hold, one can deduce that Q holds at loop exit.

Another reformulation of such an axiomatic semantics is given by the so-called *weakest pre-condition calculi* originally due to Dijkstra. This setting is indeed the basis of the Wp plugin of Frama-C, described in Chap. 4. In this setting, the semantics of **while** loops is given by the following rule for computing the minimal precondition to guarantee that a given post-condition Q holds:

$$\text{WP}(\texttt{loop invariant}\ I\ \texttt{while}\ (c)\ \{\, b\, \}, Q) = \\ I \wedge \forall \mathbf{w}.\ (I \wedge c \Rightarrow \text{WP}(b, I)) \wedge (I \wedge \neg c \Rightarrow Q)$$

where **w** denotes the write effects of the loop body.

1.4.6.2 Inductive Invariants

What we want to emphasize here is that the semantics of invariants expressed by Hoare logic or by weakest precondition is *stronger* than the one defined by the blocking operational semantics: the former are necessarily *inductive* invariants. It is a fact that there exist invariants that are valid with respect to the blocking operational semantics, but are not inductive. A toy example is as follows.

```
void f(void) {
  int x = 0;
  int y = 0;
  /*@ loop invariant y > 0 ==> x == y;
    @ loop assigns x,y;
    @*/
  while (y < 10) {
    x += 1;
    y += 1;
  }
}
```

1 Formally Expressing What a Program Should Do: The ACSL Language

Since during the execution, the values of x and y are identical at the start of each loop iteration, the invariant holds. Yet, it is not preserved inductively, in particular by the first iteration when y==0.

This situation is not uncommon. A less toy example is

```
int a[10];
void f(void) {
  int i = 0;
  /*@ loop invariant 0 <= i <= 10;
    @ loop invariant \forall int j;
    @    0 <= j < i ==> a[j] == \spsat(a[j],Pre)+1;
    @ loop assigns i,a[0..9];
    @*/
  while (i < 10) {
    a[i] += 1;
  }
}
```

The second loop invariant is valid during execution, but it is not inductive: it cannot be shown to be preserved by an arbitrary loop iteration. This is because it does not state that the values in array a for indices larger or equal to i are unchanged since the beginning of the execution of the loop. This invariant should be strengthened to make it inductive, for example by adding

```
/*@ loop invariant \forall int j;
  @     i <= j < 10 ==> a[j] == \spsat(a[j],Pre);
```

or, alternatively, by adding a precise **loop assigns** clause as

```
/*@ loop assigns i,a[0..i-1];
```

which would subsume the additional loop invariant above.

Notice finally that this subtle difference between operationally valid invariants and inductive invariants also appears for other kinds of ACSL annotations, because of potential weaknesses of function contracts. A toy example is as follows.

```
int x;

/*@ requires x < INT_MAX;
  @ assigns x;
  @ ensures x > \old(x);
  @*/
void f(void) { x += 1 }

int main(void) {
  x = 0;
  f();
  //@ assert x == 1;
}
```

During execution of main, all ACSL annotations, both in the contract of f and the assertion in the body of main, hold. The contract of f is indeed valid. Yet, a proof of the assertion using a Hoare logic approach or a WP-style calculus will fail, because the post-condition of f is too weak, it should be strengthened, for example with x == \old(x)+1. On the other hand, the Eva plugin (see Chap. 3) is able to prove the assertion valid.

1.4.7 Specifying Termination

An aspect not yet considered so far is the *termination* of programs. The meaning of annotations is defined independently of termination of the underlying programs. Yet, it might be useful to additionally specify that a given program or part of a program will not run forever. There are two potential sources of infinite executions: loops and recursive calls.

1.4.7.1 Termination of Loops

To state that a loop is terminating in ACSL, one must provide a *loop variant*, that is some logic term (of type **integer**) whose value is strictly decreasing between each loop iteration. The general shape is the following

```
1  //@ loop variant t;
2  while (c) { body }
```

meaning that the logic term t, denoting an integer, evaluated at the same execution point as the loop invariant, is strictly decreasing between two iterations. To ensure that termination is implied, t must also be always non-negative, except possibly once, at the very last iteration. The reason it works is that such a variant induces a descending chain of values $t_0 \succ t_1 \succ t_2 \succ \ldots$ corresponding to the values of term t at each iteration. If the underlying ordering relation $\succ$ is well-founded, that is, has no infinite chains, then the iteration cannot loop for infinitely long. In this precise case, the underlying ordering is $x \succ y \equiv x > y \wedge x \geq 0$, which is well-founded on the set $\mathbb{Z}$ of mathematical integers. Notice the fact that y is not required to be non-negative, explaining the sentence above that the variant t does not need to be non-negative at the last loop iteration.

Loop variant
In the following C code

```
1  //@ requires step > 0;
2  void f(int step) {
3    int x = 0;
4    //@ loop variant x;
5    while (x > 0) { ... ;x -= step; }
6  }
```

the loop variant is valid: x decreases at each step because step is positive, and x remains non-negative, except possibly at the end of the last iteration.

1 Formally Expressing What a Program Should Do: The ACSL Language   47

Notice that, in a more general case although rarely used, it is possible to provide termination measures of some other type than integers, or for some other ordering relation than the default one [1, Sect. 2.5.2].

Probably the most common specification trap with respect to termination is to state specifications that appear trivially valid, just because specifying termination was omitted. A typical example could be

```
int a[10];
void f (void) {
  int i = 0;
  while (i < 10) { a[i] += 1; }
  //@ assert <whatever>;
}
```

where one forgets to increment the loop index i. Since the loop is never terminating, the assertion is correct whatever the given proposition stated. A supposedly valid variant could have been

```
//@ loop variant 10 - i;
```

which is not valid, but becomes valid if the missing increment of i is added in the loop body.

Exercise 19 variant clause
Among the following, which loop variant clauses are valid?

```
void f(void) {
  int i = 10;
  /*@ loop variant V1: 10-i;
    @ loop variant V2: 10+i;
    @ loop variant V3: i;
    @ loop variant V4: i+5;
    @*/
  while (i > -10} {
    i -= 7;
  }
```

1.4.7.2 Termination of Recursive Function Definitions

To state that a recursive definition of a C function, or a set of mutually recursive definitions, cannot produce any infinite executions, ACSL provides the extra clause **decreases** within function contracts. Let us assume a general code shape as follows.

```
/*@ decreases v_f;
  @*/
int f(...) {
  ...
  f(..)  // recursive call to f
```

```
6      ...
7      g(..)   // mutually recursive call to g
8      ...
9    }
10
11   /*@ decreases $v_g$;
12     @*/
13   int g(...) {
14     ...
15     f(..)   // mutually recursive call to f
16     ...
17     g(..)   // recursive call to g
18     ...
19   }
```

For this template, the meaning of **decreases** is that when execution reaches respectively:

- line 5, v_f is strictly smaller than \at(v_f, Pre), which itself is non-negative
- line 7, v_g is strictly smaller than \at(v_f, Pre), which itself is non-negative
- line 15, v_f is strictly smaller than \at(v_g, Pre), which itself is non-negative
- line 17, v_g is strictly smaller than \at(v_g, Pre), which itself is non-negative

Exercise 20 decreases clause

Among the following program points, which are validating their corresponding **decreases** clauses?

```
1    int x;
2
3    /*@ decreases x;
4      @*/
5    void f(void) {
6      if (x >= 10) {
7        x -= 13;
8        f();    // program point P1
9      }
10     else {
11       if (x >= 0) {
12         x += 1;
13         f();    // program point P2
14       }
15       else {
16         if (x >= -5) {
17           x = -x-11;
18           f();    // program point P3
19         }
20       }
21     }
22   }
```

Additionally, propose another variant measure that would make the decreases clause valid, thus ensuring the universal termination of f.

1.5 Describing Memory

Since the beginning of this chapter, our programs do not involve pointers. However, in C, users make extensive use of pointers and mutation. C does not guarantee by typing that pointers can always be dereferenced by users, thus, before accessing to some memory location through a pointer, one has to be sure that the access is valid. Therefore, ACSL provides built-in predicates and functions to help describe memory shape or validity of pointers.

1.5.1 Structure of Memory Blocks

In C, an object represents a region of data storage in the execution environment. Each created variable or array is an object, as well as the memory pointed to by a pointer returned by an allocating function (like malloc). The correspondence between the type specifications given in the source code and the memory region depends on the size and alignment constraints of the system on which the program will be executed. For the remainder of this section, we will assume a fairly standard 64-bits architecture, i.e.:

- sizeof(short) == _Alignof(short) == 2,
- sizeof(unsigned) == _Alignof(unsigned) == 4, and
- sizeof(void*) == _Alignof(void*) == 8.

Let us now consider the following declaration:

```
struct S {
  char c ;
  int  i ;
} ;

struct S array[2];
```

which can be graphically represented as in Fig. 1.2.

A memory block is characterized by its base address and its length. The base address is the address of the declared object (for arrays, the address of the first element of the array) or the pointer returned by an allocating function. The functions **\base_addr** and **\block_length** are used to respectively get the base associated to a pointer and the length of the block. The **\offset** function returns the distance in bytes between a pointer and its base. These functions are illustrated in

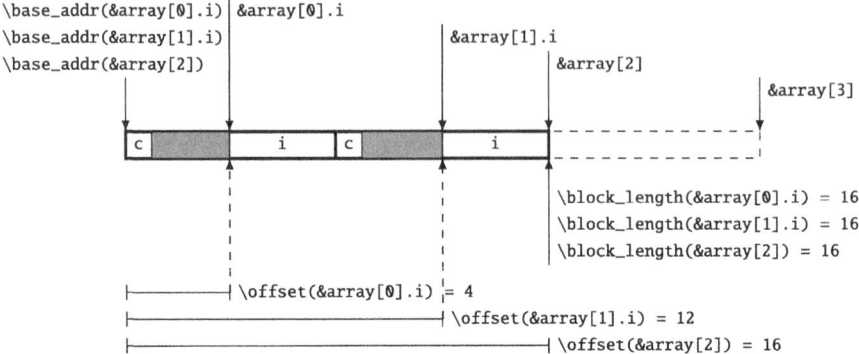

Fig. 1.2 ACSL memory blocks functions

Fig. 1.2. It is worth noticing that these functions refer to the block. Thus, the locations &array[0].i, &array[1].i and &array[2] share the same base, which is &array[0], and the same block length, which is 16, and their offset is computed with respect to the same base. Finally, it should be noted that, as mentioned in Sect. 1.3.4, these functions always return an integer, even if applied to a pointer that does not refer to an existing memory block, e.g., NULL or &array[3]. In that case, the returned value is unspecified.

Exercise 21 Memory-related predicates
In the following program:

```
struct X {
  int*      p ;
  short     s ;
  unsigned  u[2] ;
};
int main(void){
  struct X a[3] = {0};
  unsigned *p = a[1].u ;
  p ++ ;
}
```

what are the respective values of each of the following terms?

- \base_addr(&a[2].u[1])
- \offset(&a[2].u[1])
- \block_length(&a[2].u[1])
- \base_addr(p)
- \offset(p)
- \block_length(p)
- \base_addr(a[0].p)

- \offset(a[0].p)
- \block_length(a[0].p)

1.5.2 Validity of Memory Locations

The following program swaps two values pointed to by the parameters p and q.

```
1 /*@ assigns *p, *q ;
2   @ ensures *p == \old(*q) && *q == \old(*p);
3   @*/
4 void swap(int* p, int* q){
5   int tmp = *p ;
6   *p = *q ;
7   *q = tmp ;
8 }
```

This program conforms to its specification. However, it may not be safe. Indeed, since the program reads and writes values pointed to by p and q, these pointers have to be **\valid** memory locations. **\valid** expresses that a memory location can be safely read or written.

```
1 /*@ requires \valid(p) && \valid(q);
2   @ assigns *p, *q ;
3   @ ensures *p == \old(*q) && *q == \old(*p);
4   @*/
5 void swap(int* p, int* q);
```

Notice that the validity of a pointer is related to its type. Thus, here, it means that an **int** can be safely read or written. The type is also used when specifying a range of valid locations:

```
1 /*@ requires \valid(p + (0 .. n-1));
2   @ assigns *(p + (0 .. n-1)) ;
3   @ ensures \forall integer i ; 0 <= i < n ==> p[i] == 0 ;
4   @*/
5 void zero(int* p, unsigned n){
6   for(unsigned i = 0 ; i < n ; i++) p[i] = 0 ;
7 }
```

where **\valid**(p + (0 .. n-1)) refers to the set of **int** memory locations from p+0 to p+(n-1).

Taking **char** as a reference type, **\valid**(p) with p of type **char*** means that the byte at location p can be safely read or written. Ignoring any alignment constraint, we can derive the following property for any other type:

$$\forall p; \text{\valid}(p) \Leftrightarrow \text{\valid}((\text{char}*)p + (0..\text{sizeof}(*p) - 1))$$

A function might not need to write a memory location and that same memory location might be readable only. Thus, there exists a variant of **\valid** for this, which is **\valid_read**, that behave essentially the same except that the memory might not be writable. Note that a **\valid** memory location is always a **\valid_read** (that is, $\forall p.\valid(p) \Rightarrow \valid\_read(p)$). According to the semantics of C, C objects declared with the **const** qualifier are **\valid_read**, but not **\valid**.

Our swap function has another potential problem. Sometimes, while reading a memory location is authorized, the value might not be usable because it has not been initialized previously. Depending on the situation, it might cause *undefined* or *unspecified* behavior. While the last is not as bad as the former, it is generally not an intended behavior, thus ACSL does not distinguish these cases. One can specify that a memory location is initialized using the **\initialized** predicate.

```
/*@ requires \valid(p) && \valid(q);
  @ requires \initialized(p) && \initialized(q);
  @ assigns *p, *q ;
  @ ensures *p == \old(*q) && *q == \old(*p);
  @*/
void swap(int* p, int* q);
```

It is also possible in C to create entirely invalid pointers. Ideally, a pointer should always be NULL or point to some location in an object or one element past an object. In Fig. 1.2, &array[0].i, &array[1].i and &array[2] are valid pointer objects (even though &array[2] cannot be dereferenced), while &array[3] is entirely invalid. This idea is denoted by the predicate **\object_pointer**:

```
void example(void){
  int i = 0 ;
  //@ check \object_pointer(&i) ;
  //@ check \object_pointer(1+&i) ;
  //@ check ! \object_pointer(2+&i) ;
}
```

The predicate **\object_pointer** can be defined as follows:

$\forall p: \object\_pointer(p) \Leftrightarrow (\valid\_read(p) \vee \valid\_read(p-1) \vee p = \null)$

Exercise 22 Validity of memory-related formulas

Assuming 64 bits pointers, which of the following checks are valid?

```
struct X {
  int i ;
  int a[8] ;
  int *p ;
};

int main(void){
  struct X x, y = { 0 };
  struct X const z = { 0 };
```

1 Formally Expressing What a Program Should Do: The ACSL Language

```
10
11   //@ check A: \valid(&x) ;
12   //@ check B: \valid_read(&y) ;
13   //@ check C: \valid(&z.i) ;
14   //@ check D: \valid_read(y.p) ;
15
16   int* p = x.a ;
17   //@ check E: \valid_read(p+10) ;
18
19   //@ check F: \initialized(&x) ;
20   //@ check G: \initialized(&y) ;
21   //@ check H: \initialized(y.p) ;
22
23   //@ check I: \object_pointer(&x+1) ;
24   //@ check J: \object_pointer(p) ;
25   //@ check K: \object_pointer(x.p) ;
26   //@ check L: \object_pointer(z.p) ;
27 }
```

1.5.3 Specifying Memory Separation

It is often necessary to specify whether memory locations are separated or not. A simple example of this is the memcpy function that copies the content of a buffer to another. A simple implementation of this function is the following:

```
1 void* memcpy(void* dst, void const* src, size_t length){
2   for(size_t i = 0 ; i < length ; i++)
3     ((char*) dst)[i] = ((char*) src)[i] ;
4   return dst ;
5 }
```

This function copies as intended only if the buffer pointed by dst from 0 to length-1 is separated from the buffer pointed by src from 0 to length-1. Otherwise, the function could modify the original memory (and even fail to copy the original content). Thus, one has to specify memory separation, using the **\separated** predicate:

```
1 /*@ requires \valid((char*)dst + (0 .. length-1));
2   @ requires \valid_read((char*)src + (0 .. length-1));
3   @ requires \separated((char*)dst + (0 .. length-1),
4   @                    (char*)src + (0 .. length-1));
5   @ ...
6 */
7 void* memcpy(void* dst, void const* src, size_t length);
```

The **\separated**(l1, l2, ..., ln) predicate receives a list of memory locations. This expresses that all pairs of memory locations in the list are separated from each other. Just as the **\valid** predicate, the type of the parameter is used for determining the size of the memory location. Thus taking **char** as a reference type, **\separated(l1, l2)** with l1 and l2 of type **char*** states that l1 and l2 are distinct memory locations that is, either their base is different or their offset is different. From this, we derive for any other type:

$$\separated(l_1, ..., l_n) \Leftrightarrow$$
$$\forall i, j. i \neq j \Rightarrow$$
$$\separated(\text{char}*)l_i + (0..\text{sizeof}(*l_i) - 1), (\text{char}*)l_j + (0..\text{sizeof}(*l_j) - 1)$$

Exercise 23 Memory copy
The following function mimics the C++ standard library `std::copy` function:

```
void copy(const int * a, int* b, unsigned n){
  for (unsigned i = 0; i < n; ++i) {
    b[i] = a[i];
  }
}
```

it copies the content of the buffer a into the buffer b. Aliasing between the buffers a and b can be allowed (with a simple additional condition). Write a **requires** clause that guarantees that we can read the content of a (and get sensible values) to b, guaranteeing that it will indeed copy the content correctly (even if a content may be modified).

1.6 Ghost Code

Ghost code is basically C code that can observe actual code but cannot interfere with its original behavior. This code is added in annotations, thus only Frama-C can use it, the compiler ignores it like any other program comment. Therefore, when a program contains ghost code, it must not change the visible behavior of the program. In short, executing the original program or executing the program together with the ghost code should have the same behavior for any program input. As a consequence, ghost code can read real memory, but it cannot write it, it must always terminate, and it cannot change the control flow of the original program. Conversely, real code cannot refer to ghost code.

In ACSL, ghost code is provided in a code annotation block starting with the keyword **ghost**:

```
/*@ ghost
  // ...
*/
```

Ghost code is often used in verification to help program proofs, it can for example be used to record information about the memory context at some program point to reuse the information later in the proof, or to split the proof in subcases. Chapter 4 on Wp will elaborate on how to use ghost code for program proofs. For now, let us present what is ghost code, how to write ghost code in ACSL and what are its specificities in the context of the C language.

1.6.1 Ghost Memory

We can declare ghost variables at both global and local scope (including ghost formal parameters). Ghost code can read or write these variables and read real memory. It is possible to create ghost functions as well. The following code fragment shows an example of ghost code use:

```
/*@ ghost int g_function(int x){
  @    return x + 42;
  @ }
  @*/

//@ ghost int glob ;

void function(int x) /*@ ghost (int parameter) */ {
  //@ ghost int save_x = x ;
  x ++ ;
  //@ assert save_x == x - 1 ;
  //@ ghost glob = g_function(save_x) ;

  // int v = save_x ;
  //     <-- rejected: real code cannot refer to ghost
  // @ ghost x ++ ;
  //     <-- rejected: ghost code cannot write actual memory
  // g_function(x) ;
  //     <-- rejected: real code cannot call ghost function
}

void caller(void){
  function(3) /*@ ghost(4) */ ;
}
```

In C, aliasing makes ghost memory violation hard to detect at compile time. For example, the following code would change the behavior of the function:

```
void print_pointed(int* ptr){
  /*@ ghost
    @ int* q = ptr ;
    @ *q = 42;
    @*/
  printf("%d\n", *ptr); // Oops
}
```

ACSL provides a type qualifier \ghost to help detect such cases. In Frama-C, it is mandatory to have this qualifier when some pointed memory is ghost, and Frama-C rejects code when the types do not match. For example, the previous example is in fact rejected by Frama-C because on line 4, we write to a memory location that is not qualified \ghost. Changing the code to:

```
void print_pointed(int* ptr){
  /*@ ghost
    @ int \ghost * q = ptr ;
    @ // ^ error: assignment discards \ghost qualifier
    @ *q = 42;
    @*/
  printf("%d\n", *ptr);
}
```

Frama-C still rejects the program but now because we try to assign a pointer to ghost memory with the value of a pointer to normal memory. Variables declared in ghost code automatically have a ghost status, but it must be explicitly provided for pointed memory:

```
// x is in ghost memory
//@ ghost int x ;
// a is in ghost memory (thus its elements are ghost)
//@ ghost int a[10] ;
// p is a ghost pointer to non-ghost memory
//@ ghost int * p ;
// p is a ghost pointer to ghost memory
//@ ghost int \ghost * gp ;
```

Of course, since real memory cannot see ghost memory, the following declaration is ill-formed and thus rejected:

```
// a ghost pointer to normal pointers to ghost memory ?
//@ ghost int \ghost * * p ;
```

1.6.2 Interaction with Control Flow

When an **if** statement does not have an **else** part, one can add a **ghost else**, for example:

```
//@ ghost int in_if = 1 ;
if(c){
  r = 1 ;
} /*@ ghost else {
  in_if = 0 ;
} */
```

However, ghost code cannot break the existing control flow. For example, in the following program fragment:

1 Formally Expressing What a Program Should Do: The ACSL Language

```
1  int function(int x){
2    if(x){
3      return 0 ;
4    } /*@ ghost else {
5      return 1 ;
6    } */
7    return 42;
8  }
```

instead of returning 42 when x is 0, the added **ghost else** returns 1. This code is thus rejected by Frama-C, like any ghost code that changes the control flow.

For this, Frama-C syntactically checks that each non-ghost instruction has the same predecessors and successors with or without ghost code. Note that it means that ghost code that cannot semantically break the control-flow (because it is unreachable for example) can still be rejected by Frama-C.

```
1  int function(int x){
2    if(x){
3      return 0 ;
4    } /*@ ghost else {
5      if(x) return 1 ; // obviously unreachable but rejected
6    } */
7    return 42;
8  }
```

Another—non-trivial—way to break control flow is to introduce some ghost code that does not terminate. Such a code of course breaks the semantics of the program, however, except for trivial non-termination, one cannot syntactically detect such a problem. Thus, it is up to the user to ensure that all loops and function calls involved in ghost code terminate, for example using loop variants and proving termination using Wp.

1.6.3 Ghost Code Annotations

Ghost code can be enriched with ghost code annotations just as normal code. These annotations start with /@ and end with @/, then any keyword available in normal annotations (except **ghost**) are available. This is in particular useful to give contracts to ghost functions, or provide loop invariants, in order to verify properties about the ghost code itself. For instance, if we define a ghost `threshold` function similar to the one presented in Exercise 4, we can give it a contract like this:

```
1  /*@ ghost
2    /@ ensures ResultBound: min <= \result <= max;
3    @ ensures NoThresholdCase: min <= n <= max ==> \result == n;
4    @/
5    int threshold(int min, int max, int n) {
6      if (n < min) { return min; }
7      if (n > max) { return max; }
8      return n;
9    }
10 */
```

1.7 Global Annotations

In the same way as we define functions in C in order to be able to reuse code fragments across the whole application, it is possible to define in ACSL logic functions and predicates, that can be used in C function contracts and code annotations. These definitions occur in *global annotations* that are treated as a C *external-declaration*. In other words, they cannot be written inside a C function definition or between a contract and the corresponding declaration or definition.

1.7.1 Defining Predicates and Functions

The simplest way for defining a logic function or a predicate is to give a direct definition. Let us first see an example of a logic function definition.

```
1 #include <limits.h>
2
3 /*@ logic integer double_acsl(integer x) = 2 * x; */
4
5 /*@ requires INT_MIN / 2 <= x <= INT_MAX / 2;
6     ensures \result == double_acsl(x);
7 */
8 int double_c(int x) { return 2 * x; }
```

In this example, we define on line 3 a logic function double_acsl, which simply returns the double value of its argument. We then use it on line 6 in the post-condition of C function double_c, to indicate that the \result of double_c should indeed be the same as what double_acsl computes (provided that the actual parameter respects the pre-condition on line 5, which guarantees that there is no overflow in the C part).

On using the integer type

It would be perfectly possible to define double_acsl to take as parameter and/or return a standard C **int** type, but recall that each arithmetic operation returns an (unbounded) **integer** as explained in Sect. 1.3.2.2, which would need to be explicitly converted to an **int**. Hence, we would write:

```
1 /*@ int double_acsl(int x) = (int) (2 * x);
2     @*/
```

Moreover, calling double_acsl over an integral constant would also require such a conversion, as in double_acsl((int)2) == 4 (the result of the call is implicitly converted to an **integer** to do the comparison with 4). Finally, although the conversion has a completely defined semantics in ACSL (in case the **integer** does not fit, we take the unique **int** that is equal to it modulo

$2^{\texttt{sizeof(int)}}$), this can lead to counter-intuitive facts. For instance, here, the following equality would hold:

```
1  double_acsl((int)INT_MAX) == -2
```

Arguably, this is not something you would expect from a function called `double_acsl`.

For these reasons, it is usually better to use unbounded **integer** in all logic definitions, and resort to C integral types only when dealing with C symbols in function contracts or code annotations.

The code sample below presents the definition of an ACSL predicate, `positive`, and its use in the post-condition of the C function `filter` to state that the result of `filter` is non-negative.

```
1  /*@ predicate positive(integer x) = x >= 0; */
2
3  /*@ ensures positive(\result); */
4  int filter(int x) {
5    return x >= 0 ? x : 0;
6  }
```

Boolean functions and predicates
ACSL has a **boolean** type, and, in practice, there is little difference between a function returning a Boolean and a predicate. In fact, we can perfectly rewrite the previous example with a Boolean function.

```
1  /*@ logic boolean positive(integer x) = x >= 0; */
2
3  /*@ ensures positive(\result); */
4  int filter(int x) {
5    return x >= 0 ? x : 0;
6  }
```

Technically, Frama-C normalizes the post-condition the following way:

```
1  /*@ ensures positive(\result) == \true; */
2  int filter(int x);
```

but the end result is strictly equivalent. The main difference between predicates and Boolean logic functions lies in the fact that predicates can be defined inductively, as will be shown in Sect. 1.7.3, which is not the case for logic functions (whether Boolean or not).

Logic functions can be defined recursively, as shown by the example below, defining factorial:

```
/*@ logic integer fact(integer n) =
        n <= 0 ? 1 : n * fact(n-1);
*/
```

Predicate definitions can also be recursive. A recursive definition for a predicate is_fact(n, f) that holds if and only if f is the factorial of n is the following:

```
/*@ predicate is_fact(integer n, integer f) =
        (n <= 0 && f == 1) ||
        (n > 0 && f % n == 0 && is_fact(n-1,f/n));
*/
```

Notice that *inductive definitions*, as introduced later in Sect. 1.7.3, provide an alternative way to recursively define predicates, sometimes more expressive because it allows a form of non-determinism.

> **! Recursive definition and termination**
> **Frama-C** does not check that the recursive calls always terminate. It is thus extremely important to carefully review recursive definitions to ensure that this is indeed the case. Indeed, having non-terminating definitions can lead to inconsistencies. For instance, after the following (ill-formed) definition of p:
>
> ```
> /*@
> predicate p(integer x) = !p(x);
> */
> ```
>
> it is possible to let Wp (see Chap. 4) prove **\false** through a sequence of lemmas, as shown below:
>
> ```
> /*@
> lemma wrong: !p(0);
> 
> lemma wrong2: p(0);
> 
> lemma wrong3: \false;
> */
> ```

Finally, it is also possible in a global **ACSL** annotation to state *lemmas*, i.e., properties that must be proved true. Note however that in practice such lemmas only concern the deductive verification plugin Wp, and the other plug-ins will usually neither be able to check them nor to make use of them. As an example, we can state that our two definitions of factorial are equivalent:

```
/*@ lemma fact_is_fact:
        \forall integer n, f;
            is_fact(n, f) <==> f == fact(n);
*/
```

Exercise 24 Array queries

1. (Searching elements) Given an array a of **char** and two indices low and high, write a logic function search that returns the smallest index i between low (included) and high (excluded) where a[i] == 0. If a does not contain 0 within these bounds, return high.
   Hint: remember that an ACSL array is infinite: a[i] exists for any i without restriction.
2. (Positive elements) Given an array a of **char** and two indices low and high, write predicates positive1 and positive2 that hold if and only if all elements of a between indices low (included) and high (excluded) are (strictly) positive:

   - positive1 is recursive and does not use quantifiers,
   - positive2 is not recursive, and contains a universal quantification.

3. (Lemmas about positive and search) Write a lemma that states that your two predicates positive1 and positive2 are equivalent, and another one that states that if one of them holds, then search, applied to the same arguments, will not find an element of a equal to 0.

1.7.2 Referring to Memory States

By definition, when we are defining a logic function or a predicate, we are not at a precise point during the program execution. Therefore, in principle, we cannot refer to any memory location. Since this situation would make logic functions and predicates more or less useless, ACSL allows them to take one or more *logic label(s)* as parameter. It is then possible to use the \at construction (see Sect. 1.4.2) to evaluate a term at a particular program point.

Let us start with the most common usage, namely using a single label. A classical example consists of defining a predicate that states that a given memory block is valid and contains **int**s sorted in increasing order:

```
/*@ predicate sorted_expanded{L}(int* a, integer length) =
     \valid_read{L}(a + (0 .. length - 1)) &&
     \forall integer i; 0 <= i < length - 1 ==>
       \at(a[i],L) <= \at(a[i+1],L);
*/
```

As can be seen above, the logic label L is introduced by {L}. It is then used in the body of our predicate sorted_expanded for every term that can only be evaluated at a given program point. This is in particular the case for the built-in predicate \valid_read, which gets passed {L} as a specific parameter. Similarly, we use \at to state explicitly that we intend to evaluate a[i] and a[i+1] at the program point denoted by L.

While the formal semantics of ACSL requires such logic label parameters, writing them explicitly can quickly become cumbersome. Hence, it is possible to omit them and have them inferred implicitly. For instance, the sorted predicate defined below is exactly the same as sorted_expanded, but all references to the program point where it is evaluated are kept implicit:

```
1  /*@ predicate sorted(int* a, integer length) =
2       \valid_read(a + (0 .. length - 1)) &&
3       \forall integer i; 0 <= i < length - 1 ==> a[i] <= a[i+1];
4  */
```

Similarly, a logic function or predicate that takes a logic label as parameter can of course be used in a contract or in code annotation. In that case, if it is intended to be evaluated at the current program point of the annotation, we can pass the label **Here**, mentioned in Sect. 1.4.2. However, since this is the most frequent situation, **Here** can stay implicit.

```
1  void test(void) {
2      int a[] = {0, 1};
3      //@ assert init_a: sorted(&a[0],2);
4  L:
5      a[0] = 2;
6      //@ assert final_a: sorted{L}(&a[0],2) &&
7          !sorted{Here}(&a[0],2);
8  }
```

The small example above shows this distinction. In the assertion init_a, we simply use sorted, indicating implicitly that we want to evaluate the predicate at the current point. On the other hand, in the assertion final_a, we pass explicitly labels to sorted, first to state that a was sorted when reaching the C label L, and second to state that a is not sorted anymore at **Here**, the current point where the assertion is evaluated.

It is also possible to define logic functions or predicates that deal with several memory states. For that, they need to declare multiple logic labels, e.g., {L1,L2}. In such a definition, it is of course not possible to have an implicit instantiation of the logic labels, since the ACSL typechecker has no way to decide which of the logic label parameters should be used at a given place. A classical example of such a multi-state predicate is the swap predicate below, which indicates that the content of pointers a and b has been swapped between L1 and L2.

```
1  /*@ predicate swapped{L1,L2}(int* a, int* b) =
2       \at(*a,L1) == \at(*b,L2) &&
3       \at(*a,L2) == \at(*b,L1);
4  */
```

It can then be used for instance in the contract of a C function swap which performs such an exchange. The pre-conditions require that the pointers are valid and do not

1 Formally Expressing What a Program Should Do: The ACSL Language   63

overlap. The post-condition then ensures that the `swapped` predicate holds between the **Pre** state (the state just before entering the function's body) and the **Here** state, which, for a post-condition, is the state just before returning to the caller.

```
/*@ requires validity: \valid(a) && \valid(b);
    requires separation: \separated(a,b);
    assigns *a,*b;
    ensures swapped{Pre,Here}(a,b);
*/
void swap(int *a, int *b) {
  int tmp = *a;
  *a = *b;
  *b = tmp;
}
```

Exercise 25 Array Update

1. (Two-states Predicate) Write a predicate `array_update` that takes as arguments the address a of a memory block containing **int**s, its number of elements, the index i of an element, and an **int** value v, as well as two logic labels, and holds if and only if all elements of a are the same in the two memory states, except potentially for the one at i, which is equal to v in the second state.
2. (Specifying Array Assignment) Given the `array_update` as defined above, Replace L1, L2, A1, A2, A3, A4 with appropriate arguments so that the assertion holds:

```
void f() {
  int a[2] = { 0, 1 };
L:
  a[0] = 2;
  //@ assert array_update{L1,L2}(A1,A2,A3,A4);
}
```

1.7.3 Inductive Predicates

As mentioned in Sect. 1.7.1, predicates can also be defined *inductively*. Such a definition takes the following form:

```
inductive P(parameters) {
  case case_1:
    \forall x_1, ..., x_n;
      Cond_1(x_1, ..., x_n) ==> P(args_1);
  case case_2:
```

```
     \forall y_1, ..., y_m;
       Cond_2(y_1, ..., y_m) ==> P(args_2);
  ...
}
```

where each **case** describes one of the conditions where P holds. The semantics of such a definition is that P(args) holds if *and only if* it can be deduced by successive applications of the different **case**s defining P. Another way to express this semantics is that P is the smallest (in the sense: the one that is false the most often) predicate satisfying each clause of the definition. In the general case, such a smallest predicate may not exist, hence introducing such a definition would produce an inconsistent logic context. In order to avoid inconsistencies, the conditions Cond_1, Cond_2, ..., must only contain *positive* occurrences of P itself, i.e., P cannot occur under a negation.

Inductive definition of factorial
It is perfectly possible to give an inductive definition to the factorial predicate defined in Sect. 1.7.1, as is done below.

```
/*@ inductive ind_fact(integer n, integer f) {
  case zero: \forall integer n; n <= 0 ==> ind_fact(n,1);
  case succ: \forall integer n, f; n > 0 && ind_fact(n-1,f) ==>
    ind_fact(n,f*n);
}
*/
```

The zero case tells us the ind_fact(0,1) holds. From that and the succ case, we can first derive ind_fact(1,1), then ind_fact(2,2), and so on.

In addition, inductive predicates, and the cases defining them, can of course refer to specific memory states. A classical example of such an inductive definition amounts to stating that the content of an array a at a given program point is a permutation of the content of the same array at another program point. For that, we start by refining our swap predicate from Sect. 1.7.2 into swap_array, which states that we have swapped the two elements of an array a at indices i and j, leaving all other elements untouched.

```
/*@ predicate swap_array{L1,L2}
    (int* a, integer i, integer j, integer l) =
    0 <= i < j < l &&
    (\forall integer k; 0 <= k < l && k != i && k != j ==>
      \at(a[k],L1) == \at(a[k],L2)) &&
    swapped{L1,L2}(a+i,a+j);
*/
```

On top of this definition, we then define a permut predicate as follows.

```
/*@ inductive permut{L1,L2}(int* a, integer length) {
  case Refl{L}: \forall int* a, integer l; permut{L,L}(a,l);

  case Swap{L1,L2}:
```

1 Formally Expressing What a Program Should Do: The ACSL Language

```
      \forall int* a, integer i, j, l;
        swap_array{L1,L2}(a,i,j,l) ==> permut{L1,L2}(a,l);

    case Trans{L1,L2,L3}:
      \forall int* a, integer l;
        permut{L1,L2}(a,l) && permut{L2,L3}(a,l) ==>
        permut{L1,L3}(a,l);
    }
    */
```

The predicate `permut` takes two logic labels as arguments, as well as a pointer a to the start of the array and the array's `length`. The first case, `Refl`, takes only one label as argument and states that if the two states in which we compare the content of a are in fact the same, then `permut(a,l)` holds (the permutation being the identity). The second case, `Swap`, uses our `swap_array` predicate to tell that if only two elements of the array have been swapped between L1 and L2, then we indeed have a permutation. Finally, we take the transitive closure of these transformations with the `Trans` case, which states that if there's a permutation of a between L1 and L2 and another one between L2 and L3, then `permut{L1,L3}(a,l)` holds.

Exercise 26 Sum of elements
Define an inductive predicate sum such that sum{L}(a,i,j,s) holds if and only if s is the sum of all elements of a between indices i (included) and j (excluded).

1.7.4 Axiomatics

Finally, it is possible to define logic functions and predicates axiomatically, by just providing a prototype and a set of *axioms*, glued together inside an **axiomatic** block. Going back to our factorial example, it can be axiomatized as such:

```
/*@ axiomatic Ax_fact {

  logic integer ax_fact(integer n);

  axiom ax_fact_0: \forall integer n; n <= 0 ==> ax_fact(n) == 1;

  axiom ax_fact_S:
    \forall integer n; n > 0 ==> ax_fact(n) == n * ax_fact(n-1);

  }
  */
```

The semantics of such an axiomatic block is that it introduces in the logic the given logic functions and predicates, with *any possible interpretation that satisfies the axioms*. The axioms can be arbitrary formulas, offering thus large freedom on the way axioms are written (unlike, say, the clauses of inductive definitions, which must respect positivity constraints). There is a very important caveat though, such an axiomatic could be *inconsistent*, a trap that is discussed in details in Sect. 1.7.5 below.

Axioms can of course take as parameters logic labels in case the predicate or logic function they define reads some memory locations. As an example, we can rewrite the `permut` predicate seen above as an axiomatic definition as follows:

```
/*@ axiomatic Ax_permut {
  predicate ax_permut{L1,L2}(int* a, integer length)
  reads \at(a[0 .. (length - 1)],L1),
        \at(a[0 .. (length - 1)],L2);

  axiom ax_refl{L}: \forall int* a, integer l; ax_permut{L,L}(a,l);

  axiom ax_swap{L1,L2}:
    \forall int* a, integer i, j, l;
      swap_array{L1,L2}(a,i,j,l) ==> ax_permut{L1,L2}(a,l);

  axiom ax_trans{L1,L2,L3}:
    \forall int* a, integer l;
      ax_permut{L1,L2}(a,l) && ax_permut{L2,L3}(a,l) ==>
      ax_permut{L1,L3}(a,l);
}
*/
```

The axioms themselves are very similar to the cases of the inductive definition. The declaration of the predicate is slightly different, though. Namely, the **reads** clause in the declaration of the `ax_permut` predicate gives the set of memory locations that may have some influence on the truth value of the predicate. Here, we need to read all elements of a, at both L1 and L2 states.

1.7.5 The Danger of Inconsistent Axiomatizations

The fact that the axioms stated in an axiomatization are not constrained at all opens an important source of possible mistakes. The stated axioms may of course not well formalize the intended meaning of the logic function considered, and as such must be reviewed carefully as any other ACSL annotation. But the additional risk is that a set of axioms may be *inconsistent*, meaning that there exist no interpretation of the logic function that satisfies the axioms. An example could be seen on the factorial axiomatization above, if we imagine a small mistake in writing:

```
axiom ax_fact_S :
  \forall integer n; n >= 0 ==> ax_fact(n) == n * ax_fact(n-1);
```

that is writing n >= 0 instead of n > 0 in the premise. In that case, using the axiom with 0 for the value of n implies

```
ax_fact(0) == 0 * ax_fact(0-1) == 0
```

1 Formally Expressing What a Program Should Do: The ACSL Language   67

which contradicts the first axiom. It should be clear that a direct definition of factorial like the one from Sect. 1.7.1 is less error-prone and thus preferable. A rule of thumb is that **one should never specify a logic function with an axiomatic block if there is another way of doing it**. If an axiomatic block is the only possible way to define the wanted logic symbols, it is recommended, in addition to a very careful manual review, to use tools that can detect inconsistencies, such as the so-called *smoke tests* of the Wp plugin (See Chap. 4).

Among the multiple traps associated to axiomatic blocks, there is a quite specific one related to **reads** clauses for logic functions involving memory. Not declaring all the memory locations upon which a logic function depends can introduce unexpected inconsistencies, as shown by the toy example below in which we try to define a predicate specifying that the element pointed to by a pointer is zero.

```
/*@ axiomatic ReadsTrap {
  @
  @   predicate points_to_zero(int *p);
  @
  @   axiom A{L}: \forall int *p;
  @      points_to_zero(p) <==> (\spsat(*p,L) == 0);
  @ }
  @*/
```

Since the predicate is declared without any **reads** clause, it is supposed to depend only on the value of p but not on the value pointed to by p. The inconsistency can be exposed in a program like

```
/*@ requires \valid(q);
  @*/
void f(int *q) {
   *q = 0;
   //@ assert points_to_zero(q);
   *q = 1;
   //@ assert !points_to_zero(q);
}
```

where the two assertions are valid consequences of the axiom A, and yet since the predicate only depends on q which is not changed, both points_to_zero(q) and its negation are valid. A consistent axiomatic block for such a predicate would declare the predicate as

```
predicate points_to_zero{L}(int *p) reads \at(*p,L);
```

Of course, as our rule of thumb above suggests, it would have been less error-prone to directly define the predicate as

```
predicate points_to_zero{L}(int *p) = (\at(*p,L) == 0);
```

where there is no way to forget to give a label as parameter to the predicate: otherwise the label L in \at(*p,L) would be unbound and the Frama-C kernel would immediately signal an error.

1.8 Conclusion

To summarize, ACSL is a feature-rich formal specification language, primarily aimed at expressing function contracts and assertions using the full power of first-order logic. It also provides the auxiliary annotations that are needed for using deductive verification techniques (see Chap. 4), notably loop invariants and loop assigns. Moreover, the use of ghost code can play an important role in building an abstract model of the original program. Finally, the definition of logic functions and predicates allows developers to structure their specifications by naming the most important concepts of their systems.

This chapter is only a (not so) brief introduction to ACSL, and it only aims to introduce the most important concepts of the language, notably the ones that are needed to understand the rest of this book. For a more in-depth presentation of ACSL, we refer the interested reader to two tutorials on both ACSL and Wp [2, 16].

Finally, it should be noted that while ACSL forms the *lingua franca* understood by all Frama-C plugins, it does not mean that a user has to express all the annotations they want to verify about their C code in the form of ACSL annotations. First, the Rte plug-in[8] can automatically generate all assertions that need to be verified to ensure that the code is safe. Moreover, a growing set of plug-ins is dedicated to generating ACSL annotations from domain-specific languages in order to let users express more easily certain kinds of properties for which manually writing traditional function contracts and assertions would be extremely tedious and error-prone. Chapter 10 describes several of them, notably for expressing temporal logic properties, relational properties, or security (integrity or confidentiality) properties.

Acknowledgements We gratefully thank the anonymous reviewers of the preliminary versions of this chapter, whose remarks greatly helped to improve its contents.

1.9 Answers to Exercises

To start with, let us give the answers to the introductory example of Fig. 1.1: the first, the second and the fifth post-conditions are invalid, whereas the third and fourth post-conditions are valid. The explanations can be found in this chapter, in particular in Sect. 1.2.3.2 for the first three post-conditions, and Sect. 1.3.1 for the last two. Some may argue that none of the post-conditions can be valid since the execution of the function may exhibit an integer overflow (that is, when x is the maximum value of the **int** type. But in fact this is not the case: as seen in Sect. 1.2.3.3, it is possible that a post-condition is valid even in presence of undefined behaviors.

---

[8] https://frama-c.com/fc-plugins/rte.html.

Answer to Exercise 1
There are of course several possible solutions. Here is one of them

```
/*@ requires x < y;
  @ ensures ResultInBetween: x <= \result && \result < y;
  @ ensures ResultCloseToMiddle:
  @     y - \result == \result - x ||
  @     y - \result == \result - x + 1;
  @*/
int in_middle(int x, int y) { ... }
```

The second post-condition implies that the result must be $(x + y)/2$, even it does not explicitly use the division operator.

Notice that another solution like

```
    ensures ResultUnique: \result == x;
```

would be a perfectly valid answer, since we never informally require, in this exercise, that the result should be the middle value!

Answer to Exercise 2
A first rule can be given to state that evaluation of a conditional statement must first evaluate the condition itself to a value. A possible rule is:

$$\frac{(C, e) \to (C', e')}{(C, \texttt{if}\,(e)\,i_1\,\texttt{else}\,i_2) \to (C', \texttt{if}\,(e')\,i_1\,\texttt{else}\,i_2)}$$

Two additional rules state which branch of the conditional must be taken. It is worth remembering here that in C, an expression is considered as false when it is zero and true otherwise. So the rule when the condition is true is:

$$\frac{v \neq 0}{(C, \texttt{if}\,(v)\,i_1\,\texttt{else}\,i_2) \to (C, i_1)}$$

and the rule when the condition is false is:

$$\overline{(C, \texttt{if}\,(0)\,i_1\,\texttt{else}\,i_2) \to (C, i_2)}$$

The evaluation of a **while** loop can be elegantly given by a unique rule by remarking an equivalence with a conditional, as follows.

$$\overline{(C, \texttt{while}\,(e)\,i) \to (C, \texttt{if}\,(e)\,\{\,i;\,\texttt{while}\,(e)\,i\,\})}$$

Answer to Exercise 3

The declaration of a local variable of type **int** has the form int $x = e$;. The execution must proceed by evaluating e first, expressed by a rule like

$$\frac{(\Sigma_1, \Pi_1, e_1) \rightarrow (\Sigma_2, \Pi_2, e_2)}{(\Sigma_1, \Pi_1, (\text{int } x = e_1;)) \rightarrow (\Sigma_2, \Pi_2, (\text{int } x = e_2;))}$$

and then a second rule can be given when the initializer is reduced to a value v as follows

$$\overline{(\Sigma, \pi \cdot \Pi, (\text{int } x = v;)) \rightarrow (\Sigma, (\{x = v\} \cup \pi) \cdot \Pi, ())}$$

meaning that the variable x is added in the local context with the value v.

Answer to Exercise 4

1. `NoThresholdCase` is valid, but `ResultBound` is not. For example when n=0, min=1 and max=0, the result would be 1 (first test is true) but 1 is larger than 0. A suitable pre-condition is

```
/*@ requires WellFormedThreshold: min <= max;
```

2. When n is not in the interval min..max initially, then any result would validate the post-conditions. Suitable extra post-conditions are

```
/*@ ensures LowerThreshold: n < min ==> \result == min;
  @ ensures HigherThreshold: n > max ==> \result == max;
```

Answer to Exercise 5

`Post1` holds, `Post2` does not necessarily hold since y in a post-condition is the same as in the pre-condition, `Post3` holds.

1   Formally Expressing What a Program Should Do: The ACSL Language

Answer to Exercise 6
The first, second, and fourth **assigns** clause are invalid because the execution of f for example in a context where all variables are zero would result in a context where y is one, meaning that it is wrong to forget y from the **assigns** clause. The two other clauses are valid. Notice that the code of f may overflow, but this does not affect the conformity of the contract (Definition 1.4).

Answer to Exercise 7
The function f is correct: for any value of x satisfying the **requires** clause, it executes safely (without integer overflow) and returns a value that, under the condition that the **assumes** clause holds, satisfies the **ensures** clause. The function g is also correct, since the value 7 satisfies the **requires** clause of f, and it returns 17 which ensures its own post-condition.

If one exchanges the formulas in the **requires** and **assumes** clause, then the function f is not correct anymore because its execution may expose an integer overflow. Yet, it still conforms to its contract since its post-condition is required to hold when both the **requires** and the **assumes** clauses hold. The function g is not safe anymore since its execution violates the pre-condition of f. Yet, it conforms to its contract anyway, since none of its executions is safe.

Answer to Exercise 8
Two possible alternatives:

```
    \forall size_t i; a <= i < b-1 ==> t[i] < t[i+1]
```
and
```
    \forall size_t i,j; a <= i < j < b ==> t[i] < t[j]
```

The second is indeed a logical consequence of the first, though proving this fact would need reasoning by induction. The second alternative is somehow more precise since it gives more information, and it is usually a better form in practice.

Answer to Exercise 9
With **integer**: all three propositions hold.
With **int**: the first proposition holds, because the multiplication is performed as a mathematical one, without overflow, so the result is indeed always non-negative. The second proposition, however, does not hold. In fact, INT_MIN usually has no opposite in the range of machine **int**s. Indeed, on the overwhelming majority of machines currently in existence (i.e., using two's complement and no trap representation), -INT_MIN is strictly speaking an undefined behavior as far as the C standard is concerned. It may or may not evaluate to INT_MIN itself, but in any case there is no value for y in the range of machine **int**s that would give INT_MIN+y == 0 for an addition in mathematical integers. The third formula holds with type **int** because even when x is INT_MIN, the term -x is evaluated in mathematical integers, giving a non-negative value.

Answer to Exercise 10
It is not safe since addition may overflow, which is an undefined behavior. But it conforms to its contract since any safe execution will return a result that is equal to the mathematical addition. When replacing **int** by **unsigned int**, the situation is reversed. It is safe since addition on unsigned machine integers has a well-specified behavior, that is it wraps around when result is larger than UINT_MAX. But it does not conform to its contract because the post-condition tells the result is equal to the mathematical addition which is wrong in case of overflow.

Answer to Exercise 11
Assertion R pretends there is no real number r whose square is 3, it is wrong, $r = \sqrt{3}$ invalidates that assertion. That number is not exactly representable in type **double** or **float** (only rational numbers are representable), hence assertions D and F are valid, since the square is computed using the exact real arithmetic. Assertions DD and FF are trickier, because it could be possible that there is a representable number that, when multiplied by itself and rounded, gives 3. This is indeed the case in the single format: The representable number 0x1.bb67aep+0 squared and rounded[9] gives 3, but it is not the case for the double format: the number 0x1.bb67ae8584caap+0 multiplied by itself gives 0x1.7fffffffffffffp+1, and its successor 0x1.bb67ae8584cabp+0 gives 0x1.8000000000001p+1. Hence, the assertion DD is valid but assertion FF is not.

---

[9] Using the default rounding mode nearest-ties-to-even.

1 Formally Expressing What a Program Should Do: The ACSL Language

Answer to Exercise 12

$$\frac{\Sigma, \Pi \models t \Rightarrow v \quad \Sigma, \{x = v\} \cdot \Pi \models \varphi}{\Sigma, \Pi \models (\texttt{\textbackslash let}\ x\ =\ t; \varphi)}$$

$$\frac{\Sigma, \Pi \models t \Rightarrow \mathit{true} \quad \Sigma, \Pi \models \varphi_1}{\Sigma, \Pi \models ((t)?\varphi_1 : \varphi_2)}$$

$$\frac{\Sigma, \Pi \models t \Rightarrow \mathit{false} \quad \Sigma, \Pi \models \varphi_2}{\Sigma, \Pi \models ((t)?\varphi_1 : \varphi_2)}$$

Answer to Exercise 13
The first proposition is not valid, the two others are.

Answer to Exercise 14
This program is unsafe because it may access to y without initializing it. The assertion A1 makes uses of the underspecified value of y, but whatever that value is, the formula y == y is valid. The assertion A2 involves a specified value for y, yet it is not valid since x may be equal to 1. The assertion A3 is not valid either, since y is underspecified and may be equal to x. Assertion A4 takes place after an undefined behavior of the code, so it is indeed valid by definition. The post-condition E1 is valid, since even if the value of **\result** is underspecified, it must be equal to itself. The post-condition E2 is naturally valid. The third post-condition E3 is valid too, since all executions of f with x <= 0 expose an undefined behavior. And indeed, the last post-condition E4 is valid! Any safe execution of f ends in a program state where x is positive!

Answer to Exercise 15
The safety of this code depends on some architecture-dependent behavior. If the compilation and the underlying execution architecture is assumed to be strictly compliant to the IEEE-754 norm for floating-point arithmetic, then this code is indeed safe, since floating-point division always returns a well-defined value, such as infinite values when the divisor is zero, or even NaN when dividing zero by zero. The first post-condition holds, as it is an instance of reflexivity of equality. The second post-condition does not necessarily hold: the predicate

\eq_float denotes the IEEE-compliant equality, whose value is in fact false whenever one of the compared term is NaN, even when comparing to itself. Since the \result can be NaN, the post-condition is not valid.

Answer to Exercise 16

The values are 3, 2, 6, 4 and 8.

Answer to Exercise 17

```
void f(void) {
  int x = 0;
  //@ loop invariant 0 <= x <= 9;
  while (++x < 10);

  x = 0;
  //@ loop invariant 0 <= x <= 10;
  while (x++ < 10);

  x = 0;
  //@ loop invariant 0 <= x <= 9;
  do {} while (++x < 10);

  x = 0;
  //@ loop invariant 0 <= x <= 10;
  do {} while (x++ < 10);

  //@ loop invariant 0 <= x <= 9;
  for(x=0; ++x < 10; );

  //@ loop invariant 0 <= x <= 10;
  for(x=0; x++ < 10; );
}
```

B1 can be 9 since invariant is checked before the side effect of the condition. Then when x is 9, the invariant is checked, x gets incremented, the condition evaluates to false, the loop exits, and the value of x is indeed 10 afterwards. The second loop is similar, except that it exits only when x is 10 before the condition, and thus 11 after the loop. B3 can be 9 since invariant is not required to hold when the loop exits.

Answer to Exercise 18

```
void f(void) {
  int x = 0;
  //@ loop invariant 0 <= x <= 9;
  while (1) { if (++x >= 10) break; }

  x = 0;
  //@ loop invariant 0 <= x <= 10;
  while (1) { if (x++ >= 10) break; }

  x = 0;
  //@ loop invariant 0 <= x <= 9;
  do { if (++x >= 10) break;} while (1);

  x = 0;
  //@ loop invariant 0 <= x <= 10;
  do { if (x++ >= 10) break;} while (1);

  //@ loop invariant 0 <= x <= 9;
  for(x=0; ; ){ if (++x >= 10) break;}

  //@ loop invariant 0 <= x <= 10;
  for(x=0; ; ){ if (x++ >= 10) break;}
}
```

Notice that the answers are indeed exactly the same as the previous exercise. It emphasizes the fact that generally speaking, a loop can always be transformed into the same loop without condition but with a **break** statement in the body instead. It also illustrates again the semantics of **do..while** loop not requiring the invariant to hold at exit.

Answer to Exercise 19

V1 is wrong (it does not decrease), V2 is OK, V3 is wrong (it becomes negative already at the penultimate iteration), V4 is OK.

Answer to Exercise 20

P1 is OK ; P2 is not OK ; P3 is not OK, because even if it decreases, the previous value is negative.

Notice that finding an appropriate variant for this recursive function is indeed tricky, because of the increment by one. A possible solution is

```
    (x > 10) ? x+1 : (x >= 0) ? 13-x : x+5
```

Answer to Exercise 21

The base of &a[2].u[1] is the address of the first element of the array, thus (**void***) a (which is also &a). p points to the same array, thus it has the same base: (**void***)a. Finally, since the array is 0-initialized, the value of a[0].p is NULL, whose base is NULL.

In **struct** X, two padding bytes are added between s and u, and four padding bytes are added after u. Thus, **sizeof**(**struct** X) is **sizeof**(**int***) + **sizeof**(**short**) + 2 + 2***sizeof**(**unsigned**) + 4, that is 24 bytes.

The offset of &a[2].u[1] is 2***sizeof**(**struct** X) + **sizeof**(**int***) + **sizeof**(**short**) + 2 + 1***sizeof**(**unsigned**) = 64. Initially, the pointer p points to &a[2].u[0] incrementing it moves it to &a[2].u[1] whose offset is 64. The offset of a[0].p which is NULL is 0.

The length of the block associated to &a[2].u[1] is the size of the array a, equal to 3***sizeof**(**struct** X) that is 72 bytes. p refers to the same memory block, so its block length is the same. The size of the block pointed by NULL is unspecified.

Answer to Exercise 22

A is true as x is a writable structure. B is also true since y is a writable (thus readable) structure. C is false since z is declared **const**, its content is not writable. y.p is the NULL pointer, thus D is false. E is true, p points out of x.a but it still points to x.p (it points at ((**char***) &x.p)+4).

F is false, the structure has been declared but not initialized. G is true as y is 0 initialized. As y.p is NULL, we cannot prove H (but it does not really matter, since we can't evaluate *(y.p) anyway).

I is true, we point just after x which is a valid pointer. J is also true, for the same reason E is true. K is false because the pointer x.p is uninitialized, thus invalid. L is true because z.p is null.

Answer to Exercise 23

The **requires** are the following:

```
/*@ requires \valid_read(a + (0..n-1));
  @ requires \initialized(a + (0..n-1));
  @ requires \valid(b + (0..n-1));
  @ requires \separated(a + (0..n-1), b) || b == a ;
  @*/
```

That is:

- the content of a must be readable and correctly initialized,
- the content of b must writable (and large enough for the n values),
- b must be exactly a or non overlapping with the buffer a.

The last condition guarantees that we either keep a unchanged or that we do not erase values that have not been copied yet.

Answer to Exercise 24

Question 1:

```
/*@ logic function search(char a[], integer low, integer high) =
      high <= low ? high :
        a[low] == 0 ? low :
        search(a,low+1,high);
*/
```

Question 2:

```
/*@
    predicate positive1(char a[], integer low, integer high) =
      high <= low || (a[low] > 0 && positive1(a,low+1,high));

    predicate positive2(char a[], integer low, integer high) =
      \forall integer i; low <= i < high ==> a[i] > 0;

*/
```

Question 3:

```
/*@
    lemma positive_equiv:
      \forall char a[], integer low, high;
        positive1(a,low,high) <==> positive2(a,low,high);

    lemma positive_no_zero:
      \forall char a[], integer low, high;
        positive1(a,low,high) ==> search(a,low,high) == high;

*/
```

Note that actually proving these lemmas requires an inductive argument, which makes them too difficult for most of the automated provers used by Frama-C. However, they can easily be verified with Wp's interactive tactics (see Chap. 4).

Answer to Exercise 25

Question 1

```
/*@ predicate
    array_update{Before,After}
      (int* a, integer length, integer i, int v) =
      \at(a[i],After) == v &&
      \forall integer j; 0 <= j < length && i != j ==>
        \at(a[j], Before) == \at(a[j], After);
*/
```

Question 2

```
void f() {
  int a[2] = { 0, 1 };
  L:
  a[0] = 2;
  //@ assert array_update{L,Here}(&a[0],2,0,(int)2);
}
```

Answer to Exercise 26

```
inductive sum{L}(int* a, integer i, integer j, integer s) {
  case Empty: \forall int* a, integer i,j; i >= j ==> sum(a,i,j,0);
  case Elem: \forall int* a, integer i,j,s; i < j && sum(a,i,j-1,s)
  ==> sum(a,i,j,s+a[j-1]);
}
```

The first case indicates that the sum of an empty set of elements is 0. For the Elem case, we have at least one element, namely j-1, and if we know that the sum of the elements from i to j-1 is s, then the sum of the elements up to j is s+a[j-1].

References

1. Baudin P, Cuoq P, Filliâtre JC, Marché C, Monate B, Moy Y, Prevosto V (2020) ACSL: ANSI/ISO C specification language, version 1.16. https://frama-c.com/html/acsl.html
2. Blanchard A (2020) Introduction to C program proof with Frama-C and its WP plugin. Tutorial, CEA List. https://allan-blanchard.fr/frama-c-wp-tutorial.html
3. Blazy S, Leroy X (2009) Mechanized semantics for the Clight subset of the C language. J Autom Reason 43(3):263–288. http://hal.inria.fr/inria-00352524/
4. Bobot F, Filliâtre JC, Marché C, Paskevich A (2015) Let's verify this with Why3. Int J Softw Tools Technol Trans (STTT) 17(6):709–727. https://doi.org/10.1007/s10009-014-0314-5. http://hal.inria.fr/hal-00967132/en. http://toccata.lri.fr/gallery/fm2012comp.en.html
5. Boldo S, Marché C (2011) Formal verification of numerical programs: from C annotated programs to mechanical proofs. Math Comput Sci 5:377–393. https://doi.org/10.1007/s11786-011-0099-9. http://hal.inria.fr/hal-00777605
6. Burdy L, Cheon Y, Cok D, Ernst M, Kiniry J, Leavens GT, Leino KRM, Poll E (2004) An overview of JML tools and applications. Int J Softw Tools Technol Trans
7. Chalin P (2005) Reassessing JML's logical foundation. In: Proceedings of the 7th workshop on formal techniques for java-like programs (FTfJP'05). Glasgow, Scotland
8. Chalin P, Kiniry JR, Leavens GT, Poll E (2006) Beyond assertions: advanced specification and verification with JML and ESC/Java2. In: de Boer FS, Bonsangue MM, Graf S, de Roever WP (eds) Formal methods for components and objects. Springer, pp 342–363
9. Cheon Y, Leavens GT (1994) The Larch/Smalltalk interface specification language. ACM Trans Softw Eng Methodol 3(3):22–153. https://doi.org/10.1145/196092.195325
10. Cok DR (2014) OpenJML: software verification for Java 7 using JML, OpenJDK, and Eclipse. In: Dubois C, Giannakopoulou D, Méry D (eds) Proceedings 1st workshop on formal integrated development environment. Electronic proceedings in theoretical computer science, vol 149, pp 79–92. https://doi.org/10.4204/EPTCS.149.8
11. Cok DR, Kiniry J (2004) ESC/Java2: uniting ESC/Java and JML. In: Barthe G, Burdy L, Huisman M, Lanet JL, Muntean T (eds) CASSIS. Lecture notes in computer science, vol 3362, pp 108–128. Springer
12. Ellison C, Rosu G (2012) An executable formal semantics of C with applications. In: Principles of programming languages, POPL'12, pp 533–544. Association for Computing Machinery, New York, NY, USA. https://doi.org/10.1145/2103656.2103719
13. Filliâtre JC (2003) Verification of non-functional programs using interpretations in type theory. J Funct Program 13(4):709–745
14. Filliâtre JC, Marché C (2004) Multi-prover verification of C programs. In: Davies J, Schulte W, Barnett M (eds) 6th International conference on formal engineering methods, vol 3308. Lecture notes in computer science. Springer, Seattle, WA, USA, pp 15–29
15. Floyd RW (1967) Assigning meanings to programs. In: Schwartz JT (ed.) Mathematical aspects of computer science. Proceedings of symposia in applied mathematics, vol 19, pp 19–32. American Mathematical Society, Providence, Rhode Island
16. Gerlach J (2020) ACSL by example. Tutorial, Fraunhofer Fokus. https://github.com/fraunhoferfokus/acsl-by-example
17. Greenaway D, Lim J, Andronick J, Klein G (2014) Don't sweat the small stuff: formal verification of C code without the pain. In: Programming language design and implementation, PLDI'14, pp 429–439. Association for Computing Machinery, New York, NY, USA. https://doi.org/10.1145/2594291.2594296
18. Hatcliff J, Leavens GT, Leino KRM, Müller P, Parkinson M (2012) Behavioral interface specification languages. ACM Comput Surv 44(3). https://doi.org/10.1145/2187671.2187678
19. Herms P (2013) Certification of a tool chain for deductive program verification. Thèse de doctorat, Université Paris-Sud. http://tel.archives-ouvertes.fr/tel-00789543
20. Hoare CAR (1969) An axiomatic basis for computer programming. Commun ACM 12(10):576–580 and 583

21. International Organization for Standardization (ISO): ISO/IEC 9899:2011 – Programming languages – C. http://www.open-std.org/jtc1/sc22/wg14/www/docs/n1548.pdf
22. Kosmatov N, Marché C, Moy Y, Signoles J (2016) Static versus dynamic verification in Why3, Frama-C and SPARK 2014. In: Margaria T, Steffen B (eds) 7th International symposium on leveraging applications of formal methods, verification and validation (ISoLA). Lecture notes in computer science, vol 9952. Springer, Corfu, Greece, pp 461–478. https://doi.org/10.1007/978-3-319-47166-2_32. https://hal.inria.fr/hal-01344110
23. Krebbers R, Wiedijk F (2015) A typed C11 semantics for interactive theorem proving. In: Certified programs and proofs, CPP'15. Association for Computing Machinery, New York, NY, USA, pp 15–27. https://doi.org/10.1145/2676724.2693571
24. Leavens GT (1996) An overview of Larch/C++: behavioral specifications for C++ modules. Springer, pp 121–142. https://doi.org/10.1007/978-0-585-27524-6_8
25. Leavens GT, Baker AL, Ruby C (2000) Preliminary design of JML: a behavioral interface specification language for Java. Technical report. Iowa State University, pp 98-06i
26. Leavens GT, Cheon Y, Clifton C, Ruby C, Cok DR (2003) How the design of JML accomodates both runtime assertion checking and formal verification. Technical Report 03-04, Iowa State University
27. Leavens GT, Leino KRM, Poll E, Ruby C, Jacobs B (2000) JML: notations and tools supporting detailed design in Java. In: OOPSLA 2000 Companion, Minneapolis, Minnesota, pp 105–106
28. van Leeuwen J (ed) (1990) Handbook of theoretical computer science. Formal models and semantics, vol B. Elsevier, Amsterdam
29. Leino KRM, Nelson G (1998) An extended static checker for Modula-3. In: Koskimies K (ed) Compiler construction. Springer, pp 302–305
30. Marché C, Paulin-Mohring C, Urbain X (2004) The Krakatoa tool for certification of Java/JavaCard programs annotated in JML. J Logic Algebr Program 58(1–2):89–106
31. McCormick JW, Chapin PC (2015) Building high integrity applications with SPARK. Cambridge University Press
32. Meyer B (1992) Eiffel: the language. Prentice Hall
33. Meyer B (1997) Object-oriented software construction. Prentice Hall

Chapter 2
The Heart of Frama-C: The Frama-C Kernel

André Maroneze, Virgile Prevosto, and Julien Signoles

Abstract This chapter provides an overview of the Frama-C distribution, including its main plugins that are covered in depth by other chapters. It mainly focuses on the Frama-C kernel and the main services that it offers to the user. This includes notably passing proper arguments to launch Frama-C and drive an analysis, controlling the parsing and code normalization phases of the analysis, and visualizing results.

Keywords Parsing · Type-checking · C semantics · Program analysis · Combination of analyses

This chapter presents the Frama-C kernel and provides an overview of the Frama-C distribution, plug-ins included, covered in this book. By itself, most pieces of information provided by the kernel are usually computed by any common C compiler. Indeed, the strength of Frama-C comes from its plug-ins: each of them is a code verification tool, a program transformation tool, or a code analysis tool that helps validate or verify C code. However, the Frama-C kernel is at the heart of Frama-C, since it allows new analyzers to be added to the framework through new plug-ins, provides many features automatically shared by all of them, and services that ease their collaborations for implementing specific code verification methodologies. Therefore, understanding what the kernel provides and how to use it is necessary to use cleverly any plug-in.

This chapter does not assume any particular background about Frama-C, but it is *not* a replacement for its user manual [12]. In particular, it does not present as many details as the manual and does not exhaustively present all the ways to parameterize Frama-C: any serious usage of Frama-C certainly benefits from reading the Frama-C user manual in addition to this book chapter, as well as reading the book chapter(s) and the manual(s) of the relevant plug-in(s). Similarly, this chapter describes the kernel from an end-user point of view, explaining which kind of information is readily available with the current Frama-C distribution. Readers interested in extending Frama-C via the kernel API can refer to Chap. 7 instead. However, this chapter

A. Maroneze (✉) · V. Prevosto · J. Signoles
Université Paris-Saclay, CEA, List, Palaiseau, France
e-mail: andre.maroneze@cea.fr

provides a gentle introduction to using the whole platform: reading it should be enough for simple usage and for getting a taste of what Frama-C provides.

This chapter is organized as follows. Section 2.1 explains how to run Frama-C from scratch. Section 2.2 details the Frama-C plug-ins covered in this book and how they may interact with the kernel and each other. Section 2.3 presents how to parse a C source code with Frama-C. Section 2.4 explains how the code is normalized by the kernel. Section 2.5 shows how users can properly drive Frama-C. Finally, Sect. 2.6 introduces how users may visualize the results computed by the analyzers.

2.1 Running Frama-C

2.1.1 Invoking Frama-C

Frama-C can be obtained from https://frama-c.com/, or, more commonly, be installed through a package manager. Docker images are also available via the Docker Hub.[1] We refer the reader to https://git.frama-c.com/pub/frama-c/blob/master/INSTALL.md for more information about the possible installation options. Once it is installed, the simplest way to invoke Frama-C is via the command line, like most code analyzers:

frama-c [options] C_source_files

Each source file passed to Frama-C is parsed by the kernel that generates an intermediate representation, named the Cil [33] Abstract Syntax Tree (Cil AST, or AST for short). Depending on which analyzers have been invoked, one or several analyses take place. Some plug-ins can modify the AST *during* parsing (e.g., Variadic specializes variadic function calls, so that other plug-ins do not need to handle them), but most of them only start after the parsing stage is finished. Plug-ins can transform the code, creating new ASTs, which are then passed on to other analyses, and so on. Finally, Frama-C can *save* its internal state, storing it on a disk file, so that later executions can *load* it instead of performing again all the analysis steps that led to it.

Frama-C also has a powerful Graphical User Interface: frama-c-gui. It is invoked in exactly the same way as the command-line version, except that it opens a GUI (developed in Gtk3[2]). In recent Frama-C releases (from 25-Manganese), a newer GUI, Ivette, is also present. Ivette is based on a more modern framework, namely Electron/React,[3] and should entirely replace frama-c-gui at some point. For all the examples presented in this section, both frama-c-gui and Ivette operate in the same way. Unless mentioned otherwise, the reader may assume any one of these two interfaces when "GUI" is mentioned.

---

[1] https://hub.docker.com/u/framac.

[2] https://www.gtk.org.

[3] https://www.electronjs.org.

2 The Heart of Frama-C: The Frama-C Kernel

Each invocation mode has its strengths and weaknesses. The command line offers a more precise control, better repeatability, and better error handling; the GUI version provides option discovery and some unique interaction modes (source code navigation, result exploration, etc.). In theory, the GUI can entirely replace the command line. In practice, however, it is often better to *start* with the command line, especially to sort out parsing issues and their error messages, and *then* use the GUI to see results or perform interactive analyses. This can be done by running the command line, saving results, and then loading them directly in the GUI, as exemplified in Sect. 2.5.3.

Note that Frama-C can also be used in a CI pipeline, for instance via a *GitHub Action*[4] for Frama-C. Frama-C supports SARIF[5] output via its MDR plug-in. It is thus possible to run Frama-C remotely after each commit and see the results in your IDE, without ever having to install or run Frama-C locally. However, the initial setup of an analysis (which is rarely trivial for realistic code bases) and the amount of information available in SARIF (much lesser than what the Frama-C GUI can provide) make it a much less compelling experience. Thus, invoking Frama-C via a CI/CD pipeline should only be envisaged as a *complement* to locally running it, rather than a *replacement*.

2.1.2 Brief Overview of **Frama-C**'s Command Line

Frama-C's command line is simple at its heart: arguments are either source code, options, or arguments to options. For example, `frama-c -help` shows help information, `frama-c file.c` parses `file.c` (without doing anything else), and `frama-c -main a -eva file1.c file2.c` parses files `file1.c` and `file2.c`, sets the main function (analysis entrypoint) to function `a` (`-main` expects an argument), and runs the Eva analyzer. Files and options can be interchanged, with a few exceptions, notably `-then` and its siblings, that allow for sequencing several analyses over a single Frama-C run (see Sect. 2.5.2).

In practice, however, industrial-scale code analysis is all but simple: users want to customize *everything* they can. For instance, they would like to add/remove warnings, change analysis hypotheses (e.g., "unsigned integer overflows are forbidden, treat them as errors",[6] or "just assume `malloc` never fails"), or make the analysis precise for *this* function, but fast for *that* one. Just as compilers like Gcc and Clang have dozens of options, so too do Frama-C and its plug-ins. Options are grouped in categories and listed by plug-in: kernel-related options are available with `-kernel-help`, and options related to a specific plug-in with `-<plugin>-help`. For instance, to know

---

[4] https://github.com/frama-c/github-action-eva-sarif.
[5] SARIF [21] is a standard format for static analysis results supported by GitHub as well as IDEs such as Visual Studio.
[6] In the C standard, *unsigned* arithmetic overflow has a perfectly defined semantics, hence, by default, it does not lead to an alarm by Frama-C.

the options declared by the Wp plug-in, one would use -wp-help. Usually, plug-in options are similarly prefixed by their short name (e.g. -eva-precision is an option of the plug-in Eva), though a few exceptions exist for historical reasons.

More information about the main kernel options allowing to drive a complete sequence of analyses and/or saving intermediate results on disk can be found in Sect. 2.5.

2.2 Plug-ins

On top of the Frama-C kernel, plug-ins provide the main Frama-C features: analyzing source code written in C in order to compute meaningful information for validating and verifying code properties.

Section 2.2.1 introduces one key feature provided by the Frama-C kernel: plug-in collaboration. Section 2.2.2 presents the Frama-C plug-ins that are relevant for this book, and Sect. 2.2.6 shows how to load them in practice.

2.2.1 Plug-in Collaboration

No program analysis technique is perfect by nature: most program analysis problems are *undecidable* as shown by Rice's theorem [37]. In other words, it is impossible to create a tool capable of solving them in an exact and sound manner, in finite time. However, some approaches and tools give better results for particular kinds of properties or programs than others. To benefit from the strengths of the different techniques, Frama-C promotes analyzer collaboration. It can be used to decompose verification work and comes in two different flavors: *sequential* and *parallel* collaborations.

Sequential collaboration consists in using the result of an analyzer as input to another one. It can also consist in generating annotated C code that encodes a verification problem in such a way that it can be understood by another analyzer. Such collaborations are the genesis for the *expressiveness* and *support* plug-ins respectively presented in Sects. 2.2.3 and 2.2.4. The former are usually used *before* a verification plug-in in order to generate additional code and ACSL annotations to be analyzed. The latter can be used either *before* a verification plug-in or *afterward*. When used before, they infer or hide some pieces of information. When used *afterward*, they synthesize some results in a user-friendly way.

Parallel collaboration consists in using several analyzers to verify program properties, each analyzer verifying a subset of properties. For instance, absence of undefined behaviors can be verified by Eva, while functional properties can be proved by Wp. Eventually, the few remaining properties may be checked at runtime by E-ACSL. Frama-C ensures the consistency of partial results emitted by the analyzers, and summarizes what has been verified so far and what remains to be checked.

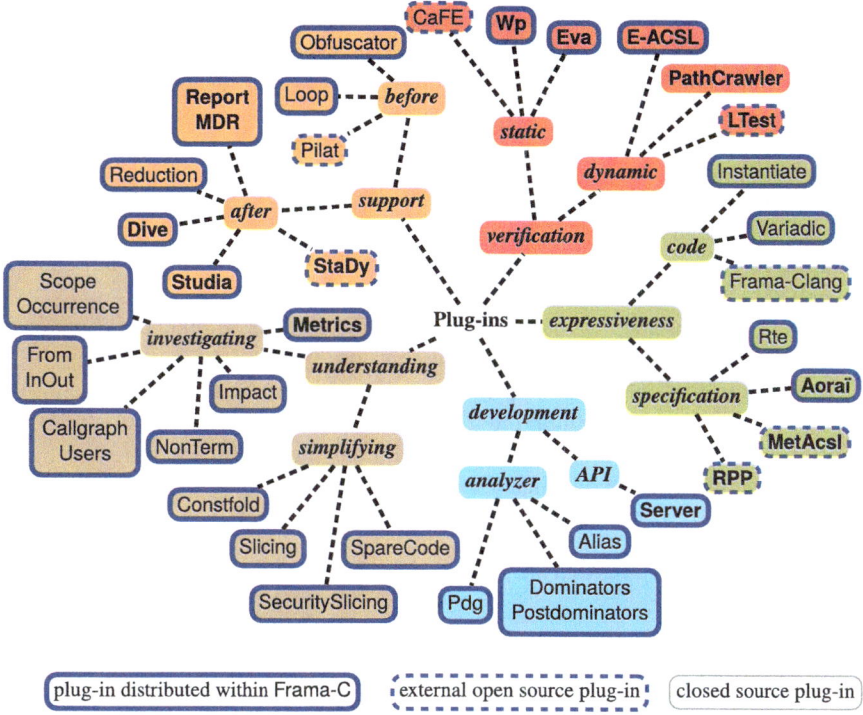

Fig. 2.1 Frama-C plug-ins covered in the book

Both kinds of collaboration are illustrated on concrete examples in the next sections. More advanced examples are also provided in other chapters of the book, in particular in Chap. 7 for sequential combinations using expressiveness plug-ins, Chap. 8 for sequential combinations using support plug-ins, and Chap. 9 for parallel combinations.

2.2.2 Plug-in Gallery

Figure 2.1, updated from previous works [6, 39], shows a large set of plug-ins of interest. It includes all the plug-ins that are part of the open source distribution **Frama-C 28.0-Nickel**, released on November 30th, 2023, as well as a few other plug-ins. They are either open source, but distributed outside the mainstream distribution, or closed source, but still available for academic usage.

This section briefly introduces these plug-ins. Our goal is only to give the readers a quick insight of what **Frama-C** contains, but not to give a comprehensive explanation of what the plug-ins provide and how they work. The next chapters in this part of the book are dedicated to the most important plug-ins, while the second and third parts of the book provide a lot of explanations, examples and concrete use cases about several other plug-ins. The interested reader may also refer to their dedicated webpage on https://frama-c.com for their user manuals. The plug-ins are organized in five main categories, presented below, roughly following their order of importance for a new user of Frama-C.

- **Verification plug-ins** verify program properties, either *statically* (i.e., at compile-time), or *dynamically* (i.e., at run-time).
- **Expressiveness plug-ins** extend the variety of *code* or *annotation* that can be handled by the framework.
- **Support plug-ins** are plug-ins that ease the verification tasks, either *before* or *after* using one (or several) of the verification plug-ins.
- **"Understanding" plug-ins** help the user comprehend the code under scrutiny, either by rewriting and *simplifying* it, or by computing extra information relevant for *investigating* it, often in combination with the graphical interface.
- **Development plug-ins** are meant to be used by other plug-ins, either for their *analysis' results*, or for handling the Frama-C *API* in a specific way.

For each category, plug-ins that are presented in-depth in this book will be introduced with a **bold face**.

2.2.2.1 Verification Plug-ins

The verification plug-ins are the most important ones, since verifying program properties is the main task of Frama-C. There are three static analysis plug-ins that only look at the source code without running the program.

- CaFE [17] (CaRet Frama-C's Extension) verifies temporal logic properties over sequences of C function calls by means of model checking [20, 36]. They are expressed in the CaRet language [1], an extension of LTL [35] (Linear Temporal Logic).
- **Eva** (Evolved Value Analysis) computes over-approximations of the possible values of each program variable at each program point by abstract interpretation [14, 15]. Its primary usage consists in detecting undefined behaviors. It is one of the key Frama-C plug-ins, covered in depth in Chap. 3.
- **Wp** (Weakest Precondition) verifies C code with respect to ACSL annotations through deductive verification [19]. It is one of the key Frama-C plug-ins, covered in depth in Chap. 4.

There are also three dynamic analysis plug-ins that are based on concrete program executions.

- **E-ACSL** generates an instrumented program p' from an annotated program p in order to verify the ACSL annotations of p at run-time. It is one of the key Frama-C plug-ins, covered in depth in Chap. 5.

- **PathCrawler** is a white-box test case generation tool based on several coverage criteria (e.g., MC/DC[7]). While not open source, an online version is freely available online at http://pathcrawler-online.com [28]. It is one of the key Frama-C plug-ins, covered in depth in Chap. 6.
- **LTest** is covered in Chap 9. It provides different services for test automation [3, 5, 29, 31], based on the notion of (hyper)labels [4, 30]. It has in fact 4 components, namely:
  - LGenTest for generating test cases, based on PathCrawler,
  - LAnnotate, which offers label generation for several standard coverage criteria,
  - LReplay,[8] that runs an existing test suite and provides coverage estimation,
  - and LUncov, for detecting uncoverable test objectives.

  LAnnotate, LUncov, and LReplay are available under the LGPL license.

2.2.3 Expressiveness Plug-ins

Expressiveness plug-ins are also quite important, because they allow the user to verify more complex pieces of code or specifications than Frama-C does by default. Those focusing on code help scaling up or target other programming languages.

- Frama-Clang [11] takes as input a C++ program and generates an equivalent C program. This way, every Frama-C plug-in may analyze C++ code. The necessary information is kept in order to convert results and user messages emitted from Frama-C back to the original C++ code. Frama-Clang is currently still a prototype that handles only a subset of C++ code, including exceptions, templates, classes and inheritance. It also supports a subset of ACSL++ [10], an extension of ACSL for specifying properties over C++ code.
- Instantiate replaces some widely-used libc functions taking **void** * arguments (typically, memory manipulation functions such as `malloc` or `memcpy`) by Frama-C built-ins in order to allow some plug-ins (typically, Wp) to analyze their calls more precisely.
- Variadic replaces each variadic function call[9] by a non-variadic function call by means of a monomorphisation technique [43].[10] Such a technique transforms a generic (polymorphic) function into a set of instantiated (monomorphic) functions, according to its calling contexts. This way, the other plug-ins do not need to take

---

[7] Modified Condition and Decision Coverage [2].

[8] Technically, LReplay is not a plug-in, but an independent tool.

[9] A variadic function is a function with a variable number of arguments (e.g., `printf`).

[10] The authors of the referenced paper [43] named this technique *"code specialization"* or *"polymorphism removal"*. It appears that the name *"monomorphisation"* becomes popular a bit later when introduced in the SML optimizing compiler MLton (http://mlton.org).

care of variadic function calls. The provided transformation is so useful for most Frama-C analyzers that it is set by default.

Other plug-ins allow the user to verify complex program specifications, beyond ACSL properties. Aoraï, MetAcsl, and RPP are covered in more details in Chap. 7, while Rte is illustrated in several chapters.

- **Aoraï** takes as inputs a C program together with an automaton which specifies acceptable sequences of function calls. It generates a C program annotated with ACSL specifications and/or ghost code that encodes the automaton and its transitions in the input code. With respect to the CaFE plug-in, Aoraï doesn't do any verification by itself. Instead, it relies on the verification plug-ins to check the generated specifications.
- **MetAcsl** allows the user to express system-level properties in an extension of ACSL, named Hilare. From a user code and such specifications, it generates a new code containing ACSL annotations that can be verified by standard verification tools like E-ACSL or Wp. If so verified, it means that the code complies with the original Hilare specifications.
- **RPP** (Relational Properties Prover) extends ACSL with relational properties that involve several functions or several calls to the same functions. These extra properties are then translated into additional code and axiomatics in order to be verified by other means, e.g. E-ACSL or Wp.
- Rte [23] (RunTime Errors) generates ACSL annotations for different kinds of undefined behaviors including validity of pointers before dereferencing them, array accesses, and arithmetic overflows. Verifying these annotations (e.g. with E-ACSL or Wp) guarantees the absence of such undefined behaviors.

2.2.4 Support Plug-ins

Support plug-ins aim at simplifying the use of verification plug-ins. Some of them work prior to the verification task:

- **Loop** approximates the number of loop iterations in order to automatically infer Eva's slevel parameters (see Chap. 3).
- **Obfuscator** obfuscates a program by replacing any name (variable name, type name, etc.) by a meaningless name while preserving program semantics. It also generates a separate dictionary linking the generated names to the original ones, that allows decoding the obfuscated code, when known.
- **Pilat** [18] (Polynomial Invariants through Linear Algebra Tool) automatically generates polynomial loop invariants by relying on linear algebra techniques. Loop invariants are ACSL properties of particular interest for Wp (see Chap. 4) and other verification tools that benefit from summaries of loop semantics.

Some other plug-ins work afterward, to help users interpret the raw results of the analyzers:

- **Dive** operates on Eva results. It generates a dependency graph of the values that contribute to the value of a memory location at a given program point. Navigating in this graph from an imprecise location helps to understand where the imprecision introduced by Eva comes from. It is mainly used through Ivette and covered in more details in Chap. 8.
- **MDR** (MarkDown Report) and **Report** output in various textual formats (e.g., CSV or JSON) a synthesis of analysis results. Both are covered in Sect. 2.6.
- **Reduction** generates ACSL annotations from Eva's internal information: for each statement, an ACSL assertion describing the possible values (as computed by Eva) of the locations modified within the statement. This plug-in is intended to make Eva results easier to exploit, but is currently very experimental.
- **StaDy** explains proof failures of Wp by means of testing through PathCrawler. It is covered in depth in Chap. 9.
- **Studia** traces, in the GUI, for a given location at a given program point, the possible points where this location was written to, according to Eva analysis. It is presented with more details in Chap. 8.

2.2.5 *"Understanding" Plug-ins*

Such plug-ins help grasp the code under scrutiny. Some of them operate in a read-only fashion, highlighting meaningful parts of the code in the GUI to help *investigation*.

- **Callgraph** and **Users** respectively compute the callgraph of a C program and the function callees inside its functions. If Eva has already been computed, function pointers are taken into account when computing the callgraph, while Users automatically runs Eva if not already computed.
- **From** and **InOut** respectively compute the functional dependencies and different kinds of inputs and outputs of a function by relying on Eva's results. For instance, InOut may compute that the inputs (resp. outputs) of some function f is x and *p (resp. *p and t[0..9], meaning the first 10 cells of the array t), while From can indicate that *p and t[0..8] are computed from (the original value of) *p, and t[9] is computed from x.
- **Impact** shows statements whose semantics may depend on the semantics of a given statement [32]. It relies on both Eva's and Pdg's (see Sect. 2.2.5.1) results.
- **Metrics** [9] shows several purely syntactic metrics and a few metrics related to the coverage ratio of Eva (see also Chap. 8).
- **Occurrence** and **Scope** respectively display in the GUI, for a selected left-value,[11] its occurrences, and scoping information (e.g, its potential definition points

---

[11] A left-value is an expression that can be written through an assignment (e.g., a variable).

from a particular use). Both rely on Eva to take into account pointer aliasing. Unlike Dive and Studia, whose usage is intrinsically related to Eva's results, Occurrence and Scope are more syntactical in nature. Still, the similarities between these plug-ins allow the user to try different approaches, depending on code characteristics; e.g., Scope can be more precise, but Studia scales better.

- NonTerm relies on Eva to warn the user about definitely non-terminating pieces of code.

Other plug-ins rewrite the code, *simplifying* it with respect to some criteria.

- Constfold is a program transformation that tries to replace each constant expression—that is, expressions that evaluate to the very same value for every program execution—by that constant value. It relies on the over-approximation of values computed by Eva to decide which expressions can be safely propagated: it is safe only if the over-approximation is a singleton.
- Slicing implements a program transformation technique that removes program constructs that are not involved for preserving a given semantic criterion (e.g., preserving the semantics of some statements) [44]: the generated program is equivalent to the original one with respect to this criterion, but is shorter.
- SecuritySlicing implements a specific slicing algorithm whose criterion is to preserve confidentiality of data [32].
- SpareCode is a specific slicing whose criterion is to preserve the final state of the program. Thus, it will remove dead code, but also assignments whose value is never read afterward (e.g., the first assignment to x in x=0; x=1;).

2.2.5.1 Development Plug-ins

Development plug-ins are of (almost) no interest to the end user. However, they aim to be used when developing other plug-ins as explained in Chap. 7, and many plug-ins presented above do take advantage of them. First, some of them implement a specific analysis.

- Alias computes aliasing and points-to information based on a variant of the Steensgaard algorithm [42]. It aims to be much faster than Eva but less precise when computing this particular data.
- Dominators and Postdominators respectively compute the dominator and the postdominator sets of a statement in a control flow graph [22].
- Pdg computes an over-approximation of the Program Dependence Graph [34] (shortly, PDG) by relying on From's and Eva's results.

Second, one specific plug-in provides an API to other plug-ins and/or tools.

- Server is a request server, which is usable when implementing a client that considers Frama-C as a server. Notably, it is used by Ivette, the Electron/React GUI introduced in Sect. 2.1.1.

2.2.6 Loading Plug-ins

When run, one of the first actions of the **Frama-C** kernel is to load all the plug-ins that it finds in its default plug-in directory. It means that the code of the plug-in is loaded from the disk and linked to the **Frama-C** kernel, and becomes available to the end-user.[12] This directory's location is printable through the option -print-plugin-path. The list of the loaded plug-ins is available through the option -plugins. If you want **Frama-C** to not load any plug-ins by default, you should use option -no-autoload-plugins.

Typical Setting

Here is the default plug-in directory for a typical Linux setting:
```
1 $ frama-c -print-plugin-path
2 ~/.opam/default-opam-switch/lib/frama-c/plugins
```

The printed directory includes all the plug-ins distributed within **Frama-C**, as well as any additionally installed plug-in:
```
1 $ frama-c -plugins
2 Alias Analysis   Lightweight May-Alias Analysis (-alias-h)
3 Aorai            verification of behavioral properties
4                  (experimental) (-aorai-h)
5 <...>
```

Loading a Single Plug-in

It is possible to precisely control which plug-ins are loaded by combining -no-autoload-plugins and -load-plugin. First, with -no-autoload-plugins, no plug-ins are loaded:
```
1 $ frama-c -no-autoload-plugins -plugins
2 <no output>
```

Then, the following command loads **Obfuscator**.
```
1 $ frama-c -no-autoload-plugins -load-plugin obfuscator
      -plugins
2 Obfuscator            <description>
```

If any, the dependencies of the loaded plug-ins are automatically loaded (and so, they must be in the searched paths). For instance, **Wp** actually depends on plug-ins **Eva**, **Rte** and **Server**:

---

[12] Yet, it usually has no effect if no option or **GUI** action for this plug-in is activated.

```
1 $ frama-c -no-autoload-plugins -load-plugin wp -plugins
2 Eva                     <description>
3 Rtegen                  <description>
4 Server                  <description>
5 WP                      <description>
```

2.3 Parsing Annotated C code

What we usually denote by *parsing* C in Frama-C is in fact a combination of several steps: preprocessing, lexing, parsing, typing, linking of source files and incorporating ACSL annotations. This section presents these steps, as well as several aspects (such as architecture-dependent features and standard library support) related to the complex process that starts with a set of source files and ends with an AST, ready to be used by sophisticated analyzers. Fig. 2.2 provides an overview of this process, whose main steps are described in the rest of this section.

2.3.1 Preprocessing C and Build Tools

Frama-C does not (yet) perform modular parsing (akin to separate compilation): it needs all the code to be analyzed to be given in the command line (or #include'd via

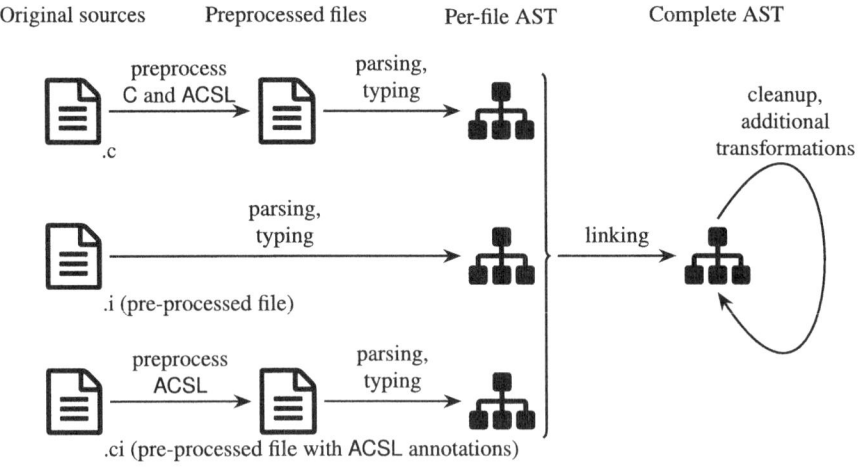

Fig. 2.2 Source files processing steps in Frama-C's kernel

2 The Heart of Frama-C: The Frama-C Kernel

preprocessor directives). It uses the file extension to identify which kind of source[13] it is:

- .c or .h for non-preprocessed sources and headers;
- .i for preprocessed files;
- .ci for preprocessed files containing ACSL annotations that need to be preprocessed.

Besides .ci, which is Frama-C-specific, the others are commonplace C conventions.

Frama-C does not have its own preprocessor: it expects to find one in the system (usually Gcc or Clang). There are several options for controlling preprocessing: the most common one is -cpp-extra-args, which passes its arguments to the preprocessor. For instance, -cpp-extra-args="-DEXTRA_FEATURE-I../my_includes".[14]

It can also be specified on a per-file basis, with -cpp-extra-args-per-file:

```
frama-c -cpp-extra-args-per-file="a.c:-DFILE_A" \
        -cpp-extra-args-per-file="b.c:-DFILE_B" \
        a.c b.c
```

The entire preprocessing command can be changed via -cpp-command. Finally, another possibility is to rely on the information contained in a *Json Compilation Database* file,[15] a *de facto* standard format for holding compilation instructions that is used by several tools including e.g. Clang, Build EAR,[16] and CMake.[17] More information about using these tools for generating such a database can be found in the user manual [12, Sect. 13.2.4]. Once a database has been created for the project under analysis, it can be given to Frama-C through the —json-compilation-database option, which takes as argument the path to the file.

An additional layer added by Frama-C at this step is the handling of *ACSL annotations*. For a compiler, ACSL annotations are simply comments, usually ignored or preserved as such, but without any preprocessing. For Frama-C, however, they are a full-fledged language; and, since the user might want to use preprocessing symbols in their annotations (e.g., ensures \result != EOF; to indicate that a given function returns a value that is different from the standard library end-of-file symbol), the comments *also* need to be preprocessed. Typically, Frama-C invokes the C preprocessor twice, once for "normal" preprocessing, and once for "logical preprocessing" of annotations. This is often transparent to the end-user, but in case of problems, these steps may show up unexpectedly. In particular, this second step

---

[13] Frama-C plug-ins are able to extend the Frama-C front-end to allow parsing of other file types, e.g. Frama-Clang does so for .C/ .cpp files, associated with C++ sources.

[14] Note the use of the -<option>=<value> syntax, instead of -<option> <value>. If <value> starts with a hyphen, the latter form would be parsed as a different option name.

[15] https://clang.llvm.org/docs/JSONCompilationDatabase.html.

[16] https://github.com/rizsotto/Bear.

[17] https://cmake.org/cmake/help/latest/variable/CMAKE_EXPORT_COMPILE_COMMANDS.html.

explains why .ci files exist for Frama-C: they had their first preprocessing stage completed, but not the second one.

2.3.2 From Preprocessed Code to Typed and Unified AST

After preprocessing has been sorted out (or if the user already provided .i files), Frama-C can finally start building its AST. This follows the classical stages of a compiler's front-end, namely lexing, parsing, and type-checking the resulting AST. Frama-C inherited these stages from Cil [33], an OCaml library for manipulating C code on top of which Frama-C was originally built [16].[18] Such a process is not far from what heavy-duty compilers do: it handles several language variants (e.g. some pre-C99 syntax, for maximum compatibility), ubiquitous compiler extensions, some semantic verifications, and several steps of code normalization, necessary to simplify code handled by the plug-ins. This step also detects non-standard code being unsupported by Frama-C. Some rigor and discipline via the usage of compilation flags such as -pedantic can help detect and prevent such cases before involving Frama-C, but there are always a few which pass through the cracks.

Finally, for each source file beyond the first, Frama-C *merges* its Cil AST to the existing one. This step, similar to the *linking* stage in a compiler, ensures that types, definitions and annotations are all consistent between them. Due to the deep semantic nature of Frama-C's analyses, this step ensures that interprocedural analyzers obtain all the information they may need. From the user's point of view, this step can reveal some issues, e.g. multiple definitions of the same function among different sources, global variable initialized multiple times, or discrepancy between an extern declaration in a header file and the actual definition (notably, declaring a pointer but defining an array is a pattern that seems to occur regularly in the wild).

2.3.3 Kernel and Plug-in Transformations and Cleanup

After the previous step, the main AST has been prepared by Frama-C, but it is still not finished: some plug-ins may operate on it, to simplify the work for others. For instance, Variadic is a plug-in that specializes variadic function calls and definitions for simplifying the work of other plug-ins that do not have to deal with variadic functions. For standard library variadic functions Variadic will also add ACSL specifications, allowing others to verify properties related to them.[19]

---

[18] Frama-C forked from Cil at its very beginning; the original Cil GitHub repository (https://github.com/cil-project) is no longer maintained since 2016, but a more recently maintained fork exists at https://github.com/goblint/cil.

[19] In the case of printf, scanf, and their relatives, the specification generation is dependent on the fact that the format string is a literal.

The Frama-C kernel also offers options for applying syntactic transformations that can impact the AST: loop unrolling (-ulevel), constant folding (-constfold) and switch transformation (-no-keep-switch), just to name a few.

Finally, the cleanup stage is usually one of the last steps the kernel performs on the AST.[20] It consists in removing unused types and declarations, according to a few options. It is worth noting that Frama-C's normalization steps (detailed in Sect. 2.4) can create several identifiers that are useless to the user (e.g., temporary variables, symbols from included headers of the standard library, leftovers after merging). Without such a cleanup stage, the final code would be much more polluted and less readable. The only downside of this step is when the user actually *wants* some of this information to remain present in the code. In this case, kernel options such as -keep-unused-types and -keep-unused-specified-functions, help prevent such syntactic elements from being removed. Such a situation occurs e.g. if these seemingly unused symbols are in fact meant to be used by a plug-in performing some kind of code instrumentation at analysis stage.

2.3.4 Supporting Several C Standards, and Non-Standard Extensions

Frama-C intends to be useful for different industrial applications. They often contain old legacy code, or modern non-standard features. Supporting the wide variety of C code in existence is a Herculean effort. However, Frama-C provides such support on a *best-effort* basis and extends it as new case studies arrive. The primary standards supported by Frama-C are, currently, ISO C99 [26] and ISO C11 [27], but they are not *entirely* supported. For instance, complex numbers (via the libc header complex.h, or the _Complex keyword) are not supported. A table listing the unimplemented features is available in the *Compliance* chapter of Frama-C's user manual [12].

Generally speaking, ISO C standard makes a distinction between *undefined behaviors* and *unspecified behaviors*. Quoting the standard [27, Sect. 3.4.3], the former is "[a] behavior, upon use of a nonportable or erroneous program construct or of erroneous data, for which this International Standard imposes no requirements". On the other hand, the latter is [27, Sect. 3.4.4] "[a] behavior where this International Standard provides two or more possibilities and imposes no further requirements on which is chosen". Frama-C is usually meant to raise an alarm when there is a possibility for the program under analysis to exhibit undefined behavior, which is always an error. Unspecified behaviors, which can lead to portability issues, can also be flagged. A certain number of Frama-C's options control the emission of these alarms and are detailed in Sect. 2.5.1.

Frama-C also supports some legacy code, despite often emitting warnings related to non-portability, such as implicit type definitions as int (from K&R, pre-C89).

---

[20] Technically, some transformations can be done after the cleanup phase. However, this point concerns mainly plug-ins developers and is beyond the scope of this chapter.

Finally, Frama-C supports a fair amount of compiler extensions, which are in practice necessary for parsing many code bases. Besides *attributes* for several kinds of declarations, Frama-C supports some syntactic sugar such as pre-C99-based VLAs[21] (using zero-sized arrays), empty structs, and empty initializers. Frama-C also supports a certain number of Gcc builtins and extensions, as well as a few MSVC constructions.

By default, some Gcc-specific features are "always on" (accepted as if they were standard features), but others are only enabled when using a Gcc-based *machdep*, as explained in Sect. 2.3.6. This avoids accepting too much "non-portable" code by default, which helps users to be aware that their code may stop working in the future.

2.3.5 Standard C Library, a.k.a. libc, and POSIX

The C programming language standard includes a sizable number of headers, types and functions that are an integral part of the language. Without them, writing portable programs would be *much* more difficult. Their presence offers both requirements and opportunities for code analysis: while their support requires some considerable amount of effort, their presence in the standard enables the definition of several *semantic* properties concerning them. For instance, dynamic memory allocation can be specified in terms of malloc and free. This allows some plug-ins such as Eva to define *built-in* mechanisms to handle these omnipresent functions.

Not all standard library functions are equally useful; some headers such as stdio.h, stdlib.h, string.h and math.h are present in almost every application written in C, while others such as wchar.h are much less frequent. Their support in Frama-C tends to be correlated to their frequency and importance in the kinds of programs handled by Frama-C users.

Frama-C's support of the standard library comes at different levels:

1. At the lowest level, Frama-C offers a portable version of the libc headers, with type definitions and function prototypes. This ensures that programs calling such functions and using such types are parseable.
2. For several functions, Frama-C's standard library also offers ACSL specifications of varying detail: some specify only very basic input/output requirements (e.g. assigned global variables), while others specify several possible behaviors, with preconditions and postconditions;
3. Finally, for a few functions, Frama-C also offers a *stub*, that is, a model implementation that complements and/or replaces ACSL specifications. For a small part of these, the stubs have been annotated and their specification has been proven with Wp, turning them into *reference implementations*.

Note that ACSL specifications and stubs are not intrinsically better or worse; some plug-ins can take full advantage of sophisticated specifications, while others obtain

---

[21] Variable-Length Arrays.

more precise results thanks to stubbing. The interested reader should refer to the dedicated chapter for each plug-in in this book, or to their respective user manuals, in order to know how to specifically handle the standard library for each of them.

The Frama-C standard library is not limited to the headers and functions in the C standard; it also incorporates several functions defined in POSIX [24], which are often present in general-purpose libraries and applications. It also contains a few non-standard (neither C nor POSIX) functions. Their inclusion allows parsing more code, extending the reach of Frama-C. Adding specifications improves precision of analyses and, as a bonus, serves as ACSL usage examples.

Importantly, the Frama-C standard library does *not* provide the same soundness guarantees than the rest of the platform, that is, the Frama-C kernel and the plug-ins distributed with Frama-C. Indeed, due to the huge amount of specifications needed, plus the fact that there are several possible choices in terms of abstraction level (e.g., should functions manipulating streams include all of their implementation details, such as current position, or should they abstract away most of their internal data?), verifying all the specifications of the Frama-C library would require extensive work. The usefulness of this work is uncertain: on the one hand, safety-critical certification often includes verification of the actual implementation of library functions, to ensure they match the system's behavior; and on the other hand, the abstraction level chosen for Frama-C's specifications might not match the one needed by the user, minimizing its utility. Therefore, we focus on providing a broader range of specifications to help users overcome the initial steps, on a best-effort basis. We also accept requests and contributions concerning library functions, to foster the community around a useful set of specifications.

Finally, due to the parsing difficulties mentioned in Sect. 2.3.4, and the fact that most libc implementations incorporate huge amounts of non-portable code (e.g., adding non-standard fields to structs, using assembly code, performing system-dependent optimizations), *mixing Frama-C's libc with the system's libc is very likely to fail*. One should therefore strive to ensure that either *all* included libc headers come from Frama-C's library, or none of them do. Practical experience shows that even a single included mixed header is likely to lead to a cascade of inclusions of system-dependent headers (e.g., including libc's string.h leads to including bits/libc-header-start.h, which leads to including features.h, etc.) and a plethora of parsing errors, leading users astray. Frama-C allows disabling the use of its own library (-no-frama-c-stdlib) to handle such situations.

To conclude, the standard C library is essential for handling real-world code, and Frama-C tries to support it as much as possible to help users focus on other issues. However, the complexity of the matter leads to imperfect solutions. Being aware of them helps fix errors when they arise.

2.3.6 Architecture-Dependent Features: *Frama-C's* machdep

Even "portable C" is not really *that* portable: the C standard specifies several limits and sizes, but it leaves *much* of it up to the implementation to decide: the size (in bits) of type int, the signedness of type char, the underlying type used for size_t, etc. All these aspects are actually dependent on the architecture. Another important source of disagreements is the compiler used: whether Gcc, Clang, MSVC, or some proprietary C compiler is being used can lead to different behaviors and availability of certain extensions. Finally, the operating system also has an influence on the program's behavior, especially concerning libc.

To take into account all of this variability and maintain a semantic equivalence with the code as it will be executed in the target machine, Frama-C uses the notion of *machdeps* (for "machine-dependent" parts): virtual architectures that decide many machine-dependent behaviors. The user can, of course, define their own machdep.

The default machdep in Frama-C is x86_64, which corresponds to a "generic" 64-bit x86 Linux architecture, on which ints are 4 bytes long, some Gcc-specific features are not available, and most libc constants are defined according to a chosen Linux 64-bit implementation. Another machdep is gcc_x86_64, a variant of x86-64 that enables Gcc-specific behaviors. The machdep x86_32 is a 32-bit variant, and so on.

Choosing the right machdep is essential for correctness of the analysis with respect to the target platform. It also influences whether some code can be parsed or not. For instance, if it uses Gcc-specific syntax or features, a non-Gcc-based machdep might prevent it from being parsed. Option -machdep allows setting it. Any user can get the list of available machdeps by giving -machdep help to Frama-C.

Machdeps are defined in YAML files. Frama-C's *share* directory contains a machdeps subdirectory with the set of default machdeps, as well as a script, make_machdep.py, which allows automatic creation of a machdep file, given the existence of a C11-compatible (cross-)compiler for the target architecture. If no such compiler is available, the user can manually fill in the required fields to create their own machdep, and load it with -machdep < file.yaml >.

2.3.7 *ACSL:* a Lingua Franca, *But Not Equally Well-Spoken*

ACSL is the formal specification language of Frama-C, presented in details in Chap. 1. It allows communications between plug-ins: while each Frama-C plug-in may have its own data structures and internal representations of the program under analysis, ACSL offers a common ground, via its properties and annotations, to ensure plug-ins can exchange precise semantic information between them and complement each other.

In practice, however, things are usually not as simple as that. First, it is important to notice that not everything that is defined in the ACSL Specification Language Ref. [7]

is actually *implemented* in Frama-C: the ACSL Implementation Ref. [8] indicates in red the features not yet implemented in a given Frama-C release. Besides, note that what the Frama-C *kernel* supports in terms of ACSL constructs is not necessarily what a *given plug-in* supports: each plug-in may or may not support some features, usually documenting its own limitations. In parallel collaboration, if a plug-in is not able to verify some properties, it might be possible to use another one to do so. Conversely, in sequential collaboration, inter-plug-ins communication will obviously be restricted to the ACSL features that are supported by both plug-ins.

2.4 Normalized Code Representation

As mentioned in Sect. 2.3.2, the main AST on which Frama-C operates is a normalized representation of C (and ACSL) constructs, in which many implicit operations appear explicitly as AST nodes. This section describes the main features of this representation. First, it is worth noting that this AST is still valid C, in the sense that the -print option of Frama-C outputs a C file that should be accepted by any C compiler, and, at least for strictly C 99 compliant programs, have an equivalent semantics as the original code. However, the pretty-printer activated by -print does not give an exact view of the normalized AST, but tries to revert some of the transformations done during the normalization in order to provide the user with a more readable result. Option -print-as-is outputs code that is closer to the actual normalized AST: this can be helpful to better understand the main syntactic differences between the original code and its Frama-C representation.

2.4.1 Side Effects and Sequencing

Perhaps the most important difference with ISO C lies in the fact that the normalized AST makes a strict distinction between *instructions*, which modify the state of the program (except of course for the do-nothing Nop instruction, representing a null statement, i.e., a single ;), and *expressions*, which are pure computations, without any side effects. In order to achieve this separation, some temporary variables can be introduced to store the result of intermediate computations.

Making explicit this intermediate computation steps amounts to fixing an evaluation order for the original C expressions. This normalization might thus hide undefined or unspecified behaviors related to proper sequencing of side effects, in the sense that the pretty-printed code, with its intermediate variables, will always be properly sequenced.

Sequencing of Side Effects

Consider for instance the following C function containing the (in)famous i = i++ + ++i statement at line 3.

```
1 int f() {
2   int i = 0;
3   i = i++ + ++i;
4   return i;
5 }
```

It is normalized by Frama-C as follows.

```
1 int f(void)
2 {
3   int tmp;
4   int i = 0;
5   tmp = i;
6   i ++;
7   i ++;
8   i = tmp + i;
9   return i;
10 }
```

However, in order to be able to identify such undefined behaviors, the resulting AST conveys internally more information than the pretty-printed compound statement. In fact, the internal Frama-C representation takes the form of an *unspecified sequence*, which associates to each statement in the block the lists of location that are written and read, as well as function calls (as the callee might itself have side effects). Such information can be displayed in Frama-C's output through the -unspecified-access option.

Displaying Unspecified Accesses

Assuming the previous function is part of file f.c, one gets the following output.

```
1 $ frama-c -unspecified-access f.c
2 [kernel] f.c:3: Warning:
3   Unspecified sequence with side effect:
4   /*    <-   */
5   tmp = i;
6   /* i <-   */
7   i ++;
8   /* i <-   */
9   i ++;
10  /* i <- tmp */
11  i = tmp + i;
```

The literal syntactic format of the output should be considered as an internal representation artifact, and should not be relied upon for further analyses (instead, a proper plug-in directly accessing the corresponding AST nodes, as described in Chap. 7, should be preferred). Hence, we will not detail it here. Broadly speaking, each read/write information is displayed as a comment before the instruction it relates to, with the written location on the left handside of the arrow, and read ones on the right handside. One can remark that the very first instruction is deemed not to access anything. In fact, this is a generated instruction storing a result in an intermediate variable. These are deemed to always be properly sequenced (in other words we assume that the normalization is correct and focuses on the undefined behaviors that exist in the original code), hence the empty lists. We also clearly see three write accesses to i, that are unsequenced, meaning that the original expression contains an undefined behavior.

Among the main analysis plug-ins of Frama-C, currently only Eva can take advantage of the information contained in unspecified sequences to display proper warnings, provided the option -unspecified-access is activated. Consequently, the rest of this section is Eva-specific and can be safely ignored by readers who are not interested in this plug-in.

Eva only checks proper sequencing at intra-procedural level, i.e. it does not warn if a side effect inside a callee interferes with a side effect in the caller. Continuing with our example, the alarm emitted by Eva in this case is the following.

```
[eva:alarm] example.c:3: Warning:
  undefined multiple accesses in expression. assert
      \separated(&i, &i);
[eva] done for function f
[eva] example.c:3: assertion 'Eva,separation' got final status
      invalid.
```

The pretty-printed code also has a slightly different shape:

```
/*@ assert Eva: separation: \separated(&i, &i); */
{ /* sequence */
  tmp = i;
  i ++;
  i ++;
  i = tmp + i;
}
```

namely, as Eva has emitted an assertion on the unspecified sequence itself, it gets materialized in the code in the form of a block with a specific /* sequence */ comment on top of it.

However, with the following code, where we have a write access in incr and a read access in f on the same memory location that are indeterminately sequenced, no alarm is emitted by Eva.

```
int incr(int *p) {
  return (*p)++;
}

void f() {
  int i = 0;
  i = incr(&i) + i;
}
```

It should be noted however, that this situation is merely an unspecified behavior: as there is a sequence point when entering incr and another one when it returns, the read and the write access are always separated by a sequence point, even though we do not know the relative order in which they will occur: there are only two values possible for i at the end of f. This is in contrast with the previous example, where the accesses were completely unsequenced, which leads to an undefined behavior, after which the state of the program can be anything.

2.4.2 Making Implicit Conversions Explicit

Another important normalization step lies in the fact that all implicit conversions that might occur during the evaluation of an expression, in particular for arithmetic operands, are made explicit by the addition of casts to the destination type. While in most cases such conversions have no consequences on the final value of the computed expressions, since implicit conversions usually go from a type of smaller conversion rank to a type of higher conversion rank, the peculiarities of the C standard can sometimes lead to unexpected results, in particular when arithmetic promotion is involved.

Implicit Conversion Made Explicit

Consider the following function.

```
void f() {
  unsigned short s = 1 << 15;
  short s1 = s * s;
}
```

Although the result of s * s exceeds the range of a **short** (at least in Frama-C's default x86_64 architecture), there is no arithmetic overflow *per se*. Indeed, if we take a look at the normalized version of the code, we can see that the multiplication is done in the **int** type, and it is the result of this operation that is converted in **short**.

```
void f(void)
{
  unsigned short s = (unsigned short)(1 << 15);
```

```
4    short s1 = (short)((int)s * (int)s);
5    return;
6  }
```

Strictly speaking, conversions to a signed integer type of lesser rank of a value that is outside the range of this type are not undefined behavior. They are merely implementation-defined, even though an implementation may choose to raise a signal instead of returning a result. For this reason, Frama-C, as will be detailed in Sect. 2.5.1, and its plug-ins do not activate the corresponding alarm by default: it must be explicitly done through the -warn-signed-downcast option. For instance, if this option is passed to Frama-C, Eva raises the following alarm, while it does not without the option.

```
1  [eva:alarm] examples/implicit_conversion.c:3: Warning:
2    signed downcast. assert (int)((int)s * (int)s) <= 32767;
```

Similarly, some care must be taken with integer constants. Indeed, it must be recalled that in C there is no negative integer constant: -123 is in fact the result of applying the unary minus operator to the constant 123. Again, this is mostly harmless, but can lead to different types than expected. In particular, -2147483648 is in fact a **long int** (for an architecture with 32-bit **int**), and attempting to store it into an int leads to a cast in the normalized code, such as **int** C = (**int**)(-2147483648);. In any 2-complement architecture, which is the case for all architectures supported by Frama-C, this conversion always succeeds, and returns -2147483648 as an **int**. Hence, this code is considered correct by any Frama-C analyzer (e.g., Eva emits no alarm), even in presence of -warn-signed-downcast.

2.4.3 Logical Operators

In C, logical AND (**&&**) and OR (**||**) operators, as well as the ternary conditional operator (...?...:...), have a particular status among expressions, in the sense that the execution flow of the program changes depending on the value of the first operand. This stands in contrast with the fact that expressions in the normalized AST are supposed to be pure computations that do not modify the internal state of the program, much less its execution path. Hence, these operators are normalized in the form of **if** statements. For instance, x++ || x-- is represented as the following statements, with the result ultimately available in temporary variable tmp_1.

```
   int tmp_1;
   int tmp;
   tmp = x;
   x ++;
   ;
   if (tmp) tmp_1 = 1;
   else {
     int tmp_0;
     tmp_0 = x;
     x --;
     ;
     if (tmp_0) tmp_1 = 1; else tmp_1 = 0;
   }
```

2.4.4 Loops

The normalized AST only has a single node for representing loops, without any condition for exiting the loop: everything is done through jump statements. For instance, if we take a simple **for** loop as in:

```
void f() {
  int S = 0;
  for(int i = 0; i < 10; i++) {
    S+=i;
  }
}
```

the normalized code, obtained with the -print-as-is option will be the following one.

```
void f(void)
{
  int S = 0;
  {
    int i = 0;
    while (1) {
      if (i < 10) ; else break;
      S += i;
      i += 1;
    }
  }
  return;
}
```

If we do not use -print-as-is, the pretty printer detects that the exit condition is in fact the condition of the **if** that is the first statement of the loop's body, and output a slightly more compact version there: **while**(i<10) S+=i; i+=1; .

Additionally, we can remark that the normalization has introduced a new block (lines 4–11). It denotes the scope of the index i of the original **for** loop.

2.4.5 Return Statement

For most analyses, it is easier to have functions that have only a single exit node. Hence, in the normalized AST, all functions have exactly one **return** statement, at the end of the function. If a function has several **return** statements in the original code, they are replaced by an assignment to a temporary variable, followed by a jump to the now unique **return**. This normalization step also takes care of functions in which the execution flow may reach the last } of the function's body, with an ACSL assertion indicating that this must not happen for functions whose return type is not **void**. This constraint is slightly stronger than what the C standard mandates (strictly speaking, the behavior is undefined only if the caller is trying to use the (absent) return value), but this lets Frama-C keep the assertion local to the function.

Normalization of Return Statements

Consider the following file.

```
1  int several_return(int c) {
2    if (c < 0) return 0; else return 1;
3  }
4
5  int no_return_non_void(int c) {
6    if (c < 0) return 0;
7    // missing return for non-void function
8  }
9
10 void no_return_void(int c) {
11   if (c < 0) return;
12   //missing return for void function
13 }
```

Its normalized version as printed by -print is the following:

```
1  int several_return(int c)
2  {
3    int __retres;
4    if (c < 0) {
5      __retres = 0;
6      goto return_label;
7    }
8    else {
9      __retres = 1;
10     goto return_label;
11   }
12   return_label: return __retres;
13 }
14
15 int no_return_non_void(int c)
16 {
17   int __retres;
18   if (c < 0) {
19     __retres = 0;
20     goto return_label;
```

```
21    }
22    /*@ assert missing_return: \false; */ ;
23    __retres = 0;
24    return_label: return __retres;
25  }
26
27  void no_return_void(int c)
28  {
29    if (c < 0) goto return_label;
30    return_label: return;
31  }
```

For several_return, we see that both branches now contain a jump to a newly created **return** statement at the end of the function. In no_return_non_void, we only have a single **return**, but inside a conditional branch: it is again replaced by a **goto** to a new **return** statement at the end of the function's body. Furthermore, as was mentioned above, line 22 contains an ACSL assertion assessing that no execution should reach this point (since this would mean that in the original code we would have reached the outermost closing brace } of a function supposed to return a non-**void** value). The dummy assignment to __retres on the following line will thus never be executed. Finally, no_return_void has a very similar normalization, except that we do not have the ACSL assertion, since in the case of a void return type, we can leave the body of the function without an explicit **return**.

2.4.6 Assembly Code

Frama-C has no means to handle assembly code, which must somehow be stubbed by some equivalent C code in order to have accurate analyses. For inline assembly blocks that follow the GNU extended syntax, in which the memory locations that are written and read by the assembly instructions are explicitly mentioned, the normalized AST generates a minimal ACSL contract surrounding the assembly block, containing a single **assigns** clause analysis plug-in about the parts of the program's state that are modified by the assembly instructions. Note that, of course, this generated clause assumes that the outputs and inputs mentioned in the assembly block are correct with respect to the actual assembly code.

GNU Extended ASM

We can consider the example provided in Gcc documentation [41, Sect. 6.47.2], where the asm statement is basically doing dst = src; dst++;.

```
1
2    int src = 1;
3    int dst;
4
5    asm ("mov %1, %0\n\t"
6         "add $1, %0"
7         : "=r" (dst)  // outputs
8         : "r" (src)); // inputs
9
10   printf("%d\n", dst);
11 }
```

Based on the output specification given on line 6 and the input specification on line 7, the code as printed by Frama-C is the following, with an assigns clause indicating the dst is modified, with a dependency on the value of src.

```
1    int dst;
2    int src = 1;
3    /*@ assigns dst;
4        assigns dst \from src; */
5    __asm__ ("mov %1, %0\n\t"
6             "add $1, %0" : "=r" (dst) : "r" (src));
7    printf("%d\n",dst); /* printf_va_1 */
8    return;
```

2.5 Driving Frama-C

As mentioned in Sect. 2.1.2, Frama-C offers a rich set of command-line options in order to drive analyses that are tailored to the user's specific verification needs. While many options are in fact defined by the plug-ins themselves for their own purpose, and thus out of scope for this chapter, this section focuses on the main options that are introduced by the kernel itself and have an impact on the whole platform. The complete list of kernel options together with a short documentation can be obtained with frama-c -kernel-help.

2.5.1 Handling Frama-C Warnings

An important group of options governs the alarms that might be emitted by the various analysis plug-ins. These options concern mainly situations that are not strictly speaking undefined behaviors in the ISO standard sense (i.e., unspecified or implementation-defined behaviors), or for which, while formally being undefined behavior, a given behavior is prevalent among most if not all current implementations (e.g., for overflows in signed operations). In this latter case, the alarm is activated by default, but can be explicitly turned off if desired.

We already presented the -unspecified-access in Sect. 2.4.1 for checking proper sequencing of memory accesses. Even though unsequenced accesses are undefined behaviors, the option must be explicitly activated, mostly because very few plug-ins are able to raise the corresponding alarms or verify that no problem can occur. Many options deal with integer operations and types. First, it is possible to decide whether to raise an alarm for an operation whose result is out of range for its result type and for a conversion from an integer type to another (of lesser rank) when the initial value cannot fit in the targeted type. This can be done separately for signed and unsigned types, resulting in 4 options: -warn-signed-overflow, -warn-unsigned-overflow, -warn-signed-downcast, and -warn-unsigned-downcast. Only the first one is activated by default, as the other ones are not undefined behaviors (even though a conversion to a signed type of a value that cannot be represented in it is implementation defined and might result in raising a signal). Strictly speaking, the unsigned operations are perfectly well defined in C, but it is frequently the case that an unsigned value wrapping around should be treated as an error, hence the options. Similarly, two options indicate what to do when left- or right-shifting a negative value (-warn-left-negative-shift and -warn-right-negative-shift). The former is activated by default, as this is an undefined behavior, while the latter, being an implementation defined behavior, is not. Finally, a specific option exists for the _Bool type. Namely, -warn-invalid-bool, activated by default, leads to an alarm whenever an attempt is made to store a value other than 0 or 1 into a location whose type is _Bool.

Floating-point operations are assumed to respect the IEEE-754 standard [25], as is the case for nearly every architecture currently in use. However, Frama-C offers the option -warn-special-float to deal with infinity and NaN values. It has three possible values as arguments. Its default value, non-finite, indicates that any computation resulting in either an infinity or NaN should raise an alarm. The value nan instructs Frama-C to raise an alarm only for NaNs, while the value none uses the full range of floating-point values. While computations including NaNs and infinities are perfectly well-defined in IEEE-754, their semantics, especially for NaNs, is quite weird, and many programs are not well prepared to deal with them, hence the default choice made by Frama-C to warn when they might appear.

The last set of options in this category is related to pointer and array accesses. First, -warn-pointer-downcast is similar to the downcast options for integer types, except that it concerns the case where the source type is a pointer type, and the target is not large enough to hold all possible values, meaning that a conversion back to pointer could lead to a different value than the original one. It is activated by default. Second, -warn-invalid-pointer leads to raising an alarm as soon as an invalid pointer value is created (e.g. p+2 if p points to a single value). By default, an alarm is raised only whenever such a value is actually used. Finally, for multidimensional arrays or arrays used within structures, Frama-C (more precisely, its analyzers) warns by default if an index is out of bounds for the inner array, even though the standard accepts such access provided the location is still inside the outer array or structure (assuming there is no issue with padding bytes). To avoid alarms related to such accesses, option `-unsafe-arrays` can be used.

> **-unsafe-arrays usage**
>
> We consider the following code, where we attempt to access field `x` of structure `s` either by considering that it is another cell of array `s.a`, or through pointer arithmetic from the base address of `s.a[0]`.
>
> ```
> 1  extern struct S {
> 2    int a[4];
> 3    int x;
> 4  } s;
> 5  int main() {
> 6    s.x = 2;
> 7    int *p = &s.a[0];
> 8    if (s.a[0] >= 0) return s.a[4];
> 9    else return *(p+4);
> 10 }
> ```
>
> If we use `-unsafe-arrays`, Eva will consider both `s.a[4]` and `*(p+4)` the same way: since we are making an access within the bounds of object `s`, everything is fine. Without this option, the default behavior will be to emit an alarm for `s.a[4]`, as we are *not* within the bounds of `s.a`. On the other hand, p points to a memory block that can handle 5 ints, and `*(p+4)` is still fine, even without `-unsafe-arrays`.

Finally, for really low-level code, it is often the case that some memory accesses must be done at an absolute address. By default, Frama-C emits an alarm for such accesses, but it is possible to specify that a range of integer values is a valid target for memory accesses with the option `-absolute-valid-range` nnn-mmm, where the bounds are given as integer literals in decimal or hexadecimal (with prefix 0x), octal (with prefix 0o), or binary (with prefix 0b) format.

2.5.2 Sequencing Analyses

A Frama-C run is separated in a number of steps that are executed sequentially. First, the plug-ins are loaded (see Sect. 2.2.6). Then, if a `-load` option has been specified, the requested state is loaded from disk (see Sect. 2.5.3). After that, regular kernel and plugin options are parsed, and the analyses are launched. If several analyses are activated at once, it is unspecified in which order they will be executed. Yet, as mentioned in Sect. 2.2.1, it is often the case that a verification task is done by chaining several runs of different plug-ins, in a particular sequence. In order to achieve that, option `-then` and its derivatives must be used. They introduce a separation in the command line between the options that come before and the ones that are after. For this reason, when using -then, it is important to remember to put the names of the source files, the pre-processing options, and/or the -load option (if any), at the beginning of the command line, or at least before the first -then option.

In addition, some plug-ins put their results into a new *project*. From an user point of view, a project is an AST together with associated information (notably property statuses as described in Sect. 2.6.1), and can be referred to by its name. A Frama-C session can have any number of projects, and each of them is independent from the others. A default project is created at the launch of Frama-C and, by default, all analyses operate on it. The -then option does not change the current project, so that it will not be very useful after the run of a plug-in that creates a new project. To run analyses on another project, two options can be used. First, -then-on prj-name indicates that subsequent analyses must be done on the project named prj-name. Second, -then-last tells Frama-C to use the most recently created project. Note that neither of them changes the default project: an occurrence of -then makes Frama-C operate on the old project. This can be changed through the -set-project-as-default option. Finally, a variant of -then-last is -then-replace, which removes the old project from memory and makes the new one the default.

As an example, the following command line performs numerous actions. First, it parses two files. Second, it instruments the result with MetAcsl.[22] Third, it generates annotations for runtime errors (including signed downcasts) with Rte on the new project, making it the default. Fourth, it calls Slicing on the new default project to focus on the generated assertions in some specific functions. Last, it calls Wp on the project named "sliced-export" and containing the sliced result.

Complex Chaining of Analyses on the Command Line

```
frama-c -cpp-extra-args="-DFEATURE=A" file1.c file2.c \
        -then -meta \
        -then-last -set-project-as-default \
                -rte \
                -warn-signed-downcast \
        -then -slicing-project-postfix="-export" \
                -slicing-project-name="sliced" \
                -slice-assert="f1,f2,f3" \
        -then-on sliced-export -wp
```

---

[22] Assuming MetAcsl is properly installed, since it is distributed as an open-source plug-in outside of the standard distribution of Frama-C.

2.5.3 Saving to and Loading from Disk Previous Results

Some operations in Frama-C can take quite some time, and it is thus often useful to store some intermediate results on disk in order to retrieve them afterward. For this purpose, -save file saves all currently existing projects in file, while -load file loads the projects stored in file. Indeed, since the loading step is performed quite early, only a default, empty project exists at that point, and it is simply replaced by the information contained inside file. Note that parts of the loaded file might be unknown to the actual Frama-C run, notably if the set of plug-ins that have been loaded differs with respect to the run which created the file. In that case, these parts, which are referred to as states by Frama-C, are ignored, and a warning giving the number of ignored states is emitted.

Sample Makefile for Visualizing Eva Results

It is possible to decompose an analysis with Eva in three main step:

1. parsing source files
2. launching the analysis itself
3. investigating the result in the GUI

The first step needs only be redone when the source files change, while the second one can be activated more often, as various parameters are tuned to increase the precision of the analysis. Finally, the results of a single run are often visualized several times. The following simple Makefile, inspired from the one generated by the Frama-C scripts described in Chap. 8, ensures that the results of the first two steps are saved on disk and only remade as needed. With this Makefile, typing make first remakes the analysis (resp. the parsing) if eva.sav (resp. parsing.sav) is out-of-date. Then, it launches the GUI.

```
SOURCES=file1.c file2.c file3.c
PARSE_ARGS=-cpp-extra-args="-DFEATURE=A"
EVA_ARGS=-eva-precision 3

default: gui

parsing.sav: $(SOURCES)
    frama-c $(SOURCES) $(PARSE_ARGS) -save $@

eva.sav: parsing.sav
    frama-c -load $< -eva $(EVA_ARGS)

gui: eva.sav
    frama-c-gui -load $<
```

2.6 Visualizing Analyses Results

This section introduces first in Sect. 2.6.1 the notion of *property statuses* that summarizes which properties have been proven (or not) by Frama-C. Then, Sect. 2.6.2 presents a few ways to generate convenient reports about what Frama-C and its analyzers have computed.

2.6.1 Property Statuses

The primary goal of Frama-C consists in checking *properties* of C code thanks to its verification plug-ins, mainly Eva, Wp and E-ACSL. Properties are usually expressed through ACSL annotations, which can come from different sources. For instance, an annotation can be:

- manually written by the user;
- automatically generated by a Frama-C expressiveness plug-in;
- emitted as a consequence of an alarm by Eva or Rte.

Most of the time, there is a one-to-one correspondence between an ACSL annotation and a property, even if it is not always the case.

Visualizing Properties of Function gethostname

The function gethostname defined in the file unistd.h of the C standard library is specified as follows in the Frama-C standard library:

```
1 #include "__fc_define_size_t.h"
2 #include "limits.h"
3
4 extern volatile char __fc_hostname[HOST_NAME_MAX];
5
6 /*@ requires name_has_room: \valid(name + (0 .. len-1));
7   @ assigns \result, name[0 .. len-1]
8     \from indirect:__fc_hostname[0 .. len], indirect:len;
9   @ ensures result_ok_or_error: \result == 0 || \result ==
        -1; */
10 extern int gethostname(char *name, size_t len);
```

It is not our goal in this chapter to precisely explain this specification (see Chap. 1 for the meaning of the different clauses involved in this function contract), but the `requires` and the `ensures` clauses correspond to exactly one property each, while the `assigns` clause actually introduces three different properties. These properties are explicitly shown in the panel *Properties* at the bottom of the Frama-C GUI, as illustrated by the screenshot of Fig. 2.3.

2 The Heart of Frama-C: The Frama-C Kernel

```
 0 /*@ requires name_has_room: \valid(name + (0 .. len - 1));
 @     ensures result_ok_or_error: \result ≡ 0 v \result ≡ -1;
 @     assigns \result, *(name + (0 .. len - 1));
 @     assigns \result
         \from (indirect: __fc_hostname[0 .. len]),(indirect:len);
 @     assigns *(name + (0 .. len - 1))
         \from (indirect: __fc_hostname[0 .. len]),(indirect:len);
   */
   extern int gethostname(char *name, size_t len);
```

```
unistd.h
885
886 extern volatile char __fc_hostname[HOST_NAME_MAX];
887
888 /*@
889   requires name_has_room: \valid(name + (0 .. len-1));
890   assigns \result, name[0 .. len-1]
891        \from indirect:__fc_hostname[0 .. len], indirect:len;
892   ensures result_ok_or_error: \result == 0 || \result == -1;
893 */
894 extern int gethostname(char *name, size_t len);
895
```

Information	Messages (0)	Console	Properties (5)	Values	Red Alarms	WP Goals		
Refresh	Function	File	Kind				Status	Consolidated Status
⊘ Current function	gethostname	unistd.h	requires name_has_room: \valid(name + (0 .. len - 1))				○	no verification attempted
▼ Status	gethostname	unistd.h	ensures result_ok_or_error: \result ≡ 0 v \result ≡ -1				⊘	unverifiable but considered VALID; requires external review
⊘ Valid	gethostname	unistd.h	assigns \result, *(name + (0 .. len - 1))				⊘	unverifiable but considered VALID; requires external review
Valid under hyp.	gethostname	unistd.h	assigns \result \from (indirect: __fc_hostname[0 .. len]), (indirect: len);				⊘	unverifiable but considered VALID; requires external review
⊘ Unknown	gethostname	unistd.h	assigns *(name + (0 .. len - 1))				⊘	unverifiable but considered VALID; requires external review
⊘ Invalid			\from (indirect: __fc_hostname[0 .. len]), (indirect: len);					
Invalid under hyp.								
⊘ Considered valid								

Fig. 2.3 Properties for the contract of function gethostname shown in the Frama-C GUI

Each Frama-C's verification plug-in may verify code properties expressed through ACSL annotations. In some cases, though, the verified property can stay implicit and not be materialized by an ACSL annotation. This is notably the case for Eva: an alarm (hence an ACSL assertion) is only emitted if Eva cannot guarantee the absence of undefined behavior at a given program point. Conversely, if no alarm is raised for a given instruction, then we know that it is free of undefined behavior.

Annotations Generated by Eva

Consider the function next defined as follows:

```
1 int next(int x) {
2    return x+1;
3 }
```

It is possible to analyze this function with Eva through the kernel option -main and to save the result in a file next.sav as follows.

```
1  $ frama-c next.c -eva -main next -save next.sav
2  [kernel] Parsing next.c (with preprocessing)
3  [eva] Analyzing a complete application starting at next
4  [eva] Computing initial state
5  [eva] Initial state computed
6  [eva:initial-state] Values of globals at initialization
7
8  [eva:alarm] next.c:2: Warning:
9     signed overflow. assert x + 1 ≤ 2147483647;
10 [eva] done for function next
11 [eva] ====== VALUES COMPUTED ======
12 [eva:final-states] Values at end of function next:
13    __retres ∈ [-2147483647..2147483647]
14 [eva:summary] ====== ANALYSIS SUMMARY ======
15 ----------------------------------------------------------------
16    1 function analyzed (out of 1): 100% coverage.
```

```
17    In this function, 2 statements reached (out of 2): 100%
          coverage.
18    ----------------------------------------------------------------
19    No errors or warnings raised during the analysis.
20    ----------------------------------------------------------------
21    1 alarm generated by the analysis:
22           1 integer overflow
23    ----------------------------------------------------------------
24    No logical properties have been reached by the analysis.
25    ----------------------------------------------------------------
```

The output shows that Eva raises an alarm (because x+1 may overflow if x is equal to 2147483647, the biggest int in Frama-C's default *machdep*). Additionally, it generates an associated ACSL assertion that states explicitly what are the safe values of x. It is clearly visible in the output message starting with [eva:alarm]. It is also visible when loading the saved file and pretty printing the code in the GUI or as follows.

```
1  $ frama-c -load next.sav -print
2  /* Generated by Frama-C */
3  int next(int x)
4  {
5    int __retres;
6    /*@ assert Eva: signed_overflow: x + 1 ≤ 2147483647; */
7    __retres = x + 1;
8    return __retres;
9  }
```

For each property, a verification plug-in may generate three different *validity statuses* (or, shortly, *statuses*): either the property is *true*, or it is *false*, or it is *unknown* because the plug-in is not able to conclude whether the property is true or false. Additionally, when the property is false, a plug-in may also weaken this status by stating that it is false *only if* the corresponding program point is reachable. Indeed, most Frama-C plug-ins are not able to prove that a particular statement is necessarily reachable from some program's inputs.

Statuses Emitted by Eva

Consider the following program.

```
1  /*@ assigns \result \from n; */
2  extern int no_code(int n);
3
4  int main(void) {
5    int a = no_code(0);
6    int x = 0;
7    /*@ assert x == 0; */
8    /*@ assert a == 0; */
```

```
 9    /*@ assert x == 1; */
10    return 0;
11 }
```

When running **Eva** on it and displaying the results in the **GUI** as follows, one gets the screenshot of Fig. 2.4.

```
1 $ frama-c-gui -eva status.c
```

This screenshot shows that the first assertion is proven valid by **Eva**, the third one is proven invalid assuming that the property is reachable, while **Eva** has not been able to conclude for the second assertion since it has no information about the behavior of function no_code on which the value of a in this assertion depends. Each status is textually shown in the *Property* panel of the GUI, while colored bullets provide a convenient way to visualize them: green (o) for *valid*, orange (o) for *unknown*, red (●) for *false*, and a mix of red and orange (●) when *false if reachable*.[23]

During the verification process, some other statuses may happen. First, a blue circle (o) means that no plug-in attempted to verify this property. Second, a mix of green and blue (●) denotes a property that is assumed to be valid but should be verified by external review, which is typically the case of function postconditions for which no implementation is provided (in particular functions from **Frama-C** standard library). Third, a mix of green and orange (●) means that a plug-in successfully verified the considered property P under the hypothesis of (at least) another unproven property. Currently, **Wp** is the only plug-in that generates some hypotheses when proving properties. Therefore, this last status usually happens when (and only when) **Wp**

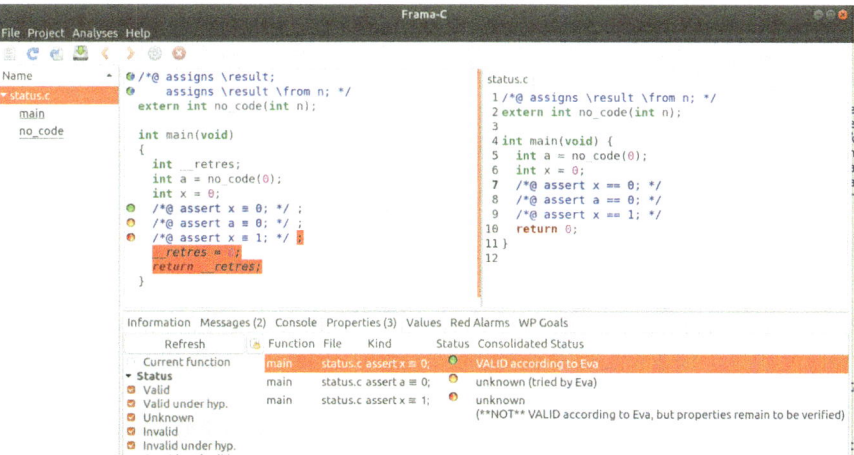

Fig. 2.4 Validity statuses emitted by Eva

proves some but not all properties of a particular function. As soon as some other plug-ins prove all the necessary unproven properties, the status of P is automatically updated to valid (◉). Until this happens, however, it is important to keep in mind that P *cannot* be considered "quasi-valid": by construction, Frama-C has no clue about the validity of the hypotheses on which the validity of P itself depends.

Statuses when Combining Plug-ins

Consider the following program for which verification engineers would like to demonstrate both the absence of runtime errors and a functional property expressed by the last assertion at line 26 stating that the content of array a is always non-negative at the end of the execution.

```
1  /*@ assigns \result \from n, m;
2    @ behavior even:
3    @   assumes m % 2 == 0;
4    @   ensures \result >= 0;
5    @ behavior odd:
6    @   assumes m % 2 != 0;
7    @   ensures n >= 0 ==> \result >= 0;
8    @   ensures n < 0 ==> \result <= -1; */
9  int pow(int n, unsigned int m);
10
11 int sum(int x) {
12   int n = 0, i;
13   for(i = 1; i <= x; i++) n += i;
14   return n;
15 }
16
17 int main(void) {
18   int i, n = 0;
19   unsigned int a[100];
20   n = sum(10);
21   /*@ loop invariant 0 <= i <= n ;
22     @ loop invariant \forall integer k; 0 <= k < i ==> a[k]
                >= 0;
23     @ loop assigns i,a[0..n-1]; */
24   for(i = 0; i < n; i++)
25     a[i] = pow(i, i);
26   /*@ assert \forall integer k; 0 <= k < n ==> a[k] >= 0; */
27 }
```

For this verification task, they would like to use the Wp plug-in, and so they wrote a few additional annotations for specifying the possible behaviors of function pow for which no implementation is provided, and for specifying the invariants of the for-loop in function main. For the purpose of this example, it is not necessary to understand the meaning of these annotations and why they are required, but the interested reader may refer to Chap. 1 for their meaning and to Chap. 4 that explains how to verify properties with Wp.

When running plug-in **Wp** on this example and displaying the results in the GUI as follows, they get the screenshot of Fig. 2.5.

```
$ frama-c-gui -wp -wp-rte mix.c
```

It shows that the postconditions of function `pow` are assumed valid (⬤), while some properties have been successfully proven but relies on other unproven properties (⬤). These properties are the one whose status is still unknown (⬤). It would be possible to add additional **ACSL** annotations to complete the proof with **Wp**, which is left as an exercise for the readers familiar with deductive verification, but the engineers decide to complete the proof with **Eva**, which requires no additional **ACSL** annotations. For reaching their goal, they just set the **Eva**'s `slevel` parameter to 100^{24} and click on button Run in the **Eva**'s panel of the **GUI**. They instantaneously get the screenshot of Fig. 2.6, which shows that their verification goal is reached since all the expected properties are proven (⬤). Note that the postconditions of function `pow` are assumed valid by the **Frama-C** kernel (⬤) since they cannot be proven by any mean without providing a function body.

It is worth noting that the verification engineers still need to take care of the implementation of function `pow`, which has not been verified by **Frama-C**. In particular, they have no evidence that it would satisfy its **ACSL** contract. One way of doing so would be to run the **E-ACSL** plug-in after running **Wp** and **Eva**: it would generate an instrumented version of the code example that will monitor at runtime any implementation of `pow` that will be linked to it (see Chap. 5 for details about **E-ACSL**).

A few other statuses are also possible. Three of them mix green, orange or red with black in order to respectively denote a valid (⬤), unknown (⬤), or invalid (⬤) status for a property in a piece of dead code. From a theoretical point of view, these statuses are equivalent to valid (because any property in dead code is valid for all execution paths reaching it), but distinguishing them is useful in some industrial verification methodologies. The notion of dead code is mostly used by the **Eva** plug-in, and refers to code that cannot be reached from the main entry point.

Lastly, the status of a property can also be inconsistent (⬤), meaning that something wrong happens when combining plug-ins: one proves that the property is valid, while another proves that it is invalid. In practice, it can happen either if there is a critical soundness bug in one of the involved plug-ins or in the **Frama-C** kernel,[25] or if users incorrectly use them (for instance, they break one of the soundness hypothesis indicated in the analyzer's manual).

---

[25] Should such a bug occur, it should be reported immediately on the **Frama-C** bug tracking system at https://git.frama-c.com/pub/frama-c.

Fig. 2.5 Wp has only partially proven a program

Fig. 2.6 Eva has completed the Wp's partial proof

For the interested reader, [13] shows the backstage by explaining how the Frama-C kernel automatically computes the correct status for each property, according to the analyzers' results.

2.6.2 Generating Reports

Skimming through the output of a Frama-C run or inspecting the statuses of the properties in the GUI is not the only way to retrieve information about the results of the analyses that have taken place. There are various ways to select and present the points that are relevant for the verification task at hand.

2.6.2.1 Redirecting Selected Messages

First of all, a typical Frama-C run generates quite a lot of messages, coming from the kernel as well as from the various plug-ins involved at some point. Not all messages have the same importance for a given verification task, and it is possible to have each message of a given category (**f**eedback, **r**esult, **w**arning, **u**ser error, **i**nternal error) and a given plug-in (or of the kernel) be copied into a file. The same file can be used for several categories and plug-ins. For instance, the following command line creates a file `fc-warnings.txt` that holds the warnings emitted by either the kernel or Eva during the analysis of `file.c`: `frama-c -kernel-log w:fc-warnings.txt -eva-log w:fc-warnings.txt -eva file.c`, but no other message.

2.6.2.2 Audit Information

In addition to verifying that ACSL properties are properly validated, it is also important to check that Frama-C has been called with an appropriate set of parameters. In order to be able to review these parameters, Frama-C offers an *audit mode*, in the form of option `-audit-prepare file`, which puts information about parameters that are relevant for the correctness of the analysis in `file`, in the form of a JSON object. This mode is currently only supported by the Eva plug-in.

Information Produced by Audit Mode

If we go back to the `status.c` file from Sect. 2.6.1, the command `frama-c -audit-prepare fc-audit.json -eva status.c` produces a JSON object following the structure below.

```
{
  "eva": {
    "correctness-parameters": {
      "-absolute-valid-range": "",
      ...
      "-lib-entry": "false",
      "-main": "main",
      "-safe-arrays": "true",
      "-unspecified-access": "false",
      ...
      "-warn-signed-overflow": "true",
      ...
    },
    "warning-categories": {
      "enabled": [
        "*", "alarm", "builtins", "builtins:missing-spec",
        ...
      ],
      "disabled": [
        "garbled-mix", "invalid-assigns",
            "loop-unroll:auto",
        "loop-unroll:missing", "loop-unroll:missing:for",
        "loop-unroll:partial", "malloc:weak"
      ]
    }
  },
  "kernel": {
    "warning-categories": {
      "enabled": [
        "*", "CERT", "CERT:EXP", "CERT:EXP:46", "CERT:MSC",
            "CERT:MSC:37",
        "CERT:MSC:38", "acsl-extension", "annot",
            "annot:missing-spec",
        ...
      ],
      "disabled": [
        "CERT:EXP:10", "acsl-float-compare",
            "file:not-found",
        "ghost:already-ghost", "transient-block"
      ]
    }
  },
  "sources": { "status.c":
      "86c29d85c6b5d9074ef3ecff0ada0369" }
}
```

We can see the status of the options that have an influence on the alarms produced by Eva, as well as the status (enabled or not) of the warning messages for both Eva and the kernel, and a hash of the content of each source file considered in the analysis. The JSON format makes it easy to use scripts to perform some checks. For instance, with the `jq` utility, we can easily verify that the signed overflow alarms were activated:

```
1 $ jq
     '.eva."correctness-parameters"."-warn-signed-overflow"'
     fc-audit.json
2 "true"
```

The main feature of the audit mechanism is to allow replaying the analysis with the same parameters as was done when creating the audit file: the user then submits this file to the auditor, who can then re-run the analysis themselves and check that the results are identical. With the `-audit-check` option, Frama-C will emit a warning if this is not the case.

Options Mismatch between Command Line and Audit File

Continuing the previous example, if one tries to disable `-warn-signed-overflow` with respect to what stands in the JSON object, then a warning is emitted by the kernel.

```
1  $ frama-c -audit-check fc-audit.json examples/status.c -eva
       -no-warn-signed-overflow
2  ...
3  [kernel:audit] Warning:
4    correctness parameter -warn-signed-overflow: expected
         value true, but got false
5  [eva] Analyzing a complete application starting at main
6  ...
7  [eva:summary] ====== ANALYSIS SUMMARY ======
8  --------------------------------------------------------------
9  1 function analyzed (out of 1): 100% coverage.
10 In this function, 5 statements reached (out of 7): 71%
       coverage.
11 --------------------------------------------------------------
12 Some errors and warnings have been raised during the
       analysis:
13   by the Eva analyzer:     0 errors     0 warnings
14   by the Frama-C kernel:   0 errors     2 warnings
15 --------------------------------------------------------------
16 0 alarms generated by the analysis.
17 --------------------------------------------------------------
18 Evaluation of the logical properties reached by the
       analysis:
```

```
19      Assertions            1 valid      1 unknown      1 invalid
                      3 total
20      Preconditions         0 valid      0 unknown      0 invalid
                      0 total
21      33% of the logical properties reached have been proven.
22      ----------------------------------------------------------------
23      [kernel] Warning: warning audit treated as deferred error.
            See above messages for more information.
24      [kernel] Frama-C aborted: invalid user input.
```

The auditor is still responsible for defining which parameters are appropriate for a given system; the -audit-* options help to automate manual checking steps, but do not replace human expertise.

2.6.2.3 Generating Report

It is also possible to obtain a summary of the status of each property in the current project through the use of the Report plug-in, which is fully documented in Frama-C's user manual [12, Chap. 10]. The most basic way to use the plug-in is to simply ask for a textual output after an analysis has taken place.

Textual output of Report

Consider again the example of Sect. 2.6.1. If we only launch Eva on file mix.c before asking for a report through the command line frama-c mix.c -eva -then -report, we obtain the following output.

```
1       ----------------------------------------------------------------
2       --- Properties of Function 'pow'
3       ----------------------------------------------------------------
4
5       [ Extern   ] Post-condition for 'even' (file examples/mix.c,
            line 4)
6                    Unverifiable but considered Valid.
7       [ Extern   ] Post-condition for 'odd' (file examples/mix.c,
            line 7)
8                    Unverifiable but considered Valid.
9       [ Extern   ] Post-condition for 'odd' (file examples/mix.c,
            line 8)
10                   Unverifiable but considered Valid.
11      [ Extern   ] Assigns nothing
12                   Unverifiable but considered Valid.
13      [ Extern   ] Froms (file examples/mix.c, line 1)
14                   Unverifiable but considered Valid.
15      [ Valid    ] Default behavior
16                   by Frama-C kernel.
17      [ Valid    ] Behavior 'even'
```

```
18                by Frama-C kernel.
19 [  Valid   ] Behavior 'odd'
20                by Frama-C kernel.
21
22 ------------------------------------------------------------
23 --- Properties of Function 'sum'
24 ------------------------------------------------------------
25
26 [     -     ] Assertion 'Eva,signed_overflow' (file
      examples/mix.c, line 13)
27                tried with Eva.
28
29 ------------------------------------------------------------
30 --- Properties of Function 'main'
31 ------------------------------------------------------------
32
33 [     -     ] Invariant (file examples/mix.c, line 21)
34                tried with Eva.
35 [     -     ] Invariant (file examples/mix.c, line 22)
36                tried with Eva.
37 [     -     ] Assertion (file examples/mix.c, line 26)
38                tried with Eva.
39 [     -     ] Assertion 'Eva,index_bound' (file
      examples/mix.c, line 25)
40                tried with Eva.
41
42 ------------------------------------------------------------
43 --- Status Report Summary
44 ------------------------------------------------------------
45      3 Completely validated
46      5 Considered valid
47      5 To be validated
48     13 Total
49 ------------------------------------------------------------
```

This textual output is human-readable, but if one wants to perform additional transformations, it is also possible to ask **Report** to generate a **CSV** file (or more precisely a tab-separated values file) that conveys the same information as above. It can then be imported into one's favorite spreadsheet application or statistical computation tool. On the same example as above, using `frama-c mix.c -eva -then -report-csv status.csv`, we obtain in `status.csv` an output similar to Fig. 2.7.

```
1  file    line  func  property kind      status       property
2  mix.c   1     pow   from clause        Cons.valid   assigns \result \from n, m;
3  mix.c   4     pow   postcondition      Cons.valid   \result ≥ 1
4  mix.c   7     pow   postcondition      Cons.valid   \old(n) ≥ 0 ==> \result ≥ 1
5  mix.c   8     pow   postcondition      Cons.valid   \old(n) < 0 ==> \result ≤ -1
6  mix.c   9     pow   assigns clause     Cons.valid   assigns \nothing;
7  mix.c   13    sum   signed_overflow    Unknown      n + i ≤ 2147483647
8  mix.c   21    main  loop invariant     Unknown      0≤i≤n
9  mix.c   22    main  loop invariant     Unknown      ∀ integer k; 0≤k<i ==> a[k] ≥ 1
10 mix.c   25    main  index_bound        Unknown      i < 100
11 mix.c   26    main  user assertion     Unknown      ∀ integer k; 0≤k<n ==> a[k] ≥ 1
```

Fig. 2.7 CSV file for property statuses

In fact, Report itself can perform some classification tasks. Moreover, it can be used to classify not only properties, but also warnings or errors emitted during the analysis. Note however that in the latter case, Report processes the warning as they are emitted, so it should *not* be separated from the main analysis by a -then option. The classification is driven by a JSON file containing a list of rules. Each rule is a JSON object whose fields can be separated into two parts: first, a pattern describing the events that are matched by the rule, and second the treatment that should be done for matching events.

For determining if a warning matches, we can specify a plugin field (which can be the kernel itself), and potentially a category. Finally, we can give an OCaml regular expression (regexp) [26] in the warning field. Conversely, to match a property of a given status, we simply put a regexp in a field named after the status (unknown, valid, or invalid), without plugin or category field.

Regarding the output of the event, the main point is the action we want to perform on it: SKIP removes it entirely from the report, INFO reports it as informational content, REVIEW indicates that someone should review what happened, and ERROR marks the fact that the event should be treated as an error (making Frama-C end up with an error after having output the record). It is also expected that each class of events gets a name through the classid field. Finally, it is possible to change the default title and description of the event, potentially using the information contained in the matching regex.

Example of Classification File for Report

To make things more concrete, we use a classification file that only marks for review the alarms emitted by Eva (that are warnings of category alarm), and use the first part of the message, which is the short name retained by Eva to describe the alarm, as the title, prefixed by Alarm:. We will also separate these kinds of warning in two classes, named respectively eva-alarm and eva-misc through the classid field. If other warnings are emitted by Eva, we keep them, but only as INFO. Finally, any unknown property is ignored, but in case an invalid property is discovered, we'll raise an ERROR.

---

[26] There are some slight variations with respect to standard POSIX regexp. See https://v2.ocaml.org/releases/5.1/api/Str.html for a complete description of the format.

2 The Heart of Frama-C: The Frama-C Kernel

```
[{ "plugin": "eva" ,
   "category": "alarm" ,
   "warning": "\([^.]*\).",

   "classid": "eva-alarm",
   "title": "Alarm:_\1" },

 { "plugin": "eva" ,
   "action": "INFO",
   "warning": ".*",
   "classid": "eva-misc" },

 { "unknown": ".*",
   "action": "SKIP" },

 { "invalid": ".*",
   "action" : "ERROR" }]
```

With this classification file, we run Eva on mix.c with the following command in order to store the classified events as a JSON array into report.json.

```
$ frama-c mix.c \
          -eva \
          -report-classify \
          -report-rules report-driver.json \
          -report-output report.json
```

The content of this file is shown below.

```
[
 { "classid": "eva-misc", "action": "INFO",
   "title": "The builtin pow will not be used for function
       pow of incompatible type.",
   "descr": "The builtin pow will not be used for function
       pow of incompatible type.\n(got: int (int n,
       unsigned int m)).",
   "file": "examples/mix.c", "line": 9 },
 { "classid": "eva-alarm", "action": "REVIEW",
   "title": "Alarm: signed overflow",
   "descr": "signed overflow. assert n + i \226\137\164
       2147483647;",
   "file": "examples/mix.c", "line": 13 },
 { "classid": "eva-alarm", "action": "REVIEW",
   "title": "Alarm: loop invariant got status unknown",
   "descr": "loop invariant got status unknown.",
   "file": "examples/mix.c", "line": 21 },
 { "classid": "eva-alarm", "action": "REVIEW",
   "title": "Alarm: loop invariant got status unknown",
   "descr": "loop invariant got status unknown.",
   "file": "examples/mix.c", "line": 22 },
 { "classid": "eva-alarm", "action": "REVIEW",
```

```
19      "title": "Alarm: accessing out of bounds index",
20      "descr": "accessing out of bounds index. assert i <
            100;",
21      "file": "examples/mix.c", "line": 25 },
22    { "classid": "eva-alarm", "action": "REVIEW",
23      "title": "Alarm: assertion got status unknown",
24      "descr": "assertion got status unknown.", "file":
            "examples/mix.c",
25      "line": 26 }
26  ]
```

2.6.2.4 Generating a SARIF file

As mentioned in Sect. 2.1.1, Frama-C is also able to output its result in the Static Analysis Result Interchange Format [21] (SARIF). SARIF is a standard format managed by OASIS, and comes in the form of a JSON schema. As its name suggests, it is meant to offer a common way for static software analysis tools to present their results, including the parameters that have been used, the problems that have been found, and the properties that have been validated. It is in particular used in many continuous integration pipelines, including in many GitHub repositories.[27]

SARIF support is handled by the MDR plug-in. A SARIF output can be obtained with the following command: frama-c -eva mix.c -then -mdr-gen sarif -mdr-out eva.sarif. The format itself is a bit verbose. Hence, we do not give the full content of the resulting eva.sarif file here, but focus on the most important components below.

SARIF File as generated by Frama-C

```
1  { ...
2    "runs": [
3      {
4        "tool": {
5          "driver": {
6            "name": "frama-c",
7            "fullName": "frama-c-25.0 (Manganese)",
8            "version": "25.0 (Manganese)",
9            ...
10         }
11       },
12       "invocations": [
13         {
```

---

[27] See https://docs.github.com/en/code-security/code-scanning/integrating-with-code-scanning/sarif-support-for-code-scanning.

```
14      "commandLine": "frama-c mix.c -eva -no-unicode -then
           -mdr-gen sarif -mdr-out eva.sarif",
15         ...
16      "exitCode": 0, "executionSuccessful": true }
17    ],
18    "originalUriBaseIds": {
19      "FRAMAC_SHARE": ...
20      "FRAMAC_LIB": ...
21      "FRAMAC_PLUGIN": ...
22      "PWD": ...
23    },
24    "artifacts": [
25      { "location": { "uri": "mix.c", "uriBaseId": "PWD" },
26      "roles": [ "analysisTarget" ], "mimeType":
           "text/x-csrc" }
27    ],
28    "results": [
29      { "ruleId": "signed_overflow", "kind": "open",
           "level": "none",
30      "message": ...
31      "locations": ...
32      },
33      { "ruleId": "index_bound", "kind": "open", "level":
           "none", ... },
34      { "ruleId": "user-spec", "kind": "pass", "level":
           "none", ... }
35    ],
36    "taxonomies": [
37      { "name": "frama-c",
38      "rules": [
39        { "id": "user-spec",
40        "shortDescription": { "text": "User-written ACSL
             specification." } },
41        { "id": "signed_overflow",
42        "shortDescription": { "text": "Integer overflow or
             downcast." } }, {
43        "id": "index_bound",
44        "shortDescription": { "text": "Array access out of
             bounds." } }
45      ], "contents": [ "nonLocalizedData" ] }
46    ] }
47  ]
48 }
```

After the (omitted) header indicating which SARIF version is in use, we start by finding generic information about Frama-C itself, including its version, and the options that have been activated for this run. Then, we have the list of source files that have been considered (only mix.c in this case). The main part of the file consists in the list of results, i.e. properties with their status (kind in the vocabulary of SARIF).

Each property has its category, called `ruleId` in SARIF. These `ruleId` are specific to Frama-C. Their description is given directly in the generated file, in the last field of the main object, `taxonomies`.

2.7 Conclusion

This chapter has presented the main features of the Frama-C kernel. They include how to prepare a C source code, how to extend it with ACSL annotations, how to drive Frama-C, and how to visualize its results. The chapter has also presented Frama-C's core plug-ins and the main ways to make them collaborate, which is a key feature of the framework.

Since it does not focus on any particular plug-in, even if some examples use a few of them, this chapter is not enough by itself for analyzing a C source code with Frama-C in any useful way. However, understanding the common set of concepts introduced here is definitely necessary in order to use correctly and efficiently any Frama-C plug-in, in particular those covered in the next chapters of this book.

Frama-C is an evolving tool, and so is its kernel: its development was initiated in 2004 and is still ongoing. Even if its software architecture has evolved over time [38] and might still evolve in the future, Frama-C tries to be backward compatible as much as possible: most features that were present 10 or 15 years ago are still here, even if new functionalities are added on a regular basis. In particular, one might expect that the Frama-C kernel will offer in the future native support for new key features that will directly benefit Frama-C plug-ins. First, the Frama-C kernel might natively support several programming languages of different natures. It would allow Frama-C's analyzers to deal with code written in various languages (e.g., Java and Rust). Second, it might provide support for incremental analyses that would allow the user to quickly analyze a revision of the program with respect to an original version that would have already been analyzed by Frama-C. This should help integrate Frama-C into CI/CD toolchains. Last, Frama-C might take benefit from parallelism in order to speed up the analyzers themselves, but also their combination capabilities offered by the kernel. The recent support of shared memory parallelism in a multicore setting by the OCaml compiler [40], the language in which most of Frama-C is written (see Chap. 7), should help.

References

1. Alur R, Etessami K, Madhusudan P (2004) A temporal logic of nested calls and returns. In: International conference on tools and algorithms for the construction and analysis of systems (TACAS)
2. Ammann P, Offutt J (2008) Introduction to software testing, 1 edn. Cambridge University Press (2008)

3. Bardin S, Chebaro O, Delahaye M, Kosmatov N (2014) An all-in-one toolkit for automated white-box testing. In: Tests and proofs. TAP 2014. Lecture notes in computer science, vol 8570. Springer Verlag, York, United Kingdom, pp 53–60. https://doi.org/10.1007/978-3-319-09099-3_4
4. Bardin S, Kosmatov N, Cheynier F (2014) Efficient leveraging of symbolic execution to advanced coverage criteria. In: 2014 IEEE seventh international conference on software testing, verification and validation. IEEE Computer Society, Cleveland, OH, United States, pp 173–182. https://doi.org/10.1109/ICST.2014.30
5. Bardin S, Kosmatov N, Marcozzi M, Delahaye M (2021) Specify and measure, cover and reveal: A unified framework for automated test generation. Sci Comput Program 207:102641. https://doi.org/10.1016/j.scico.2021.102641, https://www.sciencedirect.com/science/article/pii/S0167642321000344
6. Baudin P, Bobot F, Bühler D, Correnson L, Kirchner F, Kosmatov N, Maroneze A, Perrelle V, Prevosto V, Signoles J, Williams N (2021) The dogged pursuit of bug-free C programs: the Frama-C software analysis platform. Commun ACM
7. Baudin P, Filliâtre JC, Marché C, Monate B, Moy Y, Prevosto V ACSL: ANSI/ISO C Specification Language. http://frama-c.com/acsl.html
8. Baudin P, Filliâtre JC, Marché C, Monate B, Moy Y, Prevosto V ACSL: ANSI/ISO C specification language version 1.18. Implementation in Frama-C 26.0. http://frama-c.com/acsl.html
9. Bonichon R, Yakobowski B Frama-C's metrics plug-in. https://www.frama-c.com/download/frama-c-metrics-manual.pdf
10. Cok DR ACSL++: ANSI/ISO C++ specification language. https://github.com/acsl-language/acsl
11. Cok DR Frama-Clang plug-in user manual. https://www.frama-c.com/download/frama-clang-manual.pdf
12. Correnson L, Cuoq P, Kirchner F, Maroneze A, Prevosto V, Puccetti A, Signoles J, Yakobowski B Frama-C user manual. https://www.frama-c.com/download/frama-c-user-manual.pdf
13. Correnson L, Signoles J (2012) Combining analyses for C program verification. In: Formal methods for industrial case studies (FMICS)
14. Cousot P (2021) Principles of abstract interpretation. MIT Press
15. Cousot P, Cousot R (1977) Abstract interpretation: a unified lattice model for static analysis of programs by construction or approximation of fixpoints. In: Symposium on principles of programming languages (POPL)
16. Cuoq P, Signoles J, Baudin P, Bonichon R, Canet G, Correnson L, Monate B, Prevosto V, Puccetti A (2009) Experience report: OCaml for an industrial-strength static analysis framework. In: International conference of functional programming (ICFP)
17. De Oliveira S (2018) Finding constancy in linear routines. PhD thesis, Université Paris-Saclay
18. de Oliveira S, Bensalem S, Prevosto V (2016) Polynomial invariants by linear algebra. In: International symposium on automated technology for verification and analysis (ATVA'16)
19. Dijkstra EW (1975) Guarded commands, nondeterminacy and formal derivation of programs. Commun ACM
20. Emerson EA, Clarke EM (1980) Characterizing correctness properties of parallel programs using fixpoints. Autom Lang Program
21. Golding FMC, L.J.G. (ed.) (2020) Static analysis results interchange format (SARIF) Version 2.1.0. OASIS standard. https://docs.oasis-open.org/sarif/sarif/v2.1.0/sarif-v2.1.0.html
22. Georgiadis L, Tarjan RE, Werneck RF (2006) Finding dominators in practice. J Graph Algorithms Appl
23. Herrmann P, Signoles J Annotation generation: Frama-C's RTE plug-in. http://frama-c.com/download/frama-c-rte-manual.pdf
24. IEEE (2018) IEEE Std 1003.1-2017 the open group base specifications, Issue 7, 2018 edition. USA
25. IEEE Standard for Floating-Point Arithmetics (2019). Standard 754-2019, IEEE. https://ieeexplore.ieee.org/document/8766229

26. ISO/IEC JTC 1/SC 22 (1999) ISO/IEC 9899:1999 programming languages - C. Standard 9899:1999, International standard organisation. https://www.iso.org/standard/29237.html
27. ISO/IEC JTC 1/SC 22 (2011) ISO/IEC 9899:2011 programming languages - C. Standard 9899:2011, International Standard Organisation. https://www.iso.org/standard/57853.html
28. Kosmatov N, Williams N, Botella B, Roger M, Chebaro O (2012) A lesson on structural testing with PathCrawler-online.com. In: International conference on tests and proofs (TAP)
29. Marcozzi M, Bardin S, Delahaye M, Kosmatov N, Prevosto V (2017) Taming coverage criteria heterogeneity with LTest. In: 2017 IEEE international conference on software testing, verification and validation (ICST) (2017), pp 500–507. https://doi.org/10.1109/ICST.2017.57
30. Marcozzi M, Delahaye M, Bardin S, Kosmatov N, Prevosto V (2017) Generic and effective specification of structural test objectives. In: Proceedings of the 10th IEEE international conference on software testing, verification and validation (ICST 2017), pp 436–441. https://doi.org/10.1109/ICST.2017.57
31. Martin T, Kosmatov N, Prevosto V, Lemerre M (2020) Detection of polluting test objectives for dataflow criteria. In: Dongol B, Troubitsyna E (eds) Integrated formal methods, pp 337–345. Springer International Publishing, Cham. https://doi.org/10.1007/978-3-030-63461-2_18
32. Monate B, Signoles J (2008) Slicing for security of code. In: International conference on trusted computing and trust in information technologies (TRUST)
33. Necula GC, McPeak S, Rahul SP, Weimer W (2002) CIL: intermediate language and tools for analysis and transformation of C programs. In: Horspool RN (ed) Compiler construction, 11th international conference, CC 2002, held as part of the joint European conferences on theory and practice of software, ETAPS 2002, Grenoble, France, April 8-12, 2002, proceedings, Lecture notes in computer science, vol 2304. Springer, pp 213–228. https://doi.org/10.1007/3-540-45937-5_16
34. Ottenstein KJ, Ottenstein LM (1984) The program dependence graph in a software development environment. In: Symposium on practical software development environments. https://doi.org/10.1145/800020.808263
35. Pnueli A (1977) The temporal logic of programs. In: Symposium on foundations of computer science (FCS)
36. Queille JP, Sifakis J (1982) Specification and verification of concurrent systems in CESAR. In: International symposium on programming (ISP)
37. Rice HG (1953) Classes of recursively enumerable sets and their decision problems. Trans Am Math Soc 74:358–366. https://doi.org/10.1090/s0002-9947-1953-0053041-6
38. Signoles J (2015) Software architecture of code analysis frameworks matters: the Frama-C example. In: Workshop on formal integrated development environment (F-IDE)
39. Signoles J (2018) From static analysis to runtime verification with Frama-C and E-ACSL. Habilitation Thesis
40. Sivaramakrishnan KC, Dolan S, White L, Jaffer S, Kelly T, Sahoo A, Parimala S, Dhiman A, Madhavapeddy A (2020) Retrofitting parallelism onto OCaml. In: International conference on functional programming. https://doi.org/10.1145/3408995
41. Stallman RM, the GCC Developer Community (2022) Using the GNU compiler collection For GCC version 12.2.0. Free software foundation. https://gcc.gnu.org/onlinedocs/gcc-12.2.0/gcc/
42. Steensgaard B (1996) Points-to analysis in almost linear time. In: symposium on principles of programming languages (POPL). https://doi.org/10.1145/237721.237727
43. Tolmach AP, Oliva D (1998) From ML to Ada: strongly-typed language interoperability via source translation. J Funct Prog
44. Weiser M (1984) Program slicing. Transactions of software engineering

Chapter 3
Abstract Interpretation with the Eva Plug-in

David Bühler, André Maroneze, and Valentin Perrelle

Abstract This chapter provides an overview of the Eva plug-in of Frama-C, a static analyzer based on abstract interpretation, intended to automatically prove the absence of runtime errors in critical software. It aims at giving users a good understanding of how Eva works, by describing the theoretic principles underlying its analysis and by detailing its most important features. More practically, it also explains how to use Eva, set up an analysis and exploit its results.

Keywords C semantics · Program analysis · Abstract interpretation · Runtime errors

3.1 Introduction

The Eva plug-in is a static analyzer intended to automatically prove the absence of runtime errors in embedded critical software. One of its greatest successes is the verification of safety-critical software used in nuclear power plants [21]. Over the years, Eva has been improved to perform faster and more accurate analyses on more diverse programs and application contexts. In general, it can analyze a code base of tens of thousands of lines of code in a few minutes.

While the original goal is to detect all possible undefined behaviors in a C program, a great deal of properties about analyzed programs are obtained as a byproduct: possible ranges of values and addresses for variables, code reachability, call graphs, and so on. These properties are useful for other Frama-C plugins which need to resolve pointer accesses or get a list of callers. This also opens up completely different usage scenarios for Eva, as these properties can be visualized in a graphical interface

D. Bühler (✉) · A. Maroneze · V. Perrelle
Université Paris-Saclay, CEA, List, Palaiseau, France
e-mail: david.buhler@cea.fr

A. Maroneze
e-mail: andre.maroneze@cea.fr

V. Perrelle
e-mail: valentin.perrelle@cea.fr

for code audit, either for safety or cybersecurity purposes. The user can browse the analyzed code, display the inferred ranges for any variable, review the list of emitted alarms, highlight dead code, and more.

3.1.1 Main Features

The Eva analysis is fully context-sensitive: it starts at the *main* entry point and analyzes the entire program; all calls are inlined and all functions are interpreted separately for each call context.

Once started, Eva is entirely automatic; it does not require manual proof or existing specifications, although some code annotations may guide the analysis and lead to more precise results. However, many parameters are available to finely configure its behavior beforehand, and the accuracy of results and analysis time highly depend on this configuration step.

The Eva analysis is *sound*: it captures all possible behaviors of the program execution, and is thus able to detect all undefined behaviors that might happen in any execution—in the class of undefined behaviors supported by the analysis.

If the Eva analysis emits no warning, then the analyzed program is guaranteed to be free of these undefined behaviors. However, this strong property comes at a cost: false alarms may be issued on correct code. In the general case, the exact set of every possible behavior of any program execution cannot be computed in a reasonable time. To ensure the soundness of its analysis in a finite time, Eva *over-approximates* the possible behaviors of the program, and emits an alarm whenever it fails to prove the absence of undefined behaviors. Thus, each alarm may either reveal a real bug in the analyzed program, in which case the program can never be proved to be correct, or be due to the imprecision of the over-approximation made by Eva, in which case this is a false alarm that could be removed with a more precise analysis. A more precise analysis usually requires more analysis time to complete, or more engineer time to configure.

The historical target of Eva was safety-critical embedded software. As such, it fully supports the subset of C 99 commonly used in embedded software. Dynamic allocation is also supported but often leads to imprecise results. Eva is increasingly being used in non-embedded code, in contexts related to cybersecurity, such as cryptographic and network libraries [9], where absence of runtime errors leads to absence of several kinds of security vulnerabilities, such as buffer overflows. This transition requires new features, which are constantly being added to the plug-in.

3.1.2 First Example

The very simple C code from Fig. 3.1 computes the first hundred terms of the Collatz sequence from the integer parameter n, and puts them in an array. The first element of

Fig. 3.1 First C example for an Eva analysis

```
1  #define MAX 100
2
3  int compute_next(int x) {
4    return (x % 2 == 0) ? x / 2 : 3*x + 1;
5  }
6
7  void collatz(int n) {
8    int t[MAX];
9    t[0] = n;
10   for (int i = 0; i < MAX; i++) {
11     t[i+1] = compute_next(t[i]);
12   }
13 }
```

the array is n, and the following elements are computed by applying compute_next to the previous value: if it is even, divide it by two; if it is odd, triple it and add one.

Let us run a first Eva analysis on this file:

```
frama-c collatz.c -eva -main collatz
```

This command starts an Eva analysis from the collatz function: -eva enables Eva, and -main collatz instructs Frama-C to consider the collatz function as the main entry point of the program. On such a simple program, the analysis is immediate.

The main result of this analysis consists in the alarms reporting the possible undefined behaviors that may occur in any execution of this program, as established by Eva. Here, the analysis raises 4 alarms, as stated in the summary printed at the end of its execution:

```
4 alarms generated by the analysis:
     2 integer overflows
     1 access out of bounds index
     1 access to uninitialized left-values
```

Each of these alarms may reveal a real issue with the code, or be a false alarm caused by some analysis imprecision. We should thus examine these alarms and investigate their origin. They can be seen as warnings in the complete log of the analysis:

```
[eva:alarm] collatz.c:4: Warning:
  signed overflow. assert -2147483648 <= 3 * x;
[eva:alarm] collatz.c:4: Warning:
  signed overflow. assert 3 * x <= 2147483647;
[eva:alarm] collatz.c:11: Warning:
  accessing uninitialized left-value. assert \initialized(&t[i]);
[eva:alarm] collatz.c:11: Warning:
  accessing out of bounds index. assert (int)(i + 1) < 100;
```

Fig. 3.2 First C example corrected

```
1  #include "limits.h"
2  #define MAX 100
3
4  int compute_next(int x) {
5    return (x % 2 == 0) ? x / 2 : 3*x + 1;
6  }
7
8  int collatz(int n) {
9    int i, t[MAX];
10   t[0] = n;
11   if (n < 0)
12     return -1;
13   for (i = 0; i < MAX-1; i++) {
14     int v = t[i];
15     if (v == 1) break;
16     if (v > INT_MAX / 3) return -1;
17     t[i+1] = compute_next(v);
18   }
19   return i;
20 }
```

Let us review these alarms, and try to determine whether they report real bugs in the program or are false alarms:

- The first two alarms warn about possible arithmetic overflows in the operation 3*x of the compute_next function. This fault may actually happen, as the code does not check the range of its entry n: for very large values, the result of 3*x would exceed the range of mathematical integers that can be represented by the C type **int**.
- The third alarm warns about the access to an uninitialized memory when reading t[i]: as this array is a local variable, it is not initialized at the beginning of the function execution, and manipulating uninitialized memory is an undefined behavior according to the C standard. Here however, when a loop iteration reads t[i], it has always been initialized by the previous loop iteration (or the assignment t[0] = n for the first iteration). So this should be a false alarm.
- The last alarm warns about the access t[i+1], where the index i+1 must be strictly less than the array size 100. We have here a classical error in array manipulation: the last iteration of the loop tries to write t[MAX], one past the array length, as it writes t[i+1] when i is MAX-1. The loop exit condition should thus be i < MAX-1 instead of i < MAX.

Figure 3.2 presents a corrected version of the first code. The collatz function now stops if the argument is negative, or if a value of the computed sequence is large enough to lead to an arithmetic overflow when applying compute_next to it. It returns −1 if it stops early, and the first index for which the sequence reaches the value 1 otherwise. Finally, the loop exit condition has been fixed to avoid the out-of-bounds write to t.

If we run Eva on this corrected version, with the same command as before, the analysis now raises 2 alarms, that we know to be false alarms. We will now see how to better configure the Eva analysis to avoid these false alarms.

```
[eva:alarm] collatz.c:5: Warning:
  signed overflow. assert -2147483648 <= 3 * x;
[eva:alarm] collatz.c:14: Warning:
  accessing uninitialized left-value. assert \initialized(&t[i]);
---------------------------------------------------------------
2 alarms generated by the analysis:
    1 integer overflow
    1 access to uninitialized left-values
```

The default configuration of Eva, that we have used until now, is designed to be as fast as possible while remaining sound, and is very imprecise as a consequence.

Eva provides a wide range of options to finely adjust and improve the precision of the analysis, at the cost of longer analysis times. Some of the most crucial analysis parameters are detailed and explained in the following sections. On large and complex case studies, the analysis must often be thoroughly configured to achieve actionable results (by removing most false alarms) in a satisfactory analysis time. However, a comprehensive configuration of the analysis often requires advanced knowledge of both the analyzer and the analyzed code.

For simpler uses or as a first approach, Eva also provides the -eva-precision parameter, which allows for a quick setup of the analysis between 12 pre-determined analysis modes, from 0 (fastest but imprecise) to 11 (most precise but very slow). For user experience reasons, the Eva default configuration settings favor the speed of the analysis rather than its accuracy.

Let us apply the analysis with different values for -eva-precision:

```
frama-c collatz.c -eva -main collatz -eva-precision 2
```

With -eva-precision 2, Eva only emits one alarm. With -eva-precision 5 or higher, Eva does not emit any alarm. As the precision parameters have no impact on the analysis soundness, this guarantees that the remaining alarms were indeed false alarms. The program is now proved to be free of the undefined behaviors detected by Eva. Conversely, Eva always raises at least three alarms on the initial code, regardless of the precision used: only the false alarm about the uninitialized array can be removed with a higher analysis precision.

We can also see how the analysis time increases with -eva-precision, even on such a small program: while the analysis takes less than a second with low precision, it takes a few seconds with -eva-precision 10 or 11.

Finally, we can change the value of MAX in the analyzed source code of Fig. 3.2 and see the effect on the analysis results. With **#define MAX 10**, -eva-precision 2 is enough to get no alarm instantly. With **#define MAX 1000**, -eva-precision 11 is required to get the same result in a few tens of seconds. For even higher values of MAX, the complete proof of the program would be out of reach of Eva: at least one false alarm would then be raised and should be proved by other means.

3.1.3 Overview

The sections of this chapter can be read in any order. Nevertheless, reading Sect. 3.2 before the others might be enlightening as it goes through the theoretic principles underlying the analyzer. Sections 3.3 and 3.4 cover the concrete implementation of this theory in Eva and how to take advantage of the available features. The former describes the *abstract domains* implemented in Eva—each domain inferring some kind of properties that hold in all program executions, such as a set of possible values for each variable. The latter details one of the most efficient techniques to improve the precision of the results by cleverly partitioning the analysis. The two last sections are more practical and discuss the setup of an analysis (Sect. 3.5) and the scope of the results as well as their exploitation (Sect. 3.6).

Readers who cannot wait to play with Eva can focus on the introductory and *experiment* subsections in Sects. 3.2 and 3.4, and on Sects. 3.5 and 3.6. Sections 3.2 to 3.4 aim at giving users a good understanding of *how* the analyzer works; this knowledge is often useful to better interpret its results, and to leverage advanced features to achieve more precise analyses, able to prove more complex properties on large programs.

3.2 Abstract Interpretation

The underlying theory behind Eva is *abstract interpretation* [7, 18, 23, 28], founded during the late seventies, which aims to infer properties about programs without the need to execute them or to enumerate the set of their possible executions. This theory has been improved over the years and is still an active research subject that led to several practical implementations [6, 8, 10, 14, 15, 26].

To illustrate abstract interpretation principles, let us take the very simple program in Fig. 3.3, limited to a single, harmless function. We assume we want to prove the absence of errors—like arithmetic overflow—in this program. Several obvious arguments can be used, but let us start by a very naive approach and enumerate the *reachable states* of this program. A program state here can be seen as a pair: the current value of i and the current value of j. A state is said to be *reachable* if there is an execution from some initial state at the beginning of the program that leads to this state at some point.[1]

At the beginning of the program, the initial state is (*Uninitialized*, *Uninitialized*). After the first initialization, the only reachable state is (0, *Uninitialized*) and then (0, 10) after the second one. Inside the loop, each control point is reachable by several

---

[1] The current program location could be considered as part of a state but, by convention, we will not include it in the state and choose instead to talk about the set of reachable states *at a program location*.

3 Abstract Interpretation with the Eva Plug-in

Fig. 3.3 An illustrating example

```
1  void f(void) {
2    int i = 0;
3    int j = 10;
4    while (i < 10) {
5      i = i + 1;
6      j = j - 1;
7    }
8  }
```

states. They can be enumerated while following the execution: the state $(1, 9)$ is reached after one iteration, the state $(2, 8)$ after two iterations and so on. At the end, we can conclude that 11 states are reachable at the loop head:

$$(0, 10), (1, 9), (2, 8), (3, 7), (4, 6), (5, 5), (6, 4), (7, 3), (8, 2), (9, 1), (10, 0)$$

It follows that the last one, $(10, 0)$, is the only state reachable at the end of the function.

In other words, one can enumerate the reachable states by simulating each possible execution and by collecting each state encountered. This naive approach reaches its limit quickly: the number of states may be too large to be represented on a finite memory machine—and may even be unbounded if, for instance, the program uses unbounded integers or memory allocation. If we replace the loop bound 10 by one billion, the set of states increases accordingly and its computation, even on this simple program, may take a while.

To solve this first issue we can use a more compact representation of the set of reachable states. Instead of enumerating each reachable state, let us describe the set in a more *abstract* way by using a short description based on intervals:

$$i \in [0, 10], j \in [0, 10]$$

While not equivalent, this description has the advantage of remaining short even if we change the loop bound.

Interestingly, we can compute this abstract description directly, without having to compute the actual set of reachable states. The idea is similar to the previous naive algorithm. We compute the interval representation from the start of the program by following the execution, statement after statement. We call this process the *propagation* of intervals.

For the sake of brevity, we will adopt a diagram notation to describe the propagation of intervals. These diagrams show how statements—edges in the diagram—modify the intervals. We associate intervals to each node of the diagram, such that the intervals at the target of an edge are obtained by modifying the intervals at the source of that edge according to the statement labeling the edge. The diagram for the first two initializations would be the following:

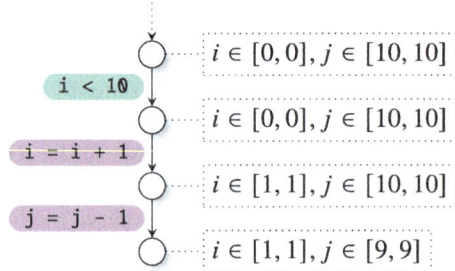

Note that in order to give a meaning to an uninitialized variable, we consider that such a variable can have any value. We can continue the propagation inside the loop with the following color code: dataflow instructions in pink and control-flow instructions (tests, loops) in green.

Since we do not want to enumerate each reachable state, we will use a different strategy. We can say that the states reachable at the loop head are the states coming from the loop entry and the state coming from previous iterations of the loop. In terms of intervals, this means that the interval of each variable at the loop head should be the union of their intervals at the loop entry and at the end of the previous iterations of the loop, in our case:

$$i \in [0, 0] \cup [1, 1], j \in [10, 10] \cup [9, 9]$$

which we can rewrite as:

$$i \in [0, 1], j \in [9, 10]$$

We will note this union with a ⊔ node in the diagram.

3 Abstract Interpretation with the Eva Plug-in

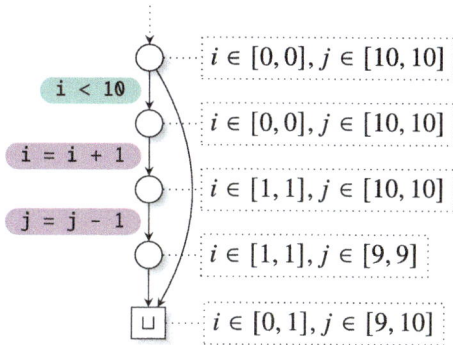

We could obviously continue for a second iteration of the loop and from these two new intervals, ending up with $i \in [0, 2]$ and $j \in [8, 10]$. Iteration after iteration, we would compute successive intervals until we reach the actual intervals of the variables i and j.

But again, this method would not scale easily, as a simple change of bounds in the loop would make the computation too expensive. This is why abstract interpretation introduces a second idea: using extrapolation. When we first entered the loop, we had $i \in [0, 0]$; after one propagation through the loop we had $i \in [0, 1]$. Based only on this information, we can, at best, assume that the variable i is increasing. With a coarse inductive reasoning, let us say that it will increase indefinitely[2] and consider from now on that the actual interval of variable i is $i \in [0, +\infty[$. Similarly, let us say that j is infinitely decreasing and that its interval of values is $j \in]-\infty, 10]$. This operation is called a *widening* and is noted ∇:

$$[0, 0] \nabla [0, 1] = [0, +\infty[$$

Note that, while these intervals are very imprecise, the two assertions "i is greater than or equal to 0" and "j is less than or equal to 10" are perfectly true. We will be able to recover some precision though, as we will now see.

We can restart the propagation through the loop with these new coarse intervals as shown by the following diagram:

---

[2] Note that we are handling *mathematical* (infinite) integers here, not machine integers. Later, we will address the verification of machine integer overflows.

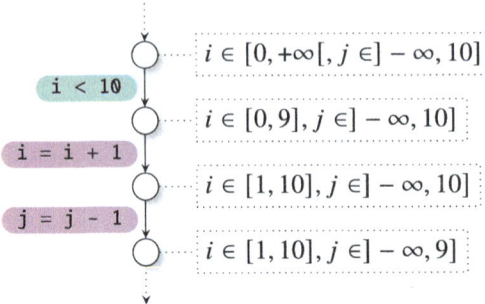

One important point about this iteration is that when we go through the loop condition i < 10 we know that only values in the interval [0, 9] are possible for i. We cannot say anything about j, however, as the loop condition does not mention it.

A second important point is that the intervals we get at the end of this iteration are strictly smaller than the ones we had before propagating the intervals through the loop body. This proves that these latter intervals are not only a loop invariant but also an *inductive invariant*—a loop invariant is a property that is true for each loop cycle, and an inductive loop invariant means that the property is true when entering the loop and if it is true at cycle n, then the same property is true at cycle $n + 1$. Indeed, the fact that the propagation of these intervals through the loop leads to smaller intervals is a sufficient condition. This can be seen with an induction on the trace—history of states along the possible execution paths starting at the beginning of the analysis—of a reachable state σ:

- Base case: if the execution never goes through the loop, then σ must match the intervals we use at loop entry, which are included in our invariant.
- Induction step: if the execution goes through at least one loop iteration, then σ is obtained after one loop iteration from a previous state σ' which, by induction, satisfies the invariant. Then, the propagation of the intervals through the loop iteration shows that σ must satisfy the intervals we get at the end of the propagation. Since these intervals are included in the invariant, σ also satisfies the invariant.

This is the third and last idea we needed to introduce abstract interpretation: the propagation of abstractions through loops can be stopped as soon as it leads to a formula that is *stronger* (that is, produces smaller intervals) or equal after the iteration than before.

To finish the computation of the set of reachable states, we must propagate the invariant intervals up to the end of the function, only keeping states for which $i < 10$ does not hold.

3 Abstract Interpretation with the Eva Plug-in

While the final intervals are not very satisfying, they can be improved either by refining the widening operator or by various strategies for recovering precision after a widening that we will not discuss here.

In fact, the imprecision at the end of the program is not a problem if we are only interested in finding execution errors, as i and j are not used after the loop. They are used inside the loop though, where we have a precise interval for i but not for j. This means we can prove the absence of arithmetic overflow on i = i + 1; but not on j = j - 1;.

To solve this problem, intervals are not enough; we need to enrich our interval abstraction. On this example, it is enough to add the information $i + j = 10$:

$$i \in [0, 9] \wedge j \in]-\infty, 10] \wedge i + j = 10$$

which is equivalent to

$$i \in [0, 9] \wedge j \in [1, 10] \wedge i + j = 10$$

and thus we can prove that j = j - 1; does not overflow.

To prove this equality we will use the same method, propagating it through the program statements like we did with the interval. We start the analysis of the program where no equality holds. The equality is true as soon as the two variables are initialized and we can prove that it remains true by propagating the equality through the loop as before:

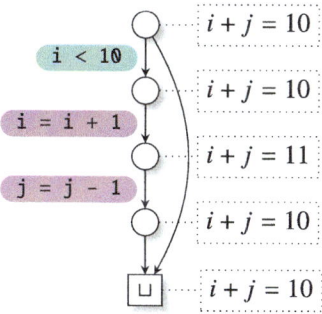

This propagation is straightforward here, though more mathematical reasoning might be useful in the general case [16]. Since the equation is true before the loop and is preserved by the loop, we can conclude it is an invariant.

To summarize, we have seen three abstract interpretation core principles:

- we use efficient abstract representations for sets of states and we propagate these abstractions through the program;
- we use a widening operator to converge rapidly to an inductive invariant; and
- we know we obtained an invariant as soon as a propagation through a loop leads to a stronger property.

The choice of which abstraction to use depends widely on the kind of program we want to analyze. In practice, we define *abstract domains* (mathematical representation of the content of variables—see Sect. 3.3) which are specialized to handle a finely delimited set of abstractions and we choose a suitable combination of abstract domains for the targeted program.

These principles will be formalized in the following sections. Before that, we will see how to experiment with Eva to confront the theory and the implementation.

3.2.1 Experiment with Eva

While Eva results can be seen with a graphical interface, it does little to help understand how these results have been obtained. It is possible however to visualize the propagation of abstractions. Frama-C provides a special builtin function Frama_C_show_each that can be used to display the properties inferred about its parameters each time the analyzer propagates through this builtin, even if an invariant has not been reached yet.

Let us change slightly the previous example to include a call to Frama_C_show_each as in Fig. 3.4a.

If we run Eva on this example using the command

```
frama-c -eva experiment.c -main f
```

it will output several lines including the contents listed in Fig. 3.4b. Some of these outputs should be familiar, as through Frama_C_show_each we observe some of the intervals we have seen earlier. We also see that the analyzer raised a (false) alarm about the potential overflow of j = j - 1;. Finally, the analyzer gives the final intervals inferred for i and j at the end of the functions.

However, some outputs do not match the theory:

1. The interval for i never reaches $[0, +\infty[$. This is because the widening operator used here identified 10 as a potential candidate to bound an interval invariant. So instead of widening to $[0, +\infty[$, it tried $[0, 9]$ first.
2. There seem to be four iterations before actually widening the interval for i. There is indeed a delay before Eva applies a widening operator. Moreover, once it has applied a first widening, it only applies following widenings every two iterations. The delay and period of widenings can be adjusted with the command-line parameters -eva-widening-delay and -eva-widening-period respectively.
3. There seem to be even more iterations before the intervals for j are widened. Eva does not apply a widening to all of the variables at first, but uses heuristics to choose when and to which variables it must be applied.

To conclude the experiment we can try to get rid of the alarm and prove that no overflow can happen. By default, only the interval abstract domain is activated. To infer the relationship i+ j = 10 we can enable another abstract domain. Among Eva's abstract domains that can infer equalities, the gauges domain (see Sect. 3.3.4)

3 Abstract Interpretation with the Eva Plug-in

```
1  void f(void) {
2      int i = 0;
3      int j = 10;
4      while (i < 10) {
5          Frama_C_show_each(i, j);
6          i = i + 1;
7          j = j - 1;
8      }
9  }
```

(a) Program with call to Frama_C_show_each builtin

```
1   [eva] experiment.c:5: Frama_C_show_each: {0}, {10}
2   [eva] experiment.c:4: starting to merge loop iterations
3   [eva] experiment.c:5: Frama_C_show_each: {0; 1}, {9; 10}
4   [eva] experiment.c:5: Frama_C_show_each: {0; 1; 2}, {8; 9; 10}
5   [eva] experiment.c:5: Frama_C_show_each: [0..9], {7; 8; 9; 10}
6   [eva] experiment.c:5: Frama_C_show_each: [0..9], {6; 7; 8; 9; 10}
7   [eva] experiment.c:5: Frama_C_show_each: [0..9], [1..10]
8   [eva] experiment.c:5: Frama_C_show_each: [0..9], [0..10]
9   [eva] experiment.c:5: Frama_C_show_each: [0..9], [-1..10]
10  [eva] experiment.c:5: Frama_C_show_each: [0..9], [-2..10]
11  [eva] experiment.c:5: Frama_C_show_each: [0..9], [-2147483648..10]
12  [eva:alarm] experiment.c:7: Warning:
13       signed overflow. assert -2147483648 ≤ j - 1;
14  [eva] done for function f
15  [eva] ====== VALUES COMPUTED ======
16  [eva:final-states] Values at end of function f:
17       i ∈ {10}
18       j ∈ [-2147483648..10]
```

(b) Output of the above program

Fig. 3.4 A basic experiment with Eva

is suitable for this program, as it can infer linear relationships between loop variables, like i + j = 10. We complete the command-line as follows:

```
1  frama-c -eva experiment.c -main f -eva-domains gauges
```

This time, the output of this command, presented in Fig. 3.5, shows that the intervals inferred for j remain precise inside the loop; therefore, the alarm is no longer emitted.

```
1  [eva] experiment.c:5: Frama_C_show_each: {0}, {10}
2  [eva] experiment.c:4: starting to merge loop iterations
3  [eva] experiment.c:5: Frama_C_show_each: {0; 1}, {9; 10}
4  [eva] experiment.c:5: Frama_C_show_each: {0; 1; 2}, {8; 9; 10}
5  [eva] experiment.c:5: Frama_C_show_each: [0..9], {7; 8; 9; 10}
6  [eva] experiment.c:5: Frama_C_show_each: [0..9], {6; 7; 8; 9; 10}
7  [eva] experiment.c:5: Frama_C_show_each: [0..9], [1..10]
8  [eva] experiment.c:5: Frama_C_show_each: [0..9], [1..10]
9  [eva] done for function f
10 [eva] ====== VALUES COMPUTED ======
11 [eva:final-states] Values at end of function f:
12    i ∈ {10}
13    j ∈ [0..10]
```

Fig. 3.5 Results with the gauges domain

3.2.2 Soundness

We want our analyses to be *sound*, in the sense that if we state that an execution error can never happen, then it has to be true. To formalize this notion of soundness, we use the notion of a program state that we will call a *concrete state*.

A concrete state σ can be seen as a function from a set of variables $\mathcal{V}$ to the set of all possible values $\mathbb{V}$. In the previous example, we had $\mathcal{V} = \{i, j\}$ and we could consider that the possible values are integers, i.e., $\mathbb{V} = \mathbb{Z} \cup \{Uninitialized\}$. While this is clearly not general enough to describe the semantics of C code, it can be extended to handle dynamic allocation, bit-precise addresses, floating-point numbers and pointers. In this section, we will keep things simple and use this formalism to describe the set S of reachable states at the loop head in our example:

$$S = \left\{ \begin{bmatrix} i \mapsto 0 \\ j \mapsto 10 \end{bmatrix}, \begin{bmatrix} i \mapsto 1 \\ j \mapsto 9 \end{bmatrix}, \begin{bmatrix} i \mapsto 2 \\ j \mapsto 8 \end{bmatrix}, \dots \begin{bmatrix} i \mapsto 9 \\ j \mapsto 1 \end{bmatrix} \right\}.$$

From the definition of concrete state and a language specification, we can derive the notion of *execution error*. For instance, an execution error occurs each time the execution encounters a division and the divisor evaluates to zero in the current concrete state.

An *abstract state* is used to simplify a set of possible concrete states—whose memory representation would quickly fill the computer memory—into a single mathematical representation. These are our abstractions for the memory at a given program point. The abstract interpretation framework does not have strong requirements[3] on these abstractions and the scientific literature proposes a large choice of abstract domains suitable for program analysis as well as ways to combine them. For an interval abstract domain, the abstract state $S^\#$ representing the previous set of concrete states would be:

---

[3] The original formalism assumed a complete semi-lattice structure, but we will not need such a structure in this book.

3 Abstract Interpretation with the Eva Plug-in

$$S^\# = \begin{bmatrix} i \mapsto [0..9] \\ j \mapsto [1..10] \end{bmatrix}$$

Note that an abstract state does not represent a concrete state but a *set* of concrete states. Formally, an abstract state in interval domains can be seen as a function which maps variables to intervals. Intervals themselves can be defined as a pair with a lower bound and an upper bound, both possibly infinite.[4] This can be formally expressed by:

$$S^\# : \mathcal{V} \to (\mathbb{Z} \cup \{-\infty\}) \times (\mathbb{Z} \cup \{+\infty\})$$

We map each abstract state to the set of concrete states it represents using a *concretization* function γ. In our example:

$$\gamma\left(\begin{bmatrix} i \mapsto [0..9] \\ j \mapsto [1..10] \end{bmatrix}\right) = \left\{ \begin{matrix} \begin{bmatrix} i \mapsto 0 \\ j \mapsto 1 \end{bmatrix}, & \begin{bmatrix} i \mapsto 1 \\ j \mapsto 1 \end{bmatrix}, & \cdots & \begin{bmatrix} i \mapsto 9 \\ j \mapsto 1 \end{bmatrix}, \\ \vdots & \vdots & & \vdots \\ \begin{bmatrix} i \mapsto 0 \\ j \mapsto 10 \end{bmatrix}, & \begin{bmatrix} i \mapsto 1 \\ j \mapsto 10 \end{bmatrix}, & \cdots & \begin{bmatrix} i \mapsto 9 \\ j \mapsto 10 \end{bmatrix} \end{matrix} \right\}$$

For an abstract state to be a sound approximation of a program semantics at a given point, its concretization may contain more than the states that are actually reachable—but not less. This is the case in our example, since $S \subseteq \gamma(S^\#)$ but $\gamma(S^\#)$ also contains many more concrete states than S.

The concretization function γ of an abstract state $S^\#$ returns the set of concrete states σ—maps from variables to values—that satisfy its mathematical representation. Hence, an interval abstraction that provides two bounds $[l^v, u^v]$ for any variable v, induces this concretization function γ:

$$\gamma(S^\#) = \left\{ \sigma \;\middle|\; \forall v \in \mathcal{V}.\; l^v \leq \sigma(v) \leq u^v \text{ where } (l^v, u^v) = S^\#(v) \right\}$$

The definition of γ is fundamental to guarantee the soundness. Hence, an abstract interpretation based analyzer should ensure the three parts below:

1. Program assignments and tests must be correctly interpreted.[5] We will check this via their denotational semantics [25, 29], which is a way of defining the effects of the execution of a statement s as a mathematical function on concrete states, noted $[\![s]\!] : S \to S$. $[\![s]\!](\sigma)$ is then the concrete state resulting from the execution of statement s on the concrete state σ. For instance, the semantics of the assignment i = i + 1 is the function $[\![i = i + 1]\!]$, which maps a concrete

---

[4] As in the introduction, we consider that uninitialized values are represented by the interval $(-\infty, +\infty)$.

[5] In practice, to interpret a real-world programming language, more than assignments and tests would have to be considered. Function calls are a good example of something that would need to be added, but are much more complex to handle.

state σ to the resulting state after the instruction, i.e., a state almost equal to σ but mapping i to the previous value of i plus one:

$$[\![\mathtt{i = i + 1}]\!](\sigma) = \sigma[i \mapsto \sigma(i) + 1]$$

Tests appear in loops and if-then-else statements, and are also denoted by a function which maps concrete states to a truth value in $\{\mathit{true}, \mathit{false}\}$. To define the semantics of the previous program, we would need to define $[\![\mathtt{i < 10}]\!]$ and symmetrically $[\![\mathtt{i \geq 10}]\!]$ for states leaving the loop. For the former we would have:

$$[\![\mathtt{i < 10}]\!](\sigma) = \begin{cases} \mathit{true} & \text{if } \sigma(i) < 10 \\ \mathit{false} & \text{otherwise.} \end{cases}$$

We extend the semantics of assignments and tests from states to sets of states, by applying the semantics to each state they contain. Formally, for an assignment a and a test t:

$$[\![\mathtt{a}]\!](S) = \{[\![\mathtt{a}]\!]\sigma \mid \sigma \in S\}$$
$$[\![\mathtt{t}]\!](S) = \{\sigma \mid \sigma \in S \wedge [\![\mathtt{t}]\!](\sigma) = \mathit{true}\}$$

In order to perform an abstract interpretation, one must define an abstract version $[\![\mathtt{a}]\!]^\#$ and $[\![\mathtt{t}]\!]^\#$ of this semantics for each assignment a and each test t. We call $[\![\mathtt{a}]\!]^\#$ and $[\![\mathtt{t}]\!]^\#$ **abstract transfer functions** in contrast to $[\![\mathtt{a}]\!]$ and $[\![\mathtt{t}]\!]$ which are called *concrete transfer functions*. In our case, we could define

$$[\![\mathtt{i = i + 1}]\!]^\#(S^\#) \stackrel{\text{def}}{=} S^\#[i \mapsto (l+1, u+1) \text{ where } (l, u) = S^\#(i)]$$

In this example, the abstract transfer function performs exactly as the concrete transfer function. This means that, if we look at the concrete states by applying γ before and after the application of $[\![\mathtt{i = i + 1}]\!]^\#$, the concrete set of states we obtain are also mapped by the concrete transfer function.

$$\forall S^\#. [\![\mathtt{i = i + 1}]\!](\gamma(S^\#)) = \gamma([\![\mathtt{i = i + 1}]\!]^\#(S^\#))$$

But, in general, we will not be so lucky. For instance, instruction i = i * 2 has no exact abstract transfer function using intervals. We allow, however, the result of an abstract transfer function to represent *more* than what the corresponding concrete transfer function would lead to. In mathematical language, an abstract transfer function $[\![\mathtt{f}]\!]^\#$ is sound if and only if

$$\forall S^\#. [\![\mathtt{f}]\!](\gamma(S^\#)) \subseteq \gamma([\![\mathtt{f}]\!]^\#(S^\#))$$

3 Abstract Interpretation with the Eva Plug-in

2. The union of abstract states must be correctly computed. Unions happen each time there is a merge of control flows, e.g., after if-then-else statements, at loop heads or at labels targeted by a goto statement. We note $S_1^\# \sqcup S_2^\#$ the union of two abstract states $S_1^\#$ and $S_2^\#$ and we call $\sqcup$ a **join operator**. For intervals, a possible join operator would be to take the union of intervals—the minimum of their lower bounds, and the maximum of their upper bounds—for each variable v in the program.

$$S_1^\# \sqcup S_2^\# \stackrel{def}{=} v \in \mathcal{V} \mapsto (min\{l_1, l_2\}, max\{u_1, u_2\}) \text{ where } \begin{cases} (l_1, u_1) = S_1^\#(v) \\ (l_2, u_2) = S_2^\#(v) \end{cases}$$

Like abstract transfer functions, the join operator may introduce new concrete states that are not reachable, but it can not remove concrete states of its operands. That means that the following property holds in any situation:

$$\gamma(S_1^\#) \cup \gamma(S_2^\#) \subseteq \gamma(S_1^\# \sqcup S_2^\#)$$

3. The inclusion of one abstract state into another must be correctly tested. This inclusion operator decides whether to stop the propagation or not. If, in a loop, the abstract state after the body of the loop is included in the abstract state before the body of the loop, then the analysis has locally reached a fixpoint, meaning that no new concrete states can be discovered with new loop iterations. Hence, the analysis stops analyzing the loop, even if the execution may loop forever.
The **abstract inclusion operator** is noted $\sqsubseteq$. For intervals, we have an exact inclusion operator, which tests interval inclusion for each variable v:

$$S_1^\# \sqsubseteq S_2^\# \stackrel{def}{\Leftrightarrow} \forall v \in \mathcal{V}. l_1^v \geq l_2^v \wedge u_1^v \leq u_2^v \text{ where } \begin{cases} (l_1^v, u_1^v) = S_1^\#(v) \\ (l_2^v, u_2^v) = S_2^\#(v) \end{cases}$$

In the general case, whenever $S_1^\# \sqsubseteq S_2^\#$ holds, their corresponding concrete sets of states must be included one another:

$$S_1^\# \sqsubseteq S_2^\# \Rightarrow \gamma(S_1^\#) \subseteq \gamma(S_2^\#)$$

The converse is not needed though: if the operator fails to see an inclusion, this will delay the end of the propagation algorithm, but that will not harm its soundness.

Let us summarize these soundness properties into our first definition.

Definition 3.1 (*Sound abstract domain*) An abstract domain with concretization function γ is **sound** if its join operator $\sqcup$, its inclusion operator $\sqsubseteq$ and all its abstract transfer functions $[\![f]\!]$ of each C instruction satisfy

$$\forall \llbracket \mathtt{f} \rrbracket. \ \forall S^{\#}. \llbracket \mathtt{f} \rrbracket (\gamma(S^{\#})) \subseteq \gamma(\llbracket \mathtt{f} \rrbracket^{\#}(S^{\#})) \tag{3.1}$$

$$\gamma(S_1^{\#}) \cup \gamma(S_2^{\#}) \subseteq \gamma(S_1^{\#} \sqcup S_2^{\#}) \tag{3.2}$$

$$S_1^{\#} \sqsubseteq S_2^{\#} \Rightarrow \gamma(S_1^{\#}) \subseteq \gamma(S_2^{\#}) \tag{3.3}$$

Now that we know how to do a sound analysis, the next step is to ensure that it will terminate.

3.2.3 Algorithm Termination

The termination of abstract interpretation depends on the right choice of a widening operator. As stated before, the propagation of abstract states may never terminate without it. It might be the case if we consider programs with unbounded integers, for instance, or more generally with dynamic allocation: the set of reachable states may be infinite and the propagation algorithm may never converge to an over-approximation of this set.

As a reminder, the widening operator, noted $\triangledown$, tries to guess what would be the inferred abstract value after a possibly unbounded number of iterations. It is an operator which takes two arguments: the abstract state before the propagation through the loop and the abstract state after this propagation. Since the difference between the two states is one loop iteration, the operator can observe the effect of an iteration on an abstract state and use this to extrapolate.

Informally, the widening we used on the example can be described in the following way. It is an operator which compares two abstract states, variable by variable, and checks if the variable inferred interval is growing. More precisely, we extract the interval for each variable in the two abstract states, and compare the two interval bounds: if the lower bound is decreasing, we set the lower bound of the result to $-\infty$, and if the upper bound is increasing, we set the upper bound of the result to $+\infty$.

$$X_1^{\#} \triangledown X_2^{\#} \stackrel{\text{def}}{:} v \in \mathcal{V} \mapsto (l, u) \text{ where } \begin{cases} l = & l_1 \text{ if } l_1 \leq l_2 \text{ else } -\infty \\ u = & u_1 \text{ if } u_1 \geq u_2 \text{ else } +\infty \\ (l_1, u_1) = & X_1(v) \\ (l_2, u_2) = & X_2(v) \end{cases}$$

Note that when a bound is tighter on the right argument, we keep the bound of the left argument. This is important as we will now see.

Widening must not only accelerate the convergence of the propagation to an invariant but must also ensure that the propagation will terminate in a finite number of steps. Each time we interpret a loop iteration i, we start from a state $X_i^{\#}$ at loop head and we end with a state $Y_i^{\#}$ at the end. If $Y_i^{\#}$ is not included in $X_i^{\#}$ we need to continue

3 Abstract Interpretation with the Eva Plug-in

the propagation for another iteration but we apply a widening $W_i^\# = X_i^\# \triangledown Y_i^\#$ before, which provides us a new state $W^\#$ to start the next iteration. We repeat the process until we reach an inclusion.

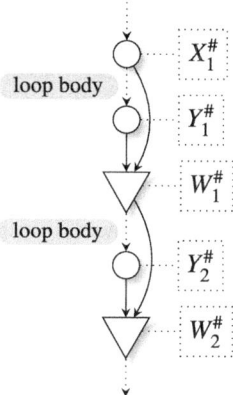

Thus, we have a sequence of states $Y_1^\#, Y_2^\#, Y_3^\#, \ldots$ obtained by the propagation through the loop body, and a sequence of states $W_1^\#, W_2^\#, W_3^\#, \ldots$ linked by the widening operator:

$$W_i^\# = W_{i-1}^\# \triangledown Y_i^\#$$

To start these sequences, we can say that $W_0 = Y_0 = X_1$. The analysis terminates if at some point we have $Y_i^\# \sqsubseteq W_{i-1}^\#$. We require this to happen after a finite number of iterations, independently of how the abstract domains are implemented. In other words, we require that, whatever the sequence $(Y_i)_i$ may be—even a meaningless sequence—applying widenings will always lead to an inclusion.

Definition 3.2 (*Widening*) A widening operator $\triangledown$ is a binary operator such that, for any sequence $(Y_i^\#)_{i \geq 0}$, there exists an iteration n such that the sequence $(W_i^\#)_{i \geq 1}$ defined by

$$\begin{aligned} W_0^\# &= Y_0^\# \\ W_i^\# &= W_{i-1}^\# \triangledown Y_i^\# \end{aligned}$$

verifies $Y_n^\# \sqsubseteq W_n^\#$.

Counterintuitively, the definition of widening is unrelated to the soundness of the analysis; you can choose any widening operator and still have a sound analysis. The widening can only impact the convergence and the precision of the analysis. The widening is usually defined as an over-approximation of the join operator, satisfying $X^\# \sqcup Y^\# \sqsubseteq X^\# \triangledown Y^\#$. This is the case of the widening that we defined on intervals. However, this is neither required to ensure the soundness nor the termination of the analysis [19], and thus our formal widening definition is kept minimal.

We can check that the widening we defined for intervals follows this definition. The idea of the proof is that once an interval bound reaches $-\infty$ or $+\infty$, it will stay infinite. It is implied by the definition and this is the very reason why we always keep the bound of the left operand even when the right bounds are tighter. At each step i, either the inclusion test $X_i^\# \sqsubseteq W_i^\#$ succeeds, or at least one interval is growing between $X_i^\#$ and $W_i^\#$. This means that at least one finite interval bound will become infinite after the widening. Since the number of variables $|\mathcal{V}|$ is bounded, so is the number of interval bounds in an abstract state. Thus, the number of widening states changing a bound is necessarily lower than $2 \times |\mathcal{V}|$, after which all interval bounds will be infinite, that is, an abstract value including any other abstract value.

Widening was the last ingredient we needed to build a powerful static analysis framework. The Abstract Interpretation theory also introduces narrowing operators to optimize some of the results of an analysis. In Eva, we choose to use progressive widening with thresholds instead of narrowing, since they are more customizable by the user (new widening thresholds can be added with ACSL annotations `widen_hints`) and since they avoid removing warnings that would be emitted by a too coarse widening before the narrowing.

The next section will describe the actual abstract domains implemented in Eva.

3.3 Abstract Domains in Eva

This section presents the main abstract domains currently available in the Eva analyzer. These are enabled by providing a list of domain names to the `-eva-domains` parameter. A list of the available domains, together with a short description of each one, can be displayed with `-eva-domains help`.

Among them, the historical *cvalue* domain is of paramount importance for the analysis precision: it infers the possible values of all C variables at any program point, and is especially useful to handle assignments and dereferences through pointers. It has been the first abstract domain of Eva and is still the only one enabled by default. Other abstract domains have then been introduced over the years to overcome some limitations of the *cvalue* domain and improve the analysis precision on some codes; their only downside is that they increase the analysis time.

We start by describing the abstractions of scalar values (integer, floating-point and pointer values) used by many domains, including the main *cvalue* domain.

3.3.1 Abstractions of C Values

The abstract representation of a set of scalar C values (integer, floating-point or pointer values) plays a central part of the Eva analysis: they are a *lingua franca* between all abstract domains to interact and exchange information about the variables and C expressions of the analyzed program, as detailed in Sect. 3.3.6.

3 Abstract Interpretation with the Eva Plug-in   151

Moreover, being one of the main results of an Eva analysis, the set of possible values inferred for each variable or expression at any program point is made available to several other Frama-C plugins through the Eva API. The user can also inspect them in great detail via the Frama-C graphical interface. Figure 3.6 presents the syntax used to express the different value abstractions available in Eva to the user, both in textual logs and in the graphical interface.

A set of *integer values* can be represented either as a concrete set or as an interval with a congruence information defined below and illustrated in the three first lines of Fig. 3.6:

- Concrete sets are implemented by sorted arrays and represent exactly their elements. They are used for small integer sets up to a given limit, and are converted into intervals once they exceed it. This limit is 8 by default, and can be changed with the -eva-ilevel parameter.
- An integer interval is a pair of two values *min* and *max* in $\mathbb{Z} \cup \{+\infty; -\infty\}$, such that $min \leq max$. A congruence information is encoded by two natural integers *rem* and *mod* of $\mathbb{N}$, such that $0 < mod$ and $0 \leq rem < mod$. Such an interval, with such a congruence, represents the set of integers between *min* and *max* included and congruent to *rem* modulo *mod*.

Eva maintains a canonical representation of these integer abstractions. The lower (respectively upper) bound of the interval is always refined up (resp. down) to the nearest integer validating the congruence. An interval with no congruence information has $mod = 1$ and $rem = 0$ (which is not printed). Any interval representing less than the ilevel limit is converted back into a concrete set.

A set of *floating-point values* is always represented by a floating-point interval, whose bounds can be finite or infinite—see lines 4-5 of Fig. 3.6. A singleton floating-point value f is represented as the interval $[f..f]$. These intervals, augmented with an additional boolean to express the possibility that the value may be a NaN, implement the floating-point semantics defined by the IEEE-754 standard.

Finally, *pointer values* are represented as maps from the base address of allocated (local, global, or dynamically allocated) variables to integer abstractions (small concrete sets or intervals) representing the possible offsets of the pointed memory locations, expressed in bytes. The last line of Fig. 3.6 shows such an example.

These value abstractions are directly used by the main *cvalue* domain to represent an over-approximation of all the memory content of a C program during its execution.

3.3.2 The Cvalue Domain

Cvalue is a non-relational domain representing separately the set of possible values of each program variable at any program point.

For integer, floating-point and pointer variables, it simply uses the abstractions described in the previous section to represent scalar values. In particular, pointer

```
{2; 3; 5; 7; 11; 13; 17; 19}      any one of these integers
[2..23]        any integer between 2 and 23 included
[0..90],0%3         any integer between 0 and 90 and congruent to 0 modulo 3
{3.14159274101}          exactly this floating-point number
[1.1 .. inf]        any number between 1.1 and infinity included
[-0.75 .. 2.5] ∪ {NaN}        any number between -0.75 and 2.5, or NaN
{{ &x; &a + [0..36],0%4; &s + {12} }}       Value of a pointer that may point to variable x, or to one of the first ten elements of an int32_t array a, or to a field of a structure s at offset 12.
```

Fig. 3.6 Examples of the C value abstractions used by Eva

values are able to precisely represent the set of addresses a pointer may point to. This allows the *cvalue* domain to infer alias information of pointers during the analysis. *Cvalue* is currently the only Eva domain featuring an alias analysis, and the other Eva domains rely on it to soundly interpret assignments and dereferences involving pointers, without having to track themselves alias information about pointers.

The C standard also defines *indeterminate* values: any access to such a value is an undefined behavior. The initial value of uninitialized local variables is indeterminate. The value of a pointer becomes indeterminate if the pointed object is out of scope. To emit alarms on accesses to indeterminate values, the *cvalue* domain also stores two additional booleans for each scalar value, to track the possibility that the stored value may also be uninitialized or a dangling pointer.

Besides scalar variables, the *cvalue* domain also needs to accurately describe the possible values stored in aggregate types, such as arrays and structs. This requires the use of a proper C memory model.

A Low-Level Untyped Memory Model

A hypothesis of the Eva analysis is that C variables are absolutely *separated*: pointers cannot jump from one variable to another via pointer arithmetic. Therefore, the *cvalue* domain represents separately the possible values of each program variable. However, this assumption does not hold for the cells of an array or the fields of a struct, which are *not* separated: a pointer can freely jump from one to another.

Generally speaking, the C standard exposes both a low-level and a high-level view of memory objects. Their values can be manipulated through the effective type of structured objects (such as arrays, structs and unions), but they can also be interpreted as an array of unsigned characters, and can always be read or written through untyped byte manipulations.

Moreover, C unions allow mixing values of different types in the same memory location, as illustrated by the code presented in Fig. 3.7, where the function write_byte_union reads and writes respectively the second and third integer bytes of a floating-point value. Right below it, the function write_byte_ptr performs the same operations using only pointer conversions.

3 Abstract Interpretation with the Eva Plug-in

Fig. 3.7 Examples of C low-level operations

```
1  union float_bytes {
2    float f;
3    char bytes[4];
4  };
5
6  float write_byte_union (float f) {
7    union float_bytes x = { .f = f };
8    char c = x.bytes[1];
9    x.bytes[2] = c;
10   return x.f;
11 }
12
13 float write_byte_ptr (float f) {
14   char c = *((char *)(&f) + 1);
15   *((char *)(&f) + 2) = c;
16   return f;
17 }
18
19 void main (void) {
20   float f = 0.75;
21   float f1 = write_byte_union(f);
22   float f2 = write_byte_ptr(f);
23 }
```

The *cvalue* domain is able to accurately represent all C memory objects, as well as all the low-level manipulations previously described. To this end, the memory of a variable is modeled as an untyped array of bits, which can be arbitrarily read, written and re-interpreted as any C type. This memory model is efficiently implemented through a data structure called *offsetmaps*.

Offsetmaps

Offsetmaps [3] are basically maps from ranges of bits to value abstractions, able to represent the possible values of any sequence of bits of a C object. For instance, a struct with two int fields containing the abstract values v_1 and v_2 respectively, can be represented, on a 32-bit architecture, by:

$$< 0 - 31 > \to v_1 \qquad < 32 - 63 > \to v_2 \qquad (3.4)$$

As it is untyped, this offsetmap could also model any object with the same memory representation, such as an array of two 32-bit elements v_1 and v_2, or even one 64-bit variable accessed through pointer arithmetic as two adjacent 32-bit values. In the *cvalue* domain, the value abstractions of offsetmaps are those described in the previous section.

An offsetmap contains successive bit ranges of any size. To ensure a compact representation, contiguous bit ranges bound to the same value are automatically fused as one bigger range containing one repeating value. Values are stored with their size in bits; a range larger than the size of its value represents a repeating value.

Fig. 3.8 Assignments of an array

```
1  volatile unsigned int nondet;
2
3  void main (void) {
4    int t[16];
5    for (int j = 0; j < 16; j++) {
6      t[j] = nondet % 10;
7    }
8    t[10] = 42;  // strong-update
9    int i = nondet % 15;
10   t[i] = -1;   // weak-update
11 }
```

Let us take the code of Fig. 3.8 as example. After line 7, each of the 16 cells of the array t may have any value between 0 and 9 inclusive. For 32-bit integers, this is simply represented as:

$$< 0 - 511 > \rightarrow [0..9](\text{size } 32) \quad (3.5)$$

Any update of an offsetmap automatically fuses adjacent ranges when possible, and splits existing ranges when needed. At line 8 of the same example, writing the value 42 in the 10th cell of the previous array leads to this new offsetmap:

$$\begin{aligned} < 0 - 319 > &\rightarrow [0..9](\text{size } 32) \\ < 320 - 351 > &\rightarrow \{42\}(\text{size } 32) \\ < 352 - 511 > &\rightarrow [0..9](\text{size } 32) \end{aligned} \quad (3.6)$$

Writing the previous value back in the 10th cell would revert the offsetmap to its previous state.

Updates of an offsetmap can affect a set of possible addresses, in which case the precise address being written is unknown. When the address being written is exactly known, a *strong update* overwrites the previous value with the new one. This was the case at line 8 of our example, where the previous interval [0..9] at range 320−351 was replaced by the singleton {42}.

On the other hand, when the address is imprecise and represents several possible memory locations, a *weak update* joins the previous value with the new one at each possible memory location: as the exact address being written is unknown, each affected range may have the new value, or keep the old one if the update happened elsewhere. This is the case at line 10, where one of the first 15 cells of t is written, depending on the unknown value of the argument x, while the other cells are unchanged. The resulting offsetmap is the following:

$$\begin{aligned} < 0 - 319 > &\rightarrow [-1..9](\text{size } 32) \\ < 320 - 351 > &\rightarrow \{-1; 42\}(\text{size } 32) \\ < 352 - 479 > &\rightarrow [-1..9](\text{size } 32) \\ < 480 - 511 > &\rightarrow [0..9](\text{size } 32) \end{aligned} \quad (3.7)$$

This result can be seen in the final values of array t printed at the end of the Eva analysis of the code of Fig. 3.8, with parameter -eva-precision 1:

```
1  [eva:final-states] Values at end of function main:
2    t[0..9] ∈ [-1..9]
3    [10] ∈ {-1; 42}
4    [11..14] ∈ [-1..9]
5    [15] ∈ [0..9]
```

Finally, assignments to a very large number of non-contiguous ranges in an offsetmap may be over-approximated—writing the new value at more locations than required in order to improve performance.

Thanks to offsetmaps, the *cvalue* domain is able to efficiently represent the set of possible values of any C variables, including nested arrays, structs and unions. Furthermore, it is able to interpret any assignments of such objects, including those that would not match the C type of assigned variables. Nevertheless, the *cvalue* domain comes with some strong limitations.

Limitations

Since offsetmaps explicitly model the bit sequence that constitutes each C object, the *cvalue* domain depends on the hardware platform that the analyzed code will run on, which determines how C objects are encoded in memory. The Frama-C parameter -machdep defines the target platform, including endianness and the size of each C type. x86_64 is used by default; -machdep help provides a list of currently supported platforms. The results of an analysis are only valid for the selected architecture, and Eva is unable to assess the portability of a program between various architectures.

Another limitation of the *cvalue* domain comes from the abstractions used to represent scalar values: some sets of concrete values cannot be precisely represented as intervals and must be over-approximated by the domain. For instance, on a pattern such as **if (y != 0) return x/y;** where the variable y may initially have any value in $[-1024..1024]$, the *cvalue* domain cannot represent this interval deprived of 0 after the condition. Thus, the interval inferred for y remains unchanged, and a false alarm is then emitted for a possible division by zero. Such an imprecision can often be mitigated by some state partitioning techniques presented in Sect. 3.4, such as the case-based and path-based reasonings.

Most importantly, the *cvalue* domain infers the possible values of each variable separately, and is therefore unable to infer relational properties between variables. It is equally unable to infer relations between the different values of an array or a struct. This is the main limitation of the *cvalue* domain, which is addressed by the relational domains presented in Sect. 3.3.4. The simple code of Fig. 3.9 illustrates this shortcoming.

In this example, the parameter x has any value in the interval [0..10]. The variable y is assigned to 2*x, and thus has a value in the interval [0..20]. At line 5, unaware of the relation between x and y, the *cvalue* domain can only infer the imprecise interval

Fig. 3.9 Limitations of the *cvalue* domain

```
1  /*@ requires 0 <= x <= 10; */
2  int main(int x) {
3      int y, i, r1, r2, r3, t[2];
4      y = x + x;
5      r1 = y - x;
6      t[0] = x;
7      t[1] = x+1;
8      r2 = t[1] - t[0];
9      i = x % 2;
10     if (t[i] < 5) {
11         r3 = t[i];
12     }
13 }
```

[−10..20] for variable r1, even though its value is always positive. Likewise, the *cvalue* domain ignores the relation between the two elements of array t at lines 6 and 7, only inferring intervals [0..10] and [1..11] for each value respectively. Then, at line 8, it can only infer the once again imprecise interval [−9..11] for variable r2, instead of the precise value 1. However, the *octagon* domain, presented in Sect. 3.3.4, is able to infer the required relations, and thus to improve the intervals computed for these two variables.

Finally, the *cvalue* domain can only infer precise values for exact memory locations. When a left-value may refer to several possible addresses, the *cvalue* domain can only compute its value as the join of the values stored at each possible address, which may be overly imprecise. This frequently happens for dereferences of pointers that may point to different variables, or for array accesses with imprecise offsets.

After the condition of line 10, the possible values of t[i] could be reduced to the interval [0..4], but this left-value can refer to t[0] or t[1] according to the exact value of i. Without knowing which exact location is concerned, the *cvalue* domain cannot reduce its state according to this condition, and infers the imprecise interval [0..11] for the variable r3.

This last limitation can be mitigated by using *symbolic* domains, able to infer properties on symbolic expressions instead of exact memory locations.

3.3.3 Symbolic Domains

In Eva, the *equality* domain and the *symbolic locations* domain are two symbolic domains that infer information about the values of syntactic C left-values or expressions, manipulated as opaque terms without any assumptions about their semantics. They only associate these terms to some values, which can be useful if the same syntactic terms are reused later in the program. Both domains maintain some inferred information about a term as long as the memory location on which it depends remains unchanged, and they remove such information as soon as this memory location is modified in any way. They rely on the alias analysis performed by the *cvalue* domain to remain sound on writes through pointers.

3 Abstract Interpretation with the Eva Plug-in

Precise analysis by default:	Precise analysis with the equality domain:	Precise analysis with symbolic locations:
	1 `int tmp = i < len;`	
1 `if (i < len)`	2 `if (tmp)`	1 `if (i + k < len)`
2 `t[i] = 0;`	3 `t[i] = 0;`	2 `t[i+k] = 0;`
1 `tmp = t[i];`	1 `tmp = t[i]`	
2 `if (tmp > 0)`	2 `if (tmp > 0)`	1 `if (t[i] > 0)`
3 `res = tmp;`	3 `res = t[i];`	2 `res = t[i];`
1 `tmp = d1 - d2;`	1 `tmp = d1 - d2;`	
2 `if (tmp > 0)`	2 `if (tmp > 0)`	1 `if (d1-d2 > 0)`
3 `r = sqrt(tmp);`	3 `r = sqrt(d1-d2);`	2 `r = sqrt(d1-d2);`

Fig. 3.10 Examples of code patterns handled by symbolic domains

The *equality* domain infers and propagates equalities between the values of different C expressions. When interpreting an assignment `*p = expr;`, the domain simply infers the subsequent equality, except when the assignment modifies the involved values, as with `i = i+1;`. When interpreting a condition **if**(x==y), the domain also infers the equality (except if it involves the floating-point values −0 and 0, as they compare equal while behaving differently in some operations).

The *symbolic locations* domain infers and propagates the abstract values of C left-values whose exact memory location is unknown, or of C expressions that depend on several memory locations. When interpreting a condition such as **if**(t[i]> 0 && x+y > 0), this domain infers that `t[i]` and x+y are both strictly positive, regardless of the specific values of `t, i, x` and `y`.

These domains make the analysis results less dependent on the exact syntax chosen to write the analyzed programs. The equality domain also tends to improve the analysis by inferring equalities about temporary variables introduced by the code normalization done by Frama-C for expressions with side effects. On the other hand, the symbolic locations domain overcomes a specific limitation of the *cvalue* domain, which is unable to infer precise values for left-values that do not refer to a single memory address, such as `t[i]` or `*p` when either `i` or `p` is imprecise. This is precisely what the symbolic locations domain does.

Figure 3.10 presents some typical code patterns on which these symbolic domains improve the precision of the Eva analysis. Each one has three variants: although they are semantically equivalent, the first one is always precisely analyzed, while the other two require the symbolic domains to achieve the same results.

In the first example, `t` is an array of size `len`, and the value of variable `i` may exceed this size. The left pattern is precisely analyzed, as the *cvalue* domain is able to reduce the set of possible values of `i` when interpreting the condition `i < len`: no alarm is ever raised here. The middle pattern introduces a temporary variable `tmp` for the value of `i < len`, used in the condition. The *cvalue* domain reduces the value of `tmp` but is not able to reduce `i` accordingly: an out-of-bounds alarm is then emitted. The equality domain can associate the value of `tmp` to `i < len`, allowing to reduce the value of `i`, and avoiding the alarm. The right pattern uses `i+k` as array index. If both i and k may range from 0 to $len - 1$, it is not possible to reduce their possible values based on the condition i+k< len. However, the symbolic locations domain infers the interval $[0..len - 1]$ for the expression i+k, and uses it to avoid the alarm at the next line.

In the second example, the index i may range over the whole array t, which contains positive and negative values. The left pattern is precisely analyzed by the *cvalue* domain, which reduces the value of the tmp variable and then uses it for the assignment of res, which can thus be proved to be positive. In the middle pattern, the analysis needs the equality relation between tmp and t[i] to achieve the same precision. On the right pattern, there is no temporary variable; as t[i] may point to several possible memory locations, none of them can be reduced safely. Nevertheless, the symbolic locations domain infers the value of t[i], regardless of the exact array cell it represents.

The third example requires similar reasoning to prove that the argument of sqrt is positive, and thus avoid a false alarm.

These examples show how these two symbolic domains help to overcome some limitations of the *cvalue* domain, while being very fast: they only slightly slow down the Eva analysis. However, they are unable to conduct advanced reasoning as they only use C expressions and left-values as opaque terms, regardless of their semantics. To infer more complex invariants, Eva also features some standard relational domains.

3.3.4 Relational Domains

Eva features several abstract domains that infer relational properties between the values of different program variables: the *octagon* domain [17], the *gauges* domain [27], and a binding to abstract domains provided by the Apron library [13]. A relation between some variables x and y, such as $x < y$, can lead to a more precise evaluation of the values of some expressions involving both variables (for instance, $y - x$ is positive), or to reduce the possible values of one variable according to the values of the other (for instance, if we learn that y is negative, then we can deduce that x is also negative).

Figure 3.11 presents two minimal code examples on which relational domains can improve the analysis precision. On the first code, a relation exists between variables x and y: $0 \leq y - x \leq 10$. On the second code, a relation exists between variables i, j and p: at each loop iteration, $j = 2 * i + 3$ and $p = \&t + i$. Without these relations, the interval semantics of the *cvalue* domain infers very imprecise ranges for r1 and r2 on the first code, and emits alarms on increment of j and p on the second one. Moreover, as these relations do not directly appear as syntactic expressions in the codes, they cannot be captured by the symbolic domains introduced above. Relational domains are thus required to precisely analyze these examples.

The *octagon* domain, based on [17], infers relations between integer variables as numerical constraints of the form $a \leq \pm X \pm Y \leq b$, where X and Y are program

Fig. 3.11 Examples of use of relational domains

```
1  /*@ requires -100 <= y <= 100; */
2  void main (int y) {
3      int k, x, r1, r2;
4      k = Frama_C_interval(0, 10);
5      x = y - k;
6      r1 = x + 3 - y;
7      if (y > 15)
8          r2 = x;
9  }
```

(a) Simple use of the octagon domain

```
1  void main (void) {
2      int j = 3;
3      int t[100];
4      int *p = &t[0];
5      for (int i=0; i<100; i++) {
6          j += 2;
7          *p++ = i;
8      }
9  }
```

(b) Simple use of the gauges domain

variables, and a and b are constants. Currently, this domain only infers relations between a pair of variables when both occur in the same instruction. If an instruction involves 3 variables (or more), relations are inferred for each pair of variables. On the code of Fig. 3.11a, the octagon domain infers the relation $0 \leq y - x \leq 10$ at line 4, and uses it later to:

- evaluate the expression x+3-y to the precise interval $[-7..3]$ (instead of the interval $[-207..203]$ computed by the *cvalue* domain alone).
- after the condition at line 6, reduce the possible values of x according those of y: as y has value in $[16..100]$, x (and thus r2) has value in $[6..100]$ (instead of the interval $[-110..100]$ computed by the *cvalue* domain).

The *gauges* domains, based on [27], infers affine inequalities between variables modified within a loop and the number of loop iterations. More precisely, it encodes numerical constraints of the form $a\lambda + b \leq x \leq c\lambda + d$, where x is a program variable, λ is a special counter representing the number of iterations of the current loop, and a, b, c, d are constants. It is able to infer such constraints on integer variables and on the offset of pointers, but only on local variables (and not global variables). Moreover, no relation is inferred outside of loops. The gauges domain can improve the analysis results on finite loops that increase multiple variables in an affine way. However, it is not useful on fully unrolled loops, or if the relation between variables is not affine (for instance computing a square). Figure 3.11b presents a classic example of a loop that the gauges domain can precisely analyze: by default, the analyzer emits alarms for a possible overflow of y and an invalid memory access through p. The gauges domain infers that $y = 2 * i + 5$, so y does not exceed 203, and that the

```
1  #include "__fc_builtin.h"
2
3  #define SIZE 1024
4  extern char buffer[SIZE];
5  unsigned char r1, r2, r3;
6
7  int read_buffer (char *buffer, int index, int size) {
8    unsigned char msg_len;
9    int msg_read = 0;
10   while (index < size) {
11     msg_len = buffer[index];
12     if (size - index - 1 < msg_len)
13       break;
14     if (msg_len >= 1) r1 = buffer[index+1];
15     if (msg_len >= 2) r2 = buffer[index+2];
16     if (msg_len >= 3) r3 = buffer[index+3];
17     index += msg_len + 1;
18     msg_read++;
19   }
20   return msg_read;
21 }
22
23 void main (void) {
24   int index = Frama_C_interval(0, SIZE);
25   int nb = read_buffer(buffer, index, SIZE);
26 }
```

Fig. 3.12 A more complex example for the octagon and gauges domains

offset of pointer p does not exceed i, so all memory accesses are valid: no alarm is emitted with the gauges domain.

Figure 3.12 introduces a more complex C code on which both the octagon and the gauges domains are required to remove false alarms from the analysis. In this code, the read_buffer function reads messages in the array buffer. A message consists of several consecutive characters, the first one is its length, i.e., the number of other characters to be read in the message. For each message read from index index, the function:

- reads the first character: it is the length l of the message.
- stops if l characters cannot be read in the array.
- reads up to three characters, if the length l allows it.
- increments index by the message length plus one, to read the next message.
- increments the variable msg_read, which counts the number of messages read in the loop.

The default analysis of this code raises several alarms: according to Eva, the dereferences of buffer could be out of bounds, and the increment of variable msg_read could overflow. Enabling both octagon and gauges domains via the analysis parameter -eva-domains octagon,gauges removes these false alarms:

- The octagon domain infers the relation between index and msg_len after line 13: $index + msg\_len \leq size - 1$, where size is a constant. So if $msg\_len \geq 1$, then

3   Abstract Interpretation with the Eva Plug-in

$index + 1 \leq size - 2$, and `buffer[index+1]` is a valid dereference at line 14. The same reasoning applies to the dereferences at line 15 and 16.
- The gauges domain infers the relation $msg\_read \leq index$. As the value of `index` remains smaller than `size` in the loop, so does `msg_read`, and its increment at line 18 cannot overflow.

Finally, Eva provides a binding to the abstract domains of the Apron library [13], including octagons, linear equalities and polyhedra. This binding is an experimental proof-of-concept that should only be used on small examples; it may result in very long analyses and massive memory usage on large case studies.

The current implementation of these relational domains in Eva shares some limitations:

- They are able to infer relations between integer values or pointer offsets, but not between floating-point variables.
- While the octagon domain is able to infer relations between any integer C left-values, the gauges domain and the Apron binding can only infer relations between scalar variables; they do not infer properties on array values, structure fields or pointer dereferences.
- These domains are intraprocedural by default: the inferred relations are local to the current function and are lost when entering a function call or returning to its caller. The `-eva-octagon-through-calls` parameter can be used to propagate relations inferred by the octagon domain through function calls, resulting in a more precise but slower analysis.

3.3.5   Inferring New Properties

All domains presented above infer properties on the sets of possible values for program variables during the program execution, and are primarily intended to prove alarms or assertions about these variables values. The *numerors* and *taint* domains address different challenges by inferring new kinds of properties.

The *numerors* domain [12] targets floating-point computation, and aims at computing the difference between computations in the floating-point semantics and their equivalent in mathematical reals. For each floating-point variable and expression of a program, the domain computes over-approximations of:

- the real value, computed with the mathematical semantics;
- the floating-point value, computed according to the IEEE 754 standard;
- the absolute error: the possible difference between the real and the floating-point value;
- the relative error: the absolute error divided by the real value.

The numerors domain is experimental and should only be used on toy examples. It also has strong limitations: it does not support loops, and only infers floating-point errors for scalar variables, while ignoring arrays, structs and unions.

The *taint* domain performs a taint analysis, which is a data-dependency analysis tracking the impact an external user may have on the program execution when supplying some data. The initial taint must be specified by the user with ACSL annotations //@ taint< lvalues>;. Such annotations are hypotheses of the taint analysis, and thus have no logical status. The taint is then propagated through the Eva analysis, which computes an over-approximation of the set of tainted memory locations at each program point.

This taint analysis can be used to find alarms that are relevant in a cybersecurity context: if an attacker can modify the values on which an alarm depends, then they may be able to use the related undefined behavior to their advantage. Conversely, alarms that only depend on untainted memory locations could never be exploited by an attacker. When the taint domain has been enabled for an Eva analysis, the graphical interface highlights tainted values and statements propagating the taint, and allows the user to list the alarms related to tainted values.

The ACSL predicate \tainted can also be used to verify that some values are *not* tainted. Indeed, the domain only infers over-approximations of the tainted locations: while locations in the over-approximations may be tainted, all other locations are proven to be safe.

All these domains are also described in the Eva user manual [5], Sect. 6.7.

3.3.6 Communication Between Abstract Domains

Many of the abstract domains provided by Eva have been presented above. As the verification of a program often requires properties inferred by different domains, the abstract interpretation framework provides some standard ways to combine abstract domains and their semantics. Most of them allow abstract domains to interact with each other during the analysis, exchanging information in order to reach a more precise approximation of the concrete program semantics.

In this spirit, Eva features a generic communication system between abstract domains, based on abstractions of C values. Its design, illustrated in Fig. 3.13, relies on the separation between *state abstractions* and *value abstractions*:

- An abstract *state* represents a set of possible concrete memory states at a given point of a program execution. State abstractions are the standard domains described above. They are propagated through the analyzed program using the abstract transfer functions that correctly approximate the effect of C statements on their abstract states, as stated by Sect. 3.2.2. State abstractions also provide queries able to produce abstract *values* for some variables or expressions.
- An abstract *value* represents the set of possible C values of one variable or expression in a given abstract state. The main value abstractions currently used in Eva

3 Abstract Interpretation with the Eva Plug-in 163

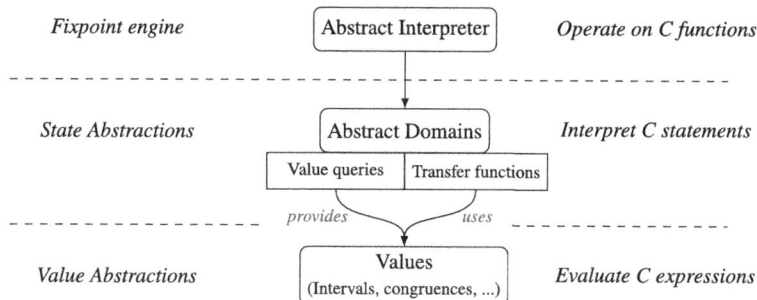

Fig. 3.13 Architecture of Eva abstractions

are those described in Sect. 3.3.1, and are able to precisely represent C values of integer, floating-point and pointer types. They provide an abstract semantics of arithmetic operations on C expressions (additions, shifts...), and are used as a communication interface between state abstractions.

Eva features a cooperative evaluation of expressions in a product of abstract states, in which each domain can provide abstract values for any variable or sub-expression on which they have inferred some properties. If several domains provide different abstract values for the same variable or expression, these values are intersected.

When interpreting a statement, all involved expressions are evaluated into value abstractions, using values provided by each domain. Then, all computed value abstractions are made available for the state transfer functions, to model the effect of the statement as precisely as possible. As these value abstractions have been computed cooperatively, information may flow from a domain to another, without direct exchanges between them.

Let us illustrate this communication between domains on a minimal example, the interpretation of assignment *p = 2 * (x -y) in the following product of states: the *cvalue* domain has computed that integer variables x, y and z are bounded by [0..10], and that pointer p points to z, while the *octagon* domain has inferred the relations $x \geq y$ and $x \geq z$.

$$\left\{ \begin{array}{l} x \in [0..10] \\ y \in [0..10] \\ z \in [0..10] \\ p \in \{\&z\} \end{array} \right\} \times \left\{ \begin{array}{l} x \geq y \\ x \geq z \end{array} \right\}$$

Eva starts by evaluating abstract values for the right expression 2*(x-y). First, it queries the domains for the values of x and y; the *cvalue* domain provides the interval [0..10] for each. Then, interval semantics is used to compute the value [−10..10] for x-y, and the *octagon* domain uses its first relation to also provide the value [0..∞] for this same expression; both values are intersected, leading to the more precise interval [0..10]. Once again, interval semantics is used to compute the value [0..20] for 2*(x-y), that no domain can improve.

Eva also evaluates the value of pointer p, whose location is assigned by the statement; only the *cvalue* domain provides a relevant value for a pointer, which here is the singleton &z. Finally, each domain interprets the statement *p = 2 * (x-y), using all these computed values. The *cvalue* domain can directly use the interval computed for the right expression as the new value of z (which is more precise than the interval that the *cvalue* domain could have computed on its own). The *octagon* domain uses the value of the assigned pointer p to remove the relation about the pointed variable z, as it may not hold anymore after the assignment. This leads to the precise and sound product of states:

$$\begin{Bmatrix} x \in [0..10] \\ y \in [0..10] \\ z \in [0..20] \\ p \in \{\&z\} \end{Bmatrix} \times \{x \geq y\}$$

As shown in this example, the communication of pointer values is crucial to ensure the soundness of all Eva domains, as only the *cvalue* domain infers the alias information required to properly interpret pointer assignments and dereferences.

The Eva architecture and its communication system between domains are described in great detail in [2, 4]. Value abstractions are extensible, should new domains require more expressiveness in their communications. However, a current limitation of this system is that domains can only obtain information from each other during the interpretation of C statements: in particular, there is no communication during the join or widening operations, where domains often lose some precision.

Finally, both state and value abstractions are easily extensible within Eva, and new abstractions are regularly implemented to tackle new verification challenges, and to further improve the precision of the analyses. However, abstract domains are not the only lever available to reach a better precision, and the next section presents how *state partitioning* can achieve this goal without adding new domains.

3.4 State Partitioning

During our journey to banish false alarms from the analysis, we may find out that even with the activation of all abstract domains, some imprecision remains. To get rid of these false alarms, Eva also proposes a set of tools based on *state partitioning* [22] that we will describe in this section and that we will illustrate on the example in Fig. 3.14.

In this toy example, we use a function parameter to index an array. The index is offset by three, such that any input parameter x between -3 and 3 is valid. However, when we analyze this piece of code with Eva, two alarms are emitted.

Fig. 3.14 An example of program where several partitioning strategies can improve the analysis precision

```
1  int abs(int x) {
2    if (x >= 0)
3      return x;
4    else
5      return -x;
6  }
7
8  int t[7] = { 1, 2, 5, 9, 5, 2, 1 };
9
10 int f(int x) {
11   if (abs(x) <= 3)
12     return t[x + 3];
13   else
14     return -1;
15 }
```

```
1  [eva:alarm] partitioning.c:12: Warning:
2    accessing out of bounds index. assert 0 ≤ (int)(x + 3);
3  [eva:alarm] partitioning.c:12: Warning:
4    accessing out of bounds index. assert (int)(x + 3) < 7;
```

Eva does not seem able to prove that the index x+3 is inside the array bounds, that is $0 \leq x + 3 < 7$, although it derives from the preceding test abs(x) <= 3. The reason why this simple inequality is not inferred is not completely obvious. To be rigorous, the test condition only states a property about abs(x), and not a property about x. To be able to deduce the latter from the former, we need to prove another property about the result of abs: either $abs(x) = x$ when $x \geq 0$, or $abs(x) = -x$ otherwise. Combined with $abs(x) \leq 3$, this property implies the correctness of the array access.

The equality domain (see Sect. 3.3.3) seems suitable to prove the property about the return value of abs(x). If x can only have positive values, this domain can easily prove abs(x) == x. Conversely, if x can only have negative values, it can prove abs(x) == -x. But what if x can both have positive and negative values? In that case, no equality is true for every possible value of x, and the domain will not infer any.

Depending on the context, there are several ways to solve this problem. Let us first suppose that the function f is called inside a loop such that the value of its parameter x can only have one possible value at each loop iteration. This is the case in the following example.

```
1  void g(void) {
2    int outputs[10];
3    for (int i = 0 ; i < 10 ; i ++)
4      outputs[i] = f(i-5);
5  }
```

In this case, we can apply a first technique called *loop unrolling* during the analysis of the loop. Instead of looking for a loop invariant, the analyzer will compute a distinct abstract state for each iteration of the loop. Instead of one state, we will have a set of inferred states. Functions f and abs are then interpreted multiple times, once for each iteration of the loop of g. At each iteration, the parameter x has only one possible value, and the equality domain can infer the proper equality between x and abs(x). We fall into the case where Eva can prove the correctness of the array accesses in function f, and it will compute the precise output value that is stored in the array outputs in function g. To unroll a specific loop with Eva, we can use a program annotation just before the given loop.

```
void g(void) {
  int outputs[10];
  //@ loop unroll 10;
  for (int i = 0 ; i < 10 ; i ++)
    outputs[i] = f(i-5);
}
```

This loop-unrolling technique would work on this example as long as the set of possible values of the f's parameter x are all positive or all negative at each iteration. But even if Eva has to interpret the function f with a wider range of values for x, including both positive and negative values, it is still possible to achieve a maximal precision on this example. To be able to benefit from the equality domain, we must *split* the analysis of abs into two cases :

- when $x \geq 0$ where we can prove that the result is equal to x and
- when $x < 0$ where we can prove that the result is equal to $-x$.

This is the second technique available to Eva. Similarly to loop unrolls, a program annotation can be used, and the array access can be proven valid with the use of the equality domain.

```
int f(int x) {
  //@ split x < 0;
  if (abs(x) <= 3)
    return t[x + 3];
  else
    return -1;
}
```

These two techniques will be described more into details in the following section along with three other techniques. But before that, let us look into the theory of state partitioning.

3.4.1 Experiment with State Partitioning

We can use the mechanism described in Sect. 3.2.1 to witness the internal representation of state partitioning. The Frama_C_show_each builtin functions will print the computed abstract values once for each individual state of the partition. Let us look at the following artificial example.

3   Abstract Interpretation with the Eva Plug-in

```
1  /*@ requires 0 ≤ x < 200; */
2  void main(int x) {
3    //@ split x < 100;
4    //@ loop unroll 3;
5    for (int i = 0 ; i < 3 ; i++) {
6      Frama_C_show_each(i, x);
7    }
8  }
```

It combines a two cases split with a loop unrolling of 3 iterations, thus we expect to have $2 \times 3 = 6$ abstract states for the loop. Indeed, if we analyze this example with Eva, we will have these 6 outputs:

```
1  [eva] partitioning-prop.c:6: Frama_C_show_each: {0}, [0..99]
2  [eva] partitioning-prop.c:6: Frama_C_show_each: {0}, [100..199]
3  [eva] partitioning-prop.c:6: Frama_C_show_each: {1}, [0..99]
4  [eva] partitioning-prop.c:6: Frama_C_show_each: {1}, [100..199]
5  [eva] partitioning-prop.c:6: Frama_C_show_each: {2}, [0..99]
6  [eva] partitioning-prop.c:6: Frama_C_show_each: {2}, [100..199]
```

Which give us, for each computed abstract state, the values of i and x.

To be perfectly precise, note that these outputs do not occur simultaneously during the analysis. Obviously, the two states of the first iteration have to be computed before the next iteration can be analyzed. Each analysis iteration of the loop adds two abstract states, and propagates only these two new states. There is an easy confusion here between the loop iterations and the analysis iterations as these two coincide, but it is possible to build examples where this is not the case.

3.4.2   Theory

Both introduced techniques share a common idea: instead of inferring a single abstract state for a control point of our program, we allow the analyzer to infer a set of abstract states. We will have one abstract state per iteration when unrolling loops, and one abstract state per case when splitting.

An abstract state can often be described by a formula on program variables, e.g., $x \in [0, 3]$. A set of abstract states would be described by the disjunction of all formulas describing individual abstract states, e.g., $x \in [0, 3] \vee x \in [-3, -1]$.

For termination and performance reasons, this set must remain bounded and not be too large. For each technique, we ensure that there is a limit on the number of abstract states that will be created. For instance, Eva will always enforce a bound on the number of unrolled iterations and on the number of split cases. Multiple partitioning instructions can be used together, multiplying the number of abstract states considered.

Note that, despite the name, this set of abstract states is not a partition in the mathematical sense, since there is no guarantee that two states of the partition do not overlap. However, whenever Eva detects that a state is included in another state previously propagated, it stops the propagation as there is no point in recomputing the same result.

3.4.2.1 Partitioning Key

Each state in the partition is associated to a partitioning key. This key retains the information about why this state is separate from the others. In the example of Sect. 3.4.1, the key would be a pair (*case*, *iteration*) identifying, for each state, whether it corresponds to the case $x < 100$ or to the case $x \geq 100$, and in which iteration of the loop it can appear.

The partitioning key is updated each time we encounter a *partitioning directive*. Loop unrolling will update the key when entering and exiting the loop by extending or shrinking the key with a new field for the loop and will update this field at the end of each iteration to keep track of the iteration count. This means that a loop unrolling annotation induces three partitioning directives. As for split annotations, they will update the key by extending it with a new field where we can store the evaluation of the split expression for the associated state.

If two states have distinct keys, they must be kept separate in the partition. But if two states have the same key, they must be joined together. Consider for instance, that we are analyzing an if-then-else statement with a partition of several states and that this statement does not itself contain any partitioning directive. The then and the else branch will both be analyzed with each state of the partition. At the end of the conditional structure, these states must be joined together and we have now twice the number of states. We could continue the analysis with this larger set, but instead we will join states which have the same partitioning key. Actually, since there were no partitioning directives, each key has been preserved during the analysis of the statement, and we should now have, for each key, two states: one that comes from the then branch and one from the else branch.

With this in mind, a state partition can be seen as map from partitioning keys to abstract states. Note that, unfortunately and for the time being, there is no way to request Eva to print the partitioning key of the states it propagates; this key is only stored internally.

3.4.2.2 Partitioning

State partitioning can be seen as an *abstract domain functor*, that is, a transformation of an abstract domain to a more powerful abstract domain. Typically, such a functor could also be applied to a product of abstract domains, so that all domains benefit from the transformation. To define it, we have to formalize the abstract states of the augmented domain, the concretization function, the operators $\sqsubseteq, \sqcup, \nabla$ on these states, and the abstract semantics.

Fortunately, these definitions are pretty straightforward. Let us say that we are given an abstract domain $\mathcal{D}$ and a set of implementation-defined partitioning keys $\mathbb{K}$.

3 Abstract Interpretation with the Eva Plug-in

Our partitioned domain will contain partitioned states $X^\#$ which will be sets of states from the original domain associated to a partitioning key, that is, a set of pairs from $\mathbb{K} \times \mathcal{D}$. Thus our state partitioning functor $\mathcal{SP}$ is defined as the powerset $\mathcal{P}(\mathbb{K} \times \mathcal{D})$:

$$\mathcal{SP}(\mathcal{D}) = \mathcal{P}(\mathbb{K} \times \mathcal{D})$$

To enforce the uniqueness of keys, we can reduce partitioned states with a *unique* function, that joins together states $x^\#$ which share the same partitioning key k.

$$unique(X^\#) = \left\{ \left(k, \bigsqcup_{(k,x^\#) \in X^\#} x^\# \right) \mid k \in \mathbb{K} \right\}$$

Each abstract state inside the partition represents reachable states. In other words, the set of reachable states can be seen as the union of what each abstract state represents. Formally, we extend the concretization function γ from $\mathcal{D}$ to $\mathcal{SP}(\mathcal{D})$.

$$\gamma(X^\#) = \bigcup_{(k,x^\#) \in X^\#} \gamma(x^\#)$$

It follows from the two previous definitions that *unique* performs an over-approximation.

From these definitions, we can lay down the last required definitions.

- The inclusion of two partitioned states is defined key-wise from the inclusion of $\mathcal{D}$.[6]
$$X^\# \sqsubseteq Y^\# \Leftrightarrow \forall (k, x^\#) \in X^\#. \exists (k, y^\#) \in Y^\#. x^\# \sqsubseteq y^\#$$

- The join of two partitioned states is defined as the reduced union of partitions.
$$X^\# \sqcup Y^\# = unique(X^\# \cup Y^\#)$$

- The widening of two partitioned states is defined key-wise. If a key k exists only on one of the partitioned states, the pair $(k, s^\#)$ from this state is kept in the result.
$$\begin{aligned}X^\# \triangledown Y^\# = &\left\{ (k, x^\# \triangledown y^\#) \mid (k, x^\#) \in X^\# \land (k, y^\#) \in Y^\# \right\} \\&\cup \left\{ (k, x^\#) \mid (k, x^\#) \in X^\# \land \nexists y^\#.(k, y^\#) \in Y^\# \right\} \\&\cup \left\{ (k, y^\#) \mid \nexists x^\#.(k, x^\#) \in X^\# \land (k, y^\#) \in Y^\# \right\}\end{aligned}$$

- We extend every semantic function f to abstract states by applying them to each state of the partition.

---

[6] This choice is not optimal in terms of precision, but drastically reduces the number of inclusion tests to be done. However, Eva regains precision by using other mechanisms to detect inclusion between partitioned states during the propagation, but the details are beyond the scope of this book.

$$f(X^\#) = \left\{ (k, f(x^\#)) \mid (k, x^\#) \in X^\# \right\}$$

3.4.3 Loop Unrolling

While loop unrolling is a method to optimize compiled programs, it is used in program analysis to improve the precision of the inferred invariants. The idea is to locally avoid (or delay) the use of widenings (presented in Sect. 3.2) that guess one inductive state representing all iterations of a given loop at once. Instead, the analysis interprets separately each iteration of the loop, and infers a set of distinct states, one for each iteration. Though it seems that it could only be used when the number of loop iterations is bounded, we will see that this is not necessarily the case. In the previous Fig. 3.14 example, a useful invariant is the property

$$0 \leq x + 3 \leq 6$$

valid at line 12. When unrolling the loop, instead of proving the invariant for all possible values of x at once, we prove it individually for each value of the inputs array.

In Eva, this is done either manually, using a program annotation, or automatically by using a built-in heuristic. The annotation is tied to a loop and can set a limit on the number of unrollings. In the previous example, we used:

```
//@ loop unroll 10;
```

As stated before, this limit is necessary to ensure the termination of the analysis. It can be any expression that dynamically evaluates to a constant and can be omitted (but this only means that Eva will use a default limit).

It does not have to match the exact number of loop iterations in the concrete execution of the program:

- If the given limit is greater than the number of loop iterations, the loop will be fully unrolled. The analyzer will naturally find that, at some iteration, it is not possible to reach further iterations The analysis of the loop will stop, fully completed and without any computation overhead.
- If the given limit is smaller than the number of loop iterations, only the initial iterations will be unrolled. After the required number of unrollings, the analyzer will detect that further iterations are still possible and will proceed to find a loop invariant for the remaining iterations in a standard way—using widenings. This feature allows the unrolling of unbounded loops, which might be useful if, for instance, the first iterations of this loop perform initialization steps of some kind.

Loop unrolling might be very expensive, in particular if the limit is high (requiring thousands of iterations or more). The time cost of an unroll annotation is linear on the actual number of performed unrollings. Counterintuitively, it might actually *speed the analysis up* for loops with a small number of iterations (a few dozen at most), as the interpretation of a loop iteration on a precise state is often faster than the tedious process of invariant inference.

Also for this reason, but mainly because loop unrolling performs very well at removing false alarms, Eva proposes a heuristic to choose automatically which loops to unroll: if the loop has a sufficiently small number of iterations, always unroll it. It may or may not be useful to unroll each of these loops, but focusing on a small limit will ensure this has minimal impact on performance on a worst case scenario. This heuristic is controlled by the option -eva-auto-loop-unroll n, disabled by default and that must be given an integer n, the limit of the heuristic: if the algorithm can detect that a loop has fewer iterations than n, it will be unrolled. In practice, using values around 100 for the limit of -eva-auto-loop-unroll usually gives good results without slowing down the analysis too much.

This heuristic is based on a simple algorithm that will use the inferred properties of the program when reaching a loop to do a simple semantic analysis of the loop. If the algorithm fails to detect that a loop can be unrolled with a small number of iterations, the manual annotation can still be used. Conversely, if the algorithm chooses to unroll a loop that is useless and too expensive to unroll, this unrolling can be manually disabled with a **loop unroll 0;** annotation.

In the previous example, loop unrolling not only helps with the proof of array accesses, but it also improves the precision of the analysis about the contents of the array outputs. Indeed, thanks to unrolling, Eva manages to prove that the array is initialized at the end of the loop.

For the same reason, in this chapter's introducing example of Collatz series, Fig. 3.1, unrolling the loop allows proving the initialization of the array. Thus, the access to t[i] can be proven safe and this eliminates the false alarm about array initialization. Note that -eva-precision implies an increasing value for automatic loop unrolling with higher precision levels; this is why the initialization alarm disappears when the precision level is high enough.

Proving array properties, like in these two examples, is in fact one of the most common uses of loop unrolling.

3.4.4 Case-Based Reasoning

Case-based reasoning is probably the second most used technique when trying to eliminate false alarms. It effectively allows inferring disjunctive invariants. For instance, in the Fig. 3.14 example, the interesting invariant would be the following formula.

$$x \geq 0 \wedge \text{result} = x \bigvee x < 0 \wedge \text{result} = -x$$

```
 1  /*@ ensures 0 ≤ \result < i;            1  /*@ ensures 0 ≤ \result < i;
 2      assigns \result \from               2      assigns \result \from
        \nothing; */                            \nothing; */
 3  int rand(int i);                         3  int rand(int i);
 4                                           4
 5  void f() {                               5  void f() {
 6    int t[10] = {0};                       6    int t[10] = {0};
 7    int i = 0;                             7
 8    //@ dynamic_split i;                   8    for (int i = 0; i < 10;) {
 9    while (i < 10) {                       9      t[i] = rand(10*(i+1));
10      t[i] = rand(10*(i+1));              10      if (10*i <= t[i]) {
11      if (10*i <= t[i]) {                 11        //@ split i;
12        i++;                              12        i++;
13      }                                   13      }
14    }                                     14    }
15  }                                       15  }

         (a) dynamic_split                          (b) split
```

Fig. 3.15 An example of unrolling alternatives with split and dynamic_split

Since most of the abstract domains available only use conjunctive formulas, each time one needs a disjunction it is likely that they will also need case-based reasoning. As we have seen in the introduction of this section, this can be done using the split annotation.

```
 1  //@ split x < 0;
```

This annotation does not only accept boolean expressions, but also any integer expression and even ACSL predicates. It splits the analysis into as many cases as there are possible values for the expression. For the previous split, this means *at most* two cases. Since Eva computes the possible values for the split's expression, there may be less than two cases if the expression is always true or always false. For integer splits, there may be more cases, but for termination reasons, Eva will check that the number of cases does not exceed a configurable limit, set by the option -eva-split-limit.

3.4.4.1 Static and Dynamic Splits

What if, in our example, the value of x does change after the annotation? This is likely to alter the evaluation of the expression. However, the split annotation only changes the partitioning when it is encountered, it does not have any effects on future partitions. But, if we really need to reevaluate the split whenever there is a change in the involved variables, there is another annotation that will maintain the split during the analysis: the dynamic_split annotation. It will imply the re-evaluation of the expression or the predicate after each instruction.

For instance, take the example of Fig. 3.15a. We have a loop that cannot be strictly unrolled since its number of iterations might be unbounded. However, the analyzer will find that there is only a finite and small set of possible values for i, and we can split the analysis for each of these values. Since the split is dynamic, the analyzer will update the partitioning each time the value of i is incremented, and will always have a separate abstract state for each value of i. This allows the analysis to infer the distinct intervals [0..9], [10..19], [20..29] and so on for each cell of the array t at the end of function f, instead of the less precise interval [0..99] for the whole array that is inferred by default.

Note that this trick can only work if the declaration of i is before the dynamic_split annotation: the split expression must only contain declared variables at the point of insertion.

3.4.4.2 Splits Lifetime

Splits can be expensive. Ideally, we want to limit the case-based reasoning to a local part of the code, as minimal as possible. By default, a split is stopped at the end of the functions in which it is declared. However, it is possible to end the case-based reasoning earlier with a merge annotation, like this one:

```
//@ merge i;
```

This annotation is symmetric to split and must be followed by the same expression. This effectively allows several case-based reasoning to be done at the same time without requirements about the order they start and end. It has some drawbacks though, as this means several splits can not use the same expression; otherwise, we would not know which split to end. Instead, when the same split expression is encountered multiple times, the split is refreshed: the expression is re-evaluated, the old split is dropped and the new split replaces it. To illustrate this, we can use this mechanism as another way to precisely analyze the loop of Fig. 3.15b: the split is reevaluated each time the value of i is incremented, and we obtain the same partitioning (and the same result) as with the dynamic split in Fig. 3.15a. Note that if several assignments modified the value of i, it would be necessary to add a split annotation for each one, while only one dynamic_split is required.

While a split can be carried through calls, as we already said, they end when the function they are declared in returns, by default. Sometimes we need to carry the case-based reasoning through returns as well. When this is needed, we can use the option -eva-interprocedural-splits with care: as soon as this option is set, every split not associated to a merge will carry until the end of the analysis, which may last way longer.

Note that the merge annotation has a limitation: it cannot be used inside another function deeper in the call tree than the corresponding split. In other words, any called function between the split and the merge must have returned. This limitation comes from intricate implementation reasons and is tied to caching and performances [30].

3.4.5 Path-Based Reasoning

A special instance of case-based reasoning is the reasoning on paths of the program. We might want to do this kind of analysis when we need to infer properties which might differ depending on which branches have been taken before. If we consider our abs example in Fig. 3.14, it falls in this category. Indeed, we are reasoning on cases $x \geq 0$ and $x < 0$ which correspond exactly to the program traces taking the branch then and else in the function abs.

Of course, it is possible to just use split annotations just before an if-then-else each time we want to distinguish between states that go into different branches. But this requires a manual intervention to identify favored if and annotate them. On the other hand, a naive automatic partitioning on paths would be time prohibitive.

Eva proposes a global parameter -eva-partition-history n which builds a state partitioning based on paths but only considering the last branching taken in a trace. More accurately, it is focused on *join nodes*, i.e., control points where several control flows join: the end of an if-then-else or of a switch, a for, while or do-while's loop head, a goto's target, etc. When the analyzer reaches a join node, it would normally compute the abstract domain's join of each incoming state of all the incoming branches. Instead, we keep states from all these branches separated, until the next join node, or until the next n join nodes if n > 1. In other words, we delay abstract domains' join application. The number of propagated states is generally exponential in n. Internally, the partitioning key stores which incoming branches were taken to reach the last n join nodes.

This solution enables local path-based reasoning, fully automatically, with a high time cost, however. In our experience, this option is only useful in rare occasions.

Since the cost of path partitioning is high, it is not carried through function returns by default, similarly to splits. Again, when necessary, an option -eva-interprocedural-history can be used to enable interprocedural path partitioning.

Let us put this all together to solve the problem of example Fig. 3.14. We need to activate path partitioning, through function returns and we also need the equality domain. So, here is a combination of working parameters:

```
-eva-partition-history 1 -eva-interprocedural-history
   -eva-domains equality
```

3.4.6 Brute-Force Partitioning

The last partitioning technique available in Eva follows a very simple principle: it keeps abstract states separated as long as possible, until their number becomes too high. When there are too many states, the analyzer starts to join them as it would have done if the technique were disabled. It can be used to reproduce loop unrolling and path partitioning.

This technique can be enabled with the parameter -eva-slevel n where n is the limit to the number of abstract states that can be computed at each control point before joining them. It can be defined individually for each function with -eva-slevel-function or locally with annotations. By default, this partitioning is dropped when a function returns, but can be expanded using -eva-split-return.

While -eva-slevel n has been for a long time the only way to partition the set of states, it has many drawbacks compared to the other techniques and is now discouraged.

- It does not allow fine-tuning. For instance, one can not simply unroll a loop without also differentiating paths inside the loop with this technique.
- It might be very hard to guess what a suitable limit would be. To estimate the number of propagated states, one would have to know the analyzer in depth, and do possibly long calculations by hand. Moreover, a higher -eva-slevel does not always imply more precision; increasing this parameter may push the control points where the join—due to the excess number of states—happens to unwelcome places. Thus, in practice, the limit is often chosen at random and adjusted until the desired behavior is obtained (e.g., a loop is completely unrolled).
- Small modifications to the analyzed code or to the analysis parameters may require updating the -eva-slevel or -eva-slevel-function parameters in an unexpected manner.

3.5 Setting Up an Analysis

As seen in the introduction to this chapter, setting up a simple analysis with Eva is immediate: given a list of sources and a main function, -eva -eva-precision <N> is enough to get the analysis going. In more realistic (and complex) scenarios, however, some preparation is essential to iterate effectively and obtain the best results.

3.5.1 Code Availability, Stubs, Specifications, Builtins

First of all, it is important to remember that Eva is a *whole-program analysis*: it requires code or specifications for *every* called function. This includes library functions and drivers. In the absence of such functions, Eva will assume some arbitrary conditions which are inherently *unsound*, e.g., if a driver written in assembly (thus without C code) modifies some global variable (e.g., errno), Eva will not be aware of it and will assume such changes never happen. This can lead to some code being considered as unreachable, leading to an incomplete analysis. Eva duly emits a warning whenever this happens: *Neither code nor specification for function f.* The user *must* pay attention to such warnings and provide a solution:

- add the missing code to the list of sources to be parsed by Frama-C. For instance, add a forgotten .c to the command-line, or the sources of an external library used by the application;
- use the corresponding Eva builtin for the missing function, if such a builtin exists (only for a few select standard library functions, mostly in math.h and string.h; the *Primitives* chapter of the Eva user manual [5] details them);
- provide an ACSL specification for the function (including a mandatory **assigns** clause), with a sound approximation of the memory zones it modifies (return value, outgoing parameters, global variables)[7];
- provide a *C stub*, that is, some substitute C code that emulates (an over-approximation of) the behavior of the actual code.

Any of the above solutions will provide the information required by Eva to soundly compute the effects of the previously missing function.

It is thus essential that, before running a precise and long analysis, the user has properly set up the sources, stubs and specifications for the functions to be analyzed. Chapter 2 contains a dedicated section that provides instructions for how to parse complex C code.

3.5.2 Main Function

The initial context of the analysis performed by Eva requires some special consideration. For instance, a command-line application in C typically uses the prototype int main(int argc, char *argv[]), where argv contains the command-line options given by the user. However, the initial context considered by Eva is the following:

- objects with static storage duration (e.g., global variables) are initialized to zero[8];
- extern scalar objects are assumed to contain a non-deterministic value (e.g., a signed char variable has the interval [CHAR_MIN, CHAR_MAX], while a double variable has any finite or non-finite value, including NaNs[9]);

For extern pointer objects, Eva creates a *context* composed of freshly-created bases, whose height and depth are given, respectively, by -eva-context-depth and -eva-context-width. For instance, if a program contains a variable extern int *p, the default initial context for Eva will assume that p can point either to the *NULL* base (used for scalar values) or to a freshly-created base S_p, which corresponds to

---

[7] The *ACSL Quick Guide for Eva* appendix of the Eva user manual [5] provides some concise information about the parts of the ACSL language that are the most relevant for configuring an Eva analysis, including a few examples of **assigns** clauses.

[8] There are options to change this behavior, such as -lib-entry. See the Eva user manual [5] for more details.

[9] The actual allowed values for floating-point variables depend on the Frama-C kernel option -warn-special-float.

3 Abstract Interpretation with the Eva Plug-in

an array of *width* objects of the type pointed by p. Note that, in particular, the initial state will never consider that p can point to an *existing* object, which may exclude real-case scenarios.

Note that neither Frama-C nor Eva enforce the constraints related to program startup for hosted environments, defined in C11 [11], Sect. 5.1.2.2.1. The user must manually generate an equivalent environment. This means that, for an analysis emulating a command-line application receiving a variable number of arguments, each of them of variable size, the user typically defines a function eva_main, which uses Frama-C builtins such as Frama_C_interval to initialize argc and argv, and then calls main.

The example below emulates an application invoked from the command line with up to 5 arguments, each of them with variable size, up to 256 characters long. This does not take into account *all* possible command-line uses of an application, but it is enough in practice for most scenarios.

```
#ifdef __FRAMAC__
#include "__fc_builtin.h"
int main(int, char *[]);
int eva_main() {
  int argc = Frama_C_interval(0, 5);
  char argv0[256], argv1[256], argv2[256], argv3[256],
       argv4[256];
  char *argv[6] = {argv0, argv1, argv2, argv3, argv4, 0};
  //@ loop unroll 5;
  for (int i = 0; i < 5; i++) {
    Frama_C_make_unknown(argv[i], 255);
    argv[i][255] = 0;
  }
  return main(argc, argv);
}
#endif // __FRAMAC__
```

3.5.3 Iterative Refinement

Finally, setting up an analysis with Eva must take into account that the iterative process of running an analysis, inspecting alarms, then adding annotations and tweaking parameters before re-running, benefits greatly from a structured approach. Practical usage of Eva, both by Frama-C developers as well as external users, relies on:

- the Frama-C session mechanism (via options -save and -load);
- the use of scripts/Makefiles to store command-line options, allowing better reproducibility;
- analysis profiling (with options such as -eva-show-perf and -eva-flamegraph);
- versioning of used options as well as results (e.g., those obtained via -report-csv) to help compare evolutions and track regressions between different iterations.

Overall, fine-tuning an analysis is an engineering process that benefits from several features of the Frama-C platform. Knowing these features to leverage them is essential for long-term success. There is an ongoing effort to put this into practice in a public Git repository containing a set of case studies which serve as tests, examples and benchmarks: https://git.frama-c.com/pub/open-source-case-studies.

3.6 Results of an Eva Analysis

An Eva analysis computes an over-approximation of all possible behaviors of a program, using the abstract domains of Sect. 3.3 and the partitioning techniques of Sect. 3.4. These over-approximations allow Eva to emit an alarm for each undesirable behavior that might occur during a program execution: this is the main result of its analysis.

But the over-approximations computed by Eva also have other uses: they can prove (or disprove) some user-written assertions, they can be inspected in the Frama-C GUI to better understand the behavior of the analyzed program, and other Frama-C plugins rely on them for their own analyses (e.g., alias information).

This section lists the main undesirable behaviors currently detected by Eva, and highlights some other uses of an Eva analysis.

3.6.1 *Undesirable Behaviors Detected by Eva*

All alarms emitted by an Eva analysis are displayed in the textual log, prefixed with [eva:alarm]. The complete list of emitted alarms can also be inspected in the Frama-C GUI, and can be summarized by the Report plugin.

Due to the over-approximations made by the analysis, each alarm may thus report a real issue in the analyzed program, or be a false alarm due to an imprecision of the analysis. Each alarm is emitted as an ACSL assertion: if the assertion can be proved by other means—for instance, by another Frama-C plugin, typically Wp—then the undesirable behavior cannot happen (and it was a false alarm).

This is a list of the main undesirable behaviors currently detected by Eva:

Memory Alarms

Invalid memory access   Dereference of an invalid pointer or array access with an out-of-bounds index.

Invalid pointer arithmetic   Creation of a pointer that does not point inside an object or one past an object. This can happen when incrementing a pointer or when converting an arbitrary integer into a pointer. This alarm is not enabled by default, and is only emitted when option -warn-invalid-pointer is set. However, the dereference of such an invalid pointer always leads to an alarm.

Uninitialized memory Reading the value of a local variable that has not been initialized. No initialization alarm is emitted when copying a struct or a union, even if they contain uninitialized fields; however, reading an uninitialized field leads to an alarm.

Dangling pointers Reading the value of a pointer that points to a local variable outside its scope.

Arithmetic Alarms

Division by zero Integer division by an expression that may be zero.

Invalid shift operands Logical shift by an integer larger than the size in bits of the shifted value.

Shift of negative integers Left-shifting a negative integer is an undefined behavior and leads to an alarm by default, which can be disabled with option -no-warn-left-shift-negative. Right-shifting a negative integer is implementation-defined and does not lead to an alarm, unless the option -warn-right-shift-negative has been set.

Integer overflow Arithmetic operation computing an integer outside the range of the operation type, or downcast of a value outside the range of the destination type. By default, Eva only emits alarms for signed integer overflows, which is an undefined behavior according to the C standard [11] (Sect. 6.5, §5), but not for unsigned overflows, which have defined semantics (Sect. 6.2.5, §9). Similarly, Eva emits by default alarms for the downcasts of pointer values into integer types that cannot represent these values, which are undefined behaviors (Sect. 6.3.2.3, §6), but not for any other downcasts, which are not (Sect. 6.3.2.3, §5 and Sect. 6.3.1.3, §2 and §3). This behavior can be changed with options [-no]-warn-[un]signed-overflow, [-no]-warn-[un]signed-downcast and -no-warn-pointer-downcast.

Overflow in conversion from floating-point to integer Conversion of a floating-point value that exceeds the range of the destination integer type.

Special floating-point values Floating-point operation resulting in a NaN or an infinite value. These alarms are controlled by the -warn-special-float option: with argument non-finite, NaN and infinite values are forbidden and lead to alarms; with argument nan, only NaN are forbidden; with argument none, all floating-point values are allowed and no alarms are emitted.

3.6.2 *libc Pre-conditions*

Most C programs rely on functions of the standard C library, such as malloc, memchr, printf, pow... Although these library functions are not part of the application itself, their execution may also lead to runtime errors if they are wrongly used. The C standard defines the behavior of these functions, as well as the conditions in which they should be called. Violating their requirements results in undefined behaviors, and thus leads to arbitrary errors.

For instance, sqrt requires a positive argument, and acos and asin both require a floating-point value in the range [−1, 1]. The strlen function requires a valid pointer to a well-formed string (that is, a contiguous sequence of characters terminated by a null character). A call to memcpy(p, q, n) is only safe if p and q are both valid pointers to objects that do not overlap, and such that n bytes can be read from *q and written to *p. free can only be applied to a null pointer, or to a pointer returned by an allocation function (such as malloc) that has not already been deallocated. In these last examples, any violation of these rules may result in buffer overflows or segmentation faults.

Frama-C provides a specification for many functions of the standard C library, in the form of ACSL function contracts whose preconditions precisely encode all contexts in which these functions can be safely called. At each call to such a library function specified by Frama-C, Eva checks that the function preconditions are satisfied, and emits alarms if this is not the case.

These alarms are also shown in the textual log with the [eva:alarm] prefix:

```
[eva:alarm] file.c: Warning:
  function sqrt: precondition 'arg_positive' got status unknown.
```

They can also be seen in the Frama-C GUI and in the outputs of the Report plugin. The complete ACSL specifications used by Frama-C for the standard library can be inspected function by function in the GUI, or are available online: https://git.framac.com/pub/frama-c/-/tree/master/share/libc.

Finally, although new specifications are regularly added to Frama-C, the support of the C standard library remains incomplete: some functions have no ACSL specification yet, and C programs using them cannot be directly verified by Eva. To allow the complete analysis of such programs, the user should write ACSL specifications for all used library functions. Such specifications can be submitted through merge requests on the Frama-C public gitlab: external contributions to improve the support of the C standard library are always welcome.

3.6.3 User Assertions

In addition to the emission of alarms for the undesirable behaviors described above, Eva also tries to verify any ACSL specification written by the user in the analyzed program: assertions are evaluated at their program point, preconditions of called functions are verified at each call site, and postconditions are verified after the return statements of the given function.

As with alarms, the Eva analysis is *sound*: if a user assertion is proven, then it is guaranteed to be satisfied in any execution of the program. On the other hand, when

Eva fails to prove an assertion, either the assertion does not hold for all possible executions of the program, or the Eva analysis was too imprecise to prove it. Unfortunately, the latter case is the most common, due to the limited support of the ACSL language in Eva.

Currently, only the *cvalue* domain can evaluate ACSL properties, and properties inferred by other domains are not used to prove user-written annotations. This means that Eva is only able to prove assertions that can be expressed by the *cvalue* domain: initialization of variables, range of integers and floating-point values, validity of pointers. In addition, the partitioning techniques presented in Sect. 3.4 can also be used to prove disjunctions of properties.

But beyond that, Eva is not able to conduct complex logical reasoning. As a general rule, any properties involving logic quantifiers, external lemmas or axiomatics have very little chance to be proven by Eva. Other Frama-C plugins, such as Wp, are better suited to prove such high-level logical properties.

3.6.4 Graphical Interface

An interesting strength of the Eva analysis is also that the over-approximations computed at each program point are kept in memory and can be later examined via the Frama-C GUI. This includes all properties inferred by the domains presented in Sect. 3.3, such as the possible values of all program variables, and the possible relational properties between them.

Thus, all results of an analysis can be saved with option -save <filename>, and then loaded in the Frama-C GUI with option -load < filename>.

Then, the Frama-C GUI allows the user to:

- inspect all alarms generated by the analysis, as well as the logical status of each ACSL clause evaluated during the analysis.
- check the coverage of the analyzed C program by Eva, i.e., the set of functions and statements that have been reached by its analysis.
- show the possible values inferred by Eva for variables or expressions of any type, at any statement reached by the analysis. This includes integer and floating-point values, pointers, as well as the content of arrays, structs and unions. Values of objects referenced by pointers can also be shown with one click.
- show the inferred values separately for each callstack in which the current function can be executed.
- compare the values of some variables or expressions before and after a given statement, or between two separate statements of the same function.
- highlight all statements writing or reading a given memory location. This is especially useful to easily find where a particular value is defined or used.
- navigate through data-dependency graphs, generated by the Dive plugin from the Eva results.

These main features enable a better understanding of both the Eva analysis and of the analyzed program, which is often required to gain confidence in the program correctness. On one hand, all this information is often necessary to separate false alarms, that may sometimes be removed by improving the configuration of further analyzes, from real issues in the analyzed code that should be fixed. On the other hand, Eva's results and the Frama-C GUI can also help to browse or review a code without missing some of its most obscure behaviors.

3.6.5 API

Finally, the over-approximations computed by Eva can be used programmatically by any other Frama-C plugin through an external API.

This API allows configuring and running analyses, and to access its results once it is completed, mainly by evaluating the possible values of any variable or expression at a given program point in given callstacks. This is especially useful to evaluate which memory locations may be read or written through pointer dereferences. Many Frama-C plugins use the alias information computed by Eva for their own needs.

This interface can be seen at https://git.frama-c.com/pub/frama-c/-/blob/master/src/plugins/eva/Eva.mli.

3.6.6 Limitations

The results provided by Eva do come with a few limitations. Knowing them helps better understand unexpected results and complement them with specifications and other analyses.

3.6.6.1 Support of the C Language

Eva supports most features from C 99, including unions, bit-fields, variadic functions, pointer arithmetic, function pointers, unstructured control-flow (e.g., goto) and variable-length arrays. Eva does *not* support setjmp/longjmp functions, long double and complex types, and floating-point rounding modes other than the default (nearest-even).

Recursive calls and inline assembly codes require an ACSL specification containing **assigns** clauses describing modified memory locations. Eva relies on these specifications without being able to verify them, so the user must pay attention when writing them.

Dynamic allocation is supported, but analyses of programs making heavy use of dynamically-allocated structures such as linked lists can take a long time to converge, or produce very imprecise results. While dynamic allocation remains a challenge for static analysis in general, existing techniques to precisely represent dynamically-allocated memory, such as recency abstractions [1] or shape analysis [24] are not available in Eva currently.

Finally, Eva does not support multi-threaded programs yet; a proprietary Frama-C plug-in, named Mthread, uses Eva to analyze concurrent programs and identify shared memory zones on which race conditions may occur, taking into account all possible thread interactions. The Mthread analysis should be integrated within Eva in the near future.

3.6.6.2 Hypotheses of the Analysis

The soundness of an Eva analysis always relies on some hypotheses that the user should check to ensure the results are relevant for its use case and match their expectation.

The Eva analysis is dependent on its entry point and its initial context: values of global variables and of the entry point arguments when the analysis starts. The default behavior of Eva is to analyze complete applications, starting from the *main* function. The -main parameter can be used to specify another entry point, allowing the analysis of only a subpart of the application. Obtaining a comprehensive analysis of library code usually requires writing a generic test case: a test scenario using Frama-C builtins such as `Frama_C_Interval` to generalize all possible input values. Some parameters also change the initial context, such as -lib-entry which starts the analysis with some non-deterministic values for global variables. See Sect. 3.5.2 or Eva's user manual [5] for more details.

The results of an Eva analysis are only valid for a given architecture, specified by the kernel parameter -machdep, which defines a number of features for the low-level description of the target platform, including the endianness and size of each C type. The option -machdep help provides a list of currently supported platforms; the default is x86_64.

Eva relies on ACSL specifications for functions whose implementation is not provided. Frama-C already provides ACSL specification for many libc and POSIX functions. These specifications are part of the trusted computing base of Eva. The user must provide an ACSL specification for any other missing function; the soundness of the analyses depends on the correctness of these specifications. Eva also relies on user-written ACSL specifications to interpret recursive functions and inline assembly code.

Finally, some parameters allow the user to disable some alarms, or to change the behavior of Eva when interpreting some C constructs. For instance, parameter -no-warn-signed-overflow disables alarms about signed operations that overflow; -eva-no-alloc-returns-null instructs Eva to assume that functions allocating dynamic memory (malloc, calloc, realloc) never return a null pointer; -absolute-valid-range specifies that some absolute addresses are valid (for reading or writing). It is the user's responsibility to only enable parameters that match its use case. Once again, more details can be found in Eva's user manual [5].

3.7 Conclusion

Eva is one of the most mature tools of the Frama-C platform and is currently deployed in many industrial contexts. It has been successfully applied to verify safety-critical embedded software, especially in the nuclear industry [21], and is increasingly used for cybersecurity purposes on non-embedded codes, such as cryptographic and network libraries [9].

Eva has been considerably optimized for years to achieve scalability on large programs. The implementation of most domains rely on Patricia trees [20] that perform very well on the inclusion and join operators of the abstract interpretation framework, which is critical for the analysis performances. Most internal data-structures, including Patricia trees and offsetmaps [3] described in Sect. 3.3.2, also use intensive hash consing to improve performance and to limit the memory consumption of analyses, even when they save all abstract states computed to show precise information in the graphical interface. Finally, Eva is fully context-sensitive: it analyzes the entire program and interprets each function separately for each possible call context, which can lead to many analyses of a same function body. An on-the-fly summarization and cache mechanism [30] avoids new analyses of a function body in a context similar to a previous one, reusing instead the previous results. This technique considerably speeds up analyses of large programs.

Eva is relatively easy to set up, as it is completely automatic, fully supports the subset of C 99 used in embedded software and does not require writing ACSL specifications. Its default configuration is designed to be fast but imprecise; many parameters are available to finely tune the balance between precision and efficiency, by configuring the abstractions and their partitioning used during the analysis, as described respectively in Sects. 3.3 and 3.4. The optimal configuration depends on the complexity of the analyzed program, the properties to be proved and the analysis time required. The meta `-eva-precision` parameter can also be used for a quick setup of the analysis, from 0 (fastest but imprecise) to 11 (most precise but very slow).

Regardless of the configuration used, the Eva analysis is *sound*: it captures all behaviors that can occur during the program execution. Its main result is to emit an alarm for each possible undefined behaviors that the program may exhibit. Thus, it can be used to prove the absence of such undefined behaviors, if no alarm is emitted. However, this is usually not the case, and each alarm may reveal a real bug in the analyzed program or be a false alarm due to some analysis imprecision: some investigative work is then needed to ascertain the cause of the alarm. Each alarm is expressed as an ACSL assertion: if the assertion is broken by a program execution, then the undesirable behavior occurs. Otherwise, if the assertion can be proven to always hold, manually or by other Frama-C plug-ins, then the undesirable behavior cannot happen, and it was a false alarm.

Eva can also verify simple user-written assertions or function pre-conditions, if they are within the scope of properties inferred by its abstract domains—for instance, properties about the ranges of values of some program variables. However, the Eva support of the ACSL language is limited, and its abstract domains are not tailored to tackle arbitrary functional proofs on C programs. To verify more complex properties, other Frama-C plug-ins presented in the next chapters are more suitable. At a function level, the Wp plug-in is designed to prove that a C implementation satisfies a given ACSL specification, and it is able to perform higher-level reasoning than Eva. The Wp plug-in is also typically used to prove that alarms emitted by Eva are indeed false alarms. At a program level, the Aoraï plug-in can help verify temporal properties on a program execution, where its expected behavior is expressed as an automaton, and the MetAcsl plug-in allows specifying high-level ACSL requirements that must be satisfied at many (or even all) program points. Both Aoraï and MetAcsl generate ACSL specifications which can then be verified using other Frama-C plug-ins, such as Eva or Wp. Finally, while all these approaches are *static* analyses, the E-ACSL plug-in is a verification tool able to *dynamically* detect undefined behaviors and verify high-level properties at runtime.

Acknowledgements We thank the anonymous reviewers for their several remarks and suggestions, which contributed to improve this chapter, as well as Franck Védrine, for his invaluable help during the revision stage.

References

1. Balakrishnan G, Reps TW (2006) Recency-abstraction for heap-allocated storage. In: Yi K (ed) Static Analysis, 13th International Symposium, SAS 2006, Seoul, Korea, August 29–31, 2006, Proceedings, LNCS, vol 4134, pp 221–239. Springer. https://doi.org/10.1007/11823230_15
2. Blazy S, Bühler D, Yakobowski B (2017) Structuring abstract interpreters through state and value abstractions. In: Bouajjani A, Monniaux D (eds) Verification, Model Checking, and Abstract Interpretation - 18th International Conference, VMCAI 2017, Paris, France, January 15–17, Proceedings, Lecture notes in computer science, vol 10145, pp 112–130. Springer (2017). https://doi.org/10.1007/978-3-319-52234-0_7
3. Bonichon R, Cuoq P (2011) A mergeable interval map. Stud Inform Univ 9(1):5–37. https://rbonichon.github.io/papers/rangemaps-jfla11.pdf
4. Bühler D (2017) Structuring an abstract interpreter through value and state abstractions: EVA, an evolved value analysis for frama-C. (Structurer un interpréteur abstrait au moyen d'abstractions de valeurs et d'états : Eva, une analyse de valeur évoluée pour Frama-C). PhD thesis, University of Rennes 1, France. https://tel.archives-ouvertes.fr/tel-01664726
5. Bühler D, Cuoq P, Yakobowski B (2023) The Eva Plug-In. http://frama-c.com/download/frama-c-eva-manual.pdf
6. CodePeer. https://www.adacore.com/static-analysis/codepeer. Accessed 06 Nov 2023
7. Cousot P, Cousot R (1977) Abstract Interpretation: a unified lattice model for static analysis of programs by construction or approximation of fixpoints. In: Proceedings of the 4th ACM SIGACT-SIGPLAN Symposium on Principles of Programming Languages, pp 238–252
8. Cousot P, Cousot R, Feret J, Mauborgne L, Miné A, Monniaux D, Rival X (2005) The ASTRÉE analyzer. In: Programming Languages and Systems: 14th European Symposium on Programming, ESOP 2005, Held as Part of the Joint European Conferences on Theory and Practice of Software, ETAPS 2005, Edinburgh, UK, April 4–8, 2005. Proceedings 14, pp 21–30. Springer
9. Ebalard A, Mouy P, Benadjila R (2019) Journey to a RTE-free X. 509 parser. In: Symposium sur la sécurité des technologies de l'information et des communications (SSTIC 2019)

10. Infer. https://fbinfer.com/. Accessed 06 Nov 2023
11. ISO (2011) ISO/IEC 9899:2011: Programming languages—C. International Organization for Standardization, Geneva, Switzerland
12. Jacquemin M, Putot S, Védrine F (2018) A reduced product of absolute and relative error bounds for floating-point analysis. In: Podelski A (ed.) Static Analysis - 25th International Symposium, SAS 2018, Freiburg, Germany, August 29–31 (2018), Proceedings, LNCS, vol 11002, pp 223–242. Springer. https://doi.org/10.1007/978-3-319-99725-4_15
13. Jeannet B, Miné A (2009) Apron: a library of numerical abstract domains for static analysis. In: Bouajjani A, Maler O (eds) Computer Aided Verification, 21st International Conference, CAV 2009, Grenoble, France, June 26–July 2. Proceedings, LNCS, vol 5643, pp 661–667. Springer. https://doi.org/10.1007/978-3-642-02658-4_52
14. Jourdan JH, Laporte V, Blazy S, Leroy X, Pichardie D (2015) A formally-verified C static analyzer. ACM SIGPLAN Notices 50(1):247–259
15. Journault M, Miné A, Monat R, Ouadjaout A (2019) Combinations of reusable abstract domains for a multilingual static analyzer. In: Chakraborty S, Navas JA (eds) Verified Software. Theories, Tools, and Experiments - 11th International Conference, VSTTE 2019, New York City, NY, USA, July 13–14, 2019, Revised selected papers, LNCS, vol 12031, pp 1–18. Springer. https://doi.org/10.1007/978-3-030-41600-3_1
16. Karr M (1976) Affine relationships among variables of a program. Acta Inf 6(2):133–151
17. Miné A (2006) The octagon abstract domain. High Order Symb Comput 19(1):31–100. https://doi.org/10.1007/s10990-006-8609-1
18. Miné A (2017) Tutorial on static inference of numeric invariants by abstract interpretation. Found Trends Program Lang 4(3–4):120–372. https://doi.org/10.1561/2500000034
19. Monniaux D (2009) A minimalistic look at widening operators. Higher-Order Symb Comput 22(2):145–154. https://doi.org/10.1007/s10990-009-9046-8. https://hal.science/hal-00363204. Online version Dec 2009, paper version 2010
20. Okasaki C, Gill A (1998) Fast mergeable integer maps. In: ACM SIGPLAN workshop on ML, pp 77–86
21. Ourghanlian A (2015) Evaluation of static analysis tools used to assess software important to nuclear power plant safety. Nuclear Eng Technol 47(2):212–218, https://doi.org/10.1016/j.net.2014.12.009. https://www.sciencedirect.com/science/article/pii/S1738573315000091. Special Issue on ISOFIC/ISSNP2014
22. Rival X, Mauborgne L (2007) The trace partitioning abstract domain. ACM Trans Program Lang Syst 29(5):26. https://doi.org/10.1145/1275497.1275501
23. Rival X, Yi K (2020) Introduction to static analysis. MIT Press. https://hal.science/hal-02402597
24. Sagiv S, Reps TW, Wilhelm R (2002) Parametric shape analysis via 3-valued logic. ACM Trans Program Lang Syst 24(3):217–298. https://doi.org/10.1145/514188.514190
25. Scott DS, Strachey C (1971) Toward a mathematical semantics for computer languages, vol 1. Oxford University Computing Laboratory, Programming Research Group Oxford
26. Sparrow. http://ropas.snu.ac.kr/sparrow/. Accessed 06 Nov 2023
27. Venet A (2012) The gauge domain: scalable analysis of linear inequality invariants. In: Madhusudan P, Seshia SA (eds) Computer Aided Verification - 24th International Conference, CAV 2012, Berkeley, CA, USA, July 7–13 2012 Proceedings, LNCS, vol 7358, pp 139–154. Springer. https://doi.org/10.1007/978-3-642-31424-7_15. https://doi.org/10.1007/978-3-642-31424-7_15
28. Wilhelm R (2022) Principles of abstract interpretation: By patrick cousot, pp 1–819. MIT Press. ISBN 9780262044905. Reviewed by Reinhard Wilhelm. Formal Aspects Comput 34(2):1–3. https://doi.org/10.1145/3546953
29. Winskel G (1993) The formal semantics of programming languages: an introduction. MIT Press, Cambridge, MA, USA
30. Yakobowski B (2015) Fast whole-program verification using on-the-fly summarization. In: Workshop on tools for automatic program analysis

Chapter 4
Formally Verifying that a Program Does What It Should: The Wp Plug-in

Allan Blanchard, François Bobot, Patrick Baudin, and Loïc Correnson

Abstract This chapter presents how to prove ACSL properties of C programs with the Wp plug-in of Frama-C using *deductive verification* and SMT solvers or Proof Assistants. Specifically, this chapter explores the internals of the Wp plug-in, with a specific focus on how ACSL and C are encoded into classical first-order logic, including its various memory models. Then, the internal proof strategy of Wp is described, which leads to a discussion on specification methodology and proof engineering, and how to interact with the Wp plug-in. Finally, we compare Wp with a selection of other amazing systems and logics for the deductive verification of C programs.

Keywords Deductive verification · Hoare logic · Memory models · SMT solvers · Proof engineering

4.1 Introduction

In a previous chapter (see Chap. 1), the ACSL language is introduced for decorating individual C functions with formal specifications, together with potential additional code annotations and global properties stated at global level. The Wp plug-in is meant to deductively verify that functions indeed fulfill their specifications, that properties provided in code annotations are verified at corresponding program points, and that properties stated at the global level are enforced during execution.

A. Blanchard (✉) · F. Bobot · P. Baudin · L. Correnson
Université Paris-Saclay, CEA, List, Palaiseau, France
e-mail: allan.blanchard@cea.fr

F. Bobot
e-mail: francois.bobot@cea.fr

P. Baudin
e-mail: patrick.baudin@cea.fr

L. Correnson
e-mail: loic.correnson@cea.fr

Consequently, with the Wp plug-in we can:

- verify the functional correctness of our program, that is, verify that our program behaves as intended by the developer,
- verify the absence of runtime errors, because by relying on a plug-in like Rte or Eva, we can emit annotations whose properties must be verified in order to guarantee absence of runtime errors,
- and more generally, verify any program property that can be expressed in terms of an ACSL annotation, provided that the developer provides enough elements of specification to prove them, which is generally the hard part when working with the Wp plug-in.

From the C code and the ACSL specification, the Wp plug-in builds *verification conditions*, which are logic formulas that are *sufficient* to guarantee that the program behaves as expected in the specification. When *all* the generated verification conditions can be verified, the program is proved to conform to its specification. Wp is incomplete by nature: when the proof fails, it does not necessarily mean that there is a bug in the code or in the specification. Maybe we need to rework the code annotations to make Wp and its companion external solvers capable of finishing the proof.

Using Wp is definitively not a push-button process. First, we have to carefully design ACSL specifications to meet proof objectives. Then, by adding intermediate ACSL code-annotations, we have to provide hints and investigate some side properties of the program in order to reach the final proof objectives.

Several introductory books and articles are available for learning deductive verification techniques in general and Wp in particular. The goal of this Wp book chapter is to provide the interested reader with an overview of Wp's *internal* technical features. Although we tried to choose simple enough introductory examples, some topics are probably more accessible to experienced users. We strongly recommend our kind readers to refer to the Wp tutorial [11] to learn the basics and to practice with the tool.

This chapter is organized as follows. Section 4.2 introduces how deductive verification and generation of verification conditions works, through simple examples examined in all their gory details. Section 4.3 focuses on how memory is modeled for verifying programs, especially when dealing with pointers: we review the different *memory models* available in Wp and the advantages they provide together with their respective drawbacks and limitations. Then, Sect. 4.4 presents the different kinds of ACSL annotations and how they are precisely handled by the Wp plug-in: understanding Wp's internal proof strategy can be typically helpful when designing the intermediate code annotations and the proof strategy. Section 4.5 focuses on various kinds of typical errors related to formal specifications, how to deal with them and how some interesting Wp features can help in this perspective. Section 4.6 finally presents two essentials components of the Wp plug-in that make it usable in practice on real world use cases: Qed, a built-in logical formula simplifier and the TIP, an interactive tool to decompose complex verification conditions into simpler ones, for debugging or proof completion. Finally, Sect. 4.7 reviews related works and tools implementing similar or complementary deductive verification techniques: Wp is not necessarily the appropriate tool for every proof objective, although it provides yet a mature and interesting balance between expressiveness, automation and complexity. We sincerely hope that this introduction to Wp's internals will benefit to all practitioners and help them into getting expertise with the tool.

4.2 Deductive Verification of Programs

Deductive verification is a *modular technique* that aims at verifying that a program fulfills its specification. Each function of the program is verified in isolation from the other ones. For each function with its specification, the technique consists in building *verification conditions* (also called *proof obligations*) that are strong enough to guarantee that the function conforms to its specification. Those generated formulas are then submitted to external solvers to determine whether they are all verified. In this section, we introduce deductive verification with simple programs from a high-level perspective. We focus on general proof objectives, verification conditions, runtime errors and the general proof process when using Wp. Starting from a fairly naive example program, we will see that fixing all its weird corner cases is not as trivial as it seems at a first sight. This section requires no preliminary knowledge of Wp but some basic understanding of ACSL, logical formulas and first-order logic.

4.2.1 Companion Provers

As briefly mentioned in introduction, Wp relies on external provers to finally verify specifications, typically SMT (Satisfiability Modulo Theories) solvers such as Alt-Ergo [19], Z3 [39], CVC4 [4], CVC5 [3], and proof assistants, Coq [7] for instance.

The interaction with external provers is handled by the Why3 [14] platform. Hence, before using Wp, the user shall install one or several provers and configure them for being used with Why3. The examples of this chapter have been proved by using the Alt-Ergo SMT solver. The prover configuration can be easily updated with the following command:

```
> why3 config detect
```

Consult the Wp plug-in manual for further installation details.

4.2.2 An Example Program

We introduce deductive verification with an example function that computes the mathematical mean of two C integers a and b:

```
1 /*@ ensures \result == (a + b) / 2 ;
2   @ assigns \nothing;
3   @*/
4 int mean(int a, int b){
5   return (a + b) / 2 ;
6 }
```

The specification here seems to be identical to the code, however, remember that ACSL language is using mathematical integers, whereas the C code will use 64-bits integer or 32-bits integers depending on the architecture. Let us put this remark as a side note for now. We can ask Wp to verify our example program using the following command:

```
> frama-c mean.c -wp

[kernel] Parsing mean0.i (no preprocessing)
[wp] Running WP plugin...
[wp] Warning: Missing RTE guards
[wp] 2 goals scheduled
[wp] Proved goals:    2 / 2
  Qed:                2
```

Or, using the Frama-C graphical user interface (GUI):

```
> frama-c-gui mean.c -wp
```

```
/*@ ensures \result ≡ (\old(a) + \old(b)) / 2;
    assigns \nothing; */
int mean(int a, int b)
{
  int __retres;
  __retres = (a + b) / 2;
  return __retres;
}
```

| Information | Messages (1) | Console | Properties | V |

| Kind | Source | Plugin | Message |

| Global | wp | Missing RTE guards |

The mean function has been proved correct by Wp! However, notice the emitted warning regarding Rte. Before investigating this issue, let us describe how Wp managed to obtain this first result.

For the verification of the mean function, Wp has generated two verification conditions (abbreviated VC or VCs). Those formulas are also called the proof *goals*. One generated VC proves the validity of the **ensures** clause, and the other VC proves the validity of the **assigns \nothing** clause.

4 Formally Verifying that a Program Does What It Should ...

The VCs can be displayed from the command line with option -wp-print or from the GUI by switching to the "WP Goals" panel. Though, with this too simple example, we need to also turn off Wp default let-binding simplifications with option -wp-no-let, otherwise the generated formulas are trivially true. Here, we have slightly cleaned the proof obligation to focus on the important parts:

```
> frama-c mean.c -wp -wp-prover none -wp-print -wp-no-let

[wp] Warning: Missing RTE guards
------------------------------------------------------------
  Function mean
------------------------------------------------------------

Goal Post-condition (file mean0.i, line 1) in 'mean':
Assume {
  Have: (a_1 = a) /\ (b_1 = b).
  (* Block In *)
  Have: ((a_1 + b_1) / 2) = retres_0.
  (* Return *)
  Have: retres_0 = mean_0.
}
Prove: ((a + b) / 2) = mean_0.
```

Let us describe how to read those formulas. Logical variables such as a and a_1 represents the value of C variables at various program points of the function body. Notice the retres_0 variable that is associated to the temporary variable introduced by the Frama-C kernel for storing the function result. Each verification condition consists of a collection of hypotheses (Assume) followed by a formula (Prove) to be verified. The Type formula are hypotheses that constrain logical variables to the range of their corresponding C types. The Have hypotheses translate the C program into mathematical statements. Finally, the Prove formula is the translation of the ACSL formula to be verified.

When running Wp with default options, we observed that all the generated goals were proved to be valid. Indeed, after simplifications, Wp can easily conclude that the code and its specification exactly coincide in this simple example.

Of course, if the code does not conform to the specification, the Wp plug-in cannot prove it. For example, in the following code:

```
1 /*@ ensures \result == (a + b) / 2 ;
2   @ assigns \nothing;
3   @*/
4 int mean(int a, int b){
5   return (a + b) / 2 + 1;
6 }
```

The proof of the **ensures** reaches a *timeout*:

```
1 [wp] 2 goals scheduled
2 [wp] [Timeout] typed_mean_ensures (Qed 2ms) (Alt-Ergo)
3 [wp] Proved goals:    1 / 2
4    Qed:              1
5    Timeout:          1
```

meaning that the solver *did not succeed to* prove that the property is verified. Here, we know that the reason is that the program is wrong, but it is worth pointing that it is not guaranteed that the program or the property is wrong, just that the solver did not prove it.

Note also that the conformance is always computed with respect to the specification, thus the Wp plug-in does not prevent the user from writing an incorrect specification, the conformance of the following program to its (wrong) specification is proved:

```
1 /*@ ensures \result == (a + b) / 2 + 1;
2   @ assigns \nothing;
3   @*/
4 int mean(int a, int b){
5   return (a + b) / 2 + 1;
6 }
```

One shall thus be particularly careful when writing the specification. As we will present in Sect. 4.5, some techniques might help this process, but it is still the responsability of the developer to carefully design good specifications.

Let us come back to our previously proved correct mean function. Its postcondition is verified from Wp point of view, which means that, if the provided function terminates, then the final state of the program verifies its postcondition.

However, in C, when a program performs an invalid action, it can lead to an *undefined behavior* which is modeled in Frama-C by a *runtime error*. Here, Wp warns about "missing RTE (runtime error) guards". Hence, at this point, we have no guarantee that the program will only perform valid operations with respect to C standard. By default, Wp assumes that no runtime errors occur during program execution. Hence, the obtained result so far for function mean can be reformulated in the following terms: "Provided the function terminates without any runtime error, then it conforms to its ACSL specification."

4.2.3 Handling Runtime Errors

The option -wp-rte can be used to enable the verification of absence of runtime errors in addition to ACSL specifications:

```
> frama-c mean.c -wp -wp-rte

[kernel] Parsing mean0.i (no preprocessing)
[wp] Running WP plugin...
[rte:annot] annotating function mean
[wp] 4 goals scheduled
[wp] [Unknown] typed_mean_assert_rte_signed_overflow (Qed
     2ms) (Alt-Ergo)
[wp] [Unknown] typed_mean_assert_rte_signed_overflow_2 (Qed
     0.64ms) (Alt-Ergo)
[wp] Proved goals:    2 / 4
```

4 Formally Verifying that a Program Does What It Should ...

```
Qed:           2
Unknown:       2
```

Or with the **Frama-C GUI**:

```
> frama-c-gui mean.c -wp -wp-rte
```

```
  /*@ ensures \result ≡ (\old(a) + \old(b)) / 2;
      assigns \nothing; */
  int mean(int a, int b)
  {
    int __retres;
    /*@ assert rte: signed_overflow: -2147483648 ≤ a + b; */
    /*@ assert rte: signed_overflow: a + b ≤ 2147483647; */
    __retres = (a + b) / 2;
    return __retres;
  }
```

```
Information   Messages (0)   Console   Properties   Values   Red /
Kind  Source  Plugin  Message
```

With the `-wp-rte` option, **Wp** asks the **Rte** plug-in to insert new assertions for verifying the absence of runtime errors. Notice that there is no emitted warning anymore. Hence, **Wp** now generates additional verification conditions to be proved for our example to be correct. Indeed, in **C**, **int** values are bounded in the range [INT_MIN; INT_MAX], hence computing a+b might cause an overflow for large enough input values. As an illustration, here are the two additional VCs generated by **Wp** for verifying the absence of overflow for this operation (again, we slightly cleaned the VCs):

```
Goal Assertion 'rte,signed_overflow' (file mean0.i, line 5):
Assume {
  Type: is_sint32(a) /\ is_sint32(b).
}
Prove: (-2147483648) <= (a + b).

------------------------------------------------------------

Goal Assertion 'rte,signed_overflow' (file mean0.i, line 5):
Let x = a + b.
Assume {
  Type: is_sint32(a) /\ is_sint32(b).
  (* Assertion 'rte,signed_overflow' *)
  Have: (-2147483648) <= x.
}
Prove: x <= 2147483647.
```

As illustrated above, the **Wp** did not succeed in proving these VCs, which are actually wrong since the execution of `mean` will overflow for large enough input values.

The kernel option `-(no)-warn-signed-overflow` can be used to specify for all plug-ins whether **C** signed overflows are runtime errors (when the option is "on") or computed with a rounding operation (when the option is "off"). Similar options exist for unsigned overflows and other kind of runtime errors, see the Frama-C reference manual for more details. These options have a direct impact on Wp and Rte plug-ins. For instance, if we turn off the default signed overflow runtime errors:

```
> frama-c mean.c -wp -no-warn-signed-overflow -wp-rte

[kernel] Parsing mean0.i (no preprocessing)
[wp] Running WP plugin...
[rte:annot] annotating function mean
[wp] 2 goals scheduled
[wp] [Timeout] typed_mean_ensures (Qed 2ms) (Alt-Ergo 10s)
[wp] Proved goals:    1 / 2
  Qed:             1
  Timeout:         1
```

Now, -wp-rte does not emit additional assertions anymore. However, Wp generates modified VCs for the postcondition: it now introduces the mathematical function to_sint32 to encode the rounding operation actually performed in case of signed overflow, which makes the goal for the postcondition no longer provable:

```
Goal Post-condition (file mean0.i, line 1) in 'mean':
Let x = a + b. Assume { Type: is_sint32(a) /\ is_sint32(b). }
Prove: (to_sint32(x) / 2) = (x / 2).
```

We can see in the simple example the actual difference between the ACSL specification of the program and its C implementation, although they both are textually identical. Actually, we can also see that the translation of C code into mathematical formulas is heavily impacted by the machine architecture and the runtime errors configured by the user *via* Frama-C kernel options.

4.2.4 Fixing the Example Program

For now, let us come back to the potentiel signed overflows in our program. We want to fix the program so that it will compute the correct mean in every situation. Recall that such an overflowing mean is an historical bug of the Java standard library, which remained hidden during many years inside the array dichotomic-search algorithm. The bug was only revealed when Google used it on huge arrays [13]. Indeed, the summation requires little more care and can be replaced with the following code:

```
1  /*@ requires 0 < a < b ;
2    @ ensures \result == (a + b) / 2 ;
3    @ assigns \nothing; */
4  int mean_pos(int a, int b){
5    return a + (b - a) / 2 ;
6  }
```

This alternative code is correct, but only under the hypothesis that a is positive and smaller than b. This requirement is fulfilled for the dichotomy search algorithm, however a generic mean function requires to handle all possible cases.

For the sake of introducing modular proofs, we split the implementation of the fully generic mean function into three separate functions to handle the different cases to consider. First, a and b are switched with each other in case b<a. Then, four cases must be treated differently: either a==b, or 0<a<b, or a<b<0 or finally a<=0<=b. In the first case there is no mean to compute. The next two cases are handled by computing respectively a+(b-a)/2 or b-(b-a)/2, where b-a works without producing any overflow. In the later case, the operation (a+b)/2 now works without producing any overflow. The full code for the generic mean function is provided below:

```
1  /*@ requires 0 < a < b ;
2    @ ensures \result == (a + b) / 2 ;
3    @ assigns \nothing; */
4  int mean_pos(int a, int b){
5    return a + (b - a) / 2 ;
6  }
7
8  /*@ requires a < b < 0 ;
9    @ ensures \result == (a + b) / 2 ;
10   @ assigns \nothing; */
11 int mean_neg(int a, int b){
12   return b - (b - a) / 2 ;
13 }
14
15 /*@ ensures \result == (a + b) / 2 ;
16   @ assigns \nothing; */
17 int mean(int a, int b){
18   int x = (a < b) ? a : b ;
19   int y = (a < b) ? b : a ;
20
21   if(x == y) return x ;
22   if(y < 0) return mean_neg(x, y);
23   if(x > 0) return mean_pos(x, y);
24   return (x + y) / 2;
25 }
```

```
> frama-c mean2.c -wp -wp-rte

[kernel] Parsing mean2.i (no preprocessing)
[wp] Running WP plugin...
[rte:annot] annotating function mean
[rte:annot] annotating function mean_neg
[rte:annot] annotating function mean_pos
[wp] 33 goals scheduled
[wp] Proved goals:    33 / 33
  Qed:                18
  Alt-Ergo 2.2.0:     15  (4ms-7ms-19ms)
```

Now, the function mean is proved to have no runtime errors *and* to fulfill its specification in *all* cases. The internal Wp simplifier (named Qed and described in Sect. 4.6) managed to handle 18 of the 33 generated VCs, and the remaining 15 ones are easily discharged by the external Alt-Ergo prover.

Note the presence of **ACSL** contracts for every function: this is due to the modular nature of deductive verification. Each function needs its contract in order to be verified and, on function call side, *only* the contract of the function is used, not its actual code. For example, one cannot verify the function mean if mean_pos does not have a contract because when verifying mean, the **Wp** plug-in will never visit the body of mean_pos. It would only rely on its contract to determine the necessary preconditions to the call and only use its provided postconditions.

The proof actually exercises many cases and execution paths. We can see in the generated VC for the postcondition of the mean function how those different cases are encoded. After **Wp** simplifications, this generated VC is more complicated than before although still readable:

```
> frama-c mean2.c -wp -wp-prover none -wp-print -wp-rte

Goal Post-condition (file mean2.i, line 25) in 'mean':
Assume {
  Type: is_sint32(a) /\ is_sint32(b) /\ is_sint32(mean_0).
  (* Residual *)
  When: b <= a.
  If b = a
  Then { Have: mean_0 = a. }
  Else {
    Core: ((a + b) / 2) = mean_0.
    If a < 0
    Then { (* Call 'mean_neg' *) Have: b < a. }
    Else {
      If 0 < b
      Then { (* Call 'mean_pos' *) Have: b < a. }
      Else {
        Let x = a + b.
        (* Assertion 'rte,signed_overflow' *)
        Have: (-2147483648) <= x.
        (* Assertion 'rte,signed_overflow' *)
        Have: x <= 2147483647.
      }
    }
  }
}
Prove: (x / 2) = mean_0.
```

Many simplifications have been performed by **Wp**, some of which are visible in the residual When condition, which means that the other branch has been trivially pruned. We also observe a Core formula that has been factorized between different branches and a Let binding introduced to factorize duplicated terms. The Rte assertions generated by -wp-rte are also visible in the goal, although there are proved by other generated VCs.

4.2.5 Combining Wp with Other Plug-ins

Finally, let us mention than the Eva plug-in can be used in conjunction with Wp for proving the absence of runtime errors. In this case, the -wp-rte option is no longer necessary, since Eva will check the absence of runtime errors. More precisely, Eva would emit new assertions on program points where it cannot guarantee absence of runtime errors, so that Wp can try to prove them in turn. The plug-in Report is useful in this context to consolidate the results obtained by the various involved plug-ins. For instance, we can make Wp and Eva collaborate on our full-featured mean function thanks to the following command line:

```
> frama-c mean2.c -eva -lib-entry -main mean -eva-precision 2
    -then -wp -then -report
...

--------------------------------------------------------------
--- Status Report Summary
--------------------------------------------------------------
    13 Completely validated
    13 Total
--------------------------------------------------------------
```

In the detailed output from Report (not shown here), the user can see which property has been proved by which of the involved plug-in, and which properties remain unproved yet, if any.

Combining the use of Eva and Wp has been used for proving the absence of runtime errors of realistic real-life libraries, for instance the x509 certificate parser [24]. Indeed, using Wp only for proving the absence of runtime errors (using Eva or Rte) can be an interesting goal in itself. Moreover, it generally requires less specifications and intermediate code annotations than when trying to specify and prove the functional behavior of a program: the absence of runtime errors and the associated ACSL annotations *are* the only specification to prove. We call this approach "minimal contracts" and it has been successfully experimented on various use cases [28].

Finally, let us mention the MetAcsl plug-in, described in Chap. 10, which is typically used in combination with the Wp plug-in. With MetAcsl, the developer writes the specifications in a higher-level dedicated language rather than in ACSL. Then, the plug-in compiles user's high-level specifications into standard low-level ACSL code annotations that are sufficient to entail the expected properties. In the end, Wp and its companion solvers are invoked to prove the generated code-annotations, possibly using side properties provided by the developer and written in standard ACSL annotations.

Hence, even if using ACSL and Wp to specify and prove programs can be seen as a difficult process that requires a lot of expertise, Wp can also be used with a large variety of simpler and more accessible methodologies, depending on the proof objectives.

4.3 Memory Models and Related Properties

The C language is imperative by nature: variables can be written during program execution and the whole memory can be randomly modified *via* pointers. Hence, when building a logical formula representing the semantics of the code, Wp must deal with the successive values of the modified variables during the execution of the program.

In the introduction section of this chapter, we briefly mentioned this issue, and we saw that a C variable x is sometimes modeled by a collection of logical variables x_0, x_1, etc., each one modelling the value of the original variable at a specific program point. This is a general technique that works for imperative programs without pointers, but that does not work when using pointers, which are ubiquitous in C.

For modeling all kinds of C assignments, Wp combines several known methods for deductive verification of imperative programs with pointers, using *memory models*. Memory models describe how the memory behaves when some assignments are performed by the program. Wp embeds several memory models. In this section, we present these memory models, in particular the Hoare model in Sect. 4.3.1, the Typed model in Sect. 4.3.4, how WP mixes aliasing and non-aliasing in Sect. 4.3.5 and how those different models can be tweaked in Sect. 4.3.7. We will explore in details the different techniques used by Wp to model different kinds of memory operations.

The reader shall be familiar with basic usage of Wp and with the general structure of generated proof obligations as detailed in Sect. 4.2, and have some reasonable practice with C pointers. We will use the notation $v@L$ to denote "the value of variable v at program point or label L". We will also use the notation $M[l \leftarrow e]$ which denotes "the map M where the cell l is now set to value e". Memory maps like M and $M[l \leftarrow e]$ are just pure total functions from memory locations (to be further defined) to values. They are the basic ingredients to model memory in presence of pointers and aliasing and these kinds of objects benefit from very good built-in support in mainstream SMT solvers.

4.3.1 Simple Variable Assignments

We start with the case of assignment to simple variables, without pointers nor aliasing. Let us take the following example:

```
void example(int x){
  x += 1;
  x += 2;
  //@ assert x <= \at(x, Pre) + 3;
}
```

This simple program receives a variable x in input and then modifies this variables by assigning it its value plus one, and then reassigns it with its new value plus two. We then want to prove that the final value of x is no more than the original value of x *at* the **Pre** state, plus three. This specification translates to the ACSL code annotation at line 4 above. From this annotated program, Wp can generate the following verification condition (we deactivate Qed simplifications to clarify the example):

```
> frama-c local_mutation.i -wp -wp-no-let -wp-print
...
Goal Assertion (file local_mutation.i, line 4):
Assume {
  Have: x_3 = x_1.
  Have: (1 + x_3) = x_2.
  Have: (2 + x_2) = x.
}
Prove: x <= (3 + x_1).
Prover Qed returns Valid
```

Several logic variables x are generated for the C variable x. They represent the state of x at different program points of the execution. Here for example, logical variable x is the value of the C variable x at the end of the function and x_1 represents its value at the beginning of the function (\at(x, Pre)). Starting from the end of the function, Wp generates x_2. Then Wp goes to the previous instruction, and generates a variable x_3. Finally, Wp reaches the beginning of the function, where x_3 is exactly x_1. At every execution step, Wp generates intermediate assertions that relate those different values of C variable x with each others.

The basic idea is that for each variable x of the program, one can attach a collection of logic variables that represent its value at each program point. For each program point L, we denote by logical variable x@L the value of x at program point L. This is illustrated below for a single variable x and a sequence of two non-branching instructions <A> and <B>:

```
int x ;
L1:            // x@L1 : initial value
   <A> ;
L2:            // x@L2 : value of x after instruction A
   <B> ;
L3:            // x@L3 : value of x after instruction B
```

Then, for each instruction between program points L and L', Wp generates an equation relating the logical variables x@L and x@L' for each C variable x. When the C variable x is not assigned by the instruction, this equation is simply x@L = x@L'; otherwise, the generated equation models the new value of x, hence x@L', with respect to the variable values before the instruction, hence the values at program point L. This is illustrated for a single example instruction below:

```
L1:            // x@L1                     , y@L1
   x = x+y+1 ;
L2:            // x@L2 = x@L1 + y@L1 + 1 , y@L2 = y@L1
```

To handle conditional statements, all variables are duplicated along the different paths of execution, with dedicated conditional equations for each branch depending on the condition, as illustrated in the following example:

```
L1:       // value of COND at L1 is some expression E
if(COND) {
  L2:     // x@L2 = x@L1
  ...
  L3:     // x@L3 = ...
} else {
  L4:     // x@L4 = x@L1
  ...
  L5:     // x@L5 = ...
}
L6:       // x@L6 = E ? x@L3 : x@L5
```

Exercise 1 Explain the verification condition generated by Wp on the following example:

```
1  void example_if(int c, int x){
2    if(c){
3      x += 1;
4    } else {
5      x += 2;
6    }
7    //@ assert x <= \at(x, Pre) + 2;
8  }
```

Result:
```
Goal Assertion (file exercise.i, line 1):
Assume {
  Have: x_2 = x_1.
  If c != 0
  Then {
    Have: x_2 = x_3.
    Have: (1 + x_3) = x.
  }
  Else {
    Have: x_2 = x_4.
    Have: (2 + x_4) = x.
  }
}
Prove: x <= (2 + x_1).
```

Solution 1 The (logical) x is the value of x at the end of the function, and x_1 the value at the beginning of the function (\at(x, Pre)). When entering the function body, Wp generates a variable x_2 (thus its value equals x_1). In the **then** branch of the conditional, Wp generates x_3 with equals to the value x_2 of x before the conditional, then from the assignment, we relate the final value of x with 1 + x_3. In the **else** branch of the conditional, Wp generates x_4 with equals to the value x_2 of x before the conditional, then from the assignment, we relate the final value of x with 2 + x_4.

This model for variable assignments is called the Hoare memory model [30], and the technique of introducing logical variables at each program points is named the Single Static Assignment form or SSA [35, 43] since each logical variable x_i is assigned only once by an equation $x_i = e$ in the generated VCs.

4.3.2 Introducing Pointers

In C, random memory access is allowed *via* pointers. One can take the *address* of a variable x and store it into a new variable, like in p = &x. Then, reading or writing to *p is the same as reading or writing to x.

Pointers introduce the possibility of aliasing: two pointers may target the same memory location. In presence of aliasing, reasoning on program assignments becomes more subtle. Consider for example the following program:

```
1 void function(int* a, int* b){
2     *a = 4 ;
3     *b = *b + 1 ;
4 }
```

Since a and b can be aliased, one cannot simply associate logic variables to actual values of *a and *b at the different program points. Indeed, the first assignment modifies *a but might also modify *b depending on the presence of aliasing (when a == b) or not (when a != b). Thus, aliasing makes relations between logic variables hard to express in a concise and usable way.

This problem is solved using another representation of the memory, completely different from the Hoare logical memory model introduced in the previous section. Before presenting the actual model used in Wp, let us present a simpler but representative one.

A simple way to model memory is to use, at each program point, a mathematical function that assigns a value to each address, called a memory map M. Reading a value at address l in memory map M is then simply modeled by $M[l]$. Writing a value v at address l in memory map M is modeled by operation $M[l \leftarrow v]$.

Then, the entire memory is modeled by a single variable M that is modified by each assignment in the program. Now, we can apply the Hoare memory model to the unique variable M throughout the program, by introducing different logical variables $M@L$ at each program point L. Going back to our simple example above, the equations on variables $M@Li$ generated for this model are illustrated below:

```
*a = 4 ;           // M@L1 = M@L0[a <- 4]
*b = *b + 1 ;      // M@L2 = M@L1[b <- M@L1[b] + 1]
```

Finally, at then end of the function body, the values of *a and *b are respectively modeled by M@L2[a] and M@L2[b]. And indeed, these expressions get different values whether pointers a and b are aliased (equal) or not.

More generally, a pattern like $M[l \leftarrow v][l']$ is called a *read-after-write*: location l' is read after writing some value v at location l into memory map M. To reason about a read-after-write, SMT into v, or $l \neq l'$ and the write operation can be pruned to obtain $M[l']$. Hence, each read-after-write pattern introduces a split for SMT solvers, so when there are a lot of these, the time needed for the proof can rapidly increase. Moreover, exploring case $l \neq l'$ will reveal different subcases, since memory locations l are actually represented by pairs (base, offset) as we will see later in the presentation.

Although reasoning with pointer aliasing can be complicated in general, on the previous example, we can specify and prove with Wp the two possible outcomes of a and b with a single ACSL annotation:

```
void function(int* a, int* b){    // M@L0
   *a = 4 ;                       // M@L1 = M@L0[a <- 4]
   *b = *b + 1 ;                  // M@L2 = M@L1[b <- M@L1[b] + 1]
//@ check *a == *b == 5 || (*a == 4 && *b == \at(*b, Pre)+1) ;
}
```

Modulo variable renaming and some Qed simplifications, the VCs actually generated by Wp with its default memory model are the following ones, which is consistent with the model presented above:

```
> frama-c simple_alias.i -wp -wp-no-let -wp-print
...
Goal Check (file simple_alias.i, line 4):
Let x = M3[b].
Let x_1 = M3[a].
Assume {
  Have: M1 = M0 /\ b = b_1.
  Have: M1[a <- 4] = M2.
  Have: M2[b <- 1 + M2[b]] = M3.
}
Prove: (x = 5 /\ x = x_1) \/ (x_1 = 4 /\ x = (1 + M0[b_1])).
```

The simple memory model informally introduced above is actually a simplified version of the actual Wp memory model, however, it captures its main logical intuition: reasoning with aliasing pointers can be solved by introducing a memory map M and reasoning by applying the Hoare memory model on logical variable M.

We will now introduce gradually more advanced features of the actual Wp memory model currently implemented in Frama-C.

4.3.3 The General WP Memory Model

Even with the simple examples presented so far, we have silently introduced some shortcuts in our presentation. Typically, the input parameters of the functions were still modeled by a logic variable like in the Hoare model. Would any variable be aliased, *all* of its occurrences shall be modeled using the memory map M.

For instance, consider the simple instruction *p = *p + 1 where the variable p itself might be aliased (by another pointer). Such an assignment is modeled by the following equation, where M and M' are the memory maps respectively before and after the assignment, and A_p is the (constant) address where variable p is actually stored in memory:

$$M' = M[M[A_p] \leftarrow M[M[A_p]] + 1]$$

In this equation, $M[A_p]$ models the value of variable p before the assignment, and $M[M[A_p]]$ the value of expression *p.

However, it is practically not possible to use this naive memory model for all variables and all pointers: this would make **Wp** generating far too large and too complex formulas for SMT solvers. The reason is quite simple: with this naive model, each assignment now introduces *nested* occurrences of read-after-write patterns, which were already mentioned to be difficult to tackle by SMT solvers and leading to exponential blow up.

In practice, **Wp** uses a much more sophisticated memory model by combining different approaches that will be described in more details in dedicated sections. Let us briefly introduce them now.

First, **Wp** uses a prior syntactical analysis in order to decide whether a variable might be aliased or not. Hence, only potentially aliased variables would be modeled using a global memory map M, and non-aliased variables will be modeled with the Hoare memory model. This mechanism is described more precisely in Sect. 4.3.5.

Sometimes, a given pointer p is the unique, constant and temporary alias to a variable, an array element or a structure field. In such a case, it is legitimate to temporary consider the value of *p to be identical to a temporary variable v, with no alias: reads and writes through *p are then just considered as reads and writes to v and can be handled by the Hoare memory model. A typical case pattern is when a function takes pointers parameters for returning more values to the caller, which are often called *reference parameters*. Such patterns can be detected by using the so-called "Reference" memory model as described in Sect. 4.3.7.

For aliased variables, putting all values of variables into the same global memory-map does not scale for programs with large data structures and many pointers. Hence, to refine its memory model, **Wp** takes further assumptions based on the type of elementary C values being read or written, as described in further details in Sect. 4.3.4.

Another improvement to the **Wp** memory model deals with separating memory locations from each others. As mentioned in a previous example, reasoning on a memory map often leads to deciding whether two memory locations are equal or different. This is particularly important for read-after-write patterns which can be demanding for SMT solvers as already explained. To optimize branching decisions, **Wp** groups the different locations l' into memory regions: as soon as locations l and l' belong to different disjoint regions, then necessarily $l \neq l'$, hence read-after-write patterns like $M[l \leftarrow v][l']$ can be directly simplified by **Qed** and SMT solvers. This mechanism is described into more details in Sect. 4.3.6.

To be complete, the memory model is not only responsible for modelling the values of variables and pointers during the execution of the program, but it is also responsible for modelling the validity of pointers and the allocation and initialization of variables. These two topics are addressed in more details in Sects. 4.3.8 and 4.3.9, respectively.

All these optimizations are mainly responsible for the ability of **Wp** to tackle the proof of relatively complex programs and complex properties in the industrial field [16, 23]. However, those improvements also come at the price of possible restrictions that are discussed when relevant.

4.3.4 The Typed Memory Model

As introduced above, Wp makes further assumptions about memory. The so-called "Typed" memory model assumes that memory can be split into different memory regions according to the types of the considered memory locations. Thus, when a program involves pointers to integers and to floating point values, they are assumed to refer to different memory regions. Hence, they are modeled by different memory maps and different logical variables, instead of a single, shared memory map M for all kind of values.

This model originates from the work of Bornat [15] and some of its more recent improvements [32]. The original idea of Bornat comes one step further regarding separation by considering that each field of structured data are separated from each others: hence, Bornat introduces a distinct memory map for each structure field. But this model is not consistent with field aliasing: C programs might alias an integer variable and an integer field of some structure. Moreover, it is common in embedded C programs to make pointer arithmetic to access consecutive structure fields, hence accessing the structure like an array.

Although it would be possible to combine the two approaches, Wp currently only uses a type-based separation and has currently no field-based separation strategy. However, the central idea remains the same: by splitting a shared memory-map M into many disjoint memory maps, one can dramatically reduce the occurrences of read-after-write patterns, hence, the number of branching conditions to be considered by SMT solvers.

Going back to the Wp "Typed" memory model, each C basic type (**char**, **int**, **float**, ...) now has its own memory map. For compound types (including arrays and structures), the different fields are dispatched into these different memory maps. Figure 4.1 illustrates this for the presented **struct** X. Since it is composed of an array of three **int** and one **float**, modeling the memory pointed by ptr at the beginning of the function requires two memory regions: one for integers (Mi@L0) and the other for floating point values (Mf@L0).

```
1  struct X {
2    int x[3] ;
3    float y ;
4  };
5
6  void f(struct X *ptr){
7    // Mi@L0[ptr + 1]
8    ptr->x[1] = 42 ;
9    // Mf@L0[ptr + 3]
10   ptr->y = 3.14 ;
11 }
```

Fig. 4.1 Compound type value dispatching in "Typed" model

4 Formally Verifying that a Program Does What It Should ...

When reading or writing to *ptr, the "Typed" model dispatches values in the right memories according to their type. When a compound type is modified (either because of an assignment or when taking into account some **assigns** clause), Wp generates new memory-map variables for all impacted types. For example, in Fig. 4.1, on line 8, a new variable is generated for the **int** memory-map; on line 10, a new variable is generated for the **float** memory map.

Notice that in this model, Wp does not use the actual field offset (based on the size and alignment of fields) but rather the logical offset of each cell of elementary type inside the structure. For instance, in structure **struct X** below, elements of array field **x** have logical offsets 0,1 and 2 and float field **y** has logical offset 3, as illustrated in Fig. 4.1. Hence, array fields are inlined in the structure, allowing for complex patterns of field and array aliasing, contrary to the original Bornat memory model.

The price of adopting the "Typed" memory model is that Wp implicitly *assumes* that pointers to **char** and pointers to **int** respectively points to necessarily separated memory ranges. This typically corresponds to enforcing strict aliasing, plus further restrictions on union types. Notice that Wp tries to detect codes that would violate such assumptions and emits warnings accordingly (like in Sect. 4.3.7), but it is a best-effort feature that limits displayed locations to addresses of globals and to the first level of indirection of global pointers and pointers in parameters.

4.3.5 Mixing Aliasing and Non-aliasing

While our model of the memory could always be used, including on local variables and formals for example, this would lead to much more complex formulas than necessary. Indeed, it is more efficient for SMT solvers to directly deal with simple variables than with memory-maps and read-over-write patterns.

Wp optimizes verification condition generation by using simple logic variables for memory locations that cannot be aliased and thus only relies on memory-maps when necessary. This is the default memory model of Wp.

Intuitively, Wp uses syntactic heuristics for determining which memory model shall be used for modeling a given variable: if the variable's address is taken from the C code, then it is potentially aliased, hence the Typed model is chosen for this variable. Otherwise, the variable can never be aliased and the Hoare model is chosen.

Exercise 2 The following program swaps two pointed values. Since the address of a and b are not taken they can be modeled using simple logic variables, this is not the case of the pointed memories *a and *b. Taking this remark into account, apply the Hoare and Typed memory models when relevant and list the logical variables and equations generated by Wp on this example.

```
1  void swap(int* a, int *b){
2    int tmp = *a ;
3    *a   = *b ;
4    *b   = tmp ;
5  }
```

Does the assertion *a == \at(*b, Pre) && *b == \at(*a, Pre) hold at the end of the function?

Solution 2 Since none of a, b and tmp are aliased in the code, they are modeled by the Hoare model. However, values of *a and *b shall be modeled by the Typed memory model. Here, only the memory-map for **int** values in involved, denoted M here for clarity. From these observations, we get:

```
1  void swap(int* a, int *b){   // a, b, M@L1
2    int tmp = *a ;              // tmp@L2 = M@L1[a]
3    *a   = *b ;                 // M@L2  = M@L1[a <- M@L1[b]]
4    *b   = tmp ;                // M@L3  = M@L2[b <- tmp@L2]
5  }
```

We can translate the assertion into:

    M@L3[a] == M@L1[b] && M@L3[b] == M@L1[a]

Thus, after rewriting relations, we have:

        M@L1[a <- M@L1[b]][b <- M@L1[a]][a] == M@L1[b]
    && M@L1[a <- M@L1[b]][b <- M@L1[a]][b] == M@L1[a]

Let us take the expression M@L1[a <- M@L1[b]][b <- M@L1[a]][a]: either a == b, then it evaluates to M@L1[a] which equals to M@L1[b] because a == b; or a != b, then it evaluates to M@L1[b]. The first part of the conjunction is true. The same reasoning can be applied on the second part.

In practice, for dispatching variables between aliasing and non-aliasing memory models, Wp not only looks for variables whose address are taken, but also uses a more subtle strategy. Without diving into too many details, we now provide to the interested reader several insights on the Wp dispatching heuristics. In the following, we say that "variable x is aliased from A" when Wp considers that x might be aliased only from syntactical elements of A (that might be either an instruction or an annotation).

Code Aliasing. If &x appears as a *value* in a C instruction A, then x is aliased from A. However, when the & operator is immediately simplified, like in expression *(&x + k) for instance, no alias is taken.

ACSL Aliasing. If &x appears in an ACSL annotation A, then x is aliased from A, like for code aliasing. However, when A is a \valid(...), \separated(...) or similar built-in ACSL predicate, no alias is taken. This allows stating memory properties about variables without aliasing them. This is typically the case for annotations related to runtime errors and separation.

Modularity. If x is aliased from the function body of a function F, it only affects the dispatching of x for proving annotations of this function F. On the contrary, for callers of this function F, aliasing sources are only taken from the ACSL contract of F. This strategy allows for a function to temporarily take the address of a global variable, without considering that this global is aliased out of F. When some variable x is actually aliased and its address escapes from a function F, one shall explicitly mention &x in its specification (see example below).

Assigns. For variables in callers to some function F to be fairly dispatched, Wp expects F to have fully specified assigns clauses. Moreover, when assigned locations may have pointer values, an ACSL clause **assigns ... \from ...** is expected, in order to infer correctly which aliases might escape when calling F. This can result in Wp warnings regarding imprecise or missing assigns clauses.

Generally speaking, the variable dispatching heuristics of Wp are optimistic in the following sense: they try to capture common patterns where x is, in practice, not aliased from A whereas &x syntactically appears inside A. Most of the time, Wp is able to generate an ACSL hypothesis that captures the implicit assumption reflecting its optimistic guess and a warning is emitted to draw user's attention: maybe the function is proved, but only under some *extra*, possibly unwanted hypotheses, that shall be verified at call sites.

For instance, consider the following program:

```
1  int x ;
2
3  /*@ requires \valid(p) ;
4   @ assigns *p;
5   @ ensures *p == 42 && x == \old(x); */
6  void f(int* p){ *p = 42; }
```

When proving function f with Wp, we obtain the following output:

```
> frama-c -wp hyps.i
[kernel] Parsing hyps.i (no preprocessing)
[wp] 2 goals scheduled
[wp] Proved goals:    2 / 2
  Qed:             2   (0.30ms-0.68ms)
[wp] hyps.i:6: Warning:
  Memory model hypotheses for function 'function':
  /*@ behavior wp_typed:
       requires \separated(p, &x); */
  void function(int *p);
```

Here, Wp succeeds in proving that the contract is verified, including the fact that variable x is not modified. But this is obviously wrong when p==&x, i.e. when calling f(&x). Actually, the Wp dispatching infers that variable x is *not* aliased in function f. Hence, variable x is modeled by the Hoare model and *p by the Typed model, which implicitly means that &x and p are actually separated and makes Wp able to prove the contract x == \old(x).

This optimistic assumption is reflected by the warning emitted by Wp: it alerts the user on the fact that function f is correct with respect to its specification, but only under the additional hypothesis that &x and p are separated from each other.

Such a verification can be done by manual review for external code, or automatically by using the option -wp-check-memory-model that adds memory model dispatching hypotheses to function preconditions.

If the user does not want such an extra hypothesis, they shall mention explicitly that x can be aliased with *p: for instance, maybe p is equal to &x. In the above example, the ensures clause x == \old(x) would be of course no longer provable:

```
/*@ requires \valid(p) || p == &x ;
    assigns *p;
    ensures *p == 42; */
void f(int *p) { *p == 42; }
```

This time, Wp no longer emits a warning and one can soundly prove that x==42 after calling f(&x).

4.3.6 Base-Pointer Memory Regions

The "Typed" model presented above allows reasoning about programs with aliasing. However, it assumes that *all* pointers may alias, in particular they may point to any global, local or formal variable. However, even if pointers might alias any variable, it is often possible to greatly simplify aliasing. In particular, one can observe that locals, formals and global variables are indeed separated from each others.

A naive way to model this separation would be to enumerate for each pair of variables their separation hypothesis, however, this would be extremely inefficient for SMT solvers to take as input such a huge amount of properties. A better way to model such separation properties is to use the notion of pointer bases introduced in Chap. 1.

In ACSL, a pointer is formally defined as a pair (base, offset), so that pointers with different bases necessarily point to disjoint memory locations. This is exactly how pointers are encoded in the "Typed" memory model. Variable address &x is then represented by pointer (b,0), with a unique base b for each variable. This actually models that distinct variables are actually separated by definition.

As an illustration, let us consider the following example program that creates aliases on variables of different kinds:

```
int global ;

void function(int formal){
    int local ;
    int *pg, *pf, *pl ;
    pg = &global ;
    pf = &formal ;
```

```
 8    pl = &local ;
 9    //@ assert *pg == *pf && *pl == 1;
10  }
```

We can see in the generated VCs a bundle of symbols, one for each variable base address:

```
# frama-c base.i -wp -wp-no-let -wp-print
...
Goal Assertion (file base.i, line 9):
Assume {
    ...
    Have: global(G_global_23) = pg_0.
    Have: global(P_formal_26) = pf_0.
    Have: global(L_local_27) = pl_0.
}
Prove: (Mint_0[pg_0] = Mint_0[pf_0]) /\ (Mint_0[pl_0] = 1).
```

In the VCs depicted above, the logical *injective* function **global** associates unique pointer bases to variable identifiers. Formally, for any variable identifiers X and Y, we have the following property:

$$\texttt{global}(X) = \texttt{global}(Y) \implies X = Y$$

But we can do much better in this direction and obtain more separations between different *groups* of variables.

In the Typed memory model, Wp actually classifies memory locations into different regions to make further separation properties easier to detect. When generating VCs for a given function, let us make the observation that formals and local variables have *fresh* bases, i.e. they are distinct from any other location seen before. This can be encoded in a rather efficient way for SMT solvers, as follows.

First, a pure logical function `region` is introduced assigning pointer bases to distinct "Regions" represented by numbers: global variables are assigned to region 0, formals to region 1 and locals to region 2. At this point, notice that whenever $\texttt{region}(l) \neq \texttt{region}(l')$, then we can immediately deduce $l \neq l'$.

Second, each memory model in Wp is asked to assign any *stored* pointer value before a function body to be also in region 0. Hence, the `region` function efficiently encodes that formal and local variables of the function under proof are really fresh: they cannot be aliased by already stored pointers in memory before the function call, since formals and locals live in regions 1 and 2, whereas all previously allocated variables live in region 0. More precisely, Wp also includes negative regions for initial pointer values, since negative regions are reserved for string literals.

Such hypotheses of stored pointers living in non-positive regions (i.e., separated from formals and local variables) are called the Heap assumptions. To complete this presentation, let us now briefly describe how those heap assumptions are actually generated by Wp for the Hoare and Typed memory models.

In the Hoare memory model, for any variable v that contains a pointer (be it a pointer variable, an array of pointers or a compound with pointer fields or arrays), its initial pointer value is assumed to be in a negative region in the function pre-state. For instance:

```
int *p;
void f(void) {
L0: // Heap assumption: region(p@L0.base) <= 0
    ...
}
```

In the Typed memory model, the memory-map Mptr that stores the value of pointers in memory is initially assumed to only contain pointers with negative bases. This is stated thanks to the predicate framed(Mptr) and the associated axiom from the Wp standard library:

$$\forall M, l, \mathtt{framed}(M) \implies \mathtt{region}(M[l].\mathrm{base}) \leq 0$$

Then, the initial memory-map variable for pointers is assumed to be framed, for instance:

```
void f(void) {
L0: // Heap assumption: framed(Mptr@L0)
    ...
}
```

Exercise 3 Explain the verification condition generated by Wp and its proof on the following example:

```
1  int *p, *q;
2
3  //@ assigns \nothing;
4  void f(void) {
5    int x;
6    int *y = &x;
7    int **z = &p;
8    //@ check \separated(&x,p) && \separated(&x,q);
9  }
```

Verification Condition (with variables slightly renamed for clarity):

```
Goal Check (file region.i, line 8):
Let x = global(L_x).
Let p = global(G_p).
Assume {
   (* Heap *) Type: (region(q.base) <= 0) /\ framed(Mptr).
}
Prove: (x != q) /\ (x != Mptr[p]).
Prover Alt-Ergo 2.2.0 returns Valid (Qed:0.85ms) (11ms) (39)
```

Solution 3 Let us first describe how variables are modeled here. Variables x and p are modeled by the Typed memory model, since their address are explicitly taken at lines 6 and 7 in the code. Variable q, however, is modeled by the Hoare memory model, since its address is never taken.

4 Formally Verifying that a Program Does What It Should ...

Consider now initial heap assumptions: variable q contains a pointer value, hence we find the predicate region(q.base)<=0 at line 4 in the proof obligation. Aliased pointer variables are stored in memory-map Mptr of the Typed memory model, hence we find the predicate framed(Mptr) at line 4 in the proof obligation.

Consider now the proof obligations for separation checks at line 8 in the code: location &x is modeled by expression x = **global**(L_x) from the Typed model; location &p is modeled by expression p = **global**(G_p) from the Typed model; location p is then modeled by expression Mptr[p] from the Typed model; location q is simply modeled by variable q from the Hoare model.

The proof is finally tackled by SMT solvers thanks to the following observations:

- We have region(x.base)=2 since x is a local variable.
- We have region(q.base)<=0 by heap assumption, so x.base != q.base, so x != q.
- We have region(Mptr[p].base)<=0 from heap assumption framed(Mptr), thus x.base != Mptr[p].base, so x != Mptr[p].

The proof is actually straightforward for most SMT solvers.

4.3.7 Pointers as References

By default, memory to which pointers point does not have a particular validity status, that means that their allocation status is not known. However, one can provide this information. The most common way to do that is to provide it by indicating in **requires** clause that the pointer points to a global variable or that it is a **\valid** memory location. However, when dealing with pointers, some separation hypotheses are quickly necessary as well.

In some circumstances, however, pointers are used with implicit validity and separation hypotheses. The "Ref" memory model allows Wp to automatically infer them for free. In this model, Wp assumes that any pointer p only accessed with exactly a simple dereferencing *p (and no p++ or p[42], etc.) necessarily points to a valid memory location separated from any other pointer. For example, if we want to verify the swap function, we can see that Wp makes different hypotheses about memory depending on whether the model "Ref" is used:

```
/*@ assigns *a, *b ;
    ensures P: *a == \old(*b) && *b == \old(*a); */
void swap(int* a, int* b){
  int tmp = *a;
  *a = *b;
  *b = tmp ;
}
```

With the default "Typed" memory model, we obtain the following results:

```
> frama-c -wp swap.i -wp-prop P
[kernel] Parsing swap.i (no preprocessing)
[wp] Warning: Missing RTE guards
[wp] 1 goal scheduled
[wp] Proved goals:    1 / 1
    Qed:              0  (4ms)
    Alt-Ergo 2.2.0:   1  (9ms) (26)
```

Notice that, by default, this program could not be proved to have no runtime errors since there is *a priori* no reason for pointers parameters to be valid. Moreover, it turns out that the swap function fulfills its contract whether the pointer parameters are equal or separated.

Instead, let us now use the "Ref" memory model instead:

```
> frama-c -wp swap.i -wp-prop P -wp-model ref
[kernel] Parsing swap.i (no preprocessing)
[wp] Warning: Missing RTE guards
[wp] 1 goal scheduled
[wp] Proved goals:    1 / 1
  Qed:                1  (1ms)
[wp] swap.i:3: Warning:
  Memory model hypotheses for function 'swap':
  /*@
     behavior wp_typed_ref:
       requires \valid(a);
       requires \valid(b);
       requires \separated(a, b);
    */
    void swap(int *a, int *b);
```

Using the model "Ref" introduces more validity and separation assumptions. Of course, these are strong hypotheses that are not always applicable, but it can dramatically impact proof efficiency on industrial applications of Wp. Indeed, by assuming that pointers are valid and separated, Wp basically assumes that these pointers are no more than proxies to non-aliased variables. Thus, we can use the "Hoare" model instead of the "Typed" model for a lot of pointers, which is much more efficient. The effect of this optimization is already visible on this simple example: instead of realizing the proof with an SMT solver (without the "Ref" model), the proof is directly realized by Qed. Moreover, with this model, function swap can be proved to be free of runtime errors.

Let us cite also the "Caveat" memory model which is similar to "Ref" but allows reasoning on possible aliasing between reference parameters and *other* variables. This memory model historically comes from the CAVEAT [6] tool which can be considered as the ancestor of Wp before the development of Frama-C.

Such memory models ease the verification of programs with pointers at the price of implicit restrictions and constraints on input variables that shall be verified either manually or by Wp at call sites (when available).

4.3.8 Memory Allocation

Wp memory models are not only responsible for modelling read and write operations, but they are also responsible for modelling the validity of memory access since deferencing an invalid pointer is an undefined behavior according to the C standard.

For example, consider the following program:

```
void function(void){
  int *ptr ;
  {
    int x ;
    ptr = &x ;
  }
  //@ assert W: \valid(ptr) ;
  *ptr = 42 ;
  //@ assert EQ: *ptr == 42 ;
}
```

Wp can indeed prove that the assertion EQ is verified, but the write operation on line 7 is invalid since ptr points to variable x outside its validity scope: p is actually a *dangling* pointer according to C. Thus, the assertion W cannot be proved (note that the Rte plug-in could have generated such an assertion), and the program is incorrect.

In Wp, memory models are responsible for keeping track of the allocation status of memory locations. To this end, in addition to value memory variables, memory models associate logical variables that track to the validity of memory locations at each program point. In this additional mapping, instead of associating values to locations, they associate an allocation status to locations, which is updated whenever the locations are allocated or deallocated. Memory models also keep track of the size of the valid memory associated to each memory location when appropriate.

Hence, in the "Hoare" memory model, in addition to logic variable x@L for the value of variable x at program point L, we would have a logic variable vx@L for the validity of variable x at program point L. Similarly, in the "Typed" memory model, we have a memory-map Malloc@L that associates to each pointer base b at program point L the size of the memory block allocated for b. More precisely, a block of size n is registered with value n in the allocation map and non-allocated blocks with value 0.

For example, the (slightly simplified) goal associated to assertion W above is the following:

```
Goal Assertion 'W' (file allocation.i, line 7):
Assume {
  ..
  (* Block In *)
  Have: Malloc_1[L_x_26 <- 1] = Malloc_2.
  Have: global(L_x_26) = ptr_0.
  (* Block Out *)
  Have: Malloc_2[L_x_26 <- 0] = Malloc_0.
}
Prove: valid_rw(Malloc_0, ptr_0, 1).
```

When entering the function block (line 4), variable x is allocated by assigning 1 to its block size (line 5). Then, pointer ptr is assigned the address of variable x (line 6). Finally, we leave the block (line 7) and reset the block size of variable x to zero (line 8). The predicate valid_rw(M, p, n) is part of the Wp standard library and states that, in memory allocation M, the memory range starting at pointer p with size n is included in a large enough allocated block with read-and-write permissions. There is a similar predicate for read-only validity access.

4.3.9 Initialization

According to the C standard, reading an uninitialized value can have unspecified or undefined behavior (that can even lead to security vulnerabilities), in Frama-C, they are considered as runtime errors. Just like memory models are responsible for representing the validity of memory locations at some program points, they are also responsible for representing their initialization status.

Initialization is handled by associating to each location a particular variable (for the "Hoare" memory model) or a particular entry in a memory map (for the "Typed" memory model). Initialization can be handled exactly the same way as reading or writing locations: instead of reading or writing values, memory models stores boolean values indicating whether the location is initialized or not. The following example illustrates this usage of memory models:

```
1  void function(void){
2    int i ;
3    //@ check !\initialized(&i);
4    i = 42 ;
5    //@ check \initialized(&i);
6  }
```

The generated verification conditions are the following:

```
> frama-c -wp -wp-no-let initialization.i -wp-print
[kernel] Parsing initialization.i (no preprocessing)
[wp] 2 goals scheduled
...
Goal Check (file initialization.i, line 3):
Assume {
   (* Block In *) Have: (Init_i_0=false).
}
Prove: (Init_i_0=false).

...
Goal Check (file initialization.i, line 5):
Assume {
   Have: (Init_i_0=true) /\ (i = 42).
}
Prove: (Init_i_0=true).
```

At function body entrance, the initialization status of i is `false`. Then, once it has been written, its value is changed to 42, and its initialization status is now `true`.

4.4 Annotations

Code annotations play an important role when proving programs with the Wp plug-in. Typically, loops need to be equipped with loop invariants and complex programs are likely to be annotated with intermediate assertions to be proved. For this purpose, the ACSL language comes with a rich variety of code annotations that are used in slightly different way by Wp for generating verification conditions. In this section, we dive into the details of Wp proof strategy for generating VCs for those different kinds of code annotations. The purpose is to help the developer efficiently use code annotations by anticipating how they are used by Wp for generating verification conditions. Only basic knowledge about Wp is required before reading this section. It can also be read as a pragmatic presentation of ACSL *for* deductive verification.

4.4.1 Assertions

The most basic code annotations are assertions. In short, they allow to state that some property should be valid at some program point:

```
1   int a = 42 ;
2   //@ assert A: a == 42 ;
3   a++ ;
4   //@ check  C: a == 43 ;
```

There are three kinds of assertions: **assert**, **admit** and **check**. Depending on the assertion kind, Wp generates different proof goals and propagates the annotation differently for proving further code annotations that come *after* in the program control flow:

- **assert** annotations are proved *and* used as further hypotheses;
- **check** annotations are proved but are *not* used further;
- **admit** annotations are *not* proved but used as further hypotheses.

For example, the VCs generated from previous example are as follows:

```
> frama-c assert.i -wp -wp-no-let -wp-print
...
Goal Assertion 'A' (file code-annots/assert.i, line 3):
Assume { Type: is_sint32(a). (* Initializer *) Init: a = 42. }
Prove: a = 42.

...
Goal Check 'C' (file code-annots/assert.i, line 5):
```

```
Assume {
  (* Assertion 'A' *)
  Have: a_1 = 42.
  Have: (1 + a_1) = a.
}
Prove: a = 43.
```

Wp generates two goals, one for the **assert** A and the second for the **check** C. In the hypotheses of the VC generated for the **check** C annotation, the property that corresponds to the **assert** A is listed.

An **assert** annotation is useful to provide an intermediate property that is used for proving properties that follow this assertion, yet without just assuming that it is true since we have to verify it. A **check** annotation requires a verification, but it is not assumed in following goals. Thus, it is useful to verify some properties without overloading the proof context. It is also useful for "testing" a property without taking the risk of introducing an inconsistency for further properties.

An **admit** annotation is merely the introduction of an axiom and shall be used with extreme care. In an industrial context, every introduction of an **admit** annotation shall be justified and submitted to manual code review. However, it can be useful for debugging purpose. For instance, adding an `//@ admit false;` annotation allows temporarily removing some execution paths and to focus proof efforts on sub-parts of the program.

4.4.2 Function Calls

As we previously mentioned, the weakest precondition calculus is a modular analysis. Thus, when a function is called, the calculus does not rely on its source code to prove the contract of the caller. Instead, it uses its contract.

Namely, when we prove the correction of a function, we *assume* that its **requires** clauses are satisfied, and we *prove* that the **ensures** clause are satisfied. On the opposite, when a function is called, we *prove* that its **requires** clauses are satisfied, and we *assume* that its **ensures** clauses are satisfied. This approach is the so-called *design by contract*.

Of course, since the memory can be modified by the function, the **Wp** plug-in needs **assigns** specification to invalidate what may have been modified. Hence, if such an **assigns** specification is not provided, **Wp** would assume that *everything* is modified and will emit a warning.

Like assertions, there are different variants of **requires** and **ensures**. One can write **check requires/ensures** or **admit requires/ensures**. In such a case, a **check requires** on f is only verified when calling f but not when verifying f, while an **admit requires** is not verified when calling f but still admitted when verifying f. A **check ensures** is verified during the verification of f but not assumed when calling f, and an **admit ensures** is not verified but assumed after f call.

4.4.3 Loop Invariants

Formally verifying C programs requires dealing with loops. Whereas many static analysis methods typically infer loop invariants, like abstract interpretation (see Chap. 3) or model checking, deductive verification with Wp needs to get them from the user.

Verifying an invariant is done by induction on the loop iterations. First, Wp proves that the invariant is established when the loop starts. Then, assuming that it is true at the loop entrance after any number of iterations, Wp proves that the invariant is still valid after one further loop iteration. Finally, when exiting the loop body, Wp assumes that an arbitrary number of iterations has been executed before the last loop entrance, where the loop invariant can now be safely assumed by induction.

This also means that loop invariants shall contain all the necessary knowledge about the loop in order to prove subsequent code annotations or ACSL properties *after* the loop.

Let us now illustrate the proof strategy for loop invariants with the following example code:

```
unsigned i = 0u ;
                           // i@LoopEntry (=0)
/*@ loop invariant Inv: i <= 10u ;
    loop assigns i ; */
while(1)                   // i@LoopCurrent (fresh, <= 10)
{
  if(! (i < 10)) break ;
  i++ ;                    // i@LoopEnd = i@LoopCurrent+1
                           // (<= 10)
}
                           // i@AfterLoop = i@LoopCurrent
                           // (>= 10)
//@ check C: i == 10 ;
```

Here, we use the internal representation of loops according to Frama-C kernel, as explained in Chap. 2. Notice that with this way of modeling loops, the invariant must be verified each time we enter the loop (since we always enter the loop). The code is annotated by logical variables i@L that represent the value of variable i at program point L, like in the Hoare memory model as explained in Sect. 4.3.1.

The two program points **LoopEntry** and **LoopCurrent** are built-in ACSL logic labels that refer respectively to program states where the loop starts and to the beginning of the current loop iteration. Let us also introduce two other memory states for our explanation: the LoopEnd state that corresponds to the end of the current iteration and AfterLoop that corresponds to the state reached when the loop terminates.

The Wp proof strategy for verifying this program is illustrated in Fig. 4.2. This figure shows a control-flow graph of the program instructions and verification conditions generated during Wp processing. We now explain this process step by step.

At **LoopEntry**, i still holds its initial value, hence we have i@LoopEntry=0. Hence, Wp is able to prove the establishment of the invariant. At **LoopCurrent**, Wp assumes that any number of loop turns has been executed so far. Hence,

Fig. 4.2 Example control flow graph with verification and assume points

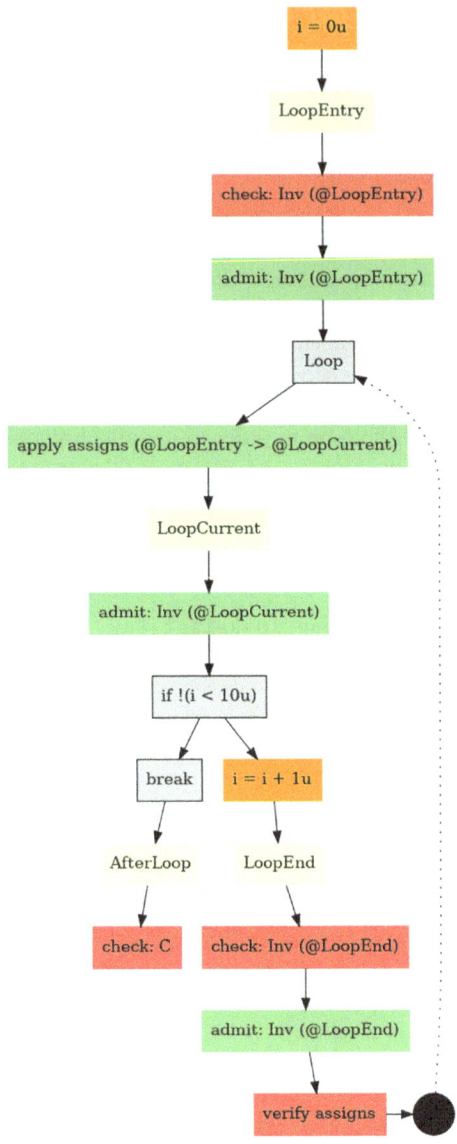

i@LoopCurrent has an arbitrary value, although the invariant is inductively assumed to hold. Thus, we still have i@LoopCurrent<=10 by induction hypothesis.

Then, in the body of the loop execution, there are two execution paths to consider: the one that continues the loop and the one that breaks the loop. Let us follow the continuing path first: condition i<10 is taken, so after increment, we can assert that i@LoopEnd<=10 so the invariant still holds. By induction, we can now assert that the loop invariant hold after an arbitrary number of loop iterations.

Finally, let us consider the loop-breaking execution path: from the **LoopCurrent** point, condition i>=10 is taken. Combined with the loop invariant hypothesis at **LoopCurrent**, we have both 10<=i and i<=10 so i=10. Hence, the proof of final check annotation C.

As illustrated above, all loop annotations can be derived into simple **check** and **admit** actions. The establishment of the invariant is verified at **LoopEntry**, it is first checked, and then can be assumed. The loop invariant is then assumed at **LoopCurrent** (which is sound: either because it is reached after proof of establishment or after proof of preservation). The proof of the preservation is verified at LoopEnd, the end of the loop body, on the edge that has been cut (thus, in the hypotheses, we have the induction hypothesis). The proof of the code that follows the loop is done at AfterLoop (thus again, here, the invariant Inv@LoopCurrent is admitted), notice that it means that we collect any code met before leaving the body of the loop, here, it is only the condition, but in a loop that execute some code before breaking, the effect of this code would also be collected.

Let us summarize how the loops are handled, and in particular why we do not need a special WP rule for it. Semantically, from a loop:

```
1 /*@ loop invariant I ;
2     loop assigns A ; */
3 while(1){
4   body (might go to AfterLoop)
5 }
6 AfterLoop:
```

Wp acts as if it was transformed into:

```
1 //@ check I (@ Loop Entry);
2 //@ admit I (@ Loop Entry);
3 {
4   // do assigns (this cannot be modeled using ACSL)
5   //@ admit I (@ Loop Current);
6   body (might go to AfterLoop)
7   //@ check I ;
8   //@ admit I ;
9   // check assigns (this cannot be modeled using ACSL)
10  goto EmptyWPState ;
11 }
12 AfterLoop:
```

where EmptyWPState is a particular state with an empty VC set. Thus, only the rules for **check** and **admit** annotations, and the handling of **assigns** annotation is needed.

Since the body is obtained via the transformation of the **Frama-C** kernel, it generally starts with a conditional that leaves the loop when the condition of the loop is false. Note that it means that proving the annotations that follow the loop can only rely on

- the code that precedes the loop,
- the knowledge of the loop invariant (I @ **LoopCurrent**),
- the last instructions executed in the body before leaving it.

When the code after loop is unreachable, the proof relies on the unreachability.

In fact, we can further detail how this computation behaves, in particular when several **loop invariant** clauses are provided. Just as assertions, loop invariants can be "checked", "admitted" or both (which is the default).

When a loop invariant appears before another, and is admitted, then it is available as a hypothesis for the verification of loop invariants that follow it. For the preservation of the invariant, a particular case applies for **check loop invariant**. Indeed, when a property is only *checked*, it does not appear as a hypothesis for goals that follow. *But*, for the property **check loop invariant** P, property P is still assumed as an induction hypothesis at label **LoopCurrent** for the preservation part of the invariant.

Consider the following example:

```
/*@      loop invariant Inv_1: P(i) ;
    admit loop invariant Adm_2: Q(i) ;
    check loop invariant Chk_3: R(i) ;
          loop invariant Inv_4: S(i) ;
    loop assigns i ; */
while(condition(i)){
   i++ ;
}
```

Hence, when establishing the invariants:

- Inv_1 is proved with only hypotheses from the code before the loop,
- Adm_2 does not require a proof,
- Chk_3 receives the same hypotheses as Inv_1 but also the fact that Inv_1 and Adm_2 are established,
- Inv_4 receives the same hypotheses as Chk_3, except Chk_3 itself since Chk_3 is just checked.

Similarly, when proving the preservation of invariants under inductive assumptions:

- Inv_1, Adm_2 and Inv_4 are assumed at **LoopCurrent**, but not Chk_3 since it is only checked (excepted for proving the preservation of Chk_3, see below).
- Inv_1 is proved at LoopEnd with only hypothesis from **LoopCurrent** and the code body,
- Adm_2 does not require a proof,

- Chk_3 receives the same hypotheses as Inv_1 but also the fact that Inv_1 and Adm_2 have been preserved, in addition to the inductive hypothesis Chk_3 itself.
- Inv_4 receives the same hypotheses as Chk_3, but no hypothesis from Chk_3 (no establishment, nor induction hypothesis and preservation) since it is just checked.

4.4.4 Writing Loop Annotations

As pointed previously, when the Wp calculus meets a loop, it cannot rely on all the iterations of the body of the loop when proving properties that comes after in the control-flow graph. Wp can only rely on the properties captured by the invariants after an arbitrary number of iterations. Thus, providing a good loop invariant is crucial when verifying programs with Wp.

Though there is no perfect guidance on how to write good loop invariants, we would like to report in the section on our experience and to provide the reader with general principles and few recipes.

The very first step when annotating loops is to write provable loop assigns. If loop assigns are incorrect, then Wp is likely to infer a wrong memory state after an arbitrary number of loop iterations at the **LoopCurrent** program point: let variable x being modified in the loop body, if x is not loop-assigned by the loop annotations, then Wp incorrectly assumes that x still hold its initial value at loop entry. Thus, it allows dummy proofs of the preservation of invariants, together with a silly loop-assigns hypothesis that cannot be proved.

Then, the second step is to bound the iterating and indexing variables. Sometimes, especially when dealing with arrays, one has to write loop invariants that bound the indexing variables for proving the loop-assigns. Hence, one often writes those loop annotations at the same time, before writing anything else.

When iterating over an array of size n, remember that array indices are likely to be in range 0..n, with n included as illustrated in previous examples, since the loop body generally *starts* with the loop exit condition.

When the loop body incrementally assigns an array, do not try to use the loop-assigns clause to characterize which parts of the array have been updated, and which parts of the array have been left unchanged so far. It is usually much easier and effective to assign the entire array at each loop turn, prove the loop-assigns, and then characterize more precisely what has been modified or left unchanged so far using dedicated loop invariants.

Write loop invariants incrementally when possible, name them and break complex properties into many smaller ones: it helps localize Wp proof failures reports, proof debugging and code maintenance.

We will now illustrate these general principles on a simple program that increments all elements of an array. Without considering runtime errors (overflows and validity of memory accesses), this function can be written and annotated in ACSL as follows:

```
1  /*@
2    assigns a[0..n-1];
3    ensures \forall integer k; 0<=k<n ==> a[k] == \old(a[k]+1);
4  */
5  void incr(int *a, int n)
6  {
7    for (int i=0; i<n; i++) { a[i]++; }
8  }
```

This program has no loop annotations and cannot be proved as it is. We gradually add loop annotations until we can prove its contract. During the process, to avoid waiting for timeouts, we temporarily comment the **ensures** clause of the function contract. We then concentrate on bounding the loop iterating variable i and proving the loop assigns:

```
1  /*@
2    assigns a[0..n-1];
3    //ensures \forall integer k; ...
4  */
5  void incr(int *a, int n)
6  {
7    /*@
8      loop invariant Lower: 0<=i;
9      loop invariant Upper: 0<=n ==> i<=n;
10     loop assigns i, a[0..n-1];
11   */
12   for (int i=0; i<n; i++) { a[i]++; }
13 }
```

```
> frama-c -wp incr-1.i
[kernel] Parsing incr-1.i (no preprocessing)
[wp] Running WP plugin...
[wp] Warning: Missing RTE guards
[wp] 7 goals scheduled
[wp] Proved goals:    7 / 7
  Qed:                6
  Alt-Ergo 2.2.0:     1 (18ms)
```

Notice the upper bound that only holds for positive n and includes n as an off-by-one of the last array index.

We try now to characterize which parts of the array has *not* been modified yet during loop iterations. Since the code incrementally modifies the array from left to right, we add the invariant RightPart at lines 12–14 which states that elements with index k such that i<=k<n are left unchanged:

```
1  /*@
2    assigns a[0..n-1];
3    //ensures \forall integer k; ...
4  */
5  void incr(int *a, int n)
6  {
7    /*@
8      loop invariant Lower: 0 <= i;
```

```
      loop invariant Upper: 0 <= n ==> i <= n;
      loop assigns i, a[0..n-1];

      loop invariant RightPart:
         \forall integer k; i <= k < n ==>
            a[k] == \at( a[k], LoopEntry);
      */
   for (int i=0; i<n; i++) { a[i]++; }
}
```

```
> frama-c -wp incr-2.i
[kernel] Parsing incr-2.i (no preprocessing)
[wp] Running WP plugin...
[wp] Warning: Missing RTE guards
[wp] 9 goals scheduled
[wp] Proved goals:    9 / 9
   Qed:               7
   Alt-Ergo 2.2.0:    2 (17ms-19ms)
```

Now we can characterize the elements of the array that are actually modified by the loop, with invariant LeftPart at lines 16–18 below:

```
/*@
  assigns a[0..n-1];
  //ensures \forall integer k; ...
*/
void incr(int *a, int n)
{
   /*@
      loop invariant Lower: 0 <= i;
      loop invariant Upper: 0 <= n ==> i <= n;
      loop assigns i, a[0..n-1];

      loop invariant RightPart:
         \forall integer k; i <= k < n ==>
            a[k] == \at( a[k]   ,LoopEntry);

      loop invariant LeftPart:
         \forall integer k; 0 <= k < i ==>
            a[k] == \at( a[k]+1 ,LoopEntry);

      */
   for (int i=0; i<n; i++) { a[i]++; }
}
```

```
> frama-c -wp incr-3.i
[kernel] Parsing incr-3.i (no preprocessing)
[wp] Running WP plugin...
[wp] Warning: Missing RTE guards
[wp] 11 goals scheduled
[wp] Proved goals:    11 / 11
   Qed:               8
   Alt-Ergo 2.2.0:    3 (16ms-19ms)
```

Notice here that invariant RightPart is necessary to prove the LeftPart: when iterating on **LoopCurrent** at index i, we increment the current value of a[i]. The RightPart invariant let us know that its value has been left unchanged since the **LoopEntry** point, hence making Wp able to prove that the element at current index k has been incremented from its initial value before the loop.

We are now finally equipped to prove the desired ACSL contract of function incr, as follows:

```
/*@
  assigns a[0..n-1];
  ensures \forall integer k; 0<=k<n ==> a[k] == \old(a[k]+1);
*/
void incr(int *a, int n)
{
  /*@
    loop invariant Lower: 0 <= i;
    loop invariant Upper: 0 <= n ==> i <= n;
    loop assigns i, a[0..n-1];

    loop invariant RightPart:
      \forall integer k; i <= k < n ==>
        a[k] == \at( a[k]    ,LoopEntry);

    loop invariant LeftPart:
      \forall integer k; 0 <= k < i ==>
        a[k] == \at( a[k]+1 ,LoopEntry);

  */
  for (int i=0; i<n; i++) { a[i]++; }
}
```

```
> frama-c -wp incr.i
[kernel] Parsing incr.i (no preprocessing)
[wp] Running WP plugin...
[wp] Warning: Missing RTE guards
[wp] 12 goals scheduled
[wp] Proved goals:    12 / 12
  Qed:                8
  Alt-Ergo 2.2.0:     4 (12ms-21ms)
```

Interestingly, the LeftPart loop invariant is quite close to the expected **ensures** clause of the function, except the upper bound which is limited to index i instead of n. This is a common pattern when dealing with loops. For more examples on how to write loop invariants for loops, the interested readers can consult the book "ACSL by Example" [29].

Exercise 4 Replay the previous example with the extended objective to get rid of Wp warnings about runtime errors. At which step of the process is it preferable to fix them?

Solution 4 The best time to look at runtime errors is after the loop-assigns clauses have been settled and proved. One generally benefits from bounded indices without being precluded by functional annotations that are usually not needed for proving the absence of RTE.

For the incr function, there are two kinds of possible RTE: array access validity and overflows when computing the increment of index i and array elements. One can use `frama-c -wp -wp-rte -print` on the program to see all of them. Hence, on a 64-bit architecture, this means elements strictly less than $2^{31} - 1$. The most minimal requirements to get rid of RTE are the two following **requires** clauses, the loop invariants being left unchanged:

```
1  /*@
2    requires \valid(&a[0..n-1]);
3    requires \forall integer k; 0<=k<n ==> a[k]+1 < (1 << 31);
4    assigns a[0..n-1];
5    ensures \forall integer k; 0<=k<n ==> a[k] == \old(a[k]+1);
6  */
7  void incr(int *a, int n) //...
```

```
> frama-c -wp -wp-rte incr-rte.i
[kernel] Parsing incr-rte.i (no preprocessing)
[wp] Running WP plugin...
[rte:annot] annotating function incr
[wp] 16 goals scheduled
[wp] Proved goals:    16 / 16
  Qed:                 9
  Alt-Ergo 2.2.0:      7 (12ms-24ms)
```

4.4.5 Termination

Termination of a program is often a desirable property. Most of the functions of a program should terminate, if not all. However, in the general case, proving that a program terminates is undecidable [46]. For this, Wp relies on some ACSL annotations that allow to provide a measure for loops and recursive functions.

A measure is intuitively an upper bound on a number of loop iterations or the depth of a recursion. It is thus a positive expression that must strictly decrease at each new loop iteration or recursive call. Generally, this expression is an expression of integer type, it is by the way recommended using such a simple measure. Wp also supports ACSL general measures for **loop variant** and **decreases** clauses, however it does not provide any way to check that such a measure is well-founded, so it should be used with extreme care.

The measure for a loop is provided as explained in Chap. 1 using the **loop variant** clause:

```
1  int i = 0 ;
2  /*@ loop invariant 0 <= i <= 10 ;
```

```
3      loop variant 10 - i ;
4      loop assigns i ; */
5   while(i < 10){
6      i++ ;
7   }
```

Notice that the verification condition generated for the **loop variant** takes benefit from the preserved **loop invariant** (except if it is a **check** of course).

The measure for a recursive function is provided as explained in Chap. 1 using the **decreases** clause:

```
1   /*@ requires n >= 0 ;
2       decreases n ; */
3   void f(int n){
4     if(n != 0){
5       f(n - 1);
6       g(n - 1);
7       h(n - 1);
8     }
9   }
10  /*@ requires n >= 0 ;
11      decreases n ; */
12  void g(int n){
13    if(n != 0) f(n-1);
14  }
15  void h(int n){}
```

In order to determine which calls require to verify the **decreases** clause, Wp computes the strongly connected components in the syntactic callgraph. These components are called *clusters* of mutually recursive functions in ACSL. This computation being syntactic, there are two main consequences:

- a function whose body is not provided is not considered to be part of any cluster,
- if a function pointer is called without a `calls` specification indicating to Wp which function might be called, Wp generates a VC that requires to prove that the call is unreachable (since it might be part of the cluster).

The verification of termination can be enforced thanks to the **terminates** ACSL clause. In a function contract, indicating **terminates** P means that the function must terminate when P is true in the **Pre** state. This is verified by the Wp plug-in by checking that all loops in the function have a **loop variant** clause, that the function has a **decreases** clause if it is recursive, and that all called functions terminate. The most common usage of the clause is to set it to **\true** (the function always terminate). One can make it a default value for Wp, and generate automatically the clause, thanks to the options: `-wp-frama-c-stdlib-terminate`, `-wp-declarations-terminate` and `-wp-definitions-terminate`.

In ACSL, when a **terminates** clause is specified, **loop variant** clauses are verified only when the condition of the **terminates** clause is satisfied. In extreme cases, if a non-terminating loop is not reachable under the **terminates** condition, the proof of loop variant is trivial because of the terminating condition. This situation is illustrated below:

4 Formally Verifying that a Program Does What It Should ...

```
//@ terminates x > 0;
void function(int x){
  if(x <= 0){
    //@ loop variant -1 ;
    while(1);
  }
}
```

The diverging loop is only entered when x<=0, which is contradictory with the terminate clause x>0. So, even if the loop variant annotation would be non-provable under normal circumstances, it is here trivially true under the terminates clause.

The same discussion applies for recursive calls. This behavior is enabled in Wp by using the option -wp-variant-with-terminates. For detailed description of the clause, the careful reader should refer to [5].

4.4.6 Lemmas and Axiomatic Definitions

In ACSL, one can declare lemmas and logical definitions, as introduced in Chap. 1. Lemmas must be proved and can then be used for helping verification. Let us see how do so with Wp.

The following logic function counts the number of occurrences of a value in an array between two bounds (upper bound excluded):

```
/*@ logic integer l_occ(int v, int* a, integer b, integer e) =
      e <= b ? 0 : (a[e-1] == v ? 1 : 0)
              + l_occ(v, a, b, e-1) ;
*/
```

One can easily use this definition to prove the following function:

```
//@ ensures \result == l_occ(v, a, 0, len) ;
int occ_easy(int v, int * a, unsigned len){
  int occ = 0 ;

  /*@ loop invariant 0 <= i <= len ;
      loop invariant occ == l_occ(v, a, 0, i) ;
      loop assigns i, occ ; */
  for(unsigned i = 0 ; i < len ; ++i){
    if(a[i] == v) occ++;
  }
  return occ ;
}
```

However, it not so easy for this slightly modified version that iterates from the end of the array instead of from its beginning:

```
//@ ensures \result == l_occ(v, a, 0, len) ;
int occ_not_so_easy(int v, int * a, unsigned len){
  int occ = 0 ;

```

```
5   /*@ loop invariant 0 <= len <= \at(len, Pre);
6       loop invariant occ == l_occ(v, a, len, \at(len, Pre)) ;
7       loop assigns len, occ ; */
8   while(len--){
9     if(a[len] == v) occ++;
10  }
11  return occ ;
12 }
```

What happened? In the first version, elements are counted by adding the current one at each iteration. Thus, for proving the preservation of the invariant, we have to prove the following formulas:

```
occ == l_occ(v,a,0,i) ==>
           occ+(a[i]==v?1:0) == l_occ(v,a,0,i+1)
```

Let us introduce the notation A(i) for test occurrence test (a[i]==v?1:0). Then we can reformulate the above formula as:

```
l_occ(v,a,0,i) + A(i) == l_occ(v,a,0,i+1)
```

which is exactly the definition of l_occ(v,a,0,i+1).

On the opposite, with the second example, we have to prove the following formula, where we denote by N the initial value of len at label **Pre**:

```
l_occ(v,a,len,N) + A(len-1) == l_occ(v,a,len-1,N)
```

Unfolding the definition in right-hand side now leads to:

```
l_occ(v,a,len,N) + A(len-1) == A(N-1) + l_occ(v,a,len-1,N-1)
```

which does not help that much.

In fact, we would have to reason by induction to finish this proof, which is out of reach of SMT solvers in general, and that would be the case for any loop using this pattern with the l_occ function. Thus, it is a good idea to help the solver by providing a lemma that explains how to count in the opposite direction and to prove it separately:

```
1 /*@ lemma opposite:
2       \forall int v, int* a, integer b, e ;
3       e > b ==>
4       l_occ(v, a, b, e) ==
5         (a[b] == v ? 1 : 0) + l_occ(v, a, b+1, e) ;
6 */
```

By adding this lemma, the proof of the second version succeeds easily. The remaining task is to prove the lemma, which indeed requires reasoning by induction, we will see how to do so in Sect. 4.6.4. Thus, the presence of the lemma implies that Wp creates a verification condition for the lemma, and uses it as a hypothesis for the other lemmas and for proving functions annotations.

Just like other code annotations, lemmas can be of different kinds: they can be only checked, in this case, they are denoted **check lemma** and they are not used for the verification of functions and other lemmas. On the opposite, they can be simply admitted, in which case they must be declared as **axiom** inside an axiomatic block (see Chap. 1).

In the previous example, the presence of the relation between the properties to verify, and the lemma was direct enough so that the added lemma was immediately used by SMT solvers. Sometimes, it is not that easy. In the following example, we have voluntarily written annotations in a way that makes the axiom hard to use for SMT solvers (notice the axiom ax_2 that relates two different memory labels as presented in Chap. 1):

```
/*@ predicate eq{L1, L2}(int* x) =
      \at(*x, L1) == \at(*x, L2) ; */

/*@
  axiomatic Ax {
    predicate P(int* x) reads *x ;
    predicate Q(int* x) reads *x ;
    axiom ax_1:
      \forall int* x ; P(x) ==> Q(x);
    axiom ax_2{L1, L2}:
      \forall int* x ; eq{L1, L2}(x) ==>
        P{L1}(x) ==> P{L2}(x);
  }
*/

/*@ assigns *x ; */
void g(int* x);

/*@ requires \separated(x, y);
    requires P(x) ;
    ensures  Q(x) ; */
void example(int* x, int* y){
  g(y);
}
```

The previous example can be verified because *x has not changed in the call to g, then, by ax_2, P(x) is still true and by ax_1 we can deduce Q(x). However, with Alt-Ergo 2.2.0, the proof fails. A way to overcome this failure is to provide the assertion:

```
//@ assert eq{Pre, Here}(x);
```

after the call to g. Usually, SMT solvers heuristics are not able to choose among all possible occurrences of \at(*x,L1) and \at(*x,L2) in order to apply the axioms and to explore all those possible combinations. By adding such an assertion explicitly in the context, the solver is able to only trigger the necessary axiom instances and to finish the proof.

During a verification process, it is common that small changes in the proof context make SMT solvers able or unable to make use of lemmas. This can be really hard to predict, assertions are a way to make it more predictable.

For verifying a lemma, Wp creates a formula that is quite close to the one expressed in ACSL. However, it still has to model the memory. For example, in the two previous examples, the lemmas and function definitions contains pointers and accesses to the pointed memory. Again, that means that in the memory that is "accessed" in the lemma, some locations are *assumed* by the Wp plug-in to be separated when proving it.

Exercise 5 Lemmas and predicate definitions can deal with several memory labels. For example, the following definition and lemma state that when no memory cell in an array has changed between two memory labels, then the number of occurrences is the same:

```
1  /*@ predicate all_same{L1, L2}(int *a, integer b, integer e) =
2      \forall integer i ;
3        b <= i <= e ==> \at(a[i], L1) == \at(a[i], L2) ;
4
5    lemma unchanged_occ{L1, L2}:
6      \forall int* a, int v, integer b, e ;
7        all_same{L1, L2}(a, b, e) ==>
8        l_occ{L1}(v, a, b, e) == l_occ{L2}(v, a, b, e);
```

This lemma, together with the assertion on lines 18–21 allows proving the **ensures** clause of the first behavior of the function change that write the value textbf at index pos of a:

```
1  /*@ requires 0 <= pos < len ;
2    assigns a[pos] ;
3    behavior same:
4      assumes new == a[pos];
5      ensures
6        l_occ{Pre}(new, a, 0, len) == l_occ(new, a, 0, len);
7    behavior different:
8      assumes new != a[pos];
9      ensures
10       l_occ{Pre}(new, a, 0, len)+1 == l_occ(new, a, 0, len);
11     ensures \let old = \at(a[pos], Pre) ;
12       l_occ{Pre}(old, a, 0, len)-1 == l_occ(old, a, 0, len);
13    complete behaviors;
14    disjoint behaviors;
15  */
16 void change(int new, unsigned pos, int *a, unsigned len){
17   a[pos] = new ;
18   /*@ assert
19     \at(a[pos], Pre) == new ==>
20     all_same{Pre, Here}(a, 0, len) ;
21   */
22 }
```

Write the body of the predicate one_change that states that all cells of a except the one at index p (that is between b and e) are the same:

```
1  /*@ predicate one_change{L1, L2}
2          (int *a, integer b, integer p, integer e) =
```

together with a lemma that allows proving the second behavior. Proving the lemma is not asked (at least for now).

Solution 5 Once the predicate is defined, the lemma consists in enumerating the cases:

```
/*@ predicate one_change{L1, L2}
        (int *a, integer b, integer p, integer e) =
        b <= p < e &&
        (\forall integer i ; b <= i < p ==>
            \at(a[i], L1) == \at(a[i], L2)) &&
        (\forall integer i ; p < i  < e ==>
            \at(a[i], L1) == \at(a[i], L2)) &&
        \at(a[p], L1) != \at(a[p], L2);

    lemma one_change_occ{L1, L2}:
        \forall int* a, int v, integer b, p, e ;
          one_change{L1, L2}(a, b, p, e) ==> (
            (v == \at(a[p], L1) ==>
                l_occ{L1}(v, a, b, e)-1 ==
                l_occ{L2}(v, a, b, e)) &&
            (v == \at(a[p], L2) ==>
                l_occ{L1}(v, a, b, e)+1 ==
                l_occ{L2}(v, a, b, e)) &&
            ((v != \at(a[p], L1) && v != \at(a[p], L2)) ==>
                l_occ{L1}(v, a, b, e) == l_occ{L2}(v, a, b, e)));
*/
```

We trigger the use of the lemma by stating what is one_change when the old value is not the same as the new one:

```
void sol_change(int new, unsigned pos, int *a, unsigned len){
    a[pos] = new ;
    /*@ assert
        \at(a[pos], Pre) == new ==>
            all_same{Pre, Here}(a, 0, len) ;
    */
    /*@ assert
        \at(a[pos], Pre) != new ==>
            one_change{Pre, Here}(a, 0, pos, len) ;
    */
}
```

Logic definitions, axioms and lemmas can be defined at different levels: at top level (same level as global definitions in C) or in axiomatic blocks. This can be used to organize these symbols. In particular, while symbols at top-level are always loaded in VCs, this is not the case for symbols that are defined in axiomatic blocks. For example, in the following code:

```
/*@ axiomatic A {
    predicate P(int* p) reads *p ;
    axiom a: \forall int* p ; *p == 42 ==> P(p);
    }

    axiomatic B {
    predicate Q(int* q) reads *q ;
```

```
 8      axiom b1: \forall int* q ; *q == 24 ==> Q(q);
 9      axiom b1: \forall int* p, *q ; *p == 24 && *q == 24 ==>
            P(p) && Q(q);
10    }
11  */
12
13  //@ ensures P(p) ;
14  void function(int *p){
15
16  }
```

when proving the **ensures**, Wp loads only the symbols of the axiomatic block A where P is defined. While, if we replace the property with Q(p), it loads B and since this block also uses P, it also loads A. This can be used to limit the quantity of symbols that Wp provides to the SMT solvers and thus avoid polluting the context with too much information.

4.4.7 Ghost Code and Lemma Functions

Ghost code is basically regular C code that only appears in annotations and can thus be seen by Frama-C but not by the compiler. Thus, this code can observe the behavior of the program but cannot change it. While the details about ghost code is left to Chap. 1, let us present how ghost code can be useful for verifying code using Wp.

Ghost code allows writing "proof carrying code". In the literature, this method is often called "auto-active verification" [36].

For example, proving the following assertion requires reasoning by induction (and ideally a lemma):

```
1  //@ requires split <= len ;
2  void f1(int* a, unsigned split, unsigned len){
3    /*@ assert \forall int v ;
4         l_occ(v, a, 0, split) <= l_occ(v, a, 0, len) ;
5    */
6  }
```

But in fact, thanks to ghost code, we can easily provide this reasoning:

```
1   //@ requires split <= len ;
2   void f2(int* a, unsigned split, unsigned len){
3     /*@ ghost
4       /@ loop invariant split <= i <= len ;
5          loop invariant
6            \forall int v ;
7              l_occ(v, a, 0, split) <= l_occ(v, a, 0, i) ;
8          loop assigns i ;
9          loop variant len - i ;
10        @/
11       for(unsigned i = split ; i < len ; ++i);
```

```
12   */
13   /*@ assert \forall int v ;
14         l_occ(v, a, 0, split) <= l_occ(v, a, 0, len) ;
15   */
16 }
```

Without changing the behavior of the original code, we are able to provide a reasoning by induction in the verification process (through the loop invariant) and this allows finishing the proof.

In fact, one can often go further and separate this kind of code into functions called "lemma functions" that acts like lemmas (providing a proof of a property), but that are manually instantiated by the user where it is needed instead of being used automatically by SMT solvers. For example, in the following code, we define a lemma function called l_occ_bound that states the bounds on the number of occurrences of a value in an array:

```
1  /*@ ghost
2    /@ requires b <= e ;
3       assigns \nothing ;
4       ensures \forall int v ; 0 <= l_occ(v, a, b, e) <= e - b;
       @/
5    void l_occ_bound(int* a, unsigned b, unsigned e){
6       /@ loop invariant b <= i <= e ;
7          loop invariant
8             \forall int v ; 0 <= l_occ(v, a, b, i) <= i - b ;
9          loop assigns i ;
10         loop variant e - i ;
11      @/
12      for(unsigned i = b ; i < e ; ++i);
13   }
14 */
```

This lemma function can then be used in the code of a function to deduce properties about an array:

```
1  void f3(int* a, unsigned len){
2    //@ ghost l_occ_bound(a, 0, len) ;
3    //@ assert 0 <= l_occ((int)42, a, 0, len) <= len ;
4  }
```

Since deductive verification is modular, we only have to prove the **requires** at call points to obtain the **ensures** *for free* (provided that we have proved the correctness of the lemma function).

Notice that since the ghost code must not change the behavior of the original code, it is mandatory to prove that any loop or call in a ghost code terminates.

This basically makes application of lemma(-function)s fully predictable since we apply them by hand when it is suitable. However, such lemma cannot be used automatically by **Wp** without manual intervention. The other main limitation is that lemma functions are just regular **C** functions: **Frama-C** does not support logic types in ghost code and a function cannot deal with several memory states that would be provided in input (thus lemmas with multiple labels cannot be handled with a simple

Exercise 6 Instead of writing a lemma to prove the function occ_not_so_easy, write a lemma function that states the same property and use it to finish the proof of the function.

Solution 6 The lemma-function is the following:

```
/*@ ghost
  /@ requires e > b ;
     assigns \nothing ;
     ensures
       l_occ(v, a, b, e) ==
         (a[b] == v ? 1 : 0) + l_occ(v, a, b+1, e) ;
  @/
  void l_occ_opposite(int v, int* a, unsigned b, unsigned e){
    /@ loop invariant b+1 <= i <= e ;
       loop invariant
         l_occ(v, a, b, i) ==
           (a[b] == v ? 1 : 0) + l_occ(v, a, b+1, i) ;
       loop assigns i ;
       loop variant e - i ;
    @/
    for(unsigned i = b+1; i < e; i++);
  }
*/
```

One can notice that the premise of our lemma appears as precondition for our function. With this lemma, we can finish the proof of occ_not_so_easy:

```
//@ ensures \result == l_occ(v, a, 0, len) ;
int occ_not_so_easy(int v, int * a, unsigned len){
  int occ = 0 ;
  //@ ghost unsigned orig_len = len ;

  /*@ loop invariant 0 <= len <= \at(len, Pre);
      loop invariant occ == l_occ(v, a, len, \at(len, Pre)) ;
      loop assigns len, occ ; */
  while(len--){
    if(a[len] == v) occ++;
    //@ ghost l_occ_opposite(v, a, len, orig_len) ;
  }
  return occ ;
}
```

In this code, in addition to the ghost call to our lemma function, we need a ghost variable to store the original value of len since we cannot use the logic construction \at in ghost code.

4.4.8 Proving Generated Annotations

As already mentioned in the introduction for simple programs, Wp can be used to just or also prove the absence of runtime errors and, more generally, to prove annotations generated by other plug-ins, like Rte, Eva or MetAcsl. Proving the absence of runtime errors often requires additional contracts, typically constraining the range of integral parameters and the validity of memory regions accessible through input pointers.

For example, in the function occ_easy illustrated above, if we ask Wp to *also* prove the absence of runtime errors, it would fail to prove the validity of the array accesses and the absence of overflow when incrementing the number of occurrences. For these proofs to succeed, we have to strengthen both the precondition and the loop invariant:

```
/*@ requires len <= INT_MAX; // added for the overflow
    requires \valid(a+(0 .. len-1)) ; // added for the access
    assigns \nothing ;
    ensures \result == l_occ(v, a, 0, len) ; */
int occ_easy(int v, int * a, unsigned len){
  int occ = 0 ;

  /*@ loop invariant 0 <= i <= len ;
      loop invariant 0 <= occ <= i ; // added for the overflow
      loop invariant occ == l_occ(v, a, 0, i) ;
      loop assigns i, occ ; */
  for(unsigned i = 0 ; i < len ; ++i){
    if(a[i] == v) occ++;
  }
  return occ ;
}
```

Exercise 7 Complete the annotations of the function occ_not_so_easy from the previous section, so that the absence of runtime errors can also be proved thanks to Wp.

```
/*@ requires len <= INT_MAX;
    requires \valid(a + (0 .. len-1)) ;
    assigns \nothing ;
    ensures \result == l_occ(v, a, 0, len) ;
*/
int occ_not_so_easy(int v, int * a, unsigned len){
  int occ = 0 ;

  /*@ loop invariant 0 <= len <= \at(len, Pre);
      loop invariant 0 <= occ <= \at(len, Pre) - len ;
      loop invariant occ == l_occ(v, a, len, \at(len, Pre)) ;
      loop assigns len, occ ; */
  while(len--){
    if(a[len] == v) occ++;
  }
```

```
16    return occ ;
17 }
```

But Wp can be also used to *only* prove the absence of runtime errors. In this approach, we generally need simpler, "minimal contracts" [28]. If we only focus on proving the absence of runtime errors, the minimal contract and code annotations for the function occ_easy are (only) the following ones:

```
1 /*@ requires len <= INT_MAX;
2     requires \valid(a + (0 .. len-1)) ;
3     assigns \nothing ;
4 */
5 int occ_easy_minimal(int v, int * a, unsigned len){
6   int occ = 0 ;
7
8   /*@ loop invariant 0 <= occ <= i <= len;
9       loop assigns i, occ ; */
10  for(unsigned i = 0 ; i < len ; ++i){
11    if(a[i] == v) occ++;
12  }
13  return occ ;
14 }
```

Exercise 8 Write a minimal contract and other code annotations for the function occ_not_so_easy.

Solution 8 The only necessary annotations and invariants for a minimal contract are the following:

```
1 /*@ requires len <= INT_MAX;
2     requires \valid(a + (0 .. len-1)) ;
3     assigns \nothing ;
4 */
5 int occ_not_so_easy_minimal(int v, int * a, unsigned len){
6   int occ = 0 ;
7
8   /*@ loop invariant 0 <= len <= \at(len, Pre);
9       loop invariant 0 <= occ <= \at(len, Pre) - len ;
10      loop assigns len, occ ; */
11  while(len--){
12    if(a[len] == v) occ++;
13  }
14  return occ ;
15 }
```

Even when Wp is used to prove annotations and contracts generated from other plug-ins, typically Rte, Eva or MetAcsl, the developer might have to add domain-specific ACSL annotations to complete the proofs. However, writing minimal contracts if often simpler that specifying the complete functional behavior of a program.

4.4.9 Annotations and Verification Conditions

Let us briefly summarize what is generated by WP (and where for each type of annotation) for each clause:

- **lemma**: 1 VC
- **axiom**: no VC (admitted with no proof)
- **ensures**: 1 VC
- **exits**: 1 VC
- **disjoint**: 1 VC
- **complete**: 1 VC
- **requires**: 1 VC for each call
- **terminates**: 1 VC for each call, 1 VC for each loop without **loop variant**
- **decreases**: 1 VC for each recursive call
- **assigns**: 1 VC for each assigned left-value
- **admit**: no VC (admitted with no proof)
- **assert/check**: 1 VC
- **loop invariant**: 2 VCs (established, preserved)
- **loop variant** (integer): 2 VCs (positive, decreasing)
- **loop variant** (general measure): 1 VC (the measure is *assumed* to be well-founded)
- **loop assigns**: 1 VC for each assigned left-value within the loop.

Recall also that, for external functions that are only specified but for which no implementation is provided, no VC can be generated for the **ensures**, **exits**, **terminates** and **assigns** clauses. Hence, those clauses are *assumed* to be verified, and no VC is generated, like for the **axiom** and **admit** clauses.

4.5 Detecting Specification Errors

When verifying a program with Wp, we write a specification, and we check that the program indeed conforms to it. As program complexity increases, its ACSL specification is likely to increase accordingly. We already illustrated such a situation in early introduction of this chapter: if both the specification and the program code are wrong in the same way, Wp proves them correct with each others.

In this section, we investigate issues related to incorrect specifications. Only basic knowledge of Wp is required for reading this section. However, we present some Wp features that are not commonly implemented in deductive verification tools and that can significantly improve confidence in specifications. Hence, we think that this is a worth reading section for both beginners and experienced practitioners.

4.5.1 Specification Threats

When proving code with Wp, we rely on some assumptions. First, of course we rely on the tool itself, in particular on all the assumptions Wp makes about the C language or the memory context. We also rely on possible abstract logic definitions, and in particular *axioms*. Code annotations also contain implicit hypotheses, typically any *admitted* annotations and contracts of functions without a body. Conversely, **requires** clause of an entry-point function can also be seen as an implicit hypothesis: maybe it is not possible to fullfill the requirement of the function from external code. Moreover, as already seen in Sect. 4.3, Wp sometimes infer extra requirements on functions, dependending on the memory model used.

Any of those implicit hypotheses might be wrong or inconsistent with each others. This can have a dramatic impact on the soundness of the verification: if inconsistent assumptions are present in a proof context, one can prove wrong properties. Actually, the property $\bot \implies \varphi$ is true for any formula φ.

Pushing forward in this direction, let us examine the following *extreme* specifications:

```
/*@ requires \true; ... */
void always (void);

/*@ requires \false; ... */
void never (void);

/*@ ensures \true; ... */
void silent (void);

/*@ ensures \false; ... */
void unsound (void);
```

The first function, always has no restriction on its precondition. This is a good property, meaning that no specific hypothesis is taken in input. On the opposite, for the second function, never, the precondition is too restrictive. Actually, the function can *never* be called from any context: all its other contracts and code annotations are proved with zero effort by Qed. But there is no added value for these proofs, since they can never be invoked.

The third function, silent has no valuable postcondition. Although there is no risk associated with such a contract, we gain no information after calling the function, which can be a problem when the function has side effects. Notice that, by default, if the function has no body, it is assumed to assign *everything*, which means that it would be very difficult to prove something related to memory after calling such a function.

The last function, unsound is the most dangerous extreme specification that we may encounter. If this function has a body and the specification is provable, it means that it never terminates normally: it always exits abnormally *or* loops infinitely. If the function has no body, the postcondition is assumed. After calling the unsound function, every code-annotation becomes provable with zero effort by Qed.

Let us also briefly mention a typical variant of such an extreme specification for code-annotations:

```
void job(void)
{
    ...
    //@ admit Unsound: \false;
    ...
}
```

In this situation, any other code-annotation after the Unsound admitted assertion above is proved valid with zero effort by Qed.

Altough we have considered here extreme cases, it can be the case where a complex collection of ACSL annotations becomes unexpectedly equivalent to \false. If ever any of the SMT solvers is smart enough to discover the inconsistency, everything is silently proved with minimal effort, but there is no value in such proofs: the inconsistency makes it unusable in practice and the specification describes only impossible situations.

Hence, tracking and avoiding such specification threats is important when using deductive verification. We now explain how to deal with such issues.

4.5.2 Testing the Specifications

Avoiding specification threats like the ones depicted above is not possible in the general case: the problem of detecting an inconsistency in specifications is proven to be *undecidable* in first-order logic.

However, we can gain confidence into our ACSL specification by applying standard testing techniques.

A first possibility is to apply Wp on unit tests of the functions, when available. In such a process, we want to check, using the existing tests, that we can indeed prove that our preconditions are satisfied in each test, and that our properties of interest from the postcondition hold. This allows for detecting if our preconditions are permissive enough for already identified use cases from our unit tests. This also allows for checking that our postconditions are strong enough to enforce properties of interest of the tested functions.

While the unit tests can already help in finding specification errors, Wp also provides an option to *systematically* track inconsistencies, in particular related to memory models, axioms and other implicit or admitted hypotheses. When option -wp-smoke-tests is set, Wp tries to prove \false at various places. Did any such proof succeed, then there is surely an inconsistency somewhere in the background or dead code, and Wp hopefully found it! Notice however that this is a best-effort search, since, even if none of those proof attempts succeed, there still might remain some inconsistencies, although Wp and the SMT solvers did not manage to find them... yet. Notice that smoke-tests timeouts can be tuned independently of normal timeouts: actually, all smoke-tests proofs shall timeout to succeed, a process that usually requires a long time to complete. Because of the incompleteness of the method, smoke tests are indeed a "testing" process, as opposed to the "proving" process of deductive verification.

Let us illustrate the Wp smoke tests process on a trivial example:

```
1 //@ requires i < 0 && i > 0;
2 void f(int i){
3
4 }
```

when one use the following command on this example, Wp indicates that the **requires** clause of the function is "doomed", meaning that it is certainly inconsistent since we can prove \false (and thus everything) using it:

```
> frama-c smoke.i -wp -wp-smoke-tests
...
[wp] [Failed] (Doomed) typed_f_wp_smoke_default_requires (Qed)
[wp] rtes/smoke.i:2: Warning: Failed smoke-test
```

Of course, here, detecting the broken **requires** is easy, since most SMT solvers will immediately prove false from $i < 0$ and $i > 0$. Even a manual review would reveal it. However, with more ellaborated theories, it is easy to introduce inconsistencies that are not easy to detect by manual review. For example, in the following code:

```
1  /*@ axiomatic Ax {
2      predicate ready(int* x);
3      axiom a{L}: \forall int* x; \valid(x) ==> ready(x) ;
4      axiom b{L}: \forall int* x; ready(x) ==> *x == 42 ;
5  } */
6
7  /*@ requires ready(x) && *x == 24 ; */
8  void g(int* x){
9      *x = 24 ;
10 }
11 //@ requires \valid(x);
12 void c(int *x){
13     *x = 24 ;
14     g(x) ;
15 }
```

It is possible to satisfy the precondition of g (see in function c), however because of the axiom b, the precondition is inconsistent. Such an inconsistency is hopefully detected by SMT solvers thanks to Wp smoke tests:

```
> frama-c smoke.i -wp -wp-smoke-tests
...
[wp] [Failed] (Doomed) typed_g_wp_smoke_default_requires (...)
[wp] rtes/smoke.i:7: Warning: Failed smoke-test
```

On the same topic, any **requires** that would break an axiom of the memory model is also detected by smoke-tests. For example, in the following code, if we use the reference model, the **requires** clause is doomed:

```
1  //@ requires ! \valid(p) ;
2  void invalid_model(int* p){
3
4  }
```

```
   > frama-c smoke.i -wp -wp-model ref -wp-smoke-tests
   ...
   [wp] [Failed] (Doomed)
       typed_ref_invalid_model_wp_smoke_default_requires (Qed)
   [wp] rtes/smoke.i:31: Warning: Failed smoke-test
```

Another feature provided by smoke tests is the systematic detection of dead code, either because the actual program indeed has unreachable code, or because of some mistake in the specification. For example, in the following code, either because of a wrong condition in the code or because of a wrong **requires** clause, the then branch of the conditional is actually unreachable:

```
1  //@ requires x < 0 ;
2  void unreachable(int x){
3    if(x >= 0){
4      x ++ ;
5    }
6    x ++ ;
7  }
```

```
   > frama-c smoke.i -wp -wp-smoke-tests
   ...
   [wp] [Failed] (Doomed)
       typed_unreachable_wp_smoke_dead_code_s12
   [wp] rtes/smoke.i:4: Warning: Failed smoke-test
```

Note that since the principle is to prove unreachability, this also applies to non-termination of loops.

To avoid false positives, the user shall explicitly introduce **assert \false** code-annotations at unreachable-by-design code points. Hence, Wp generates VCs to *prove* that the assertion hold, hence the code is trully dead, but it will *not* generate a smoke test for this program point.

Smoke tests are a good way to detect potential flaws in specifications, however one should keep in mind that if all smoke tests succeed, it does not mean that the specification has no mistakes, just that the smoke tests were not able to detect them, if any. Smoke tests might just fail as any other proof, or the mistake is hidden in place that is not covered by smoke tests. For more information about "doomed" properties or code, the careful reader can refer to [31].

4.6 Sequent Simplification and Interactive Proof

The previous sections described the generation of verification conditions (VCs) by Wp plug-in to verify that the source code complies with its ACSL specification.

During this computation and before calling a prover on the generated VCs, many simplifications and transformations are performed in order to reduce the search space for the provers. This process uses the Qed library [20] for the representation of the logical terms manipulated by the Wp plug-in.

However, the automatic theorem provers may still fail to discharge the verification conditions. This can be due to limitations in the automatic theorem provers (the demonstration is too complex, or some theories are missing), or just because the source code does not comply with the specifications (the source code or the specifications are wrong, or the specifications are incomplete). In both cases, the user can use the Wp interactive prover, also named TIP, to interactively complete or debug the proof. Alternatively, external proof assistant can also be used.

Both topics, simplifications with Qed and interactive proof with TIP are closely coupled with each other: both deal with VCs transformations, although Qed is systematically used, whereas the TIP is user-directed and efficiently powered by Qed.

We now introduce the simplifications performed by Qed at the term level, then the VCs transformations based on proof sequents transformations and, finally, the TIP interactive prover of Wp and how to use external (interactive) proof assistants. The purpose of this section is to provide the interested reader with background understanding of what is performed behind the scene to make Wp efficient in practice. It is not *necessary* to understand all these techniques, although, having a minimal background about them might help debugging proofs or writing more efficient specifications.

This section is rather technical, with non-trivial operations at logical formula level. As a matter a fact, VCs transformations are only necessary for *complex* formulas: simple cases are generally automatically discharged by Qed and SMT solvers and we never have to go into the details of the associated VCs. Hence, this section presents non-trivial examples and requires some practice with Wp and with reading complex verification conditions.

The section is organized as follows. First (Sect. 4.6.1), we review general simplifications performed systematically by Qed on every term and logical formula. Second (Sects. 4.6.2 and 4.6.3), we introduce different techniques provided by Qed internals for transforming VCs. Then (Sect. 4.6.4), we illustrate how the TIP allows the developer to interactively apply those techniques on VCs for debugging or completing its proofs. Finally (Sect. 4.6.5), we briefly overview how to use external proof assistants to discharge the most difficult VCs.

4.6.1 Normalization and Term Simplifications

The Qed library builds normalized terms efficiently from a collection of built-in smart constructors and an extensible simplifier engine.

The library defines built-in operators for first-order logic with quantifiers and some built-in theories, notably real and integer arithmetic, records and arrays. Qed also supports uninterpreted functions and functions with algebraic properties (associative, commutative and idempotent operators, injective functions, constructors, neutral and absorbent elements). Moreover, Qed offers an *API* allowing Wp to add specific predicates and logical functions (i.e. bitwise operators, list data type) with their smart constructors and rewriters.

The smart constructors have a low algorithmic complexity; they perform on-the-fly computations, partial evaluations and some algebraic simplifications in roughly constant-time or linear-time with respect to their number of arguments. They are also responsible for *normalizing* terms and predicates, with an efficient in-memory representation that maximizes sharing.

For boolean expressions, De Morgan rules are applied for normalizing expressions. For instance, the negation operator is lifted towards the root of terms as follows:

$$\neg \neg p \rightsquigarrow p$$
$$\neg (p_1 \Rightarrow p_2) \rightsquigarrow p_1 \wedge \neg p_2$$
$$\neg (p_1 \vee \cdots \vee p_n) \rightsquigarrow \neg p_1 \wedge \cdots \wedge \neg p_n$$
$$\neg (p_1 \wedge \cdots \wedge p_n) \rightsquigarrow \neg p_1 \vee \cdots \vee \neg p_n$$
$$\neg \forall x.p \rightsquigarrow \exists x. \neg p$$
$$\neg \exists x.p \rightsquigarrow \forall x. \neg p$$

Arithmetic expressions are normalized towards integer and real linear forms, like in the following example rules, where p, q, r are numerical constants:

$$p.a + q.b + r.a \rightsquigarrow (p+r).a + q.b$$
$$p.a < q.a + r \rightsquigarrow (p-q).a \leq (r-1)$$

Functions with algebraic properties are also normalized accordingly. For example, the ACSL *bitwise or* operator (connector |) is commutative, associative and idempotent. It has 0 as neutral element, -1 as absorbent element and the ACSL *bitwise negation* (connector ~) as opposite operator. So, the construct of the operation contained in the left column of the following table builds the corresponding term of the right column:

$$0 \mid a_1 \mid \ldots \mid a_n \rightsquigarrow a_1 \mid \ldots \mid a_n$$
$$-1 \mid a_1 \mid \ldots \mid a_n \rightsquigarrow -1$$
$$a \mid \sim a \mid b_1 \mid \ldots \mid b_n \rightsquigarrow -1$$
$$b \mid c_1 \mid (a \mid b) \mid c_2 \mid \ldots \mid c_n \rightsquigarrow a \mid b \mid c_1 \mid c_2 \mid \ldots \mid c_n$$

The Qed simplifications can also be extended to perform computations that cannot be delegated to automated provers. For example, the construct of the *bitwise or* operation applied on two constants returns another constant by computation: the term 0x0F | a | 0x50 | a would be actually on-the-fly normalized into 0x5F | a.

For further simplifications, the Qed *API* allows the Wp plug-in to extend the simplifier engine with *conditional* transformations, here denoted by:

$$c \vdash t_1 \rightsquigarrow t_2$$

This notation means that the terms t_1 is rewritten into t_2, provided condition c can be normalized into true. Such a conditional transformation is semantically equivalent to applying instances of the following lemma (where $x, \ldots$ stands for all free variables appearing in the rule):

$$\forall x, \ldots, \ c \implies t_1 = t_2$$

However, Qed applies conditional transformations systematically when the condition c reduces to true, a strategy that offers different trade-offs from what SMT solvers typically do when applying lemmas. The rule is applied less frequently than what SMT solvers would do, but only when the condition is *evaluated* to be true, hence pruning many unwanted branches from the search space.

Combined with the ability of Qed to perform computations in guards, conditional rules also allows implementing transformations that are actually beyond the capabilities of SMT solvers. For instance, the following rules are applied whenever a and b are numerical constants:

$$(a \ \& \ {\sim}b) \neq 0 \vdash (a \mid c_1 \mid \ldots \mid c_n) = b \leadsto \text{false}$$
$$a < 0 \land 0 <= b \vdash (a \mid c_1 \mid \ldots \mid c_n) \leq b \leadsto \text{true}$$
$$a < 0 \land b <= a \vdash b \leq (a \mid c_1 \mid \ldots \mid c_n) \leadsto \text{true}$$

Actually, there are many conditional rules on *bitwise* operators that make Qed capable of simplifying VCs generated from C code with bit-level operations. Moreover, Wp introduces a dedicated symbol $\texttt{bit\_test}(a, k)$ (associated to a Why3 theory) to reason specifically over the k-th bit of a, as defined in ACSL by generalized *bitwise* operators:

$$\texttt{bit\_test}(a, k) \equiv a \ \& \ (1 \ll k) \neq 0$$

From this definition, many conditional simplification rules are defined, that introduces $\texttt{bit\_test}(a, k)$ expressions or simplify expressions with $\texttt{bit\_test}(a, k)$ occurrences. For instance, under the condition guard $0 \leq k$, the following rules are defined:

$$a \ \& \ (1 \ll k) \neq 0 \leadsto \texttt{bit\_test}(a, k)$$
$$\texttt{bit\_test}({\sim}a, k) \leadsto \neg \texttt{bit\_test}(a, k)$$
$$\texttt{bit\_test}(a_1 \mid \ldots \mid a_n, k) \leadsto \texttt{bit\_test}(a_1, k) \lor \cdots \lor \texttt{bit\_test}(a_n, k)$$
$$\texttt{bit\_test}(a_1 \ \& \ \ldots \ \& \ a_n, k) \leadsto \texttt{bit\_test}(a_1, k) \land \cdots \land \texttt{bit\_test}(a_n, k)$$

To finish with this overview of Qed simplifications, let us mention that there are also many conditional rules regarding ACSL memory predicates and Wp memory models. For instance, in the "Typed" memory model, the expression \separated(&a.f[i], &a.f[j]) are automatically rewritten into i!=j. Similarly, an assignment to p[i] is proved by Wp to be consistent with the ACSL clause **assigns** p[0..n-1] by generating the condition p+i \in p+(0..n-1), which in turn is automatically rewritten by Qed into 0 <= i < n.

4 Formally Verifying that a Program Does What It Should ...

Hence, Qed simplifications makes Wp able to dramatically reduce the complexity of deductive verification even in the presence of complex C and ACSL constructs, including arithmetic, logical, *bitwise*, memory operators and many other ACSL built-in theories such as lists, sets, arrays, records and so on.

Let us finally mention that the Qed simplification engine can also be extended by advanced users *via* Wp drivers. The interested reader is invited to consult the Wp reference manual for further details.

4.6.2 Sequent Decomposition

The verification conditions (VCs) computed by Wp might result in huge formulas with many hypotheses and nested terms at arbitrary levels, even after the many simplifications performed by Qed as described above. Proving huge proof terms is likely to stand far beyond the capabilities of SMT solvers and Wp has to transform them further. This is where sequent decomposition and transformations are involved.

The generated VCs are encoded as *Sequents* that have the following general structure:

$$\Delta \vdash \phi$$

where ϕ is a Qed formula representing the *goal* of the VC and $\Delta = H_1, \ldots H_n$ is the list of *hypotheses* of the VC. Each hypothesis H also have a recursive structure, and can consist of:

- a Qed formula $H \equiv \varphi$
- a conditional collection of hypotheses $H \equiv \text{if}\,\varphi\,\text{then}\,\Delta^+\,\text{else}\,\Delta^-$
- a general disjunction of hypotheses $H \equiv \text{either}\,\Delta_1 \ldots \Delta_n$.

Such structured hypotheses naturally come from the structure of the original C code and ACSL annotations. Moreover, each hypothesis node H carries additional attributes, not represented here, that allows for a categorization of the formulas φ it consists of. Hence, Wp can distinguish between formulas originating from the C code and ACSL properties, and those originating from type constraints and memory model assumptions. Such information is crucial for designing efficient heuristics for sequent simplifications, as we will see below.

Mathematically, a proof sequent is naturally associated to a purely logical formula, that is actually computed (and simplified) by Qed when transmitting the VCs to external SMT solvers and proof assistants. However, keeping track of the structured hypotheses eases the development of efficient sequent transformations and heuristics.

Wp uses classical rules of Natural Deduction [27] for decomposing sequents. Typical decomposition rules transform a proof sequent into an equivalent one, without losing any information. Here are some examples of sequent decomposition rules, where φ_x denotes formula φ where free variable x has been freshly renamed:

$$\Delta \vdash (\varphi \Rightarrow \phi) \quad \rightsquigarrow \quad \Delta, \varphi \vdash \phi$$
$$\Delta \vdash \forall x.\varphi \quad \rightsquigarrow \quad \Delta \vdash \varphi_x$$
$$\Delta, (\exists x.\varphi) \vdash \phi \quad \rightsquigarrow \quad \Delta, \varphi_x \vdash \phi$$

Those simplifications allow removing a logical operator from the goal: to prove formula $\varphi \Rightarrow \phi$, one simply needs to assume φ and prove ϕ (first rule); to prove formula $\forall x.\varphi$, one simply need to prove φ for a fresh instance of variable x (second rule); finally, from hypothesis $\exists x.\varphi$, one can obtain one instance of x for which property φ hold (third rule).

Natural deduction rules also deal with conjunctions and disjunctions, but the associated rules produce *multiple* sequents from a single one. Here is the transformation rule for conjunction:

$$\Delta \vdash (\varphi_1 \wedge \varphi_2) \quad \rightsquigarrow \quad \begin{cases} \Delta \vdash \varphi_1 \\ \Delta \vdash \varphi_2 \end{cases}$$

Indeed, for proving $\varphi_1 \wedge \varphi_2$, one has to prove both φ_1 and φ_2, separately. Conversely, here is the transformation rule for disjunction:

$$\Delta, (\varphi_1 \vee \varphi_2) \vdash \phi \quad \rightsquigarrow \quad \begin{cases} \Delta\varphi_1 \vdash \phi \\ \Delta\varphi_2 \vdash \phi \end{cases}$$

That is to say, to prove goal ϕ under hypothesis $\varphi_1 \vee \varphi_2$, one has to prove it both under hypothesis φ_1 and hypothesis φ_2, separately.

Such rules are *not* systematically applied by Qed unless all-but-one of the resulting sequents can be trivially simplified. Otherwise, the simplification would produce multiple sequents from the original one, which might lead to combinatorial explosion in the number of sequents to consider. Although, the user may still apply such rules by hand using the interactive TIP, as explained in Sect. 4.6.4.

Structured hypotheses with conditionals are also subject to simplification by pruning. For instance:

$$\text{if } \varphi \text{ then } \Delta^+ \text{ else } (\ldots, \text{false}, \ldots) \quad \rightsquigarrow \quad \varphi, \Delta^+$$
$$\text{if } \varphi \text{ then } (\ldots, \text{false}, \ldots) \text{ else } \Delta^- \quad \rightsquigarrow \quad \neg\varphi, \Delta^-$$

In the first rule, the **else** branch of the conditional can be trivially simplified since it would introduce false in hypothesis, so only the **then** branch is relevant. Similarly, for the second rule, the **then** branch can be trivially simplified and only the **else** branch is relevant.

Common equations between conditional branches are also factorized, moving some equations one level up in the structure of the left-hand-side of the sequent:

$$\text{if } \varphi \text{ then } (\phi, \Delta^+) \text{ else } (\phi, \Delta^-) \quad \rightsquigarrow \quad \phi, \text{if } \varphi \text{ then } \Delta^+ \text{ else } \Delta^-$$

In practice, even if each of the transformation rules presented above only has a limited impact on the sequent, it is often the case that *many* cascading transformation rules can be applied in a fruitful way. Moreover, by decomposing complex formula into smaller ones and moving equations around, opportunities for further simplifications might occur, as investigated in the next section.

4.6.3 Sequent Simplifications

Sequent decomposition techniques depicted above are simple rules that can be applied systematically. Other sequent simplification techniques are also used by Wp, although they require much more computations and some heuristics to choose between alternatives.

Consider for instance the following sequent transformation, where $A_{a/b}$ denotes A where every occurrence of term a has been recursively replaced by term b:

$$a = b, \Delta \vdash \phi \quad \leadsto \quad a = b, \Delta_{a/b} \vdash \phi_{a/b}$$
$$x = e, \Delta \vdash \phi \quad \leadsto \quad \Delta_{x/e} \vdash \phi_{x/e}$$

The first rule leads to a logically equivalent sequent, although it can also be applied by replacing b with a, which can lead to a very different proof obligation with different Qed simplifications.

In the second rule, the initial equation $x = e$ can be safely removed, since variable x no more appear elsewhere. However, this rule might offer even more alternative choices than the first one. Consider for instance we have equation $x = y + z$: it is not clear whether to apply the rule above by replacing x with $y + z$, or by replacing y with $x - z$, or by replacing z with $x - y$.

Hence, turning equalities between terms and variables into actual substitutions is a complex part of sequent transformations that takes into account the structure of the hypotheses Δ and their attributes. Heuristics are used to choose which equations to take into account and which variables to rewrite first. They are repeatedly applied until some fixpoint is reached.

Some other simplifications try to remove hypotheses from Δ since too large proof contexts are often hard to deal with for SMT solvers. Removing a hypothesis H from a sequent requires extra care: maybe the hypothesis H is necessary to prove the goal, or maybe it is inconsistent with another hypothesis H' while the goal ϕ is not provable on its own. In both cases, removing H from the sequent would make it non-provable anymore. Wp applies some heuristics for determining some hypotheses H that can be erased without (hopefully) weakening the original sequent:

- A hypothesis $\varphi(x)$ where x appears nowhere else can be safely erased.
- Type constraints on variables that do not appear elsewhere can also be erased.
- Hypotheses from C code that do not interfere with other hypotheses or with the goal can also be removed since formulas from C code semantics are never contradictory with each others. Typically, hypotheses related to the initialization of C variables that are not mentioned in the VC are erased by this method.

Finally, let us mention the existence of *semantic* sequent simplifiers that try to infer some knowledge from hypotheses and to rewrite formulas accordingly, e.g. by using abstract interpretation domains. Typically, there is a dedicated simplifier that infers numerical bounds for arithmetic expressions[1] from constraint hypotheses, and then simplifies other relations accordingly.

Most Wp sequent transformations can be turned off by Frama-C command line options. For instance, -wp-(no)-filter turns on/off hypothesis filtering and -wp-(no)-let turns on/off variable substitutions. The interested reader is invited to consult the Wp reference manual for further details and insights on other sequent simplifications performed by Wp.

4.6.4 *The Interactive Prover (TIP)*

Deductive verification with WP is supported by SMT solvers. While they allow finishing most proofs automatically, sometimes, they fail to provide another result than "unknown". In such a case, one wants to:

- at least understand why the proof fails, in case something can be done at the C or ACSL level (for example, adding a clause, or fixing a wrong code, or helping the proof with additional annotations);
- alternatively, finish the proof interactively by transforming the proof sequent until SMT solvers can conclude, in which case replaying the proof later is an important feature.

For both purposes Wp provides an interactive prover, named TIP, available from the Frama-C GUI. This tool relies on the same machinery as the sequent simplifiers described above. Although, we will present here other kinds of sequent transformations, called *Tactics*. Contrarily to previously presented sequent transformations, Tactics require some extra information from the user to be applied. Or, they might weaken the initial proof sequent by making some choice, may fail and hence may require the user to backtrack and investigate another proof strategy. Another particularity of Tactics is that applying them often leads to *several* children sequents to be proved, which we denote by the following notation:

$$\frac{\Delta_1 \vdash \phi_1 \ \ldots \ \Delta_n \vdash \phi_n}{\Delta \vdash \phi}$$

which reads as follows: provided one can prove all the sequents $\Delta_i \vdash \phi_i$, then sequent $\Delta \vdash \phi$ is also proved. The associated theorem can be stated as follows:

$$\bigwedge_{i \leq n}(\Delta_i \Rightarrow \phi_i) \implies (\Delta \Rightarrow \phi)$$

---

[1] This Wp simplifier actually reuses the Eva abstract domains for this purpose.

For instance, the *Split* tactic can operate on a conjunction of goals:

$$\frac{\Delta \vdash \varphi_1 \ \ldots \ \Delta \vdash \varphi_n}{\Delta \vdash (\wedge_{i \leq n} \varphi_i)}$$

or on a disjunction of hypotheses:

$$\frac{\Delta, \varphi_1 \vdash \phi \ \ldots \ \Delta, \varphi_n \vdash \phi}{\Delta, (\vee_{i \leq n} \varphi_i) \vdash \phi}$$

or on a conditional:

$$\frac{\Delta, \varphi, \Delta^+ \vdash \phi \quad \Delta, \neg\varphi, \Delta^- \vdash \phi}{\Delta, \text{if } \varphi \text{ then } \Delta^+ \text{ else } \Delta^- \vdash \phi}$$

Such rules lead to children sequents that are equivalent to the initial sequent, however, they cannot be applied by Wp systematically: this would lead to an exponential number of children sequents. Hence, they shall be triggered manually by the user (or by using dedicated strategies, as explained later).

As an example of a weakening tactic, let us present the Non-Overflow tactic, here on the special case where we want to remove the conversion from an arbitrary integer a to its 32-bit unsigned representation in expression to_uint32(a) from the goal ϕ:

$$\frac{\Delta \vdash 0 \leq a < 2^{32} \quad \Delta, \vdash \phi_{\text{to\_uint32}(a)/a}}{\Delta \vdash \phi}$$

Although this transformation is correct, maybe expression a cannot be proved to be in unsigned 32-bit range, in which case the first children sequent generated by the tactic is not provable, and the user has to backtrack.

Let us now illustrate how to use the TIP on the following example:

```
1  // logic integer l_occ() = ...
2  // lemma opposite: ...
3
4  /*@ requires \valid(&a[0..len-1]) ;
5   @ ensures \result == l_occ(v, a, 0, len) ; */
6  unsigned occ_not_so_easy(int v, int * a, unsigned len){
7    unsigned occ = 0 ;
8
9    /*@ loop invariant 0 <= len <= \at(len, Pre);
10       loop invariant occ == l_occ(v, a, len, \at(len, Pre)) ;
11       loop assigns len, occ ; */
12    while(len--){
13      if(a[len] == v) occ++;
14    }
15    return occ ;
16  }
```

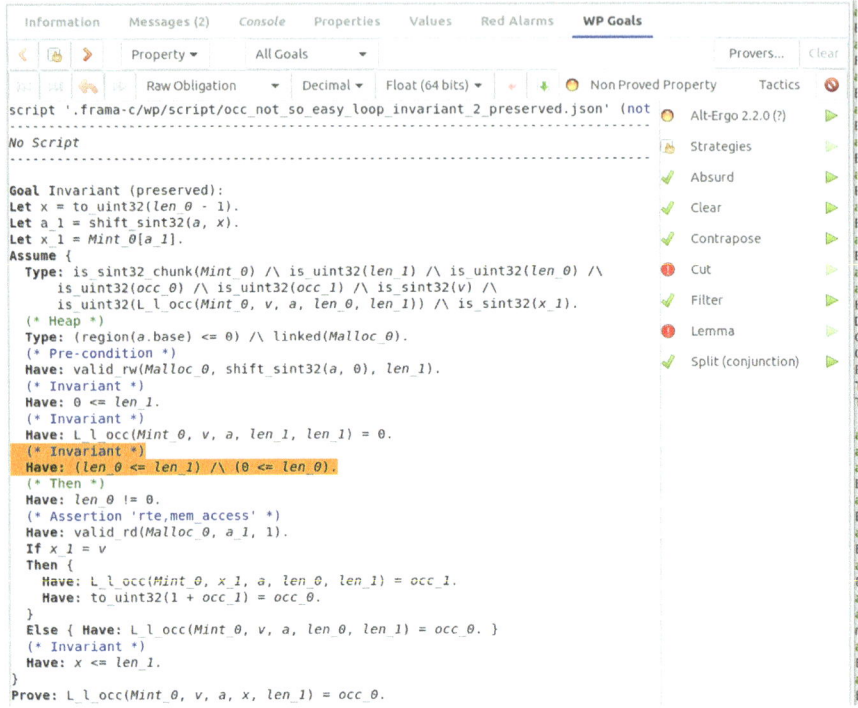

Fig. 4.3 The Wp TIP

This is basically the same example as in previous sections except that we use a type **unsigned** for the counting variable occ. Now, the loop invariant is a bit too weak to easily finish the proof.

From the Frama-C GUI, the TIP can be focused by double-clicking on a goal in the "WP goals" tab of the GUI. The interface then shows the proof obligation and the available tactics from the current selection in the goal. For example, if we run the previous example with -wp-rte, we fail to prove one of the invariants. This is illustrated in Fig. 4.3 where we can see the unproved verification condition and the tactics available after selecting the inductive hypothesis of the bounding invariant.

From this goal, let us now investigate the reasons of proof failure by applying tactics as illustrated in Fig. 4.4. First, we unfold the definition of l_occ. Then we split on the head condition of the goal, which leads to two children sequents. One branch of the condition is easily tackled by SMT solvers, while the other branch remains unproved. Now, we can split on the condition from the code. Here also, one path is easily discharged by SMT solvers. Remains the final residual path where it seems that the value of occ might potentially overflow. Applying the associated tactic to eliminate the conversion to unsigned 32-bit integer, it seems that we cannot indeed prove the absence of overflow, since there is no upper bound for occ available from the loop invariants.

4 Formally Verifying that a Program Does What It Should ...

1 Prove: L_l_occ(Mint_0, v, a, x, len_1) = occ_0.
  ✓ Definition ▷

2 Prove: (if (len_0 <= x) then 0
    else ((if (Mint_0[shift_sint32(a, x_2)] = v) then 1 else 0)
    + L_l_occ(Mint_0, v, a, x, x_2))) = occ_0.
  ✓ Split (<=) ▷

3
```
(* Assertion 'rte,mem_access' *)
Have: valid_rd(Malloc_0, a_1, 1).
If x_1 = v
Then {
  Have: L_l_occ(Mint_0, x_1, a, len_0, len_1) = occ_1.
  Have: to_uint32(1 + occ_1) = occ_0.
}
Else { Have: L_l_occ(Mint_0, v, a, len_0, len_1) = occ_0. }
(* Invariant *)
Have: x <= len_1.
```
  ✓ Split (branch) ▷

4 Prove: ((if (x_1 = Mint_0[shift_sint32(a, x_3)]) then 1 else 0)
    + L_l_occ(Mint_0, x_1, a, x, x_3)) = to_uint32(1 + x_2).
  ✓ Overflow ▷

Fig. 4.4 Investigation of a proof failure

By fixing the missing invariant (Cf. complete solution in Sect. 4.4.8), the proof is now automatically handled by SMT solvers and the TIP is no more needed.

When a proof fails, investing the goal by applying tactics as illustrated above is a good strategy to understand the reasons of the failure and fix the code or the missing annotations. However, sometimes, the proof is indeed beyond what SMT solvers are able to prove. For example, the opposite lemma in previous example can *not* be proved automatically by SMT solvers. In this case, a proof script generated by Wp can finish the job.

For example, Fig. 4.5 shows the generated goal for the opposite lemma: we want to reason inductively on integer e, using value b as base. Selecting expression e makes popups several tactics, including the "Induction" tactic. We can in turn configure the "Induction" tactic in order to use expression b as the base case. Applying the tactic results into two proof sequents that are easily tacked by SMT solvers, hence finishing the proof.

We can now ask Wp to save the script so that the proof can be reused later. To replay the proof from the Frama-C command-line, simply ask Wp to also use the prover script with option -wp-prover. During proof replay, Wp tries to slightly adapt the proof script in case the C code or ACSL annotations have been modified. There are other options to make Wp updating the scripts after replay, or not, depending on your needs. The interested reader is invited to consult the Wp reference manual where a complete workflow mixing interactive editing and batch replay is provided in details.

Finally, let us talk about "Strategies". Wp provides various heuristics that can try to automatically prove a goal by applying the various available tactics. External

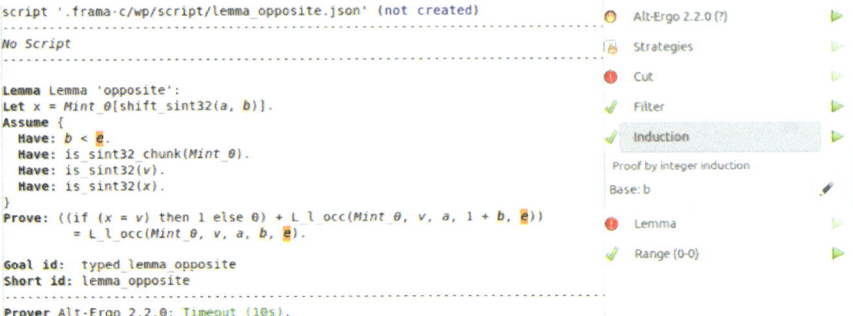

Fig. 4.5 Finishing a proof by using the induction tactic

Frama-C plug-ins can extend available strategies. Strategies are **OCaml** routines that can analyze proof sequents thanks to **Wp** programmatic *API* to infer candidate tactics instances to be explored. For example, the **Wp** "Auto Range" strategy looks for integer bounds and expressions to automatically apply the "Range" tactic. Similarly, **Wp** provides several strategies to automatically apply the many *bitwise*-oriented available tactics.

The user can combine several strategies when exploring proofs. The **Wp** proof search engine tries to apply the tactic candidates generated by each strategy, backtracking when necessary. Eventually the user might complete, cancel or adapt the generated script from the **TIP**. Notice that proof-search strategies can be also triggered from the **Frama-C** command-line by using the `tip` prover in combination with option `-wp-auto <strategy,...>`.

For even more advanced users, recall that user-defined strategies and tactics can be developed in **OCaml** as external **Frama-C** plug-ins, thanks to the **Wp** programmatic *API*. This *API* provides you with basic sequent transformers that you can safely combine with existing tactics to implement arbitrarily complex decision procedures. Hence, whatever your **OCaml** heuristic code infer, the only risk would be to generate non-provable proof sequents. This is a powerful way of extending **Wp** and **TIP** capabilities for specific domains.

4.6.5 Using External Proof Assistants

For most of the generated verification conditions, SMT solvers are sufficient for proving your **ACSL** properties. In rare cases, when **TIP** tactics cannot fulfill your needs, you might want to use full-featured proof assistant, typically the **Coq** proof assistant.

Interactive proof assistants supported by **Why3** shall be usable off-the-shelf with **Wp**. However, it is worth mentioning that it may require a significant amount of maintenance overhead on the part of the user: first, proofs are usually longer to write since they do not benefit from the SMT solvers capabilities; second, generated proof scripts are fragile with respect to even small modifications of the C source code and ACSL properties. TIP also has these problems but in a much more limited way since generated subgoals are verified using SMT solvers and because it embeds routines to reconcile scripts when the code sligthly changes.

Hence, one shall rely on interactive proof assistant only when necessary. A good practice would be to restrict their usage for ACSL lemmas and ghost lemma-functions (cf. Sect. 4.4.7), since they are generally quite stable over time and benefit from a limited proof context compared to verification conditions generated from annotated C code (especially with pointers).

To use Wp with interactive proof assistant, one shall define a Wp session from the Frama-C command-line with option `-wp-session <dir>`. The session directory is used by the TIP and by external proof assistant to store the proof scripts.

Then, proof assistants are simply activated by passing their name to option `-wp-prover`, preferably *after* SMT solvers in order to be activated only when necessary. Notice also that, by default, Wp only tries to replay existing scripts for the selected proof assistant(s). Hence, it is likely to fail on the first attempt, since there is no available script. To start *editing* proof scripts for external proof assistants, one shall use option `-wp-interactive <mode>` to choose which kind of interaction is needed. We recommend using the `fix` mode when debugging proofs, which generates a proof script if necessary, tries to check existing ones and finally launch the proof assistant editor in case of failure. When replaying a proof campaign, one shall simply use the default `batch` mode that submits `jobs` running the proof scripts without any user interaction.

Other modes are available, see `-wp-interactive` option help or consult the Wp reference manual for more details.

4.7 Related Work

In this chapter, we have described and illustrated how the Wp plug-in of Frama-C can be used to *prove* ACSL properties of C programs by Deductive Verification. For further reading and learning more proof techniques step-by-step, we strongly recommend the two following books[2]:

- "Introduction to C program proof with Frama-C and its Wp plugin" [11] is a comprehensive tutorial on the use of ACSL and Wp, it enters into the details of many aspects of the tool and gives some theoretical background on deductive verification. As already mentioned, it is a strongly recommended reading before going into the most technical parts of this chapter.
- "ACSL by Example" [29] is collection of algorithms inspired from the STL standard library of C++, adapted to C and proved with the Wp plug-in; it is extremely

---

[2] These resources are also available from the Wp pages of the Frama-C website: https://www.frama-c.com/fc-plugins/wp.html, https://www.frama-c.com/html/publications.html#wp.

useful for learning how to specify algorithms with ACSL and conducting the proofs using Wp. Although not actively maintained and probably up-to-date with recent Wp versions, it is still a reference handbook of ACSL specifications for array-based programs.

Deductive Verification of C programs is a large and active research domain with a huge amount of academic papers and practical tools. The article "Deductive Software Verification" [25] is an introductory survey with many entry points on historical foundations, scientific challenges and modern techniques available in this domain.

Briefly speaking, Wp uses weakest-precondition calculus and Hoare logic [30] with some kind of SSA translation [43] for variable assignments in absence of pointer aliasing, and a memory store *à la* Bornat [15] to handle memory assignments through pointers. To avoid the well known exponential growth of the generated formulas by such methods, Wp also uses many techniques inspired by efficient weakest-precondition calculus from Leino [35].

In this section, we would like to provide the reader with a fair overview of other tools than Wp, without any claim for completeness, and we apologize in advance for all the many valuable contributions to the domain that are not directly cited here. In the following presentation, we present different tools and platforms organized with respect to the underlying main techniques they implement, which is an arbitrary choice only guided by presentation purpose.

4.7.1 Hoare Logic

Let us start by some tools that generate Verification Conditions like Wp, but in a slightly different way. The Frama-C ancestor of Wp, an independent Frama-C plug-in named Jessie [26], unfortunately not maintained anymore, translates the C code and ACSL annotations into an equivalent, purely functional program in WhyML language for the Why platform, the ancestor of Why3. Jessie translates pointer aliasing into purely functional maps, using a memory model *à la* Bornat using field-based separation and implicit assumptions on function pointer parameters [32]. The generation of VCs reuses the weakest-precondition algorithm of Why, sent to SMT solvers. Compared to Wp, Jessie uses a more optimistic and simpler memory model than Wp at the price of few limitations on the source programs that can be modelled, especially for low-level system C code. Moreover, it does not have any simplification stage like the one offered by Qed.

The VCC verifier [21] from Microsoft is another tool that uses a similar strategy. VCC translates the C codes and specifications (written in a dedicated language similar to ACSL) into a purely logical langage and delegates the generation of VCs to the Boogie tool which uses the Z3 [39] SMT solver as a backend. VCC is quite powerful with a rich specification language and targets *parallel concurrent* C programs whereas Wp only targets single sequential programs. It supports a rich memory model with the possibility to define *frame predicates* that ease reasoning on complex data structures. However, it only works with the Z3 SMT solver.

Let us also mention here the KeY [2] platform and the GNATprove [18] tool that work similarly for annotated programs respectively written in Java and Spark2014. Worth to be mentioned, GNATprove re-uses the Why3 back-end for generating VCs and launching SMT solvers.

4.7.2 Separation Logic

Reasoning on C programs with pointers and complex data-structures, like linked-lists or trees is known to be difficult with classical Hoare program logics. Specifications are difficult to write *and* proof obligations are difficult to tackle. To overcome such difficulties, another category of tools relies on a more powerful logic named *Separation Logic*. This logic is powerful for describing concisely the memory layout of complex structures. However, proof obligations in Separation Logic cannot be handled by SMT solvers often required to decompose the proofs into tractable classical proof obligations.

In this category, let us mention RefinedC [44] and the VeriFast [33] tools. Both offer a similar workflow than Wp and VCC, but within a *Separation Logic* settings. RefinedC uses a type-directed approach to translate Separation Logic proof obligation to SMT solvers, whereas VeriFast relies more on user-directed proof script. More recently, the Gillian [38, 45] framework proposes an implementation of those techniques in a unified setting for several programming languages, notably C and JavaScript.

Another important framework based on separation logic is the VST-Floyd [17] tool, although it is based on deep embedding within a proof assistant, and we review it specifically in the next section.

These tools are ongoing research projects with promising capabilities that explore interesting and complementary trades-off between code and annotation languages restrictions, proof automation and proof scripting.

Despite its complexity, it shall be pointed out that Separation Logic is currently the only efficient way to reason on programs with complex memory separation or aliasing subtleties, typically linked lists, vectors, graphs, dynamic allocation, etc. On such class of programs, classical deductive verification with Hoare-based logic is doomed to fail in practice, including Frama-C/Wp.

4.7.3 Deep Embedding

All the aforementioned tools are based on some automated or semi-automated generation of Verification Conditions that are later discharged by external provers such as SMT solvers. Another category of tools proposes to perform deductive verification directly inside a proof assistant: both the semantic of the program and its desired properties are expressed in the (richer, higher-order) logic of the proof assistant.

This method is called *Deep Embedding*. It has the advantage of being non-restrictive on the program and its properties, since the developer benefits from all the capabilities of a proof-assistant. However, it suffers from a lack of automation: developers have to apply tactics and decompose the full proof by hand in every detail.

The L4·project [34] is an historical and well-known framework using deep embedding in the Isabelle/HOL proof assistant for developing and proving the correctness of a secured micro-kernel with a dedicated subset of C language. More recently, the VST-Floyd system goes one step beyond with a deep embedding of the C programming language into Separation Logic, which can be considered as the most expressive and fully consistent setting to deal with program logics. Moreover, the VST framework is also proved to be sound with respect to the semantics of the CompCert [37] compiler, hence providing strong guarantees that the *binary* code will also satisfy the properties that were proved at source code level. Interestingly enough, VST-Floyd provides a rich set of tactics that allows to automate many intermediate proof steps. This reduces the amount of interactive work to be done within the Coq proof assistant.

Indeed, Deep Embedding approaches are the most powerful and flexible solutions for deductive verification. However, the proof assistant context and the lack of full automation make those techniques reserved for skilled and experienced users. Like for any other deductive programming system, complex program and complex properties still require a huge amount of intermediate annotations, invariants and/or proof directives.

4.7.4 C-Extraction

Another radically different way of performing Deductive Software Verification is to *not* write the desired program directly in C, but within a more friendly, purely logical environment. Then, the program will be eventually *translated* into a C program which is (provably) equivalent to the original one. Such methods are often called *Code Extraction* and benefit from the high-level constructs of the host language which is not precluded by pointers and all the aliasing problems of C.

In this category, we shall mention the re-implementation of the GMP multi-precision library in Why3 [42] with final extraction to C code with comparable performances compared to the original library.

Probably the most advanced platform in this domain is F* [41], which is a purely functional language with composable Monads and a Dependent Higher-Order Logic with extraction to C. The F* system has been notably used to implement safe, efficient, provably robust and correct cryptographic libraries.

To complete the panorama related to C extraction, we shall also cite the B method [1] and its associated platforms, where the software is designed as abstract machines with invariants expressed in a classical first-order logic. Those abstract machines are then successively refined into more concrete machines using the same logic, until machines that are low-level enough to be directly converted to C code.

The common feature of these methods by extraction is that the finally extracted code is *correct by construction*. With the previously discussed techniques, like Wp, the C code is often already developed, without verification in mind, and verified *a posteriori*. On one hand, tools like Wp can be applied on existing projects, whereas extraction-based methods can only be applied from scratch on new projects. However, on the other hand, methods that are correct by construction benefit from a better design dedicated to verification from the very beginning.

4.7.5 *Model Checking*

Although Model Checking, Symbolic Evaluation, Deductive Verification can be regarded as separate research fields, they actually share a lot of common concepts and techniques and shall not be considered opposed to each others.

From a practical point of view, Model Checking is radically different from Deductive Verification *à la* Hoare: usually, Model Checking requires hardly any annotations from the user and focuses on implicit properties, such as absence of runtime errors or unreachability of code branches tagged as "Errors." Model Checking techniques are powerful in the sense that they *automatically infer* intermediate properties and necessary loop invariants to prove the desired properties.

Existing model checkers for C, typically CPAchecker [9] and its ancestor BLAST [8], are really efficient at automatically proving the kind of properties described above on large industrial code bases, with amazing capabilities to produce counterexamples on failure.

However, let us point out some intrinsic limitations of Model Checking techniques. They rely on automated inference of loop invariants that only exists for boolean and well-chosen arithmetic theories (roughly linear arithmetic with few extensions). To the best of our knowledge, inferring the necessary invariants for the kind properties we have proved with Wp in this chapter, or properties from "ACSL By Example" [29] for instance, remains far beyond the state-of-the-art Model Checking capabilities.

Still, improving Model Checking inference techniques and application domains is an active research field. As an illustration, memory shape analysis [22, 40] recently opened the route to more elaborate properties on complex data structures to be automatically proved.

Hence, automated Model Checking techniques and manual Deductive Verification shall be considered as complementary with each other. Practical methods have already been proposed [10] that, for instance, make Frama-C and CPAchecker interoperate effectively.

4.7.6 Choose Your Tool

As briefly and partially presented above, there are many available tools implementing various techniques for Deductive Software Verification of C programs. Each tool focuses on some specific application domain or some kind of C codes and each one has its own strength and weakness, including Wp of course.

4.8 Conclusion

In this chapter, we have presented the Wp plug-in of Frama-C. It is meant to verify that a program conforms to its specification using deductive verification techniques, and can also verify other global and local annotations making it able to also verify other properties like absence of runtime errors or global properties. We have seen how Wp generates the verification conditions associated to these properties.

Wp embeds different memory models that allows tweaking the analysis based on the knowledge we have about the verified code. It also provides additional tools like its internal simplifier Qed, its interactive proof tool TIP, or the smoke tests that allows scalability when verifying real world use cases.

As a final note, let us recall the Frama-C mantra as an open-source, extensible, *platform*: combining a large variety of analysis techniques will be always the key for successfully proving complex software. Hence, split your complex systems to be verified into smaller code units and use the good tool on each part and for each desired property!

Acknowledgements The authors would like to warmly cite the other main contributors of Wp, namely Anne Pacalet who was at the very origin of the project, Zaynah Dargaye who contributed to the early design of Wp memory models, and to thank all the Frama-C kernel developers for their support. The authors also thank the anonymous reviewers who provided numerous helpful comments and questions that allowed to greatly improve this chapter.

References

1. Abrial JR, Hoare A, Chapron P (1996) The B-Book: assigning programs to meanings. Cambridge University Press. https://doi.org/10.1017/CBO9780511624162
2. Ahrendt W, Beckert B, Bubel R, Hähnle R, Schmitt PH, Ulbrich M (eds) Deductive software verification—the KeY Book—from theory to practice. Lecture notes in computer science, vol 10001. Springer (2016). https://doi.org/10.1007/978-3-319-49812-6
3. Barbosa H, Barrett CW, Brain M, Kremer G, Lachnitt H, Mann M, Mohamed A, Mohamed M, Niemetz A, Nötzli A, Ozdemir A, Preiner M, Reynolds A, Sheng Y, Tinelli C, Zohar Y (2022) CVC5: a versatile and industrial-strength SMT solver. In: Fisman D, Rosu G (eds) Proceedings of the tools and algorithms for the construction and analysis of systems—28th international conference, TACAS 2022. Lecture notes in computer science, vol 13243. Springer (2022), pp 415–442. https://doi.org/10.1007/978-3-030-99524-9_24

4. Barrett CW, Conway CL, Deters M, Hadarean L, Jovanovic D, King T, Reynolds A, Tinelli C (2011) CVC4. In: Gopalakrishnan G, Qadeer S (eds) Proceedings of the computer aided verification—23rd international conference, CAV 2011. Lecture notes in computer science, vol 6806. Springer, pp 171–177. https://doi.org/10.1007/978-3-642-22110-1_14
5. Baudin P, Cuoq P, Filliâtre JC, Marché C, Monate B, Moy Y, Prevosto V (2020) ACSL: ANSI/ISO C specification language, version 1.16. https://frama-c.com/html/acsl.html
6. Baudin P, Pacalet A, Raguideau J, Schoen D, Williams N (2002) CAVEAT: a tool for software validation. In: Proceedings of the international conference on dependable systems and networks DSN 2002, p 537. Bethesda, USA (2002). https://doi.org/10.1109/DSN.2002.1028953
7. Bertot Y (2016) Coq in a hurry. https://cel.hal.science/inria-00001173. Lecture
8. Beyer D, Henzinger TA, Jhala R, Majumdar R (2007) The software model checker blast. Int J Softw Tools Technoly Trans 9(5–6):505–525. https://doi.org/10.1007/s10009-007-0044-z
9. Beyer D, Keremoglu ME (2011) CPAchecker: a tool for configurable software verification. In: Gopalakrishnan G, Qadeer S (eds) Proceedings of the computer aided verification—23rd international conference, CAV 2011. Lecture notes in computer science, vol 6806. Springer, pp 184–190. https://doi.org/10.1007/978-3-642-22110-1_16
10. Beyer D, Spiessl M, Umbricht S (2022) Cooperation between automatic and interactive software verifiers. In: Proceedings of the software engineering and formal methods—20th international conference, SEFM 2022. Lecture notes in computer science, vol 13550. Springer, pp 111–128. https://doi.org/10.1007/978-3-031-17108-6_7
11. Blanchard A (2020) Introduction to C program proof with Frama-C and its WP plugin. https://allan-blanchard.fr/frama-c-wp-tutorial.html
12. Blanchard A, Loulergue F, Kosmatov N (2019) Towards full proof automation in Frama-C using auto-active verification. In: Badger JM, Rozier KY (eds) Proceedings of the NASA formal methods—11th international symposium, NFM 2019. Lecture notes in computer science, vol 11460. Springer, pp 88–105. https://doi.org/10.1007/978-3-030-20652-9_6
13. Bloch J (2006) Nearly all binary searches and mergesorts are broken. https://ai.googleblog.com/2006/06/extra-extra-read-all-about-it-nearly.html
14. Bobot F, Filliâtre JC, Marché C, Paskevich A (2011) Why3: shepherd your herd of provers. In: Boogie 2011: first international workshop on intermediate verification languages. https://hal.inria.fr/hal-00790310
15. Bornat R (2000) Proving pointer programs in hoare logic. In: Proceedings of the mathematics of program construction—5th international conference, MPC 2000. Lecture notes in computer science, vol 1837. Springer, pp 102–126. https://doi.org/10.1007/10722010_8
16. Brahmi A, Delmas D, Essoussi MH, Randimbivololona F, Atki A, Marie T (2018) Formalise to automate: deployment of a safe and cost-efficient process for avionics software. In: Embedded real time software and systems—9th european congress, ERTS 2018
17. Cao Q, Beringer L, Gruetter S, Dodds J, Appel AW (2018) VST-Floyd: a separation logic tool to verify correctness of C programs. J Autom Reason 61(1–4):367–422. https://doi.org/10.1007/s10817-018-9457-5
18. Comar C, Kanig J, Moy Y (2012) Integrating formal program verification with testing. In: Embedded real time software and systems—6th European congress, ERTS 2012. Toulouse, France. https://hal.archives-ouvertes.fr/hal-02263435
19. Conchon S, Coquereau A, Iguernlala M, Mebsout A (2018) Alt-Ergo 2.2. In: SMT workshop: international workshop on satisfiability modulo theories. https://hal.inria.fr/hal-01960203
20. Correnson L (2014) Qed. Computing what remains to be proved. In: Proceedings of the NASA formal methods—6th international symposium, NFM 2014. Lecture notes in computer science, vol 8430. Springer, pp 215–229. https://doi.org/10.1007/978-3-319-06200-6_17
21. Dahlweid M, Moskal M, Santen T, Tobies S, Schulte W (2009) VCC: contract-based modular verification of concurrent C. In: 31st international conference on software engineering, ICSE 2009. Companion volume. IEEE. https://doi.org/10.1109/ICSE-COMPANION.2009.5071046
22. Dams D, Namjoshi KS (2003) Shape analysis through predicate abstraction and model checking. In: Proceedings of the verification, model checking, and abstract interpretation, 4th international conference, VMCAI 2003. Lecture notes in computer science, vol 2575. Springer, pp 310–324. https://doi.org/10.1007/3-540-36384-X_25

23. Djoudi A, Hána M, Kosmatov N (2021) Formal verification of a JavaCard virtual machine with Frama-C. In: Huisman M, Pasareanu CS, Zhan N (eds) Proceedings of the formal methods—24th international symposium, FM 2021. Lecture notes in computer science, vol 13047, pp 427–444. Springer. https://doi.org/10.1007/978-3-030-90870-6_23
24. Ebalard A, Mouy P, Benadjila R (2019) Journey to a RTE-free X.509 parser. Symposium sur la sécurité des technologies de l'information et des communications (SSTIC)
25. Filliâtre J (2011) Deductive software verification. Int J Softw Tools Technol Trans 13(5):397–403. https://doi.org/10.1007/s10009-011-0211-0
26. Filliâtre J, Marché C (2007) The Why/Krakatoa/Caduceus platform for deductive program verification. In: Damm W, Hermanns H (eds) Proceedings computer aided verification, 19th international conference, CAV 2007. Lecture notes in computer science, vol 4590. Springer, pp 173–177. https://doi.org/10.1007/978-3-540-73368-3_21
27. Gentzen G (1935) Untersuchungen über das logische Schließen I. Mathematische Zeitschrift 39:176–210
28. Gerlach J (2019) Minimal Contract Hoare-Style Verification versus Abstract Interpretation. Technical report, Fraunhofer FOKUS. VESSEDIA Project
29. Gerlach J (2020) ACSL by Example. Tutorial, Fraunhofer FOKUS. https://github.com/fraunhoferfokus/acsl-by-example
30. Hoare CAR (1969) An axiomatic basis for computer programming. Commun ACM 12(10):576–580. https://doi.org/10.1145/363235.363259
31. Hoenicke J, Leino KRM, Podelski A, Schäf M, Wies T (2009) It's doomed; we can prove it. In: Proceedings of the formal methods—2nd workd congress, FM 2009. Lecture notes in computer science, vol 5850. Springer, pp 338–353. https://doi.org/10.1007/978-3-642-05089-3_22
32. Hubert T, Marché C (2007) Separation analysis for weakest precondition-based verification. In: HAV 2007—heap analysis and verification, pp 81–93. https://hal.inria.fr/hal-03630177
33. Jacobs B, Smans J, Philippaerts P, Vogels F, Penninckx W, Piessens F (2011) VeriFast: a powerful, sound, predictable, fast verifier for C and Java. In: Bobaru MG, Havelund K, Holzmann GJ, Joshi R (eds) Proceedings of the NASA formal methods—3rd international symposium, NFM 2011. Lecture notes in computer science, vol 6617. Springer. https://doi.org/10.1007/978-3-642-20398-5_4
34. Klein G, Andronick J, Elphinstone K, Heiser G, Cock DA, Derrin P, Elkaduwe D, Engelhardt K, Kolanski R, Norrish M, Sewell T, Tuch H, Winwood S (2010) seL4: formal verification of an operating-system kernel. Commun ACM 53(6):107–115. https://doi.org/10.1145/1743546.1743574
35. Leino KRM (2005) Efficient weakest preconditions. Inf Proc Lett 93(6):281–288. https://doi.org/10.1016/j.ipl.2004.10.015
36. Leino KRM, Moskal M (2010) Usable auto-active verification. http://fm.csl.sri.com/UV10/
37. Leroy X (2009) Formal verification of a realistic compiler. Commun ACM 52(7):107–115. https://doi.org/10.1145/1538788.1538814
38. Maksimovic P, Ayoun S, Santos JF, Gardner P (2021) Gillian, Part II: real-world verification for JavaScript and C. In: Silva A, Leino KRM (eds) Proceedings of the computer aided verification—33rd international conference, CAV 2021. Lecture notes in computer science, vol 12760. Springer, pp 827–850. https://doi.org/10.1007/978-3-030-81688-9_38
39. de Moura L, Bjørner N (2008) Z3: an efficient SMT solver. In: Proceedings of the tools and algorithms for the construction and analysis of systems—14th international conference, TACAS 2008, vol 4963, pp 337–340. https://doi.org/10.1007/978-3-540-78800-3_24
40. Nicole O, Lemerre M, Rival X (2022) Lightweight shape analysis based on physical types. In: Finkbeiner B, Wies T (eds) Proceedings of the verification, model checking, and abstract interpretation—23rd international conference, VMCAI 2022. Lecture notes in computer science, vol 13182. Springer, pp 219–241. https://doi.org/10.1007/978-3-030-94583-1_11
41. Protzenko J, Zinzindohoué JK, Rastogi A, Ramananandro T, Wang P, Béguelin SZ, Delignat-Lavaud A, Hritcu C, Bhargavan K, Fournet C, Swamy N (2017) Verified low-level programming embedded in F*. Proceedings of the 22nd international conference on functional programming, ICFP 2017, vol 1, pp 17:1–17:29 (2017). https://doi.org/10.1145/3110261

42. Rieu-Helft R (2019) A Why3 proof of GMP algorithms. J Form Reas 12(1):53–97. https://doi.org/10.6092/issn.1972-5787/9730
43. Rosen BK, Wegman MN, Zadeck FK (1988) Global value numbers and redundant computations. In: Ferrante J, Mager P (eds) Conference record of the 15th annual ACM symposium on principles of programming languages, POPL 1988. ACM Press, pp 12–27. https://doi.org/10.1145/73560.73562
44. Sammler M, Lepigre R, Krebbers R, Memarian K, Dreyer D, Garg D (2021) RefinedC: automating the foundational verification of C code with refined ownership types. In: Freund SN, Yahav E (eds) Programming language design and implementation—42nd international conference, PLDI 2021. ACM, pp 158–174. https://doi.org/10.1145/3453483.3454036
45. Santos JF, Maksimovic P, Ayoun S, Gardner P (2020) Gillian, Part I: a multi-language platform for symbolic execution. In: Donaldson AF, Torlak E (eds) Proceedings of the programming language design and implementation—41st international conference, PLDI 2020. ACM, pp 927–942. https://doi.org/10.1145/3385412.3386014
46. Turing AM (1937) On computable numbers, with an application to the entscheidungsproblem. In: Proceedings of the London mathematical society, vol s2-42(1), pp 230–265. https://doi.org/10.1112/plms/s2-42.1.230
47. Volkov G, Mandrykin M, Efremov D (2018) Lemma functions for Frama-C: C programs as proofs. In: Proceedings of the 2018 Ivannikov ISPRAS open conference (ISPRAS-2018), pp 31–38. https://doi.org/10.1109/ISPRAS.2018.00012

Chapter 5
Runtime Annotation Checking with Frama-C: The E-ACSL Plug-in

Thibaut Benjamin and Julien Signoles

Abstract Runtime Annotation Checking (RAC) is a lightweight formal method consisting in checking code annotations written in the source code during the program execution. While static formal methods aim for guarantees that hold for any execution of the analyzed program, RAC only provides guarantees about the particular execution it monitors. This allows RAC-based tools to be used to check a wide range of properties with minimum intervention from the user. Frama-C can perform RAC on C programs with the plug-in E-ACSL. This chapter presents RAC through practical use with E-ACSL, shows advanced uses of E-ACSL leveraging the collaboration with other plug-ins, and sheds some light on the internals of E-ACSL and the technical difficulties of implementing RAC.

Keywords Runtime annotation checking · Inline monitoring · Dynamic analysis · Memory debugging

This chapter presents the E-ACSL plug-in of Frama-C. Together with Eva (see Chap. 3) and Wp (see Chap. 4), E-ACSL is one of the three main Frama-C plug-ins for verifying program properties. Contrary to those, E-ACSL does not verify properties statically, before executing the program, but it verifies them dynamically, at runtime, during concrete program executions. Therefore, it does not provide as strong guarantees as the two others, since it only checks properties for the executed runs, but not for all possible execution traces. However, it is much easier to use. In particular, it requires neither proof effort as Wp, nor complicated fine-tuning as Eva to complete the verification process. Furthermore, since dynamic verification analyzes concrete program executions, it requires no user knowledge about the runtime environment

---

T. Benjamin (✉) · J. Signoles
Université Paris-Saclay, CEA, List, Palaiseau, France
e-mail: tjb201@cam.ac.uk

J. Signoles
e-mail: Julien.Signoles@cea.fr

T. Benjamin
University of Cambridge, Cambridge, UK

(contrary to static approaches), even if it must be available in a concrete or virtual setting.

The underlying verification technique used by E-ACSL is called *Runtime Annotation Checking* [25] or *Runtime Assertion Checking* [11]. It is a lightweight formal method, which consists in verifying formal properties during program execution. As any formal method, it provides formal guarantees about its outputs. In particular, it must detect the invalid properties on the checked execution traces (*soundness*), while not modifying the functional behavior of the observed program (*transparency*) when the checked properties are all valid.

In the case of E-ACSL, the checked formal properties are a subset of ACSL annotations (see Chap. 1). However, in many use cases, E-ACSL takes advantage of Frama-C to *not* requiring writing annotations manually, but relying on other plug-ins to generate them automatically. Therefore, these cases are fully automatic: they make E-ACSL even easier to use in practice.

This chapter is organized as follows. Section 5.1 introduces runtime annotation checking. Then, Sect. 5.2 presents the subset of ACSL that can be verified at runtime. Next, Sect. 5.3 explains how to use E-ACSL in practice and what it provides. Therefore, it is probably the most important section for the working engineer. Afterwards, Sect. 5.4 provides some additional details regarding its uses. Section 5.5 gives some technical details about the underlying runtime verification techniques implemented in E-ACSL. Therefore, it targets computer scientists who would like to better understand how E-ACSL internally works.

5.1 Runtime Annotation Checking

Runtime Annotation Checking (RAC), also named Runtime Assertion Checking, is the discipline of verifying program annotations, in particular assertions, at runtime, i.e. when executing the code. It takes its roots in the late seventies as a simpler and more practical alternative than formal proof of correctness [11]. Since runtime checks were usually introduced through a dedicated *assert* language construct or a specific *assert* preprocessor macro, the wording *runtime assertion checking* became popular even if the expression *self-checking programs* was also used for a while [59]. Such runtime assertions were regularly introduced in programming languages and systems in the late seventies and during the eighties, notably through extensions of Fortran [54] and Ada [33]. They were eventually added to the C programming language through the macro `assert` and were popularized to support *defensive programming* by the programming language Eiffel [37]. Nowadays, they also support *offensive programming*.[1] Eiffel was also one of the first languages to implement Meyer's *design by contract* approach [38], which lifts assertions from statements to functions and larger components like objects and modules through the design of Behavioral Interface Specification Languages (BISL). In such languages, assertions

---

[1] https://en.wikipedia.org/wiki/Offensive_programming.

are a particular case of *annotations* expressing various code properties of interest. Since more than twenty years, RAC for BISL has been proposed for mainstream programming languages like OpenJML [12] for checking JML annotations [30] of Java code, Spec# [4] for C#, Spark2014 [36] for Ada and, more recently, Ortac [21] for OCaml. E-ACSL belongs to this line of works by checking ACSL annotations of C code.

From a research perspective, while several works have focused on the design of BISL (see Sect. 5.2), RAC as a formal verification techniques has received much less attention than the other formal code verification techniques, notably abstract interpretation [13], deductive verification [15] and model checking [18, 43]. For instance, the authors of Spec# stated that implementing RAC is straightforward in a short paragraph, while dedicating the rest of their paper to deductive verification [3]:

> The run-time checker is straightforward: each contract indicates some particular program points at which it must hold. A run-time assertion is generated for each, and any failure causes an exception to be thrown.

Indeed, the problem looks quite simple at a first glance. For instance, the ACSL assertion /*@ **assert** x == 0; */ can easily be translated to the C statement **assert**(x == 0)[2]; However, as we will explain in Sect. 5.5, the translation scheme is not always as easy, in particular when trying to be both *correct* and *efficient*, while translating complex properties. Here, being correct means that the runtime checker should not detect at runtime valid properties as invalid, and conversely. For instance, if we slightly modify the assertion to /*@ **assert** x+1 == 0; */, and assuming x of type **int**, it is not correct anymore to convert it to the C statement **assert**(x+1 == 0); because x+1 might overflow in C while it is computed over the (unbounded) set $\mathbb{Z}$ of mathematical integers in ACSL. In the general case, correctness requires to rely on an arbitrary precision arithmetic library, such as Gmp,[3] to translate the ACSL arithmetic operations, but it is not always efficient enough in practice if there are many such mathematical operations. Section 5.5.2 explains how E-ACSL solves this issue. In practice, regarding the previous example, E-ACSL generates code equivalent to **assert**(x+1 == 0L),[4] meaning that all the operations are computed over type **long** in order to be both efficient and correct.

Beyond E-ACSL, Cheon was the first researcher to consider the translation of formal annotations to executable code as a research problem, in the context of JML [10]. A few others [20, 27, 31, 41] target more specific research questions, described later in this chapter. Also, a quite important topic is the problem of checking memory properties such as **\valid**(_), which is the purpose of memory debuggers (e.g., AddressSanitizer [45]). However, since such properties are specific to languages in which memory management is delegated to programmers (typically, C), they are only relevant for E-ACSL among the existing RAC tools (see Sect. 5.5.3).

---

[2] The **assert** macro is part of assert.h, a file of the C standard library.
[3] http://gmplib.org.
[4] Assuming a standard architecture in which **sizeof(int)** < **sizeof(long)**.

More generally, RAC may be seen as a particular case of Runtime Verification (RV), a lightweight verification technique with an active research community that aims at verifying formal properties at runtime [23]. In particular, RAC tools generate monitors, which check properties during concrete runs (i.e., *online* monitors) and are directly embedded in the code under scrutiny (i.e. *inline* monitors) [19]. However, non-RAC RV tools usually focus more on temporal and liveness properties (properties regarding events emitted during an execution) [32] than BISL-like state properties (properties about data at a given program point, or set of program points). Therefore, many RV research results are not directly applicable to RAC, even if they are an important source of inspiration.

RAC tools are usually evaluated according to four main criteria:

- *Expressiveness*: The more formal properties a RAC tool is able to check, the better. When the tool is used to detect vulnerabilities in a program, the term *precision* is also often used to quantify the amount of detected vulnerabilities.
- *Efficiency*: Running a monitor on top of a program requires additional time and memory resources, that may depend on the size of the monitored program. For a monitored program to be usable in practice, it is critical to limit the time and memory overheads induced by the verification.
- *Transparency*: The instrumented program should perform the same operations as the original one. In particular, the presence of annotations should not interfere with the behavior of the original program, beyond interrupting the execution when detecting an invalid property.
- *Soundness*: The instrumented program should check the annotations accurately, i.e. it should interrupt the program exactly when an assertion is violated. Some related challenges are discussed in Sect. 5.5. Transparency and soundness together are sometimes referred to as *correctness*.

Transparency and soundness are formal properties: either the monitoring is transparent (resp. sound), or it is not. Conversely, expressiveness and efficiency are quantitative measures: they are useful to compare tools or different versions of the same tool. In general those two axes tend to work against each other since runtime verification tools often implement a trade-off that gives up expressiveness for efficiency (or conversely, sometimes).

5.2 An Executable Subset of ACSL

The ACSL specification language (see Chap. 1) is dedicated to state properties about C code. It is shared by all Frama-C plug-ins. However, it was designed with static analysis in mind and, more precisely, deductive verification, such as provided by plug-in Wp (Chap. 4), but E-ACSL operates differently as it checks the properties at runtime through RAC. It leads to some issues in this context. In particular, some ACSL properties are not suited for RAC. For this reason, the E-ACSL plug-in only relies on a subset of ACSL, also called E-ACSL, which stands for "Executable ACSL". This section provides a short presentation of the E-ACSL specification language. Its complete specification is available in its reference manual [50]. For the time being, the E-ACSL plug-in only supports a subset of this specification language [51], which becomes larger and larger from one version of Frama-C to the other.

5.2.1 Expressiveness of the E-ACSL Language

The E-ACSL specification language is a strict subset of the ACSL specification language. In particular, it does not contain the annotations that cannot be dynamically evaluated in finite time at runtime. It aims at being agnostic as to which tool uses it. The main restrictions are the following.

- Quantifications must be bounded: Checking a quantified predicate at runtime requires looping over the quantified domain. Thus quantifying over an infinite domain may result in a non-terminating loop. For this reason, the E-ACSL language only allows for quantification over finite domains, that is, types with finite number of elements (such as **int**), or *guarded* quantifications over integer types, where every quantified variable x is subject to a condition of the form t1 <= x < t2, where t1 and t2 are two (non-necessarily constant) terms. In practice quantifying over large types (like all the elements of **int**) is likely to induce a prohibitive overhead.
- Sets must be finite: As for quantifications, ranging over infinite sets yield non-terminating loops, hence only the finite sets are part of the E-ACSL language. In particular, any set defined by comprehension (i.e., a set of the form $\{x|P(x)\}$) must be bounded in a similar way to quantifications.
- No **terminates** clauses: When a function does not terminate, the program point stating its termination is unreachable. Therefore if this program point is reached, it implies that the function has already terminated and there is nothing more to check. Hence the E-ACSL specification language does not contain any **terminates** clause.
- \at constructs on C left values: The \at construct is part of the language as long as, for each call \at(t,id) and every C left-value x that contributes to the definition of a logic variable involved in a term t, the equality \at(x,id) == \at(x,Here) holds, i.e. the value of x must be equal between labels id and **Here**. Indeed, the value of a logic variable is always locally defined at program point **Here** and constant. If this value depends on the current value of a C variable, it cannot be computed at the program point id occurring earlier in the execution flow. Since checking equalities of C values is undecidable in the general case, it is left to tools to implement practical sound criteria (e.g., either fully automatic but partial, complete but with the help of the user, or based on runtime checks) to decide which \at constructs are correct.

Supported and unsupported references

In the following example, the first assertion is supported, while the second one is not since the guard defining the logic variable u depends on n whose value is modified between labels L1 and L2 : the value of n at L2 would be unknown when evaluating u at L1.

```
main(void) {
   int m = 2;
   int n = 7;;
   L1:
   n = 4;
   L2:
   /*@ assert
     @ \let k = m + 1;
     @ \exists integer u; 9 <= u < 21 &&
     @ \forall integer v; -5 < v <= (u < 15 ? u + 6 : k) ==>
     @    \at(n + u + v > 0, L1); */ ;

   /*@ assert
     @ \let k = m + 1;
     @ \exists integer u; n <= u < 21 && // [u] depends on
       [n]
     @ \forall integer v; -5 < v <= (u < 15 ? u + 6 : k) ==>
     @    \at(n + u + v > 0, L1); */ ;
   return 0
}
```

- Real numbers: Real numbers are part of the E-ACSL specification language, but most real numbers are not representable at runtime with infinite precision. For this reason, the E-ACSL language does specify that real number shall either be supported up to a certain precision, or raise undefinitive verdicts when necessary. It is left to the tool interpreting the E-ACSL language to choose a sound way of handling real numbers. At the time of writing, the E-ACSL plugin implements no support for real numbers beyond rational numbers.
- Inductive predicates and axiomatics: In the general case, they cannot be checked at runtime, but a subset can. However, characterizing precisely which inductive predicates and axiomatics are checkable at runtime is an open research problem. For this reason, the E-ACSL language contains both of them but does not specify a way to handle them: the choice of a heuristics to determine which of those are supported is left at the discretion of each tool. An experimental support for a small subset of them has been recently introduced in the E-ACSL plug-in.

5.2.2 Semantics of the E-ACSL Language

Besides the few ACSL constructs removed from the E-ACSL language, there are also some statements whose meaning is slightly different from ACSL in order to reflect the specifics of RAC.

- Undefinedness: in a dynamic approach, one may encounter undefined constructs, such as the mathematical term $\frac{1}{0}$. Translating and executing those constructs to C code usually leads to undefined behaviors, which breaks the guarantees provided by the verification of subsequent assertions. Consider for instance the predicate 1/0 == 1/0. In ACSL, this predicate is valid by reflexivity of equality. If it were not, the underlying logic would be inconsistent: proving invalid properties as being valid would become possible and even trivial in most cases. This property is critical in practice for theorem provers which are used by plug-in Wp (see Chap. 4). However, from a RAC perspective, this predicate cannot be evaluated without leading to undefined behaviors. This is issue was named the *undefinedness problem* by Cheon [10]. To solve it, the E-ACSL specification language relies on the strong validity principle [26], which has been proposed for RAC by Chalin [9]: only properties which are both valid in ACSL *and* defined are valid in E-ACSL. In practice, the E-ACSL plug-in adds guard conditions to prevent executing assertions containing undefined behaviors. For instance, for the assertion /*@ **assert** x/y == z; */, the guard /*@ **assert** y != 0; */ is generated, thus when y is 0, the guard assertion is violated and the program stops before running into an undefined behavior.
- Lazyness of operators: the E-ACSL specification language states that the logic operators such as && and ==> are lazy. This allows to reduce the number of undefined predicates. Indeed, in the predicate x != 0 && 1/x == 5 in the case where x is 0, the first operand evaluates to *False* and, by lazyness, the second member is never evaluated. In this case, even though this predicate contains the undefined term 1/x, it evaluates to *False* since it prevents the evaluation of that term. Without the lazyness, this predicate would be undefined when x is 0. Additionally, being lazy is computationally more efficient: the second operand is evaluated only when it is needed to compute the result.

Despite these differences between the languages ACSL and E-ACSL, they allow for a sound and effective collaboration between E-ACSL and the other Frama-C plug-ins. This property is stated more precisely as Conjecture 3.1 in [49], and although it is not yet proven at the time of writing, it has been observed in practical applications involving such a collaboration that the interoperability behaves as expected, for instance between E-ACSL and Eva [40] or E-ACSL and Wp [29, 44].

5.3 How to Use E-ACSL Plug-In

This section is dedicated to a presentation of the E-ACSL plug-in for practical purposes. It is intended for people who want to get started with using E-ACSL to verify programs.

5.3.1 What is *E-ACSL*, Basic Usage

E-ACSL is the RAC tool of Frama-C. It relies on its own specification language, also named E-ACSL and introduced in Sect. 5.2. At its core, the principle is fairly straightforward: it consists in verifying E-ACSL assertions during a program's execution, as if they were the **assert** macros of C.

Monitoring an annotated C program

Consider the following annotated C program, which contains an invalid access to an array, and annotations that verify the array accesses. For further references, this program is assumed to be in a file named example.c.

```
1  #include<stdio.h>
2  #include<stdlib.h>
3
4  int main () {
5    int *a, *b;
6    a = (int*) malloc (10 * sizeof(int));
7    b = (int*) malloc (3 * sizeof(int));
8    for(int i = 0; i <= 10; i++){
9      //@ assert(i < 10);
10     a[i] = i;
11   }
12   printf("Done!\n");
13   return 0;
14 }
```

Compiling this program as a normal C program (i.e. without monitoring with E-ACSL) using Gcc 12.2.1 and the option -fsanitize=undefined, and monitoring it with E-ACSL yield the following

Without monitoring	With monitoring
1 Done!	1 file.c: In function 'main' 2 file.c:9: Error: Assertion failed: 3 The failing predicate is: 4 i < 10. 5 With values at failure point: 6 - i: 10

Without monitoring, the program does not detect the invalid array access and performs an unsafe operation, then continues the execution (hence printing "Done!" in the end). With the monitoring, the assertion is checked at every step through the loop. The last one where i has value size makes it false. The monitor detects that the assertion is failed and reports the failure with the concrete values of i and size, and aborts the execution. The unsafe array access is never performed, nor the printing of the "Done!" message.

The rest of this section is dedicated to a presentation of the basic usage of E-ACSL allowing the user to go from the original program example.c all the way to the execution of the program with monitoring. Any subsequent reference to a file named example.c is assumed to be the above file.

E-ACSL Workflow.

The primary use of E-ACSL is not through the Frama-C graphical interface, but through the command line. The process is close to using a C compiler. It can be decomposed in three steps:

1. Run E-ACSL on the C file of interest to transform all the E-ACSL annotations into executable C code. During this step, a new C program is generated, where all the initial assertions are translated to C code.
2. Run a C compiler on the new program to get an instrumented executable. In particular, it requires to link the E-ACSL Runtime library (RTL) to the generated code, as well as the external libraries it depends on (typically, the Gmp library[5]).
3. Run the instrumented executable in order to detect and report invalid monitored ACSL properties as in the previous example in addition to the normal program behavior. By default, execution stops after the detection of the first invalid property.

---

[5] https://gmplib.org/.

The last step is just like running any other executable on the computer. The rest of this section focuses on the various ways to achieve the first two steps along with the options associated.

Compiling E-ACSL Annotations.

We describe step 1 of the above workflow. It consists in calling Frama-C with the E-ACSL plug-in on file example.c of the running example as follows:

```
$ frama-c -e-acsl example.c
```

This command opens example.c in Frama-C and then runs E-ACSL on the default project. E-ACSL creates a new Frama-C project containing the instrumented version of example.c. However this command does not specify what to do with this project, so it is eventually discarded, when Frama-C stops. Therefore, any practical use of E-ACSL requires to extend this command in order to specify what to do with the resulting project thanks to the Frama-C option -then-last (see Chap. 2).

Generation of a monitored program

The command below calls Frama-C and E-ACSL on file example.c, and stores the generated program into file example_monitored.c.

```
$ frama-c -e-acsl file.c \
    -then-last -print -ocode example_monitored.c
```

Compiling the Monitored Program.

We describe step 2 of the workflow, assuming that step 1 has been successfully performed to produce a file example_monitored.c containing the monitored program. Step 2 consists in compiling the file with a standard C compiler, such as Gcc:

```
$ gcc -c example_monitored.c
```

However, a program monitored by E-ACSL uses external libraries, such as the Gmp library, as well as its own RTL, in order to monitor the memory state of the executions, and the library to soundly perform assertion checking and abort the program execution. These libraries must be linked to example_monitored.c to produce executable code. Therefore, the required command line is quite tedious to write and is not explained here. Instead, E-ACSL provides a wrapper script called e-acsl-gcc.sh which performs the appropriate call to a C compiler (Gcc by default). Using this script to compile and link file example_monitored.c is highly recommended.

Compiling the generated code

The following command produces a new executable called example_monitored, obtained by compiling and linking the file example_monitored.c.

```
$ e-acsl-gcc.sh example_monitored.c -C -Oexample_monitored
```

When running this executable, it aborts and reports a failed assertion like in the running example.

Workflow Simplification.

The e-acsl-gcc.sh script provides several features to customize the process (see the E-ACSL user manual [52] or the e-acsl-gcc.sh man page for details). For instance, it allows the user to change the C compiler (e.g., Clang is also supported), provide flags to Frama-C, the C compiler or the linker, or provide paths to the Frama-C and Gcc executables. It also provides features to streamline more advanced uses of E-ACSL. One of them allows the user to perform steps 1 and 2 of the workflow at the same time.

Streamlining the workflow with e-acsl-gcc.sh

Running the following command on file example.c:

```
$ e-acsl-gcc.sh file.c -c -oexample_monitored.c -Oexample
```

produces three outputs: the file example_monitored.c, and two executables example and example.e-acsl. Executable example is uninstrumented: it is the very same as one obtained by direct compilation of example.c. Executable example.e-acsl is obtained by compiling and linking example_monitored.c with Gcc. It corresponds to the executable example_monitored of the previous example. Running both executables produce the results presented at the beginning of this section. Omitting the -O argument produces executables named a.out and a.out.e-acsl and omitting the -o argument produces a file named a.out.frama.c

5.3.2 Checking Undefined Behaviors

E-ACSL is commonly used in combination with another Frama-C plug-in called Rte. The purpose of this plugin is to generate ACSL annotations ensuring that this code will not hit an undefined behavior. These annotations can then be checked at runtime by E-ACSL. This particular use of E-ACSL is quite simple, since it does not require the user to write any formal annotations. The resulting program behaves the exact same way as the original program, except that it will abort when reaching an undefined behavior (supported by Rte). This is one of the most accessible usages of formal methods since it requires very little intervention from the user.

Detecting an incorrect array access

Consider the following program, which performs an invalid access to an array. The size of the array is stored in a variable, so that Gcc does not report the invalid access at compilation time.

```
int main()
{
   int size = 3;
   int p[size];
   for (int i = 0; i <= 3; i++)
      p[i] = 0;
   return 0;
}
```

Running the Rte analyzer and compiling the result with E-ACSL can be done with the following command, and executing the resulting executable yields the result:

```
$ e-acsl-gcc.sh -c -Omonitored-app --rte=all file.c
$ ./monitored-app.e-acsl
file.c: In function 'main'
file.c:6: Error: Assertion failed:
    The failing predicate is:
    rte: mem_access: \valid(p + i).
    With values:
    - rte: mem_access: \valid(p + i): 0
    - sizeof(int): 4
    - i: 3
    - p: 0x7fffbe6f0a00
Aborted
```

The instrumented program detects the incorrect access to the array and exits before this undefined behavior actually happens. It also provides useful information about the involved values. In particular, it shows that i is equal to 3, which should help the developer to debug their program.

Which undefined behavior should be tracked by Rte is customizable and documented in its user manual [24].

5.3.3 Checking Other Properties

This section explains how it is possible to check properties at runtime with E-ACSL, beyond manually-written annotations and undefined behaviors through the Rte plug-in. First, Sect. 5.3.3.1 explains how to automatically check requirements for some of the most important libc functions. Then, Sect. 5.3.3.2 presents how to detect format string vulnerabilities at runtime. Finally, Sect. 5.3.3.3 introduces how to verify high level properties beyond ACSL by combining E-ACSL to other Frama-C plug-ins.

5.3.3.1 Security-Critical Libc Functions

Many libc functions have requirements that any developer must take care of when writing C code. In particular, several of them are security-critical: violating them may introduce security vulnerabilities. For instance, the Heartbleed vulnerability[6] discovered in 2014 in OpenSSL is a buffer overflow that may occur because of an incorrect call to function memcpy provided in the libc's string library.

To circumvent this issue with E-ACSL, it is possible to use the Frama-C version of the C standard library (see Chap. 2) or to manually add the necessary preconditions on the relevant libc function. Preferably, the user can also use option -e-acsl-replace-libc-functions, named --libc-replacement if called through e-acsl-gcc.sh: these options generates efficient checks for several security-critical functions of the libc.[7]

Preventing memcpy's Buffer Overflow

Consider the following C program memcpy.c.[8] It takes a buffer size as argument and some string from the standard input and copies this string up to the buffer size in variable username. For doing so, it uses the libc's function memcpy.

```
1 #include <stdio.h>
2 #include <string.h>
3 #include <stdlib.h>
4 #include <unistd.h>
5
6 int main(int argc, char *argv[])
7 {
8   if (argc > 1) {
9     int i;
10    char username[8] = {0};
```

---

[6] https://en.wikipedia.org/wiki/Heartbleed.

[7] The E-ACSL user manual [52] provides the exhaustive list of supported functions.

[8] This example is freely inspired by a Stack Overflow's thread: https://security.stackexchange.com/questions/143625/problem-finding-a-vulnerability-in-memcpy.

```
11    char msg[2048] = {0};
12    /* size in hexadecimal */
13    long size = strtol(argv[1], NULL, 16);
14    /* the user types the command he wants */
15    i = read(STDIN_FILENO, msg, sizeof(msg)-1);
16    memcpy(username, msg, size);
17  }
18  return 0;
19 }
```

Unfortunately, this program is not correct when trying to use it with a large size and an empty input string.

```
1 $ ./a.out 0xffffffff
2
3 Segmentation fault
```

Fortunately, the incorrect call to memcpy can be properly detected if the program is monitored by E-ACSL as follows:

```
1 $ e-acsl-gcc.sh --libc-replacement -c memcpy.c
2 ./a.out.e-acsl 0xffffffff
3
4 memcpy: unallocated (or insufficient) space in source space
5 Aborted
```

5.3.3.2 Format String Vulnerabilities

E-ACSL allows the user to check that the arguments of `printf`- and `scanf`-like functions are consistent with the provided (non-necessarily constant) format string. Such inconsistencies are well-known security vulnerabilities in C code.

Since there is no way to specify the absence of such vulnerabilities with ACSL (and so with the E-ACSL specification language), this behavior is hardcoded in E-ACSL and only activated when the option `-e-acsl-validate-format-strings` is provided on the Frama-C command line.

This option is shortened to `--validate-format-strings` when called through e-acsl-gcc.sh.

Detecting Format String Vulnerabilities at Runtime

Consider the following program.

```
1 #include <stdio.h>
2
3 int main(void) {
4   printf ("is %d really an int?", "foo");
5   return 0;
```

```
6 }
```

E-ACSL detects that its call to printf is incorrect in the following way.

```
1 $ e-acsl-gcc.sh -c --validate-format-strings printf.c
2 $ ./a.out.e-acsl
3 printf: directive 1 ('%d') expects argument of type 'int'
      but
4 the corresponding argument has type 'char*'
5 Aborted
```

5.3.3.3 High-Level Properties

Similarly to combining E-ACSL with Rte in Sect. 5.3.2, E-ACSL can be combined with any Frama-C plug-in that generates ACSL annotations which are in the subset supported by E-ACSL in order to check advanced non-functional properties. Such plug-ins include Aoraï, MetAcsl, and RPP. Plug-in Aoraï allows the user to specify acceptable sequence of function calls. Plug-in MetAcsl allows the user to express system-level properties through an extension of ACSL named Hilare. Plug-in RPP is dedicated to so-called relational properties, which involve several functions or severals calls to the same function. These plug-ins are covered in depth in Chap. 10.

Correct Sequences of Function Calls

Consider the C program of Fig. 5.1a that initializes an array a, and searches some values inside a. Figure 5.1b describes an automaton in the Ya syntax that specifies the valid behaviours for this program: after having called function main (state a), it only accepts the programs that calls function init exactly once, then may call function find, possibly several times, before ending the execution. By combining Aoraï and E-ACSL, it is possible to check at runtime that this program in file aorai.c satisfies its specification aorai.ya as follows.

```
1 e-acsl-gcc.sh -F "-aorai-automata aorai.ya" \
2   --then-last -c aorai.c
```

This command first runs Aoraï on the given program and specification thanks to option -F that allows e-acsl-gcc.sh to give an extra argument to Frama-C. It generates a new project with the necessary code and ACSL annotations for verifying that the former is compliant with the latter. This project is given in turn to E-ACSL via the option --then-last, which is processed as usual via the option -c. The instrumented binary generated by E-ACSL checks the expected property. For instance, if the call to function

```
1  void init(int *a, int n)
2  {
3    for(int i = 0; i < n; i++)
4      *(a+i) = i;
5  }
6
7  int find(int *a, int n, int k)
8  {
9    for(int i = 0; i < n; i++)
10     if (a[i] == k) return i;
11   return -1;
12 }
13
14 int main()
15 {
16   int a[10];
17   init(a, 10);
18   find(a, 10, 5);
19 }
```

(a) Initialize then Search a Value in an Array.

```
1  %init: a;
2  %accept: ok;
3  %deterministic;
4  a:  { CALL(main) } -> b;
5  b:  { init() } -> c;
6  c:  { find()+ } -> ok;
7  ok: { RETURN(main) } -> ok;
```

(b) Specification through a Ya Automaton.

Fig. 5.1 A program and its specification for plug-in Aoraï

init at line 17 is omitted, E-ACSL detects an error at runtime because the modified program does not match the specification anymore, and aborts the execution:

```
1  $ ./a.out.e-acsl
2  : In function 'main'
3  :0: Error: Precondition failed:
4      The failing predicate is:
5      \false.
6  Aborted
```

5.4 Effective Usage of E-ACSL Plug-in

This section presents some specific features of E-ACSL, what guarantees it provides, how it compares with others tools and how to integrate it in software development life cycle. Advanced users shall refer to the E-ACSL user manual [52] for additional details.

typing	mathematical reals, beyond rational numbers
terms	labeled functions
predicates	unguarded quantifications over bounded types
	quantifications over pointers and enums
	comparisons of unions and structures
	labeled predicates
clauses	decreases clauses
	assigns clauses
	reads clauses
annotations	loop assigns
	lemmas

Fig. 5.2 Most important not-yet-implemented features

5.4.1 Fine-Tuning *E-ACSL*

Supported Features

The E-ACSL plug-in currently supports a large subset of its specification language presented in Sect. 5.2 [51]. At the time of writing, most commonly used features have been implemented. Figure 5.2 shows features that are still lacking support in the current implementation.

How to customize property checking

E-ACSL uses a custom function written in C and called `__e_acsl_assert` as a replacement for the C **assert** macro to check the assertions provided. Using this function not only allows E-ACSL to provide a rich feedback, it also gives control over the behavior of E-ACSL upon annotation checking. The script e-acsl-gcc.sh provides a few ways of customizing this behavior. They are introduced below and illustrated with the running example when necessary.

- Option `--fail-with-code=CODE` allows to change the behavior upon assertion failure from raising an ABORT signal to exiting the program with a return code CODE.

Exiting with exit code 125

This example shows how to exit with error code 125 in case of failure detected by E-ACSL.

```
1 $ e-acsl-gcc.sh example.c -c -Oexample --fail-with-code=125
2 $ ./example.e-acsl
3 example.c: In function 'main'
4 example.c:8: Error: Assertion failed:
5     The failing predicate is:
6     i < size.
```

```
 7      With values at failure point:
 8       - size: 10
 9       - i: 10
10 $ echo $?
11 125
```

- Option `--keep-going` provides a way to report assertion violations without interrupting the program. It provides a way to report several defects at once, but it should be used with care when the assertion would have prevented running an undefined behavior. The behavior of E-ACSL (and the behavior of the original program) is undefined as soon as the monitor code has undefined behavior. In particular, any subsequent check of ACSL annotations is not necessarily sound.

Continuing the execution after a failed assertion

Using option `--keep-going` on the running example allows the instrumented program to print its final message.

```
 1 $ e-acsl-gcc.sh example.c -c -Oexample --keep-going
 2 $ ./example.e-acsl
 3 example.c: In function 'main'
 4 example.c:8: Error: Assertion failed:
 5     The failing predicate is:
 6     i < size.
 7     With values at failure point:
 8      - size: 10
 9      - i: 10
10 Done!
```

- Option `--no-assert-print-data` implies that E-ACSL does not report the values of the involved variables at the point of failure. This option provides fewer feedbacks to the end-user, but the instrumented program is executed more efficiently.

Omitting the values at the point of failure

The following shows a few Shell commands along with their results, to make the example monitored program not print the data at the point of failure.

```
 1 $ e-acsl-gcc.sh example.c -c -Oexample
     --no-assert-print-data
 2 $ ./example.e-acsl
```

5 Runtime Annotation Checking with Frama-C: The E-ACSL Plug-in

```
3 example.c: In function 'main'
4 example.c:8: Error: Assertion failed:
5    The failing predicate is:
6       i < size.
7 Aborted
```

- Option `-external-assert=<filename>` allows the users to overload the function `__e_acsl_assert`, part of the E-ACSL RTL, and use their custom function to check annotations. The custom function must be declared in a C file, and respect the following signature

```
1 void __e_acsl_assert(int predicate, __e_acsl_assert_data_t *
     data)
```

More details about the `__e_acsl_assert_data_t` structure can be found in the E-ACSL User Manual [52].

Logging the failed assertion in a file

Consider the following file `custom_assert.c`, which defines an `__e_acsl_assert` function that does not interrupt the program when a failed assertion is met, but instead creates a log file `example.log` where the assertion is reported.

```
1  #include <stdio.h>
2  #include <e_acsl.h>
3
4  void __e_acsl_assert(int predicate, __e_acsl_assert_data_t
       * data) {
5    if (!predicate){
6      FILE *logfile;
7      logfile = fopen("example.log", "a");
8      fprintf (logfile,
9              "%s in file %s at line %d in function %s is
                 invalid.\n \
10   The failed predicate was: '%s'.\n",
11             data->kind,
12             data-> file ,
13             data-> line ,
14             data->fct,
15             data->pred_txt);
16     fclose (logfile);
17   }
18 }
```

One can require the monitored program to use this custom function and execute the result as follows:

```
1 $ e-acsl-gcc.sh basic.c -c -Oexample
      --external-assert=custom_assert.c
2 $ ./example.e-acsl
3 Done!
4 $ cat example.log
5 Assertion in file example.c at line 8 in function main is
     invalid.
6 The failed predicate was: 'i < size'.
```

Practical use-case of custom behavior upon error

Customizing assertion checking in E-ACSL can be used to integrate smoothly E-ACSL in larger projects. For instance one could write a function __e_acsl_**assert** that writes an output in specific a data format interoperable with other tools, such as xml, csv, json, yaml or sarif. Preventing E-ACSL from stopping the running program and performing a less invasive action at a point of failure is commonly done in embedded or server programs, for which one does not want the execution to stop but might instead be interested in logging the violated assertion. This has been done for instance in the implementation of the CURSOR method [40] in order to execute security counter-measures whenever weaknesses are detected.

5.4.2 Guarantees Offered by *E-ACSL* and Comparison with Other Tools

This section aims at giving an assessment of E-ACSL performance according to the four criteria presented in Sect. 5.1. First, E-ACSL is expected to be both transparent and sound. Any defect of soundness or transparency in E-ACSL is considered a bug. However, there is currently no formal proof that E-ACSL respect those two properties, even if some parts have been formalized for a subset of C and ACSL, such as monitoring of memory properties [34, 35, 57, 58] and checking arithmetic properties [7, 28]. Its implementation has been tested on home-made test cases, open source benchmarks and industrial case studies.

At the time of writing, E-ACSL is the only RAC tool for C program, but it is not the only tool that checks properties on C code at runtime (referred to as *dynamic tools*). Typically, another kind of dynamic tools are memory debuggers, which aim at finding memory vulnerabilities. Contrary to RAC tools, memory debuggers do not require the user to write any formal annotation. These tools detect frequent vulnerabilities in C code, such as incorrect memory accesses or incorrect array indexes. Importantly, they are unsound: they can choose to not detect some (infrequent) incorrect memory accesses because it would be too costly. E-ACSL may be used as a sound memory debugger, when used in combination with the Rte plugin, as explained in Sect. 5.3.2. Memory debuggers are the only available tools with which a comparison with

E-ACSL makes sense for assessing expressiveness and efficiency. The evaluations presented here are part of benchmarks run between 2016 and 2018 on memory debuggers including E-ACSL for finding memory vulnerabilities [56, 58]. They only evaluate a particular use of E-ACSL, whose scope is wider.[9]

Evaluating E-ACSL Expressiveness.

The detection power of E-ACSL with respect to standard security defects of C programs has been evaluated [56] on the NIST's SARD-100 test suite[10] and the Toyota ITC benchmark [48].[11] The results are respectively presented in Figs. 5.3 and 5.4. For these security defects, E-ACSL is compared to several Google's sanitizers—namely AddressSanitizer [45] (also known as ASan), MemorySanitizer [53], ThreadSanitizer [46] and UndefinedBehaviorSanitizer[12] (also known as UBSan)— seen as a single tool and Runtime Verification Inc.'s RV-Match [22], an automatic debuger based on the $\mathbb{K}$ framework [16, 17] whose objective is to find as many C undefined behaviors as possible. Our comparison does not include other tools whose detection capabilities on these benchmarks are lower.

Overall, these tests show that some security defects (e.g. SQL injection) are out of the scope of these tools and not detected. Currently, E-ACSL detects no concurrency defects, even if an experimental support for concurrency code has been recently implemented. Additionnally, E-ACSL does not try to detect incorrect function pointers (notably contained in Toyota ITC's "pointer-related" category), stack overflows (notably contained in Toyota ITC's "stack-related" category) and overflows through bitfield values (notably contained in the Toyota ITC's "numeric" category). Beside those limitations, E-ACSL performs better than Google's sanitizers (all included). It can be explained because E-ACSL extensively detects *block-level properties*, which are properties related to the bounds of memory blocks, while Google's sanitizers only implement ad-hoc strategies, typically for detecting off-by-one (but not off-bytwo). RV-Match performs quite accurately as well. Interestingly, all the tools are complementary: each tool is able to detect at least one defect that the other tools cannot detect, because of limitations (e.g., no detection of some block-level properties) and different focuses (e.g., no concurrency check).

Evaluating E-ACSL Efficiency On Memory Properties.

E-ACSL has several mechanisms to reduce the runtime overhead induced by monitoring, both in terms of time and memory usage. The two most notable sources of overhead are the use of arbitrary precision arithmetic and the monitoring of the memory properties. For the former, a custom type system is implemented in E-ACSL to determine whether an arithmetic property can safely use machine integers. This type system is presented in Sect. 5.5.2. No other tool uses such a mechanism for RAC on

---

[9] The definition of the CWEs can be found at https://cwe.mitre.org/.
[10] https://samate.nist.gov/SRD/view.php?tsID=100.
[11] https://github.com/Toyota-ITC-SSD/Software-Analysis-Benchmark.
[12] https://clang.llvm.org/docs/UndefinedBehaviorSanitizer.html.

Weakness[12]	E-ACSL	Google's Sanitizers	RV-Match
Non-memory Defects			
CWE-078: Command Injection	– (0/6)	– (0/6)	– (0/6)
CWE-080: Basic XSS	– (0/5)	– (0/5)	– (0/5)
CWE-089: SQL Injection	– (0/4)	– (0/4)	– (0/4)
CWE-099: Resource Injection	– (0/4)	– (0/4)	– (0/4)
CWE-259: Hard-coded Password	– (0/5)	– (0/5)	– (0/5)
CWE-489: Leftover Debug Code	– (0/1)	– (0/1)	– (0/1)
Memory Defects			
CWE-121: Stack Buffer Overflow	✓(11/11)	91% (10/11)	91% (10/11)
CWE-122: Heap Buffer Overflow	✓(6/6)	✓(6/6)	✓(6/6)
CWE-416: Use After Free	✓(9/9)	✓(9/9)	✓(9/9)
CWE-244: Heap Inspection	– (0/1)	– (0/1)	– (0/1)
CWE-401: Memory Leak	✓(5/5)	80% (4/5)	60% (3/5)
CWE-468: Pointer Scaling	50% (1/2)	50% (1/2)	50% (1/2)
CWE-476: Null Dereference	✓(7/7)	✓(7/7)	✓(7/7)
CWE-457: Uninitialized Variable	✓(4/4)	75% (3/4)	✓(4/4)
CWE-415: Double Free	✓(6/6)	✓(6/6)	67% (4/6)
CWE-134: Format String	✓(8/8)	– (0/8)	– (0/8)
CWE-170: String Termination	✓(5/5)	✓(5/5)	✓(5/5)
CWE-251: String Management	✓(5/5)	✓(5/5)	✓(5/5)
CWE-391: Unchecked Error	– (0/1)	– (0/1)	– (0/1)
Concurrency Defects			
CWE-367: Race Condition	– (0/4)	– (0/4)	– (0/4)
CWE-412: Unrestricted Lock	– (0/1)	– (0/1)	– (0/1)
Overall	67% (67/100)	56% (56/100)	54% (54/100)

Fig. 5.3 Detection results of E-ACSL, Google's sanitizers and RV-Match over SARD-100 test suite

Defect Type	E-ACSL	Google's Sanitizers	RV-Match
Dynamic Memory	94% (81/86)	78% (67/86)	94% (81/86)
Static Memory	✓(67/67)	96% (64/67)	✓(67/67)
Pointer-related	56% (47/84)	32% (27/84)	99% (83/84)
Stack-related	35% (7/20)	70% (14/20)	✓(20/20)
Resource	99% (95/96)	60% (58/96)	98% (94/96)
Numeric	93% (100/108)	59% (64/108)	98% (106/108)
Miscellaneous	94% (33/35)	49% (17/35)	71% (25/35)
Inappropriate Code	– (0/64)	– (0/64)	– (0/64)
Concurrency	– (0/44)	73% (32/44)	66% (29/44)
Overall	71% (430/604)	57% (343/604)	84% (505/604)

Fig. 5.4 Detection results of E-ACSL, Google's sanitizers and RV-Match over Toyota ITC benchmark

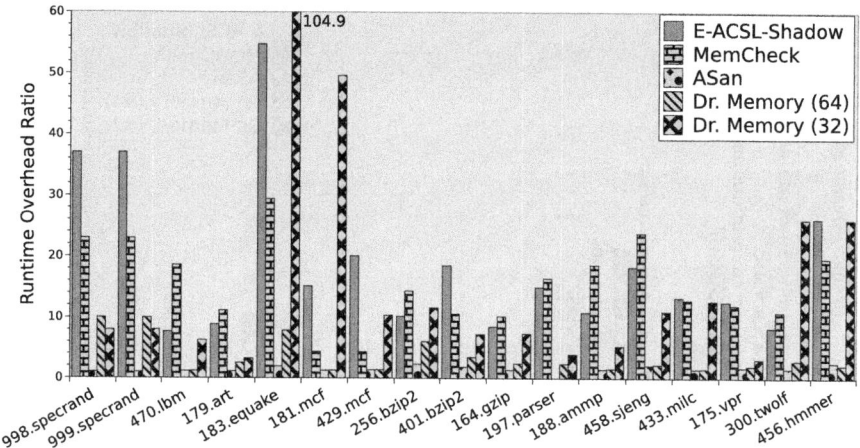

Fig. 5.5 Runtime overhead of E-ACSL, AddressSanitizer, MemCheck and Dr. Memory on SPEC CPU programs

C programs, hence it is not possible to assess its efficiency against alternatives, even if it has been shown efficient enough for some concrete use-cases [2, 40, 58].

For the monitoring of memory properties, E-ACSL uses a *memory model*, which is a dedicated memory space where E-ACSL registers the status of various memory spaces available to the program. More details on E-ACSL memory model are presented in Sect. 5.5.3. Figure 5.5 shows, on the vertical axis, the time overhead of E-ACSL against other tools that monitor memory properties, and Fig. 5.6 presents the memory overheads of the same cases. The numbers were obtained on the SPEC CPU benchmarks[13] containing 17 C programs (horizontal axis) ranging from 74 to 36,037 lines of code. For each program, our evaluation compares E-ACSL to efficient memory debuggers, namely AddressSanitizer [45], MemCheck [47] embedded in Valgrind [39], and Dr. Memory [8]. In particular, it includes neither all the Google's sanitizers (but only ASan) since we focus here on efficiency, nor RV-Match that does not scale (e.g., it overflows or does not provide any result for the largest test cases). In this evaluation, E-ACSL was used in conjunction with the Rte plugin of Frama-C in order to generate the assertions to be checked.

E-ACSL has a runtime overhead averaging at 19 times, and a memory overhead averaging 2.48 times the normal execution without instrumentation. It represents a slightly lower memory overhead, but a higher runtime overhead than the average of the other tools. However, the different tools do not have exactly the same detection capabilities on this benchmark, as summarized in Fig. 5.7. Since E-ACSL is not just a memory debugger, it supports the monitoring of non-memory properties, such as arithmetic properties, which are not in the scope of the other memory debuggers. Read-only properties check that read-only memory locations (typically, string literals) are never written to. They are also not checked by the other memory

---

[13] https://www.spec.org/cpu/.

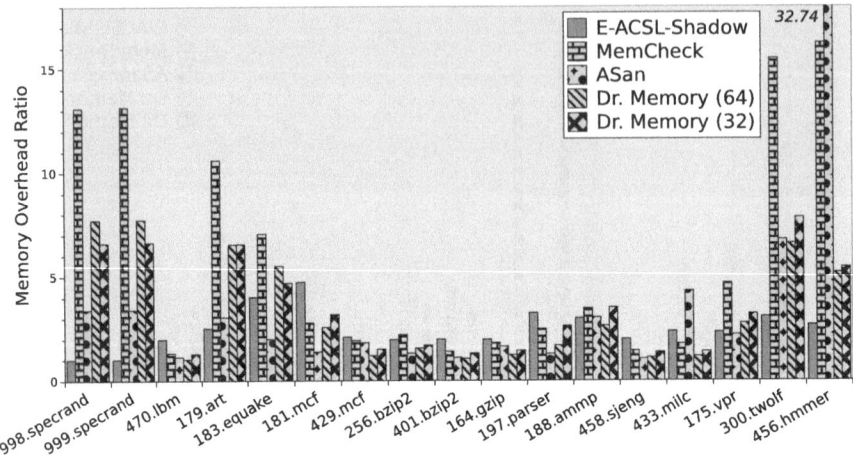

Fig. 5.6 Memory overhead of E-ACSL, AddressSanitizer, MemCheck and Dr. Memory on SPEC CPU programs

Properties \ Tool	E-ACSL	AddressSanitizer	MemCheck	Dr. Memory 32/64	
Heap Tracking	+	+	+	+	+
Stack Tracking	+	+	−	+	+
Global Tracking	+	+	−	+	+
Allocation	+	+	+	+	+
Initialization	±	−	+	+	−
Pointer Initialization	+	−	−	−	−
Bounds Check	+	±	±	±	±
Arithmetic Overflow	+	−	−	−	−
Read-only	+	−	−	−	−
Block Properties	+	−	−	−	−
Heap Leak	±	+	+	+	+

Fig. 5.7 Properties tracked during experimentation

debuggers. As already explained, E-ACSL also detects block-level properties. More details about this evaluation are provided in [58].

To sum-up, considering that E-ACSL is generally more precise than the other tools, as it is able to find more vulnerabilities and supports the monitoring of more properties, it is overall a competitive tool, with a focus on precision at the cost of a slightly more important overhead.

5.4.3 How to Integrate **E-ACSL** in Software Development Life Cycle

E-ACSL can be used during development, during a validation and verification (V&V) step, or when the software has been deployed and is being used. When used during development, it usually integrates to continuous integration (CI) in order to validate that no regression has been introduced. When used during a V&V campaign, it can be part of test campaigns. When used during operations, it is usually associated to defensive or recovery mechanisms that are activated when E-ACSL detects invalid properties.

Using E-ACSL in CI or a V&V campaign requires to have a test suite executable on a concrete or virtual environment. However, E-ACSL is agnostic with respect to such a test suite: it can be manually written and used for non-regression, unit or integration testing, or automatically generated (e.g., through mutation testing or fuzzing [60]). In such a context E-ACSL is often combined with Rte to automatically detect undefined behaviors in the test cases: many undefined behaviors are detected this way that would remain undectable without E-ACSL and Rte because unobservable. For instance, the assertion in the first example of Sect. 5.3.1 can be automatically generated by Rte and so detected invalid by E-ACSL, while the out-of-bound access to the array is undetectable without monitoring.

If the test cases are manually written, it is also possible to add ACSL annotations (e.g., assertions or function's requirements) by hand for detecting functional issues. In critical contexts in which formal proof with Wp (see Chap. 4) is also a cost-effective option, E-ACSL and Wp can be used together to prove some properties and check the others at runtime: such a combination lowers the verification cost. It is up to the user to choose the best trade-off between tests and proofs, even if some methodologies can be applied to decide whether the combined verification campaign is extensive enough according to some coverage criteria [29].

Using E-ACSL during operations allows the system to apply defensive, mitigation or recovery mechanisms when an issue is detected at runtime. For instance, Dassault Aviation experimented applying security counter-measures in such a case [40]. This usage relies on the customization capability of the `__e_acsl_assert` function, as explained in Sect. 5.4.1. It also fully depends on the operational runtime constraints of the system: for embedded systems, the runtime and/or memory overheads of E-ACSL is not acceptable for operational uses. However, combining Eva (see Chap. 3) with E-ACSL significantly lowers this overhead, as reported for instance by Dassault Aviation [40], since only the alarms remaining after having run Eva are monitored at runtime. Applying security-oriented measures when using E-ACSL can also be used in combination with fuzzing during pen-testing campaigns.

We do not detail here how to do such methodological integrations practically (e.g., how to integrate E-ACSL into an existing build system): it is not E-ACSL-specific but shared by any (combination of) Frama-C plug-in(s), and so it is covered by Chap. 2 that explains in particular how to drive Frama-C for dealing with complex C code.

5.5 Design and Technical Solutions for E-ACSL

This section presents some design choices and technical details about how E-ACSL works in order to check the formal properties of interest at runtime. First, Sect. 5.5.1 briefly introduces its internal software architecture. Then, it presents how E-ACSL deals with two important research questions regarding RAC tools for the C language: Sect. 5.5.2 explains on how E-ACSL checks arithmetic properties efficiently, and Sect. 5.5.3 focuses on memory property checking. They are the two topics which most research works on E-ACSL have been invested in.

Yet, they are not the only interesting research questions about RAC. Specifically, Lehner [31] focused on the problem of translating JML's **assignable** clauses, which are similar (yet slightly different) from ACSL's **assigns** clauses (see Chap. 1 for details), while Kosiuczenko [27], Petiot et al. [41] and, more recently, Filliâtre and Pasciutto [20] addressed the translation of the **\old(_)** keyword, which is a particular case of **\at**. Currently, the E-ACSL tool does not support **assigns** clause, while the construct **\at** is only partially supported. Fully supporting both constructs is future work.

5.5.1 Software Architecture

From a technical point of view, E-ACSL is a compiler that takes as input C source code with ACSL annotations and outputs new C source code in which the necessary pieces of C code have been introduced in order to monitor the logical annotations.

E-ACSL delegates to the Frama-C kernel the first steps of the compilation process that parses, type-checks, normalizes, and links the source files in order to generate an Abstract Syntax Tree (AST) in the internal representation of Frama-C, as well as the very last step that pretty prints—or runs any other Frama-C analysis on—the generated AST (see Chap. 2 for details about these steps). E-ACSL itself focuses on translating the input AST a into a new one that inserts the generated monitor into a. Figure 5.8 presents this part of the compilation process.

Globally, it contains three main stages. The first one translates the input AST to a new one, the E-ACSL's internal AST, that is ready for integrating the monitor. First, this stage prepares the source (box "AST preparation") code to enforce a few invariants that ease the monitor generation. For example, after this step, each labeled statement is a block of code in which new statements can easily be injected. Second, it integrates the .h header files of the E-ACSL RTL (box "RTL headers") to the input AST (box "RTL linking"), so that its type and function declarations can be used by the generated monitor. For example, these files include the declaration of the function **__e_acsl_assert** that is called to generate the verdict when evaluating a logical property.

Once the internal AST on which E-ACSL really works is generated, the second stage consists in performing several static analysis steps before generating the

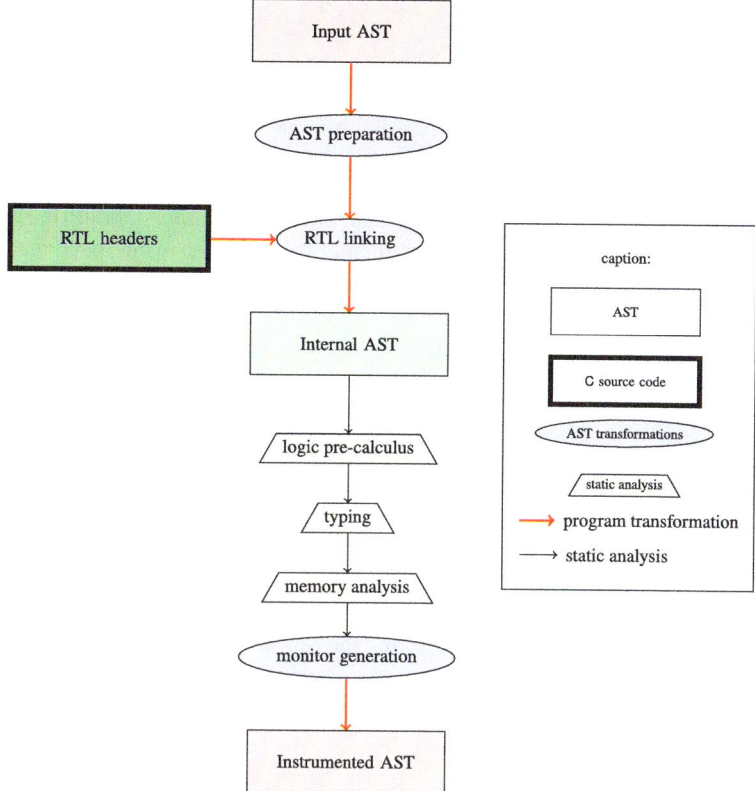

Fig. 5.8 E-ACSL architectural scheme

monitor. As a whole, this stage helps to optimize the generated code for the monitored properties, while remaining sound. First, it pre-computes several pieces of information that help the other analysis steps and the monitor generation (box "logic pre-calculus"). For example, it computes the lower and upper bound terms of each quantified variables. Second, **E-ACSL** applies a dedicated type system that allows it to generate sound efficient code for arithmetic properties, as explained in Sect. 5.5.2 (box "typing"). Third, **E-ACSL** applies a dedicated memory dataflow analysis that allows it to optimize the monitor dedicated to memory properties, as explained in Sect. 5.5.3 (box "memory analysis").

The last stage consists in generating the monitor by injecting the necessary pieces of C code into the AST (box "monitor generation"). It relies on the pre-established invariants and the static analysis steps to generate sound and efficient monitors for each logical property.

5.5.2 Arithmetic Properties

Checking arithmetic properties both soundly and efficiently is not as straightforward as it may seem. A direct translation is unsound in general since, in C programs, an operation on machine integers may overflow. When the integers are signed, this leads to an undefined behavior which makes the translation unsound, and when they are unsigned, the behavior is defined and correspond to a computation modulo the bound, which does not translate accurately the mathematical operation in the specification. In both cases, the translation becomes unsound when an arithmetic operation overflows.

Generation of an unsound monitor

For the following input program

```
int main() {
  short x = 168;
  long y;
  scanf("%ld", y);
  //@ assert x/(y+1) <= 200;
  return 0;
}
```

a naive strategy consists in generating the following monitor, by translating the E-ACSL primitive **assert** into a usual **assert** in C

```
int main() {
  short x = 168;
  long y;
  scanf("%ld", y);
  assert (x/(y+1) <= 200);
  return 0;
}
```

But this translation is unsound, since the computation y+1 may overflow if the user enters the maximal number representable with a **long**.

Using the Gmp Library

E-ACSL solves this problem by using the Gmp library that provides facilities to operate over integer and rational numbers of arbitrary size. The semantics of the integers defined by the library Gmp is that of mathematical integers, so they can be used to translate annotations. However, the generated code becomes an order of magnitude more complex: all the variables of the type mpz (the type of integers provided by the Gmp library) have to be allocated and initialized with the appropriate mpz_init_* functions and freed with mpz_clear, while each intermediate step of computation needs to be stored in an intermediate variable. This additional complexity comes at a runtime cost since the time and the memory overhead is much higher than using machine integers. Therefore, although using the Gmp library systematically allows

5 Runtime Annotation Checking with Frama-C: The E-ACSL Plug-in

$$\frac{}{\vdash cst :: [cst, cst]} \qquad \frac{x : \tau}{\vdash x :: [\min \tau, \max \tau]}$$

$$\frac{\vdash t :: [a, b] \quad \vdash u :: [c, d] \quad \diamond \neq / \text{ or } 0 \notin [c, d]}{\vdash t \diamond u :: [\min\{a \diamond c, a \diamond d, b \diamond c, b \diamond d\}, \max\{a \diamond b, a \diamond c, b \diamond c, b \diamond d\}]}$$

$$\frac{\vdash t :: [a, b] \quad \vdash u :: [c, d] \quad 0 \in [c, d]}{\vdash t/u :: [-\max\{|a|, |b|\}, \max\{|a|, |b|\}]}$$

Fig. 5.9 A few rules for interval inference

E-ACSL to be correct, it may prevent scalability because of prohibitive overheads in practical applications.

Using the Gmp Library Only When Necessary

In practice, most ACSL properties actually involve integers small enough to be stored and computed over machine integers. Therefore, reducing the overhead induced by the Gmp library can be achieved by restricting the calls to this library to the necessary cases. For instance, relying on Gmp for computing x + 1 when x is a **char** is a waste of valuable computational resources. E-ACSL conciliates soundness and efficiency by using a dedicated type system that discriminates an integer type in which the computation is sound among the machine integers and the Gmp integers and rational numbers. This typing step performed prior to code generation relies on interval arithmetic. The rest of this section is dedicated to a brief presentation of this mechanism, described in more details in [28].

The aim of this type system is to restrict computations to machine integers when it is sound to do so. Since soundness is more critical than efficiency, the analysis computes an over-approximation and may assign a larger type than necessary. Still this analysis drastically reduces the overhead induced by using the Gmp library, and makes it a usable tool for industrial-scale applications.

Interval Inference

The mechanism for discriminating when to call the Gmp library is separated in two distinct steps, which are performed during the phase labeled "typing" in Fig. 5.8. The first step determines, for every term of the program, an over-approximating interval containing all the values the term may evaluate to. The second step assigns a C type to a term according to the interval computed during the first step. Figure 5.9 presents a few rules of the interval arithmetics that is implemented in E-ACSL in order to compute an interval in which the term ranges.

The first two rules infers the interval of a constant or a C variable. The former is a singleton containing its value, while the latter is the interval containing all the values that fit in the the type τ of the variable. The next rule describes the interval of an arithmetic operation, which is the convex envelope of the operation on the bounds on the interval. This only works when the operations are monotonous, and thus it does not apply for the division (and the modulo, which is not shown here), if the denominator ranges in an interval containing 0. In this case the infered interval is unbounded as specified by the last rule: Since the system is designed to work with integers and there is a separate mechanism to handle the division by 0, this rule assumes that the denominator is not 0, and thus belongs to the set $[c, -1] \cup [1, d]$.

The rules presented in Fig. 5.9 are simplified. In the actual implementation, the inference happens in an environment that records the range of the bound logic variables. The typical constructs that introduce such bindings are quantifiers and calls to logic functions and predicates. The interval inference in those cases is out of the scope of this book, but has been detailed in related papers [6, 28].

Derivation of the interval inference of a term

Continuing the example presented at the beginning of this section, the following shows the derivation trees for interval inference of the term x/(y+1) that appears in the assertion, for an architecture giving the specified bounds to the types **short** and **long**.

$$\frac{x : \text{short}}{\vdash x :: [-2^{15}, 2^{15} - 1]} \quad \frac{y : \text{long} \quad \vdash y :: [-2^{63}, 2^{63} - 1] \vdash 1 :: [1, 1]}{\vdash y+1 :: [-2^{63} + 1, 2^{63}]}$$
$$\vdash x/(y+1) :: [-2^{15}, 2^{15}]$$

Typing Rules

The C type into which a term is translated is computed using the result of the interval inference. The basic intuition is to simply use the smallest type containing the interval infered for each term. However, this intuition is too simplistic: considering for instance the term x+1 where x was infered to be in the interval $[-\infty, \infty]$, even though the best possible type that contains the constant 1 is the type **char**, the generated code actually needs to convert this constant to type mpz (the types of integers provided by the Gmp library) in order to perform the addition. The E-ACSL type system solves these issues by working in a lattice of types, with a subsumption and a coercion rule. The former is standard in type theory [42], while the latter is ad-hoc to ours. The order relation in the lattice is denoted $\preceq$: $\tau \preceq \tau'$ means that every integer value representable in τ is also representable in τ'. For the sake of simplification, the type mpz is assumed to be able to represent every integer[14] and is thus denoted $\mathbb{Z}$. The link between interval inference and the type system is provided through an operator Θ that computes the smallest C type containing a given interval. It is omitted here and defined in [28]. Figure 5.10 presents the few typing rules needed to translate the basic arithmetic operations. The result of the interval inference for a term t is denoted $\kappa(t)$ in order to simplify the rules.

---

[14] This assumption is not strictly satisfied: the Gmp actually provides an upper bound to the representable integers, but in practice this upper bound is rarely reached.

5 Runtime Annotation Checking with Frama-C: The E-ACSL Plug-in

$$\dfrac{}{\vdash cst : \Theta(\kappa(cst) \cup [\min\text{int}, \max\text{int}])}\,[\text{Cst}] \qquad \dfrac{}{\vdash x : \Theta(\kappa(x))}\,[\text{Var}]$$

$$\dfrac{\vdash t : \tau' \quad \tau' \leq \tau}{\vdash t : \tau}\,[\text{Sub}] \qquad \dfrac{\vdash t : \tau' \quad \tau < \tau' \quad \Theta(\kappa(t)) \leq \tau}{\vdash t : \tau}\,[\downarrow]$$

$$\dfrac{\tau = \Theta(\kappa(t \diamond u) \cup \kappa(t) \cup \kappa(u)) \quad \vdash t : \tau \quad \vdash u : \tau}{\vdash t \diamond u : \tau}\,[\text{Op}]$$

$$\dfrac{\tau = \Theta(\kappa(t) \cup \kappa(u)) \quad \vdash t : \tau \quad \vdash u : \tau}{\vdash t \triangleleft u : \text{int}}\,[\text{Cmp}]$$

Fig. 5.10 A few rules for type inference

Rule [Cst] types a logic constant, whose type is at least **int**, in accordance to the C 99 promotion rule.[15] The rule [Var], combined with the interval inference for variables entails that a C variable preserve its type. The rules [Sub] and $\downarrow$ are the subsumption and the coercion rules. The former states that if a type can represent an element then so do all the larger types, and the latter states that an element of a type may be downcasted to a smaller type when the inferred interval fits into this type. The rule [Op] is the rule for typing arithmetic operations, which requires the type to be the maximal among all the operands involved and the result in order to soundly deal with non-monotonic operators. Finally, the rule [Cmp] types a comparison operator. Since its result is always a Boolean value at runtime, there is no need for an interval inference in this case. Due to the presence of the rules $\downarrow$ and [Sub], there is no uniqueness in the derivation tree. The strategy implemented in E-ACSL is to use these rules only when necessary.

> **Derivation of the interval inference of a term**
>
> Continuing the previous example, the following shows a derivation tree for typing the inequality x/(y+1) <= 200.
>
> $$\dfrac{\dfrac{\vdash x : \text{short}}{\vdash x : \mathbb{Z}} \quad \dfrac{\dfrac{\vdash y : \text{long}}{\vdash y : \mathbb{Z}} \quad \dfrac{\vdash 1 : \text{int}}{\vdash 1 : \mathbb{Z}}}{\vdash y+1 : \mathbb{Z}}\,[\text{Op}]}{\dfrac{\vdash x/(y+1) : \mathbb{Z}}{\vdash x/(y+1) : \text{int}}}\,[\text{Op}] \quad \dfrac{\text{int} \prec \mathbb{Z} \quad \kappa(x/(y+1)) \leq \text{int}}{\vdash 200 : \text{int}}\,[\downarrow]$$
>
> $$\dfrac{}{\vdash x/(y+1) <= 200 : \text{int}}\,[\text{Cmp}]$$

Here, the unlabeled rules all correspond to axioms or to the subsumption rule [Sub]. The results of the interval inference previously computed impose some of the intermediate computation to take place in type $\mathbb{Z}$ denoting Gmp

---

[15] Section 6.3.1.1 of the ISO C Standard 1999. Technical report (1999).

integers. The application of the rule [↓] indicates a downcast from a Gmp to an **int**, where the interval inferences guarantee that the result does not overflow in the new type, hence doing it is safe.

Translating Using Types

The last step for handling arithmetic properties in E-ACSL is the generation of C code with the same semantics as the annotation, using the result of the type system to guide the C types during the translation. The strategy for translating a predicate that has already been typed is to follow the derivation tree and add a downcast for every use of the rule [↓], and an upcast for every use of the rule [SUB]. The type system guarantees that all the downcasts generated this way are safe.

Translation of an assertion using the type system

Once again, continuing the example for the predicate with the previously computed typing:

```
1  short x = 168;
2  long y;
3  scanf("%ld", y);
4  /* compute x /(y +1) with GMP integers */
5  mpz_t _x, _y, _cst_1, _add, _div;
6  int _div2 , _and;
7  mpz_init_set_si(_x, x );
8  mpz_init_set_si(_y, y );
9  mpz_init_set_si(_cst_1, 1) ;
10 mpz_init( _add ); mpz_add(_add, _y, _cst_1);
11 mpz_init( _div ); mpz_tdiv_q(_div, _x ,_add);
12 /* soundly downcast the result of the division from GMP to
       int;
13    it corresponds to the application of the coercion rule.
     */
14 _div2 = mpz_get_si(_div);
15 int c = _div2 <= 200;
16 /* register the data for reporting the assertion */
17 {
18   __e_acsl_assert_data_t data = ... ;
19   ...
20 }
21 __e_acsl_assert(c, data);
22 /* de-allocate the allocated GMP variables for integers */
23 mpz_clear(_x); mpz_clear(_y); mpz_clear(_cst_1);
24 mpz_clear(_add); mpz_clear(_div);
```

In this example, the constant 1 is directly computed in the type mpz_t, rather than first generated and then coerced. This heuristics allows E-ACSL to avoid generating unnecessary variables and casts.

5.5.3 Memory Properties

Memory properties are quite specific to programming languages that allow the developper to have a precise control over the program's memory operation, as it is typical in C. For that reason, E-ACSL is the only runtime annotation checker that supports checking memory properties such as \valid at runtime, as far as we know. In particular, no BISL and associated RAC tools listed in Sect. 5.1 support memory properties.

Verifying such properties at runtime requires having a deep knowledge of the structure of the program memory. For instance, it requires knowing whether some memory address pointed to by some pointer p is writable by the program (otherwise, the predicate \valid(p) would not hold), or whether it has been initialized at some moment during the execution in order to check \initialized(p). Such dynamic verifications are actually performed by a category of tools named *dynamic memory analyzers*, also called *memory debuggers*, such as AddressSanitizer [45] or Valgrind [39]'s MemCheck [47]. These tools are specialized in detecting memory errors such as buffer and heap overflows, and accesses to uninitialized data in the program memory before they really occur. They are also fully automatic and do not require the user to write any specification. When used as explained in Sect. 5.3.2, E-ACSL is also a dynamic memory analyzer, even if it can also verify other properties as already explained. This section is an introduction to the challenge of verifying memory properties in E-ACSL. The interested reader may refer to [49, Sect. 3.3.4] for a more complete overview of the topic and to [58] for additional technical details.

Different Kind of Memory Models

Dynamic memory analyzers rely on a *runtime memory model* to perform its analysis, i.e. a way for the analyzer to register and use information about the state of the memory during the program's execution. E-ACSL and all those tools use the same approach to implement their memory models, namely *memory shadowing*. It consists in associating addresses from program memory to the necessary values stored in a disjoint memory space, called shadow memory. Accessing the shadow memory from the program memory is performed in constant time, just by adding some fixed offset from the program memory. Traditional memory shadowing techniques, typically used in AddressSanitizer [45] and MemCheck [47] only allow to store a small amount of information, typically one or two bit(s) of information for each byte or bit of the program memory. Such techniques are very efficient, but can only detect bit- or byte-level properties (e.g. whether it has been properly allocated). In particular, it fails to detect block-level properties, which require to have information about memory block boundaries and sizes.

Consider for instance the following code snippet containing a buffer overflow at the second line

```
1 char a[1], b[4];
2 a[1] = '0';        /* buffer overflow */
```

Variables a and b are local, usually allocated on the stack one after the other. Hence at line 2, even though the access to a overflows, it may reach a valid address in the memory block of b, which has been properly allocated. Traditional memory shadowing fails to detect this error, since it does not contain the information that the access a[1] crosses the boundaries of the memory block of a, even if ad-hoc techniques such as AddressSanitizer's red-zoning may detect particular patterns: in the above example, the off-by-one is detected, but no error would be detected if a[1] is replaced by a[3] at Line 2.

E-ACSL Memory Model

E-ACSL relies on a custom shadow memory model [58]. Contrary to traditional memory shadowing techniques, E-ACSL is able to handle block-level memory properties. It actually uses two different mechanisms to monitor overflows in the heap and in the stack, but they share a key idea consisting of splitting the shadow representation into segments, and using a *meta-segment* to encode information about the other segments, as shown on examples below. They offer a good trade-off between efficiency and expressiveness, as already illustrated by Figs. 5.3 and 5.4. Their C implementations are part of the E-ACSL's RTL, which is linked against the code generated by E-ACSL, as explained in Sect. 5.3.1.

To further improve efficiency of the generated monitors, E-ACSL also implements a dedicated dataflow analysis [34]. The purpose of this analysis is to determine (an over-approximation of) which variables should be monitored. For instance, there is no need to monitor by shadow memorying the memory block corresponding to some variable x if it is provably not involved in any memory property, either directly (e.g., \valid(&x)) or indirectly through aliasing (e.g., \valid(&y) with x and y being aliased). This avoids registering information about the memory blocks corresponding to variables which will never be accessed.

Heap Shadow Encoding

The heap shadow encoding assumes that there is a common size s at least twice as big as the number of bytes to represent a physical address (so in a 64-bit architecture, $s \geq 64/8 \times 2 = 16$). All the memory blocks are assumed to be aligned at boundary divisible by s. This ensures no overlap between memory blocks, and also allows to find the base address of some segment from any address a with a single operation $a - (a \bmod s)$. The model also assumes s bytes of padding between blocks in order to ensure enough space before each block for a special segment, called the meta-segment.

Assuming $s = 16$ from now on, Fig. 5.11 shows an example of the E-ACSL encoding of a heap memory block B of 40 bytes starting at physical address 0x100. The block B is split into three 16-bytes segments starting at addresses 0x100, 0x110 and 0x120. Its meta-segment is just before the first segment, and the last $16 \times 3 - 40 = 8$ bytes of the last segment are unallocated.

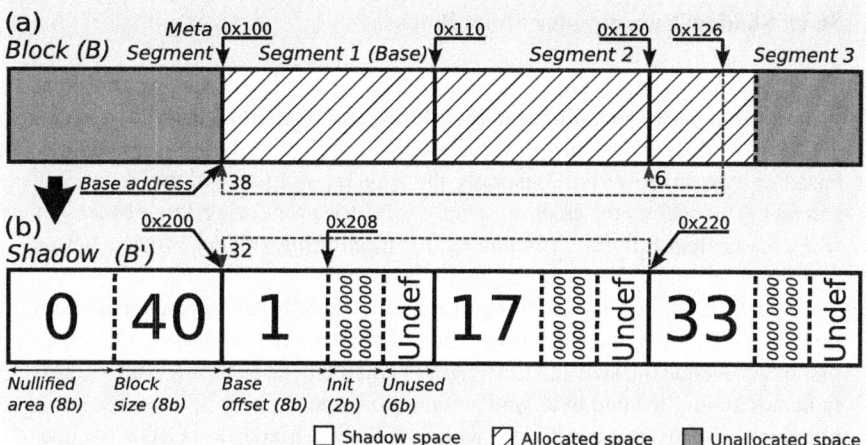

Fig. 5.11 Example of E-ACSL heap shadow encoding

The shadow segment of block B is block B'. Assuming an offset of 0x100 bytes between the program memory and the shadow region, it starts at address 0x100 + 0x100 = 0x200 and has the same structure as block B: it is split into three 16-byte segments with an additional meta-segment before its first one. The first 8 bytes of the meta-segment are nullified, and its last 8 bytes encode the size of block B (here 40 bytes). Each of the other segments is divided into three parts. The first 8 bytes of each of these segments encode the offset of the base address of the segment to the base address of the block, incremented by one. These values are each non-zero, allowing to distinguish the meta-segment from the other ones, and indicate that at least the first byte of the corresponding segment has been allocated. The 16 following bits (2 bytes) encode the initialization status of each of the bytes of the corresponding segment in B. The ith bit is set to 1 when the ith byte of the corresponding segment is initialized. The rest of each of the segment (6 bytes) is unused. This region might be used in later version of E-ACSL to monitor additional pieces of information (e.g., related to concurrency or security properties).

This shadow encoding allows to retrieve the necessary information for monitoring ACSL memory properties. For instance, it is possible to know whether the single byte at address 0x126 is initialized. Its base segment starts at the address 0x126 − 0x126 mod 16 = 0x120. Thus the shadow segment representing this segment starts at 0x220. Hence the initialization status of the byte at address 0x126 is given by the sixth bit starting from address 0x220 + 8 = 0x228. Similarly, its base address (i.e., the first address of the block containing it, which is B) can be retrieved by reading the 8 bytes starting from 0x220, so here 33, and computing 0x120 − 33 + 1 = 0x100, while the size of the block is stored in the last half of the meta-segment, which starts at address 0x200 − 8 = 0x1F8.

Stack Shadow Encoding for Short Blocks

The blocks in the stack are often small and unaligned, so applying the same strategy as the heap encoding leads to adding a lot of padding and thus to large memory overheads. For this reason, E-ACSL uses a slightly different strategy based on two shadow stores, namely the *primary shadow* and the *secondary shadow*. The secondary shadow is only used for monitoring large blocks of size greater than 8 bytes. It is unused for monitoring smaller blocks. Let us first focus on small blocks, since their shadow encoding is easier.

Figure 5.12 shows the encoding of a small 4-byte block starting as address 0x100. Each byte of the stack is encoded by a byte in the primary shadow. The first 6 bits encode the size and the length of small blocks as follows: the value 0 indicates an unallocated byte, and a non-zero value encodes all possible combinations of pairs (Size, Offset), with $1 \leq \text{Size} \leq 8$ and $0 \leq \text{Offset} < \text{Size}$. There are exactly $\Sigma_{i=1}^{8} = 36$ such combinations, and the number between 1 and 36 are chosen to represent those values. The other values between 37 and $2^6 - 1 = 63$ are unused for small blocks. The seventh bit of the shadow byte encodes the initialization status of the monitored byte, and the last bit is left unused.

Assuming that there is an offset of 0x100 between the stack and the primary shadow, retrieving the information about the byte 0x102 which is part of a short memory block can be done by accessing to the byte at address 0x202. The first 6 bits of this byte contain the value 9, which encodes a size of 4 bytes and an offset of 2 bytes. Thus, the address 0x102 is part of a 4-bytes block starting at address $0x102 - 2 = 0x100$. Moreover, the seventh bit of the byte at address 0x202 is 1, indicating that the byte starting at address 0x102 is initialized.

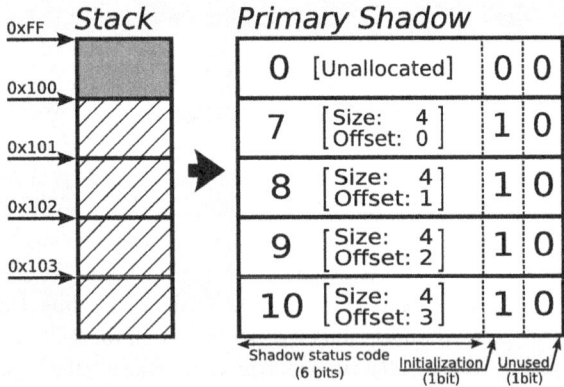

Fig. 5.12 Example of E-ACSL stack shadow encoding for a small 4-byte block

Stack Shadow Encoding for Large Blocks

Figure 5.13 shows a long block B of 18-bytes starting at address 0x100 in the stack. Its shadow encoding uses both the primary and the secondary shadow. The offset between the block and its encoding in the primary (resp. secondary) shadow is assumed to be 0x100 (resp. 0x200).

The primary shadow block monitoring block B starts at 0x100 + 0x100 = 0x200, and its secondary shadow block starts at 0x100 + 0x200 = 0x300. The initialization bit for a byte in a large block is at the place in the primary shadow as for a short block. Furthermore, the encoding of the first 6 bits in the primary shadow allows us to distinguish a long block from a short block: a value between 48 and 63 denotes a long block, while, as already explained, a value between 1 and 36 denotes a short blocks (the values between 37 and 47 are unused). For a long block, this value decremented by 48 encodes the offset from the corresponding shadow segment base address in the secondary shadow. Like in the heap encoding, the secondary shadow is partitioned into 8-byte segments, but the last segment may be smaller than the others. In this case, it is zeroified. If not zeroified, the first 4 bytes of each segment is used to store the size of the block, and the last 4 bytes are used to encode the offset between the start of the segment and the start of the block. Any additional byte in the segments is left unused.

Consider for instance the byte at address 0x10F, the corresponding byte in the primary block is at address 0x20F. The seventh bit of this byte is 0 indicating that the byte has not been initialized. The first 6 bits of this byte contain the value 55, which encodes an offset of $55 - 48 = 7$ to the start of the

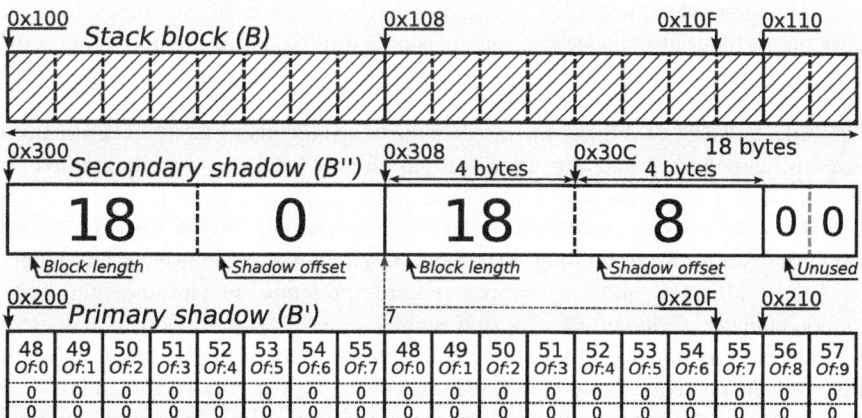

Fig. 5.13 Example of E-ACSL stack shadow encoding for a large 18-byte block

segment in the secondary shadow. Therefore, the corresponding segment in the secondary shadow starts at address 0x30F − 7 = 0x308. The first 4 bytes of this segment contain the value 18 indicating that the byte is part of an 18-bytes block. The last 4 bytes contain the value 8, indicating that the first segment monitoring this block starts at 0x308 − 8 = 0x300. Thus the base address of the block containing 0x10F is 0x100. From here, one can easily deduce the offset of 0x10F, which is 0x10F − 0x100 = 15.

5.6 Conclusion and Perspectives

In this chapter, we presented E-ACSL, the Frama-C plug-in dedicated to runtime annotation checking of C programs. Thanks to this lightweight formal methods, E-ACSL allows the user to check ACSL annotations at runtime. These annotations may be written manually by the end-user or generated automatically by other Frama-C plug-ins, typically Rte for preventing undefined behaviors. Therefore, it is easier to use compared with the other verification techniques provided by Frama-C, even if it provides fewer guarantees since it only checks some concrete execution traces and not all of them. It can also be used in combination with these techniques. For instance, combining E-ACSL and Eva may help to detect undefined behaviors efficiently at runtime, while combining E-ACSL and Wp lowers the verification cost of program proof. Therefore, E-ACSL can be integrated, alone or in combination with Eva or Wp, within CI, V&V or testing campaigns. It can also be used during operations for detecting issues at runtime, allowing the system to apply ad-hoc defensive, mitigation or recovery measures if it happens.

This chapter also presented a few challenges for implementing this technique in order to generate efficient and sound monitors and how E-ACSL solves them. In particular, it focused on the generation of code for arithmetic and memory properties.

Perspectives

Currently, E-ACSL is still less mature than the two main other Frama-C verification plug-ins, namely Eva and Wp, even if it has already been successfully applied on several small- to medium-size concrete use cases [2, 14, 40, 44, 55].

Only a subset of the specification language is currently supported [51]. While supporting some of the missing constructs only requires manpower, a few others are more challenging and lead to open research problems. For instance, the latter include support of properties over real numbers and a good support of inductive predicate and axiomatic definitions. Each of them are not soundly monitorable in finite time in the general case. Similarly, supporting the \at construct would require to copy the whole program memory in the general case, even if partial solutions already exist [20, 27, 41]. E-ACSL currently implements its own solution, which needs to be improved both from a theoretical and a practical point of view. Potential solutions for properties which may be non-monitorable include a smart use of partial verdicts [5] in runtime annotation checking, which would allow the tool to answer "I don't know" at runtime when necessary.

Another axis of development consists of improving efficiency of the generated code. First, the current generated code (e.g., calling Gmp function for arithmetic operations) prevents several standard compiler optimizations, such as constant propagation or common sub-expression eliminations [1], which would be easy to do internally, before generating the code. More specifically but importantly, using Rte for checking undefined behaviors leads to generating code of quadratic complexity (instead of linear) in some rather frequent cases. Modifying the code generation scheme is possible, yet not straightforward. From preliminary experiments with code manually optimized, we can expect to reduce the time overhead by about a factor of two in many cases.

Last, the E-ACSL RTL has strong requirements about the underlying system (e.g., it does not work on MacOS) and supporting a new system is expensive. Furthermore, the time and memory overhead may be prohibitive for operational uses in constrained environments (e.g., embedded systems). A way to solve these issues might be to run the monitor on a different address space, e.g. typically on a remote server. This technique is known as *outline monitoring* [19]. Designing an outline runtime annotation checker has never been tried, as far as we know.

References

1. Aho A, Ullman J, Sethi R, Lam M (2006) Compilers: principles, techniques, and tools. 2nd edn. Addison Wesley
2. Barany G, Signoles J (2017) Hybrid information flow analysis for real-world C code. In: International conference on tests and proofs (TAP). https://doi.org/10.1007/978-3-319-61467-0_2
3. Barnett M, Fähndrich M, Leino KRM, Müller P, Schulte W, Venter H (2011) Specification and verification: the Spec# experience. Commun ACM
4. Barnett M, Leino KRM, Schulte W (2004) The Spec# programming system: an overview. In: International conference on construction and analysis of safe, secure, and interoperable smart devices (CASSIS)
5. Bauer A, Leucker M, Schallhart C (2007) The good, the bad, and the ugly, but how ugly is ugly? In: International workshop on runtime verification, revised selected papers. https://doi.org/10.1007/978-3-540-77395-5_11
6. Benjamin T, Signoles J (2023) Abstract interpretation for efficient runtime assertion checking of recursive definitions. In: International conference on tests and proofs (TAP)
7. Benjamin T, Signoles J (2023) Formalizing an efficient runtime assertion checker for an arithmetic language with functions and predicates. In: Symposium on applied computing (SAC)
8. Bruening D, Zhao Q (2011) Practical memory checking with Dr. Memory. In: International symposium on code generation and optimization (CGO)
9. Chalin P (2007) A sound assertion semantics for the dependable systems evolution verifying compiler. In: International conference on software engineering (ICSE). https://doi.org/10.1109/ICSE.2007.9
10. Cheon Y (2003) A runtime assertion checker for the Java Modeling Language. Ph.D. thesis, Iowa State University

11. Clarke LA, Rosenblum DS (2006) A historical perspective on runtime assertion checking in software development. SIGSOFT Softw Eng Notes. https://doi.org/10.1145/1127878.1127900
12. Cok DR (2011) OpenJML: JML for Java 7 by extending OpenJDK. In: International symposium on NASA formal methods (NFM). https://doi.org/10.1007/978-3-642-20398-5_35
13. Cousot P, Cousot R (1977) Abstract interpretation: a unified lattice model for static analysis of programs by construction or approximation of fixpoints. In: Symposium on principles of programming languages (POPL)
14. Dieumegard A, Garoche PL, Kahsai T, Taillar A, Thirioux X (2015) Compilation of synchronous observers as code contracts. In: Symposium on applied computing (SAC)
15. Dijkstra EW (1975) Guarded commands, nondeterminacy and formal derivation of programs. Commun ACM
16. Ellison C (2012) A formal semantics of C with applications. Ph.D. thesis, University of Illinois
17. Ellison C, Roşu G (2012) An executable formal semantics of C with applications. In: Symposium on principles of programming languages (POPL)
18. Emerson EA, Clarke EM (1980) Characterizing correctness properties of parallel programs using fixpoints. In: Automata, languages and programming
19. Falcone Y, Krstic S, Reger G, Traytel D (2021) A taxonomy for classifying runtime verification tools. Int J Softw Tools Technol Transf. https://doi.org/10.1007/s10009-021-00609-z
20. Filliâtre JC, Pasciutto C (2022) Optimizing prestate copies in runtime verification of function postconditions. In: International conference on runtime verification (RV)
21. Filliâtre JC, Pascutto C (2021) Ortac: runtime assertion checking for OCaml (tool paper). In: International conference on runtime verification (RV)
22. Guth D, Hathhorn C, Saxena M, Roşu G (2016) RV-match: practical semantics-based program analysis. In: Computer aided verification (CAV)
23. Havelund K, Goldberg A (2005) Verify your runs. In: Verified software: theories, tools, experiments (VSTTE)
24. Herrmann P, Signoles J Annotation generation: frama-C's RTE plug-in. http://frama-c.com/download/frama-c-rte-manual.pdf
25. Huisman M, Wijs A (2023) Runtime annotation checking. In: Concise guide to software verification, texts in computer science. Springer, Berlin. https://doi.org/10.1007/978-3-031-30167-4
26. Konikowska B, Tarlecki A, Blikle A (1988) A three-valued logic for software specification and validation. In: VDM '88 VDM—the way ahead. https://doi.org/10.1007/3-540-50214-9_19
27. Kosiuczenko P (2010) An abstract machine for the old value retrieval. In: International conference on mathematics of program construction (MPC). https://doi.org/10.1007/978-3-642-13321-3_14
28. Kosmatov N, Maurica F, Signoles J (2020) Efficient runtime assertion checking for properties over mathematical numbers. In: International conference on runtime verification (RV). https://doi.org/10.1007/978-3-030-60508-7_17
29. Le VH, Correnson L, Signoles J, Wiels V (2018) Verification coverage for combining test and proof. In: International conference on tests and proofs (TAP)
30. Leavens GT, Baker AL, Ruby C (1999) JML: a notation for detailed design, Chap. 12. Springer, Berlin, pp 175–188 (1999). https://doi.org/10.1007/978-1-4615-5229-1_12
31. Lehner H (2011) A formal definition of JML in Coq and its application to runtime assertion checking. Ph.D. thesis, ETH Zurich
32. Leucker M, Schallhart C (2009) A brief account of runtime verification. J Log Algebr Program
33. Luckham DC, Von Henke FW (1985) An overview of anna, a specification language for Ada. IEEE Softw
34. Ly D, Kosmatov N, Loulergue F, Signoles J (2018) Soundness of a dataflow analysis for memory monitoring. In: Workshop on languages and tools for ensuring cyber-resilience in critical software-intensive systems (HILT). https://doi.org/10.1145/3375408.3375416
35. Ly D, Kosmatov N, Loulergue F, Signoles J (2020) Verified runtime assertion checking for memory properties. In: International conference on tests and proofs (TAP). https://doi.org/10.1007/978-3-030-50995-8_6

36. McCormick JW, Chapin PC (2015) Building high integrity applications with SPARK. Cambridge University Press
37. Meyer B (1988) Eiffel: a language and environment for software engineering. Syst Softw. https://doi.org/10.1016/0164-1212(88)90022-2
38. Meyer B (1992) Applying "Design by contract". Computer
39. Nethercote N, Seward J (2007) Valgrind: a framework for heavyweight dynamic binary instrumentation. In: Conference on programming language design and implementation (PLDI)
40. Pariente D, Signoles J (2017) Static analysis and runtime assertion checking: contribution to security counter-measures. In: Symposium sur la Sécurité des technologies de l'Information et des communications (SSTIC). publis/2017_sstic.pdf
41. Petiot G, Botella B, Julliand J, Kosmatov N, Signoles J (2014) Instrumentation of annotated C programs for test generation. In: International conference on source code analysis and manipulation (SCAM). https://doi.org/10.1109/SCAM.2014.19
42. Pierce BC (2002) Types and programming languages. MIT Press
43. Queille JP, Sifakis J (1982) Specification and verification of concurrent systems in CESAR. In: International symposium on programming (1982)
44. Robles V, Kosmatov N, Prevosto V, Rilling L, Gall PL (2019) Tame your annotations with MetAcsl: specifying, testing and proving high-level properties. In: International conference on tests and proofs (TAP). https://doi.org/10.1007/978-3-030-31157-5_11
45. Serebryany K, Bruening D, Potapenko A, Vyukov D (2012) Addresssanitizer: a fast address sanity checker. In: Annual technical conference (ATC). https://doi.org/10.5555/2342821.2342849
46. Serebryany K, Potapenko A, Iskhodzhanov T, Vyukov D (2011) Dynamic race detection with LLVM compiler. In: International conference on runtime verification (RV)
47. Seward J, Nethercote N (2005) Using valgrind to detect undefined value errors with bit-precision. In: Annual technical conference (ATC)
48. Shiraishi S, Mohan V, Marimuthu H (2015) Test suites for benchmarks of static analysis tools. In: International symposium on software reliability engineering workshops (ISSREW)
49. Signoles J (2018) From static analysis to runtime verification with Frama-C and E-ACSL. Habilitation Thesis, Université Paris-Sud
50. Signoles J (2022) E-ACSL: Executable ANSI/ISO C specification language. Version 1.18
51. Signoles J (2022) E-ACSL Version 1.18. implementation in Frama-C Plug-in E-ACSL. Version 25.0
52. Signoles J, Desloges B, Vorobyov K (2022) E-ACSL user manual
53. Stepanov E, Serebryany K (2015) MemorySanitizer: fast detector of uninitialized memory use in C++. In: International symposium on code generation and optimization (CGO)
54. Stucki LG, Foshee GL (1975) New assertion concepts for self-metric software validation. In: International conference on reliable software (ICRS)
55. Védrine F, Jacquemin M, Kosmatov N, Signoles J (2021) Runtime abstract interpretation for numerical accuracy and robustness. In: International conference on verification, model checking, and abstract interpretation (VMCAI)
56. Vorobyov K, Kosmatov N, Signoles J (2018) Detection of security vulnerabilities in C code using runtime verification. In: International conference on tests and proofs (TAP). https://doi.org/10.1007/978-3-319-92994-1_8
57. Vorobyov K, Kosmatov N, Signoles J, Jakobsson A (2017) Runtime detection of temporal memory errors. In: International conference on runtime verification (RV). https://doi.org/10.1007/978-3-319-67531-2_18
58. Vorobyov K, Signoles J, Kosmatov N (2017) Shadow state encoding for efficient monitoring of block-level properties. In: International symposium on memory management (ISMM). https://doi.org/10.1145/3092255.3092269
59. Yau SS, Cheung RC (1975) Design of self-checking software. In: International conference on reliable software (ICRS)
60. Zeller A, Gopinath R, Böhme M, Fraser G, Holler C (2019) The fuzzing book. CISPA and Saarland University, Saarbrücken

Chapter 6
Test Generation with PathCrawler

Nicky Williams and Nikolai Kosmatov

Abstract Structural testing allows validation engineers to ensure that all parts of the program source code are activated (or covered) by the executed tests. The parts of code to be covered are determined by the choice of a coverage criterion. Automated test generation tools can be used effectively to generate test inputs satisfying a selected coverage criterion. This chapter presents PathCrawler, an automatic test generation tool for structural testing of C code, available as a plug-in of the Frama-C verification platform. We present the structural testing approach and describe the method implemented in PathCrawler and its usage in practice. The freely available online test generation service PathCrawler-online is presented as well. Finally, we discuss the place of testing in software verification and validation with respect to other Frama-C plug-ins and give an overview of various research, industrial and teaching activities using PathCrawler.

Keywords Test generation · Structural testing · Test coverage criteria · PathCrawler · PathCrawler-online

6.1 Introduction

Testing remains a widely used technique for software validation. Amongst other testing approaches, structural testing allows the validation engineer to ensure that all parts of the program source code were activated (or covered) by the executed tests at least once, and thus increases the level of confidence in the program under test. Various structural coverage criteria [1, 24] have been proposed in the literature. Existing standards for critical software [8], for instance, in avionics and automotive software, require a certain level of coverage, depending on the degree of criticality

N. Williams
Université Paris-Saclay, CEA, List, Palaiseau, France
e-mail: nicky.williams@cea.fr

N. Kosmatov (✉)
Thales Research & Technology, Palaiseau, France
e-mail: nikolaikosmatov@gmail.com

of the software. Achieving the required coverage manually can be a tedious and error-prone task. Automated test generation tools can be used effectively to generate test inputs corresponding to a selected coverage criterion.

This chapter presents PathCrawler [4, 22, 23], an automatic test generation tool for structural testing of C code, available as a plug-in of the Frama-C verification platform. We start by briefly presenting the structural testing approach with its goals and limitations in Sect. 6.2. Section 6.3 presents the overall purpose of PathCrawler and describes its inputs and outputs. Section 6.4 describes the test generation technique implemented by PathCrawler and illustrates it on an example. Section 6.5 explains in more detail how to use PathCrawler in practice. The online test generation service PathCrawler-online [13] is presented in Sect. 6.6. Then we describe the place of PathCrawler in a general verification and validation process with respect to other Frama-C plug-ins in Sect. 6.7. Finally, Sect. 6.8 presents various research, industrial and teaching activities using the tool and concludes the chapter.

6.2 Structural Testing

Structural testing is a testing technique in which the selection of tests has to satisfy a structural coverage criterion. A *structural coverage criterion* is based on the structure of an artifact of the tested software, which for PathCrawler is the C source code. Some elements of the structure are selected, and the coverage criterion requires that each of these elements must be *covered* (that is, activated) by at least one test. For instance, in *branch coverage*, the selected elements are branches in the source code. A branch is a decision (inside a conditional statement, a loop, etc.) that leads the control flow from the branch point to one of two different instructions depending on the result of evaluation of a condition to true or false. A set of tests satisfies the branch coverage criterion (also called *decision coverage*) if each reachable branch in the code is taken by at least one test. Similarly, a set of tests satisfies the *instruction coverage* criterion if each reachable instruction in the source code is executed by at least one test. As another example, a set of tests satisfies the *path coverage* criterion if each feasible (that is, executable) path of the program is activated by at least one test. Other structural coverage criteria are related to conditions in the code. Some conditions in the code are multiple (or compound): they contain several simple conditions (or subconditions) linked by logical operations. There exist coverage criteria that target such simple conditions inside compound conditions (*condition coverage*), or all possible combinations of such conditions (*multiple condition coverage*), or both conditions and decisions (*condition-decision coverage*). *Modified condition-decision coverage* (*MC/DC*) is a much-used structural source code coverage criterion based on certain combinations of conditions in the same compound condition. An interested reader can refer to [1, 24] for a complete presentation of coverage criteria.

Structural source code coverage criteria are motivated by the simple idea that if there is a part of the code that is not activated by any test and if there is a bug which is located, wholly or partially, in that part of the code, then the tests cannot reveal

Fig. 6.1 C function which should return the maximum of two integers

```
1  int max(int x, int y) {
2      int result = y;   // error, should be: x
3      if (x <= y)
4          result = y;
5      return result;
6  }
```

Fig. 6.2 C function with unreachable branch

```
1  int infeasible(unsigned int x, unsigned int y){
2      int result = 0;
3      if (x < 100)
4        if (y < 100)
5          if ((x * y) > 20)
6            if (x == 1)
7              if (y == 1)
8                result = 1; /* unreachable */
9      return result;
10 }
```

the bug. In particular, if a branch is not taken by any test then no test can reveal a bug situated in the branch itself or in any instructions which can only be reached through the branch. However, a test may cover the branch but still not reveal the bug. By covering not only individual branches but whole execution paths, we can more rigorously test the different behaviours of the code and potentially reveal more bugs.

Instruction coverage is a relatively weak structural source code coverage criteria: it requires fewer tests and has a weaker ability to detect bugs, for instance, compared to branch coverage. In particular, instruction coverage cannot detect bugs in empty branches. Consider the (buggy) function in Fig. 6.1, supposed to return the maximum of its two arguments. All instructions of this function are covered by a single test in which x is less than y, e.g., with a call max(4,5), but this test does not cover the empty else branch of the conditional at line 3. With an additional test where x is greater than y, e.g., with a call max(6,4), branch coverage is achieved and the bug is revealed.

A coverage criterion is considered to be weaker than another one if the first one is always satisfied whenever the second one is satisfied. The example in Fig. 6.1 illustrates why branch coverage is stronger than instruction coverage. Detailed relationships between different criteria can be found in [1].

Not all execution paths are feasible and tests can only cover feasible paths. Similarly branch testing can only cover reachable branches, i.e., branches which belong to at least one feasible path. To illustrate this, consider the example of Fig. 6.2. The condition in line 7 is inconsistent with the conditions in lines 3–6 so the unique path to line 8 is infeasible and the true branch of condition in line 7 is unreachable.

Path coverage is stronger than branch coverage. The number of tests needed to cover all reachable branches is less than or equal to the total number of branches in the code which may be high in real-life software. Typically, it increases linearly with the size of the code. As for the number of tests needed to cover all feasible paths, it is exactly the number of feasible paths. The number of theoretical paths increases very

```
1  int member(int x, int array[5]){
2    for (int i = 1; i < 5; i++){ // error, should start with i = 0
3      if (array[i] == x)
4        return 1;
5    }
6    return 0;
7  }
```

Fig. 6.3 C function which should return 1 if x is in array and 0 if not

rapidly with the number of branches, and can be exponential. The feasible paths of real-life functions are often too numerous to be able to perform path testing.

Code coverage criteria have another limitation. They cannot detect *missing paths*, i.e., bugs in which part of the specification has been forgotten in the implementation. Figure 6.3 shows an example of this. Given an integer and an array, function member is supposed to return 1 if the integer is an element of the array. The loop should treat all elements of the array but mistakenly skips the element at index 0. Path testing will produce tests of all the implemented paths through the loop but will probably fail to reveal that there is no path which can detect that x = array[0]. Indeed, the arbitrary input values chosen by the test generator to ensure coverage of the different paths may reveal this bug by chance but only structural coverage of the behaviours in the specification can ensure its detection.

Another limitation of structural coverage criteria, and indeed of all test criteria, is that they only solve the problem of how to select test input values. To validate the program under test, a test engineer also needs an *oracle*: a way to decide (and produce a *verdict*) whether the results of execution with those test inputs are correct or not. It is an issue which is orthogonal to that of the test criterion but at least as important in determining how many bugs can be detected.

Structural testing does ensure that all parts of the structure are tested to some extent. This is why it is often imposed in code standards [8] such as DO-178C for avionics [9] or ISO 26262 for road vehicles.[1] Another advantage of structural testing is that test inputs can be automatically generated, using an automated structural analysis, to satisfy the coverage criterion. This is what PathCrawler does.

6.3 PathCrawler: What it Does

PathCrawler automatically generates test inputs achieving structural coverage of a given function in a program written in ANSI/ISO C. The test set will guarantee coverage of all feasible coverage objectives under the conditions detailed below. Moreover, if the user supplies an oracle, PathCrawler will also record and report the verdicts of the oracle. In this chapter, we present only selected PathCrawler options, the user can list all options using the -pc-help option.

---

[1] See https://www.iso.org/standard/43464.html.

Let f be the function under test. The inputs to PathCrawler are the following.

- The name of the tested function f. As for other Frama-C plug-ins, this is main by default, and in this case PathCrawler will suppose that the initial value of all uninitialized global variables is zero. As PathCrawler is usually used for unit testing, the Frama-C options -main f -lib-entry must be used to name the function under test and to indicate that non-constant global variables do not necessarily have their initial value when f is executed.
- The Gcc-compilable files containing the complete C source code of the function under test f and all other functions (including library functions, except functions for dynamic memory allocation) which may be directly or indirectly called by f (see Sect. 6.5.1 for advice on stubbing called functions). The files must not include assembly code.
- The PathCrawler command-line options, such as those defining the coverage criterion. By default, the criterion is all feasible paths in the tested function and all called functions but options can be used to select another coverage criterion, limits to the criterion (such as different timeouts, the maximal number of loop iterations in the generated tests or whether the criterion applies to called functions) or extensions in the form of additional coverage objectives (such as assertion violations or runtime errors). The coverage options are reviewed in Sect. 6.5.4.
- The *test context*, which is defined by certain command-line options and a *precondition* which applies to the input values of all the generated tests. A default precondition is constructed by PathCrawler but the user can choose to extend it, see Sect. 6.5.3 for more detail.
- Finally, the user can provide an oracle in the form of C code or annotate the code with assertions, as we explain in Sect. 6.5.5.

PathCrawler outputs are produced in the form of xml files, converted to html for display, and include the following data.

- A set of test-cases, where each case contains:
  - the input values,
  - the covered path,
  - concrete and symbolic (i.e., in terms of input values) output values or a detected error (runtime error, assertion violation, etc.), and
  - the oracle verdict if an oracle is supplied.

- Any uncovered coverage objectives and the reason they were not covered, i.e., because demonstrated to be infeasible or unreachable or because of a timeout of the constraint solver, as well as the unresolved constraint satisfaction problem, in the form of a *path predicate* (also called *path conditions* or *path constraints*).
- Other statistics about the test generation session, such as execution time, and the termination status (normal, timeout or fault).

The generated test-cases are guaranteed to cover all feasible test objectives (according to the selected criterion) in the tested function, provided that:

- test generation terminates normally within time and memory limits set by the user,[2] and
- the source code does not contain certain constructs which PathCrawler does not treat well yet, notably:
  - **volatile** variables,
  - pointers of type **void*** on input to the tested function,
  - pointers to functions on input to the tested function,
  - recursive structures on input to the tested function.

6.4 PathCrawler: How it Works

PathCrawler is based on a test generation method that is often called *concolic* testing, or Dynamic Symbolic Execution (DSE) [6]. It combines *symbolic execution* [11] of the program, in which the program is analyzed with symbolic (that is, unknown) values of variables, and its concrete execution, in which the compiled binary code is executed in a usual way, with concrete values of variables.[3] The method contains two major phases.

In the first phase, called the *analyzer*, PathCrawler extracts the possible inputs and the semantics of the tested function, f, and instruments the source code in order to create a test driver. The *test driver* executes the instrumented version of f to trace the execution path for a given test-case. The analyzer is implemented in OCaml as a Frama-C plug-in. The possible inputs include the formal parameters of f and the non-constant global variables. A test-case will provide a value for each possible input of f.

The second phase, called the *generator*, generates test-cases for f which respect the given precondition and satisfy the selected coverage criterion. Implemented in the Prolog language in the ECLiPSe constraint logic programming environment[4] and based on the COLIBRI constraint solver [7],[5] the generator combines concrete execution of the instrumented code to recover the execution path followed by the test and symbolic execution to extract the path constraints. By default, the paths of f are explored in a depth-first search. Let us describe (a simplified version of) the PathCrawler test generation method in more detail.

We use abbreviated path notation where we write branches only. We can denote an execution path by a sequence of branches, e.g., $+a, -b, -c, +d$, where a, b, c, d designate the positions (here, line numbers) of control points in conditional or loop

---

[2] The user can set timeouts for constraint resolution of each solver request and for the whole test generation session globally, as well as memory limits (see the command-line options).

[3] The term *concolic* indicates a combination of two techniques, *conc*rete and symb*olic* execution.

[4] See http://www.eclipse-clp.org.

[5] See https://smt-comp.github.io/2020/system-descriptions/COLIBRI.pdf.

statements. A branch is denoted by the control point position preceded by a "+" if the condition is true, and by a "−" otherwise. For instance, +4, +7 denotes the path that enters into the loop on line 4 (since the condition on line 4 is satisfied), then enters into the conditional statement on line 7 (and, therefore, exits the function at the return statement on line 8). The mark "⋆" after a branch indicates that the other branch has already been explored (as explained in detail below). For conciseness, we will omit commas and write, for example, the path $+a, -b, -c, +d$ as $+a -b -c +d$.

The generator relies on the test driver to trace the execution path for a generated test-case. The generator's main loop is rather simple: given a partial program path π, the main idea is to symbolically execute it as constraints. A solution of the resulting constraint solving problem will provide a test-case exercising a path starting with π. Then the trick is to use concrete execution of the test-case on the instrumented version to obtain the complete path. The partial paths are explored in a depth-first search.

For symbolic execution of a program as constraints, PathCrawler maintains:

- a memory state of the program at each moment of symbolic execution. It is basically a mapping associating a value to a symbolic name. The symbolic name is the name of a variable or array element or struct field, with a basic type. The value is either a constant or a logical variable.
- the current partial path π in the program. When a test-case is successfully generated for the partial path π, the remaining part of the path it activates is denoted by σ.
- a constraint store with the constraints added by the symbolic execution of the current partial path π.

The method contains the following steps:

- **Initialisation**: Create a logical variable for each input and associate it with this input. Set initial values of initialized variables. Add constraints for the precondition. Let the initial partial path π be empty. Continue to Step 1.
- **Step 1**: Let σ be empty. Symbolically execute the partial path π, that is, add constraints and update the memory according to the instructions in π. If some constraint fails, continue to Step 4. Otherwise, continue to Step 2.
- **Step 2**: Call the constraint solver to generate a test-case, that is, concrete values for the inputs, satisfying the current constraints. If it fails, go to Step 4. Otherwise, continue to Step 3.
- **Step 3**: Run the test driver with traced execution of f on the test-case generated in Step 2 to obtain the complete execution path. The complete path must start with π. Save the remaining part into σ. Continue to Step 4.
- **Step 4**: Let ρ be the concatenation of π and σ. Try to find in ρ the last unmarked branch, i.e., the last branch without a "⋆" mark. If ρ contains no unmarked branch, exit. Otherwise, if $+r$ (or $-r$) is the last unmarked branch in ρ, set π to the subpath of ρ before $+r$ (resp., $-r$), followed by $-r_\star$ (resp., $+r_\star$), i.e., the negation of the last unmarked branch in ρ marked as already processed, and continue to Step 1.

Fig. 6.4 C function for binary search in arrays of length 4

```
1  int bsearch(int a[4], int elem) {
2    int low = 0;
3    int high = 3;
4    while (low <= high) {
5      int mid = low + (high-low)/2;
6      int midVal = a[mid];
7      if (midVal == elem) {
8        return 1;
9      } else if (midVal > elem) {
10       high = mid-1;
11     } else {
12       low = mid+1;
13     }
14   }
15   return 0;
16 }
```

Notice that Step 4 chooses the next partial path in a depth-first search. It changes the last unmarked branch in ρ to look for differences as deep as possible first, and marks a branch by a "$\star$" when its negation (i.e., the other branch from this node in the tree of all execution paths) has already been fully explored. For example, if $\rho = +a -b_\star -c -d +e_\star$, the last unmarked branch is $-d$, so Step 4 takes the subpath of ρ before this branch $+a -b_\star -c$, and adds $+d_\star$ to it to obtain the new partial path $\pi = +a -b_\star -c +d_\star$.

We will use as a running example the C function shown in Fig. 6.4. To simplify the example, we limit the array size to 4, and the domain of elements to [0, 100]. The function bsearch takes two parameters: an array a of four integers $\in$ [0, 100] and an integer elem $\in$ [0, 100]. Given that a is sorted in ascending order, the function returns 1 if elem is present in a, or 0 otherwise. We assume the precondition and oracle (see Sect. 6.5) are provided, and focus on the generation of test data.

Figure 6.5 shows how our method proceeds on this example. The empty path is denoted by ε. In the state (1), we see that the Initialisation step associates a logical variable X_i to each input, i.e., to each element of a and to elem, and posts the precondition $\langle Pre \rangle$ to the constraint store. Here, $\langle Pre \rangle$ denotes the constraints: $X_0, \ldots, X_4 \in [0, 100]$ and $X_0 \leq X_1 \leq X_2 \leq X_3$. Since the initial memory associations of State (1) for inputs remain unchanged, they are denoted by [...] in the following states to avoid repetitions. As the original prefix π is empty, Step 1 is trivial here and adds no constraints.

The result of application of Steps 2 and 3 is shown in the box Test-Case 1. Step 2 consists of choosing a first test case. In Step 3, we retrieve the complete path traced during the concrete execution of Test-Case 1, and obtain $\sigma = +4 -7 -9 +4 -7 +9 -4$. (Note that when a path executes several iteration of the same loop, the loop condition appears several times, each time it is evaluated.)

Step 4 sets $\rho = +4 -7 -9 +4 -7 +9 -4$ and, therefore, the new path prefix $\pi = +4 -7 -9 +4 -7 +9 +4_\star$ by negating the last not-yet-negated branch. Now, Step 1 symbolically executes this path prefix as constraints for unknown inputs, and the resulting state is shown in (2). Let us explain this execution in detail. First, the

6 Test Generation with PathCrawler

(1) Memory	Constraints
a[0] ↦ X_0	$\langle Pre \rangle$
a[1] ↦ X_1	
a[2] ↦ X_2	
a[3] ↦ X_3	
elem ↦ X_4	
$\pi = \epsilon$	

Test-Case 1
$X_0 = 4$
$X_1 = 16$
$X_2 = 42$
$X_3 = 71$
$X_4 = 23$
$\sigma = +4 -7 -9 +4 -7 +9 -4$

⇝ →

(2) Memory	Constraints
[...]	$\langle Pre \rangle$
low ↦ 2	$0 \leq 3, X_1 \neq X_4,$
high ↦ 1	$X_1 \leq X_4,$
mid ↦ 2	$2 \leq 3, X_2 \neq X_4,$
midVal ↦ X_2	$X_2 > X_4, 2 \leq 1$
$\pi = +4 -7 -9 +4 -7 +9 +4_\star$	

⇝ infeasible →

(3) Memory	Constraints
[...]	$\langle Pre \rangle$
low ↦ 2	$0 \leq 3, X_1 \neq X_4,$
high ↦ 3	$X_1 \leq X_4,$
mid ↦ 2	$2 \leq 3, X_2 \neq X_4$
midVal ↦ X_2	$X_2 \leq X_4$
$\pi = +4 -7 -9 +4 -7 -9_\star$	

Test-Case 2
$X_0 = 26$
$X_1 = 41$
$X_2 = 47$
$X_3 = 99$
$X_4 = 99$
$\sigma = +4 +7$

⇝ ...

Fig. 6.5 Depth-first generation of all-paths test-cases for bsearch, where → denotes application of Steps 4, 1 and ⇝ denotes application of Steps 2, 3. State (1) corresponds to the state after the Initialisation and Step 1

execution of lines 2 and 3 adds low ↦ 0 and high ↦ 3 into the memory. The conditional expression at line 4 is interpreted as a constraint $0 \leq 3$ after replacing the variables by their current values in the memory map. The assignments of lines 5 and 6 add mid ↦ 1 and midVal ↦ X_1, X_1 being the current symbolic value of a[1]. The conditional expressions at lines 7 and 9—that must be both false—give the constraints $X_1 \neq X_4$ and $X_1 \leq X_4$, since X_4 is the symbolic value associated to elem. At line 12, the memory is updated with low ↦ 2. Line 4 adds the trivial constraint $2 \leq 3$. Lines 5 and 6 update the memory map with mid ↦ 2 and midVal ↦ X_2, thus overwriting their previous values. Because of the minus sign, the expression at line 7 is negated and gives the constraint $X_2 \neq X_4$. Line 9 posts the constraint $X_2 > X_4$. Line 9 changes the value of high to 1 in the memory. Finally the last conditional node +4 gives the false constraint $2 \leq 1$, so the path prefix is infeasible. The last constraint obviously fails, which is detected by our solver at the propagation step while posting the constraint, and Step 1 continues directly to Step 4. The intermediate states are not detailed in Fig. 6.5.

We are now going from (2) to (3) in Fig. 6.5. Step 4 computes the complete path $\rho = +4 -7 -9 +4 -7 +9 +4_\star$. As $+4_\star$ means that its negation has already been explored, the new prefix π is $+4 -7 -9 +4 -7 -9_\star$. Next, Step 1 symbolically executes this partial path. It can be done from the initial state (1). However, in practice, backtracking allows us to come back to the closest intermediate state (here, the state just before $X_2 > X_4$ was posted by the previous symbolic execution), from which we can reach the current path prefix in a minimal number of steps. Instead, symbolic execution of the path posts the constraint for the last branch -9, that is, the constraint $X_2 \leq X_4$, and stops. Next, Step 2 generates Test-Case 2. Step 3 computes the full path and sets σ to $+4+7$. Step 4 computes $\rho = +4-7-9+4-7-9_\star+4+7$, deduces the new prefix $\pi = +4-7-9+4-7-9_\star+4-7_\star$, and so on.

The method exits when all paths have been explored, so that the generated test-cases cover all feasible paths. The reader will find applications of this method to other examples in [12, 22, 23]. The PathCrawler method is related to other testing tools using symbolic execution, for instance, KLEE [5], SAGE [10] and PEX [18] (see [6] for a survey).

6.5 Using PathCrawler in Practice

Automatic testing tools allow huge savings but they do not exonerate the user from thinking carefully about what they want testing to achieve. To successfully use PathCrawler, the user must provide not only the full source code (with C stubs for missing functions), but also must specify the *test context* and the *oracle*. This demands a different mindset from that used for manual unit testing.

6.5.1 Stubbing

PathCrawler rigorously explores the code of the function under test and its callers. If the tested code calls undefined functions then PathCrawler will abort test generation after listing the signatures of these functions. In the case of undeclared functions, it is the supposed signature that will be listed before PathCrawler aborts. The Frama-C libc definitions cannot be used because they may depend on internal Frama-C functions and variables (the Frama-C -no-frama-c-stdlib option must be set when using PathCrawler) and in any case, most of them are defined in ACSL which cannot be used by PathCrawler to generate path conditions. Moreover, PathCrawler does not propose its own libc stubs because, as we explain below, standard stubs that will suit all purposes are very difficult to define.

This means that the user must provide either the correct definitions in C or else C stubs for all called functions. If a stub is used instead of the real definition then the effect of the stub on test generation results is the user's responsibility. Indeed, a stub which is too simple may cause PathCrawler to decide that a path is infeasible when this is not the case when the real function is used. For example, if a stub systematically returns zero and a branch in the code tests the returned value then PathCrawler will conclude that the branch for a non-zero return value is unreachable. If the code beyond this branch outputs an error message or halts the program then PathCrawler will be unable to generate a test which activates this behaviour.

6 Test Generation with PathCrawler

```
1  int * _arrayIntStubOutputs; // additional array for stub calls
2  unsigned int _arrayIntStubOutputs_i; // current index
3
4  int stub(int x, int y){
5    pathcrawler_assume(pathcrawler_dimension(_arrayIntStubOutputs)
         > _arrayIntStubOutputs_i);
6    return _arrayIntStubOutputs[_arrayIntStubOutputs_i++];
7  }
8
9  int tested_function(int x){
10   _arrayIntStubOutputs_i = 0;
11   if (stub(x, x+1) == 0)
12     return x;
13   else
14     return x+stub(0,1);
15 }
```

Fig. 6.6 C function calling minimal stub

In order to test different return values of function stubs, the user can define a new global variable for each return value (and be sure to use the -lib-entry option as explained in Sect. 6.3). **PathCrawler** considers the new global variable as an extra input to the function and attributes a concrete value for each testcase. In the example in Fig. 6.6, **PathCrawler** generates two tests, one in which _arrayIntStubOutputs[0] == 0 to cover the true branch at line 11 and one in which _arrayIntStubOutputs[0] != 0 to cover the false branch. Note that the use of the array _arrayIntStubOutputs allows for successive stub calls. The call to pathcrawler_assume is necessary if the user does not know the upper limit on the number of calls in a single execution path but it causes inefficient test generation and so it is better to use a constant dimension for the array _arrayIntStubOutputs if possible.

In the example in Fig. 6.6, the behaviour and output of the stub do not depend on the input values and this is often unsatisfactory. If it is the printf function which is being stubbed, for example, then this sort of stub will only be sufficient if the test oracle does not check the printed output and if the return values are not used in the subsequent code.

malloc, calloc and realloc are the only library functions that are internally modelled by **PathCrawler** and do not need to be stubbed. However, **PathCrawler**'s model for malloc never returns a null pointer if the requested size is greater than zero so **PathCrawler** cannot construct tests to check the results of memory running out. Indeed, it is very difficult to test what happens when some function calls malloc in a situation in which there is not enough memory. It can however be simulated using an additional global array (like on lines 1–2 in Fig. 6.6) in the following way. The user replaces the calls to malloc by another function, say, stubbedMalloc, and then defines stubbedMalloc so that it tests an element of the global array (_arrayIntStubOutputs[_arrayIntStubOutputs_i++] if we reuse the notation of Fig. 6.6) in order to decide whether to call malloc (i.e., the

PathCrawler model) or return a null pointer. Note, however, that unlike real cases where memory might run out, this behaviour for `stubbedMalloc` does not depend on the number of previous calls to `malloc` or the size of the allocated memory.

To take another example, if the `strlen` function is stubbed and the return value is used in the tested function, then returning an arbitrary, incorrect, value for the length of the real input string could cause PathCrawler to wrongly decide that a path containing a call to `strlen` is infeasible. Moreover, in many cases, the value returned by a call to one stub provides the input value for a call to another stub.

As a final example, consider a stub for the `random` function. Does the user want to generate tests in which each call to `random` produces a different value but the values are not necessarily randomly distributed, do they want to test the unlikely situation in which two calls could produce the same value or do they need true random distribution of the returned value to correctly test the calling function?

The above examples demonstrate that stubbing is important and that it is very difficult to define standard stubs for library code that will serve all test purposes and suit all tested functions. The user must create suitable stubs for the test context if code coverage is to be maximized and correctly reported.

6.5.2 Constant Inputs

PathCrawler must find a concrete value for all inputs of basic type which may be read by the tested function. To efficiently execute test-cases one after another in the same process, the PathCrawler test harness contains a loop in which each iteration assigns values to input variables before calling the function under test. However, such inputs (formal parameters, global variables or their fields, elements, results of dereferencing, etc.) may be qualified in C as **const** and the C compiler does not allow the PathCrawler test harness to directly assign a value to **const** inputs in order to run the test-case. This is why PathCrawler does not consider **const** values to be inputs. If the tested function, f, happens to be `main` (see Sect. 6.3), then the user must just ensure that all **const** inputs of `main` are initialized. If f is not `main` and has one or more **const** inputs then the user must choose between the following solutions for each of them:

1. ensure that the input is assigned before calling f, by testing a *wrapper* function which indirectly assigns this value, using a pointer, in order to bypass the compiler check, and then calls f (this solution is preferable if the input has a single, known, possible value) or
2. modify the source code to remove the **const** qualification of this input for the purpose of testing.

6.5.3 Precondition

The test context is defined by certain command-line options and by the *precondition*. The precondition will be respected by the input values in all generated tests. The default precondition

- limits the values of each input variable to those that respect its declared type,
- respects the declared dimensions of formal parameters which are arrays, and
- ensures that input pointers are either NULL or point to a single element, unless certain command-line options[6] are used.

PathCrawler supposes that the possible input variables of the tested function include the formal parameters and all global variables declared in the source code and gives an input value to

- each non-constant input variable and dereferenced input pointer with a basic type,
- each non-constant element or field or dereferenced element or field with a basic type which belongs to an input variable or dereferenced input pointer and
- each unknown dimension of an array containing input or output variables.

Indeed, in order to run the test-cases, PathCrawler needs to decide the values of the dimensions of not only the arrays containing input values but also those containing output values.

The PathCrawler analyzer (see Sect. 6.4) generates the default test context which the user should check and modify if necessary in order to

- ensure that the test input values are realistic, i.e., belong to the set of possible values which the tested function is supposed to treat, and
- in some cases, prevent generation of an unnecessary number of test-cases.

Consider the example in Fig. 6.4: if the input array is not sorted in ascending order then the algorithm may give the wrong result. The implementation does not check that the array is sorted so the test result may be wrong not because the algorithm for binary search is incorrectly implemented but just because the test inputs do not satisfy the implicit precondition. Either this is a bug and the implementation should be changed to check that the input array is indeed sorted or else the precondition should be made explicit in the test context so that PathCrawler will only generate test-cases containing sorted arrays on input and any unexpected resuts will be due to the implementation of binary search.

The implementation in Fig. 6.4 is not very realistic because it only treats arrays of exactly four elements. A more realistic implementation, in which the length of the array is indicated by the additional formal parameter dim_a, is shown in Fig. 6.7.

---

[6] Options -pc-ptr-0-1, -pc-ptr-0-max, -pc-ptr-0-n, -pc-ptr-1-1, -pc-ptr-1-max, -pc-ptr-1-n modify the definition of the default precondition, by setting the corresponding interval as the default dimension range of all arrays with unknown dimensions which are referenced by input pointers (type frama-c -pc-help for more detail).

Fig. 6.7 C function for binary search in arrays of any length

```c
int bsearch(int *a, int dim_a, int elem) {
  int low = 0;
  int high = dim_a - 1;
  while (low <= high) {
    int mid = low + (high-low)/2;
    int midVal = a[mid];
    if (midVal == elem) {
      return 1;
    } else if (midVal > elem) {
      high = mid-1;
    } else {
      low = mid+1;
    }
  }
  return 0;
}
```

For this implementation, **PathCrawler** also considers as an additional input value, as explained above, the actual length (or dimension, referred to as dim(a)) of the input array pointed to by a. The possible values of dim(a) are therefore included in the test context. If the path coverage criterion is chosen and the range of values for dim(a) is too large, then test generation will time out before full coverage is reached (but after generating a great many cases) because the number of possible iterations of the loop, and so the number of possible execution paths, depends on this range. The user may judge that it should be possible to detect any bugs on relatively small arrays and either limit the range of values for dim(a) in the test context or else use the -pc-iter-limit option to limit the number of iterations in the generated cases.

If **PathCrawler** is now run on the implementation in Fig. 6.7 with a test context restricted, as explained above, to relatively small, sorted, input arrays, then a new problem will be revealed. **PathCrawler** will generate test-cases in which the parameter value indicating the expected length of a (that is, dim_a) and the number of elements effectively allocated for array a in the test-case (that is, dim(a)) are unrelated. In this case, a memory access runtime error is likely to be generated, and reported by **PathCrawler**, when a test is generated in which dim_a is larger than dim(a) and a is accessed with an index which is smaller than dim_a but not smaller than dim(a). Indeed, the implicit precondition for this new implementation includes the relation that dim(a) is greater than or equal to dim_a. The implementation cannot check this relation and **PathCrawler** cannot guess it, so to avoid reporting a false bug, the test context must be extended to ensure this relation in the test-cases generated by **PathCrawler**. Note, however, that in cases such as this example, where both the actually allocated array length, dim(a), and the assumed array length, dim_a, are variable and a variable index is also used (in line 6 of Fig. 6.7), the user is advised to define a precondition in which dim(a) is exactly equal to dim_a (and not greater than or equal), in order to simplify the test generation problem and reveal any bugs leading to an array access beyond dim_a.

```
1  void precond_bsearch(int *a, int dim_a) {
2    pathcrawler_assume(dim_a == pathcrawler_dimension(a));
3    for (int i = 1; i < dim_a; i++) {
4      pathcrawler_assume(a[i-1] <= a[i]);
5    }
6  }
7
8  int bsearch_with_precond(int *a, int dim_a, int elem) {
9    precond_bsearch(a, dim_a);
10   return bsearch(a, dim_a, elem);
11 }
```

Fig. 6.8 C function `bsearch_with_precond` for testing `bsearch` function of Fig. 6.7 with C precondition

Finally, note that another memory access bug caused by an integer overflow in line 3 will be revealed if **PathCrawler** is allowed to generate a test-case in which dim(a) is indeed greater than or equal to dim_a but dim_a is the minimum possible negative C int value! This unlikely situation becomes more realistic in the setting where an attacker tries to use some (unusual) argument values to provoke an out-of-bound memory access that can lead to a security vulnerability. This example illustrates how **PathCrawler** can detect such situations.

The user can define the precondition in either of the following ways, of which the first is the most efficient for test generation:

- by changing the default test context file created by **PathCrawler** in Prolog (the detailed description of its format can be found in the tool manual);
- by coding the precondition in C, using calls to `pathcrawler_assume`, and applying **PathCrawler** to the combination[7] of the precondition and the tested function as shown in Fig. 6.8.

6.5.4 Coverage Options

As well as the precondition, the test context also includes the command-line coverage options, which we summarize here and whose detailed description is obtained by typing `frama-c -pc-help`.

The basic coverage options are the following:

```
-pc-all-branches,
-pc-branches-func <function_name>,
```

---

[7] In this case, the coverage criterion must apply to the tested function only, and assuming that the `-pc-ptr-0-n` option is used to limit the range of dimensions of a to 0..4, the resulting options are `-pc-ptr-0-n 4 -main bsearch_with_precond -pc-paths-func bsearch`.

```
-pc-all-mcdc,
-pc-mcdc-func <function_name>,
-pc-all-paths,
-pc-paths-func <function_name>.
```

They select branch, MC/DC or path coverage and define whether coverage extends to the tested function and all functions it calls, directly or indirectly, or is just restricted to a single function. Restricting coverage to a single function is useful to avoid unnecessary exploration of complex library functions. As shown in the example of Fig. 6.8 above, it can also be used when the precondition is coded in C. In this example, the tested function is defined as `bsearch_with_precond` and the `-pc-paths-func` option is used to ensure full path coverage of just the `bsearch` function. In this case, within the code of the precondition itself (`precond_bsearch`), the only paths covered are those that are necessary for full coverage of `bsearch`.

Supplementary or alternative coverage objectives can be defined in order to make PathCrawler try to generate tests specifically to provoke the corresponding bugs, rather than relying on the bugs being revealed by tests generated in order to cover the code structure:

- option `-pc-cover-asserts` to make PathCrawler try to generate tests provoking assertion violations;
- options `-pc-cover-int-overflow` and `-pc-cover-int-underflow` to make PathCrawler try to generate tests provoking overflows and underflows;
- option `-pc-cover-rt-errors` to make PathCrawler try to generate tests provoking runtime errors.

Moreover, any calls to `pathcrawler_label` in the text constitute an additional implicit coverage objective.

If none of the above options are set by the user then the default option is `-pc-all-paths`, unless the tested function includes calls to `pathcrawler_assert` or `pathcrawler_label`, in which case just these objectives are covered. Note that the `-pc-stop-when-assert-violated` option can be used in conjunction with `-pc-cover-asserts` to stop test generation as soon as a single assertion violation is found.

PathCrawler will try to cover all feasible objectives defined by the coverage options and it will achieve this if test generation terminates normally (as reported by the final session status) and without any reports of constraint solving timeouts. However, because the number of feasible paths in the tested code may be enormous, because constraint solving is NP-complete or because resources are limited, full coverage may not be achieved.[8]

---

[8] The following options allow the user to explicitly control certain resources, thereby effectively defining possible limits to coverage: `-pc-session-timeout`, `-pc-resol-timeout`, `-pc-harness-timeout`, `-pc-eclipse-options`, `-pc-global-stack-threads`.

The following options restrict the exploration of the execution paths during test generation, with the consequent risk of restricting coverage:

- `-pc-iter-limit` to limit the number of loop iterations,
- `-pc-recur-limit` to limit the recursion depth,
- `-pc-suffix-length` to limit the maximal path suffix length until which an exhaustive exploration is performed.

6.5.5 Oracle

The role of the oracle in the test process is to determine whether the verdict of a test-case is *success* (i.e., no bug detected) or *failure*. PathCrawler generates the input values of the test-cases and the user can just inject these values into their own test process, notably if this is performed on a different platform to that used for test generation. However, PathCrawler runs all the test-cases it generates, on the test generation platform, in order to recover the covered path, and so it gives the user the option of providing an oracle so that PathCrawler can manage the whole test process, applying the oracle after each test-case is run and including the verdict in the information it outputs on the test-case.

The oracle examines the input values generated by PathCrawler for the test-case and the effective output values produced when the test-case is run in order to decide the verdict. It is important that the user understands the difference between the oracle and the precondition. The oracle does not need to check that the precondition is satisfied by the input values because PathCrawler ensures that this is always true. The PathCrawler analyzer generates a default oracle in the form of a C function with the default verdict of *unknown* for all cases (see Fig. 6.9 for the default oracle for the example in Fig. 6.7). The user must replace the definition of this function (but without changing its signature), using the dedicated PathCrawler macros to produce `success` or `failure` verdicts. Note that in the case of variables which may have both input and output values, i.e., global variables or formal parameters with pointer types, such as a in the example in Fig. 6.7, two variables are used by the oracle. The variable with the original name (e.g., a) holds the output value and the input value is stored in another variable with the same name prefixed by `Pre_` (e.g., `Pre_a`). The signature of the oracle function includes the formal parameters of the tested function (including the return value if any) and also the additional variables

Fig. 6.9 Default oracle for bsearch

```
void oracle_bsearch(
    int *Pre_a, int *a, int dim_a, int elem,
    int pathcrawler__retres__bsearch)
{
    pathcrawler_verdict_unknown();
}
```

```
1  void oracle_bsearch(
2    int *Pre_a, int *a, int dim_a, int elem,
3    int test_result)
4  {
5    if(test_result<0 || test_result>1){
6      pathcrawler_verdict_failure();
7      return;
8    }
9    pathcrawler_verdict_success();
10   return;
11 }
```

Fig. 6.10 Simple partial oracle for bsearch

```
1  void oracle_bsearch(
2    int *Pre_a, int *a, int dim_a, int elem,
3    int test_result)
4  {
5    int i;
6    int correct_result = 0;
7
8    for(int i=0; i<dim_a && correct_result==0; i++){
9      if(a[i] == elem)
10       correct_result = 1;
11   }
12   if(correct_result!=test_result){
13     pathcrawler_verdict_failure();
14     return;
15   }
16   pathcrawler_verdict_success();
17   return;
18 }
```

Fig. 6.11 Better oracle for bsearch

named with a Pre_ prefix for the input values of any formal parameters of pointer type. Any global variables needed by the oracle must be accessed by the user using the same naming convention.

The ability of the test process to detect bugs obviously depends greatly on the quality of the oracle. With PathCrawler's default oracle only run-time errors will be detected. Consider the example of a simple oracle for bsearch in Fig. 6.10 which just checks whether the output value is either 0 or 1. This would detect certain bugs but miss other bugs which would be detected by comparison with an alternative implementation of search in an array, such as the oracle shown in Fig. 6.11. In this example, the oracle is an implementation of search with exactly the same results but which is less efficient. Sometimes the user has a previous implementation which can serve the purpose of an oracle. In fact, it is often necessary to reimplement the function, at least partially, in order to provide a good oracle. This can be avoided for some functions, for example the oracle of a sorting function just has to check whether the result is ordered and contains the right elements.

6 Test Generation with PathCrawler   323

While the oracle of Fig. 6.11 is more precise, it is still incomplete. Indeed, a buggy implementation of `bsearch` can modify the array a, say, by writing `elem` in all its elements and then returning 1. The search function is not supposed to modify the array, and the result of this buggy implementation will obviously be wrong if the initial array does not contain `elem`. Such bugs will not be detected because the oracle of Fig. 6.11 reads the values of a after the execution of the tested function and not before. The complete oracle is shown in Fig. 6.12. It checks in addition whether the input array is modified by the implementation and distinguishes wrong and invalid results: **PathCrawler** outputs the line number of the failure verdict in the oracle so that different reasons for failure can be distinguished.

Note that the user can choose to use assertions in the source code instead of (or as well as) the oracle. Without the `-pc-cover-asserts` coverage option, bugs will be revealed by any assertions that happen to be violated by the tests which are generated in order to satisfy the chosen coverage criterion. However, with this option set, **PathCrawler** will actively try to find input values which cause assertions in the tested function to be violated. If full coverage is achieved without timeouts and without finding any assertion violations then **PathCrawler** effectively proves all the properties expressed by the assertions.

Similarly to the use of assertions and the `-pc-cover-asserts` option, the user can apply **PathCrawler** to the combination of the tested function and the oracle, as in Fig. 6.13, where `cross_check` is now the tested function. This technique is sometimes called *cross-checking* and ensures that **PathCrawler** actively tries to find input values which violate the oracle and, if none are found, effectively proves that the oracle is satisfied. Note that cross-checking, and to a lesser extent the assertion coverage mode (activated by the `-pc-cover-asserts` option), increase the complexity of test generation and so the risk that full coverage cannot be achieved in a reasonable time.

6.5.6 *Testing Faulty Code*

The point of testing is to detect bugs and so the user is quite likely to apply **PathCrawler** to faulty code! This section gives some indications how to efficiently detect bugs in such cases. If any test-cases provoke a runtime error or a failure verdict, the user is advised to first check their precondition and oracle by inspecting the input and output values of these test-cases. If their input values belong to the desired test context (e.g., in the example in Fig. 6.7, the value of `dim(a)` is a reasonable array dimension, identical to the value of `dim_a`, and a is correctly sorted) and their output values do not satisfy the desired oracle (e.g., in the example in Fig. 6.7, the returned value is neither 0 nor 1 or does not correspond to the presence of `elem` in a or if a is modified by the implementation), then the user can study all the information on the failing test-cases (or all the cases which fail on the same line in the oracle) and compare them to the correct test-cases. Sometimes, the failing cases may be the only ones covering a particular branch or with a particular input value (e.g., a NULL

```c
void oracle_bsearch(int *Pre_a, int *a, int dim_a, int elem, int
    test_result) {
  int correct_result = 0;
  int modif = 0;

  for(int i=0; i<dim_a; i++){
    if(a[i] == elem)
      correct_result = 1;    // check if elem present in a
    if(a[i] != Pre_a[i])
      modif = 1;             // check if bsearch modified a
  }
  if(correct_result==0 && test_result!=0){
    if(test_result!=1){
      pathcrawler_verdict_failure();  /* implementation returned
          invalid result when elem not in a */
      return;
    }
    pathcrawler_verdict_failure();  /* implementation returned
        wrong result when elem not in a */
    return;
  }
  if(correct_result==1 && test_result!=1){
    if(test_result!=0){
      pathcrawler_verdict_failure();  /* implementation returned
          invalid result when elem in a */
      return;
    }
    pathcrawler_verdict_failure();  /* implementation returned
        wrong result when elem in a */
    return;
  }
  if(modif!=0){
    pathcrawler_verdict_failure();  /* implementation modified a
        */
    return;
  }
  pathcrawler_verdict_success();
  return;
}
```

Fig. 6.12 Complete oracle for bsearch

pointer), which can indicate an erroneous code instruction. Otherwise, the user can select a single failing test-case for detailed analysis and debugging.

Figure 6.14 shows an implementation of binary search with a bug in line 7, where condition `midVal != elem` is wrongly used instead of `midVal == elem`. Table 6.1 shows typical test-cases generated for this version[9] with a test context assuming that array a has between 0 and 4 elements. Oracle verdicts are generated

---

[9] Unless the deterministic generation mode is set (with the -pc-deter option), the selection of test input values is partly random so PathCrawler will not always generate the same test input values on each run, and will not always cover test objectives in the same order.

6 Test Generation with PathCrawler

```c
int spec(int *Pre_a, int *a, int dim_a, int elem, int* modif) {
  int result = 0;
  for(int i = 0; i < dim_a; i++){
    if(Pre_a[i] == elem)
      result = 1;
    if(a[i] != Pre_a[i])
      *modif = 1;
  }
  return result;
}

void cross_check(int *Pre_a, int *a, int dim_a, int elem){
  for(int i = 0; i < dim_a; i++){
    Pre_a[i] = a[i];
  }
  int test_result = bsearch(a,dim_a,elem);
  int modif = 0;
  int correct_result = spec(Pre_a,a,dim_a,elem,&modif);
  pathcrawler_assert(test_result == correct_result && !modif);
  return;
}
```

Fig. 6.13 Cross-checking `bsearch`, where `cross_check` is the new function under test

```c
int bsearch(int *a, int dim_a, int elem) {
  int low = 0;
  int high = dim_a - 1;
  while (low <= high) {
    int mid = low + (high-low)/2;
    int midVal = a[mid];
    if (midVal != elem) { // Bug: should be midVal == elem
      return 1;
    } else if (midVal > elem) {
      high = mid-1;
    } else {
      low = mid+1;
    }
  }
  return 0;
}
```

Fig. 6.14 Implementation of `bsearch` with a bug in line 7

using the complete oracle of Fig. 6.12, where each failure is reported with the line number of the failure reporting macro in the oracle. The user might notice that none of the test-cases cover branch +9 and find the bug this way, or else just examine more closely the simplest failing test-case, comparing its inputs and the executed path.

Now consider the implementation with a bug in line 9 shown in Fig. 6.15 and Table 6.2 showing typical test-cases generated for it. They all succeed! In fact, this is a subtle example of the missing-path problem defined in Sect. 6.2: PathCrawler covers the paths but does not reveal the bug.

However, the user might consider that for both these implementations, the test-case inputs are not very useful. Indeed, unless the `-pc-random` option is set,

Table 6.1 Tests generated for bsearch function of Fig. 6.14 with a bug in line 7

Verdict	dim_a	a[0]	a[1]	a[2]	a[3]	elem	Return	Path
Success	0					0	0	−4
Fail line 24	1	0				0	0	+4, −7, −9, −4
Fail line 24	2	0	0			0	0	+4, −7, −9, +4, −7, −9, −4
Fail line 24	4	0	0	0	0	0	0	+4, −7, −9, +4, −7, −9, +4, −7, −9, −4
Success	4	0	0	0	699510381	0	1	+4, −7, −9, +4, −7, −9, +4, +7
Success	2	0	1180171200			0	1	+4, −7, −9, +4, +7
Fail line 16	1	0				1745242961	1	+4, +7

6 Test Generation with PathCrawler

```
1  int bsearch(int *a, int dim_a, int elem) {
2    int low = 0;
3    int high = dim_a - 1;
4    while (low <= high) {
5      int mid = low + (high-low)/2;
6      int midVal = a[mid];
7      if (midVal == elem) {
8        return 1;
9      } else if (midVal < elem) { // Bug: should be: midVal > elem
10       high = mid-1;
11     } else {
12       low = mid+1;
13     }
14   }
15   return 0;
16 }
```

Fig. 6.15 Implementation of bsearch with a bug in line 9

PathCrawler uses heuristics to choose all input values which are not dimensions.[10] These heuristics start by trying certain values, such as zero, which often reveal bugs. However, this is not always appropriate and in the case of binary search, the actual value of elem and the elements of a are irrelevant (except, possibly, in the case of certain bugs) and the result of the heuristic is to set all these values to either zero or some non-zero number, chosen within the vast range of the C int type. There is very little chance of a test-case with a fortuitous occurrence of elem in a, independently of the path constraints. Moreover, bsearch only has two possible return values so the chances of a fortuitous success are high! Table 6.3 shows the results of generation for the implementation shown in Fig. 6.15 when the chances of fortuitous occurrence of elem in a are maximized by using the precondition to restrict the possible values of both elem and the elements of a to the 5 possible values in the range 0..4 and setting the -pc-random option. The test-case inputs are more varied and by chance, rather than thanks to path coverage, a bug is detected.

6.6 PathCrawler-online

Since 2010, PathCrawler has an online version, PathCrawler-online,[11] which offered the first online service for automatic structural unit testing of C programs. This section presents its main features.

---

[10] Dimension values are always chosen to be as small as possible with the given path constraints.
[11] Available at http://pathcrawler-online.com:8080/.

Table 6.2 Tests generated for bsearch function of Fig. 6.15 with a bug in line 9

Verdict	dim_a	a[0]	a[1]	a[2]	a[3]	elem	Return	Path
Success	0					0	0	−4
Success	1	0				0	1	+4, +7
Success	1	0				1473093489	0	+4, −7, +9, −4
Success	3	0	0	0		217239785	0	+4, −7, +9, +4, −7, +9, −4
Success	1	0				−1928518485	0	+4, −7, −9, −4
Success	2	0	0			−247127362	0	+4, −7, −9, +4, −7, −9, −4
Success	4	0	0	0	0	−1960667394	0	+4, −7, −9, +4, −7, −9, +4, −7, −9, −4

6 Test Generation with PathCrawler   329

Table 6.3 Tests generated for bsearch function of Fig. 6.15 with a context which restricts and randomizes values

Verdict	dim_a	a[0]	a[1]	a[2]	a[3]	elem	Return	Path
Success	0					0	0	−4
Success	1	0				0	1	+4, +7
Success	1	2				3	0	+4, −7, +9, −4
Success	2	1	4			0	0	+4, −7, −9, +4, −7, −9, −4
Success	1	3				0	0	+4, −7, −9, −4
Success	3	0	0	0		3	0	+4, −7, +9, +4, −7, +9, −4
Fail line 24	4	0	1	1	2	0	0	+4, −7, −9, +4, −7, −9, +4, −7, −9, −4

6.6.1 Inputs of **PathCrawler-online**

PathCrawler-online takes the following input data:

- an archive with complete compilable C source code, possibly in several files,
- name of the function under test,
- name of the main file,
- test parameters including a precondition and a test coverage strategy: coverage of the set of all feasible paths (*all-path*) or of their subset with a limited number of loop iterations (*k-path*),
- a C function with an oracle.

The user can also select one of the predefined examples. An online interface offers the user the possibility to conveniently customize test parameters and the oracle. Alternatively, the user can submit customized test parameters (in XML format) and a customized oracle (in a C function) within the submitted archive.

6.6.2 Ouputs of **PathCrawler-online**

During a test generation session, PathCrawler-online generates test-cases and a coverage report for the submitted C program. The results of a test session include:

- test session statistics (cf. Fig. 6.16) and coverage information (cf. Fig. 6.17),
- test-cases (cf. Fig. 6.18) along with test drivers,
- the list of the explored paths with their details.

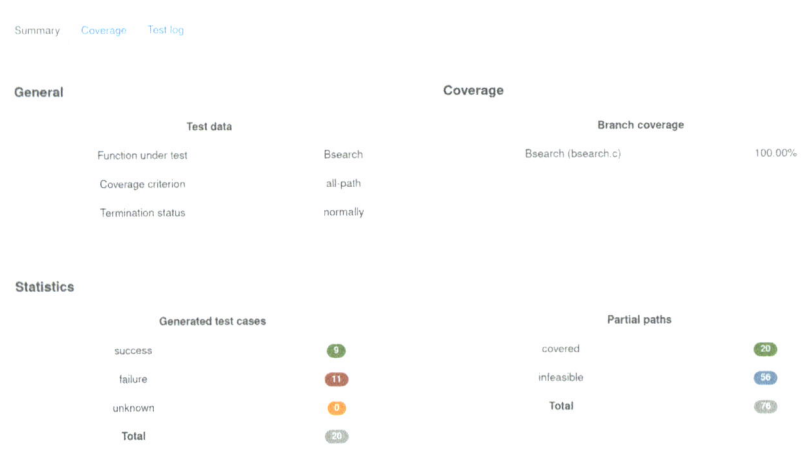

Fig. 6.16 Statistics generated by PathCrawler-online for a variant of a binary search function

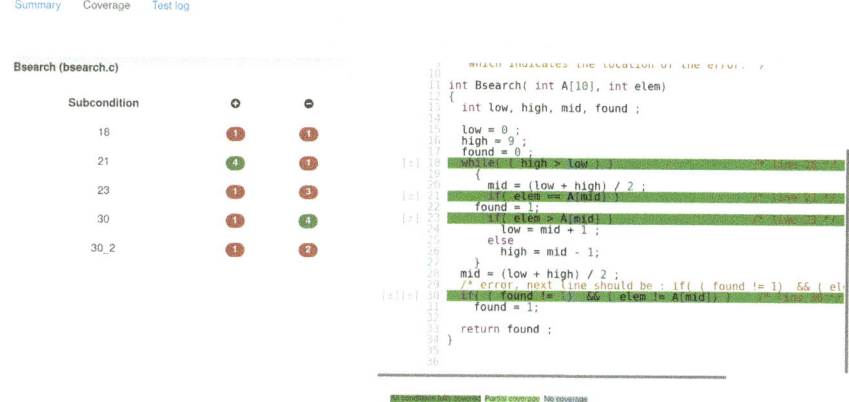

Fig. 6.17 Coverage overview generated by PathCrawler-online for a variant of a binary search function

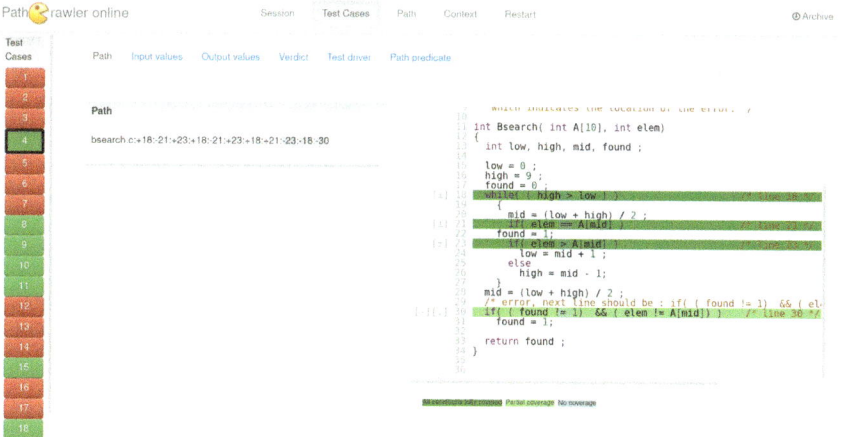

Fig. 6.18 A test-case generated by PathCrawler-online for a variant of a binary search function

The results are generated in HTML format which can be displayed in the browser, while test drivers are ready-to-compile C files which can be used to run the test-cases.

The generated test session statistics include test session duration, the number of explored paths (indicating the number of paths of each status, e.g., covered, interrupted by a timeout, infeasible), the number of generated test-cases with different oracle verdicts (e.g., failure, success, violating a user assertion, unknown). Branch coverage achieved by the generated test-cases is also computed.

Each test-case (cf. Fig. 6.18) provides *input values*, *output values* and the *oracle verdict*. Moreover, it includes the program *path* activated by the test-case, as well as *path prefix*, *symbolic outputs* and *path predicate*. The explored paths list indicated the status of each path and its path predicate.

6.6.3 Benefits and Limitations

PathCrawler-online takes advantage of the multi-core architecture of the server to run several test sessions in parallel, but each test session is run sequentially. Since PathCrawler-online is often simultaneously used by dozens of users (in particular, by students in Software Testing courses), the server restricts the resources which can be used by a user during one test session. These resources include memory, disk space and execution time.

PathCrawler-online demonstrates the paradigm of *Testing as a Service* (TaaS). TaaS brings various benefits, both to the user and to the tool provider (who becomes a test-service provider). First, TaaS removes the need for the user to install the testing tool. Instead, the test-service provider installs it on their servers. Only the user interface may need to be integrated into different software development environments, so the porting effort is considerably reduced for the test-service provider. The test-service provider can keep a record of test sessions which revealed bugs in the tool, which facilitates their replay and bug fixing. The service provider can rapidly update the tool when necessary, and these updates can be made transparent for the user. Finally, the user can use the testing service on demand and does not need to amortize the costs of the software licence or extra infrastructure to run the tests. However, an important limitation to the usage of TaaS can be confidentiality of the code, which is sent to the server.

6.7 The Place of Structural Testing in the Verification and Validation Process

Each program analysis technique is most effective when performing a specific function. This section compares PathCrawler with other core Frama-C plug-ins presented in the previous chapters: Eva, Wp and E-ACSL (see Chaps. 3, 4 and 5), and discusses the place of testing, and in particular structural testing, in a verification and validation process based on static analysis (such as performed by the Eva plug-in), proof (such as Wp) and structural testing (such as PathCrawler).

Static Frama-C plug-ins Eva and Wp allow the user to prove that some properties are verified, but they cannot usually confirm the presence of bugs when the analysis returns an *unknown* verdict for a property since some alarms can be *false positives*. PathCrawler can reveal bugs with a *witness* for each bug in the form of the test inputs, which can help to understand and correct the bug. In this sense, PathCrawler is similar to E-ACSL, whose purpose is to check annotations at runtime and report possible failures. However, E-ACSL only reports failures on the provided test inputs, whereas PathCrawler is able to generate the test input values respecting a given structural criterion. In this way, PathCrawler can demonstrate that the covered parts of code are not unreachable. Eva and Wp can detect the presence of unreachable code (also called *dead code*), but they are usually unable to prove its absence. In the general case, E-ACSL and PathCrawler, unlike Eva and Wp cannot prove the absence of bugs. However, when sufficient coverage can be achieved, E-ACSL and PathCrawler can also confirm the absence of certain bugs and PathCrawler can demonstrate the presence of unreachable code. For example, Fig. 6.2 shows an example of a situation where Eva may be unable to detect the dead code (in line 8 of Fig. 6.2) due to complex conditions, but PathCrawler does when it tries to cover the true branch of condition in line 7.

Let us compare the relative advantages of these techniques with respect to certain criteria.

Type of bug: Let us distinguish functional properties (as expressed by user assertions or, in PathCrawler, an oracle), runtime errors (of which PathCrawler currently treats fewer than Eva or E-ACSL) and unreachable code. Eva can show the absence of runtime errors, prove some simple functional properties expressed by user assertions (which it is able to reason about), and detect unreachable code but it usually cannot confirm the presence of these bugs or prove the absence of unreachable code. Wp can also prove the presence of unreachable code, show the absence of runtime errors (thanks to annotations automatically generated by the Rte plug-in) and prove a wide range of functional properties expressed by ACSL annotations. Eva and Wp can both report a proof failure for a property but in this case the property status remains unknown and the presence of a bug is not confirmed. PathCrawler can confirm, with a witness, violation of functional properties (expressed by user assertions or the oracle) and certain runtime errors. If full branch coverage of the tested function is achieved, it can also prove the absence of unreachable code. Full branch coverage of the oracle when performing cross-checking (see Sect. 6.5.5) proves the absence of oracle violations. Similarly, if the test objectives include assertion violations and certain run-time errors, achieving full coverage is a way to prove their absence.

Difficulty of use: Eva can often prove the absence of certain runtime errors without the help of user annotations, but some experience with the tool is needed to create a suitable context and to tune the analysis for a particular code. Wp can sometimes prove the absence of runtime errors or unreachable code without assistance from the user (thanks to smoke tests and Rte) but usually the user must provide annotations. To prove functional properties, Wp requires annotations in the form of function contracts, loop invariants or other proof-guiding annotations (assertions, lemmas). For this reason, Eva often demands a certain expertise and Wp almost always does. Structural testing can sometimes reveal runtime errors or prove absence of unreachable code with no user intervention but usually needs the user to provide a precondition and cannot detect functional bugs without user assertions or an oracle. However, the precondition and oracle can be written in C rather than ACSL, making the task easier for some users. Similarly, all the plug-ins need stubs for library functions: these can always be written in C or alternatively, for Eva and Wp, their ACSL contracts can be provided or, for some libc functions, the predefined ACSL specifications can be used.

Knowledge of the source code: The user can provide pre- and postconditions for all three plug-ins or an oracle for PathCrawler without any knowledge of the implementation of the target function, but the additional user annotations or specific analysis options which may be needed by Eva or Wp can require a deeper understanding of the source code of the implementation. In any case, a good understanding of the source code is needed to analyze failures of Wp proofs or fix bugs or alarms identified by Eva or by testing.

Program size: Eva is able to treat relatively large programs starting from the main entry point, while Wp must be used in a modular way, so the size of the program is mainly limited by the capacity of the user to provide annotations for all functions. PathCrawler can only achieve full coverage on relatively small (or structurally simple) programs, but it can still be used on larger functions, to obtain partial coverage.

Confidence in the results: The user can be confident in Eva's proof of absence of runtime errors or assertion violations but Eva's alarms to indicate possible bugs must be checked by the user because the bug may not really exist. In theory, the user can also be totally confident in a Wp proof but only if there are no mistakes in any annotations (for instance, initial conditions, modeling or environment-related assumptions, axioms), so the user may want to use testing for additional confirmation of their consistency. Any bug found by PathCrawler certainly exists, unless the user made a mistake in the oracle or stub (or precondition in the case of unreachable code). Full structural coverage is necessary for PathCrawler to prove the absence of unreachable code, assertion violations or runtime errors. Regarding functional properties specified by the oracle (when cross-checking is not performed) and the runtime errors not covered by PathCrawler, if no bugs are detected by the generated tests then the result of structural testing is just a degree of confidence which depends on the coverage criterion, the effective coverage and the quality of the oracle. Note that structural coverage criteria used frequently (branch, MC/DC, ...) focus on control flow and are poor at detecting some types of bug, such as computation bugs.

Help in debugging: When an alarm from Eva or a proof failure in Wp indicates a bug then to confirm its existence and correct the bug, the user must examine the source code with the help of any additional information from the tool. Eva enables the user to examine the source code annotated with value ranges. Wp provides the proof obligations, but does not provide a counterexample. If a test generated by PathCrawler detects a bug then the user is sure that it exists and can use the test inputs, or other information provided by PathCrawler, to help debug (see Sect. 6.5.6). To justify the presence of unreachable code, PathCrawler can supply the unsatisfiable path predicates.

Because of the above considerations, Eva is often deployed at the start of the verification and validation process to rapidly indicate absence or possible presence of runtime errors with a minimal user intervention.

PathCrawler can then be run on each module, with just the help of a precondition, to test for runtime errors, or with the help of an oracle or assertions too, to test for functional bugs. If full coverage is achieved by PathCrawler and no bugs are detected, then this may provide sufficient confidence. Otherwise, the test results can be used to justify the effort needed for a full proof with Wp if the user knows the implementation and has sufficient expertise.

If static analysis with Eva is inconclusive or proof with Wp fails, the user can pursue with these plug-ins, adding code annotations or trying some advanced analysis options. As an alternative, PathCrawler can be used (with the help of a precondition) to try to find a witness exhibiting each failure in a given module, by trying to cover all the assertions expressing the expected properties. A witness can often be generated even when full coverage is not achieved. If several failures were recorded, it is more efficient to run PathCrawler once on each module, with all the failing assertions, than once for each assertion.

6.8 Conclusion

In this chapter, we have presented PathCrawler, a mature tool for automated structural unit testing which allows the user a large degree of control and outputs detailed information on the results of test generation, as well as on the execution of the tests on the test-generation platform.

Development of PathCrawler started in 2002 and it was conceived as a tool to ensure full coverage of safety-critical code subjected to test standards. At around the same time, a similar tool, CUTE [17], was being independently developed. The first article on CUTE was published after that on PathCrawler [22] but it coined the term *concolic* which can also be applied to PathCrawler. However, PathCrawler has always differed from similar test generators in two ways: by using Constraint Logic Programming, instead of a SAT solver, to solve the path predicate, and by aiming for exhaustive coverage in order to satisfy standards.

Further development of PathCrawler included its integration into the Frama-C platform and making it freely available online in the form of the PathCrawler-online [13] test server. PathCrawler-online was designed for the evaluation and teaching of automated structural testing and it has been in constant, and increasing, use ever since. It is currently used for teaching in several universities, for example: Paris-Saclay, CentraleSupélec, Créteil, Évry, Bordeaux, Grenoble, Toulouse and TU Graz. A tutorial [21] and lesson examples [14] have been published to support its use in teaching.

In the course of its development, PathCrawler has also been used in numerous research projects,[12] to test code from different application domains, to support more coverage criteria or C code features [12, 20], to measure effective worst-case execution time (*WCET*) [19], and as a symbolic execution engine by researchers verifying control software [15, 25].

PathCrawler has also been used to help verification engineers to diagnose proof failures after a static analysis plug-in (like Eva) or a deductive verification plug-in (like Wp) [7, 16]. For a given unproven annotation, the StaDy plug-in instruments C code so that PathCrawler can be used to try to generate a counterexample indicating the reason of the failure, or to show—when no counterexample is found (possibly, after a partial program exploration)—that the unproven annotation is likely to hold.

Finally, the LTest toolset [2] relies on generic test objectives (called *labels*) to provide several testing services, in particular, applying static analysis techniques for detection of infeasible test objectives. LTest also extends test generators such as PathCrawler to a large range of structural coverage criteria and even proprietary industrial criteria, as it was done in joint work with Mitsubishi Electric R&D Centre Europe (MERCE), a research center of Mitsubishi Electric [3].

References

1. Ammann P, Offutt J (2008) Introduction to software testing, 1st edn. Cambridge University Press
2. Bardin S, Chebaro O, Delahaye M, Kosmatov N (2014) An all-in-one toolkit for automated white-box testing. In: Proceedings of the 8th international conference on tests and proofs (TAP 2014), held as part of STAF 2014, LNCS, vol 8570. Springer, pp 53–60. https://doi.org/10.1007/978-3-319-09099-3_4
3. Bardin S, Kosmatov N, Marre B, Mentré D, Williams N (2018) Test case generation with PathCrawler/LTest: how to automate an industrial testing process. In: Proceedings of the 8th international symposium on leveraging applications of formal methods, verification and validation. Part IV. Industrial Practice. (ISOLA 2018), LNCS, vol 11247. Springer, pp 104–120. https://doi.org/10.1007/978-3-030-03427-6_12
4. Botella B, Delahaye M, Ha SHT, Kosmatov N, Mouy P, Roger M, Williams N (2009) Automating structural testing of C programs: experience with PathCrawler. In: Proceedings of the 4th international workshop on the automation of software test (AST 2009), part of the 31st international conference on software engineering (ICSE 2009). IEEE, pp 70–78. https://doi.org/10.1109/IWAST.2009.5069043
5. Cadar C, Dunbar D, Engler D (2008) KLEE: unassisted and automatic generation of high-coverage tests for complex systems programs. In: Proceedings of the 8th USENIX symposium on operating systems design and implementation (OSDI 2008). USENIX Association, pp 209–224
6. Cadar C, Godefroid P, Khurshid S, Pasareanu CS, Sen K, Tillmann N, Visser W (2011) Symbolic execution for software testing in practice: preliminary assessment. In: Proceedings of the 33rd international conference on software engineering, (ICSE 2011). ACM, pp 1066–1071. https://doi.org/10.1145/1985793.1985995

---

[12] See the full list of publications on http://pathcrawler-online.com/ in the About menu.

7. Chebaro O, Cuoq P, Kosmatov N, Marre B, Pacalet A, Williams N, Yakobowski B (2014) Behind the scenes in SANTE: a combination of static and dynamic analyses. Autom Softw Eng 21(1):107–143. https://doi.org/10.1007/s10515-013-0127-x
8. Esposito C, Cotroneo D, Silva N (2011) Investigation on safety-related standards for critical systems. In: Proceedings of the first international workshop on software certification, (WoSoCER 2011). IEEE Computer Society, pp 49–54. https://doi.org/10.1109/WoSoCER.2011.9
9. Gigante G, Pascarella D (2012) Formal methods in avionic software certification: the DO-178C perspective. In: Proceedings of the 5th international symposium on leveraging applications of formal methods, verification and validation. Applications and case studies (ISoLA 2012), LNCS, vol 7610. Springer, pp 205–215. https://doi.org/10.1007/978-3-642-34032-1_21
10. Godefroid P, Levin MY, Molnar DA (2012) SAGE: whitebox fuzzing for security testing. Commun ACM 55(3):40–44. https://doi.org/10.1145/2093548.2093564
11. King JC (1976) Symbolic execution and program testing. Commun ACM 19(7):385–394
12. Kosmatov N (2008) All-paths test generation for programs with internal aliases. In: Proceedings of the 19th international symposium on software reliability engineering (ISSRE 2008). IEEE, pp 147–156. https://doi.org/10.1109/ISSRE.2008.25
13. Kosmatov N, Williams N, Botella B, Roger M (2013) Structural unit testing as a service with PathCrawler-online.com. In: Proceedings of the 7th IEEE international symposium on service-oriented system engineering (SOSE 2013). IEEE, pp 435–440. https://doi.org/10.1109/SOSE.2013.78
14. Kosmatov N, Williams N, Botella B, Roger M, Chebaro O (2012) A lesson on structural testing with PathCrawler-online.com. In: Proceedings of the 6th international conference on tests and proofs (TAP 2012), LNCS, vol 7305. Springer, pp 169–175. https://doi.org/10.1007/978-3-642-30473-6_15
15. Park J, Pajic M, Lee I, Sokolsky O (2016) Scalable verification of linear controller software. In: Proceedings of the 22nd international conference on tools and algorithms for the construction and analysis of systems (TACAS 2016), held as part of the European joint conferences on theory and practice of software (ETAPS 2016), LNCS, vol 9636. Springer, pp 662–679. https://doi.org/10.1007/978-3-662-49674-9_43
16. Petiot G, Kosmatov N, Botella B, Giorgetti A, Julliand J (2018) How testing helps to diagnose proof failures. Formal Aspects Comput 30(6). https://doi.org/10.1007/s00165-018-0456-4
17. Sen K, Marinov D, Agha G (2005) CUTE: a concolic unit testing engine for C. In: Proceedings of the 10th European software engineering conference held jointly with 13th ACM SIGSOFT international symposium on foundations of software engineering (FSE 2005). ACM, pp 263–272. https://doi.org/10.1145/1081706.1081750
18. Tillmann N, de Halleux J (2008) Pex-white box test generation for .NET. In: Proceedings of the 2nd international conference on tests and proofs (TAP 2008), LNCS, vol 4966. Springer, pp 134–153. https://doi.org/10.1007/978-3-540-79124-9_10
19. Williams N (2005) WCET measurement using modified path testing. In: Proceedings of the 5th international workshop on worst-case execution time analysis (WCET 2005), OASICS, vol 1
20. Williams N (2021) Towards exhaustive branch coverage with PathCrawler. In: Proceedings of the 2nd IEEE/ACM international conference on automation of software test (AST@ICSE 2021). IEEE, pp 117–120. https://doi.org/10.1109/AST52587.2021.00022
21. Williams N, Kosmatov N (2012) Structural testing with PathCrawler: tutorial synopsis. In: Proceedings of the 12th international conference on quality software (QSIC 2012). IEEE, pp 289–292. https://doi.org/10.1109/QSIC.2012.24
22. Williams N, Marre B, Mouy P (2004) On-the-fly generation of k-paths tests for C functions: towards the automation of grey-box testing. In: Proceedings of the 19th IEEE international conference on automated software engineering (ASE 2004). IEEE, pp 290–293. https://doi.org/10.1109/ASE.2004.10020
23. Williams N, Marre B, Mouy P, Roger M (2005) PathCrawler: automatic generation of path tests by combining static and dynamic analysis. In: Proceedings of the 5th European dependable computing conference (EDCC 2005), LNCS, vol 3463. Springer, pp 281–292. https://doi.org/10.1007/11408901_21

24. Zhu H, Hall PAV, May JHR (1997) Software unit test coverage and adequacy. ACM Comput Surv 29(4):366–427
25. Zutshi A, Sankaranarayanan S, Deshmukh JV, Jin X (2016) Symbolic-numeric reachability analysis of closed-loop control software. In: Proceeding of the 19th international conference on hybrid systems: computation and control (HSCC 2016). ACM, pp 135–144. https://doi.org/10.1145/2883817.2883819

Part II
Advanced Usages and Analyses

Chapter 7
The Art of Developing Frama-C Plug-ins

François Bobot, André Maroneze, Virgile Prevosto, and Julien Signoles

Abstract One of the key features of Frama-C is its extensibility. More precisely, the platform is based on a kernel, which provides the core services and datastructures that are needed for analyzing C programs, including in particular parsing C and ACSL code. Analyses themselves are then implemented by *plug-ins*, that use the kernel's API to, among other things, access the code under analysis, perform some code transformation, add ACSL annotations, and validate (or invalidate) other ACSL annotations. Furthermore, plug-ins can also export their own API to be used by other plug-ins. In this chapter, we will give an overview of Frama-C's general architecture and describe the main functionalities of the kernel, using as example a small plug-in that we build step by step during the course of the chapter.

Keywords Dataflow analysis · Abstract syntax tree · OCaml · Visitor pattern

This chapter presents how to develop new Frama-C plug-ins. Indeed, Frama-C is an extensible open source framework: anyone can contribute by developing new code analyzers and linking them to the framework by writing a plug-in. In such a context, Frama-C's kernel and the existing plug-ins can be seen as a large library for code analysis enabling the implementation of promising prototypes for research activities and the development of sophisticated and scalable tools for industrial uses at much lower cost than starting from scratch.

F. Bobot (✉) · A. Maroneze · V. Prevosto · J. Signoles
Université Paris-Saclay, CEA, List, Palaiseau, France
e-mail: francois.bobot@cea.fr

A. Maroneze
e-mail: andre.maroneze@cea.fr

V. Prevosto
e-mail: virgile.prevosto@cea.fr

J. Signoles
e-mail: julien.signoles@cea.fr

Frama-C is developed in the OCaml programming language,[1] and so are its plug-ins. OCaml is often considered a tool of choice for developing software that heavily relies on symbolic manipulations, such as compilers and code analyzers. It is one of the main reasons—but not the only one—for having chosen to develop Frama-C in this language [6]. Therefore, this chapter assumes a minimal knowledge of OCaml. If necessary, many materials for learning OCaml are available online, including the official documentation,[2] books [8, 16], and tutorials.[3] This chapter also assumes some very basic knowledge in compilation, such as the notion of Abstract Syntax Tree (AST), which can easily be learned from any introductory course in compilation, such as [2] for instance.

This chapter focuses on explaining the key programming notions that are useful and often specific to Frama-C for developing plug-ins. It does not aim at being a comprehensive guide on how to write Frama-C plug-ins. In particular, it does not replace the Frama-C Plug-in Development Guide [14] for users who would like to develop advanced plug-ins. Additionally, it does not explain how to design or efficiently implement a specific algorithm for code analysis, which is usually dependent on the underlying analysis technique (e.g., abstract interpretation [3, 12]) and is often part of an active research area. More precisely, this chapter shows the most important points that are needed for writing a new Frama-C plug-in:

- set up a compilation environment and write the code to let the plug-in register itself in the kernel;
- manipulate the AST nodes representing C and ACSL constructions;
- use the visitor mechanism for traversing the whole AST and performing code transformations;
- implement a dataflow analysis;
- create and use Frama-C's projects to perform several independent analyses on the same code base;
- use Frama-C's logging infrastructure;
- define command-line options;
- test a plug-in;
- package and distribute a plug-in.

The Frama-C platform provides its users with all the needed infrastructure to write their own analysis, either from scratch or on top of existing ones. It offers a very rich set of features but its use comes with a few challenges. The main challenge is to know which part of the kernel provides which feature. The second challenge is the complexity of C, which makes transformations of the AST and its analysis complicated, even though Frama-C provides helpers for many tasks. The third challenge is brought by Frama-C itself, since the use of mutable global states, such as the one manipulated by the project library, creates possible pitfalls. We will emphasize the

---

[1] https://ocaml.org.
[2] https://v2.ocaml.org/releases/4.14/htmlman/index.html.
[3] https://ocaml.org/docs.

7 The Art of Developing Frama-C Plug-ins

points that require some care throughout this chapter, and give a summary of the main difficulties in Sect. 7.9.

Before starting, we briefly introduce Frama-C's plug-in-based software architecture, depicted in Fig. 7.1. Each plug-in is usually a code analyzer or a source-to-source transformation. It can also be a (small) script that automatically tunes some parame-

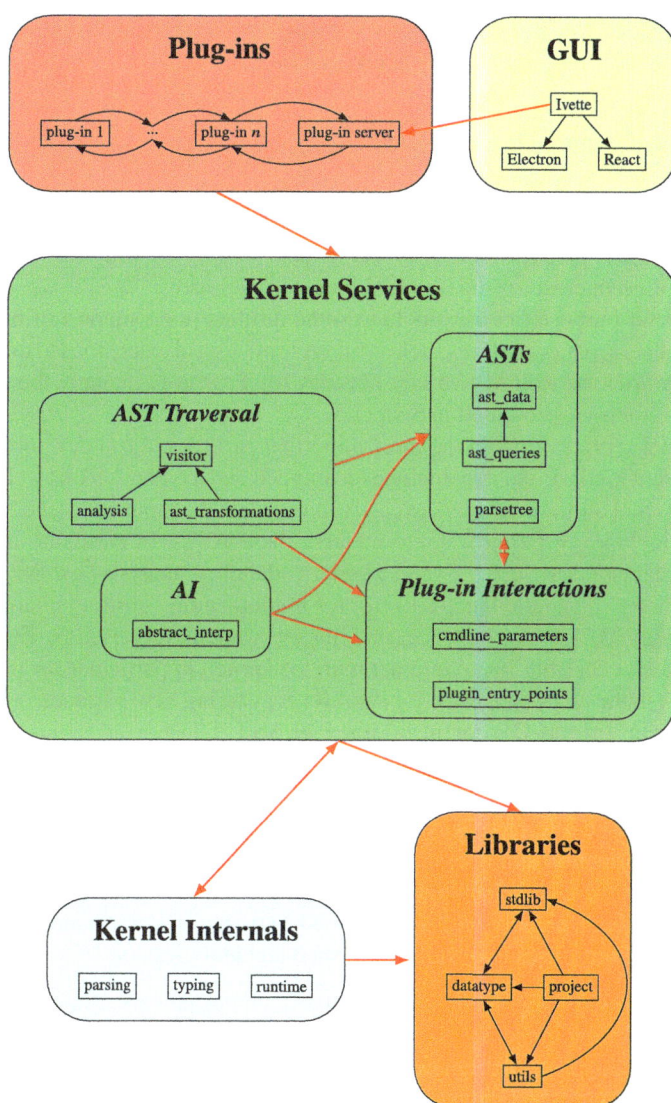

Fig. 7.1 Frama-C's Architectural Design. Red arrows are dependencies between main components, and black arrows are dependencies internal to one main component

ters of more advanced plug-ins for specific usages. The plug-ins are built on top of a *kernel* that provides a uniform setting, as well as several services, available through an API, to the plug-in developer. In addition to the kernel APIs, a plug-in may also use the API offered by other plug-ins. For instance, this chapter will illustrate how to develop a plug-in that first runs the Eva plug-in, then relies on its results to get some useful information about the program's memory values. Therefore, from a plug-in developer point of view, Frama-C is a large library that helps develop new tools handling C code.

The kernel is composed of three main parts:

- internal modules;
- some general-purpose libraries; and
- several specific services.

The kernel internal modules are in charge of managing the platform itself (e.g., initializing it) and preparing the AST that is analyzed by the plug-ins. This part of the API is of no interest for the standard plug-in developer.

The general-purpose libraries include some utilities (e.g., support of the JSON format, or generic data structures such as bit vectors), useful extensions to the OCaml standard libraries, as well as two key libraries for Frama-C, namely the *Datatype* and *Project* libraries, presented in Sect. 7.5.

The kernel services contain the most important part for plug-in developers. They are of different kinds. First, they include an untyped and a typed AST as well as associated tables that allow developers to easily manipulate the input code (see Sect. 7.4). Second, it includes different ways to traverse these ASTs, in particular a generic visitor mechanism (see Sect. 7.6) and a generic dataflow analysis (see Sect. 7.7), as well as a library of generic lattices useful for implementing abstract interpretation-based analyses [3, 12]. For instance, this library is intensively used by Eva. Third, it provides ways for plug-ins to interact with the kernel, in particular for registering the plug-ins themselves (see Sect. 7.2) and for registering new parameters available as options from Frama-C's command line (see Sect. 7.3.4).

The Graphical User Interface (GUI) is an important part of the interaction of the user with Frama-C. Indeed, some analyses results are difficult to interpret in a purely text-based interface. Any plug-in can extend the GUI in several ways, for displaying information and interacting with the users. The communication between the plugins and the GUI is handled by the plugin server as shown in Fig. 7.1. The new Frama-C GUI, namely Ivette (Integrated Verification Environment Toolkit for Trusted Executions), uses two well-established technologies:

- Electron[4] provides a graphical engine that is supported very efficiently by most modern web browsers.
- React[5] is a JavaScript framework that allows users to modify data instead of directly modifying the GUI widgets that use it.

---

[4] https://www.electronjs.org/.

[5] https://react.dev/.

The Frama-C binary communicates through sockets with the Electron application, using Remote Procedure Calls (RPCs) with a custom protocol. The Server plug-in contains the code of the server and its API can be used to define new RPCs. The OCaml side only gives access to operations and data, all the display is done in the GUI by Ivette. Since these technologies are still evolving, interested users should refer to an up-to-date documentation for concrete examples.

The rest of this chapter is organized as follows. Section 7.1 introduces a running example that is used throughout the chapter to provide concrete code snippets. Section 7.2 explains how to set up a new plug-in. Section 7.3 shows how to deal with inputs and outputs in a plug-in. Section 7.4 introduces Frama-C's AST and how to access it. Section 7.5 presents two key Frama-C libraries, namely *datatypes* and *projects*. Section 7.6 introduces Frama-C's visitor and how it helps deal with AST traversal. Section 7.7 shows Frama-C's *interpreted automata* that help develop dataflow-like analyses. Section 7.8 explains how to test, profile, benchmark, and prepare a Frama-C plug-in to be released as an open-source tool. Finally, Section 7.9 summarizes the main features offered by the kernel, as well as the most important points to be aware of when developing a plug-in.

There are numerous OCaml examples throughout this chapter, but some functions have been omitted to avoid overly long code listings. The complete source code for the plug-in that is described in this chapter can be found at https://git.frama-c.com/pub/frama-c-book-companion/-/tree/main/plugindev.

7.1 Running Example

In order to make things more concrete, let us introduce here a small analysis that we would like to implement on top of Frama-C's kernel. This section gives an overview of the information we would like to compute, together with a first set of requirements for the plug-in's implementation, which we will detail in the rest of this chapter.

7.1.1 Typestates

Typestates have been introduced by Strom and Yemini [15] as a means to keep track of the set of operations that can be performed over an object during its lifetime. It is a refinement of the static type of the object (hence the name). However, contrary to static types, typestates may change during the object lifetime, so that some kind of dataflow analysis is required to check typestates-related properties.

As an example, let us focus on a very simple API to manipulate files, as presented in Fig. 7.2. The internal representation of a file is deliberately left abstract, as we will only manipulate pointers to **struct inner_rep**. The first operation we can do is allocating a new structure via `alloc_file`. Then, the file can be opened, either for reading or writing (or both). After that, the corresponding operations can be

Fig. 7.2 Simple API for file

```
1  struct inner_repr;
2
3  typedef struct inner_repr *file;
4
5  file alloc_file(const char* filename);
6
7  void open_read(file f);
8
9  void open_write(file f);
10
11 char read(file f);
12
13 void write(file f, char c);
14
15 void flush(file f);
16
17 void close(file f);
18
19 void free_file(file f);
```

performed on the file (to keep the API simple, it is only possible to read sequentially from the start and appending characters at the end of the file). Finally, if some write accesses have been performed, data must be flushed to disk before the file can be closed. A closed file can be re-opened again (either for reading or writing), or it can be disposed of and the associated memory freed. To avoid memory leaks, all files must be freed when the program terminates its execution.

Based on this API, we can describe typestates as the nodes of a graph whose edges are the operations that are allowed on an object that is in this typestate. Figure 7.3 depicts this graph. Intuitively, the idea behind typestates is quite simple: each time the program is about to perform an operation on a file f, we have to check whether f is in a typestate that allows this operation, and update its typestate according to the graph. Finally, at the end of the execution, all files must be in an accepting state (freed in our case).

7.1.2 Requirements for MiniTypeState

Making a full-fledged typestate-checker plug-in for Frama-C would be far beyond the scope of this chapter. Instead, we focus on the main ingredients of such an analysis, and present the components of the kernel on top of which such a plug-in can be built.

First, let us explicitly narrow down the scope of MiniTypeState, our typestate checker, in order to keep the plug-in simple. Its biggest restriction is that it is only aimed at verifying the usage of the file API of Fig. 7.2. It would be possible to make it more generic, notably by using ACSL's extension mechanisms as done by the MetAcsl and RPP plug-ins (Chap. 10) to let users define arbitrary typestate graphs for their own API, but this would get us too far. Hence, we will stick to a fixed graph for a given API. We also assume that all calls to the functions in the API are

7 The Art of Developing Frama-C Plug-ins 347

Fig. 7.3 Typestates for the file API

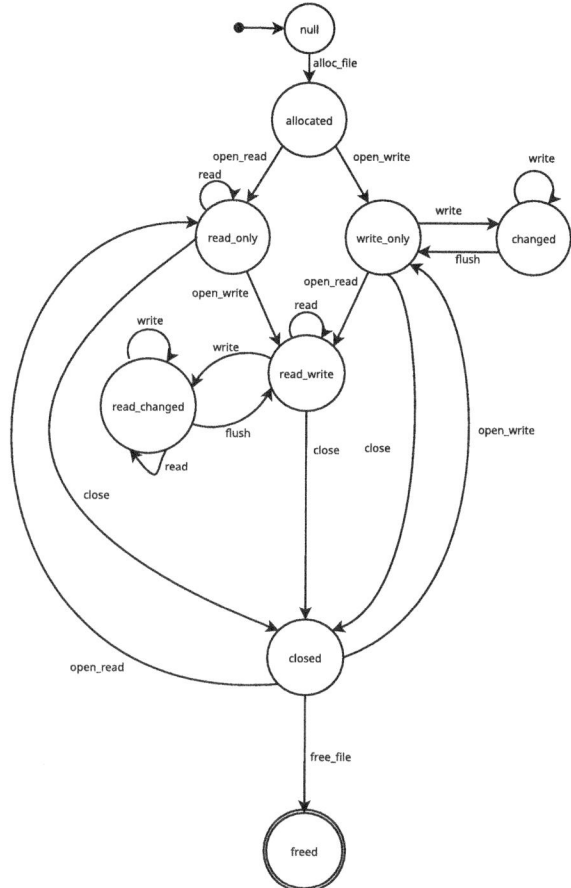

direct, i.e. that they do not occur through a function pointer. As we will see later, it is possible to check that this is the case and emit a warning otherwise, or mitigate that by relying on information from other plug-ins, e.g. **Eva** (Chap. 3) that can provide the set of all potential target functions of an indirect call.

In this context, there are two main possibilities for analyzing typestates that we will explore in turn. First, we can generate specific ACSL annotations at each call site of a function in the API, and rely on one of the main analyzers of **Frama-C**, **Eva** (Chap. 3), **Wp** (Chap. 4), or **E-ACSL** (Chap. 5) for verifying these annotations. This is described in more details in Sects. 7.4 and 7.6. Second, we can provide our own dedicated dataflow analysis[6] for verifying the API usage. This is presented in Sect. 7.7.

---

[6] Actually, it would also be possible to provide a specific domain for **Eva** to propagate typestate information together with the rest of **Eva**'s abstract state. Building **Eva** domains is presented in Chap. 3.

7.2 Plug-in Setup

Since its version 26 (Iron), Frama-C relies on the build tool Dune[7] for configuring, compiling and installing itself and its plug-ins. Dune is specialized for OCaml projects, and features predefined rules for a wide range of cases. Therefore, a Frama-C plug-in is structured as a directory containing at least the three following files, even if it usually contains several other files, in particular OCaml source files and test files:

- dune-project
- dune
- <Plugin>.ml

The dune-project file specifies: the minimal required Dune version; the fact that we use dune_site (so that we can register our project inside another one, that is, Frama-C); the fact that Dune itself will generate an opam[8] file to help with installation; the project name; and finally, the package name, synopsis, and dependencies.

```
1 (lang dune 3.2)
2 (using dune_site 0.1)
3
4 (generate_opam_files true)
5
6 (name frama-c-minitypestate)
7 (package
8  (name frama-c-minitypestate)
9  (synopsis "MiniTypeState plug-in for Frama-c")
10  (depends
11    (frama-c (>= 26.1))
12  )
13 )
```

Note that Frama-C plug-ins, by convention, have their package name starting with "frama-c-".

The dune file below specifies the plug-in name (as a library), compilation flags and package dependencies (including Frama-C's kernel and other plug-ins). By convention, the public name of a plug-in ends with .core.[9]

```
1 (library
2  (name MiniTypeState)
3  (public_name frama-c-minitypestate.core)
4  (flags -open Frama_c_kernel :standard -w -9-27)
5  (libraries frama-c.kernel)
6  (preprocess (pps ppx_deriving.show)))
7
```

---

[7] https://dune.build/.

[8] This will be used for packaging the plug-in, as shown in Sect. 7.8.2.

[9] In addition to .core, some complex plug-ins also define sub-packages with names such as frama-c-plugin.subpkg, but this is outside the scope of this chapter.

```
8  (plugin
9   (optional)
10  (name minitypestate)
11  (libraries frama-c-minitypestate.core)
12  (site (frama-c plugins)))
```

Finally, the main OCaml file, in our case MiniTypeState.ml, contains the public API of our plug-in. It is empty by default. We will decide which functions to export later.

We can add as many OCaml files as we wish. They are going to be compiled and integrated in our plugin automatically. Dependencies between files are automatically handled by Dune. For example, we can add the following options.ml file to register the plug-in in Frama-C[10]:

```
1  include Plugin.Register (struct
2    let name = "Mini-TypeState"
3    let shortname = "mts"
4    let help = "A mini-typestate plug-in"
5  end)
```

Plugin.Register is a functor that expects a structure as argument containing:

- the plug-in name (here, Mini-TypeState);
- a short name (here, mts), used in particular for prefixing the plug-in's command-line option names;
- a help message describing the plug-in.

This registers the plug-in in Frama-C's kernel, even if the plug-in currently does nothing by itself.

Whenever source files are added or modified, running dune build recompiles them. Then, in order to run Frama-C with this extra plug-in, we can run:

```
1  dune exec -- frama-c <options>
```

For now, we run dune exec - frama-c -plugins to see that our plug-in has been registered with Frama-C. The following line must appear among all other plug-ins (listed alphabetically):

```
1  Mini-TypeState  A mini-typestate plug-in (-mts-h)
```

Executing a Command Inside the Build Environment

As a side note, dune exec -- allows executing a program in the context of the current working directory, enabling, for example, locally defined plug-ins to be found. However, they are neither automatically compiled nor updated.

---

[10] We do not provide any interface file options.mli: adding it is left as an exercise for the developer.

Finally, the command `dune build @doc` allows us to generate the documentation of the API of your plug-ins using the odoc tool.[11] Through mld files, we can also generate user manuals (see odoc's documentation[12] for more information). The generated documentation can be found in _build/default/_doc.

7.3 Plug-in Inputs/Outputs

The plug-in development process is, at its core, an OCaml programming exercise, which depends on knowledge about the analysis to be performed. The following points, which are Frama-C-specific, should help plug-in developers iteratively develop and improve their plug-ins with the help of many features provided by Frama-C's kernel: the definition of the plug-in's entry point (Sect. 7.3.1); some basic debugging and pretty-printing modules (Sect. 7.3.2); Frama-C's logging facilities to parameterize output messages of various kinds and verbosity levels (Sect. 7.3.3); and a brief summary about command-line options (Sect. 7.3.4), to help handle some advanced features of the rich command line provided by Frama-C. These sections are illustrated with examples based on our MiniTypeState plug-in.

7.3.1 Entry Point

The main entry point of a plug-in is the function
`Db.Main.extend: (unit -> unit) -> unit`

Db is a module that historically contains the main plug-in APIs and entry points: it is nowadays almost exclusively used for this critical function and a few secondary plug-ins. This function is executed after Frama-C is initialized and configured. By convention, the main function of a plug-in is often called `run` or `main`. It is set up as follows, typically at the end of the main module `main.ml` of the plug-in:

```
1 let run () = assert false (* analyzer body, not yet defined *)
2 let () = Db.Main.extend run
```

In order to prevent the kernel from running the MiniTypeState plug-in each time Frama-C is run, we need to define an option `-typestate` that the user must set to run the plug-in. This is done by adding the following module definition in file `options.ml`, after the `Plugin.Register` part.

```
1 module Run =
2   False
3     (struct
4       let option_name = "-typestate"
5       let help = "Activate the mini-typestate analysis"
6     end)
```

---

[11] https://github.com/ocaml/odoc.
[12] https://ocaml.github.io/odoc/odoc_for_authors.html#doc-pages.

7 The Art of Developing Frama-C Plug-ins

The `False` functor is defined by `Plugin.Register`. It allows the developer to define a new Boolean option, unset by default (here, `-typestate`). It also defines its dual negative option (here, `-no-typestate`) to disable it once enabled.

The `run` function, entry point of this plug-in, can now be added to the `main.ml` file:

```
1 let run () =
2   if Options.Run.get () then begin
3     Options.feedback "Starting typestate analysis...";
4     (* ... analysis code goes here ... *)
5     Options.feedback "Finished typestate analysis.";
6   end
7
8 let () = Db.Main.extend run
```

This code checks if the user requested the analysis by calling `Options.Run.get`. If this is the case, we emit messages both at the beginning of the analysis and at its end. The function `Options.feedback` prints the message (see Sect. 7.3.3 for details). We can test that it works by running `dune build` to recompile, and then `dune exec -- frama-c -typestate`:

```
1 [mts] Starting typestate analysis...
2 [mts] Finished typestate analysis.
```

Remark 1 (*Advanced Entry Point Considerations*) Variadic—which performs AST transformations before other plug-ins are run—have a different entry point. Developing such plug-ins requires knowledge of the initialization stages of Frama-C, which is detailed in Section *Initialization steps* of the Plug-in Development Guide [14]. Currently, our MiniTypeState plug-in's entry point is set to run during Frama-C's main *stage*, when the main actions of most plug-ins are executed. In Sect. 7.4.3, we will set up some AST transformations that take place earlier.

Remark 2 (*Chaining plug-ins*) Dependent plug-ins, which need the result of others before they can be run, can be chained together in different ways. The simplest (for the plug-in developer) is to check and fail if the prerequisite plug-in has not been run. This requires that the user properly adds -then sequencers to Frama-C's command-line, as described in Chap. 2, and limits the flexibility of the plug-in. The other way is to programmatically run the dependent plug-in via its API.

As a general rule, avoid writing code that is evaluated "immediately", that is, via a `let () = ...` construction at top-level of a plug-in file. Such code often runs too early, before Frama-C's kernel has been able to initialize several of its data structures, leading to inexplicable errors.

7.3.2 Basic Debugging and Pretty-Printing

OCaml programmers rarely use debuggers, resorting more often to adding pretty-printing messages (typically, Format.printf) when debugging/understanding some piece of code. Frama-C plug-in development also follows this rule: several pretty-printing functions are available throughout the code to help developers quickly hack output messages to understand what is going on.

A very useful module is the Pretty_utils[13] library; it provides useful pretty-printing functions, such as pp_list that combine with elementary pretty-printers to output a list of such elements. For instance, to print a list of statements, you can use:

```
1  Format.printf "my_stmts: %a@."
2    (Pretty_utils.pp_list ~sep:"; " Printer.pp_stmt) my_stmts
```

The above example uses Format.printf. The %a is used for custom printing functions. Then, Pretty_utils.pp_list is called, with the function Printer.pp_stmt as pretty-printer for the list element type. Printer is a Frama-C module that offers pretty-printers for AST elements. Pretty-printers can be chained. For instance, one can print a list of lists of strings as follows:

```
1  Format.printf "all names: %a@."
2    (Pretty_utils.pp_list ~pre:"[" ~suf:"]" ~sep:", "
3      (Pretty_utils.pp_list ~sep:" " Format.pp_print_string))
4    my_names
```

The pre, suf, and sep arguments are optional, but often useful for readability (the first two are for example not present in Format.pp_print_list).

Some specific Frama-C modules exist to help debugging, e.g. Cabs_debug and Cil_types_debug. The latter is especially useful to understand the underlying AST components, since they are printed in "OCaml constructor" form. Note that the debug pretty-printers are configured to omit some data and to avoid excessive and cyclic outputs.

Usage of Cil_types_debug

Consider the following variable declaration in the analyzed C file:

```
1  struct st *array_of_pointers_to_struct[2];
```

Pretty-printing its type in debug form as an OCaml constructor can be done this way:

```
1  let typ = (* see Ex 1 in Sect. 4.2 *) in
2  Format.printf "AST representation of type is:@\n%a@."
3    Cil_types_debug.pp_typ typ
```

---

[13] File src/libraries/utils/pretty_utils.mli.

and would result in the following output:

```
1  AST representation of type is:
2    TArray(
3      TPtr(
4        TComp(
5          {cstruct=true; cname=st; cattr=[]})),
6      Some(
7        {enode=Const(
8            CInt64(
9              Integer(2),
10             IInt))}))
```

7.3.3 Outputting Messages

Plug-ins use the kernel's module Log[14] to emit messages, based on OCaml's Format module. The Log module signature is imported via Plugin.Register. In our example, since it is included in module Options, you can output plug-in's messages by using Options.feedback as shown in Sect. 7.3.1. Beyond function feedback, here is a summary of the existing output functions:

- debug: debugging information for developers of the plug-in;
- feedback: for general messages; equivalent to log level INFO in most logging frameworks;
- result: analysis results;
- warning: non-critical warnings; usually affect the correctness of the analysis (e.g., additional hypotheses taken into consideration);
- error: an erroneous situation, *caused by the user* (most often, some invalid input), has been detected; however, analysis will continue, reporting as many errors as possible before stopping. If this happens in a function returning a value, error requires the plug-in developer to produce a fallback value (e.g., an empty list if no valid elements were found);
- failure: also an error, but *not* caused by the user; almost always indicates a bug in the plug-in, or some unexpected violation of an internal invariant (e.g., the plug-in assumed there were no variadic functions, but such a call has been found);
- abort: a user error, but causing the analysis to stop immediately; the main difference w.r.t. error is that, for the plug-in developer, no fallback value is required;
- fatal: the internal (non-user) counterpart to abort.

Most functions are self-evident, except for the 4 error-related functions. The Table 7.1 below groups them according to whether execution halts or continues, and whether the origin of the error is due to some incorrect user input, or some internal

---

[14] Full details available in the *Plug-in Development Guide* [14], Sect. *Logging Services*.

Table 7.1 Error-reporting functions in Frama-C's logging mechanism

Error source	Continue execution?	
	Yes (Report many errors)	No (Halt immediately)
User/External input	Error	Abort
Internal (likely a bug)	Failure	Fatal

unsatisfied condition. If unsure, we recommend using abort at first, and replacing it with error when possible, supplying a meaningful default value if needed. For unimplemented features, use not_yet_implemented. When handling cases that should never arise in practice (e.g. a particular pattern in a **match ... with** that should be impossible due to semantic constraints), or checking invariants, use fatal with a meaningful error message instead of OCaml's **assert false** because it provides better debugging feedback by default.

7.3.3.1 Example of Output Messages

To illustrate the usage of the Log API on our MiniTypeState plug-in, let us revisit the initial messages in the main.ml file, with some added constraints:

- we do not want to emit the "starting/finished typestate analysis" messages by default, but only if the user requested some verbose output;
- we want to add two different execution modes: by default, a *strict* mode which fails whenever the analysis is unsure of its results, that is, whenever an indirect function call is detected; but the user can enable a *relaxed* mode that only emits a warning, without stopping the analysis;
- whenever possible, the warnings/error messages should include source location data, to help the user understand which part of their code is the culprit.

For the first requirement, we assume the user will add command-line option -mts-verbose 2 whenever they want the "starting analysis" message. The -<plugin>-verbose option is provided by Frama-C for all registered plug-ins. The code would then become:

```
1    if Options.Run.get () then begin
2      Options.feedback ~level:2 "Starting typestate
           analysis...";
3      (* ... analysis code goes here ... *)
4      Options.feedback ~level:2 "Finished typestate analysis.";
5    end
```

Alternatively, for even finer-grained control, we might want to use a specific *message key* for the starting message, and a different one for the finishing one. Here is a way to define and use them:

```
1    let start_dkey = Options.register_category "start"
```

7 The Art of Developing Frama-C Plug-ins   355

```
2 let finish_dkey = Options.register_category "finish"
3
4 let run () =
5   if Options.Run.get () then begin
6     Options.feedback ~dkey:start_dkey "Starting␣typestate␣
        analysis...";
7     (* ... analysis code goes here ... *)
8     Options.feedback ~dkey:finish_dkey "Finished␣typestate␣
        analysis.";
9   end
```

Now, to enable such messages, the user needs to add options -mts-msg-key=start and -mts-msg-key=finish (or combine both: -mts-msg-key=start,finish) to the command-line. Note that when passing dkey as argument to feedback or debug, the corresponding messages are labeled with the dkey in addition to the plug-in name. For instance, when enabled, our starting message will now read:

[mts:start] Starting typestate analysis...

For the second and third requirements (strict/relaxed modes, and source location), we assume that the user already created a module Allow_unsafe in options.ml, which creates a Boolean option. We omit the details of detecting such indirect calls (they will be presented in Sect. 7.4.1), replacing them with a comment. This is how we could implement both warning and abort messages:

```
1 (...)
2 if (* found indirect call *) then
3   if Options.Allow_unsafe.get () then
4     Options.warning ~current:true ~once:true
5       "Indirect␣call␣detected;␣results␣may␣be␣unsafe"
6   else
7     Options.abort ~current:true
8       "Indirect␣call␣detected;␣results␣would␣be␣unsafe."
```

Note the presence of current:true and once:true arguments. The former includes source code information in the message: Frama-C automatically tracks source code location during its AST visits, so that current:true will print e.g. [mts] file.c:17: Indirect call detected (...), assuming the indirect call happened at file.c, line 17. The once:true parameter prevents an identical message from being emitted multiple times. When current:true is enabled, this will only happen if the analysis passes several times over the same program point.

7.3.3.2 Automatic Switching Between Warnings and Errors

The code snippet above can be simplified and improved: since Frama-C 17, plug-in developers no longer need to add options such as Allow_unsafe and then choose between warnings and errors; instead, with option -mts-warn-key

<category>=<behavior>, plug-in developers only need to declare *warn categories*, and let the users choose what to do with them:

```
1 let wkey_indirect_call = Options.register_warn_category
      "api:indirect-call"
2 (...)
3   if (* found indirect call *) then
4     Options.warning ~current:true ~once:true
          ~wkey:wkey_indirect_call
5       "Indirect call detected; results may be unsafe"
```

register_warn_category is the counterpart of register_category, but for warning messages instead of feedback/debug. The labeled argument wkey given to the Options.warning call associates it with the specified api:indirect-call key. Like for dkey in the previous section, the messages emitted through such a call to warning will be labeled with the wkey. For instance, indirect calls will lead to an output like this:

```
[mts:api:indirect-call] file.c:17: Indirect call detected (...)
```

In addition, adding wkey allows the message to be controlled on the command line with option -mts-warn-key api:indirect-call=<behavior>, where the main behaviors are: active (by default; emit a warning), error (emit an error and continue), abort (emit an error and stop), and inactive (disable the warning entirely).

Thus, instead of a new option such as -mts-allow-api-indirect-calls, the wkey allows the user to set -mts-warn-key api:indirect-call=abort if they want to be strict. If you want the strict mode to be the default, you can simply set the default behavior for the warn category to be "abort":

```
1 let wkey_indirect_call =
2   Options.register_warn_category "api:indirect-call"
3 let () = Options.set_warn_status wkey_indirect_call Log.Wabort
```

In this case, -mts-warn-key api:indirect-call=active enables the relaxed mode. Ideally, most messages emitted by a plug-in should have a warning category, especially if some users might want to consider them as warnings/errors.

7.3.3.3 Hierarchical Categories

The name given to each category (either debug or warning) is a list of colon (:)-separated strings defining a *hierarchy* of keys. In our example, we have actually created *two* warning categories: one named api, and another named api:indirect-call. Setting the behavior of the parent also sets that of its children.

7.3.4 Command-Line Options

Frama-C has a very rich command line, and offers several facilities for plug-in developers: handling and validation of command-line options, configuration of several analysis parameters...

For instance, suppose we wanted to extend our **MiniTypeState** plug-in to handle a different API, and take a text file as input. We would want an option `-mts-input` of type `Filepath`, which would be written as follows in `options.ml`:

```
module Input =
  Filepath
    (struct
      let option_name = "-mts-input"
      let arg_name = "input"
      let file_kind = "input API"
      let existence = Fc_Filepath.Must_exist
      let help = "Name of the input file with the API
        automaton"
    end)
```

`arg_name`, `file_kind`, and `help` are just user-visible strings to help document the option. `existence` allows automatically checking that the option argument matches an existing file. If the option is not set (`Options.Input.is_set () == false`), you can assume a default value.

For a more complex example, let us assume the state transitions used for the automaton in Fig. 7.3 are not hard-coded in the plug-in, but entirely defined via the command-line, using the syntax `<f>:<src>-<dst>` to mean "function `<f>` changes the typestate from `<src>` to `<dst>`". **Frama-C** can help us check that the user entered valid function names and valid state names in the command line. We will use the `Kernel_function_map` module to define our option, `-mts-transitions`, as follows:

```
(* Predefined, hard-coded set of typestate names *)
let states = Datatype.String.Set.of_list ["null";
    "allocated"; "changed"; ...]

module Transitions =
  Kernel_function_map
    (struct
      type key = Cil_types.kernel_function
      include Datatype.String (* type of values *)
      let of_string ~key:kf ~prev:_ edge_opt =
        match edge_opt with
        | None -> abort "missing transition for '%a'"
            Kernel_function.pretty kf
        | Some edge ->
          begin
            match Str.split (Str.regexp "-") edge with
            | [src; dst] ->
              if Set.mem src states && Set.mem dst states then
                edge_opt (* valid mapping; return it *)
```

```
18              else abort "invalid transition: %s" edge
19            | _ -> abort "transition '%s' must have a single
                     '-'" edge
20         end
21     let to_string ~key:_ state = state
22   end)
23   (struct
24     let option_name = "-mts-transitions"
25     let arg_name = "f:src->dst"
26     let help =
27       Format.asprintf
28         "Map function 'f' to transition src -> dst. Valid
             state names are: %a"
29         Datatype.String.Set.pretty states
30     let default = Kernel_function.Map.empty
31   end)
```

In the above code, the Kernel_function_map functor needs to be instantiated with two modules; the first one is a datatype (see Sect. 7.5 for more details about datatypes) with two main functions: to_string and of_string, which specify how the user-provided arguments are converted into the data used by the plug-in. The of_string function in particular performs the validation.

The second module given to Kernel_function_map is the typical module used by command-line options: it contains the option name, its default value, and documentation data (arg_name and help). If the user invokes **Frama-C** with erroneous options, such as -mts-transitions invalid_function:null-allocated, or -mts-transitions foo:invalid-null, the command-line processing will abort with an error.

Adding Non-Trivial Command-Line Options

When adding non-trivial command-line options to your plug-in, it is often a good idea to find an existing option in a **Frama-C** plug-in that has a similar behavior to what you want, and then look at its source code to see how it is implemented.

7.4 AST and Associated Tables

One of the most important data structures in **Frama-C** is the AST, that describes the C program under analysis. The various nodes of the AST are described in more details in Chap. 2, which also details the various steps that are needed to transform a set of C source files into such an AST. This section focuses on the kind of information about the AST that can be obtained by the plug-ins from the kernel.

7.4.1 Accessing the Complete AST

By default, the kernel does not attempt to do any analysis. Hence, before making any query over the code, a plug-in must ensure that the AST has indeed been computed, through a call to `Ast.compute()`. This function instructs the kernel to parse all source files and fill all the tables that are directly related to the AST, as we will see in the rest of this section. Even if called multiple times, this is done only once: subsequent calls to `Ast.compute()` do not trigger new computations as long as the parameters on which the AST depends (e.g., -machdep, -cpp-extra-args, or the list of source files) do not change. It is also possible to retrieve the entire AST (triggering its computation if needed) by calling function `Ast.get()`.

Untyped AST

It is worth noting that, as explained in Chap. 2, Frama-C's AST proposes a normalized version of the code, in which some syntactic constructions present in the original source file might have been transformed into a semantically equivalent representation to ease the analyses. If a plug-in needs to access the exact syntactic representation[15] of the source code for some reason, this can be done by calling function `Ast.Untyped.get()`. It returns a list of untyped ASTs, one for each input file. However, the kernel offers far fewer functions to operate on untyped ASTs. Furthermore, the list itself is only available when working in projects that have been created by parsing actual code (as opposed to, e.g., being the result of a code transformation from a previous project). For these reasons, the usage of the untyped AST should be strictly reserved to cases where very specific syntactic information is needed. In particular, we do not use it for MiniTypeState.

Having access to the AST as a whole is mainly useful to iterate over its globals. For instance, to check that no function from the `file` API has its address taken in the code (so that we can be sure that they cannot be the target of an indirect call), as mentioned in Sect. 7.1.2, we can write the following piece of code.

Iterating Over the AST

```
1 let api_list = [ "alloc_file"; "open_read"; ... ]
2
3 let wkey = Options.register_warn_category
       "api-indirect-call"
4
5 let check_func = function
```

---

[15] Modulo the preprocessing phase, since Frama-C parses the source code after the preprocessor has run.

```
6    | GFun ({ svar = v }, _) | GFunDecl(_,v,_) ->
7      if List.mem v.vorig_name api_list && v.vaddrof then
8        Options.warning ~wkey
9          "Address of function %s is taken; results may be
              unsafe"
10         v.vorig_name
11   | _ -> ()
12
13 let check () = Cil.iterGlobals check_func (Ast.get())
```

Our check function iterates over every toplevel node of the current AST, i.e., each global declaration or definition. It applies check_func on them, which in turn considers each declaration or definition of a function, and emits a warning (using a specific category, as in Sect. 7.3.3) in case the name of (the variable representing) this function matches one of the functions of the API and its vaddrof field is true, meaning that the address of the function is taken. To check the name, we use the vorig_name field, as vname might be normalized, in particular to avoid name collisions. It is extremely unlikely that Frama-C decides to rename a global variable, at least if it is not **static**, but when looking for the name of locals or formals, it is better to be aware of the potential difference between v.vname and v.vorig_name.

7.4.2 Accessing a Specific Node

Not all syntactic operations result in an iteration over the whole AST, though. In fact, there are many cases where one wants to focus on specific nodes. Frama-C's kernel offers a number of functions in its API to perform such queries, mostly in Globals, Kernel_function, and Cil modules. More precisely, module Globals contains functions to retrieve the information associated with a given name (variable, function, or type), and to add or remove global symbols from the kernel's internal tables. Module Kernel_function allows developers to query about some components of a given function. Finally, module Cil and its siblings, notably Cil_builder and Logic_const, offer miscellaneous functions for manipulating all components of the AST. It is beyond the scope of this chapter to present all the content of these modules in detail. Instead, we focus on the most important ones through a few examples.

First, going back to MiniTypeState, imagine that we want to retrieve all the statements where a function of the API is called. This could be done by the following snippet.

Retrieving Callsites

```
1 let call_sites name =
2   try
```

```
3      let kf = Globals.Functions.find_by_name name in
4      let call_sites =
          Kernel_function.find_syntactic_callsites kf
5      name, call_sites
6   with Not_found ->
7      Options.fatal
8        "invariant␣broken:␣function␣%s␣not␣found"
9        name
10
11 let api_call_sites () = List.map call_sites api_list
```

The `call_sites` function takes as argument a function name, and retrieves the associated `kernel_function`, i.e. the Frama-C structure that stores all the information about the corresponding function. Then, it computes the list of syntactic call sites for this function, each of them being a pair composed of a `kernel_function` representing the caller, and the exact statement (the `stmt` type of Frama-C's AST) corresponding to the call. Finally, `call_sites` simply returns a pair composed of the `name` and the computed list. The function is then used to create an association list[16] mapping each name in the API to the list of its syntactic call sites.

Exercise 1 (*Retrieving a Global Variable*) Using the `Variables` module of `Globals`, and the `varinfo` type that describes a C variable in the AST, complete the definition of `typ` in the example of Sect. 7.3.2.

7.4.3 AST Transformations

Up to now, we have only looked at existing AST nodes. It is also possible to perform some changes, such as adding or removing nodes, either globals, in which case the `Globals` table must be updated accordingly, or statements inside a function, in which case the control-flow graph of the function (see Sect. 7.4.4) must be updated.

---

[16] Since there are few functions in the API, we don't really need to use a more efficient datastructure. For larger mappings, Sect. 7.5.1 will show how to define maps and hash tables.

! Living in Harmony With Other Plug-ins

Note that changing the AST potentially invalidates analysis results that might have already been computed on the original AST. It is thus strongly suggested to perform such transformations through one of the following mechanisms:

- when operating on a brand new project (see Sect. 7.5.2), just created by the plug-in;
- using a copy visitor (see Sect. 7.6);
- or registering an AST transformation, as discussed now.

As mentioned in Sect. 7.4.1, **Frama-C**'s (typed) AST is obtained from the untyped AST through a normalization phase, which includes type-checking the code. It is possible for a plug-in to customize this normalization by registering an AST transformation. Such a transformation has four main components:

- its *category*, i.e. basically a name that must be registered through the function `File.register_code_transformation_category`;
- its *dependencies*, i.e., a list of parameters that impact the transformation: if one of them changes, the kernel will automatically know that the AST must be recomputed;
- the list of transformations that must occur before or after the one being registered, identified by their categories;
- the transformation itself, i.e., a function that takes an AST as argument and transforms it in place, returning nothing.

Moreover, there are two kinds of AST transformations, depending on whether they are run before or after the clean-up phase that takes care of filling the tables related to the AST (notably, global symbol tables mentioned in Sect. 7.4.2 and annotation tables introduced in Sect. 7.4.5), as well as removing various unused symbols. AST transformations operating before clean-up obviously cannot query the associated tables for retrieving a given symbol. On the other hand, if they create some new global symbols, they do not need to worry about adding them in the appropriate table since the kernel will take care of them afterwards.

There are several steps that must be done for tracking typestates. Only some of them are presented here. The other ones require unintroduced concepts: they will be described later, in the appropriate sections. In summary, we want to perform the following transformations to the AST:

- add an enumerated type whose tags represent the typestates;
- add a formal parameter to each function of the API that represents the typestate of the corresponding `file` (see Sect. 7.6);

7 The Art of Developing Frama-C Plug-ins

- add an ACSL contract for each function of the API describing how the typestate should evolve (see Sect. 7.4.5);
- for each each `file` variable, create a fresh variable tracking its typestate[17] in the AST (see Sect. 7.6); and
- update the calls to each function of the API to match our transformation of the prototype (see Sect. 7.6).

For the first one, we choose to perform the transformation before clean-up: this way, we do not have to worry about adding information in `Globals.Types` tables, which contain information about the AST's type definitions. Hence, we just have to add a global AST node representing the definition of the **enum** type. This is done by the function below.

Adding a New enum Type Into the AST

```
let typestates_names =
  [ "null"; "allocated"; "read_only"; "write_only";
      "changed";
    "read_changed"; "read_write"; "closed"; "freed"; ]

let mts_enum =
  File.register_code_transformation_category
      "MiniTypeState_Enum"

let add_enum_ts ast =
  if Options.Run.get () then
    begin
      let loc = Cil_datatype.Location.unknown in
      let enum = mk_enum "__typestates" in
      let create_item i ts =
        let item = mk_item loc enum i ts in
        Typestates_tags.add ts item;
        item
      in
      let tags = List.mapi create_item typestates_names in
      enum.eitems <- tags;
      Typestates_enum.set enum;
      ast.globals <- (GEnumTag (enum,loc)) :: ast.globals
    end

let () =
  File.add_code_transformation_before_cleanup
    mts_enum add_enum_ts
```

As mentioned above, we start by registering a new transformation category. The chosen name can be arbitrary, except that it must not conflict with another category.

---

[17] Our plug-in doesn't attempt to handle arrays of `file` or `struct` with `file` fields. See Exercise 6 in Sect. 7.6.3.

Hence, we prefix it with the plug-in's name. The transformation itself is implemented by the add_enum_ts function. First, it creates an enuminfo record called enum, with a currently empty list of tags. Then, add_enum_ts maps all names in typestates_names to enumitem records associated with enum. In addition, we keep track of the association between each typestate name and its tag, as well as a reference to the enuminfo itself in tables Typestates_tags and Typestates_enum respectively. Their definition is provided in Sect. 7.5.2, where we explain how a plug-in's internal state is managed by Frama-C's kernel. Then, we update enum with the final list of tags, and add a GEnumTag node to the AST. Finally add_enum_ts is registered as a code transformation that must be performed before clean-up.

We also want to mark the API functions with the FC_BUILTIN attribute. They will then be treated as special functions by the kernel. In particular, they will not be removed during the clean-up phase, as would be the case by default for plain prototypes without contracts that are never called. This is done by the following transformation.

Marking the Functions of the API as Built-ins

```
1 let api_func_add_attr f =
2   f.vattr <- Cil.addAttribute (Attr ("FC_BUILTIN",[]))
        f.vattr
3
4 let update_global =
5   function
6   | GFunDecl(_,f,_) ->
7       if List.mem f.vname api_list then api_func_add_attr
            f
8   | _ -> ()
9
10 let update_api_func ast = Cil.iterGlobals ast
        update_global
11
12 let mts_formals =
13   File.register_code_transformation_category
        "MiniTypeState Formals"
14
15 let () =
16   File.add_code_transformation_before_cleanup
17     ~after:[mts_enum] mts_formals update_api_func
```

Namely, we add the attribute to the list of attributes attached to the varinfo representing the function f and apply this operation to all function declarations by iterating over globals, as done in Sect. 7.4.1. Finally, we register this operation as a new code transformation that must occur after registering the enumeration type through mts_enum.

7.4.4 Using the Control-Flow Graph

The kernel, together with two very basic plug-ins called **Dominators** and **Postdominators**, can give useful information tied to the Control-Flow Graph (CFG) of each function in the program. The CFG is a key element of many semantic analyses [1, 11]. Basically, **Frama-C**'s CFG is a graph where nodes are statements (i.e., values of type `stmt`). There is an edge between two statements `s1` and `s2` of function `f` if and only if `s2` might be a direct successor of `s1` in some execution of `f`. The CFG is intra-procedural: if `s1` is in fact a call to `g`, `s2` will be the statement occurring directly after the call in `f`, and not the first statement of `g`.

While Sect. 7.7 presents *interpreted automata*, which provide a convenient way for building dataflow analyses [11], basic information about the CFG itself can be retrieved through the `Dominators` and `Db.Postdominators` modules.[18] As their names suggest, they respectively provide the *dominators* of a statement `s` in a function `f`, i.e., the statements that are necessarily executed before `s` in any call to `f` and its *postdominators*, i.e., the statements that are always executed after `s` has been reached in any call to `f`.

As an example of `Db.Postdominators` usage, imagine a coding rule stating that each function making a call to `alloc_file` must open (either for reading or for writing) the result afterwards, as a sanity check to avoid allocating resources that would remain unused. A quick—albeit unsafe since it does not take into account potential redefinition or aliasing of the left-value holding the allocated `file`—way to achieve that would be to check that, for each call to `alloc_file`, there is a postdominator calling `open_read` or `open_write`. Using the `call_sites` function from Sect. 7.4.2, this can be done the following way.

Checking Allocated Files are Opened

```
1  let wkey = Options.register_warn_category "alloc-open"
2  exception Alloc_val_not_stored
3
4  (* access to standard functions about Lval and Stmt *)
5  open Cil_datatype
6
7  (* check whether [stmt] is a call to "open_read" or
      "open_write"
8     with [lv] as argument. *)
9  let is_open_call lv stmt = match stmt.skind with
10   | Instr
11     (Call (_,
12       { enode = Lval (Var f,NoOffset) },
13       [ { enode = Lval lv' } ], _)) ->
14     (f.vname = "open_read" || f.vname = "open_write") &&
```

---

[18] Note that `Db.Postdominators`'s values are references to functions, hence they must be dereferenced before being called.

```
15        Lval.equal lv lv'
16      | _ -> false
17
18   (* Assume that [stmt] is a direct call to "alloc_file" in
         [kf]
19      and emits a warning for each call to "alloc_file" with
             no
20      subsequent calls to "open_read" or "open_write". *)
21   let check_one_call (kf, stmt) =
22      let source = fst (Stmt.loc stmt) in
23      try
24        let lv = match stmt with
25          | Instr (Call (lv,_,_,_)) -> lv
26          | _ -> raise Alloc_val_not_stored
27        in
28        !Db.Postdominators.compute kf;
29        let pds = !Db.Postdominators.stmt_postdominators kf
             stmt in
30        let exist_call = Stmt.Hptset.exists (is_open_call lv)
             in
31        if not (exist_call pds) then begin
32          Options.warning ~wkey ~source
33            "alloc_file without open in function %a"
34            Kernel_function.pretty kf;
35        end
36      with
37      | Db.PostdominatorsTypes.Top ->
38        Options.warning ~wkey ~source
39          "call to alloc_file in a path that never returns
             from function %a"
40          Kernel_function.pretty kf
41      | Alloc_val_not_stored ->
42        Options.warning ~wkey ~source
43          "result of call to alloc_file is ignored in
             function %a"
44          Kernel_function.pretty kf
45
46   let check_all () =
47      let _, alloc_calls = call_sites "alloc_file" in
48      List.iter check_one_call alloc_calls
```

Function is_open_call returns true if the given statement is a call to either open_read or open_write taking as argument the appropriate left-value. Function check_one_call takes as argument a call to alloc_file and the kernel function in which it appears for checking the coding rule. First, it retrieves the left-value in which the result of alloc_file is stored, and raises an exception, later converted into a warning if this result is ignored by the caller. Then, it fills the table of post-dominators for kf by calling !Db.Postdominators.compute. This function does not return a value, but fills a table that is then queried by

!Db.Postdominators.stmt_postdominators. It is worth noting that, once the table is filled for a given kf, a second call to !Db.Postdominators.compute will not trigger any new computation. This behavior is quite standard in Frama-C and explained in more details in Sect. 7.5.2. Finally, we check whether there is a statement for which is_open_call lv is true among the post-dominators and emit a warning if it is not the case. The Top exception might be raised when retrieving the post-dominators if the call is in a path that never returns from the current function (e.g., in an infinite loop). In that case, we emit another warning. Finally, we iterate check_one_call over all the call sites of alloc_file.

7.4.5 Accessing ACSL Annotations

The modules presented in the previous section define accessors for AST nodes corresponding to C code. However, it is also often necessary to retrieve information about ACSL annotations, or their validity status according to the analyses that have been launched on the current project. This is done through the functions defined in the Annotations, Property, and Property_status modules. Module Annotations contains the functions that retrieve the properties corresponding to ACSL annotations, be they global declarations or tied to specific program points in the code. A property often directly corresponds to an ACSL clause or declaration (e.g., an assertion), but is sometimes a strict subset of it (e.g., for complex **assigns** clauses). Module Property mostly defines the data structures needed to unambiguously identify each ACSL property by its position in the code. It is (only) required when using module Property_status, which allows for checking the validity status of each registered property.

Section 7.4.3 has introduced an enumerated type for identifying typestates. We now continue the transformation by adding a proper contract to each function in the API, based on the typestate graph of Fig. 7.3. Basically, each function has a pre-condition stating that the file must be in a typestate that is the source of a transition labeled with this function's name, and a post-condition giving the new typestate according to the active transition. Those annotations could then be used by the plugins Eva, Wp, or E-ACSL, for example.

For instance, the function open_write can be called when the file is in one of the three states allocated, closed, or read_only. In the first two cases, the file will end up in state write_only, while, in the last case, the file will be in state read_write after the call to open_write. Hence, we would like to generate the following contract for open_write.

Typestates-Based ACSL Contract for open_write

```
1 /*@ requires *f__ts == allocated ||
```

```
2       *f__ts == closed ||
3       *f__ts == read_only;
4
5     ensures \at(*f__ts,Pre) == allocated ==> *f__ts == write_only;
6     ensures \at(*f__ts,Pre) == closed    ==> *f__ts == write_only;
7     ensures \at(*f__ts,Pre) == read_only ==> *f__ts == open_write;
8
9     assigns *f \from *f; /* supposed to be given by the API. */
10    assigns *f__ts \from *f__ts;
11  */
12  void open_write(file f)/*@ ghost (enum __typestates *f__ts) */;
```

Note that we assume that the first **assigns** clause is directly given in the header file that declares the prototypes. The rest of the contract deals with the typestate itself and is generated by our code transformation.

Assuming that we have an association list `transitions` that maps each function name to a list of pairs of state names representing the transitions, our transformation for adding pre-conditions is the following:

Code Transformation Adding a Function Contract

```
1  (* emitter taking responsibility for generated
       annotations. *)
2  let emitter =
3    Emitter.create "MiniTypeState"
4      [Emitter.Code_annot; Emitter.Funspec]
5      ~correctness:[] ~tuning:[]
6
7  (* creates an ACSL equality of the form '*ts_prm ==
       ts_name'
8     indicating that object 'prm' is in typestate 'ts_name'.
9  *)
10 let is_ts ts_prm ts_name =
11   let open Cil_builder.Exp in
12   let v = var ts_prm in
13   let cst = TConst (LEnum (Typestates_tags.find ts_name))
         in
14   let ts = term (Logic_const.term cst (Ctype
         ts_prm.vtype)) in
15   mem v == ts
16
17 (* creates a contract for function 'f' whose possible
       transitions
18    are given by 'trans'. 'f' is a function of the API. *)
19 let add_one_contract (f,trans) =
20   let kf = Globals.Functions.find_by_name f in
21   let loc = Kernel_function.get_location kf in
```

7 The Art of Developing Frama-C Plug-ins

```
22    (* assumes that we already created the ghost parameter
         holding
23       the current typestate of the given file, and that it
           is the
24       last parameter of 'f'. *)
25    let ts_prm = Extlib.last (Kernel_function.get_formals
         kf) in
26    let possible_ts =
27      List.map (fun (pre,_) -> is_ts ts_prm pre) trans
28    in
29    (* pre-condition: 'ts_prm' must be equal to one of the
         sources
30       of the transitions. *)
31    let requires =
32      Cil_builder.Exp.(cil_ipred ~loc (logor_list
           possible_ts))
33    in
34    Annotations.add_requires emitter kf [requires];
35    (* post-conditions: for each pair of states in the
         'trans'
36       list, we generate an implication indicating that if
37       'ts_prm' is equal to 'pre' before the call, it will
38       be equal to 'post' when 'f' returns. *)
39    let mk_ensure (pre,post) =
40      let pred =
41        Logic_const.pimplies ~loc (
42          Logic_const.pat ~loc
43            (Cil_builder.Exp.cil_pred ~loc (is_ts ts_prm
               pre),
44            Logic_const.pre_label),
45          (Cil_builder.Exp.cil_pred ~loc (is_ts ts_prm
             post)))
46      in
47      Normal, Logic_const.new_predicate pred
48    in
49    let ensures = List.map mk_ensure trans in
50    Annotations.add_ensures emitter kf ensures;
51    (* assigns clause: we add the content of 'ts_prm'
52       to the existing list of assignments. Its new value
53       only depends on its original one.
54    *)
55    let ts_prm_deref = Cil_builder.Exp.(mem (var ts_prm)) in
56    let assign = Cil_builder.Exp.(cil_iterm ~loc
           ts_prm_deref) in
57    (* Note that although the underlying ACSL terms can be
58       identical, the 'from' identified_term needs to be
59       physically distinct from the 'assign' one, as they
60       represent two different properties.
61    *)
62    let from = Cil_builder.Exp.(cil_iterm ~loc
           ts_prm_deref) in
```

```
63    let assign_from = Writes [ assign, From [from] ] in
64    (* we assume that other assigned locations are given
65       in the C+ACSL declarations of the API. If this is
66       not the case, the assigns clause will be kept
67       empty (i.e. anything can be assigned by 'f').
68    *)
69    Annotations.add_assigns ~keep_empty:true
70        emitter kf assign_from
71
72 let add_api_contracts _ast =
73    if Options.Run.get () then
74      List.iter add_one_contract transitions
75
76 let mts_contracts =
77    File.register_code_transformation_category
78      "MiniTypeState Contracts"
79
80 let () =
81    File.add_code_transformation_after_cleanup
82      mts_contracts add_api_contracts
```

This transformation relies on the fact that we have added a parameter to each of these functions, representing the typestate of the corresponding file. This will actually be done by the transformation described in Sect. 7.6, which will be duly registered as taking place *before* mts_contracts. Indeed, modifying the prototype of the functions implies modifying also their calling points, which is best done with a *visitor*, whose description deserves a section of its own. Also, our transformation is run *after* the kernel's tables have been filled. Therefore, those tables can be used to retrieve some needed information, e.g., the node representing a kernel function from its name through function Globals.Functions.find_by_name. As a drawback, we also have to take care to fill those tables with the information we want to add.

First, adding annotations requires registering an emitter, that is used by the kernel to track which plug-in has generated which annotations and/or emitted a validity status. An emitter is restricted in the type of information it can actually emit. Here, the "MiniTypeState" emitter is able to emit either code annotations or components of function contracts. Furthermore, we must give the list of parameters that may impact the generated annotations or statuses, with the correctness and tuning lists. In our case, these lists are empty since our analysis depends on no parameters.

Then, the code uses the Cil_builder module, which helps create AST fragments both for C (expressions and statements) and ACSL (terms and predicates). Here it helps create ACSL terms. In particular, the is_ts function creates an equality between the typestate parameter of a function and a given typestate, such as *f__ts == allocated in the contract of function open_write above. Thanks to the function and operators defined in Cil_builder, the OCaml code used for creating the equality, namely mem v == ts is actually quite close to what a user would write as an ACSL annotation.

The function add_one_contract takes as arguments the name of a function in the API and the list of associated possible typestate transitions. It assumes that the API function exists in the AST (otherwise, Globals.Functions.find_by_name would raise a fatal error) and that the last parameter of the function is the one conveying typestate information. First, we create a **requires** clause in the form of a disjunction of all possible typestates when the function is called, corresponding to lines 1–3 of the open_write example. This clause is added to the annotation table using our emitter. Then, an **ensures** clause is created for each possible transition, with the form shown at lines 5–7 in the contract of open_write above. Again, these clauses are added to the annotation table. Finally, we build a From clause indicating that the content of the typestate parameter is updated during the call and that its new value depends on its original one. Function add_assigns will append this location to the list of potentially modified locations in the existing contract. Finally, we iterate add_one_contract over the transitions mapping, and register that as a code transformation.

> **! Warning**
>
> Note that the argument keep_empty:true given to Annotations.add_assigns indicates that this clause will not be added if the contract of the function still contains the default clause (WritesAny in Frama-C's AST). Passing false to keep_empty would mean that we might end up with an incorrect **assigns** clause if an API function is not provided with an **assigns** clause summarizing its side effects on the original code, as we would change the clause from WritesAny to Writes[assigns, ...], effectively asserting that the only side effect of the function is the one we just generated.

Exercise 2 (*Adding Extra Annotations*) Complete function add_one_contract to generate an ACSL behavior for each possible transition, where the assumes clause states that we are in the initial typestate of the transition, and the ensures clause states that we are in the final typestate.

Tip: In order to check the clauses that you have generated, you can use Options.debug (see Sect. 7.3.3) and some functions from the Printer module (see Sect. 7.3.2) to pretty-print your formulas notably pp_funspec, pp_behavior, or pp_predicate.

7.5 Datatype and Project Libraries

Datatypes and projects are two key Frama-C libraries, which can be safely ignored at first glance, in most cases, when writing a plug-in. However, projects *must* be taken into account in order to be fully compatible with several services offered by the kernel, such as options -save and -load (see Chap. 2), or for writing program transformers, while datatypes are convenient in many cases and required by the project library. Section 7.5.1 introduces datatypes, while Sect. 7.5.2 focuses on the project library.

7.5.1 Datatype Library

In Frama-C, a *datatype* is a module that implements the signature Datatype.S, which provides a monomorphic type τ and some usual values over it, such as functions equal, compare, hash, and pretty for testing equality, comparing, hashing, and pretty printing values of type τ. Most datatypes implement a more precise signature, named Datatype.S_with_collections, that includes sets over values of type τ, and maps and hash tables whose keys are values of type τ.

The module Datatype itself provides datatypes for most standard OCaml types, such as Datatype.Bool for type bool, Datatype.Int for type int, Datatype.String for type string, and Datatype.Integer for type Integer.t representing the type of unbounded mathematical integer in Frama-C.[19]

Datatypes are also provided for most Frama-C's types. For instance, module Kernel_function (see Sect. 7.4) implements Datatype.S_with_collections, while module Cil_datatype provides datatypes for most AST types, such as e.g., Cil_datatype.Stmt for type stmt denoting the C statements or, for type term denoting ACSL terms, Cil_datatype.Term. Therefore, for all these types, it is straightforward to use the functionalities associated with datatype, such as comparing them or creating tables over them.

Using Datatypes

The example of Sect. 7.4.4 uses several datatypes. First, it uses the functions Kernel_function.pretty and Cil_datatype.Lval.equal for printing C functions and comparing left values, respectively. It also uses operations specific to stmt, namely Cil_datatype.Stmt.loc and Cil_datatype.Stmt.Hptset.exists, that are not necessarily provided for other types.

The Datatype library allows any developer to create datatypes for their own monomorphic types. Creating such datatypes is useful when storing values in project-compatible states (see Sect. 7.5.2) or when providing a new data structure in a plug-in API in a Frama-C-consistent style. In the general case, creating a new datatype

---

[19] Frama-C's module Integer is an extension of module Z from the Zarith library provided at https://github.com/ocaml/Zarith/.

requires applying the functor Datatype.Make, which provides no collections such as sets, or Datatype.Make_with_collections, which also provides collections.[20] In addition to some expected functions such as equal, these functors also require a few values used for consistency checks or when loading values previously saved on disk.[21] For simplicity, one can provide the desired set of functions by including either module Datatype.Undefined or Datatype.Serializable_undefined: the former does not permit serializing values of this type, while the second one serializes and unserializes it as a standard OCaml value.

Creating Datatypes

In the running example, we can create a sum type for the different kinds of states, which are the vertices of the graph in Fig. 7.3.

```
type state =
  | Null | Allocated | Closed | Freed
  | Read_only | Write_only | Changed | Read_write |
    Read_changed
```

To provide a datatype for type state, which includes collections and the necessary functions for them but no other functions, you can write the following code:

```
module State_datatype =
  Datatype.Make_with_collections
    (struct
      (* the type corresponding to the datatype *)
      type t = state
      (* a unique required name for the built datatype *)
      let name = "state"
      (* any non-empty list of values for this type,
         used for internal consistency checks. *)
      let reprs = [ Null ]
      (* define nothing else, but ... *)
      include Datatype.Undefined
      (* ... equality, compare, and hash by their standard
         values *)
      let equal (x:t) y = x = y
      let compare (x:t) y = Stdlib.compare x y
      let hash (x:t) = Hashtbl.hash x
    end)
```

---

[20] In a future version of Frama-C, datatypes will be automatically generated in most cases through ppx extensions.

[21] This mechanism allows Frama-C to customize how to unserialize OCaml values, which is particularly useful for hashconsed values [4, 5]. There is no need to customize the serialization process, which is always the default OCaml one.

The Datatype library also provides many functors that help developers to implement datatypes for monomorphic versions of polymorphic types. Following the above example, the module

Datatype.List(Datatype.Pair(State_datatype)(State_datatype))

implements a datatype for type (state*state) list, i.e. list of pairs of state values.

7.5.2 Project Library

As explained in Chap. 2, a project groups together a source code and the associated parameters and results for each analyzer. During a Frama-C session, several projects may exist simultaneously, and the user may choose to move from the current project to another one in order to compute and/or display analysis results for different pieces of code. Such a feature is particularly useful when using a program transformation tool, such as the Slicing plug-in, since it allows the user to inspect or analyze differently the program before the transformation and the program afterward.

From a developer perspective, this behavior is provided by the Project library. It implements a notion of *current project*, accessible when necessary through Project.current (), on which the kernel and each plug-in works by default. For instance, Ast.get (), introduced in Sect. 7.4.1, returns the AST that is *associated with the current project*. It is the same when getting the value of parameters, such as Run.get () for option -typestate introduced in Sect. 7.3.4. If the current project has changed, the expression Ast.get () (resp. Run.get ()) returns the AST (resp. the value of option -typestate) in the new current project, so not the same one as the previous call.

Plug-in developers should never change the current project themselves. Instead, the function Project.on should be used for applying a given closure to a given project when needed and restoring the context of the former project once done. More precisely, a global context switch is automatically performed by the project library when the current project changes: all the *internal states* of Frama-C are automatically (and efficiently) updated [13]. To do so, these states need to be registered inside the project library. This is automatically done for all the relevant states of the kernel and existing plug-ins and all the parameters built by applying a functor as explained in Sect. 7.3.4,[22] such as parameter -typestate.

However, any other state needs to be registered by the developers, typically the state storing the result of their analyzer. This is usually done by using the functors from the State_builder module.

---

[22] Except the few parameters explicitly prevented from being registered through a call to Parameter_customize.do_not_projectify.

Project-Compliant Reference

The functor `State_builder.Ref` implements a project-compliant reference and may be used as follows in order to create a reference over an `int`.

```
1  module R =
2    State_builder.Ref
3      (Datatype.Int) (* the datatype for the stored value *)
4      (struct
5        let name = "R" (* name of the state ; must be
                          unique *)
6        let dependencies = [] (* explained latter *)
7        let default () = 0 (* default value *)
8      end)
```

The signature of the resulting module R allows the developer to use this reference in the current project. For instance, the expression `R.get ()` is the value stored in the reference and `R.set 1` sets it to 1. Since the reference is project-compliant, its content is automatically updated by the project library when the project is changed. Therefore, there is actually one reference for each AST, but that is fully transparent for the developers: by default, they always access the value corresponding to the current project.

In addition to functors implementing standard mutable data structures in module `State_builder`, Frama-C also provides mutable data structures over AST values in module `Cil_state_builder`. For instance project-compliant hash tables with statements as keys are available through the `Cil_state_builder.Stmt_hashtbl` functor.

Quite importantly, since the project library is aware of the whole internal state of Frama-C (plug-ins included), it is also in charge of saving and loading each internal state from/to the disk. For example, the internal state corresponding to module R above is automatically saved on disk on `-save` and consistently loaded on `-load`. It would not be the case for a simple reference created by, say, `ref 0`.

Last, the project library includes a notion of *dependency* between states. It allows the library to automatically keep them consistent when some operations are performed. For instance, anyone can clear one state to restore its default value and automatically also clear all its dependencies without specifying them explicitly.

Hash Table Associating Tags With Typestates Names

As mentioned in Sect. 7.4.3, we need to maintain a mapping from the typestate names to the corresponding tags of the `enum` type created by our first AST transformation. This mapping can be easily implemented by means of a hash table as shown below.

```
module Typestates_tags =
  State_builder.Hashtbl
    (Datatype.String.Hashtbl)
    (Cil_datatype.Enumitem)
    (struct
       let name = "Mts_ast_transform.Typestates_tags"
       let dependencies = []
       let size = 9
     end)

let () = Ast.add_monotonic_state Typestates_tags.self
```

The only peculiarity of this state is that, instead of registering a dependency on the AST, we use a function that indicates that it is *monotonic* with respect to the AST: if a new transformation simply add new globals, this table will not need to be recomputed. It will only be automatically reset when globals are removed from or modified in the AST.

Exercise 3 (*Reference to* enuminfo) Write a definition for a Typestates_enum module that holds a reference to the enuminfo representing typestates as mentioned in Sect. 7.4.3.

Memoizing Analysis Results

Consider the running example: as mentioned in Sect. 7.1.2, one possible way to implement typestate verification is to provide our own dataflow analysis. It might be quite inefficient on large C code, though. In such cases, Frama-C promotes memoization [10], a programming technique that prevents recomputing what has already been computed. It requires storing the result of the analysis in a state. In our case, we need to store, for each program statement of type stmt and for each C variable denoting a file of (OCaml) type varinfo, the current vertex of type state in the typestate graph. In short, it requires a (mutable) datatype implementing the following mapping: stmt → varinfo → state. It can be implemented by using a hash table whose keys are stmt to a map whose keys are varinfo and values are state. Once chosen, creating this data structure in a project-compliant way is rather simple:

```
module Result =
  (* a hashtbl whose keys are statements ... *)
  Cil_state_builder.Stmt_hashtbl
    (* ... and values are maps of varinfos to states *)
    (Cil_datatype.Varinfo.Map.Make(State_datatype))
    (struct
```

```
7        let name = "MiniTypeState.Result" (* name of the
                state *)
8        let dependencies = [ Ast.self ] (* it depends on
                the AST *)
9        let size = 97 (* initial size of the hash table *)
10      end)
```

During the analysis, it is possible to add a binding (`stmt, vi, state`) in this table by using the following line of code, if we assume that it does not already exist for simplicity.

```
1  Result.add stmt (Cil_datatype.Varinfo.Map.singleton vi state)
```

Assume now we consider a more evolved plug-in in which the user can configure whether the final state `freed` is mandatory; said otherwise, whether the function `free_file` must be called eventually. One could add a Boolean parameter in module `Options` of our plug-in to let the user set this mode or not as follows.

```
1  module Mandatory_freed =
2    True(struct
3      let option_name = "-mandatory-freed"
4      let help = "calling function 'free_file' is mandatory"
5    end)
```

In this configuration, the behavior of the dataflow algorithm would differ according to the value of this parameter. Therefore, one should indicate that the `Result` table depends on this flag (in addition to depending on the AST), by modifying line 8 in its definition as follows.

```
1      let dependencies = [ Ast.self; Options.Mandatory_freed.self
                ]
```

Thanks to this single modification, the `Result` table is automatically reset each time someone modifies the value of `-mandatory-freed` (either the user or another plug-in or the kernel), in addition to being reset each time the AST is changed.

! State_builder.Register

Even if the project library provides a functor `State_builder.Register`, it is highly *not* recommended to apply it directly because its inputs invariants are usually tricky to establish. Instead, one should prefer one of the other functors, also provided in module `State_builder` (e.g. `Ref`, `Hashtbl` seen in this section). They provide OCaml mutable data structures compliant with the project library.

7.6 Visitor

7.6.1 Overview of the Visitor Mechanism in Frama-C

The visitor pattern [7] aims at providing developers with a convenient way to iterate over a complex data structure. Frama-C implements such a visitor in the form of a class, which has a method for each type of node in the AST. Its API is described in detail in the plugin development guide [14, Sect. 4.17]. We present here only its most important features through a few examples.

Each method takes as input a node of the corresponding type τ and returns a value of type τ Cil.visitAction, which tells the visitor what to do over its children (notably, stop the visit right there, continue normally, or apply a function over the resulting node after its children have been visited themselves). It should also be noted that the visitor's interface is imperative: during the visit, information can only be transferred through references or mutable instance variables.

Moreover, the visitor comes in two flavors: *inplace* visit, where the nodes of the current AST are manipulated directly, and *copy* visit, where a copy of the current node is presented to the methods. Finally, the classes defined in the Visitor module are meant to operate on the normalized AST. In particular, during the visit, they retrieve the annotations from the given tables of the Annotations module (see Sect. 7.4.5).

An inplace visitor typically inherits from Visitor.frama_c_inplace, the default inplace visitor that only traverses each node of the AST without doing anything else. Section 7.6.2 describes how to use such a visitor to observe some AST nodes throughout the code, while Sect. 7.6.3 explains how to build a code transformation pass with a visitor. Finally, Sect. 7.6.4 gives an overview of the copy visitor, which can be used to create new AST, e.g. to create a new project as explained in the previous section.

7.6.2 Observing the AST with an Inplace Visitor

The simplest usage of a visitor is to collect information while traversing the whole AST, without modifying the AST nodes. As an example, we can create a small visitor that will count the number of calls to a function of the API that occurs within the AST.

Counting Function Calls

The visitor itself goes through each Call node and increments its variable count when there is a (direct) call to a function in the API. It can be noted that the normalization done by Frama-C removes nested calls (by introducing temporary variables). More generally, given the definition of the OCaml types constituting the AST, no Call nodes can be under an `instr` or an `expr`. Hence, there is no need to continue the visit below these nodes, and we use SkipChildren to speed up the visit.

```
class count_api_visitor =
  object
    inherit Visitor.frama_c_inplace
    val mutable count = 0
    method get_count = count
    method! vinst = function
      | Call(_,{enode = Lval (Var vi,NoOffset)},_,_)
          when List.mem vi.vorig_name api_list ->
          count <- count + 1;
          Cil.SkipChildren
      | _ -> Cil.SkipChildren
    method! vexpr _ = Cil.SkipChildren
  end
```

Then, we define a simple state to hold the result (see Sect. 7.5.2).

```
module Count_API_calls =
  State_builder.Int_ref
    (struct
      let name = "Mts_ast_transform.Count_API_calls"
      let dependencies = [ Ast.self ]
      let default () = 0
    end)
```

Finally, we define a function that creates a new instance of count_api_visitor and uses it over the whole AST. After the visit, we call the method get_count to retrieve the result, and we memoize it thanks to Count_API_calls.

```
let nb_api_calls () =
  if not (Count_API_calls.is_computed ()) then begin
    let ast = Ast.get () in
    let vis = new count_api_visitor in
    Visitor.visitFramacFileSameGlobals
      (vis:>Visitor.frama_c_visitor) ast;
    Count_API_calls.set vis#get_count;
    Count_API_calls.mark_as_computed ()
  end;
  Count_API_calls.get ()
```

Furthermore, a visitor does not need to visit the whole AST. Namely, for each node type, there is a function `Visitor.visitFramac_type`, which takes as argument a node of the corresponding type and a visitor, and performs a visit starting from this node. For instance, we might also want to count the number of calls for a single function as follows.

```
let nb_api_calls_kf kf =
  let vis = new count_api_visitor in
```

```
3    ignore
4      (Visitor.visitFramacKf (vis:>Visitor.frama_c_visitor)
            kf);
5    vis#get_count
```

Note that, in that case, we need to ignore the result of `visitFramacKf`, which returns a `kernel_function`. In the former function, `visitFramacFileSameGlobals` is a specialized version of `visitFramacFile` that can be used when the visitor does not modify the list of global declarations of the AST.

Exercise 4 (*Memoization*) Using the appropriate `Cil_state_builder` module, modify the function `nb_api_calls_kf` so that it won't visit the same `kf` twice.

7.6.3 Creating a Code Transformation Pass with an Inplace Visitor

It is also possible to use a visitor to implement a code transformation that would then be registered as explained in Sect. 7.4.3. To illustrate this use case, we present a partial implementation for the visitor that would perform the code instrumentation needed to perform MiniTypeState analysis. The instrumentation is partial in the sense that it only considers cases where all expressions of type `file` are variables. Extending it for handling arrays of `file` or fields in structures or unions of type `file` would add quite some complexity and is irrelevant to present the visitor itself. It is thus left as an exercise for the reader.

Code Instrumentation for MiniTypeState

Before defining the visitor class itself, we need to introduce a few auxiliary functions. First, we define a pair of functions, namely `is_file_type` and `is_file_ptr_type` for finding variables of type `struct inner_rep`, or a pointer to it, respectively.

```
1  let is_file_type t =
2    match Cil.unrollTypeDeep t with
3      | TComp(ci,_) when ci.cname = "inner_repr" -> true
4      | _ -> false
5
6  let is_file_ptr_type t =
7    match Cil.unrollTypeDeep t with
8      | TPtr(TComp(ci,_),_) when ci.cname = "inner_repr" ->
            true
9      | _ -> false
```

7 The Art of Developing Frama-C Plug-ins

We also need the function `init_state_var` that sets the fields of a new auxiliary (ghost) `varinfo` holding the current typestate of a `file` variable present in the original code. In case the associated variable is a formal parameter of a function, the auxiliary variable holds a pointer, so that the function can modify the typestate during its execution.

```
let init_state_var is_formal vi =
  let ty =
    if is_formal then ptr_to_ts_enum ()
    else TEnum(Typestates_enum.get(),[])
  in
  vi.vlogic_var_assoc <- None;
  vi.vtype <- ty;
  vi.vghost <- true;
  vi.vname <- vi.vname ^ "__ts";
  vi.vformal <- is_formal
```

Since we are again registering a code transformation made right after parsing, we use an inplace visitor to work directly on the original AST. The visitor relies on a mapping table between the variables of type `file` and the associated typestate. Since this table is meant to be used only in the visitor, it is a plain hash table that is not explicitly registered as a Frama-C state.

Method `type_state_var` is used on variable declarations to decide whether an associated typestate should be generated or not. Note that we check whether a mapping has already been established (in which case we reuse the existing `varinfo`). Indeed, the visit may inspect the same variable more than once, as there can be several instances of the same global declaration in the AST.

```
class instrumentation =
  object(self)
    inherit Visitor.frama_c_inplace

    val correspondence =
      Cil_datatype.Varinfo.Hashtbl.create 43

    method private type_state_var is_formal vi =
      if is_file_ptr_type vi.vtype then begin
        let res =
          Cil_datatype.Varinfo.Hashtbl.find_opt
            correspondence vi
        in
        match res with
        | None ->
          let vi_new = Cil_const.copy_with_new_vid vi
          in
          init_state_var is_formal vi_new;
          Cil_datatype.Varinfo.Hashtbl.add
            correspondence vi vi_new;
```

```
20          Some vi_new
21        | Some vi_new -> Some vi_new
22      end
23    else None
```

We then find the visitor methods that actually check for variable declarations. We start with local variables. They appear in the blocals field of the block in which they are in scope. Method vblock ends with the DoChildren tag, which tells the visitor that the children of the block must be visited as well: we can have nested block, and, more importantly, we also have to visit call instructions in order to add the appropriate typestate parameters when needed.

```
1   method! vblock b =
2     let new_vars =
3       List.filter_map (self#type_state_var false)
          b.blocals
4     in
5     b.blocals <- b.blocals @ new_vars;
6     let loc =
7       Kernel_function.get_location (Option.get
          self#current_kf)
8     in
9     let local_init (vi:varinfo) =
10      let cst =
11        Cil_types.Const (CEnum (Typestates_tags.find
            "null"))
12      in
13      let init =
14        AssignInit (SingleInit (Cil.new_exp ~loc cst))
15      in
16      let instr = Cil_types.Local_init (vi,init,loc) in
17      Cil.mkStmtOneInstr ~ghost:true ~valid_sid:true
          instr
18    in
19    b.bstmts <- List.map local_init new_vars @ b.bstmts;
20    Cil.DoChildren
```

Method vfunc merely updates the list of all variables local to the functions, so that slocals field will be consistent with the blocals fields of the blocks after the visit. It also indicates to the kernel that the CFG of the function must be recomputed once the transformation is done, as we may have added new instructions (namely the initialization of the typestate variables).

Method vglob_aux is tasked with building mapping for global variables and formals of function (including the functions of the API themselves, as mentioned in Sect. 7.4.5). For global declarations, if the variable is not of type file, we return SkipChildren, instructing the visitor not to descend further in this node. Namely, there is nothing left to do for us in the children of a global variable declaration, so that we can skip this part of the AST. On the other hand, if we create a typestate variable, we use the ChangeTo constructor, to tell the visitor that we have now two globals instead of one, and that there is no need to visit their children (should we need to visit the children of our new nodes, the `ChangeDoChildrenPost` can do that, as well as applying a post-processing function on the final result).

```
1   method! vfunc b =
2     let new_vars =
3       List.filter_map (self#type_state_var false)
            b.slocals
4     in
5     b.slocals <- b.slocals @ new_vars;
6     File.must_recompute_cfg b;
7     Cil.DoChildren
8
9   method! vglob_aux g =
10    match g with
11    | GVarDecl (vi,loc) ->
12      (match self#type_state_var false vi with
13       | None -> Cil.SkipChildren
14       | Some vi_new ->
15         Cil.ChangeTo [g; GVarDecl(vi_new,loc)])
16    | GFunDecl(spec,f,loc)
17      (* The AST may contain several prototypes of
            the
18         same function, or a prototype and a
            definition.
19         We only perform the transformation once,
            when g
20         is the last global node declaring the
            function. *)
21      when Ast.is_def_or_last_decl g ->
22      let kf = Globals.Functions.get f in
23      let vars = Kernel_function.get_formals kf in
24      let new_vars =
25        (* special case for alloc_file, which
             returns a
26           newly allocated file. *)
27        if f.vname = "alloc_file" then
28          create_state_formal ()
29        else
30          List.filter_map
31            (self#type_state_var true) vars
32      in
33      let forms = (vars @ new_vars) in
34      (* For pure prototypes (defined functions
35         are treated in the the GFun case below)
36         updating the formals is not trivial,
37         and won't be detailed here.
38       *)
39      self#update_fundecl loc spec f forms;
40      Cil.DoChildren
```

```
41      | GFun(f,_) ->
42        let new_vars =
43          List.filter_map
44            (self#type_state_var true) f.sformals
45        in
46        Cil.setFormals f (f.sformals @ new_vars);
47        Cil.DoChildren
48      | _ -> Cil.SkipChildren
```

The last part of the visitor is dedicated to add the necessary ghost arguments to each function call in the code itself. For that, the vinst method inspects Call and Local_init nodes, the latter being a specific case where the call is used to initialize a local variable, and whose exact implementation is again left as an exercise. We distinguish two cases for Call: the alloc_file function itself, and all the other functions.

Indeed, the function alloc_file is special in the sense that its file element is not an argument but the value it returns. In order to find the appropriate typestate auxiliary variable, we thus have to use the left-value storing the result of the call and not the list of arguments as we do for the other functions. Apart from this point, the mechanism is the following: we check that the actual argument (or receiver of the returned value) is indeed a variable, and retrieve its typestate in the mapping table. We then take the address of this auxiliary variable as actual argument for the ghost formal parameter that has been generated previously.

```
1   method! vinst i =
2     match i with
3       | Call(d,({enode = Lval(Var f,NoOffset)} as
              ef),args,loc)
4         when f.vname = "alloc_file" ->
5         let e = self#add_ts_binding loc d in
6         let args_new = args @ [ e ] in
7         Cil.ChangeTo [Call(d,ef,args_new,loc)]
8       | Call(lv,f,args,loc) ->
9         let (_,formals,_,_) =
10           Cil.splitFunctionType (Cil.typeOf f)
11        in
12        let rec aux formals args new_args =
13          match formals, args with
14            | [], _ -> new_args
15            | (_,t,_)::formals, arg::args ->
16              let new_args =
17                if is_file_ptr_type t then begin
18                  self#add_ts_binding_expr loc arg
19                  :: new_args
20                end else new_args
21              in
22              aux formals args new_args
23            | formals, [] ->
```

```
24                      (* The type is already of the new arity
                        *)
25                      assert List.(length formals=length
                           new_args);
26                      new_args
27                    in
28                    let new_args =
29                      List.rev (aux (Option.get formals) args [])
30                    in
31                    Cil.ChangeTo [Call(lv,f,args @ new_args,loc)]
32                 | Local_init(v,ConsInit(f,args,Plain_func),loc) ->
33                    (* ... *)
34                    Cil.SkipChildren
35                 | _ -> Cil.SkipChildren
36    end
```

The treatment of the calls is based on method `add_ts_binding`, which must find the typestate-holding variable corresponding to each `file` argument given to the callee. This is mostly done by using the correspondence table filled by `vglob_aux` and `vblock` above.

```
1     method private add_ts_binding loc e =
2       match e with
3       | Some (Var v, NoOffset) ->
4         let v_new =
5           Cil_datatype.Varinfo.Hashtbl.find
              correspondence v
6         in
7         (if v_new.vformal then
8            Cil_builder.Pure.(cil_exp ~loc (var v_new))
9          else
10           Cil_builder.Pure.(cil_exp ~loc (addr (var
              v_new))))
11      | _ ->
12        (* Left as an exercise: modify the auxiliary
13           variable creation to handle arrays of
              files,
14           and struct fields with file type. *)
15        Cil_builder.Pure.(
16        cil_exp
17          ~loc (cast (of_ctyp (ptr_to_ts_enum()))
              zero))
18
19    method private add_ts_binding_expr loc e =
20      match e.enode with
21      | Lval lv -> self#add_ts_binding loc (Some lv)
22      | _ -> self#add_ts_binding loc None
```

Finally, as hinted at the beginning of the example, we define a function `instrumentation` that applies a new instance of our class on a whole AST and registers it as a code transformation. This function indicates that it has changed the AST (adding variables and instructions, and modifying call instructions), so that all the internal states that depend on the AST ought to be cleared. Furthermore, the transformation registration indicates that instrumentation must occur before contract generation, since as said previously the latter relies on the fact that the auxiliary variables have been generated.

```
let instrumentation file =
  if Options.Run.get () then begin
    Options.feedback "Add typestate instrumentation...";
    let vis = new instrumentation in
    Visitor.visitFramacFile vis file;
    Ast.mark_as_changed ()
  end

let mts_instru =
  File.register_code_transformation_category
    "MiniTypeState Instrumentation"

let () =
  File.add_code_transformation_after_cleanup
    ~before:[mts_contracts]
    mts_instru instrumentation
```

Once we have added our instrumentation function to the set of code transformations known to Frama-C's kernel, there is no need for the plug-in to call it explicitly: it will be part of the standard AST creation process (of course, the transformation itself will only be triggered when Options.Run is set).

Exercise 5 (*Missing* Call *Case*) There is a third possibility of finding file variables in a function call that is not considered in the visitor above and imply extending both vinst and vglob_aux. Which is it? How would you implement it?

Exercise 6 (*Visitor Extension*) Extend the visitor so that the instrumentation can handle arrays of file, fields in struct and union that have type file, and pointers to file.

7.6.4 Creating a New AST with a Copy Visitor

A copy visitor is used to build a deep copy of the AST (or of the fragment rooted at the node where the visit starts) and usually inherits from either Visitor.framac_copy or Visitor.framac_refresh. Both take as argument the destination project of the resulting AST (or AST fragment). In particular, a copy visitor can be used to create a new project by applying some program transformation over the current one, as done by many code transformation plug-ins in Frama-C. The refresh visitor, in addition to copying nodes, creates fresh identifiers where needed (i.e. for `varinfo`, `stmt`, `identified_predicate`, ...). This can notably be used to create copies of a fragment of the AST in the current project, as is done for instance by the kernel for inlining function definitions when the `-inline-calls` option is used.

In the case of a copy, some care must be taken to ensure that the tables of the new project are filled with nodes corresponding to the new project. For that, the functions from `Visitor_behavior` module maintain the mapping between nodes from the original project to the new one, and conversely, for each type in the AST that has an identifier (notably `varinfo`, `stmt`, `compinfo`). More specifically, `Visitor_behavior` contains the following submodules, which themselves have a function for each type of node that is handled:

- `Get` for retrieving the new node corresponding to the original one given as argument.
- `Get_orig` for retrieving the original node corresponding to the new one given as argument.
- `Memo` is similar to `Get` but creates a new node if no binding already exists.
- `Set` for adding a binding from an original node to a new one.
- `Set_orig` for adding a binding from a new node to an original one.
- `Reset` for clearing all existing bindings.

The behavior associated with a given visitor is accessible through the `behavior` method. The base visitor takes care of registering the nodes during the visit, so that `Set` and `Set_orig` are only needed for the nodes in which a transformation actually occurs.

Furthermore, switching projects back and forth along the visit would have an impact on efficiency on a large AST. Hence, operations on the tables of the new projects (for instance to add annotations or create a new global) should be put in a queue of actions accessible through the method `get_filling_actions`. These actions are played at the end of the visit, so that we switch from the current project to the new one only once.

A typical call to `get_filling_actions` is of the form:

```
let new_stmt =
  Visitor_behavior.Get.stmt self#behavior orig_stmt
in
let my_assertion = create_assertion orig_stmt in
let add_new_annot s =
  Annotations.add_code_annot my_emitter s my_assertion
in
Queue.add
  (fun () -> add_new_annot new_stmt)
  self#get_filling_actions
```

where the code of `create_assertion` must take care of always using references to nodes from the *new* project by using functions from `Visitor_behavior.Get` when necessary.

Visitor Actions and Copy

SkipChildren is fine for an inplace visitor as shown in the example above, but for a copy visitor, it is important to remember that the copy is done on the fly as each node gets visited, so that SkipChildren would leave some nodes shared between the original AST and its copy, which can later lead to hard-to-fix issues. In that case, the JustCopy action should be preferred: all the descendants of the current node will be readily copied, without any other transformation.

7.7 Interpreted Automata

As already explained in Sect. 7.4.4, Frama-C provides a CFG. It can be used to build a dataflow analysis. Frama-C historically provides such generic dataflow analyses in modules Dataflows and Dataflow2. However, it now provides a higher-level representation, named *Interpreted Automaton*. Using it is recommended instead of using the old-fashioned modules.

An interpreted automaton, provided in module Interpreted_automata, is a graph of type automaton, made of vertices of type vertex and edges of type vertex edge. Compared to using the CFG directly, an interpreted automaton relies on a smaller number of statements' kinds, a precise labelling of each transition, additional vertices at useful program points, and native handling of annotations. In short, interpreted automata are easier to use.

Section 7.7.1 presents interpreted automata, while Sect. 7.7.2 implements a dataflow analysis for MiniTypeState using an interpreted automaton.

7.7.1 Automaton Description

The function get_automaton generates an interpreted automaton for a given kernel function. This function is memoized: calling it twice on the same function generates the same automaton. The function Compute.get_automaton is not memoized. The argument annotations specifies whether annotations, such as asserts, should be added directly in the graph as vertices (loop variants and invariants are always added).[23]

The vertices of an automaton denote the control points inside the function, each of them represented by a unique integer. Some additional information is also attached to

---

[23] Only the first function is memoized since it is often more useful in practice.

each vertex. In particular, it helps locate the control point in the code. The edges of an automaton denote transitions from one control point to the other through statements:

```
type 'vertex transition =
  | Skip
  | Return of exp option * stmt
  | Guard of exp * guard_kind * stmt
  | Prop of 'vertex annotation * stmt
  | Instr of instr * stmt
  | Enter of block
  | Leave of block

and guard_kind = Then | Else
```

One step of the execution of the program can be executed by following an edge in the graph. Each kind of edge specifies when it can be executed and how the state of the program is modified. The Skip transition does not change the state of the program and can always be executed. The other transitions are labeled by the statement or the block of statements they denote, and possibly other information they carry. The Return transition is labeled by the optional expression returned by the function. The Guard transition denoting if and switch statements, can be followed if the given expression is true or false according to the guard kind. The Prop transition denotes an ACSL predicate in a code annotation such as an assertion or a loop invariant (see Chap. 1). The field labels in the annotation type maps the ACSL logic labels to the corresponding vertex of the graph. The Instr transition directly denotes an instruction from the AST. The Enter and Leave transitions respectively denote entering and leaving blocks.

```
type 'vertex annotation = {
  kind: assert_kind;
  predicate: identified_predicate;
  labels: 'vertex labels;
  property: Property.t;
}
```

7.7.2 Using an Automaton to Implement a Dataflow Analysis

A dataflow analysis [9, 11] makes use of the CFG (see Sect. 7.4.4) to attach information to each statement of the program. There are several kinds of dataflow analyses: they may go forward or backward and can over-approximate ("may"-analysis) or under-approximate ("must"-analysis) the computed information.

A forward over-approximating analysis starts from the information tied to the entry point (the very first statement of the program) and proceeds by propagating this information along the edges of the CFG through *transition functions*, until a fixpoint is reached. For nodes with more than one predecessor, information flowing from all the predecessors is merged: the resulting state of the analysis contains at least as many pieces of information as all the predecessors. A backward analysis proceeds in the same way, but starts from the return statement[24] and propagates information from successors to predecessors. An under-approximating analysis differs from an over-approximating analysis when merging since its resulting state must only contain pieces of information that are shared by all its predecessors (or successors for a backward analysis).

Dataflow Analysis for MiniTypeState

To illustrate how to use interpreted automata for dataflow analyses, we now implement the mini-typestate analysis as a forward over-approximating analysis based on interpreted automata. For this purpose, we need to apply the functor Interpreted_automata.ForwardAnalysis to a structure Domain implementing the state of our analysis and how to propagate states. For MiniTypeState, this state maps a typestate to each variable of type file.

```
1  module State =
       Cil_datatype.Varinfo.Map.Make(State_datatype)
2
3  module Domain = struct
4    type t = State.t
5
6    let join = Cil_datatype.Varinfo.Map.union
7      (fun varinfo v1 v2 ->
8        if not (State_datatype.equal v1 v2) then
9          Options.warning
10           "Variable %a has two possible states %a and %a: 
              unsound analysis"
11           Cil_datatype.Varinfo.pretty varinfo pp_state v1
              pp_state v2;
12        Some v1)
13
14   let widen t1 t2 =
15     if State.equal t1 t2 then None (* fixpoint reached *)
16     else Some (join t1 t2)
17
18   let transfer trans t = ... (* defined latter *)
19 end
```

The function join is used for merging states coming from two predecessors. Here, for simplicity, our analysis only supports programs whose predecessors have the same value (for instance, opening a file for reading in both branches of a conditional). The widen function is used for handling loops in a finite number of steps. Here, since our domain is finite, we can directly use the join function.

We now need to implement the transfer function that computes the resulting over-approximating state from a state t after having executed a transition

---

[24] As explained in Chap. 2, in the normalized AST of Frama-C, each function has exactly one return statement.

trans. The main case of the analysis is function call. It is handled separately in the function transition.

Thanks to this function, implementing the transfer function is easy.

```
1  let transfer trans t = match trans with
2  | Skip
3  | Return _
4  | Guard (_,_,_)
5  | Prop _
6  | Instr ((Asm _ | Skip _ | Code_annot _), _) ->
7    Some t
8  | Instr (Call((None | Some (Var _,NoOffset)) as lv,
9    {enode = Lval (Var varinfo,NoOffset)},args,loc), _) ->
10   let lv = match lv with
11     | None -> None
12     | Some (Var vi,NoOffset) -> Some vi
13     | Some _ -> assert false (* absurd because of lv
         shape *)
14   in
15   begin match Globals.Functions.get varinfo with
16     | exception Not_found ->
17       Some t (* unsound since it is an unknown call *)
18     | kf ->
19       let name = Kernel_function.get_name kf in
20       transition ~source:(fst loc) lv name args t
21   end
22 | Enter {bscoping = true; blocals} ->
23   let t =
24     List.fold_left
25       (fun acc vi ->
26         (* Initialize to Null each variable of type
            file* *)
27         if Mts_ast_transform.is_file_ptr_type vi.vtype
            then
28           Cil_datatype.Varinfo.Map.add vi Null acc
29         else
30           acc)
31       t
32       blocals
33   in
34   Some t
35 | Enter _ | Leave _ -> Some t
36 | Instr (Set (_, _exp, _loc), _) ->
37   Some t (* unsound if exp points to a file *)
38 | Instr (_instr, _stmt) ->
39   Some t (* unsound since it is an unknown instruction *)
```

The function transition handles function calls. It returns None when the analysis reaches only forbidden states. In the case of this analysis, the interesting function calls assign only one variable, named varinfo. For example, the call modifies this variable through its returned value in the case of `alloc_file` and through its argument in the case of `open_read`. The auxiliary function `apply` factorizes the lookup of the abstract value of variable `varinfo` in the state before the transition and the creation of the state after the transition by replacing the value of `varinfo` with its new state v instead of s.

```
let transition ~source lval name args state =
  let apply varinfo f =
    match Cil_datatype.Varinfo.Map.find_opt varinfo state with
    | None ->
       Options.fatal ~source "Variable %a has no state"
         Cil_datatype.Varinfo.pretty varinfo
    | Some s ->
       begin match f s with
       | `Result v ->
          Some (Cil_datatype.Varinfo.Map.add varinfo v state)
       | `Error_unapplicable ->
          Options.warning ~source
            "Variable %a has the forbidden state %a for %s
              call"
            Cil_datatype.Varinfo.pretty varinfo pp_state s
              name;
          None
       end
  in
  match lval, name, args with
  | Some varinfo, "alloc_file", [_] ->
     apply varinfo
       (function Null -> `Result Allocated
               | s -> `Error_unapplicable)
  | None, "open_read", [{enode = Lval (Var varinfo,
      NoOffset)}] ->
     apply varinfo (function Allocated -> `Result Read_only
                           | Write_only -> `Result Read_write
                           | s -> `Error_unapplicable);
  | ...
  | _, name, _ ->
     Options.warning ~source
       "Unknown or unsupported call %s: unsound analysis"
       name;
     Some state
```

The `Options.fatal` should never be executed since each `varinfo` of type `file` is initialized when entering in its scope.

Exercise 7 (*Missing Cases for the* `transition` *Function*) Add the cases for handling functions `open_write` and `read` in the same way as `open_read` in the implementation of `transition`.

Once implemented, the dataflow analysis can be run by modifying the main function `run` of our plug-in as follows.

```
module DataflowAnalysis =
  Interpreted_automata.ForwardAnalysis(Domain)
```

```
2
3  let run () =
4    if Options.RunDataflow.get () then begin
5      Options.feedback "Starting typestate dataflow
           analysis...";
6      (* getting the main function *)
7      let main, _ =
8        try Globals.entry_point ()
9        with Globals.No_such_entry_point ->
10         Options.abort "No entry point for the analysis."
11     in
12     ignore
13       (DataflowAnalysis.fixpoint
14         main
15         Cil_datatype.Varinfo.Map.empty (* initial state *));
16     Options.feedback "Completed typestate dataflow analysis."
17   end
18
19 let () = Db.Main.extend run
```

Exercise 8 (*Add Warning for the Unsound Cases*) The `transfer` function handles some cases unsoundly. Add warnings so that users can be warned in such cases. The following function could be used: `Options.warning source:(fst loc)` "msg"

Exercise 9 (*Handling a Set of Type States*) Change the `Domain` module in order to assign to each `varinfo` a set of typestates instead of a single one. This removes the assumption made when merging states in our example.

Hint: you can use module `State_datatype.Set` to manipulate sets whose elements are of type `State_datatype.t`.

7.8 Plug-in Packaging

A plug-in should not be limited to its OCaml code: tests, easy installation, and clear documentation are essential for the proper use and maintenance of the plug-in. This section presents some tools used by Frama-C developers to facilitate these aspects of plug-in development: testing is presented in Sect. 7.8.1; packaging and installation in Sect. 7.8.2.

7.8.1 Testing

Testing a plug-in is a very important part of its development. Frama-C provides several features to support non-regression testing. We present here how to structure the plug-in's tests and how to use Frama-C's frama-c-ptests tool, included in the Frama-C distribution. This tool, combined with Dune, facilitates running and managing tests. Note that, since the move to Dune as primary build system for Frama-C, it is also possible to use *Cram tests*[25] instead of frama-c-ptests.

7.8.1.1 A Simple Test File for Frama-C-Ptests

For frama-c-ptests, test files are simply C files with a special header (a multiline C comment) containing one or more directives:

Test Case for the MiniTypeState Plug-in

Here we provide a single test file containing several test directives.

```
1   /* run.config
2      PLUGIN: minitypestate eva
3      STDOPT: +"-typestate -eva"
4      STDOPT: +"-main main_error -typestate -eva"
5   */
6
7   struct inner_repr;
8   typedef struct inner_repr *file;
9
10  file alloc_file(const char* filename);
11  void open_read(file f);
12  void open_write(file f);
13  char read(file f);
14  void write(file f, char c);
15  void flush(file f);
16  void close(file f);
17  void free_file(file f);
18
19  int main() {
20    file f;
21    f = alloc_file("foo.ml");
22    open_read(f);
23    open_write(f);
24    return 0;
25  }
26
27  int main_error() {
28    file f;
29    file g;
30    f = alloc_file("foo.ml");
31    g = f;
32    read(f);
33    return 0;
34  }
```

---

[25] https://dune.readthedocs.io/en/stable/tests.html#cram-tests.

The run.config annotation is an indication to frama-c-ptests that this test uses the *default configuration*. Even if useless here, different configurations can be used to re-run batches with a different set of default arguments, or to split tests into subsets.

Each following line starts with a directive, which modifies the environment for the following test cases or defines a new test case. Thus, a single .c file can define several test cases, with different options. Here, STDOPT: means "add the following to the set of options run by default by any test". For instance, the first directive means adding options -typestate -eva. PLUGIN: indicates which plugin to load for the tests. Loading only the plugins needed by the test avoids spurious changes of oracles due to unrelated plugins. Other directives exist that are described in the Plug-in Developer Manual [14].

7.8.1.2 Running Tests

Assuming the plug-in directory contains a tests subdirectory with all the test files, running frama-c-ptests from the plug-in directory automatically creates Dune files that can then be run with dune build @ptests. Running this command will execute the tests, compare them to expected outputs (in directory oracle) and output the difference, if any. An empty output thus means that all tests succeeded.

Initially (and whenever a test case is added), we need to create new oracles, by running frama-c-ptests -create-missing-oracles. This will create a oracle| subdirectory inside the tests, and populate it with empty files. Then, we can run dune build @ptests, which will output something similar to the following:

```
1 File "tests/minitypestate/oracle/test.2.res.oracle", line 1,
      characters 0-0:
2 (...)
3 +[kernel] Parsing test.c (with preprocessing)
4 +[mts] Starting typestate dataflow analysis...
5 +[mts] Finished typestate dataflow analysis.
```

We manually inspect that the results are the ones we expected, and *promote* them to become the new oracles, with dune promote. We then run frama-c-ptests -remove-empty-oracles to clean useless files, and we can commit the new oracles to version control.

Whenever we change the test case, or the plug-in code, we can check it for regressions: running dune build @ptests, inspecting the new oracles, promoting them if needed.

For instance, if we modify the MiniTypeState instrumentation visitor to add calls to Cil.SkipChildren in some nodes, to speed up the analysis, we may inadvertently introduce a bug. Running dune build @ptests provides a quick way to check if such regressions were introduced (assuming enough test coverage, of course).

Note that each test file contains any number of test cases, and each test case contains two or more oracles: *stdout*, *stderr* (often empty), and possibly other files produced by test directives.

Also note that frama-c-ptests tracks dependencies between test cases and source components (plug-ins, test input files, ...), using Dune to manage them and to avoid re-running tests if none of their dependencies has been modified.

The full list of frama-c-ptests directives, as well as a tutorial on plug-in testing, are presented in the Frama-C Plug-in Developer Manual [14]. For practical usage examples, you can also *grep* the tests directories in Frama-C's source distribution.

7.8.1.3 Checking AST Invariants

AST-transforming plugins are especially prone to breaking subtle invariants that Frama-C and its visitors rely on. While it is very hard to take into account all possible mistakes that users can make (e.g., forgetting to copy some AST nodes when switching between projects), some structural checks can be enabled by adding the option -check to the beginning of Frama-C's command line. These additional verifications can be costly and are therefore disabled by default, but frama-c-ptests enables them during tests, and plug-in developers are strongly recommended to do so when developing and debugging their own transformations. Additional structural checks are added to this option on a regular basis.

7.8.2 Distribution

Now that MiniTypeState is written and tested it would be nice to be able to distribute it to users. Packaging systems help the user by dealing with the dependencies of a project, creating the installation environment and executing the compilation instructions. A Frama-C plug-in is similar to any other OCaml library, so it can use any existing OCaml distribution framework (opam,[26] Nix,[27]...). This section uses opam, but only the last part is specific to it.

First, Sect. 7.8.2.1 looks at the optional creation of a configure step. Then Sect. 7.8.2.2 shows how our plugin can use external libraries or can be used as an external library of another plugin. Section 7.8.2.3 explains how to handle external files needed during execution. Section 7.8.2.4 shows how to prepare a plugin to be used as an external library. Finally, Sect. 7.8.2.5 describes how to install it.

---

[26] https://opam.ocaml.org/.
[27] https://nixos.org/.

7.8.2.1 Configure

The build system Dune automatically detects all the needed libraries and executables. Therefore, a specific configure step is not needed before compilation. However, users who compile from source would appreciate having a simple way to check that everything is available. Unfortunately, Dune does not yet have builtin features for displaying a nice summary of the required packages. For an advanced plug-in, we could rely on a autoconf-generated configure script. Here, since our plug-in only requires Dune, a simple script does the job.

```
#!/bin/sh

if ! [ -x "$(command -v dune)" ]; then
  echo 'Error: dune is not installed.'
  exit 1
else
  echo 'dune found.'
fi

echo "Availability of OCaml libraries:"
dune build @frama-c-configure
```

The last command executes the alias frama-c-configure, whose sole purpose is to print the availability of each dependence. The rule for this alias must be added to the dune file of MiniTypeState as follows[28]:

```
(rule
 (alias frama-c-configure)
 (deps (universe))
 (action (progn
          (echo "MiniTypeState:"
                %{lib-available:frama-c-minitypestate.core}
                "\n")
          (echo " - Frama-C:"
                %{lib-available:frama-c.kernel} "\n")
 )
 )
)
```

The list must be extended when dependencies are added, which makes this an untractable configure solution for large plug-ins, but good enough for plug-ins with few dependencies.

7.8.2.2 Use of External Libraries

Using external libraries in a plug-in is straightforward: just add them to the list of libraries of the MiniTypeState library stanza in the Dune file (currently (libraries frama-c.kernel)). One can also use other Frama-C libraries as dependencies. For example, the Eva API (see Chap. 3) can be used by adding the library frama-c-eva.core.

---
[28] The universe dependency indicates to always rerun this rule when requested.

7.8.2.3 Use of Additional Files During the Execution of the Analysis

It could be necessary to use files provided by the plug-in during the analysis, such as specific headers, or data files that are too large to be added to the binary. Such files have to be declared for installation in the Dune file of the plug-in.

Installing MiniTypeState's .h File

Instead of writing OCaml code that generates the enum type __typestates as described in Sect. 7.4.3, it is also possible to write this type definition in a file minitypestate.h and merge it with the new project before applying the code transformation.

```
(install
  (package frama-c-minitypestate)
  (section (site (frama-c share)))
  (files
    (minitypestate.h as minitypestate/minitypestate.h)
))
```

This Dune stanza copies the file minitypestate.h to a subdirectory of the same name, inside Frama-C's share directory, which is where Frama-C stores additional files. Each plug-in has its own subdirectory. For instance, getting the filename of the installed version of minitypestate.h in MiniTypeState's OCaml code can be done through Plugin.Share.get_file "minitypestate.h".

7.8.2.4 Plug-in Used as an External Library

The plugin MiniTypeState is already defined as a library in the Dune file. Therefore, any other plugin defined in the same Dune project can use it. Moreover, once installed, it can be used as any other library or any other Frama-C plug-in.

For example, one can add the file MiniTypeState.ml, exporting chosen types and functions, and documenting them:

```
(** Check if the type is the type of files handled by the
    plugin *)
let is_file_type = Mts_ast_transform.is_file_type

(** Check if the type is a pointer to a file *)
let is_file_ptr_type = Mts_ast_transform.is_file_ptr_type

(** List of functions handled by the plugin *)
let api_list = Mts_ast_transform.api_list
```

Running dune build @doc will generate documentation (in HTML format) in directory _build/default/_doc.

7.8.2.5 Installing MiniTypeState

With opam, the dependency information and compilation instructions for the package frama-c-minitypestate are in a file frama-c-minitypestate.opam. Since this file contains boilerplate instructions, it is handy to generate it automatically. The generation is triggered by the line (generate_opam_files true) in the dune-project file.[29] If Git is used, the plug-in can be installed directly from source by using the following command:

```
opam pin <url of the minitypestate repository>
```

The local directory can also be used directly.

For simplifying the installation process, the last step consists in adding the plug-in to the central opam repository. The tool dune-release can automate the release process.[30] It requires some information and an authentication token from Github that should be specified following the instructions from dune-release help files. Then the following subcommands can be executed in order:

1. `dune-release tag <VERSION>` creates an annotated tag in the local repository;
2. `dune-release distrib` checks the project for common package defects and creates the archive;
3. `dune-release publish distrib` creates the release in the Github repository;
4. `dune-release opam pkg` creates locally the opam package file;
5. `dune-release opam submit` creates a new branch and adds the opam file created previously to opam-repository, which contains descriptions for all opam packages. It also opens a merge request in the Github repository of opam-repository.

Once the merge-request has been merged by the opam repository maintainers, the package of your new Frama-C plug-in is directly available to the users.

7.9 Conclusion

Frama-C provides a large API with several specific features, e.g., projects, copy visitors, or interpreted automata, which can be complicated to grasp. This chapter has provided a general overview that should help OCaml developers[31] with basic

---

[29] This might be handled automatically by Dune in future releases.

[30] Currently, uploading the source code archive of the plugin is only possible for plug-ins hosted on Github.

[31] Even if this chapter assumes knowledge of OCaml for understanding it, several developers with no expertise in functional languages in general and in OCaml in particular have already developed Frama-C plug-ins. Basically, they have just practiced a bit by reading and applying some OCaml tutorials before developing their own plug-ins.

knowledge in program analysis to enter into this framework to develop their own plug-ins. However, it does not pretend to cover everything developers should know: they are invited to refer to the dedicated manual [14] and to the API documentation for details. As explained in introduction and seen throughout this chapter, one usually faces several challenges when developing a Frama-C plug-in. Even Frama-C experts must be careful. Our main advices would be:

- Choose with care when your program transformation should be applied: before or after Frama-C's clean-up phase. If executed before, Frama-C's global tables cannot be queried; if executed after, it is your responsibility to update them according to your program transformation.
- Once the AST has been created, you cannot modify it in-place: you should do it in a new project, either by using a copy-visitor or after having duplicated it.
- Duplicating AST nodes must perform deep copies.
- When manipulating two ASTs of two different projects, be careful to not mix up AST nodes: a node of project p_1 cannot be used in project p_2 without having been (deeply) copied.

The development of Frama-C is still ongoing and so, some code snippets presented here might become incompatible with future versions of Frama-C. However, we do think that the core set of features presented here should still be present in the coming years. Indeed, most of them already exist with almost no changes since more than ten years. New kernel services might also be added in the future, while new plug-ins (with their API) are added in the public distribution on a regular basis.

As a summary, Frama-C's expanding codebase aims at being a tool of choice for any researcher who would like to quickly prototype a new analyzer for C code or any engineer who needs to build a powerful code analyzer embedded in a dedicated mechanized verification methodology at lower costs.

References

1. Allen FE (1970) Control flow analysis. In: Proceedings of a symposium on compiler optimization. Association for computing machinery, New York, NY, USA. https://doi.org/10.1145/800028.808479
2. Appel AW (2004) Modern compiler implementation in ML. Cambridge University Press
3. Cousot P (2021) Principles of abstract interpretation. MIT Press
4. Cuoq P, Doligez D (2008) Hashconsing in an incrementally garbage-collected system: a story of weak pointers and Hashconsing in Ocaml 3.10.2. In: Workshop on ML. https://doi.org/10.1145/1411304.1411308
5. Cuoq P, Doligez D, Signoles J (2011) Lightweight typed customizable unmarshaling. In: Workshop on ML
6. Cuoq P, Signoles J, Baudin P, Bonichon R, Canet G, Correnson L, Monate B, Prevosto V, Puccetti A (2009) Experience report: OCaml for an industrial-strength static analysis framework. In: International conference of functional programming (ICFP)
7. Gamma E, Helm R, Johnson R, Vlissides J (1994) Design patterns: elements of reusable object-oriented software. Addison-Wesley

8. Hickey J, Madhavapeddy A, Minsky Y (2021) Real world OCaml, 2nd ed. O'Reilly (2021). https://dev.realworldocaml.org/
9. Kildall GA (1973) A unified approach to global program optimization. In: Proceedings of the 1st annual ACM SIGACT-SIGPLAN symposium on principles of programming languages, POPL '73, pp 194–206. Association for Computing Machinery. https://doi.org/10.1145/512927.512945
10. Michie D (1968) Memo functions and machine learning. Nature
11. Nielson F, Nielson HR, Hankin C (2015) Principles of program analysis. Springer
12. Rival X, Yi K (2020) Introduction to static analysis: an abstract interpretation perspective. MIT Press
13. Signoles J (2009) Foncteurs impératifs et composés: la notion de projet dans Frama-C. In: Journées Francophones des Langages Applicatifs (JFLA). In French
14. Signoles J, Antignac T, Correnson L, Lemerre M, Prevosto V, Plug-in development guide. http://frama-c.com/download/frama-c-plugin-development-guide.pdf
15. Strom RE, Yemini S (1986) Typestate: a programming language concept for enhancing software reliability. IEEE Trans Softw Eng 12(1)
16. Whitington J (2013) OCaml from the very beginning. Coherent Press

Chapter 8
Tools for Program Understanding

André Maroneze and Valentin Perrelle

Abstract Frama-C can operate as an *exploration* framework: several plug-ins and scripts have been created to enable tool-assisted code exploration and understanding. Unassisted reviews of C code require deep expertise and knowledge of many of C's pitfalls; Frama-C should help overcome many of its quirks, to help ensure that "all bugs are shallow". Whether you want some simple metrics and an overview; or you need to quickly evaluate code written by others; or you seek deep understanding about the origin of a complex bug, several Frama-C plug-ins are available to help you. This chapter presents these plug-ins and provides pointers for further details.

Keywords Code metrics · Visualization · Dependency analysis · Makefile

8.1 Introduction

Maurice Wilkes, British computer scientist, builder and designer of the first computer with an internally stored program (EDSAC), stated in his autobiography, *Memoirs of a Computer Pioneer* [3]:

> (...) the realization came over me with full force that a good part of the remainder of my life was going to be spent in finding errors in my own programs.

Most programmers are used to debugging their code, especially after having just written them. However, practitioners of formal verification and developers of code analysis tools often need to read code that *others* have written, sometimes with very little documentation, little time, or both. Time spent *writing* code is often an order of magnitude less than time spent *debugging* it. Tools which help understand the code are very precious indeed.

A. Maroneze (✉) · V. Perrelle
Université Paris-Saclay, CEA, List, Palaiseau, France
e-mail: andre.maroneze@cea.fr

V. Perrelle
e-mail: valentin.perrelle@cea.fr

One of the strengths of having a comprehensive code analysis platform such as Frama-C is that it allows having different views about the code: syntactic, semantic, static, dynamic. Reasoning about the code with minimal influence from the execution environment provides a more abstract, "purer" view of its intentions. After undefined behavior happens, a debugger, or `printf` statements will show concrete values exhibited by the machine, which can be garbage, non-deterministic, or "just seem right". Frama-C will halt earlier, allowing the user to get to the actual source of the issue faster.

Whenever you are confronted with new code, or even code written by yourself (in C, the programmer is often their own nemesis), these plug-ins should help you more quickly grasp what the code means. Obviously, they cannot show the *intention* of the code; but they should speed up understanding how it is seen by the compiler and by the C standard. When combined with the ingenuity of the reader, they augment their capability, hopefully allowing true *understanding* to emerge.

Section 8.2 describes Ivette, the new Frama-C graphical interface; Sect. 8.3 presents a code metrics plug-in, Metrics; Sects. 8.4 and 8.5 present the Studia and Dive plug-ins, both used to navigate and understand the origins of alarms emitted by Eva; finally, Sect. 8.6 presents a toolbox called *analysis scripts*, with scripts that complement Frama-C's plug-ins before, during and after analyses.

Disclaimer: this chapter deals with some recent, sometimes experimental plug-ins and tools. The presentation here intends to be generic and adapt to changes in upcoming Frama-C versions, but be aware that some features and command line examples may change substantially; please check for updates in the documentation of your Frama-C release.

8.2 Ivette: A New Graphical Interface

Frama-C, while being essentially a command-line tool, has been complemented by a graphical user interface (GUI), `frama-c-gui`, since its first release. This graphical interface (described in the Frama-C User Manual [2]), provides several essential features:

- Visualization of the normalized AST seen by Frama-C, along with its mapping w.r.t. the original source code, and including generated ACSL annotations (e.g., for alarms raised by analyses);
- Code navigation features (jump to definition, find uses, highlight occurrences), including, under certain conditions, through function pointers;
- Logical properties and their statuses (e.g., which assertions have been proven, and by which plug-in);
- Plug-in-specific tools, such as a value viewer for Eva (displaying the value of *any* expression, at any program point) and an interactive proof assistant for WP.

Several other features complement these (saving/loading sessions, alternating between projects, navigation history, etc.), providing a very powerful and effective interface for program analysis and understanding.

However, the technical framework upon which frama-c-gui was built did not evolve as quickly as other graphical frameworks, and the emergence of reactive-based tools with web-based technologies quickly outpaced it. There are thousands of JavaScript libraries and React components that offer features such as source code highlighting and navigation, interactive graphs, customizable tables, etc. Being able to benefit from these developments is one of the reasons a new Frama-C interface has been developed. Called Ivette, this new GUI replaces the aging Frama-C GUI with modern components and enables implementing several new modes of visualization and exploration to help users better make use of Frama-C.

As of this writing, Ivette does not yet *completely* replace frama-c-gui; in particular, it does not currently allow parametrizing analyses using the graphical interface, and does not have all plug-in-specific components—namely, Wp's proof assistant, TIP, the *Interactive Proof Transformer* (acronym in French). However, as will be seen in this chapter (and in others, such as Chap. 3), Ivette already offers replacements for several plug-ins, as well as new features that do not exist in the traditional GUI.

Ivette is currently distributed either as *download-on-demand* (when running ivette for the first time, it will download its JavaScript dependencies via NPM, install itself and then run), or as AppImage (Linux)/Package Bundle (macOS).

Replacing frama-c with ivette in a command with several options will have the effect of starting Ivette and then pass the command-line options to an underlying Server process (Server is a Frama-C plug-in handling communications between the Ivette client and a Frama-C instance). This process will allow Frama-C to compute analyses and send their results to the graphical front-end, both *during* and *after* processing of analyses. Some interactions with the graphical interface can also trigger new computations, which are sent back to the Frama-C process, and so on.

Figure 8.1 shows a screenshot of Ivette, just as an illustration of the overall view of the interface. Unlike the previous frama-c-gui, whose layout was mostly fixed, Ivette allows selecting which components are visible, and arranging their panels freely. A predefined set of *Views* is available, with the most common usage scenarios such as: source code navigation, visualization of Frama-C properties, exploration of Eva results, etc. Each view has a default set of associated *components*, which are the individual panels seen in the interface: *AST*, *Source Code*, *Properties*, *Inspector*, *Call Graph*, etc. A given component can be present in several different views.

Due to its rapidly-evolving state, we do not present here the details of the current components and views of Ivette. Interested users will find information about it in upcoming Frama-C releases. Many components are similar to those present in frama-c-gui, or self-explanatory.

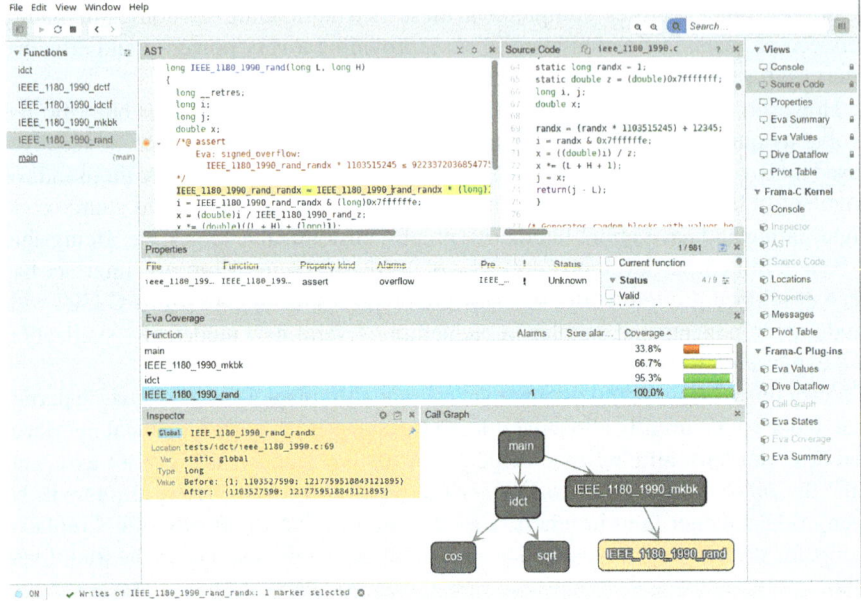

Fig. 8.1 Screenshot of Ivette, illustrating the overall appearance of the graphical interface

8.3 Metrics

Metrics is a syntax-oriented plugin, which computes several kinds of code metrics: number of statements, cyclomatic complexity, code coverage, etc. Its main advantages w.r.t. external tools, such as cloc, are:

- Metrics knows about Frama-C's internal representations; in particular, it can compute metrics both at the *Cabs* and at the *Cil* levels; the former is a more low-level view of the syntax tree, while the latter is the usual intermediate representation in Frama-C;
- Most code metrics tools ignore comments or treat them uniformly; Metrics knows about ACSL annotations and can count them;
- Metrics can include or exclude files related to Frama-C's standard library;
- When applied after other plug-ins, such as Eva, Metrics can compute some semantic information, such as function reachability and code coverage w.r.t. a specific entrypoint.

Metrics supports a few different output formats (text, HTML or JSON) and has some specific features, such as estimating the size of local variables (useful for stack size estimation), or listing unused files.

Metrics is a relatively underdeveloped plug-in, for a few reasons: first, most users of the Frama-C platform are interested in *semantic* properties and analyses; one will rarely consider Frama-C just for measuring syntactic metrics. Second, most

8 Tools for Program Understanding

Pivot Table								
Table ▼		Domain ▼	*Directory* ▼	Extension ▼	Filename ▼	Names ▼	Plugin ▼	Status ▼
Count ▼	⋮ ↔	Kind ▼	Node ▼					

Function ▼	Kind			code					
	Node								Totals
	Function	assignment	break	call	goto	if	loop	return	
	IEEE_1180_1990_dctf	28	8	2		13	8	1	60
	IEEE_1180_1990_idctf	28	8	2		13	8	1	60
	IEEE_1180_1990_mkbk	4	2	1		2	2	1	12
	IEEE_1180_1990_rand	6						1	7
	idct	45	8	2		16	8	1	80
	main	171	43	24	9	72	43	1	363
	Totals	282	69	31	9	116	69	6	582

Fig. 8.2 Pivot table in Ivette

users have specific metrics that they want to track, with specific requirements for grouping and filtering; for instance, should metrics be grouped by function, by file, or by directory? Should **Frama-C**'s standard library be included? Handling all possible combinations is costly. Third reason—and this is, paradoxically, a *good* reason for having an underdeveloped plug-in—is that, thanks to **Frama-C**'s visitors (see Chap. 7), most metrics can be easily computed by creating a custom visitor and counting the elements of interest, in a way that is very precisely suited to the user's needs.

Still, for occasional usage and a quick overview, **Metrics** does provide useful data, and can also serve as practical example and starting point for more specific calculations. For more details about **Metrics**, consult its user manual [1].

General Metrics via Pivot Tables

One way to solve the issue of different grouping and filtering needs, as mentioned above, is to use a *pivot table*: such tables, familiar to users of spreadsheets, display tabular data by aggregating and filtering lines and columns by simple drag and drop gestures. Figure 8.2 shows an example of the pivot table available in **Ivette**[1] (see Sect. 8.2 for details about **Ivette**).

In the figure, the table shows one function per row and one syntactic node kind per column. The intersections show the *count* function, that is, how many occurrences of each kind of statement happen at each function.

The headers can be dragged and dropped in either position, and are hierarchically organized: in Fig. 8.2, for instance, columns are organized by *kind* (here, only *code* is selected), and then by *node*. Names in italics indicate that some values are *filtered*: each header is also a button, and clicking on it allows filtering values, as shown in Fig. 8.3.

---

[1] This pivot table component is currently based on an external library, `react-pivottable`.

Fig. 8.3 Heading filter popup in a pivot table

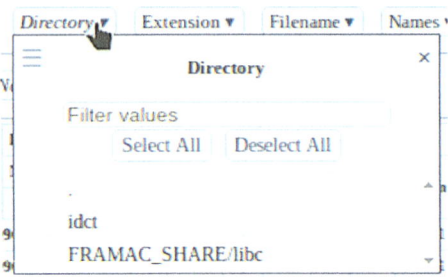

The user can also sort values according to totals, for instance to focus on rows and columns with more occurrences.

Besides syntactic elements, the pivot table can display counts related to *program properties*, that is, logical statements associated to specific program points, such as ACSL annotations and alarms. For instance, it is possible to count how many properties with *unknown* status there are per function. Finally, the pivot table provides information about messages (e.g., number of warnings).

The main downside of pivot tables is that, due to their versatility, they do not provide information in an immediately ready-to-use format; the user must choose the rows, columns and filters.

The user interface of the pivot table is likely to evolve to accommodate more features, such as drill-down (click to list the actual data rows in maximum detail), links to navigate to the source code, and common predefined table configurations.

8.4 Studia

The Studia plug-in is used to find the sources of alarms in a complex Eva analysis. It computes, for a given left-value, the set of statements reading from or writing to it. It is directly integrated in the graphical interface (both GUI and Ivette) and available via some context menus related to the AST. It is a simple plug-in used to help alleviate some of the exploratory work related to understanding the possible origins of alarms.

When inspecting the results of an Eva alarm, we often need to understand where an alarm comes from, to help us decide whether it is a *true* alarm or the result of an over-approximation. For instance, Fig. 8.4 shows a very short code extract with an initialization alarm related to variable tmp_8. Right-clicking on it and selecting *Studia: select writes* will ask Studia to compute and display the statements possibly writing to that left-value, which will highlight the statement tmp_8 = q;. Indeed, the value of tmp_8 trivially depends on the contents of q. The interesting step is the next one: repeating the operation on q will fill the *Locations* tab with 4 statements, as shown in Fig. 8.4. Note that the locations are presented in the bottom panel, and

8 Tools for Program Understanding

```
}
q = p;
while (q < & buf[ilen] - 1) {
  unsigned char *tmp_8;
  { /* sequence */
    tmp_8 = q;
    q ++;
    /*@ assert Eva: initialization: \initialized(tmp_8); */
    pad_count += (size_t)((int)*tmp_8 != 0);
```

Locations			3/4
		Writes of q	
#	Function	Statement	
1	rsa_pkcs1_decrypt	rsa.c:534	
2	rsa_pkcs1_decrypt	rsa.c:539	
3	rsa_pkcs1_decrypt	rsa.c:553	
4	rsa_pkcs1_decrypt	rsa.c:558	

Fig. 8.4 Locations computed by the Studia plug-in, as seen on Ivette: the bottom panel provides interprocedural navigation for the source code viewer (upper panel)

clicking on them navigates the AST (upper panel) to center on the selected statement, while also highlighting the other related statements.

These statements are those which may have written to the memory location q. In some cases they can be trivially found by the user via manual inspection, but as the code grows, it becomes increasingly important to have a tool automatically track this information.

Note that, as is often the case with Eva, these constitute an *over-approximation*, that is, there may be extra statements (which have not in fact written to that location), but there are no other statements which may have written to it.

The user can then click on the statements and inspect the variables to find out the cause of the alarm. This iterative process may continue for several steps before the root cause is found.

It is important to note that Studia does *not* perform a dataflow analysis when computing the statements; this has the advantage of maintaining very good performance even on very large programs. The downside is that the over-approximation mentioned before is even more noticeable; for instance, in the sequence x = 1; y = x; x = 2; y = x;, if we ask Studia to show the writes to x in the last statement, it will also include x = 1;, even if we can see in practice that its value is always overwritten before reaching this point.

To summarize, Studia is a small plug-in integrated into the graphical interface whose purpose is to help investigate sources of alarms emitted by Eva. It helps understanding what happens in Eva, and by consequence, the program. It also provides features for the Dive plug-in, described below.

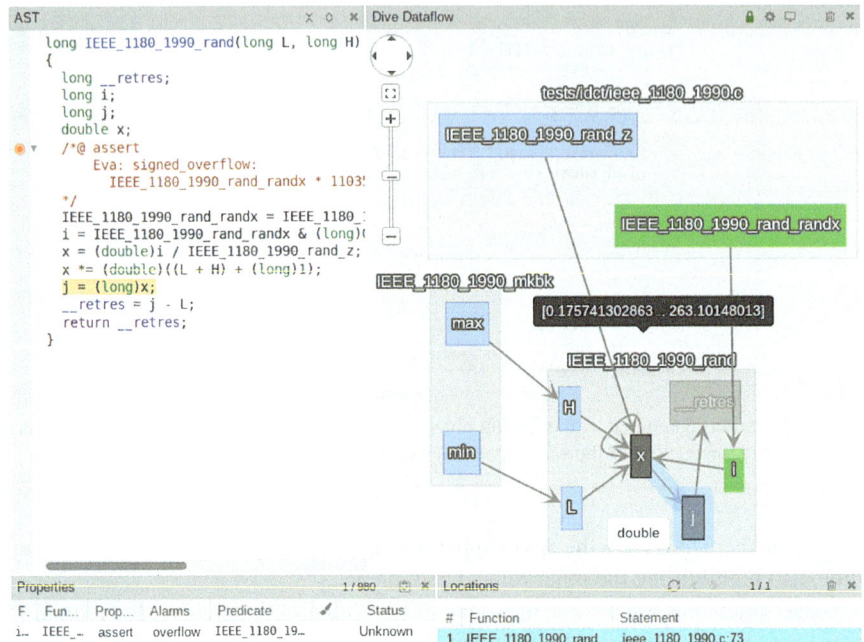

Fig. 8.5 Dive plug-in

8.5 Dive

The Dive plug-in provides visualization of data dependencies. It is intended to help the user "dive" into the origin of imprecisions during an analysis with Eva. It is based, among others, on Studia. It proposes some automation of the iterative process described in Studia.

This plug-in operates via the Ivette graphical user interface (Sect. 8.2), in a highly interactive and dynamic manner which is not easily conveyed by screenshots; nevertheless, the screenshot in Fig. 8.5 displays the *Dive Dataflow* view, with its main components: the source code AST, the Dive graph, the list of Properties and the list of Locations.

When an expression is selected in Frama-C's AST (x in the figure), Dive calls Studia to compute the data dependencies affecting its location and displays them in a graph (*Dive Dataflow*), with arrows pointing from assigned locations to read locations. This process can be iterated via user requests for further exploration (internally, Dive combines dependencies from locations to statements, and then from statements to locations, according to the information given by Studia and by Eva).

A color code indicates whether the values in a location were precisely (blue or green) or imprecisely (gray or red) computed by Eva. This allows, for instance, quickly discarding locations which are precise when searching the origin of an imprecision.

The user can click on any node in the graph and the AST view will navigate to the corresponding statements, interprocedurally, to allow the user to inspect the expressions and look for possible causes. This helps to find which functions can benefit from extra analysis time. The user can then re-parametrize Eva to focus on such functions, and hopefully obtain fewer alarms. Or, at least, obtain a more precise view of what happens with the variables.

Dive is a sophisticated tool for iterating through Eva analyses. It improves upon the usage of Studia, offering richer information which does not replace the user, but enhances their exploration, with graphical and interactive components. Note that Dive is still in an experimental state, and some aspects of its interactivity are subject to further development. Thus, they are not detailed here.

8.6 Analysis Scripts

When starting a new analysis with Frama-C, a series of steps are necessary before Frama-C's kernel is able to successfully parse the entirety of the code: the user must provide the whole set of source files, along with necessary preprocessing flags and an appropriate machdep.[2] Only then can Frama-C parse the code, build its AST, and provide it to the analyzers which perform the desired verifications and transformations.

This step is often overlooked by the users, who expect Frama-C to operate like many syntactic tools which require little more than a source file name. The deep semantic analyses that Frama-C is able to perform require extensive information about the program; sometimes even more than a compiler needs. After all, a compiler relies on the operating system and hardware to supply some information during runtime. Frama-C does not execute the code, so it cannot rely on existing object files, loader conventions, operating system implementation-defined behaviors, etc.

Besides initial setup, there are other tasks to be performed after an analysis takes place, or during the iterative process of refinement between successive analyses. They are sometimes better served by small scripts rather than custom Frama-C plug-ins: preparing Makefiles, opening results in a browser, monitoring the progress of analyses, etc.

In order to help users overcome the initial parsing step, as well as offering several auxiliary services, Frama-C provides a set of scripts, accessible via the `frama-c-script` command. These scripts help provide insights on new code bases, and offer extra comprehension features about existing ones. We do not detail all commands here, but we present some of them to illustrate what kind of help can be provided. The *Frama-C User Manual* [2] contains a chapter about analysis scripts

---

[2] See Chap. 2 for information about *machdeps*.

with details about their practical application, which complements the information given here.

On a technical note, the scripts mentioned here are mostly written in Python, while a few use bash. They are mainly directed towards Linux users, though macOS users should be able to run them as well, e.g., via Homebrew.

8.6.1 Motivating Example

This section provides a concrete example of the difficulties in handling non-trivial code bases. It may seem somewhat contrived, but it is inspired by real situations encountered during case studies, simplified into a few source files. If you already have some experience building C code, you may skip this section.

In this example, we handle a set of *counters* related to an *upvote* system (e.g., a button). Each counter starts at zero and, whenever an upvote event arrives, the counter is incremented.

We assume our code includes a *main* function which simulates the generation of some events and prints the final value of each count. The simplest form of the code is displayed below.

```
#include <stdio.h> // for printf()
#include <stdlib.h> // for rand()
#define RANDOM_COUNTER (rand() % 10)

static int counters[10];

void upvote(int c);

void main() {
  // simulate 100 voters
  for (int i = 0; i < 100; i++) {
    upvote(RANDOM_COUNTER);
  }
  // print counters
  for (int i = 0; i < 10; i++) {
    printf("counter %d: %d upvotes(s)\n", i, counters[i]);
  }
}

void upvote(int c) {
  counters[c]++;
}
```

This is straightforward both for a C compiler and for Frama-C: simply run it with the file name. Either the compiler or Frama-C will call the C preprocessor, which will process the #include directives, parse the code, link, and produce the final executable (in the case of the compiler) or some analysis result (in the case of Frama-C).

In practice, however, C code is rarely this simple. First, let us assume that the number of counters is not hard-coded, but defined via a preprocessor macro, N. The following lines must be modified:

```
#define RANDOM_COUNTER (rand() % N)
static int counters[N];
void main() {
    (...)
    for (int i = 0; i < N; i++) {
        (...)
}
```

Now, in order to parse the code, we need to provide the value of the preprocessor macro. Gcc/Clang and POSIX-compatible compilers use option -D; Microsoft's cl.exe uses option /D; etc.

Frama-C does not accept such option directly, the main reason being that Gcc accepts arguments with or without spaces (e.g., -D N=10 and -DN=10 are both accepted), but Frama-C's command line module cannot parse the latter. To avoid confusion and accidental usage, Frama-C emits a warning if -D is used, referring to its option -cpp-extra-args. Thus, in order to parse the above code, Frama-C requires the following command line (assuming you have a Gcc-compatible compiler):

```
frama-c -cpp-extra-args="-D N=10" simplest.c
```

Let us keep making the code more complex. Now a counter, instead of being a simple int, is a typedef that depends on the target architecture: an unsigned long long in a 32-bit architecture, or an unsigned long in a 64-bit architecture.[3] Such code uses implicitly-defined macros such as __amd64__ to detect the current architecture. In a 64-bit Gcc (assuming an x86 architecture), we can use option -m32 to compile to a 32-bit architecture. In Frama-C, one should prefer using *machdeps* (detailed in Chap. 2) to handle these architectural constraints. This means adding a new option to our command line, -machdep x86_32.

The next step consists in compiling two different variants of the same code: we will call them *simple* and *deluxe*. Our previous counter is the simple one, while the deluxe counter has a different API: instead of a single upvote, the upvote function takes an extra argument which sets the amount of votes to be increased. Of course, we want to reuse as much code as possible, so we will factor the common code (typedef, function prototypes, helper macros) in a header.h file. This file will be included by both simple.c and deluxe.c. The code of header.h is the following:

```
#include <stdio.h>  // for printf()
#include <stdlib.h> // for rand()

#ifdef __amd64__
typedef unsigned long counter;
#else
```

---

[3] Standards-compliant code would simply use uint64_t from <inttypes.h>, but in practice much code is *not* standards-compliant.

```
7  typedef unsigned long long counter;
8  #endif
9
10 #ifdef DELUXE
11 static void upvote(counter c, unsigned count); // Allow multi-upvote
12 #else
13 static void upvote(counter c);
14 #endif
15
16 // Simulate counters and upvotes
17 #define RANDOM_COUNTER (rand() % N)
18 #define RANDOM_AMOUNT (rand() % 10)
```

Note that the prototype for upvote is controlled by a conditional macro DELUXE: if it is defined, then the "deluxe" prototype is used, otherwise the simple one. Also note that the prototypes are declared as static, to ensure they are internal to each compilation unit; this will be useful later.

Both of our .c files will include this header; but to complicate matters, they will do so via #include "header.h", without a directory, while the file itself will lie in a directory include. Thus, another POSIX option, -I (or /I in Microsoft's cl.exe), will be necessary to guide the preprocessor. Here is the code of deluxe.c, under these conditions:

```
1  #define DELUXE
2  #include "header.h"
3
4  static counter counters[N];
5
6  void main() {
7    for (int i = 0; i < 100; i++) {
8      upvote(RANDOM_COUNTER, RANDOM_AMOUNT);
9    }
10   for (int i = 0; i < N; i++) {
11     printf("counter %d: %lu upvotes(s)\n", i, counters[i]);
12   }
13 }
14
15 void upvote(counter c, unsigned count) {
16   counters[c] += count;
17 }
```

First, we define DELUXE inside the file itself, then include header.h. This ensures the right prototype for upvote is defined. The auxiliary macros are used to simulate the upvotes in the main function. The code can be either compiled by Gcc or parsed with Frama-C:

```
1  gcc -D N=10 -I include deluxe.c
2  frama-c -cpp-extra-args="-D N=10 -I include" deluxe.c
```

The code for simple.c is very similar, except for the #define DELUXE line, and the RANDOM_AMOUNT argument to upvote.

8 Tools for Program Understanding

In practice, much C code is in the form of libraries, and does not necessarily contain a main function. We will now add another .c file, main.c, which will contain calls to main_deluxe and main_simple:

```
1  void main_deluxe(void);
2  void main_simple(void);
3
4  int main() {
5  #ifdef DELUXE
6    main_deluxe();
7  #else
8    main_simple();
9  #endif
10 }
```

Note that we need the function prototypes, otherwise the compiler (and Frama-C as well) will complain about implicit function declarations. We also need to rename the main functions in our other C files.

Now that we have multiple C sources for a single application, we must take into account different possibilities offered by *separate compilation*. With option -c, it is possible to compile a source file without *linking* it. This results in an .o object file which will be later linked to other files to produce the final executable. Frama-C cannot read .o files, so here the command lines for the compiler and for Frama-C diverge. As a reminder of the different possibilities, let us consider what we can do with simple.c and main.c:

- we can compile both files into an executable; Frama-C can parse both of them, perform its linking process, and analyze the resulting full program;
- we can compile only simple.c; Frama-C can parse it, but it will not have a main function, thus some whole-program analyses such as Eva cannot run;
- we can compile only main.c; Frama-C can parse it, and it will have a main function, but the analysis will lack the code of function main_simple and thus will be very imprecise (and possibly *incorrect*, since Frama-C will try to *guess* the effects of the nonexistent code, after emitting a warning).

Here, we believe the reader can already see why preparing the sources for a proper Frama-C analysis is non-trivial. However, we can still add an extra layer on top of it. Let us consider the following Makefile:

```
1  all: simple deluxe
2
3  simple: simple.o main.c
4          cc $^ -o $@
5
6  simple.o: simple.c
7          cc -D N=10 -I include $^ -c -o $@
8
9  deluxe: deluxe.o main.c
10         cc -D DELUXE $^ -o $@
11
12 deluxe.o: deluxe.c
13         cc -D N=10 -I include $^ -c -o $@
```

This Makefile uses separate compilation and produces two main executables: simple and deluxe. The compilation steps are very similar for both variants, except for -DDELUXE when compiling the main function of deluxe. Remember that each C file is parsed independently, thus the #define DELUXE line in the beginning of deluxe.c does *not* apply to the compilation of main.c.

Now comes the fun part: what if we wanted our simple and deluxe versions to be disjoint in terms of sizes of N? That is, we will add the following to the beginning of deluxe.c:

```
#if N < 100
#error N too small (must be >=100)
#endif
```

And the same thing, but with condition if N > 10, in simple.c. This prevents accidentally creating a too small deluxe version, or a too large simple version. Finally, now that we have two very distinct programs, we want to be able to test both in a single application, with a test.c file very similar to our main.c file, but which includes *both* versions of main:

```
void main_deluxe(void);
void main_simple(void);

int main() {
  main_deluxe();
  main_simple();
}
```

Here is a simple Makefile that can compile these three applications (deluxe, simple, and test):

```
all: simple deluxe test

simple: simple.o main.c
        cc $^ -o $@

simple.o: simple.c
        cc -D N=10 -I include $^ -c -o $@

deluxe: deluxe.o main.c
        cc -D DELUXE $^ -o $@

deluxe.o: deluxe.c
        cc -D N=100 -I include $^ -c -o $@

test: test.c simple.o deluxe.o
        cc $^ -o $@
```

Note the use of -D N=10 for simple and -D N=100 for deluxe. This redefinition of N *forces* us to use separate compilation; it is not possible to compile test using a single call to the compiler.

Situations such as the one above led to the creation of a specific Frama-C option, -cpp-extra-args-per-file, to enable handling different preprocessing options on a per-file basis, as well as the support for *JSON Compilation Databases* (JCDBs, see Sect. 8.6.3). These options allow **Frama-C** to parse code produced by what still remains a moderately simple Makefile. Using one of these options, it is possible to parse and fully analyze the `test` application, but generating the full command line for **Frama-C** is far from automatic. Even if we do have a JCDB, it will likely contain entries for both `test.c` and `main.c`. If the user simply tries to parse every C file in the directory, **Frama-C** will find two definitions for function `main` and halt. It is still up to the user to know which files should be parsed. In more complex Makefiles, the same C file may be compiled with different preprocessor flags, for different targets, preventing any kind of automatic processing.

To sum up, the following facts all contribute to the complexity of setting up an analysis with **Frama-C**:

- C has a preprocessor accessible via command-line options;
- C has separate compilation, but **Frama-C** needs the entire set of source files to perform a thorough analysis;
- much architecture-specific information (including compiler extensions and non-portable features) is implicit inside the build toolchain, so that users are not even aware of it;
- there is no "default build system" for C (even if Makefiles are very often used), which means much of the build process is "hidden away" from the user;
- even with external tools, such as **JSON** Compilation Databases, it is not possible to fully automatize the build process due to missing linking information.

These factors justify the existence of some of the scripts described in the next sections: since parsing is one of **Frama-C**'s first steps, it is sometimes necessary to rely on external tools in order to help the user obtain a first successful parse.

8.6.2 Heuristics-Based Scripts

In order to overcome the initial hurdle of getting a set of working parsing commands, some scripts operate on a set of syntactic heuristics. Their result is not guaranteed in any way to reflect the actual source code (e.g., some comments or conditional preprocessor directives might lead the heuristics astray), but in practice, the information they provide is accurate enough to help in most cases. Such scripts are useful for an initial look into a new code base. `estimate-difficulty` is the main example of this class of scripts.

`estimate-difficulty`

It is hard to precisely determine, in advance, whether a given code is going to be easy or hard to analyze with **Frama-C**. However, some syntactic elements allow scripts to quickly enumerate a set of features that are likely to affect the complexity of an analysis.

This script takes as argument a set of C files, and outputs a set of messages such as "this non-POSIX header has been included", or "this standard library function seems to be used, but it has no specification in Frama-C's libc". It also warns about features that may be problematic to some plug-ins (recursion, dynamic memory allocation, inline assembly, etc.), or that simply make analyses more complex in general.

Overall, this script is the cheapest approach before trying Frama-C. Even for large code bases, execution should take no longer than a few seconds.

8.6.3 Configuring and Building

Much C software is built using CMake, GNU Autoconf and Makefile, using commands such as ./configure && make && make install. The configure step probes the system for dependencies, such as external libraries, and can disable optional features. Customizing this step for Frama-C can improve the chances of obtaining a readily-analyzable code. The make step (or cmake, in the case of CMake) can also be used to extract information concerning the build, so that Frama-C can reuse it later to more easily parse the code.

frama-c-script offers a configure wrapper which tries to emulate a minimal system, with as few installed components as possible, which might help to analyze programs with optional components. In practice, configure scripts vary widely and this step is very hard to automate, leading this tool to fail more often than one might expect.

A more useful script is build, which relies on a build_commands.json[4] file. Such a file is similar in principle to a compile_commands.json, also called JSON Compilation Database[5] (JCDB), which is a JSON file storing the commands used to compile the sources, with all the arguments given to the compiler. This file can be produced by build-related tools, such as cmake (via variable CMAKE_EXPORT_COMPILE_COMMANDS) and Build EAR.[6] However, while JCDBs only store individual commands, a build_commands.json also stores the relations between *sources* and *targets* (.o files, libraries, or executables), including link commands. This information allows Frama-C to automate, most of the time, the list of sources and preprocessing flags to be used in order to obtain a given application. Thus, what the build script does is read a build_commands.json and produce a GNUmakefile which follows the *analysis template* provided by Frama-C. This template, detailed in the next section, is the recommended way to run analyses for realistic code bases, especially for iterative refinement of whole-program based analyses such as Eva.

---

[4] Note that, currently, build_commands.json files are not automatically produced by any tool; the Frama-C team intends to release such a tool in the nearby future to help automate this step.

[5] https://clang.llvm.org/docs/JSONCompilationDatabase.html.

[6] https://github.com/rizsotto/Bear.

8.6.4 Analysis Template

The analysis of a realistic code base with plug-ins such as Eva—that is, whole program analysis, and with a wide array of tuning options—is better performed as a series of discrete steps, as described in the *Eva User Manual* (section *Three-step approach*): parsing, then running Eva, then viewing results in the graphical interface. There is often some back-and-forth between steps; for instance, if when running Eva, the user notices a called function which has neither code nor specification, they may need to modify the list of source files, or add some specifications. If they see in the GUI that some results are imprecise, they may want to add annotations to split analysis states. Handling this directly in the command line is inefficient and error-prone: a fully-parametrized analysis for a large code base can have dozens of lines of parameters (function-specific options, custom notification levels, etc.).

The analysis template provided by Frama-C offers a Makefile-based solution to keep track of dependencies (only reparse sources when needed, for instance), parametrization options, and saved sessions and logs. By default, these are stored in a hidden folder .frama-c in the root of the C project, to avoid polluting the original sources with Frama-C-specific files.

This analysis template can be automatically produced using the build script mentioned earlier, ideally from an existing build_commands.json file. Note that the script produces an *initial* version of the GNUmakefile; the user is expected to open the file and adjust analysis parameters (mostly related to precision), then rerun the analysis until it converges. To help during this iterative step, the following script "wraps" the make command, automatically detecting some common issues and trying to provide helpful messages when possible.

make-wrapper

This is a simple wrapper around the make command used to run Eva targets from the analysis template. For instance, instead of running make program.eva, the user should run frama-c-script make-wrapper program.eva. This will run a wrapper which calls make and, whenever it hits upon some set of predefined situations, will try to automatically provide *hints* on how the user should proceed. The most common use case is when Eva stumbles upon a function with neither code nor specification. This can make the analysis unsound, so the analysis template treats it as an error. The make-wrapper script then parses the error message and uses some auxiliary scripts to try to find its definition among unprocessed sources.

One might wonder why Frama-C does not directly offer this kind of suggestion, when such errors arise. The main reason is that such reasoning is based on contextual information (e.g., searching for unparsed files in the current directory), the presence of external tools (e.g., Makefile), and the assumption that the user wants to perform a specific kind of analysis. Following Unix philosophy, we opted for composing features, by adding a layer which operates on top of existing tools (Makefile), which themselves operate on top of Frama-C.

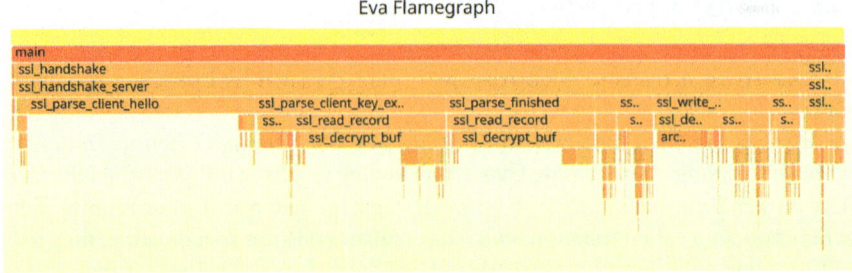

Fig. 8.6 Flame graph of an analysis with the Eva plug-in

8.6.5 Miscellaneous Utilities

A few analysis scripts are provided to help with auxiliary tasks related to external tools. The integration of these scripts with the workings of Frama-C helps save time and offers a few extra features.

flamegraph

Flame graphs[7] can be used to visualize the duration of an analysis with a long-running plug-in such as Eva, by showing a stack trace of time spent in each function, as shown in Fig. 8.6.

The analysis template (mentioned Sect. 8.6.4) sets, among other options, -eva-flamegraph. During an analysis with Eva, this option periodically outputs information to a text file. The flamegraph script then converts it to an SVG (using Brendan Gregg's Perl script) embedded in an HTML page. Via a browser, it provides an interactive visualization about which C functions consume more time during the analysis, with search and zoom features.

creduce

C-Reduce[8] is a tool that *reduces* the size of a C program according to an arbitrary criterion. It was originally developed to test compilers. In Frama-C's case, it is mostly useful to debug Frama-C crashes: by giving it a Frama-C command line and a (possibly large) C file, it can obtain a minimum example which triggers the crash. The reducing criterion is defined via a shell script, and C-Reduce applies a vast array of passes to strip parts of the code while respecting the criterion. Note that this does *not* preserve the C semantics of the original code.

This script is overall more useful to report issues with Frama-C, but it can also be used to help understand how Frama-C interacts with a given code: if a given Frama-C message seems hard to understand, it is possible to write a script to reduce

---

[7] https://www.brendangregg.com/flamegraphs.html.

[8] https://embed.cs.utah.edu/creduce/.

the program while keeping the message, and hopefully obtain something simpler that can be more easily understood.

C-Reduce requires some initial learning effort, especially to get used to writing meaningful test scripts; but in the long term, it can save a substantial amount of time reducing large programs.

8.6.6 Conclusion

In order to understand programs with Frama-C, it is first necessary to be able to parse them; several scripts presented here aid the user in this task. Later, when setting up a long analysis for deep understanding of a complex code base, a structured approach is useful. Again, the *analysis scripts* presented here offer a template as well as some helper tools. This is not entirely free—some initial investment is necessary, but in the long term it is certainly useful. Finally, after analyses are run, and further refinement is required, some plug-ins and scripts—with support from the graphical interface—help visualize, profile and understand the origin of alarms. They extend the Frama-C toolbox with several minor instruments to help the user in their enlightenment process.

References

1. Bonichon R, Yakobowski B, Frama-C's metrics plug-in. https://www.frama-c.com/download/frama-c-metrics-manual.pdf
2. Correnson L, Cuoq P, Kirchner F, Maroneze A, Prevosto V, Puccetti A, Signoles J, Yakobowski B, Frama-C user manual. https://www.frama-c.com/download/frama-c-user-manual.pdf
3. Wilkes M (1985) Memoirs of a computer pioneer. Massachusetts Institute of Technology, USA

Chapter 9
Combining Analyses Within Frama-C

Nikolai Kosmatov, Artjom Plaunov, Subash Shankar, and Julien Signoles

Abstract Combinations of analysis techniques and tools can help verification engineers to achieve their goals. The Frama-C verification platform offers a large range of possibilities for combining its analyzers with each other or with external tools. This chapter provides an overview of several combinations with different objectives and levels of maturity. First, we show how model checking and Counterexample Guided Refinement Abstraction (CEGAR) are used with value analysis and deductive verification for proving statement contracts. Runtime assertion checking can provide the user with useful information by checking annotations on selected test inputs. Next, test generation can be used to classify alarms, explain proof failures and generate counterexamples. Finally, static techniques are capable to optimize test generation by detecting infeasible test objectives and thus avoiding a test generation tool to waste time by trying to cover them.

Keywords Combined analyses · Model checking · CEGAR · Runtime assertion checking · Deductive verification · Value analysis · Test generation

---

The work of Subash Shankar was partly supported by Digito Foreign Guest Research Grant 2013-0376D and PSC-CUNY Grant 67776-00-45.

N. Kosmatov (✉)
Thales Research & Technology, Palaiseau, France
e-mail: nikolaikosmatov@gmail.com

A. Plaunov · S. Shankar
City University of New York, New York, USA
e-mail: subash.shankar@hunter.cuny.edu

J. Signoles
Université Paris-Saclay, CEA, List, Palaiseau, France
e-mail: julien.signoles@cea.fr

9.1 Introduction

Different static and dynamic analysis techniques have been proposed and successfully used in various verification projects. Typically, each technique can be efficient for some verification tasks and some kinds of target products. However, there is no magic bullet: each technique has its own limitations, for instance, related to soundness, completeness, ease of use, level of automation, capacity to scale up, capacity to exhibit a counterexample for an unproven property, capacity to handle some kinds of properties or some specific code features. Sometimes, the shortcomings of one technique can be compensated by using another one. Therefore, combinations of analysis techniques and tools can help verification engineers to achieve their goals.

The previous chapters presented core analysis techniques offered by the Frama-C verification toolset and implemented as separate analysis plug-ins. This chapter presents an overview of a few combined usages of different analyzers with different objectives and different levels of maturity. Section 9.2 shows how model checking—applied with Counterexample Guided Refinement Abstraction (CEGAR)—is used with value analysis and deductive verification. Section 9.3 illustrates how runtime assertion checking can provide the user with useful information by checking ACSL annotations on selected test inputs. Section 9.4 explains how test generation can be used to classify alarms, explain proof failures and generate counterexamples, and how static techniques can also benefit to test generation by detecting infeasible test objectives and thus avoiding a test generation tool to waste time by trying to cover them. Section 9.5 discusses some points of attention when combining analyses. Section 9.6 concludes the chapter and presents some related work.

9.2 Combinations of Model Checking with Frama-C Analyzers

Model checking is a powerful and automatic technique for verification of properties on models represented as finite state machines [15, 21, 45]. It is widely used today, especially in hardware verification. The properties being verified were traditionally written in various temporal logics, most often involving properties of the form '[something good] *eventually* occurs', or '*always* [something bad] does not occur', corresponding to liveness (reachability) and safety properties, respectively. State machines constituting the model are Boolean and finite, though they can be very large depending on the considered model checker. Consequently, the primary problem in model checking is state space explosion, which can manifest itself in either space or time complexity again depending on the considered model checker. However, the automatic turnkey nature of model checking makes it appealing for quickly dispensing with proofs for significant parts of programs, allowing the prover to concentrate on more complex proofs using other Frama-C components.

The use of model checking in program verification is not directly evident for several reasons:

- Program properties, as expressed in ACSL (and more generally in a Behavioral Interface Specification Language [32]), do not directly correspond to liveness and safety properties.
- Program variables are seldom Boolean, unlike state machines. While non-Boolean variables may be represented as Booleans (e.g., 32 Booleans for a 32-bit integer), the resulting state machines are then far too big for non-trivial programs.
- Model checking is most often used for proving properties dealing with interaction between concurrent entities, whether hardware or software, while traditional imperative programming languages are sequential with desired properties requiring knowledge of axioms about the underlying domains (e.g., axioms relating integer addition and multiplication).

Nevertheless, model checking has already successfully verified software in several cases [35].

The first of the above issues is simple to handle, as described later in Sect. 9.2.2. The most prominent technique to handle the second and third issues is based on a technique called *Counterexample Guided Refinement Abstraction* (CEGAR), and we have developed a Frama-C plug-in, CegarMC,[1] for proving ACSL statement contracts using CEGAR (future versions may also support function contracts and assertions). As state space explosion prevents all contracts from being verifiable, the intent is for the plug-in to be used in concert with other Frama-C plug-ins in an integrated verification framework.

9.2.1 The CEGAR Approach

As mentioned above, a major problem in software model checking is the disconnect between programs with non-Boolean variables and finite state machines. The most common way of addressing this disconnect is to abstract programs into Boolean programs—that is, programs where all variables are Boolean. This is done by identifying predicates on program variables (e.g., x == 0 or x != y) and/or control variables, and capturing semantics of program statements by their effects on the resulting predicates. This process, called *predicate abstraction*, was introduced by Graf and Saidi [28]. The resulting abstracted program is Boolean and may readily be translated to a relatively small finite state machine for model checking. However, information is lost in the abstraction so that while any properties that are valid in the abstracted program must also hold in the original concrete program, the converse does not hold. For a property that does not hold in the abstracted program, additional work is needed to determine whether the property is also false in the concrete program or whether the failure is due to the abstraction.

---

[1] http://www.cs.hunter.cuny.edu/~sshankar/cegarmc.html, currently for Frama-C v.21 Scandium.

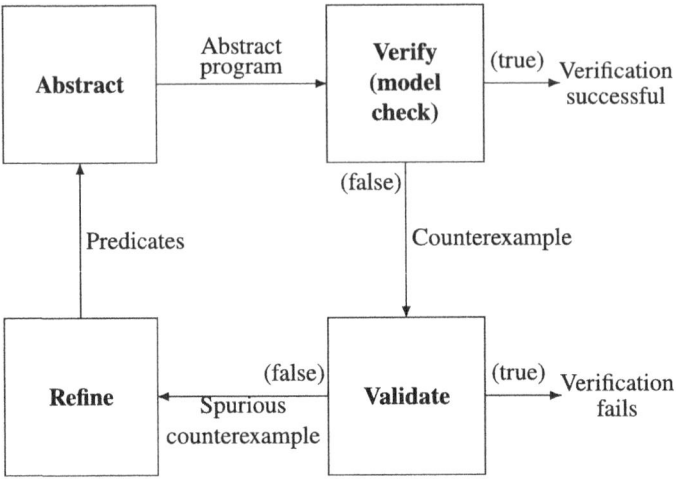

Fig. 9.1 Counterexample guided refinement (CEGAR)

We assumed above that it was possible to identify suitable predicates for abstraction, but this is in general not straightforward. A suitable initial set of predicates may be syntactically chosen to be some subset of conditions appearing in the program (including its contract), but this typically results in a too abstract program and additional predicates are needed. Model checkers provide counterexamples for failed proofs, which provide additional sources for predicates. This observation is the core of the CEGAR approach, illustrated in Fig. 9.1.

The four stages of CEGAR are:

1. **Abstract**: Given a set $\mathcal{P}$ of predicates, a predicate abstraction with respect to $\mathcal{P}$ is generated, utilizing the semantics of program statements. The resulting program is nondeterministic but Boolean.
2. **Verify**: A finite state machine corresponding to the program is constructed (as described in Sect. 9.2.2) and model checked. The number of states is manageable since an n-predicate abstraction requires only 2^n states, and normally fewer reachable states. If the verification is successful, the property is proven; otherwise a counterexample is sent to the next stage.
3. **Validate**: The counterexample is validated against the concrete program. If it is a true counterexample, the property can be concluded false. Otherwise the counterexample is spurious and the property's truth is unknown, leading to the next stage.
4. **Refine**: The counterexample is studied and additional abstraction predicates are identified, with the primary goal of distinguishing true and false counterexamples. This is generally the hardest stage though there are a number of techniques based on logic along with heuristics. Then, the process continues to the first stage, where $\mathcal{P}$ is now the extended set of predicates.

9 Combining Analyses Within Frama-C

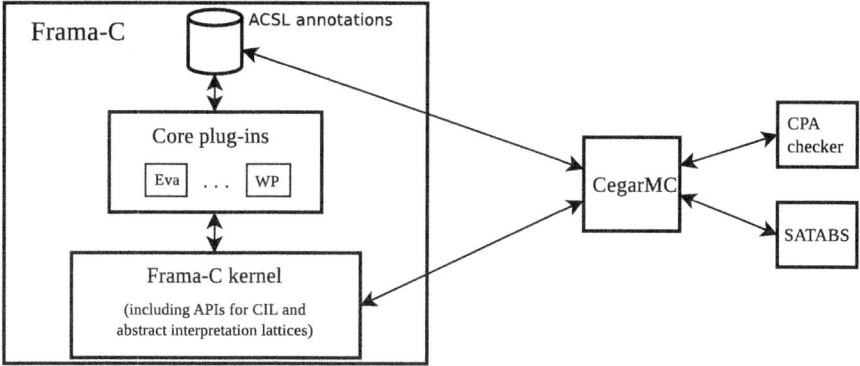

Fig. 9.2 System architecture of CegarMC

In practice, this iterative process is terminated with an indeterminate result after a timeout if not proven by then. When the result is conclusive, the number of iterations (also referred to as the number of refinements) gives an indicator of performance of the approach for a concrete program.

CEGAR was originally introduced by Clarke et al. [12], and since then has become major components of the SATABS [13] and CPAchecker [5] C verification tools. Our plug-in CegarMC verifies Frama-C statement contracts using these two tools as backends. In the case of CPAchecker, which is a configurable tool that allows for multiple model checking techniques (including some not completely sound approximations), CegarMC only supports a sound bit-precise CEGAR-based configuration (named predicateAnalysis-bitprecise). While our treatment here is from the perspective of the user, CegarMC also serves as a research tool to explore and perhaps improve the performance of CEGAR by utilizing information from other Frama-C plug-ins to improve the above stages, with refinement being particularly promising. The CegarMC plugin does not currently do any of these individual steps, as it utilizes CEGAR tools instead; however, the interested researcher may wish to explore the four stages for possible improvement if Frama-C functionalities are exploited.

9.2.2 *CegarMC Architecture*

CegarMC has already been presented in detail by Shankar and Pajela [47], so we only summarize the approach here. Figure 9.2 illustrates the CegarMC plug-in architecture. It essentially translates C programs augmented with ACSL statement contracts into a format suitable for verification by CEGAR backends. CEGAR backends (shown on the right-hand side of Fig. 9.2) use a form of C augmented with two constructs of concern here:

- Assume statements: Such a statement restricts the verification to execution paths in which the condition given in the statement's argument holds. For example, an assume(x > y) statement would not explore any executions in which x <= y at that point of the program.
- Assert statements: These are similar to C's assert, and correspond to the properties being verified.

The CegarMC plug-in constructs a program from the Cil version of the statement being verified to pass on to the CEGAR backends. Semantics of ACSL **requires** statements are captured using CEGAR assume statements, while **ensures** statements are translated to appropriately placed **assert** statements. There are additional complications imposed due to ACSL semantics dealing with abnormal terminations (see Chap. 1), and CegarMC maintains these semantics in such cases. Results are displayed using standard Frama-C GUI icons indicating statuses of properties.

An important feature is that interprocedural verification is supported by replacing called functions with their contracts (when available, and enabled by the user), instead of including the code of the function body. A resulting proof is of course then conditional on the contract of the called function, but Frama-C directly supports such dependences, thus enabling compositional verification (see Chap. 2). Such transformations alleviate the state space explosion problem prevalent in model checking since a large function may be abstracted into a much smaller contract capturing the important properties of the function. This feature also allows for support of library calls since Frama-C includes contracts for most functions of the C standard library (the so-called libc) and allows the user to add contracts to any declared functions when needed.

There are two issues that may make such an approach potentially less efficient than direct CEGAR model checking:

- The internal version of the code in Frama-C (partially transformed after parsing the original code by Frama-C, which relies on the Cil library, as explained in Chap. 2) may not be comparable to the more structured original C code (e.g., without goto statements) even though semantically equivalent;
- There are a *few* additional variables (thus states) introduced by CegarMC during the translation (see [47] for details).

However, we have evaluated the performance of CPAchecker alone against the combination with CegarMC, that we denote CegarMC⇒CPAchecker, and found no significant performance difference—in fact there is a slight improvement in some cases. Figure 9.3 shows the results of the two approaches on a collection of programs arbitrarily selected from the SV-COMP software verification competition benchmark suite.[2] These are evaluated with respect to both the total time and the number of CEGAR refinements.[3] The seemingly surprising cases where the number of CEGAR

---

[2] See https://sv-comp.sosy-lab.org/.
[3] Times are "Total time for CPAchecker" as reported by the tool. All experiments are run on a standard laptop with an Intel i5-8265U (1.6–3.9 GHz) and 16 GB RAM.

Benchmark		CPAchecker		CegarMC ⇒CPAchecker	
	# Lines	#Refinements	Time (s.)	#Refinements	Time (s.)
loops/heavy-2	10	1	1.196	1	1.183
loops/string-1	38	1	1.219	1	1.081
loops/n.c40	14	3	2.853	0	1.071
loops/terminator_02-2	22	1	1.176	1	1.140
loop-new/count_by_1	8	0	1.059	4	1.129
loop-acceleration/array_3-1	15	5	1.288	1	1.266
loop-acceleration/array_1-2	7	4	1.268	0	1.123
loop-simple/nested_3-1	15	25	9.146	25	9.144

Fig. 9.3 Performance of direct CPAchecker versus CegarMC⇒CPAchecker

iterations goes down are an artifact of the translation since contracts quantifying over entire arrays are replaced by assertion checks in each iteration of a loop. While the nature of refinements and translations makes it difficult to infer any patterns in the resulting numbers, it does not seem that CegarMC significantly affects performance.

As mentioned earlier, CegarMCcurrently supports only ACSL statement contracts, and its intended use is to quickly dispense with those proofs that do not experience state space explosion and/or for which a CEGAR approach can find suitable abstraction predicates. In such cases, no additional end-user effort is needed as model checking is completely automatic. In practice, it is but one tool in a verification toolchest and other Frama-C plug-ins will be needed to verify complete programs. Thus, CegarMC includes several flags that assist in integrating with other tools, discussed in the next two sections. The CegarMC plug-in is also intended as a research tool to develop new methods of integration (possibly automatable in the future), though we do not discuss that here.

9.2.3 *Exploiting Value Analysis*

By default, CegarMC treats statement contract semantics as follows: For a statement contract containing ACSL clauses **requires** R and **ensures** E, if the statement is executed in a state satisfying R, property E must be true upon termination of the statement (see Chap. 1 for details). However, ACSL statement contract semantics has the additional constraint that the statement contract is applicable only to actual paths in the program—i.e., in the context of the program containing the statement. If considering the context of the statement as determined by value analysis using the Eva plug-in, additional preconditions for the statement may be exploited. Consider the program in the running example of Fig. 9.4. The statement contract of lines 35–38 is proven by CegarMC by exploiting information determined by Eva. In this example, Eva can be used to determine that -99 <= x <= 199, 101 <= y <= 199, and eva%10==0 upon reaching line 34. Note that although some of these constraints

```
1   #include "__fc_builtin.h"
2   #include "limits.h"
3
4   // The loop increments/decrements x/y to maintain x+y as n,
5   // thus making the function always return false.
6   /*@
7     requires n >= 0;
8     ensures \result == 0; */
9   int foo(int n) {
10    int x = n; int y = 0;
11    /*@ loop invariant x+y == n; */
12    while (x > 0) {
13      x--;
14      y++;
15    }
16    return !(x+y == n);
17  }
18
19  int main() {
20    int base = 0; int eva = 0;
21    int x = Frama_C_interval(-99,199);
22    int y = Frama_C_interval(101,199);
23    int n = Frama_C_interval(0,INT_MAX-10)
24
25    for (int i=0; i<n; i+=10) {
26      eva = i;
27    }
28
29    // The first two conjuncts of the loop condition are
30    // the negation of the ensures. Thus, the only way to
31    // falsify the ensures is to make the third conjunct
32    // false. The verification amounts to showing that the
33    // third conjunct is true.
34    /*@
35      requires base == 0;
36      requires x >= -99 && x <= 199;
37      requires y >= 101 && y <= 199;
38      ensures x >= 100 || y <= 100;
39    */
40    while( x<100 && y>100 && (eva%10 + base%10)%10==0)
41    {
42      int flag = Frama_C_interval(0,1);
43      if ( flag ) {
44        base = foo(n);
45        x += (base%10 == 0);
46      } else {
47        x--;
48        y--;
49      }
50    }
51    return 0;
52  }
```

Fig. 9.4 Example program

may be inferrable using other techniques, the final constraint is one that would not typically be determinable without using abstract interpretation, thus enhancing the power of software model checking if done in the Frama-C setting.

The CegarMC plug-in supports this with a user-selected flag (named -mc-context). After running Eva, any uses of CegarMC with this flag set implicitly assume the results of Eva. To distinguish such proofs, they are tagged with context. Implementation-wise, information about initial values for any variables in the statement are sent to the CEGAR-based solvers. For integers, this may include ranges (e.g., $n \in [1..10]$), finite sets (e.g., $n \in \{2, 7\}$), mod fields (e.g., $n \bmod 2 = 1$), and combinations of these. Appropriate other types of abstractions as supported by Eva are used for other types. In general, adding contextual information reduces state space size for model checking; however there are cases where it can increase model checking time since these initial conditions also provide additional candidate refinement predicates for a CEGAR-based checker, and these predicates may not be helpful towards proving the desired contract thus leading to increased state space and/or refinement iterations. Thus, CegarMC also provides a flag (-mc-maxcontext) that can be set to indicate the maximum number of elements to enumerate if Eva analysis results in a finite set for a variable.[4] For example, if Eva determines that $n \in \{2, 3, 7, 11\}$ and this flag is set to 2, model checking will be based on the assumption that $2 \leq n \leq 11$ instead of enumerating all 4 values.

The use of context has two benefits:

- While contextual information may be explicitly inserted by the verification engineer into requires clauses of the statement contract, that would involve significant additional effort, especially if the statement contains many variables. It is also not clear that a typical engineer would be able to determine these preconditions to the precision of value analysis.
- Even if the statement contract is true regardless of context, the use of context may reduce the state space. In cases where the statement contract is true regardless of context, the user may call CegarMC without the context flag especially in case that improves model checker performance. But note the above discussion about how it may not necessarily improve performance in some cases due to the refinement process.

Before turning to another type of integration, some comparison of value analysis and model checking is warranted as the former can be considered equivalent to model checking in that it effectively explores the program state space but using abstract interpretation instead of state machine traversal [18]. However, there are three significant differences:

1. Eva is mostly limited to showing monadic predicates—that is predicates involving single variables. For example, it can not be used to show that x>y except in the relatively rare cases where it can independently conclude that x>a and y<=a (for some constant a). While Eva also supports some relational domains, these may be

---

[4] This flag has the same meaning as the Eva's option -eva-ilevel (see Chap. 3).

less efficient or precise than the default monadic predicates and still less general than the full set of properties that a user may wish to prove. Model checking based approaches are not limited to properties involving pre-defined abstract domains.
2. In the presence of loops, value analysis performs a fixed point computation that may lead to over-approximation for efficiency reasons. It may also be directed by the user to unroll loops a fixed number of iterations. In contrast, model checking unrolls loops as many times as needed to prove the property.
3. Even in cases where model checking and value analysis are equivalent, the use of CEGAR (as opposed to plain model checking) significantly affects performance in some cases.

However, the user is cautioned that it is not in general possible to predict when CegarMC will succeed, while Eva times are more predictable.

9.2.4 Integrating Deductive Verification

Deductive verification (based on Floyd-Hoare logic and Dijkstra's weakest precondition calculus [20, 31]) and model checking are two different approaches to proving contracts with pre- and postconditions, and Frama-C implements the former using the Wp plug-in (see Chap. 4 for detail). Deductive verification is generally more appropriate for sequential programs as it can exploit properties of arithmetic using modern SMT solvers. The presence of loops complicates the issue though since deductive methods require the user to explicitly identify loop invariants, which most software engineers find difficult. In contrast, model checking is completely automatic though it may not work in many cases.

Consider the example of Fig. 9.4 again. The previous section left unsaid that although the statement contract of line 38 is true CegarMC actually is unable to prove it (with default timeouts) due to state space explosion. Noting that the called function foo contains a function contract, it is possible to integrate Wp and CegarMC to achieve success, by abstracting foo's function body with its contract—that is, by restricting the state space to those paths in which foo returns 0. The CegarMC plug-in contains a flag (named -mc-abstractCalls) that automates this abstraction, thus reducing state space. With this flag, the statement contract is successfully proven by CegarMC though the proof is conditional on foo's function contract in line 4. The user may prove this function contract itself using Wp to complete the proof, thereby avoiding the need for an invariant on the loop in the statement contract.

Any proof dealing with multiple functions must also make independence assumptions between the caller and callee—for example, the proof in the above example is valid only if foo does not modify eva (among other variables). While this happens to be true for the example, it is conceivable that two functions interfere, for example through static global variables. To capture this, CegarMC checks for such interferences utilizing information from ACSL **assigns** clauses when provided or

```
1   /*@
2     requires R1;
3     ensures E2;
4   */                  // deduced by WP from statement contracts
5   int foo() {
6   /*@
7     requires R1;
8     ensures E1;
9   */                  // proved by CegarMC
10  S1;
11  /*@
12    requires E1;
13    ensures E2;
14  */                  // proved by WP
15  S2;
16  }
```

Fig. 9.5 A perspective of Wp-CegarMC integration for consecutive statements

explicitly traversing the code to determine modified variables otherwise.[5] Proofs are then marked accordingly if they cannot be shown independent but are otherwise verified. An important special case of this occurs when a callee contains no body and only a contract, in which case no interference conclusions can be drawn (if there is no assigns clause).

Although not directly related to deductive verification, an important consequence of function call abstraction is that it allows software model checking to cover library calls as Frama-C provides contracts for a significant number of functions in libc.[6] As most software model checkers do not directly support library calls, their integration into Frama-C is a significant enhancement.

Current versions of Wp do not support statement contracts. An additional form of Wp-CegarMC integration using the contracts of consecutive statements will be possible in the future when their support is added to Wp. Consider the abstract program with two statement contracts and one function contract shown in Fig. 9.5. Furthermore, suppose that CegarMC is able to prove the first statement contract, and the second one can be proved by Wp (thus, including the specification and proof of invariants if it has a loop). Then the function contract can trivially be proven using Wp.

---

[5] In its current version, this traversal does not take care of pointers. It is left to future work to compute a sound overapproximation in presence of pointers, e.g., by using a sound alias analysis such as provided by plug-ins Eva or Alias.

[6] CegarMC currently only supports a small ACSL subset, thus excluding library functions using other ACSL constructs, though future versions will extend it.

9.3 Lightening Static Analyses by Means of Runtime Verification

Runtime verification is a lightweight formal method that allows the user to check program properties at runtime, i.e. when the program is being executed. Even though verifying properties at runtime is used since the 1970s, in particular through program assertions such as the **assert** macro in the C programming language, it has been promoted as a research discipline at the turn of the millennium [33]. A key idea behind the creation of this research field was to deal with the state explosion problem of model checking by providing a lightweight dynamic verification technique that can be used when static model checking does not scale. However, and because of this strong inclination towards model checking, runtime verification usually focuses on verifying the very same class of properties as model checking, i.e. temporal properties expressing safety or liveness properties [37], as explained in Sect. 9.2. As a consequence, standard runtime verification tools typically do not deal with properties about one (or several) program state(s) as expressed through a Behavior Interface Specification Language (BISL) such as ACSL [24].

Nevertheless, even if less academically studied than runtime verification of temporal properties, verifying BISL-like properties at runtime is also possible. Such techniques are usually referred to as *runtime assertion checking* [14] or, more recently, as *runtime annotation checking* [34]. For ACSL annotations, it is the goal of the E-ACSL plug-in of Frama-C presented in Sect. 9.3.1. Thanks to the analysis combination facilities offered by Frama-C, this framework is also a tool of choice for using runtime verification through E-ACSL to help in reducing the verification cost of static analyzers. In particular, Sects. 9.3.2 and 9.3.3 describe combinations of runtime verification with value analysis and deductive verification, respectively. It is also possible to combine those analyses in order to verify non-ACSL properties, as demonstrated in Sect. 9.3.4.

9.3.1 Verifying ACSL Annotations with E-ACSL

This section only gives a short overview of E-ACSL, which is sufficient to understand the content of this chapter. The interested reader is referred to Chap. 5 for more technical details about this plug-in.

E-ACSL takes as input a C program annotated with ACSL annotations and generates a new C program that checks the validity of the annotations at runtime. If an annotation is invalid for the given execution trace, the generated program exits the execution[7] and reports the failure; otherwise, if all annotations are valid, its functional behavior is the same as for the initial program.

---

[7] Unless this default behavior is modified by the user.

9 Combining Analyses Within Frama-C

Fig. 9.6 Functional view of the E-ACSL plug-in

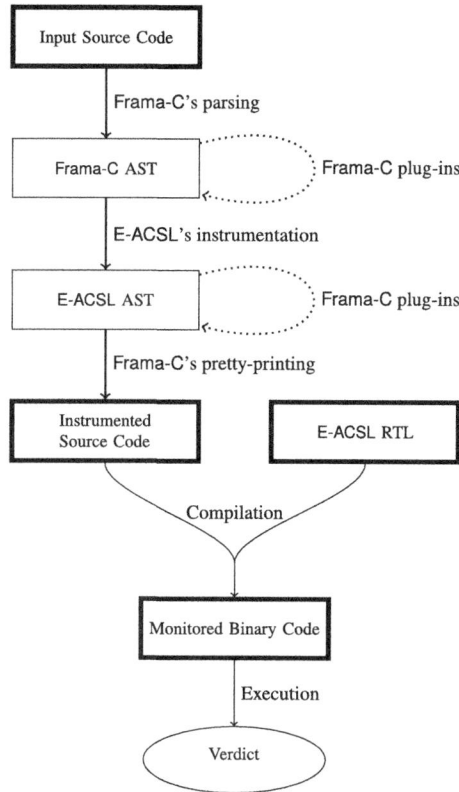

Figure 9.6 presents a functional view of E-ACSL. First, the Frama-C kernel parses the input C source files, possibly annotated with ACSL annotations, in order to generate an Abstract Syntax Tree (AST), which is the internal representation of a C program in Frama-C (see Chap. 2 for details). E-ACSL works on top of this AST and generates a new AST in which an appropriate instrumentation is added to soundly monitor ACSL annotations. Before running E-ACSL, it is possible to run any other Frama-C plug-ins, as it will be shown in the next subsections. Once a new AST has been generated by E-ACSL, it can be pretty-printed and saved into a new C source file. It is also possible to run other Frama-C plug-ins before pretty-printing the code, but this feature is rarely used.[8] The generated file should be compiled and linked using a standard C compiler (typically, Gcc) in order to generate an executable binary. This step requires to link this file to the E-ACSL RunTime Library (RTL), whose routines are called by the instrumentation introduced by E-ACSL. The

---

[8] The only known concrete example of this feature is the non-regression test suite of E-ACSL that runs the Eva plug-in on the AST generated by E-ACSL, in order to verify that it has introduced no undefined behaviors in the generated code.

generated binary embeds both the original source code and its monitor, so that its execution reports any failed property, as explained above.

In practice, since it is quite tedious to invoke Frama-C and the C compiler with the necessary options, E-ACSL comes with a convenient script e-acsl-gcc.sh for doing both steps at once, as shown in the following example.

Using E-ACSL on a Simple Program

Figure 9.7 provides a tentative answer to Euler project's problem 1,[9] which consists in finding the sum of all the multiples of 3 or 5 below 1,000. At lines 5 and 6, it contains two ACSL annotations that specify necessary conditions for calling function add adding a new number to the sum: first, the given number must be below max (the maximal acceptable value, which is 1,000 here) and, second, it must be a multiple of either 3 or 5.

Instrumenting and compiling this program with e-acsl-gcc.sh is very easy and produces two binaries, a non-instrumented one (a.out) and an instumented one capable to monitor properties at runtime (a.out.e-acsl), which can be run as shown below:

```
1 $ e-acsl-gcc.sh -c euler1.c
2 $ ./a.out
3 total = 351648
4 $ ./a.out.e-acsl
5 euler1.c: In function 'add'
6 euler1.c:5: Error: Precondition failed:
7     The failing predicate is:
8     x < max.
9     With values:
10    - max: 1000
11    - x: 1000
12 Aborted
```

The binary executable a.out is the very same as the one produced by gcc without E-ACSL instrumentation, while a.out.e-acsl is the monitored executable. It demonstrates that our tentative answer to Euler project's problem 1 is actually incorrect since it is possible that it reaches max, i.e. 1,000. Indeed, the second call to function add at line 14 in Fig. 9.7 does not check that the added value 5 * i is below 1,000. Modifying this line of code in the following way fixes this problem (say, in file euler1_fixed.c):

```
14 if (i % 3 != 0 && 5 * i < max) add(max, 5 * i);
```

Now, executing the monitored code generates no error and produces the expected output:

```
1 $ e-acsl-gcc.sh -c euler1_fixed.c
2 $ ./a.out
3 total = 233168
```

---

[9] See https://projecteuler.net/problem%3D1.

```
4 $ ./a.out.e-acsl
5 total = 233168
```

It is worth noting that this correct result (i.e., 233,168) is not the same as the wrong one (i.e., 351,648) we previously got when running the unmonitored buggy executable.

9.3.2 Combining Runtime Verification with Value Analysis

The Frama-C plug-in Eva implements value analysis by means of abstract interpretation *à la* Cousot [17, 46] (see Chap. 3 for details). Such an analysis computes an over-approximation of all the possible values of each memory location at each program point. Doing so, it also generates an *alarm* as soon as an undefined behavior could happen at runtime or, in the context of Frama-C, as soon as an ACSL annotation could be invalid. In both cases, a warning is emitted. Additionally, in the former case, Eva also generates an unproven ACSL annotation, and, in the latter case, it raises an unproven status for any possibly invalid existing annotation. It means that, if no alarm is raised, Eva guarantees that the program has no undefined behavior and all the ACSL annotations are valid. However, when an alarm is raised, the user often does not know whether the issue can indeed happen at runtime (*true alarm*), or whether it cannot occur in practice and the alarm is only raised because of a lack of precision of the analysis due to an over-approximation (*false alarm*). In this last case, the user can try to fine-tune the analyzer to trade precision against time efficiency

```
1  #include <stdio.h>
2
3  int total = 0;
4
5  /*@ requires x < max;
6    @ requires x % 3 == 0 || x % 5 == 0; */
7  void add(int max, int x) { total += x; }
8
9  void sum(int max) {
10    for(int i = 1; i < max / 3; i++) {
11      /* add multiples of 3 */
12      add(max, 3 * i);
13      /* add multiples of 5 that are not also multiples of 3 */
14      if (i % 3 != 0) add(max, 5 * i);
15    }
16  }
17
18  int main() {
19    sum(1000);
20    printf("total = %d\n", total);
21    return 0;
22  }
```

Fig. 9.7 Tentative answer to Euler project's Problem 1 (file euler1.c)

```c
1  #include <stdlib.h>
2  #include <stdio.h>
3
4  typedef struct node { int value; struct node *next; } list;
5
6  void print(list *l) {
7      list *ptr = l;
8      while (ptr) {
9          printf("%d ", ptr->value);
10         ptr = ptr->next;
11     }
12     printf("\n");
13 }
14
15 void add(list **l, int v) {
16     list *head = (list *) malloc(sizeof(list));
17     if (head) {
18        head->value = v;
19        head->next = *l;
20        *l = head;
21     }
22 }
23
24 int main() {
25     list *l = NULL;
26     for(int i = 0; i < 10; i++) add(&l, i);
27     print(l);
28     return 0;
29 }
```

Fig. 9.8 Linked list example (file list.c)

and/or memory consumption, but it requires expertise. Even worse, it is not always possible. In practice, obtaining 0 alarms on a large program is utopian. Therefore, it is up to the user to manage the remaining alarms.

The usual way to do it is by manual reviewing, as reported in [40] by EDF, the biggest French electricity provider. Each of the 153 alarms generated by Eva was reviewed in order to "(1) confirm that it was a real bug; (2) justify that the tool was not able to conclude; or (3) justify that the tool made an overapproximation" [40, Sect. 5]. However, it requires a local expertise and does not scale well. For instance, in the same case study, 995 alarms were raised by PolySpace,[10] another value analysis tool based on abstract interpretation, and EDF's "limited means" prevent them from analyzing each alarm.

Verifying the alarms generated by Eva at runtime by means of E-ACSL may come to the rescue to circumvent this issue. This combination of Eva and E-ACSL within Frama-C relies on Eva's ability to generate ACSL annotations when it raises

---

[10] https://fr.mathworks.com/products/polyspace.html.

9 Combining Analyses Within Frama-C

alarms: after running Eva, E-ACSL can in turn be used to generate monitors for the alarms expressed as ACSL annotations.

Monitoring Eva's Alarms with E-ACSL

Consider for instance the program of Fig. 9.8. This program manipulates a linked list in a trivial manner. However, Eva is not able to precisely deal with such a recursive data structure. In particular, even if it can prove that all memory accesses in function add are safe, it cannot do so for function print because it does not know the memory shape of the list. Therefore, it raises two alarms, for each of the memory accesses at lines 9 and 10, as shown below:

```
$ frama-c -eva list.c
<...>
[eva:alarm] list.c:9: Warning:
  accessing uninitialized left-value.
  assert \initialized(&ptr->value);
[eva:alarm] list.c:10: Warning:
  accessing uninitialized left-value.
  assert \initialized(&ptr->next);
<...>
 2 alarms generated by the analysis:
     2 accesses to uninitialized left-values
```

These alarms can be monitored with E-ACSL using the following command:

```
$ e-acsl-gcc.sh --frama-c-stdlib --frama-c-extra "-eva" \
    --then -c list.c
```

This command line ensures that Eva is run before E-ACSL and uses the Frama-C standard library instead of the system one. Now, executing a.out.e-acsl provides the expected output since this program is correct.

Consider now a modified version of the example of Fig. 9.8 where one swaps lines 9 and 10 (a typical student's mistake). Assume this version is stored in file buggy_list.c. If we run again the previous command for this file to generate Eva's alarms,[11] create and execute the monitored binary, we get the following output that shows a memory access error at line 10 when trying to dereference ptr->value, which is indeed NULL in the last iteration of the loop:

```
buggy_list.c: In function 'print'
buggy_list.c:10: Error: Assertion failed:
    The failing predicate is:
    Eva: mem_access: \valid_read(&ptr->value).
    With values:
    - Eva: mem_access: \valid_read(&ptr->value): 0
    - sizeof(int): 4
    - &ptr->value: (nil)
Aborted
```

---

[11] 3 alarms are now generated instead of 2.

Even though such a combination of Eva and E-ACSL provides no guarantees for all possible executions, it allows the user to check that failures do not occur for the performed runs. Therefore, if used at development and/or validation time, it helps to strengthen the software. If used at operational time, it helps to detect errors at runtime and to recover when possible. For instance, in a security setting, Dassault Aviation experimented home-made security counter-measures on a few OpenSSL's modules, where Eva's alarms corresponding to security weaknesses were confirmed as failures at runtime thanks to E-ACSL [41]. Running Eva before E-ACSL allows Dassault Aviation to get a reasonable time overhead when monitoring, because most alarms were actually removed by Eva at compile time. This use case also relies on the possibility to redefine the E-ACSL monitor's behavior when an ACSL annotation fails at runtime (see Chap. 5): instead of reporting the standard error message and aborting, the authors can implement their own security counter-measures.

When combining Eva and E-ACSL in this way, it is important to remember that only the alarms generated by Eva are monitored. In particular, nothing is done for the undefined behaviors that would occur on statements outside of the code covered by Eva. It is typically the case when Eva is only run on a subset of all the possible entries through a specific analysis context, in order to keep the analysis tractable. In other words, as usual when combining analyses, this combination assumes the union of all hypotheses made by Eva and E-ACSL. Therefore, when Eva's coverage is low, generating ACSL annotations for all Frama-C-detectable undefined behaviors before running Eva through the Rte plug-in is an option worth considering. As explained in Chap. 2, Rte is a plug-in that generates ACSL annotations when undefined behaviors may happen: if they are proven valid by another plug-in, it means that these undefined behaviors will not happen.

9.3.3 Combining Runtime Verification with Deductive Verification

As mentioned in Sect. 9.2.4, deductive verification in Frama-C is implemented by the Wp plug-in (presented in detail in Chap. 4). It aims at proving the compliance of the code to the ACSL annotations. For doing so, it generates so-called *proof obligations* (or *verification conditions*) that must be validated by external means, typically an SMT solver such as Alt-Ergo.[12] While very powerful for guaranteeing program safety and functional correctness, this technique requires strong expertise and significant efforts to be done completely. In particular, the user may need to carefully choose the way to state the required properties and to provide additional proof-guiding ACSL annotations (such as loop invariants, assertions and lemmas) in order to complete the verification process. Similarly to what has been introduced in the previous section, it is natural to combine Wp and E-ACSL in order to formally prove only a subset of the functional properties and let E-ACSL monitor the others at runtime. A typical

---

[12] See https://alt-ergo.ocamlpro.com/.

9 Combining Analyses Within Frama-C

```
/*@ requires len >= 0;
    requires \valid(a+(0..len-1));
    requires \forall integer i,j; 0 <=i<=j<len ==> a[i] <= a[j];

    assigns \nothing;

    behavior exists:
      assumes \exists integer i; 0 <= i < len && a[i] == key;
      ensures 0 <= \result < len && a[\result] == key;

    behavior not_exists:
      assumes \forall integer i; 0 <= i < len ==> a[i] != key;
      ensures \result == -1;

    complete behaviors;
    disjoint behaviors; */
int search(int* a, int len, int key) {
  int low = 0, high = len - 1;
  /*@ loop invariant 0 <= low;
    @ loop invariant high < len;
    @ loop assigns low, high;
    @ loop variant high - low; */
  while (low <= high) {
    int mid = low + (high - low) / 2;
    if (a[mid] == key) return mid;
    if (a[mid] < key) { low = mid + 1; }
    else { high = mid - 1; }
  }
  return -1;
}

#include <stdio.h>

int main() {
  int a[10];
  for(int i = 0; i < 10; i++)
    a[i] = 2 * i;
  for(int i = 0; i < 10; i++) // test all possible values
    printf("%d ", search(a, 10, i));
  printf("\n");
}
```

Fig. 9.9 Binary search example

use case consists in proving some functions with Wp and monitoring the others with E-ACSL, but it is even possible to prove some annotations of a particular function with Wp, and let E-ACSL monitor the others in the same function. In this approach, ACSL annotations proved by Wp are *not* monitored at runtime (by default), which reduces the time overhead for monitoring.

Monitoring Unproved Properties with E-ACSL

Consider for instance the program of Fig. 9.9 that implements a binary search on a sorted array. A very precise contract has been written. Assume the user has a basic knowledge in deductive verification. She might write the loop annotations on lines 19–22 quite easily. They are actually sufficient to prove program safety (i.e., absence of undefined behaviors) and all the annotations except the behavior `exists`, which is arguably the most complicated case of the specification. Even if an expert user would quickly write the missing loop invariants, they are not so easy for beginners.[13] In such a case the unproved annotation can be monitored with E-ACSL as follows:

```
1 $ e-acsl-gcc.sh --frama-c-extra "-wp -wp-rte -wp-fct
    search" \
2     --then -c search.c
3 <...>
4 [wp] Proved goals:    22 / 23
5 <...>
```

This command line ensures that Wp is run on function `search` before E-ACSL, and proves its safety properties (thanks to the option `-wp-rte`) in addition to trying to prove its annotations. Now, executing `a.out.e-acsl` to check the unproven annotations with E-ACSL reports no error since this program is correct. However, if one introduces a bug at line 25 by replacing **return** mid; by **return** key;, the same annotations are proven, but we get the following error reported by E-ACSL at runtime, since the function now incorrectly returns 2 (instead of 1) when looking for 2:

```
1 buggy_search.c: In function 'search'
2 buggy_search.c:9: Error: Postcondition failed:
3     The failing predicate is:
4     exists: 0 <= \result < \old(len) && *(\old(a) +
        \result) == \old(key).
5     With values:
6     - \old(key): 2
7     - *(\old(a) + \result): 4
8     - \old(len): 10
9     - \result: 2
10 Aborted
```

When combining proof with Wp and (monitored) tests with E-ACSL for the same function, it may be quite hard to know what is eventually validated. Indeed, the proof is usually completed when all the annotations are proved, which is not the case in this setting. A test suite is usually considered good enough when a particular structural testing criterion is fulfilled, e.g. branch coverage or MC/DC [1]. However,

---

[13] Finding them is left as an exercise (a hint is provided in the following sections).

the standard testing criteria only apply to the whole code of a function: they are not able to take into account that we only need to test a subset of the specification, and so possibly only a subset of the code. For solving this issue, Le et al. propose an algorithm for evaluating the coverage of a verification campaign that combines proofs and tests [36]. It relies on a coverage matrix that indicates, for each part of a specification paired with a code statement, whether they have been covered by the verification process (either a proof or a test case). However, this algorithm still needs more tooling and further evaluations before being used on real-world case studies.

Another way to combine runtime verification with E-ACSL and deductive verification with Wp consists in using E-ACSL *before* proving ACSL annotations with Wp. Indeed, writing ACSL specifications is error-prone. Additionally, when Wp is not able to prove a property, it is not always easy to understand what is going on. There are several reasons of errors: either the code is wrong and needs to be fixed, or the specification is wrong and needs to be fixed as well, or some previous annotation is missing (or too weak) in order to help the automatic prover (typically, a loop invariant or a callee's contract), or the property is actually beyond the scope of automatic proof with the currently existing technologies [43]. However, Wp currently gives no clue to help the user. Therefore, testing the compliance of the code to the specifications before trying to prove them can detect inconsistencies or help to gain confidence in both the code and the specifications.

Debugging Specifications with E-ACSL

Consider again the example of Fig. 9.9 and now assume that the loop invariant at line 19 has been written 0 < low instead of 0 <= low. Then, any test case monitored with E-ACSL will report a failure as follows.

```
1 $ e-acsl-gcc.sh -c buggy_contract.c
2 $ ./a.out.e-acsl
3 buggy_contract.c: In function 'search'
4 buggy_contract.c:19: Error: Invariant failed:
5    The failing predicate is:
6    0 < low.
7    With values:
8    - low: 0
9 Aborted
```

Thanks to this execution, the user can easily see that this invariant is wrong, low == 0 being a counterexample. Therefore, she can easily fix the invariant to 0 <= low before proving the program with Wp.

9.3.4 Beyond ACSL Properties

The main purpose of E-ACSL consists in checking (functional) ACSL properties at runtime. However, it can take benefit from the Frama-C plug-ins that generate such annotations in order to check other properties at runtime.

One example that has been already illustrated in Sect. 9.3.2 consists of checking Eva's alarms at runtime thanks to ACSL annotations generated by Eva. More generally, detecting undefined behaviors at runtime with E-ACSL can be automatically done by relying on plug-in Rte for generating necessary annotations. E-ACSL itself promotes this usage by integrating it through a dedicated option --rte in e-acsl-gcc.sh.

Detecting Undefined Behaviors at Runtime with E-ACSL and Rte

Consider for instance the buggy version of Fig. 9.8 (in file buggy_list.c) discussed in Sect. 9.3.2. Activating the detection of undefined behaviors with E-ACSL on this is done file as follows:

```
$ e-acsl-gcc.sh --rte all -c buggy_list.c
```

Executing the generated monitored code gives the same result as in the example of Sect. 9.3.2.

This approach can be generalized to all the plug-ins that generate ACSL annotations, as soon as the generated annotations are handled by E-ACSL. In particular, it includes the Aoraï, MetAcsl and RPP plug-ins introduced in Chap. 10, which generate ACSL annotations for system-wide and relational properties. An additional example is also provided in Chap. 5.

9.4 Combinations of Static Analysis with Test Generation

Similarly to combinations with runtime verification, static analyzers can be combined with test generation. In the context of Frama-C, such combinations typically rely on the PathCrawler test generation tool [50], based on dynamic symbolic execution [6] and presented in detail in Chap. 6. The main idea of PathCrawler is to explore execution paths of the program under test and to generate a test case for each feasible path. A key feature of such combinations is the capacity of the testing tool to automatically generate counterexamples when such counterexamples exist and the tool manages to find one. If a counterexample for an annotation is not found after a complete exploration of a representative subset of program paths (respectively, the set of all paths when it is possible), then the user gets confidence (respectively, a proof) that the annotation is always valid. The first two subsections below present an overview of combined usages of Eva and Wp with PathCrawler. Another inter-

9 Combining Analyses Within Frama-C

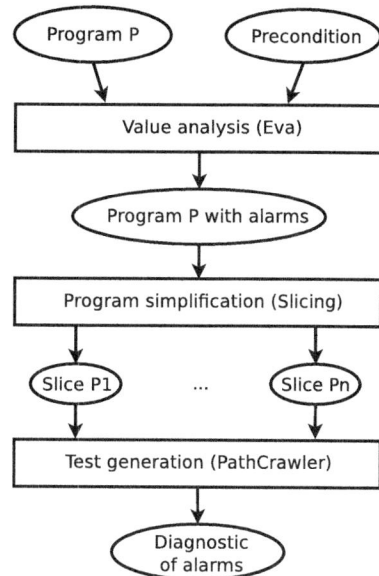

Fig. 9.10 Classification of alarms with SANTE

esting combination, presented in the third subsection, applies static analyzers Eva and Wp to identify infeasible test objectives in order to optimize the creation of test cases.

9.4.1 Test Generation to Classify Alarms Produced by Value Analysis

The goal of the SANTE (Static ANalysis and TEsting) plug-in [7, 8] (currently not supported) is to enhance the results of value analysis by testing. The SANTE method contains three main steps shown in Fig. 9.10. Given a C program P with a precondition, value analysis (with Eva[14]) first reports alarms for potential runtime errors in P and then tries to classify them as real bugs or false alarms. To do that, program simplification with the Slicing plug-in [7] is used to reduce the whole program with respect to one or several alarms. The idea of applying slicing in this case is to simplify the program by removing program statements that do not have any impact on the selected alarm(s), so that any given input will trigger an error for the target alarm (or one of the target alarms) on the initial program if and only if it will trigger an error for the same alarm on the simplified program. Slicing produces one or several slices $P_1, \ldots, P_n$. Thanks to the simplification step, the resulting programs usually contain less program paths and are easier to explore by a test generation tool

---

[14] The predecessor of Eva, called Value, was used by SANTE at the time of its creation.

Fig. 9.11 Verification of a function f containing a call to function g

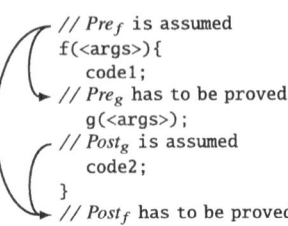

than the initial program P. Then, for each P_i, PathCrawler is run to explore program paths and to generate test cases confirming the alarms present in P_i. If such a test case activating an alarm is found, the alarm is confirmed as a *bug*. If all feasible paths were explored for slice P_i, all unconfirmed alarms in P_i are in fact false alarms. They are classified as *safe*. If PathCrawler performed only a partial exploration of all paths of P_i (e.g. due to a partial criterion like k-path, or a timeout), SANTE cannot conclude and the status of unconfirmed alarms in p_i remains *unknown*. Experiments [8] on several small real-life programs show that SANTE can be in average 43% faster than test generation alone and leaves less unknown alarms than value analysis alone. A more detailed description of the SANTE method can be found in [7, 8].

9.4.2 Test Generation to Diagnose Proof Failures

As we have mentioned in Sect. 9.3.3 above, an important difficulty of deductive verification is the need for manual debugging of proof failures by the verification engineer since proof failures may have several causes. Let us illustrate this point in more detail using Figs. 9.11 and 9.12, where for a function f we denote by Pre_f and $Post_f$, respectively, the precondition and the postcondition of f. Indeed, during the proof of function f, a failure to prove the precondition Pre_g of a callee g (see Fig. 9.11) may be due to a *non-compliance* of the code to the specification: either an error in the code code1, or an incorrect formalization of the requirements in the precondition Pre_f, or in Pre_g itself. The verification of the property can also fail because of a *prover incapacity* to realize a particular proof within a chosen timeout. It is often extremely difficult for the verification engineer to decide how to proceed: either suspect a non-compliance and look for an error in the code or in the specification, or suspect a prover incapacity and try to achieve an interactive proof with a proof assistant.

It is even more complex to analyze a failure to prove the postcondition $Post_f$ (see Fig. 9.11): in addition to a prover incapacity or a non-compliance due to errors in the pieces of code code1 and code2 or due to a wrong specification of Pre_f or $Post_f$, the failure can also result from a *subcontract weakness*: a *too weak* postcondition $Post_g$ of g that does not fully express the intended behavior of g. Notice that in this last case, the proof of g can still be successful: a weaker postcondition can still be provable. The verification engineer basically has to consider the possible reasons one

Fig. 9.12 Verification of a function f containing a loop with a loop invariant I

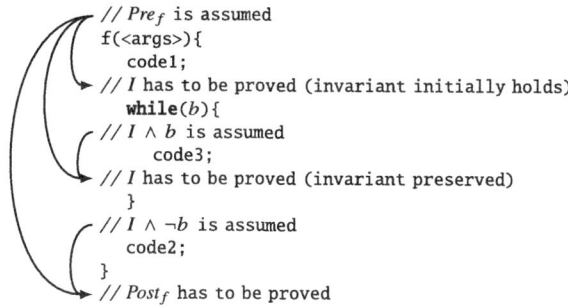

Fig. 9.13 Classification of proof failures with StaDy

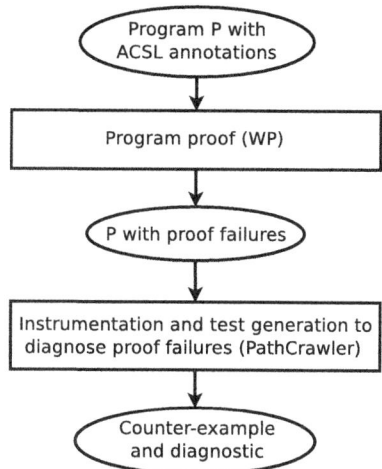

by one, and possibly initiate a very costly interactive proof. For a loop, the situation is similar (see Fig. 9.12), and offers an additional challenge: proving the invariant preservation, whose failure can be due to several reasons as well.

For instance, for function search of Fig. 9.9, the failure to prove the postcondition is an example of a subcontract weakness: the code is compliant with the specification, but it cannot be proved without a stronger loop invariant. It should state that the searched element key cannot be found outside the interval of indices low..high.

The StaDy (STAtic and DYnamic verification) plug-in[15] [43, 44] was proposed to address these issues. The motivation of this plug-in is twofold. One goal is to provide the verification engineer with a more precise feedback indicating the reason of each proof failure. A second goal is to produce—when possible—a counterexample that either confirms the non-compliance and demonstrates that the unproven annotation can indeed fail on a test datum, or confirms a subcontract weakness, showing on a test datum which subcontract (a callee or loop contract) is insufficient. The absence

---

[15] Available on https://github.com/gpetiot/Frama-C-StaDy (currently only for earlier versions of Frama-C and PathCrawler).

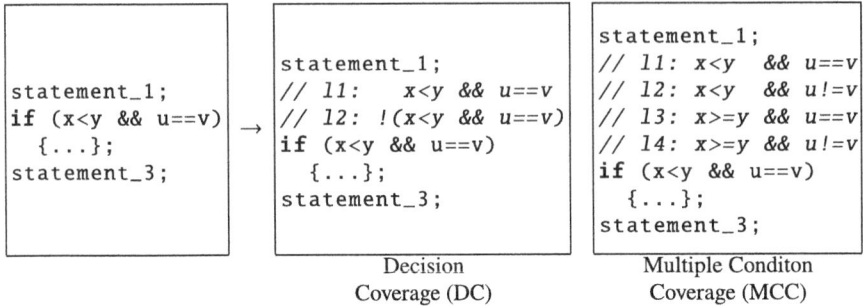

Fig. 9.14 Examples of labels encoding two common coverage criteria: decision (or branch) coverage (DC) and mutpiple condition coverage (MCC)

of counterexamples—at least after exploring a reduced program domain—suggests a prover incapacity to finish the corresponding proof.

The main steps of StaDy are shown in Fig. 9.13. A dedicated code transformation is used to translate ACSL annotations into PathCrawler preconditions and PathCrawler assume statements (for assumptions), and PathCrawler assert statements (for annotations that have to be verified). More precisely, when looking for non-compliance counterexamples, the only assumptions are the preconditions of the function under test, while all subsequent annotations become PathCrawler assert statements. In this way, the dynamic symbolic execution method of PathCrawler explores all program paths trying to violate each assertion and to generate a suitable test datum (a non-compliance counterexample).

When looking for subcontract weaknesses, the translation is more subtle: each function call is replaced by the most general code satisfying the called function contract, and each loop is replaced by the most general code satisfying the loop contract. The *most general code* is constructed as follows: variables listed in the (loop) assigns statements are assumed to receive symbolic (i.e. arbitrary) values satisfying the (loop or called function) contract. Then, running PathCrawler on the resulting code leads to generating counterexamples if the subcontracts (that is, loop or callee contracts) are too weak.

Experiments [44] show that this technique is capable to classify a significant number of proof failures within an analysis time comparable to the time of an automatic proof, including some very subtle functions, like the buggy version of OpenJDK sort function [27], in which the bug is pretty hard to exhibit because it occurs in very specific situations. A more detailed presentation of the StaDy method can be found in [43, 44].

9 Combining Analyses Within Frama-C

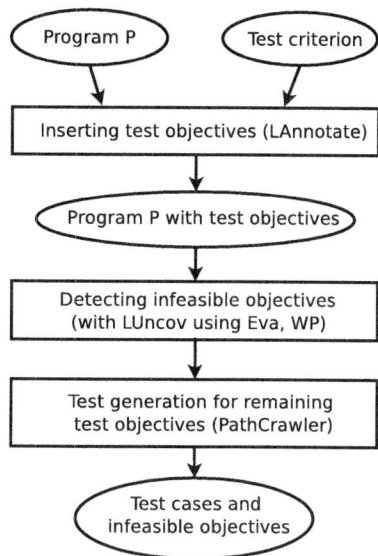

Fig. 9.15 Test generation process with LTest

9.4.3 Detecting Infeasible Test Objectives by Static Analysis

The last example of an integrative technique illustrates how test generation can in turn benefit from value analysis and deductive verification for detection of *infeasible test objectives* (or *test requirements*), that is, test objectives that cannot be covered by any execution. Infeasible test objectives have long been recognized as one of the main cost factors of software testing [49, 51]. This issue is due to the following three reasons. First, resources are wasted in attempts to improve test cases while there is in fact no hope to cover these requirements. Second, the decision to stop test generation is hard if the knowledge of what could be covered remains uncertain. Third, since identifying such infeasible test objectives is an undecidable problem [26], they require time consuming manual analysis because the test engineer must justify why she could not fulfill these requirements. In short, the effort that should be spent in testing is wasted in understanding why a given requirement cannot be covered.

Identifying infeasible test requirements with **Frama-C** was investigated in the context of the **LTest** toolset [2]. This toolset offers tools, implemented as **Frama-C** plug-ins, for several testing services: **LAnnotate** for inserting test objectives (in the form of so-called *(coverage) labels*) into the C code for a given test coverage criterion; **LUncov** for detecting infeasible test objectives; **LReplay** for replaying a given test suite to measure its coverage; and **PathCrawler** for generating test cases for coverage labels in the target program (previously added by **LAnnotate** or by the user).

Using (Coverage) Labels to Express Test Requirements

Let us illustrate how labels can be used to express test requirements associated to coverage criteria. Figure 9.14 shows a code snippet (on the left) with a conditional statement containing a conjunction of two conditions. To express test requirements for decision coverage, two labels are added (see the second code snippet): they express that the whole decision is, resp., true (then branch) and false (else branch). To express multiple condition coverage, we need to express all possible combinations of the conditions appearing in the decision. This is done by the four labels in the third code snippet.

A label is covered by a test input when the execution of the program with this test input reaches the label point and satisfies the label condition. A given test suite satisfies the coverage criterion encoded by a set of labels if and only if each label is covered by at least one test of the test suite.

An integrative technique for test generation with LTest is shown in Fig. 9.15. After inserting test objectives into a given program for a given criterion, infeasible test objectives are detected before running test generation for the remaining test objectives, that is, after excluding those that are detected to be infeasible. Let us explain the detection step in more detail.

A method to detect infeasible test requirements for several structural testing criteria was proposed by Bardin et al. [3]. The main idea of this method is that the problem of detecting infeasible requirements can be transformed into an assertion validity problem. For a test objective that requires to reach a given program point l and to satisfy a given predicate p, this test objective is infeasible if and only if the assertion //@ assert !p; inserted at program point l is valid. By using static program verification tools, such as Eva and Wp, it becomes possible to solve this problem.

This approach is generic since test requirements for various coverage criteria can be handled in the same way. It can be applied to all criteria that can be simulated by coverage labels [3]. This approach is also sound since it relies on sound static analysis techniques—if a test objective is detected as infeasible, it really is.

Experiments [3] demonstrate that static analysis can detect up to 95% of infeasible requirements for the condition coverage, multiple condition coverage and weak mutation testing criteria. Experiments also show that by identifying infeasible requirements before the test generation process we can speed up the automated test generation tool since it does not waste time trying to cover infeasible test objectives. An extension of this technique was proposed for detection of two other kinds of polluting test objectives: duplicate and subsumed objectives [38].

9.5 Points of Attention for Integrative Techniques

The previous sections presented several benefits of integrative methods. They are well supported by Frama-C, its kernel, and its specification language ACSL, which is the *lingua franca* between Frama-C's analyzers. First, when a set of ACSL annotations is proven by different plug-ins, the Frama-C kernel automatically consolidates the global results to show to the end-user which properties remain to be proven (see Chap. 2). Second, a standard approach for plug-in collaboration consists in generating ACSL annotations with one plug-in and verifying them by others. Frama-C provides smooth ways to instantiate this approach, as illustrated in Sect. 9.3, in particular through a powerful command line. Combined to the extensive nature of Frama-C, which allows any OCaml developers to extend the framework with new plug-ins (see Chap. 7), it helps design new methodologies for verifying properties that were not the original verification targets when designing Frama-C and ACSL about 15 years ago (see Chap. 10).

However, despite the benefits of integrative methods, there are several points of attention that need to be accounted for. We list some major ones here.

- **Soundness**: Considered as techniques for validating annotations, some tools provide sound results while others may be unsound, but can still be useful to provide the user with additional confidence that the annotations are likely to hold. For instance, runtime verification over selected test data or test generation after a partial exploration of program paths cannot establish that an annotation holds in all cases. In CegarMC, while the tool defaults only use sound methods, it is possible for users to tradeoff soundness for efficiency (e.g., by using a CPAchecker configuration that approximates datatypes). The overall results in such cases should be interpreted carefully.
- **Language semantics and tool limitations**: The subsets of (un)supported features for each of the combined tools can differ. For instance, runtime execution can deal more easily with complex programming language features (since the code is instrumented, compiled and executed) but has more limitations for complex annotations. Based on symbolic execution, PathCrawler may have issues in supporting some kinds of aliases and dynamic memory allocation. Value analysis has a more limited subset of supported annotations compared to deductive verification. Whereas Frama-C supports C99, SATABS and CPAchecker are based on ANSI-C and C11, respectively. Any resulting semantic issues are not handled, and CegarMC is thus not suitable for verifying any programs relying on the intricacies of a particular version. Applications of combined tools should take into account such limitations according to the tool documentation.
- **Memory model**: Any analysis of pointers (and pointer arithmetic) is dependent on the underlying memory model. Examples of memory related operations whose support can be limited in a given memory model are pointer casts and dynamic memory allocation. An integrative analysis for a given program with complex memory operations can only be as sound as the most restrictive of the tools that it interfaces with.

- **Interactivity of interface**: Dynamic techniques, such as runtime assertion checking and test generation, are often run from a terminal. Model checking is often also used in a "batch" mode due to state space explosion, while for deductive verification and value analysis many users prefer a graphical user interface. Hence, the combination with static analyzers is preferable through a GUI. Integration of the two different styles of verification in an interactive way can require specific attention to make it practical for users.

If these issues are properly taken into account, combined verification tools can provide useful solutions for software verification tasks.

9.6 Conclusion

This chapter does not have the ambition to cover all possible integrative analysis techniques. Many other techniques were proposed and evaluated for different verification tasks and tools. One example is annotation synthesis and, in particular, loop invariant generation, combined with static analysis. Daikon [23] uses dynamic analysis to detect likely invariants. Check'n'Crash [48] applies static analysis that reports alarms. It uses intraprocedural weakest-precondition computation with ESC-Java [16] rather than value analysis, so it necessitates code annotations. Next, random test generation (with JCrasher) tries to confirm the bugs. Dsd Crasher [48] applies Daikon to infer likely invariants before the static analysis step of Check'n'Crash to reduce the false alarms rate. Loop invariant synthesis was also studied by Oliveira et al. [39].

The test generation approach proposed by Engel and Hähnle [22] exploits the proof trees built by the KeY prover during a proof attempt. The approach of Ge et al. [25] is similar to SANTE. In their work, irrelevant code is excluded before dynamic analysis for CFG connectivity reasons, that is weaker than what can be removed with program slicing used in SANTE. Christakis et al. [11] propose a technique related to SANTE but assume that static analysis can be unsound. Chimento et al. [9] propose a combination of static analysis with runtime verification for Java programs. Synergy [30], Blast [4] and the approach of Pasareanu et al. [42] combine testing and partition refinement for property checking.

When SMT solvers fail on some verification conditions and provide a counter-model to explain that failure, the counter-model can be turned into a counterexample for the program under verification. This task was designed by Hauzar et al. and implemented in Spark2014 [19]. SMT models are already exploited for C programs, for instance, by the CBMC model checker [29]. Dafny has been extended for diagnosing proof failures [10], where a test generation tool is used to try to find counterexamples demonstrating non-compliance between program and specification.

This chapter illustrated several combined usages of analysis techniques and tools within the Frama-C verification platform. They show a large range of possibilities for integrative analysis methods in the platform. The design of Frama-C as an extensible toolset with a common kernel, a common annotation language and communicating analyzers makes it very suitable for designing and evaluating research prototypes and tools combining different techniques, and likely to be used for new integrative approaches in the future.

References

1. Ammann P, Offutt J (2008) Introduction to software testing. Cambridge University Press
2. Bardin S, Chebaro O, Delahaye M, Kosmatov N (2014) An all-in-one toolkit for automated white-box testing. In: Proceedings of the 8th international conference on tests and proofs (TAP 2014), held as part of STAF 2014, LNCS, vol 8570. Springer, pp 53–60. https://doi.org/10.1007/978-3-319-09099-3_4
3. Bardin S, Delahaye M, David R, Kosmatov N, Papadakis M, Traon YL, Marion J (2015) Sound and quasi-complete detection of infeasible test requirements. In: Proceedings of the 8th IEEE international conference on software testing, verification and validation (ICST 2015). IEEE, pp 1–10. https://doi.org/10.1109/ICST.2015.7102607
4. Beyer D, Henzinger TA, Jhala R, Majumdar R (2007) The software model checker Blast: applications to software engineering. Int J Softw Tools Technol Trans 9(5–6):505–525
5. Beyer D, Keremoglu ME (2011) CPAChecker: a tool for configurable software verification. In: Computer aided verification (CAV), pp 184–190
6. Cadar C, Godefroid P, Khurshid S, Pasareanu CS, Sen K, Tillmann N, Visser W (2011) Symbolic execution for software testing in practice: preliminary assessment. In: Proceedings of the 33rd international conference on software engineering, (ICSE 2011). ACM, pp 1066–1071. https://doi.org/10.1145/1985793.1985995
7. Chebaro O, Cuoq P, Kosmatov N, Marre B, Pacalet A, Williams N, Yakobowski B (2014) Behind the scenes in SANTE: a combination of static and dynamic analyses. Autom Softw Eng 21(1):107–143. https://doi.org/10.1007/s10515-013-0127-x
8. Chebaro O, Kosmatov N, Giorgetti A, Julliand J (2012) Program slicing enhances a verification technique combining static and dynamic analysis. In: Proceedings of the 27th annual ACM symposium on applied computing, software verification and testing track (SAC-SVT 2012). ACM, pp 1284–1291. https://doi.org/10.1145/2245276.2231980
9. Chimento JM, Ahrendt W, Pace GJ, Schneider G (2015) Starvoors: a tool for combined static and runtime verification of java. In: Proceedings of the 6th international conference on runtime verification (RV 2015), LNCS, vol 9333. Springer, pp 297–305. https://doi.org/10.1007/978-3-319-23820-3_21
10. Christakis M, Leino KRM, Müller P, Wüstholz V (2016) Integrated environment for diagnosing verification errors. In: Proceedings of the 22th international conference on tools and algorithms for the construction and analysis of systems (TACAS 2016), LNCS, vol 9636. Springer, pp 424–441
11. Christakis M, Müller P, Wüstholz V (2012) Collaborative verification and testing with explicit assumptions. In: Proceedings of the 18th international symposium on formal methods (FM 2012), LNCS, vol 7436. Springer, pp 132–146. https://doi.org/10.1007/978-3-642-32759-9_13
12. Clarke EM, Grumberg O, Jha S, Lu Y, Veith H (2000) Counterexample-guided abstraction refinement. In: Computer aided verification (CAV), pp 154–169
13. Clarke EM, Kroening D, Sharygina N, Yorav K (2005) SATABS: SAT-based predicate abstraction for ANSI-C. In: Tools and algorithms for construction and analysis of systems (TACAS), pp 570–574
14. Clarke LA, Rosenblum DS (2006) A historical perspective on runtime assertion checking in software development. SIGSOFT Softw Eng Notes 31(3). https://doi.org/10.1145/1127878.1127900

15. Clarke Jr EM, Grumberg O, Peled DA (1999) Model checking. MIT Press
16. Cok DR, Kiniry JR (2004) ESC/Java2: uniting ESC/Java and JML. In: Proceedings of the international workshop on construction and analysis of safe, secure and interoperable smart devices (CASSIS 2004), LNCS, vol 3362. Springer, pp 108–128
17. Cousot P, Cousot R (1977) Abstract interpretation: a unified lattice model for static analysis of programs by construction or approximation of fixpoints. In: Symposium on principles of programming languages (POPL'77), pp 238–252
18. Cousot P, Cousot R (2000) Temporal Abstract Interpretation. In: Symposium on principles of programming languages (POPL), pp 12–25. https://doi.org/10.1145/325694.325699
19. Dailler S, Hauzar D, Marché C, Moy Y (2018) Instrumenting a weakest precondition calculus for counterexample generation. J Log Algebr Meth Program 99:97–113. https://doi.org/10.1016/j.jlamp.2018.05.003
20. Dijkstra EW (1975) Guarded commands, nondeterminacy and formal derivation of programs. Commun ACM 18(8):453–457
21. Emerson EA, Clarke EM (1980) Characterizing correctness properties of parallel programs using fixpoints. Autom Lang Program 85/1980
22. Engel C, Hähnle R (2007) Generating unit tests from formal proofs. In: Proceedings of the first international conference on tests and proofs (TAP 2007), LNCS, vol 4454. Springer, pp 169–188
23. Ernst MD, Perkins JH, Guo PJ, McCamant S, Pacheco C, Tschantz MS, Xiao C (2007) The Daikon system for dynamic detection of likely invariants. Sci Comput Program 69(1–3):35–45
24. Falcone Y, Krstic S, Reger G, Traytel D (2021) A taxonomy for classifying runtime verification tools. Int J Softw Tools Technol Trans 23(2):255–284 https://doi.org/10.1007/s10009-021-00609-z
25. Ge X, Taneja K, Xie T, Tillmann N (2011) DyTa: dynamic symbolic execution guided with static verification results. In: Proceedings of the 33rd international conference on software engineering (ICSE 2011). ACM, pp 992–994
26. Goldberg A, Wang TC, Zimmerman D (1994) Applications of feasible path analysis to program testing. In: Proceedings of the international symposium on software testing and analysis (ISSTA 1994). ACM, pp 80–94. https://doi.org/10.1145/186258.186523
27. de Gouw S, Rot J, de Boer FS, Bubel R, Hähnle R (2015) OpenJDK's Java.utils.Collection.sort() is broken: the good, the bad and the worst case. In: 27th international conference on computer aided verification—27th international conference (CAV 2015), LNCS, vol 9206. Springer, pp 273–289. https://doi.org/10.1007/978-3-319-21690-4_16
28. Graf S, Saidi H (1997) Construction of abstract state graphs with PVS. In: International conference on computer-aided verification (CAV), pp 72–83
29. Groce A, Kroening D, Lerda F (2004) Understanding counterexamples with explain. In: Proceedings of the 16th international conference on computer aided verification (CAV 2004), LNCS, vol 3114. Springer, pp 453–456. https://doi.org/10.1007/978-3-540-27813-9_35
30. Gulavani BS, Henzinger TA, Kannan Y, Nori AV, Rajamani SK (2006) SYNERGY: a new algorithm for property checking. In: Proceedings of the 14th ACM SIGSOFT international symposium on foundations of software engineering (FSE 2006). ACM, pp 117–127
31. Hähnle R, Huisman M (2019) Deductive software verification: from Pen-and-Paper proofs to industrial tools, LNCS, vol 10000. Springer, pp 345–373. https://doi.org/10.1007/978-3-319-91908-9_18
32. Hatcliff J, Leavens GT, Leino KRM, Müller P, Parkinson M (2012) Behavioral interface specification languages. Comput Surv 44(3):16:1–16:58. https://doi.org/10.1145/2187671.2187678
33. Havelund K, Goldberg A (2005) Verify your runs. In: Verified software: theories, tools, experiments (VSTTE'05)
34. Huisman M, Wijs A (2023) Runtime annotation checking. In: Concise guide to software verification, texts in computer science. Springer. https://doi.org/10.1007/978-3-031-30167-4
35. Jhala R, Majumdar R (2009) Software model checking. ACM Comput Surv **41**(4), 1–54. https://doi.org/10.11455/1592434.1592438

36. Le VH, Correnson L, Signoles J, Wiels V (2018) Verification coverage for combining test and proof. In: International conference on tests and proofs (TAP)
37. Leucker M, Schallhart C (2009) A brief account of runtime verification. J Logic Algeb Program 78(5):293–303
38. Marcozzi M, Bardin S, Kosmatov N, Papadakis M, Prevosto V, Correnson L (2018) Time to clean your test objectives. In: Proceedings of the 40th international conference on software engineering (ICSE 2018). ACM, pp 456–467. https://doi.org/10.1145/3180155.3180191
39. de Oliveira S, Bensalem S, Prevosto V (2017) Synthesizing invariants by solving solvable loops. In: Proceedings of the 15th international symposium on automated technology for verification and analysis (ATVA 2017), LNCS, vol 10482. Springer, pp 327–343. https://doi.org/10.1007/978-3-319-68167-2_22
40. Ourghanlian A (2015) Evaluation of static analysis tools used to assess software important to nuclear power plant safety. Nucl Eng Technol 47(2):212–218. https://doi.org/10.1016/j.net.2014.12.009
41. Pariente D, Signoles J (2017) Static analysis and runtime assertion checking: contribution to security counter-measures. In: Symp sur la Sécurité des Technologies de l'Information et des Communications (SSTIC)
42. Pasareanu CS, Pelánek R, Visser W (2005) Concrete model checking with abstract matching and refinement. In: Proceedings of the 17th international conference on computer aided verification (CAV 2005), LNCS, vol 3576. Springer, pp 52–66
43. Petiot G, Kosmatov N, Botella B, Giorgetti A, Julliand J (2016) Your proof fails? Testing helps to find the reason. In: International conference on tests and proofs (TAP'16)
44. Petiot G, Kosmatov N, Botella B, Giorgetti A, Julliand J (2018) How testing helps to diagnose proof failures. Formal Asp Comput 30(6):629–657. https://doi.org/10.1007/s00165-018-0456-4
45. Queille JP, Sifakis J (1982) Specification and verification of concurrent systems in CESAR. In: International symposium on programming, vol 137
46. Rival X, Yi K (2020) Introduction to static analysis: an abstract interpretation perspective. MIT Press
47. Shankar S, Pajela G (2016) A tool integrating model checking into a C verification toolset. In: International symposium on model checking software (SPIN), pp 214–224
48. Smaragdakis Y, Csallner C (2007) Combining static and dynamic reasoning for bug detection. In: Proceedings of the first international conference on tests and proofs (TAP 2007), LNCS, vol 4454, pp 1–16. Springer
49. Weyuker EJ (1993) More experience with data flow testing. IEEE Trans Softw Eng 19(9):912–919. https://doi.org/10.1109/32.241773
50. Williams N, Marre B, Mouy P, Roger M (2005) PathCrawler: automatic generation of path tests by combining static and dynamic analysis. In: Proceedings of the 5th European dependable computing conference (EDCC 2005), LNCS, vol 3463. Springer, pp 281–292. https://doi.org/10.1007/11408901_21
51. Yates DF, Malevris N (1989) Reducing the effects of infeasible paths in branch testing. ACM SIGSOFT Softw Eng Notes 14(8):48–54. https://doi.org/10.1145/75309.75315

Chapter 10
Specification and Verification of High-Level Properties

Lionel Blatter, Nikolai Kosmatov, Virgile Prevosto, and Virgile Robles

Abstract The ACSL specification language allows the verification engineer to specify almost any property they might want to verify at any given point in a given C program. For some complex properties, it can sometimes be done at the price of an extremely complex encoding, which could quickly become error-prone if written manually. To facilitate this task, a certain number of Frama-C plug-ins offer dedicated specification languages, targeting different kinds of properties, that are then translated into a set of ACSL annotations amenable to verification via the core analysis plug-ins (Eva, Wp, E-ACSL). In this chapter, we present three such plug-ins. The first of them, MetAcsl, is dedicated to pervasive properties that must be checked at each program point meeting some characteristics, for instance at each write access. A single MetAcsl annotation can thus be instantiated by a very large number of ACSL clauses. Second, RPP targets relational properties. In contrast to an ACSL function contract, which specifies what happens during a single function call, relational properties express relations over several calls of potentially different functions. Finally, Aoraï is dedicated to properties over sequences of function calls that can occur during an execution.

Keywords High-level properties · Pervasive properties · Relational properties · Properties over call traces · Instrumentation · Verification

---

L. Blatter
Max Planck Institute for Security and Privacy, Bochum, Germany
e-mail: lionel.blatter@mpi-sp.org

N. Kosmatov (✉)
Thales Research & Technology, Palaiseau, France
e-mail: nikolaikosmatov@gmail.com

V. Prevosto
Université Paris-Saclay, CEA, List, Palaiseau, France
e-mail: virgile.prevosto@cea.fr

V. Robles
Tarides, Paris, France
e-mail: virgile.robles@protonmail.ch

10.1 Introduction

Behavioral interface specification languages [15] are commonly used to express the behavior of each program procedure and to specify necessary interfaces between them. ACSL (presented in detail in Chap. 1) provides such a specification language for C programs. It allows verification engineers to specify a wide range of properties they may need to verify at any given point in the considered program. Some more complex properties, involving several program points and/or several functions at the same time, can be expressed—sometimes using a specific instrumentation—by additional ACSL annotations inserted in relevant program locations. However their specification in this way often requires a rather complex or verbose encoding, which could quickly become error-prone if performed manually.

To facilitate specification and verification of such properties, the Frama-C verification platform offers a number of dedicated plug-ins, each of them supporting a specific specification language for a class of properties. The main idea is to translate specific properties into a set of ACSL annotations that can be verified by the main Frama-C analyzers, such as Eva, Wp and E-ACSL.

In this chapter, we present three such plug-ins. The first of them, MetAcsl, is dedicated to expressing pervasive properties that must be checked at all program points respecting some characteristics, for instance at each read and write access, or each function call. A single MetAcsl annotation can thus be translated into a very large number of ACSL clauses. A second plug-in, RPP, targets so-called *relational* properties. Contrary to ACSL contracts, which specify what happens during a single function call, relational properties express relations over several calls of the same or different functions. Finally, the Aoraï plug-in allows the verification engineer to write properties over sequences of function calls (i.e., over call traces) that can occur during a program execution.

10.2 Pervasive Program Properties

One useful type of high-level properties is that of *pervasive* ones. They include properties over program data in the global scope that are expected to hold for every single execution trace. While they can express a variety of program properties, those related to memory accesses are particularly relevant candidates. The specification of constraints over accesses and modifications of program memory locations can serve as a basis for such important characteristics as confidentiality and integrity.

10.2.1 Illustrative Examples of Properties

Let us imagine a low-level memory management system, as it could be found in a microkernel, holding a number of memory *pages* (i.e., blocks of data with a fixed size) that can be requested by processes to store and access their data.

10 Specification and Verification of High-Level Properties

Now let us define a notion of *confidentiality*. Assume we have several confidentiality levels represented by integers $0, 1, \ldots, n$ with $n \geqslant 1$. Each allocated page must have a confidentiality rating, that is, a confidentiality level representing how sensitive the data stored in it is. In turn, each process has an assigned confidentiality clearance, that is, a level indicating how sensitive the data it writes is assumed to be, and how sensitive the data it is allowed to access is.

The kernel may make available a set of system calls for processes to request pages, access them and modify them, for instance, with the API shown in Fig. 10.1.

Two basic guarantees that the implementation of this API should offer are the following:

1. A process must not read from a page with a confidentiality rating higher than its own clearance (to preserve the confidentiality of the data written on the page),
2. A process must not write to a page with a confidentiality rating lower than its own clearance (to prevent a leak of the process' sensitive data).

We can reasonably assume that the kernel implementing this API manages its resources in the global state as illustrated in Fig. 10.2. It defines a memory pool

```
1 struct Page; // Abstract type for memory pages
2 // Request a page
3 struct Page* page_alloc();
4 // Release a page
5 void page_free(struct Page* p);
6 // Load a whole page into a buffer
7 int page_read(struct Page* from, char* buffer);
8 // Overwrite a whole page with a buffer
9 int page_write(struct Page* to, char* buffer);
```

Fig. 10.1 Simple page management API

```
1 #define PAGE_SIZE 0x100 // Size of a page in bytes
2 #define MAX_PAGE_NB 0x1000 // Size of memory pool of pages
3 #define PROCESS_NB 1024 // Max number of processes
4
5 typedef unsigned confidentiality;
6
7 struct Page {
8   char* data;
9   confidentiality rating;
10 };
11
12 struct Page pages[MAX_PAGE_NB]; // Memory pool
13 confidentiality clearances[PROCESS_NB]; // Process -> clearance
14
15 // ID of the current calling process, assumed to be set
16 // automatically when calling the higher-level API
17 unsigned calling_process;
```

Fig. 10.2 Simple modeling of confidentiality rankings and clearances

of pages, represented by array `pages` (cf. Lines 7–12), each page having a rating and data. The clearance of each process is associated to its number (ID) by array `clearances` (Line 13). Finally, the ID of the current process is defined by variable `calling_process` (Line 17).

In this setting, the aforementioned guarantees can be trivially mapped to constraints on operations accessing and modifying the state:

1. Whenever the global state is *accessed* (that is, read), if the accessed memory address is within the bounds

$$[\text{pages[p].data}, \text{pages[p].data} + \text{PAGE\_SIZE}[$$

of a page p, then the clearance `clearances[calling_process]` of the calling process must be greater than or equal to the accessed page's rating `pages[p].rating`.
2. Similarly for memory *modifications* of a page p, but in this case the calling process' clearance `clearances[calling_process]` must be less than or equal to the written page's rating `pages[p].rating`.

With this pattern as a building block, more complex and interesting program properties can be encoded as memory transaction constraints.

10.2.2 The *MetAcsl* Plug-In

In Frama-C, the MetAcsl extension to ACSL (also called Hilare language) provided by the MetAcsl plug-in, exposes syntactical constructs to express such pervasive properties with a focus on memory constraints.

Hilare The following syntax declares a pervasive property (often also called a *High-Level ACSL Requirement* (HILARE) or a *metaproperty*):

```
1 /*@
2 meta \prop,
3   \name(NAME),
4   \targets(TARGET_SET),
5   \context(CONTEXT_TYPE),
6   CONSTRAINT;
7 */
```

where:

- NAME is an identifier indicating the name of the property;
- TARGET_SET is a set of functions where the property must hold (constructed for instance by enumerating the necessary functions, or removing some functions from the \ALL set containing all the functions of the program under analysis);
- CONTEXT_TYPE is the kind of operation which should be constrained (\writing for memory modifications, \reading for accesses, etc.);

- CONSTRAINT is an ACSL predicate expressing the property that should hold when the above conditions are met and that can use special variables to refer to the memory address undergoing an operation when applicable (see the examples below).

For example, to declare the first confidentiality property of our previous example, we can write:

```
/*@
meta \prop,
  \name(confidential_read),
  \targets(\ALL),
  \context(\reading),
  \forall unsigned p; 0 <= p < MAX_PAGE_NB ==>
    !\separated(\read,
      pages[p].data + (0 .. PAGE_SIZE - 1))
      ==> clearances[calling_process] >=
          pages[p].rating;
*/
```

We use the \reading context indicating that the constraint concerns memory accesses. It allows us to use the \read special variable in the constraint to refer to the accessed memory address. Hence, the condition !\separated(\read, pages[p].data+ (0 .. PAGE_SIZE - 1)) in our predicate encodes "an address within page p is being accessed", and the overall implication matches our initial requirement.

To express our second property, we can write a similar pervasive property using instead the \writing context and its associated \written variable as follows:

```
/*@
meta \prop,
  \name(confidential_write),
  \targets(\ALL),
  \context(\writing),
  \forall unsigned p; 0 <= p < MAX_PAGE_NB ==>
    !\separated(\written,
      pages[p].data + (0 .. PAGE_SIZE - 1))
      ==> clearances[calling_process] <=
          pages[p].rating;
*/
```

Notice that, contrary to ordinary ACSL contracts, such properties are not attached to a function or another object of the program, but are instead placed freely at the end of the program (or even in a separate file). More information and examples of various kinds of pervasive properties can be found in dedicated publications [23, 25] and the Ph.D. thesis of Virgile Robles [22]. Some security related metaproperties are also illustrated in Chap. 16. The full syntax of the Hilare language is described in MetAcsl's README.[1]

---

[1] See https://git.frama-c.com/pub/meta.

10.3 Verification of Pervasive Properties with MetAcsl

Following the usual approach of tool integration in Frama-C, the main idea behind the verification of pervasive properties with MetAcsl is to benefit from existing analyzers without re-implementing them for these new properties. For that purpose, we have to transform metaproperties into plain ACSL annotations while keeping links between the original metaproperty and the resulting ACSL annotations. These resulting ACSL annotations can then be verified using the existing analyzers.

To illustrate this approach, consider again metaproperty confidential_read given above. Let us illustrate the MetAcsl transformation for it for (a subset of operations in) function page_read given in Fig. 10.3. Since this metaproperty has all functions in its target set, it should be applied in particular to page_read.

Figure 10.4 gives the translation of the metaproperty into several assertions for the assignment on line 4 of Fig. 10.3. (The generated assertions being quite repetitive, we do not show the assertions generated for other reading operations.)

```
1  void page_read(struct Page* from, char* buffer){
2    int j;
3    for(j = 0; j < MAX_PAGE_NB; j++)
4      *(buffer + j) = *(from->data + j);
5  }
```

Fig. 10.3 Function page_read

```
1  /*@
2  assert confidential_read: meta:
3    \forall unsigned p; 0 <= p < 0x1000 ==>
4      !\separated(&from, pages[p].data + (0 .. 0x100 - 1))
5      ==> clearances[calling_process] >= pages[p].rating;
6  assert confidential_read: meta:
7    \forall unsigned p; 0 <= p < 0x1000 ==>
8      !\separated(&buffer, pages[p].data + (0 .. 0x100 - 1))
9      ==> clearances[calling_process] >= pages[p].rating;
10 assert confidential_read: meta:
11   \forall unsigned p; 0 <= p < 0x1000 ==>
12     !\separated(&j, pages[p].data + (0 .. 0x100 - 1))
13     ==> clearances[calling_process] >= pages[p].rating;
14 assert confidential_read: meta:
15   \forall unsigned p; 0 <= p < 0x1000 ==>
16     !\separated(&from->data, pages[p].data + (0 .. 0x100 - 1))
17     ==> clearances[calling_process] >= pages[p].rating;
18 assert confidential_read: meta:
19   \forall unsigned p; 0 <= p < 0x1000 ==>
20     !\separated(from->data + j, pages[p].data + (0 .. 0x100 - 1))
21     ==> clearances[calling_process] >= pages[p].rating;
22 */
23 *(buffer + j) = *(from->data + j);
```

Fig. 10.4 Assertions generated by MetAcsl for metaproperty confidential_read for the assignment on line 4 of function page_read shown in Fig. 10.3

10 Specification and Verification of High-Level Properties

```
1 /*@
2 assert confidential_write: meta:
3    \forall unsigned p; 0 <= p < 0x1000 ==>
4      !\separated(buffer + j, pages[p].data + (0..0x100 - 1))
5      ==> clearances[calling_process] <= pages[p].rating;
6 */
7 *(buffer + j) = *(from->data + j);
```

Fig. 10.5 Assertion generated by MetAcsl for metaproperty `confidential_write` for the assignment on line 4 of function `page_read` shown in Fig. 10.3

As we can see in Fig. 10.4, each reading operation of the instruction leads to a separate assertion added before the instruction. The variables and memory locations being read in the assignment on line 4 of Fig. 10.3 are the following: `from`, `buffer`, `j`, `from->data` and `*(from->data + j)`. Their addresses are, respectively, `& from`, `& buffer`, `& j`, `& from->data` and `from->data + j`. The MetAcsl translation of a metaproperty with a reading context considers each reading operation and generates an assertion for it (cf. Fig. 10.4), stating the constraint of the metaproperty where the address of the read location replaces variable `read` used in the constraint. We also say that the metaproperty is *instantiated* for all relevant operations.

Since the preprocessing phase occurs before Frama-C itself (hence before MetAcsl), the macros `MAX_PAGE_NB` and `PAGE_SIZE` are directly replaced by the corresponding constants (according to their definitions in Fig. 10.1). Since constants do not require reading any memory locations, no assertions need to be generated for macros in our example. In the cases where a macro contains reading operations, necessary assertions will be generated as well.

Consider now metaproperty `confidential_write`, which has a writing context. As above, we consider the assignment on line 4 of Fig. 10.3. This time, MetAcsl will generate the assertion shown in Fig. 10.5, added, as in Fig. 10.4, before the assignment. The only memory location being written in this assignment is `*(buffer + j)`, whose address is `buffer + j`. More generally, in MetAcsl, the instantiation of a metaproperty with a writing context considers each writing operation and generates an assertion for it, stating the constraint of the metaproperty, where the address of the written location replaces variable `\written` used in the constraint.

We illustrated the instantiation of two metaproperties separately for more clarity. In practice, several metaproperties are often provided and instantiated for the same function. In this case, necessary assertions are added to the relevant program locations one after another, sometimes leading to a large number of additional annotations. The use of MetAcsl on a real-life industrial project and the related statistics (regarding the number of the resulting annotations and their verification) are discussed in Chap. 16.

After the transformation performed by MetAcsl, the existing analyzers can verify the resulting annotations, and their results can then be interpreted for the original metaproperties. For instance, if all annotations generated by the instantiation of some metaproperty are proved by Wp, then this metaproperty is verified, and the corresponding pervasive property holds. If at least one of the annotations generated by the instantiation of some metaproperty fails at runtime (and the failure is reported thanks to E-ACSL), the corresponding pervasive property does not hold.

MetAcsl offers various options to activate a special treatment for undefined functions (based on their contracts), to activate some optimizations, to instantiate metaproperties into **check** annotations instead of **assert** annotations, to number them, etc. For more detail about the usage of MetAcsl and available features, the reader can refer to [22, 24, 25] and the tool documentation.

10.4 Relational Properties

This section presents another interesting kind of properties—those relating several calls, possibly of the same or different functions. Such properties, referred to as *relational properties*, link the results of the considered function calls and express constraints that should hold for them (possibly, under some assumptions on their input states). We first describe an example of relational properties that one might want to verify over C functions in Sect. 10.4.1. Then, we present in Sect. 10.4.2 the syntax that has been introduced by the RPP plug-in to formally specify such properties.

RPP[2] is a Frama-C plug-in allowing the user to verify relational properties and to use them as hypotheses for verification of other properties. The plug-in is based on code transformations to make relational properties amenable to verification by other Frama-C plug-ins.

10.4.1 *Example of Relational Property*

Let us consider as an illustrative example a (buggy) C function implementing a comparator, inspired by a StackOverflow question[3] for elements of the AInt data type that can be seen as a naive data type for vectors or words. The data type and the function are shown in Fig. 10.6. Type AInt is a struct composed of two fields: an array t of size 1000 to store the elements and an integer length indicating the number of elements in the vector t.

A comparator is supposed to compare two vectors[4] of type AInt by checking the lexicographic order of their elements and returning −1 (resp., 1, or 0) if the first argument is smaller than (resp., greater than, or equal to) the second argument. The

---

[2] See https://github.com/lyonel2017/Frama-C-RPP.

[3] See https://stackoverflow.com/questions/23907134/comparing-two-arrays-using-dictionary-order-in-an-array-of-arrays-java.

[4] In the illustrative example of Fig. 10.6, the parameters are directly the vector structures. In real-life code, passing vectors by pointers would provide faster function calls.

```
1  struct AInt{
2    int t[1000];
3    int length;
4  };
5
6  int compare(struct AInt o1, struct AInt o2){
7    int index = 0;
8    while ((index < o1.length) && (index < o2.length)) {
9      if (o1.t[index] < o2.t[index]) return -1;
10     if (o1.t[index] > o2.t[index]) return 1;
11     index++;
12   }
13   return 0;
14 }
```

Fig. 10.6 A (buggy) C function implementing a comparator

given implementation is incorrect since for two vectors such that one is a prefix of the other, longer, vector, the function returns 0 (so they are considered equal).

A comparator is expected to satisfy three properties, anti-symmetry (P_1), transitivity (P_2) and extensionality (P_3), expressed as follows:

$$P_1 : \forall\ s_1, s_2.\ \mathrm{compare}(s_1, s_2) = -\mathrm{compare}(s_2, s_1),$$
$$P_2 : \forall\ s_1, s_2, s_3.\ \mathrm{compare}(s_1, s_2) \geq 0 \land \mathrm{compare}(s_2, s_3) \geq 0$$
$$\Rightarrow \mathrm{compare}(s_1, s_3) \geq 0,$$
$$P_3 : \forall\ s_1, s_2, s_3.\ \mathrm{compare}(s_1, s_2) = 0 \Rightarrow \mathrm{compare}(s_1, s_3) = \mathrm{compare}(s_2, s_3).$$

These properties are relational in the sense that they relate multiple instances of the compare function. A classic ACSL function contract cannot express such properties. It might be noted that we have not mentioned another usual property of ordering relations, namely reflexivity, that can be expressed as $\forall s,\ \mathrm{compare}(s, s) = 0$. In fact, it is not a relational property *per se*, in the sense that it could be expressed as a plain ensures clause in ACSL. But more importantly, it is entailed by P_1, so that we do not need to verify it if we can prove that P_1 holds.

Notice that properties P_2 and P_3 are not satisfied by the implementation shown in Fig. 10.6. For instance, P_2 does not hold for vectors $s_1 = (1, 2), s_2 = (1), s_3 = (1, 3)$, for which the implementation returns:

$$\mathrm{compare}(s_1, s_2) = 0,\ \mathrm{compare}(s_2, s_3) = 0,\ \mathrm{compare}(s_1, s_3) = -1.$$

Similarly, P_3 does not hold for the same counterexample. Section 10.5.2 shows how to fix the implementation and how to formally express properties P_1, P_2 and P_3 using the plug-in RPP.

Fig. 10.7 Relational annotation for a C function h without parameters and return value

```
1  int y;
2
3  /*@ assigns y \from y;*/
4  void h(){
5    int a = 10;
6    y = y + a;
7    return;
8  }
9
10 /*@
11 relational R1:
12   \callset(\call(h,id1),\call(h,id2)) ==>
13   \at(y,Pre_id1)  < \at(y,Pre_id2)  ==>
14   \at(y,Post_id1) < \at(y,Post_id2);
15 */
```

10.4.2 Specification of Relational Properties in RPP

In this section we present the **relational** annotations, an extension added to the ACSL specification language to support relational specification in RPP. We first present in Sect. 10.4.2.1 the relational specification for functions without parameters and return value. Then, in Sect. 10.4.2.2 we present the extension for the general case of functions with parameters and return values. The complete grammar can be found in the RPP plug-in manual [8].

10.4.2.1 Functions Without Parameters and Return Value

Figure 10.7 shows an example of a **relational** annotation for a C function h without parameters and return value. This annotation states that h is monotonic[5] with respect to global variable y. The relational property R1 of Fig. 10.7 is introduced after the declaration of the function h, so that the name h is in scope. R1 is divided in two parts. First, on line 12 of Fig. 10.7, two function calls to h are explicitly specified in the \callset construct, using construct \call. Since we may need to refer to memory locations in either the pre- or the post-state of each call, we need to be able to make explicit references to these states. Therefore, each call has its own identifier. In the case of Fig. 10.7, we have identifiers id1 and id2 associated to each call. Each such identifier gives rise to two logic labels. For instance, for id1, label Pre_id1 refers to the pre-state of the corresponding call, and label Post_id1 to its post-state. These labels can in particular be used as label L in the ACSL term \at(e, L) indicating that the term e must be evaluated in the context of the program state at label L. In our example, we indicate that the value of global variable y is evaluated in four different states:

---

[5] We assume here for simplicity that there is no arithmetic overflow inside h.

```
1  int y;
2
3  /*@ assigns \result \from x,y;*/
4  int f(int x){
5    ...
6  }
7
8  /*@
9  relational R1:
10   \forall int x1,x2;
11   \callset(\call(f,x1,id1),\call(f,x2,id2)) ==>
12   \at(y,Pre_id1) < \at(y,Pre_id2) ==> x1 < x2 ==>
13   \callresult(id1) < \callresult(id2);
14  */
```

Fig. 10.8 Relational annotation for a C function f with a parameter and a return value

Fig. 10.9 Relational annotation for a pure C function p with a parameter and a return value

```
1  /*@ assigns \result \from x;*/
2  int p(int x){
3    ...
4  }
5
6  /*@
7  relational R1:
8    \forall int x1,x2; x1 < x2 ==>
9    \callpure(p,x1) < \callpure(p,x2);
10 */
```

- Pre_id1, the state before the execution of function h with tag id1;
- Post_id1, the state after the execution of function h with tag id1;
- Pre_id2, the state before the execution of function h with tag id2;
- Post_id2, the state after the execution of function h with tag id2.

Once we have introduced the calls and their corresponding pre- and post-states, we can state the relational property itself. Here, R1 indicates (lines 13–14 of Fig. 10.7) that if a first call to h is done in a state Pre_id1 where the value of y is strictly less than in the pre-state Pre_id2 of a second call, this will also be the case in the respective post-states Post_id1 and Post_id2.

10.4.2.2 Functions with Parameters and Return Value

An example of a relational property linking functions with parameters and return values is shown in Fig. 10.8. In this case, a relational clause is composed of three parts. In addition to the two parts previously presented, we also declare a set of universally quantified variables that will be used to express the arguments of the calls that are involved in the clause. Moreover, a new construct \callresult taking a call identifier as a parameter can be used to refer to the value returned by the corresponding call.

In the case of a *pure* function, i.e., a function without side effects and whose return value only depends on the parameters, it is possible to use the \callpure construct to denote the value returned by a pure function. An example is shown in Fig. 10.9. This allows for specifying relational properties over pure functions without having to declare the call in the \callset part, since, by definition, pure functions are independent from the global state of the program. Hence, we do not need to refer to the pre- or post-state of a call to a pure function. Nested \callpure constructs are also allowed. For a function used with \callpure, in order to ensure that this function has no side effects, RPP requires that an **assigns \result\from params** clause be specified (see Chap. 1), where params indicates a subset of the set of parameters of the function. A similar specification approach exists for defining relational properties on Java pure methods [13] in the JML specification language.

10.5 Verification of Relational Properties with RPP

This section focuses on the verification of relational properties using RPP. The plug-in has been tested with success on different benchmarks[6] in combination with Wp, E-ACSL and StaDy plug-ins [6, 7]. This section is organized as follows. Section 10.5.1 presents the code transformation performed by the RPP plug-in to make relational properties suitable for other plug-ins. Then, Sect. 10.5.2 presents an application of RPP on the comparator function introduced earlier.

10.5.1 Code Transformation

The syntax for relational properties presented in Sect. 10.4.2 enables us to deal with the states of different functions. However, this syntax is not supported by other plug-ins like Wp. Thus, the RPP plug-in performs a code and annotation transformation to make relational properties verifiable by other Frama-C plug-ins.

This transformation is composed of two parts, allowing the verification of relational properties, as well as their use as hypotheses for the verification of other annotations. To illustrate this transformation, we use function h of Fig. 10.7. Section 10.5.1.1 describes the transformation for the verification of relational properties. Section 10.5.1.2 presents the transformation for the use of relational properties.

10.5.1.1 Verification of Relational Properties

For the verification of relational properties, we apply a transformation called *self-composition* [2]. Self-composition is an approach to verify relational properties by

---

[6] See https://github.com/lyonel2017/RPP-Examples-TAP-2018.

Fig. 10.10 Self-composition for property R1 over function h of Fig. 10.7

```
1  int y_id1;
2  int y_id2;
3
4  void relational_wrapper_1(void){
5      int a_1 = 10;
6      y_id1 = y_id1 + a_1;
7      int a_2 = 10;
8      y_id2 = y_id2 + a_2;
9      /*@ assert Rpp:
10         \at(y_id1,Pre) < \at(y_id2,Pre) ==>
11         \at(y_id1,Here) < \at(y_id2,Here);
12     */
13     return;
14 }
```

reducing the verification of a relational property to the verification of a property for a newly generated function. It is constructed as a sequential composition of the considered functions' bodies where their variables are renamed to ensure memory separation. Applied to property R1 over function h of Fig. 10.7, RPP produces the code shown on Fig. 10.10.

In RPP, to perform the self-composition transformation, we generate a new function, commonly called a *wrapper*. The wrapper inlines the calls[7] occurring in the relational property under analysis, with a suitable renaming of local and global variables to avoid interferences between the calls, so that each function call operates on its own memory state, separated from the other calls. In Fig. 10.10, we can in particular notice the creation of global variables on lines 1 and 2 and local variables on lines 5 and 7. In both cases, the first variable is tied to the call id1 of relational property R1 in Fig. 10.7, and the second to the call id2.

To avoid useless creation of global variables (for variables that are not used by the function), we require that each function involved in a relational property has been equipped with a proper set of ACSL **assigns** clauses, including \from components. This constraint allows us to determine the parts of the global state that are accessed (either for writing or for reading) by the functions under analysis and that must be subject to duplication[8] In case of function h, only global variable y is read and written by the function (see line 3 of Fig. 10.7). Thus, only variable y must be duplicated.

Finally, in the spirit of calculational proofs [19], we state an assertion equivalent to the relational property (lines 9–12 in Fig. 10.10). The verification of such an assertion can be performed for example with the deductive verification tool Wp.

---

[7] Recursive functions are supported by RPP likewise: during self-composition, only the calls involved in the relational property are inlined, while the recursive calls inside their bodies are not.

[8] Note that RPP assumes that the \from clauses are correct and does not verify them. Since Wp currently does not support them, one possibility is to use MetAcsl to verify that only the provided global variables (as well as local variables) can be read by the function.

```
1  /*@
2  axiomatic Relational_axiom_1 {
3    predicate h_acsl(int y_pre, int y_post);
4
5    lemma Relational_lemma_1:
6      \forall int y_id2_pre, y_id2_post, y_id1_pre, y_id1_post;
7        h_acsl(y_id2_pre, y_id2_post) ==>
8        h_acsl(y_id1_pre, y_id1_post) ==>
9        y_id1_pre < y_id2_pre ==> y_id1_post < y_id2_post;
10 }
11 */
```

Fig. 10.11 Axiomatization of relational property R1 of Fig. 10.7

10.5.1.2 Use of Relational Properties

For common program annotations (such as pre-conditions, post-conditions, assertions, loop invariants), it is often necessary to use one annotation as a hypothesis to prove some others. The approach presented above demonstrates how to prove a relational property, but does not show how it can be used as a hypothesis to prove another property. Indeed, since relational properties are not natively supported by Wp, this requires additional efforts.

To be able to use a relational property as a hypothesis, we transform the relational annotation into a global ACSL axiomatic annotation. This axiomatic definition introduces a logical reformulation of the relational property as a lemma over otherwise unspecified predicates. We declare a predicate for each involved function that takes as parameters the relevant parts of the program states that are involved in the property: as above, we assume that appropriate **assigns ... \from ...** clauses have been provided for these functions. Figure 10.11 illustrates this transformation generated by RPP for relational property R1 of Fig. 10.7. In this example, predicate h_acsl is introduced for function h (cf. line 3), while the reformulation of relational property R1 is shown on lines 5–9, where we have four quantified variables representing the value of global variable y before and after both calls involved in the relational property. Initially, the lemma has the unknown status, and gets the valid status at the moment the corresponding assertion in the wrapper function (lines 9–12 in Fig. 10.10) is proven.

Using this reformulation, a relational property, which involves a function that calls function h, can use property R1 of Fig. 10.7. For example, consider function g shown in Fig. 10.12, calling function h. Relational property R2 states that g is monotonic. The monotonicity of g can be proved using the monotonicity of h. It is thus useful to be able to prove R2 using R1. Let us show how it can be done with RPP. Applying RPP to property R2 over function g produces the code shown in Fig. 10.13. As for the proof of property R1 in Fig. 10.10, the wrapper function inlines the calls (cf. lines 10 and 12) occurring in the relational property under analysis, and applies some renaming.

Fig. 10.12 Relational property for function g calling function h of Fig. 10.7

```
1  int y;
2
3  /*@ assigns y \from y;*/
4  void g(){
5    h();
6    return;
7  }
8
9  /*@
10  relational R2:
11    \callset(\call(g,id1), \call(g,id2)) ==>
12    \at(y,Pre_id1)  < \at(y,Pre_id2) ==>
13    \at(y,Post_id1) < \at(y,Post_id2);
14  */
```

Fig. 10.13 Self-composition for function g of Fig. 10.12

```
1  int y_id1; int y_id2;
2
3  /*@ assigns y_id1 \from y_id1;*/
4  void h_id1();
5
6  /*@ assigns y_id2 \from y_id2;*/
7  void h_id2();
8
9  void relational_wrapper_2(void){
10    l1:h_id1();
11    /*@ assert h_acsl(\at(y_id1,l1),\at(y_id1,Here)); */
12    l2:h_id2();
13    /*@ assert h_acsl(\at(y_id2,l2),\at(y_id2,Here)); */
14    /*@ assert Rpp:
15      \at(y_id1,Pre)  < \at(y_id2,Pre) ==>
16      \at(y_id1,Here) < \at(y_id2,Here);
17    */
18    return;
19  }
```

Moreover, the function calls are also renamed, which leads to the creation of function prototypes with associated contracts on lines 3–7 of Fig. 10.13. In addition, new assertions are generated after the functions calls (cf. lines 10–13). They specify that there is an exact correspondence between the original C function and its newly generated logical ACSL counterpart so that the axiomatics of Fig. 10.11 can be used. To express these assumptions, RPP currently uses assertions[9] that are considered as valid. For instance, assertion on line 11 of Fig. 10.13, states that predicate h_acsl holds with two arguments representing the values of y before and after the first execution of h.

10.5.2 Back to the Illustrative Example

Using relational annotations, we can specify anti-symmetry (P_1), transitivity (P_2) and extensionality (P_3) for the comparator function introduced in Sect. 10.4. The

---

[9] In a future version of the tool, it is planned to use **admit** clauses of ACSL (see Chap. 1) to state assumptions.

```
1  /*@ requires 0 <= o1.length && o1.length < 1000;
2   @ requires 0 <= o2.length && o2.length < 1000;
3   @ assigns \result \from o1,o2; */
4  int compare(struct AInt o1, struct AInt o2){
5    int index = 0;
6    /*@
7    loop invariant 0 <= index && index <= o2.length &&
8      index <= o1.length;
9    loop invariant \forall integer k; 0 <= k < index ==>
10     o1.t[k] == o2.t[k];
11   loop assigns index; */
12   while ((index < o1.length) && (index < o2.length)) {
13     if (o1.t[index] < o2.t[index]) return -1;
14     if (o1.t[index] > o2.t[index]) return 1;
15     index++;
16   }
17   if (o1.length < o2.length)  return -1;
18   if (o1.length > o2.length)  return 1;
19   return 0;
20 }
21
22 /*@ relational P1:
23   \forall struct AInt x1,x2;
24     \callpure(compare,x1,x2) ==
25     -(\callpure(compare,x2,x1)); */
26 /*@ relational P2:
27   \forall struct AInt x1,x2,x3;
28     (\callpure(compare,x1,x2) >= 0 &&
29     \callpure(compare,x2,x3) >= 0) ==>
30     \callpure(compare,x1,x3) >= 0; */
31 /*@ relational P3:
32   \forall struct AInt x1,x2,x3;
33     \callpure(compare,x1,x2) == 0 ==>
34     (\callpure(compare,x1,x3) ==
35     \callpure(compare,x2,x3)); */
```

Fig. 10.14 Annotated (correct) C function implementing a comparator

relational specification is shown on lines 22–35 of Fig. 10.14. It is a straightforward rewriting of the mathematical definition. Since `compare` is a pure function (cf. line 3 of Fig. 10.14), we can use the `\callpure` construct, which makes it much easier to specify the properties. In addition to the relational annotations, standard ACSL annotations are added on lines 1–3 and 6–11 for pre-conditions and a loop contract to make the properties provable.

As already mentioned in Sect. 10.4, the implementation shown on Fig. 10.6 is flawed. Lines 14–15 of Fig. 10.14 introduce a fix such that all properties P_1, P_2 and P_3 are satisfied. They can then be proved with RPP and Wp for this version of the code. As it can be expected, an attempt to prove P_2 and P_3 with the buggy version would fail.

The reader will find further examples of relational properties and a detailed description of the RPP tool in dedicated publications [6, 7], the Ph.D. thesis of Lionel Blatter [5], and the tool documentation [8]. A formalization and proof of correctness of a verification approach for relational properties in the Coq proof assistant can be found in [9, 10].

10.6 Properties Over Call Traces

We now consider another category of properties that involve several function calls: properties over call traces. Indeed, it is often the case that a set of related functions cannot be called in an arbitrary order, but that instead some constraints over the sequences of calls and/or the call stack must be met. Such constraints are difficult to express directly with simple function contracts. The Aoraï plug-in introduces a new language to express such constraints, and performs a C and ACSL instrumentation so that the usual analysis plug-ins can be used to check the corresponding properties. This section is structured as follows. First, Sect. 10.6.1 introduces an example of the kind of properties that can be handled by Aoraï, while Sect. 10.6.2 proposes a first formalization of the example in the form of an automaton. Then Aoraï itself is briefly presented in Sect. 10.6.3 and its input language in Sect. 10.6.4. The instrumentation performed by the plug-in and the verification tasks are the subject of Sect. 10.7.

10.6.1 Informal Properties

For instance, suppose that we have the small API presented in Fig. 10.15 to control the access to some sensitive data.

Fig. 10.15 An API controlling access to sensitive data

```
1  enum permission {
2    DENIED = -1,
3    READ_ONLY,
4    WRITE_ONLY,
5    READ_WRITE
6  };
7
8  /*@ assigns \nothing; */
9  enum permission
10 login(const char *user, const char *passwd);
11
12 /*@ assigns \nothing; */
13 const char *read(void);
14
15 /*@ assigns \nothing; */
16 void write(const char *data);
17
18 /*@ assigns \nothing; */
19 void logout(void);
```

We would like to verify the following properties about the way our program uses this API.

1. After three unsuccessful login attempts, the program should exit.
2. A read access can only occur after a successful login has returned READ_ONLY or READ_WRITE, without a logout in between.
3. Similarly, a write access can only occur after a successful login has returned WRITE_ONLY or READ_WRITE, without a logout in between.
4. After a successful login, login must not be called before a call to logout.
5. After a successful login, logout must be called exactly once.

10.6.2 Automaton

These properties are in fact constraints over the sequences of calls to the API that may occur during the execution of a program that uses this API. They can be expressed as an automaton where a transition is taken each time a call to a function of the API is encountered during the execution of the program. Figure 10.16 shows an automaton corresponding to the five properties described in Sect. 10.6.1.

Each state of the automaton represents a program state in which it is possible to call one or more functions of the API. Each outgoing edge corresponds to a call

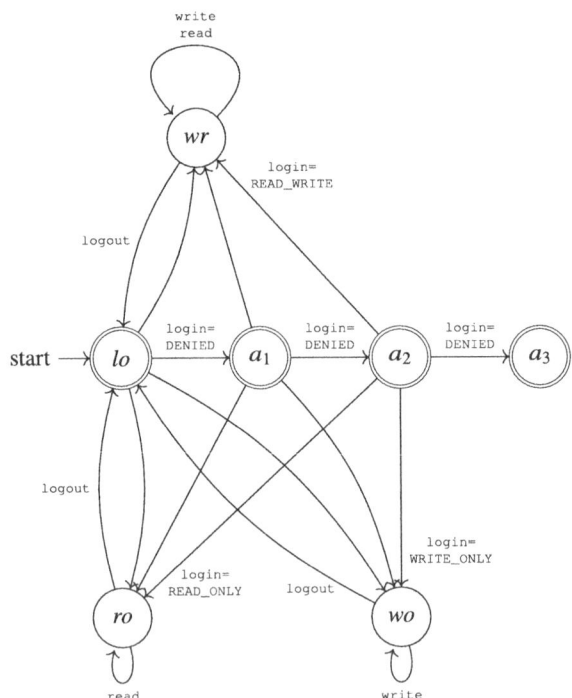

Fig. 10.16 Automaton describing possible API function calls

and is labeled with the function name, and the returned result when it is relevant. For instance, the edge from *lo* to a_1 indicates a failed login attempt (returning DENIED). The labels of the edges from *lo*, a_1 and a_2 to *ro* being identical, they are indicated only once for readability reasons. This is also the case for the labels of the edges leading to *wo* and to *wr*.

Implicitly, the program is only allowed to call a function of the API if there is an edge with the corresponding label starting at the state of the automaton that is currently active. Otherwise, we flag the call attempt as an error. At the start of the program, the automaton is in state *lo* (logged-out), where only a call to login can occur. Then, if the login succeeds, we end up in one of the three states *ro*, *wo*, or *wr*, depending on the value returned by login. In such a state, we can call read and/or write any number of times. This ensures properties 2, 3, and 4, since we know that we are in these states after a successful login, and that no logout has occurred since. Indeed, when we call logout we go back to *lo*, ensuring property 5. Since *lo* is an accepting state, the program can stop there (in fact, a run without any call to the functions of the API at all will be deemed successful as well). Alternatively, we can perform a new login. In any case, we only perform a login attempt when we are logged out, ensuring property 4. Finally, for property 1, states a_1 and a_2 represent respectively one and two consecutive failed login attempts. In a_2, we can try to call login only once. If the third attempt is DENIED as well, we end up in a_3, where we cannot call any function of the API anymore. Note that the three states a_1, a_2, and a_3 all meet property 5 (we only need to call logout after a *successful* login), thus are accepting states.

10.6.3 The Aoraï Plug-In

Within Frama-C, the Aoraï plug-in [27], which is included in the main distribution, lets users express properties over sequences of function calls in the form of an automaton similar to the one presented in Fig. 10.16. More precisely, users can write a description of the automaton in a language called Ya, which is described in the next section. Initially, Aoraï's primary input language was Linear Temporal Logic (LTL) formulas [14]. Temporal logic [21] is meant to express properties over potentially infinite traces of events (function calls in our case). However, Aoraï has always been restricted to be used on terminating programs, so that the full LTL framework and the corresponding encoding of formulas into Büchi automata did not bring any real benefit. This part of the plug-in was eventually dropped in Frama-C 26.0 Iron in order to focus on Ya automata and terminating programs.

Given a Ya automaton and a C program, Aoraï performs an instrumentation of the latter with new functions and ACSL formulas. These formulas can then be verified by the standard analysis plug-ins (Eva, Wp, and E-ACSL). Before presenting the instrumentation and the verification in detail in Sect. 10.7, let us describe the Ya language.

```
1  %init: logged_out;
2  %accept: logged_out, attempt_1, attempt_2, too_many_attempts;
3  %observables: login, read, write, logout;
4
5  logged_out: { login() {{ \result == -1 }} } -> attempt_1
6            | { login() {{ \result == 0 }} } -> logged_ro
7            | { login() {{ \result == 1 }} } -> logged_wo
8            | { login() {{ \result == 2 }} } -> logged_rw;
9
10 attempt_1:  { login() {{ \result == -1 }} } -> attempt_2
11           | { login() {{ \result == 0 }} } -> logged_ro
12           | { login() {{ \result == 1 }} } -> logged_wo
13           | { login() {{ \result == 2 }} } -> logged_rw;
14
15 attempt_2:  { login() {{ \result == -1 }} } -> too_many_attempts
16           | { login() {{ \result == 0 }} } -> logged_ro
17           | { login() {{ \result == 1 }} } -> logged_wo
18           | { login() {{ \result == 2 }} } -> logged_rw;
19
20 logged_ro: { read() } -> logged_ro
21          | { logout() } -> logged_out;
22
23 logged_wo: { write() } -> logged_wo
24          | { logout() } -> logged_out;
25
26 logged_rw: { read() } -> logged_rw
27          | { write() } -> logged_rw
28          | { logout() } -> logged_out;
29
30 too_many_attempts: -> reject;
31
32 reject: -> reject;
```

Fig. 10.17 An automaton described in the Ya language

10.6.4 The Ya Language

As said above, Aoraï's input language allows for describing simple automata whose transitions are triggered whenever a function call occurs or a **return** statement[10] is reached. The syntax is fairly simple, as can be seen in Fig. 10.17, which defines in Ya the automaton described in Sect. 10.6.2.

First, a prelude (cf. lines 1–3 in Fig. 10.17) indicates the initial state and the accepting states. We also tell Aoraï that we are only interested in observing calls to the four functions in our API, so that any other call will leave the state of the automaton unchanged.

Then, for each state of the automaton, we describe the transitions that are possible when the automaton is in this state and a call or a return event occurs. Each transition

---

[10] In the normalized AST of Frama-C, each function has exactly one return statement, at the end of the function (see Chap. 2).

10 Specification and Verification of High-Level Properties 477

can be guarded by a sequence of events, and possibly some guard that the program state must verify for the transition to be triggered. If there is no guard, the transition is always taken. More precisely, as can be seen in Fig. 10.17, this description takes the following form.

```
source_state:   { guard_1 }  -> dest_state_1
              | { guard_2 }  -> dest_state_2
              | ... ;
```

The full syntax for transition guards can be found in the Aoraï manual [27], and we only mention here what is needed to understand the meaning of our example automaton. Basically, a guard of the form `f()` indicates that we go from the source state to the destination state after `f` has been called *and* has returned. Moreover, `f` itself must not call observed functions. In addition, it is possible to add constraints to either the pre-state or the post-state of the call, with a guard of the form `{{ pre_condition }} f() {{ post_condition }}`. Conditions are written in a very small subset of ACSL: it is possible to write equalities or inequalities over globals, formals, and \result (for post-conditions only of course), and conjunctions, disjunctions and/or negations of these formulas. In our example, conditions on the returned value are used (cf. lines 5–18) to distinguish different calls to `login`. Finally, when we are in the `too_many_attempts` state, no call to any of the observable functions should occur, as per property 1. However, for technical reasons linked to its LTL past, Aoraï imposes any state to have at least one transition. Thus, we add (line 30) a transition to a `reject` state, without any guard, so that the transition will be taken whenever there is a call to any of the observable functions. Similarly, once we are in the `reject` state, we stay there forever (line 32). Since `reject` is *not* an accepting state, if we end up in this state, the code is not conforming to the automaton, as expected.

! **Deterministic Automata**

As stated in the Aoraï manual [27], it is also possible to indicate that an automaton should be considered as deterministic via the `%deterministic` directive. However, it should be noted that this applies to the normalized automaton, where call and return events are separated. In particular, the automaton of Fig. 10.16 *cannot* be marked as deterministic, since when we are at the call event for `login`, in any of the `logged_out`, `attempt_1` or `attempt_2` state, we do not know which transition will be taken. It is only at the return event that \result is known and we can discriminate on it.

10.7 Verification of Properties Over Call Traces with Aoraï

Given an automaton and corresponding C source code, Aoraï mainly works by providing an instrumentation of the original code in the form of ACSL annotations and C ghost code. The main components of this instrumentation, which is described in detail in Sect. 10.7.1, are the following. First, it contains an encoding for the current state of the automaton. Second, for each function f in the original program, Aoraï creates two (ghost) functions performing the appropriate transition each time we enter or exit f. Third, Aoraï generates ACSL contracts indicating in which state the automaton can be in the pre- and post-state of each function. Moreover, the contract of the main entry point contains an additional post-condition saying that the automaton must end up in an accepting state. Users can then use analysis plug-ins to check the instrumented code. Sections 10.7.2 and 10.7.3 present, respectively, how Wp and Eva can be used on the instrumented code. An example combining Aoraï and E-ACSL is provided in the chapter on E-ACSL (see Chap. 5).

10.7.1 Code Instrumentation by Aoraï

The -aorai-automata option takes as argument the name of a file containing the Ya description of an automaton. In presence of this option, Aoraï creates a new Frama-C project to store the instrumented code. As an example, assume[11] file aorai-example-api.c contains an implementation of the API, aorai-example.c some client code, and aorai-example.ya the Ya description of the automaton. Then, the following command line outputs the result of Aoraï's instrumentation in aorai-instru.c, thanks to options -then-last, -ocode and -print provided by the Frama-C kernel (see Chap. 2):

```
1 frama-c \
2   -aorai-automata aorai-example.ya \
3   -aorai-acceptance \
4   aorai-example.c aorai-example-api.c \
5   -then-last \
6   -ocode aorai-instru.c -print
```

Option -aorai-acceptance instructs Aoraï to output an ACSL annotation reflecting the fact that the automaton must be in an accepting state at the end of the program (i.e., at the end of the main entry point). This will be detailed below.

The full file aorai-instru.c file can be found in the companion repository. We only describe here the most important aspects of the instrumentation. First, we find a set of global (ghost) variables to store the current status of the automaton.

---

[11] The full example is available in the companion repository at https://git.frama-c.com/pub/frama-c-book-companion/-/tree/main/high-level-properties.

```
1  /*@ ghost int aorai_intermediate_state = 0; */
2  /*@ ghost int aorai_intermediate_state_0 = 0; */
3  /*@ ghost int aorai_intermediate_state_1 = 0; */
4  /* ... other intermediate states ... */
5  /*@ ghost int logged_out = 1; */
6  /*@ ghost int attempt_1 = 0; */
7  /*@ ghost int attempt_2 = 0; */
8  /*@ ghost int logged_ro = 0; */
9  /*@ ghost int logged_wo = 0; */
10 /*@ ghost int logged_rw = 0; */
11 /*@ ghost int too_many_attempts = 0; */
12 /*@ ghost int reject = 0; */
13 /*@ ghost int aorai_reject = 0; */
```

Fig. 10.18 Ghost variable declarations generated by Aoraï for the illustrative example

In our example, their declarations are shown[12] in Fig. 10.18. There is one variable for each state of the automaton. Note that, as mentioned in Sect. 10.6.4, there is a normalization phase where intermediate states might be introduced in addition to the states explicitly mentioned in the Ya file, in order for the transitions to consider a single event at a time. For instance, in our example the intermediate state variables with suffixes _0, _4 or _8 are set when login is called, respectively, from state logged_out, attempt_1 or attempt_2. In this way, we keep track of the state in which the call to login was invoked until the call result is computed. As we will see below, this information is used to determine the new state when the return value of the call is known. Indeed, depending on the original state and the return value, the new state can be different (cf. Figs. 10.16 and 10.17).

Apart from the intermediate states, another state is generated in the normalized automaton, namely aorai_reject (cf. line 13 in Fig. 10.18). It is used when a sequence of transitions in the normalized automaton representing a single transition in the original Ya automaton fails to be taken completely.

Value 0 for a variable indicates that the automaton is not in the corresponding state, and value 1 that the state is active. All variables are initialized to 0, except the one tied to the initial state (here, logged_out), initialized to 1 (cf. line 5).

Deterministic Automaton

The encoding above is needed for non-deterministic automata, since by definition several states might be active at the same time. A deterministic automaton has a simpler instrumentation: we only need a single variable, and an enumerated type with a tag for each state of the automaton.

Then, for each C function observed by the automaton, Aoraï generates two ghost functions that perform the appropriate transitions of the automaton, respectively when

---

[12] For the sake of clarity, the Aoraï output examples in this chapter may be given with some simplifications and reformatting.

```
1   /*@ ghost
2     /@ requires
3           (1 == aorai_intermediate_state ||
4            1 == aorai_intermediate_state_0 || ... 1 == reject) &&
5           0 == aorai_intermediate_state_11 && ...
6           0 == attempt_1 && 0 == attempt_2 && ... ;
7       assigns aorai_intermediate_state, ... aorai_reject,
8               attempt_1, attempt_2, logged_out, logged_ro,
9               logged_wo, logged_rw, reject, too_many_attempts;
10      assigns aorai_intermediate_state
11        \from aorai_intermediate_state, ... aorai_reject,
12              attempt_1, attempt_2, logged_out, logged_ro,
13              logged_wo, logged_rw, reject, too_many_attempts,
                  res;
14       ...
15      behavior buch_state_logged_out_out:
16        ensures 0 == logged_out;
17       ...
18      behavior buch_state_logged_ro_in:
19        assumes
20            (1 == aorai_intermediate_state_8 && res == 0) ||
21            (1 == aorai_intermediate_state_4 && res == 0) ||
22            (1 == aorai_intermediate_state_0 && res == 0);
23        ensures 1 == logged_ro;
24
25      behavior buch_state_logged_ro_out:
26        assumes
27            (0 == aorai_intermediate_state_8 || !(res == 0)) &&
28            (0 == aorai_intermediate_state_4 || !(res == 0)) &&
29            (0 == aorai_intermediate_state_0 || !(res == 0));
30        ensures 0 == logged_ro;
31    @/
32    void login_post_func(enum permission res)
```

Fig. 10.19 Extract from the contract of a ghost transition function generated by Aoraï

the C function is called and when it returns. By default, each function is equipped with a contract, even though this can be disabled with -aorai-no-generate-annotations, notably when using Eva afterwards, as explained in Sect. 10.7.3. As the contracts have to describe the value of each of the state variables in every possible case, it would be extremely tedious to reproduce them in full. Figure 10.19 presents the most important parts of the contract of function login_post_func, which describes the transitions that occur according to the automaton of Fig. 10.17 when we return from a call to login.

Function login_post_func takes a single argument, res, which is the value that login itself is about to return. Indeed, the guards of the transitions can depend on this value. Similarly, the transition functions for entering a call have the same arguments as the original function, since its parameters can be used in Ya guards.

The contract itself contains first a **requires** clause indicating which states might be active (lines 3–4) when we are about to return from a login call, and

10 Specification and Verification of High-Level Properties

which states are definitely inactive (lines 5–6). Possible active states are `reject`, in case we have already made 3 or more failed `login` attempts, and the intermediate states corresponding to the various `login` transitions, which have been activated by another function generated by Aoraï, `login_pre_func`, triggered at the beginning of the call to `login`. On the contrary, the states explicitly mentioned in the Ya automaton, as well as the intermediate states corresponding to other functions are known to be inactive at this stage.

Then, the **assigns** clause of lines 7–9 states that only the variables representing the automaton are modified. We also have a **\from** clause for each state variable, saying (lines 10–14) that their new value depends on the values of the state variables themselves, and on the value of `res`, since the guards of the transitions in Fig. 10.17 use **\result**.

Finally, for each state variable, we find one or two behaviors explaining whether their state is active or not after the transition. In case the state is surely inactive after the event, like for `logged_out` after a call to `login`, there is a single behavior stating exactly that (lines 15–16). On the other hand, if the state may be active, like `logged_ro`, we have two complementary behaviors. The first one (lines 18–23) indicates at which conditions the behavior will be activated. Namely, `logged_ro` will be active if one of the intermediate states corresponding to the transitions of lines 6, 11, and 16 in Fig. 10.17 are currently active and the corresponding guard is true. Conversely, the second behavior (lines 25–30) indicates that if none of these conditions are met, `logged_ro` will be inactive.

The body of the transition function follows the same schema as the behavior, with a sequence of **if** statements to assign the appropriate value of each state variable. It takes the form shown in Fig. 10.20. First, we declare one local variable for each state variable (line 2): they represent the activation value of the states after the transition. Then, for states that might be active, such as `logged_ro`, we check whether an activation condition is met, for instance we are in state `aorai_intermediate_state_0` and the value returned is 0 (lines 3–4). If this is the case, we set the corresponding local variable to 1, otherwise, we check the next activation condition (line 5). If no activation condition is met, the local variable is set to 0 (lines 15–16). On the other hand, if we know that the state will be inactive, we simply set the local variable to 0 (line 17). Once all local variables have been set, we can copy back their new value to the corresponding global state variables (lines 19–20).

Finally, a call to a transition function for a call or return event is inserted into the body of the corresponding C function, respectively, at the beginning of the function or just before the **return** statement. If annotation generation has not been turned off, a contract similar to the ones of the transition functions is generated for the C function as well. When option `-aorai-acceptance` is set, the contract of the main entry point additionally receives a specific **ensures** clause stating that at least one of the accepting states mentioned in the Ya description must be active at the end of the execution. In addition, all loops in the program are equipped with loop invariants stating which states might be active in all loop steps. The reader may refer

```
1  /*@ ghost void login_post_func(enum permission res)
2     { int logged_out_tmp, ..., logged_ro_tmp, ...;
3       if (aorai_intermediate_state_0 == 1) {
4         if (res == 0) logged_ro_tmp = 1;
5         else goto __aorai_label_12; }
6       else
7         __aorai_label_12:
8         if (aorai_intermediate_state_4 == 1) {
9           if (res == 0) logged_ro_tmp = 1;
10          else goto __aorai_label_11; }
11        else
12          __aorai_label_11:
13          if (aorai_intermediate_state_8 == 1)
14            if (res == 0) logged_ro_tmp = 1;
15            else logged_ro_tmp = 0;
16          else logged_ro_tmp = 0;
17      logged_out_tmp = 0;
18      ...
19      logged_out = logged_out_tmp;  ...
20      logged_ro  = logged_ro_tmp;   ...
21      return;
22    }
23  */
```

Fig. 10.20 Body of a ghost transition function generated by Aoraï

10.7.2 Analysis of the Instrumented Code with Wp

Historically, the original Aoraï implementation was meant to be used together with deductive verification tools [14]. This is in particular the reason why Aoraï generates loop annotations. It is thus perfectly possible to launch the Wp plug-in (presented in detail in Chap. 4) on the instrumented code. However, when doing so, one has to remember that the generated annotations are usually not sufficient to complete the proofs. Indeed, as usual with Wp, at least **assigns** clauses describing the memory locations of the original C code that can be assigned by each function and each loop must be provided: Aoraï only completes such clauses with the information that the state variables are modified as well. Similarly, it is most often the case that additional **requires**, **ensures**, and **loop invariant** clauses are needed to specify the behavior of some variables during the execution. The code of our example contains thus a few manually written ACSL annotations needed for Wp to complete the proofs. In the companion repository, make aorai-wp triggers the Wp verification of the instrumented code.

10.7.3 Analysis of the Instrumented Code with Eva

Recently, Aoraï gained some features that are geared towards analyzing the instrumented code with the Eva plug-in (detailed in Chap. 3). Indeed, since a Ya automaton is meant to describe all admissible sequences of calls from a given entry point, a whole-program analysis à la Eva is somehow more appropriate than a modular approach à la Wp. In particular Aoraï registers a built-in function `Frama_C_show_aorai_state` that takes any number of arguments. Like for the `Frama_C_show_each*` family of built-ins mentioned in Chap. 3, Eva will output the current abstract value of each of the arguments each time it encounters such a call. In addition, it will also output the current abstract value of the state variables, allowing to monitor the status of the automaton during Eva's analysis.

In addition, option `-aorai-instrumentation-history <n>`, which is currently only available for deterministic automata, instructs Aoraï to keep track not only of the current status of the automaton, but also of the n previous states, allowing for a finer partitioning of Eva's abstract states according to the automaton specification. Moreover, Aoraï positions Eva options and emits specific `slevel` annotations (see Chap. 3) to ensure a partitioning as precise as possible during Eva analysis. As for Wp, an example of successful Eva analysis is available in the companion repository (through `make aorai-eva`).

10.8 Conclusion

This chapter has presented three plug-ins that extend ACSL specifications abilities. MetAcsl allows users to write a few HILAREs that the tool automatically instantiates in all relevant program points. It has notably been deployed in an industrial context to generate tens of thousands of ACSL annotations, that probably could not have been handled correctly by hand (see Chap. 16). RPP is dedicated to specifying relational properties, which constitute a very active research topic, especially because of their ties with non-interference properties, which play an important role in security (more precisely for assessing confidentiality). Finally, the third tool, Aoraï, makes it possible to provide a model of the call stack of a program in the form of a Ya automaton and to verify properties over call traces.

A common characteristic of the three approaches is that complex program properties are first expressed with a dedicated specification mechanism before being translated into usual ACSL annotations and verified with other Frama-C analyzers. This is possible thanks to the design of the Frama-C platform, where ACSL annotations can be automatically generated by a plug-in and then verified by other analyzers. This methodology facilitates the design and development of analyzers for specific properties.

Specification and verification techniques for properties described in this chapter were studied in the past. We cite here only a few examples of previous research efforts.

Verification of temporal properties is mainly associated with model-checking [4]. Similarly, runtime verification of temporal properties has also been actively investigated since several years [20]. Several high-level extensions of a contract-based specification language were proposed. For example, Cheon and Perumandla [11] extended JML [17] to specify properties related to call sequences, and Trentelman and Huisman [28] adapted JML to express temporal properties. The general idea of defining a high-level concept in the global scope and then moving to the implementation is similar to the Aspect-Oriented Programming (AOP) [16] paradigm. From this point of view, metaproperties can be seen as cross-cutting concerns at the specification rather than code level, while contexts can be related to *pointcuts*, which in AOP define a set of control flow points where the code required by the concern should be added. Various theories and techniques were proposed to deal with relational properties. Relational Hoare logic can be used to show the equivalence of program transformations [3]. Cartesian Hoare logic addresses k-safety properties [26]. Self-composition [2] was used to verify relational properties by relating two execution traces. It was applied in particular to verification of information flow properties [1, 2] and properties relating two equivalent-result methods [18]. A partial support for deductive verification of relational properties was proposed in OpenJML [12].

Future work directions for MetAcsl include the design and development of mechanisms for automatic generation of metaproperties, optimizations of their instantiation into assertions, and advanced reasoning techniques (e.g., to deduce a metaproperty from some other metaproperties or annotations). Regarding RPP, future work includes proving the **\from ...** components of **assigns ... \from ...** clauses (which can be seen as non-interference properties), using verification condition generators rather than self-composition and verification of relational properties for loops. To extend and consolidate Aoraï, we plan to extend the Ya language, notably with the ability to declare and update variables within the automaton, in order to have more flexible guards.[13] Another important extension for Aoraï would be to consider other kinds of events than function calls and returns (e.g., writing to a given variable). A common (and ambitious) research perspective for the three tools is a formalization and a mechanized proof of soundness of the proposed techniques in a proof assistant, partly started in [9, 10, 22]. Another common perspective is a larger application and evaluation of these tool in various case studies. Finally, the specification and verification methodology presented in this chapter is also very likely to be used for other kinds of properties and verification tasks in the future.

References

1. Barthe G, Crespo JM, Kunz C (2016) Product programs and relational program logics. J Log Algebr Methods Progr. https://doi.org/10.1016/j.jlamp.2016.05.004
2. Barthe G, D'Argenio PR, Rezk T (2011) Secure information flow by self-composition. J Math Struct Comput Sci 21(6):1207–1252. https://doi.org/10.1017/S0960129511000193

---

[13] A limited support already exists.

3. Benton N (2004) Simple relational correctness proofs for static analyses and program transformations. In: Proceedings of the 41st symposium on principles of programming languages (POPL 2004), pp 14–25. https://doi.org/10.1145/964001.964003
4. Beyer D, Podelski A (2022) Software model checking: 20 years and beyond. In: Raskin JF, Chatterjee K, Doyen L, Majumdar R (eds) Principles of systems design: essays dedicated to Thomas A. Henzinger on the Occasion of his 60th Birthday, pp 554–582. Springer Nature Switzerland, Cham. https://doi.org/10.1007/978-3-031-22337-2_27
5. Blatter L (2019) Relational properties for specification and verification of C programs in Frama-C. PhD thesis, University Paris-Saclay. https://theses.hal.science/tel-02401884/
6. Blatter L, Kosmatov N, Le Gall P, Prevosto V (2017) RPP: automatic proof of relational properties by self-composition. In: Proceeding of the 23th international conference on tools and algorithms for the construction and analysis of systems (TACAS 2017), held as part of the european joint conferences on theory and practice of software (ETAPS 2017), LNCS, vol 10205, pp 391–397. Springer. https://doi.org/10.1007/978-3-662-54577-5_22
7. Blatter L, Kosmatov N, Le Gall P, Prevosto V, Petiot G (2018) Static and dynamic verification of relational properties on self-composed C code. In: International conference on tests and proofs (TAP), LNCS, vol 10889, pp 44–62. Springer. https://doi.org/10.1007/978-3-319-92994-1_3
8. Blatter L, Kosmatov N, Prevosto V, Le Gall P, The RPP plug-in manual. https://github.com/lyonel2017/Frama-C-RPP/blob/master/doc/rpp-manual.pdf
9. Blatter L, Kosmatov N, Prevosto V, Le Gall P (2022) Certified verification of relational properties. In: Proceedings of the 17th international conference on integrated formal methods (iFM 2022), LNCS, vol 13274, pp 86–105. Springer. https://doi.org/10.1007/978-3-031-07727-2_6
10. Blatter L, Kosmatov N, Prevosto V, Le Gall P (2022) An efficient VCGen-based modular verification of relational properties. In: Proceedings of the 11th international symposium on leveraging applications of formal methods, verification and validation. (ISOLA 2022), LNCS, vol 13701, pp 498–516. Springer. https://doi.org/10.1007/978-3-031-19849-6_28
11. Cheon Y, Perumandla A (2005) Specifying and checking method call sequences in JML. In: International conference on software engineering research and practice, pp 511–516
12. Cok DR (2011) OpenJML: JML for Java 7 by extending OpenJDK. In: International symposium on nasa formal methods (NFM). https://doi.org/10.1007/978-3-642-20398-5_35
13. Darvas A, Müller P (2006) Reasoning about method calls in JML specifications. J Object Technol 5(5):59–85
14. Groslambert J, Stouls N (2009) Vérification de propriétés LTL sur des programmes C par génération d'annotations. In: Approches Formelles dans l'Assistance au Développement de Logiciels (AFADL 2009). In French
15. Hatcliff J, Leavens GT, Leino KRM, Müller P, Parkinson M (2012) Behavioral interface specification languages. Comput Surv 44(3):16:1–16:58. https://doi.org/10.1145/2187671.2187678
16. Kiczales G, Lamping J, Mendhekar A, Maeda C, Lopes CV, Loingtier JM, Irwin J (1997) Aspect-oriented programming. In: European conference on object-oriented programming, LNCS, vol 1241, pp 220–242. Springer. https://doi.org/10.1007/BFb0053381
17. Leavens GT, Baker AL, Ruby C (1999) JML: a notation for detailed design. In: Behavioral specifications of businesses and systems, vol 523, pp 175–188. Springer. https://doi.org/10.1007/978-1-4615-5229-1_12
18. Leino KRM, Müller P (2008) Verification of equivalent-results methods. In: ESOP, LNCS, vol 4960, pp 307–321. https://doi.org/10.1007/978-3-540-78739-6_24
19. Leino KRM, Polikarpova N (2013) Verified calculations. In: Proceedings of the 5th international conference on verified software: theories, tools, experiments (VSTTE 2013), Revised selected papers, vol 8164, pp 170–190. Springer. https://doi.org/10.1007/978-3-642-54108-7_9
20. Leucker M, Schallhart C (2009) A brief account of runtime verification. J Log Algebr Progr 78(5):293–303
21. Pnueli A (1977) The temporal logic of programs. In: the 18th annual symposium on foundations of computer science (FOCS 1977), pp 46–57. IEEE Computer Society
22. Robles V (2022) Specifying and verifying high-level requirements on large programs: application to security of C programs. PhD thesis, University Paris-Saclay. https://theses.hal.science/tel-03626084/

23. Robles V, Kosmatov N, Prevosto V, Rilling L, Le Gall P (2019) MetAcsl: specification and verification of high-level properties. In: Proceedings of the 25th international conference on tools and algorithms for the construction and analysis of systems (TACAS 2019), Held as part of the European joint conferences on theory and practice of software (ETAPS 2019), LNCS, vol 11427, pp 358–364. Springer. https://doi.org/10.1007/978-3-030-17462-0_22
24. Robles V, Kosmatov N, Prevosto V, Rilling L, Le Gall P (2019) Tame your annotations with MetAcsl: specifying, testing and proving high-level properties. In: International conference on tests and proofs (TAP), LNCS, vol 11823, pp 167–185. Springer. https://doi.org/10.1007/978-3-030-31157-5_11
25. Robles V, Kosmatov N, Prevosto V, Rilling L, Le Gall P (2021) Methodology for specification and verification of high-level properties with MetAcsl. In: Proceedings of the 9th IEEE/ACM international conference on formal methods in software engineering (FormaliSE 2021), pp 54–67. IEEE. https://doi.org/10.1109/FormaliSE52586.2021.00012
26. Sousa M, Dillig I (2016) Cartesian Hoare logic for verifying k-safety properties. In: Proceedings of the 37th ACM SIGPLAN conference on programming language design and implementation, PLDI 2016, Santa Barbara, CA, USA, June 13–17, 2016, pp 57–69. https://doi.org/10.1145/2908080.2908092
27. Stouls N, Prevosto V (2023) Aoraï plug-in tutorial. https://frama-c.com/download/frama-c-aorai-manual.pdf
28. Trentelman K, Huisman M (2002) Extending jml specifications with temporal logic. In: International conference on algebraic methodology and software technology, LNCS, vol 2422, pp 334–348. Springer. https://doi.org/10.1007/3-540-45719-4_23

Chapter 11
Advanced Memory and Shape Analyses

Matthieu Lemerre, Xavier Rival, Olivier Nicole, and Hugo Illous

Abstract One of the main features of C is manual management of memory, which explains the domination of C in systems and performance-critical code. The downside of manual memory management is that designing data structures requires reasoning about custom memory invariants, and that any mistake may lead to severe cybersecurity vulnerabilities. In this chapter, we present two Frama-C plug-ins, the Codex and RMA plug-ins, respectively based on physical refinement types and separation logic, that help verify properties of C programs manipulating data structures in memory with minimal guidance from the user.

Keywords Memory safety · Shape analysis · Separation logic · Type-based analysis · Abstract interpretation

11.1 Introduction

Memory safety errors that come from programming errors in systems language such as C, are the most pervasive and severe cybersecurity errors today. In 2019, reports from both Microsoft [28] and Google [14] estimate that between 60% and 70% of the vulnerabilities that they address come from memory corruption. In addition to being

---

M. Lemerre (✉) · O. Nicole · H. Illous
Université Paris-Saclay, CEA, List, Palaiseau, France
e-mail: matthieu.lemerre@cea.fr

O. Nicole
e-mail: olivier@chnik.fr

H. Illous
e-mail: hugo.illous@gmail.com

X. Rival
Inria and Département d'informatique de l'ENS, CNRS, PSL University, Paris, France
e-mail: xavier.rival@ens.fr

O. Nicole · H. Illous
Inria and Département d'informatique de l'ENS, Paris, France

pervasive, they are also the most severe kind of vulnerabilities, as they can lead an attacker to inject code and penetrate the computer running the vulnerable program.

In addition to memory safety, many program properties in systems programs apply to values inside or flowing through the memory. Therefore, reasoning about these properties require precise reasoning about the memory. For instance, a 2014 article from Klein et al. [22] about their experience proving the seL4 microkernel explains that "There are four main categories of invariants in our proof: 1. low-level memory invariants, 2. typing invariants, 3. data structure invariants, and 4. algorithmic invariants", and furthermore that 80% of their effort went into establishing these invariants. Thus, reasoning about memory invariants is a significant part of the verification of large systems software, which is the kind of software for which the C language is generally used.

This chapter presents two different **Frama-C** plug-ins, **Codex** and **RMA**, which can be used to automatically infer and verify memory invariants, and absence of memory safety issues. Both of these plug-ins are based on abstract interpretation [9], meaning that they are automated and sound verification techniques. The goal of this chapter is to introduce both analysis techniques using a single motivating example and formal framework, to better highlight their respective strength and complementarity, and to help use both plug-ins in a combination. In a nutshell, the differences between both techniques and plug-ins is that:

- The **Codex** plug-in relies on a type-based analysis technique [32–34] which is strong at representing low-level structural invariants and code patterns, e.g., for programs using interior pointers or byte–per-byte copy. It can verify spatial memory safety in many programs, handles well unclear sharing patterns, and the analysis is easy to bootstrap using the existing C types of the program.
- The **RMA** plug-in [16] relies on a shape analysis [4] based on separation logic [5, 35] which is able to represent very precise structural invariants of the program, especially those featuring strong constraints on the separation between memory objects, and can be used to verify temporal memory safety issues such as absence of use-after-free. It is also able to infer the memory transformations [19] performed by a program (e.g., whether parts of the memory have been modified, if a list has shrunk or grown), and to use these transformations [18] to perform modular analysis of a large program.

We begin by presenting the two analysis techniques on a single shared example (Sect. 11.2). Then, we present a common abstract interpretation framework (Sect. 11.3) which is used to present the type-based static analysis technique of the **Codex** plug-in (Sect. 11.4) and the separation-logic technique of the **RMA** plug-in (Sect. 11.5). Section 11.6 finally lists additional references for readers who would want to know more about these static analysis techniques.

```
1  typedef int node_data;          /* Should be between 0 and 10. */
2  struct list_node {
3    struct list_node* next;
4    node_data data;
5  };
6
7  struct list_node*
8  list_append( struct list_node* a, struct list_node* b)
9  {
10   struct list_node* ptr = a;
11   while( ptr->next != 0){
12     ptr = ptr->next;
13   }
14   ptr->next = b;
15   return a;
16 }
```

Fig. 11.1 Running example: the list_append function

11.2 Overview

This section provides a gentle, but quite complete, introduction to the capabilities of the Codex and RMA analysis plug-ins, leaving to later sections the formal description of their specification and implementation as abstract interpretations.

11.2.1 Motivating Example

We consider the C program of Fig. 11.1. Despite its apparent simplicity, many different situations can be encountered during the execution of this program, that depends on the state at the beginning of the execution of list_append (i.e. its *pre-state*).

Figure 11.2 examines these different situations. The pre-state of list_append is composed of the memory (that we call the *heap*, even if it can also cover statically allocated memory) and of the value of the local variables used as parameters, a and b. The top of Fig. 11.2 presents different heaps h_0 to h_4, while the bottom shows 9 different executions (from a pre-state to its corresponding post-state after execution of list_append, if it exists). Note that we consider a low-level representation of memory corresponding to the result of the compilation of C, on a 32-bit machine (i.e., pointers and int values each occupy 4 bytes in memory).

The execution (1) represents the canonical execution of the program: given two separate, acyclic linked lists (heap h_0), the program creates a single acyclic linked list (heap h_1).

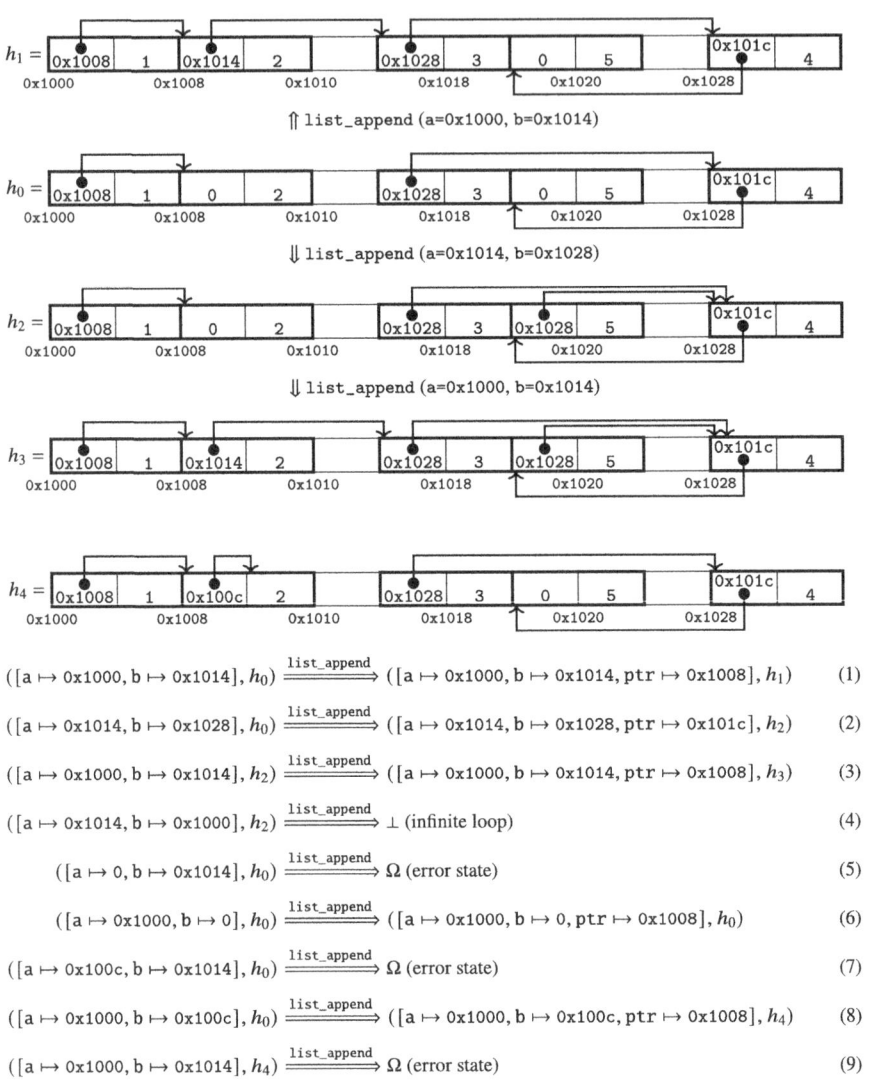

Fig. 11.2 Possible executions for list_append (with their corresponding heaps)

The pre-state of execution (2) also has h_0 as its heap, but this time the two linked lists given as an argument are not separated (i.e. b points inside the linked list starting at a). The result is that list_append creates a cycle in the linked list (heap h_2). Note that such a linked list with cycle can be appended to a acyclic linked list (this is what execution (3) does). However, passing a linked list with a cycle as the first argument (as in execution (4)) makes the program enter an infinite loop. Note that depending on the program, the possibility to manipulate linked lists with cycles may or may not be desirable.

Executions (5) and (6) correspond to special cases where a or b represent empty linked lists (i.e. their value is 0x0). In execution (5) a is an empty linked list; this results in an undefined behavior (typically a segmentation fault) when ptr is first dereferenced on line 11. In execution (6), the parameter b represents an empty linked list; in this case the memory is not modified (there is a memory write, but it does not change the heap).

The last executions correspond to *ill-typed* accesses to memory, in particular dereference of list_node* pointers that do not point to a correct list_node object. In execution (7), a does not point to the beginning of a list_node object; the execution will try to dereference the address 2, which is (generally) not a valid heap address, and will thus cause a segmentation fault. In execution (8) it is b that is ill-typed; this corrupts the linked list starting at 0x1000 (in heap h_4). The segmentation fault will then occur at the next traversal of this list (execution (9)).

To summarize, there is one expected behavior for list_append (concatenation of acyclic linked lists in execution (1)), of which concatenation with an empty list is a special case (execution (6)); some behaviors mixing linked lists with or without cycles, which may or may not be desirable (executions (2) and (3)); clearly problematic situations that lead to undefined behaviors or segmentation faults (executions (5), (7–9)), and another problematic execution leading to an infinite loop (execution (4)).

In the following, we show how both plug-ins and analyses can be used to verify the correct behavior of the function on normal executions and warn if problematic situations are possible. Both analyses will allow standard executions (1) and (6), and reject those leading to an undefined behavior (executions (5), (7–9)). But the type-based shape analysis will have a more permissive invariant that will allow situations (2–4), that the standard analysis based on separation logic will reject. Finally, the relational separation-logic-based analysis will prove that execution (6) does not modify the heap, among other transformation properties.

11.2.2 Type-Based Shape Analysis

The type-based shape analysis relies on a core invariant stating that program states are *well-typed*, for a type system that looks superficially similar, but is actually quite different from that of C. This type system features in particular subtyping and dependent types, which is necessary to verify the type safety of C programs, but prevents the types from being inferred using classical type inference. Instead, types will be inferred using abstract interpretation.

11.2.2.1 Well-Typed Heaps and States

The top of Fig. 11.3 presents the definition of two types, node_data and list_node, in this type system.

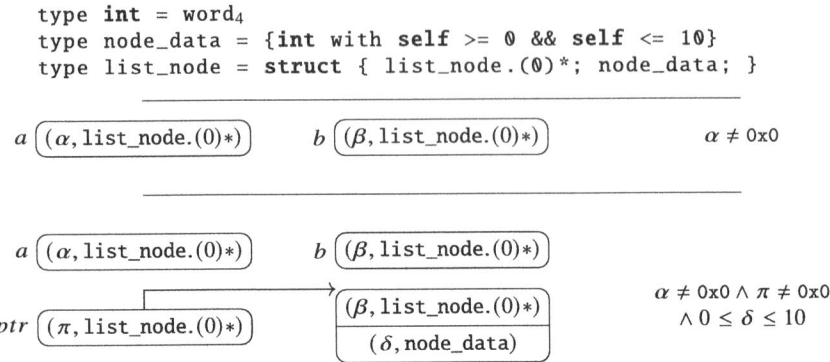

Fig. 11.3 Result of the type-based analysis of list_append : global flow-insensitive invariant (top), given pre-condition (middle) and computed post-condition (bottom)

These type definitions act as a global, flow-insensitive, invariant that *well-typed heaps* must fulfill. The definition of type node_data expresses that memory regions tagged as node_data hold a bitvector of length 4 bytes which, when interpreted as a signed integer, is between 0 and 10. The definition of type list_node says that memory regions tagged as list_node are the concatenation of two regions, the first region being tagged as list_node.(0)* (which contains addresses of list_node regions at offset 0, i.e., they are pointers to the beginning of a list_node region), and the second region being tagged as node_data (which contains integers between 0 and 10 as explained previously).

Our interpretation of these type definitions as an invariant also says that the regions holding a node_data and those holding a list_node.(0)* are separated, but also that the list_node.(0) and list_node.(4) regions are separated (i.e. a memory address cannot simultaneously contain the bytes at different offset of a list_node, which rules out h_4 in Fig. 11.2).

More precisely, we define a notion of an *allocation map* $\mathcal{A}$, labeling each address with the type of the values that it may contain. For instance, in all the heaps of Fig. 11.2, addresses from 0x1000 to 0x1007 are supposed to contain a list_node structure; this is also the case for addresses 0x1008-0x1010, 0x1014-0x101b, 0x101c-0x1024, and 0x1028-0x1030. Thus, we will write that $\mathcal{A}(\text{0x1000-0x1007}) = \text{list\_node}$, and more precisely that $\mathcal{A}(\text{0x1000}) = \text{list\_node.(0)}$ (the byte 0 of a list_node value), $\mathcal{A}(\text{0x1001}) = \text{list\_node.(1)}$ (the byte 1 of a list_node value), etc.

Using this allocation map, we can *interpret types* as sets of values, and in particular, we can interpret pointer types as set of memory addresses. For instance on Fig. 11.2, the type list_node.(0)* (i.e. the type of the pointers to addresses at the beginning of a list_node) can be interpreted as:

$$(\!|\text{list\_node.(0)}*|\!)_{\mathcal{A}} = \{0, \text{0x1000}, \text{0x1008}, \text{0x1014}, \text{0x101c}, \text{0x1028}\}$$

meaning that the possible values for this type are either the addresses of the beginning of a list_node, or 0 (representing null pointers).

The core invariant that well-typedness of the heap guarantees is that a memory region labeled with a certain type t contains a value inside the set of possible values for type t. For instance in Fig. 11.2, following the allocation map $\mathcal{A}$, addresses 0x1000-0x1003 should contain the first 4 bytes of a list_node. The first 4 bytes of a list_node correspond to the first field of the struct list_node, which is of type list_node.(0)*. Thus, addresses 0x1000-0x1003 should contain a value inside $(\!|\,\text{list\_node.(0)}*\,|\!)_{\mathcal{A}}$. In all heaps, 0x1000-0x1003 contains the value 0x1008, which satisfies the invariant. Actually, using the allocation map $\mathcal{A}$, all of $h_0 \ldots h_3$ are well-typed heaps, but h_4 is not: the addresses 0x1008-0x100b do not contain a value of type list_node.(0)* (they contain 0x100c, which is of type list_node.(4)*).

Finally, a well-typed state consists in a well-typed heap, together with a mapping from variables to types such that the values of the variables match their types. For instance in Fig. 11.2, associating both a and b to type list_node.(0)* would result in a well-typed pre-state in executions (1) to (6), but on an ill-typed pre-state for executions (7) to (9) (in (7) the value and type of a do not match, in (8) the value and type of b do not match, and in (9) the heap is ill-typed).

11.2.2.2 A Type-Based Memory Abstraction

A state abstraction is a finite description of a large or infinite set of states or, equivalently, the description of a predicate over states. Thus, abstract states can be used to represent pre and postconditions of a function, as in Fig. 11.3.

The main state abstraction in our type-based static analysis is simply an environment mapping each variable to a type, which represents all the well-typed states for which this mapping from variables to types is correct. For instance, in the precondition given in Fig. 11.3 (middle), a is given the type list_append.(0)*, and so is b, meaning that they describe states where a and b contain addresses tagged by list_append.(0). More precisely, there exists an allocation map $\mathcal{A}$ for which the pre-states in all the execution from (1) to (6) are well-typed using this mapping from variables to types. On the contrary, no such allocation map exists for the pre-states of execution (7) to (9), thus being ruled out by our precondition.

In addition to being constrained by the type, variables are also constrained numerically. This works as follows: each program variable is associated with a *symbolic value* (e.g. α for a), that denotes some numerical value at runtime; then a numerical abstract domain (intervals [9], octagons [29], polyhedra [11]) captures, at analysis time, numerical constraints on the symbolic values. For instance in Fig. 11.3, a is constrained to be different from 0: it cannot be the null pointer. Thus, the pre-state excludes the problematic execution (5).

Thus, the type-based abstraction of Fig. 11.3 (middle) represents states that exclude the most problematic executions of list_append (i.e. executions (5), (7),(8), and (9)), but includes all the other executions: actually, this abstraction represents any state where the two variables a and b points to linked list (with or without cycles), with the constraint that a is non-empty.

11.2.2.3 Retained and Staged Points-to Predicates

The type-based abstraction is sufficient to describe the correct initial states for list_append, but is not sufficient to verify that the execution starting from these correct states will not enter an error state. For this purpose, the analysis needs to remember that ptr is not null after the assignment ptr=ptr->next on line 12 of Fig. 11.1, but ptr->next loads a field in memory, of which our representation only remembers that it is of type list_node.(0)∗. To verify the program correctness, we must remember that ptr->next is non-null due to the test that we already performed in the condition of the **while**.

For this, we extend our domain with additional information using *retained points-to predicates*: when we access memory (by a load or a store), we remember the value and type of the location that we have accessed, so that it can be reused if we later load a value at that location. For instance, in Fig. 11.1, there are two accesses (on line 11 and 12) to ptr->next. Retained points-to predicates make sure that the value returned by the first load, which is constrained to be non-null, is also returned by the second load (and is thus also non-null); this step is necessary to prove that ptr is not null at the end of the loop, and the store at line 14 is valid. For another example, after line 14, our analysis retains that ptr->next contains β, the value of b (the fact that they both point to the same symbolic value implies that their value is equal); this fact is still true at the end of the analysis (Fig. 11.3).

Dually, sometimes a program might temporarily perform memory stores such that the resulting state is not well-typed. This happens for instance when performing byte-per-byte copy of pointers using memcpy: in the middle of the copy, the memory may contain two halves of pointer values that do not point to a valid memory address. For this reason, we introduce *staged points-to predicates* that can be seen as a way to "delay" memory stores until the memory can be well-typed again; i.e., staged point-to predicates can be viewed as a buffer that accumulates the writes before they are "flushed" (actually written in memory, which is when the well-typedness check is performed). The moment when the analysis decides to "flush" a write is based on a heuristic (currently, we retain the write for as long as possible, which is the end of the analysis or when there is another write that may alias with the staged write).

11.2.2.4 Strengths and Weaknesses of the Type-Based Abstraction

The type-based abstraction that we propose has numerous strengths. The base abstraction is storeless and thus very efficient: the memory does not need to be represented or propagated in the abstract domain. The domain is easy to parameterize as the types can be easily derived from the existing C types. It can be enhanced to handle more complex cases: the complete analysis includes features such as subtyping to accommodate low-level manipulation of interior pointers, dependent types, uses abstract types to represent a set of concrete types... With these extensions, the domain generally succeeds in verifying spatial memory safety of programs manipulating linked data structures. Note that despite the use of types, the verification done is a static analysis by abstract interpretation, not a rule-based type inference: if the analysis does not succeed in type-checking (which can happen on memory stores), it will emit alarms, but the rest of the analysis can proceed.

There are drawbacks, though, such as the inability to distinguish linked lists that have cycles from those who do not, or to verify the absence of temporal memory safety errors (e.g., use-after-free). In order to verify such properties, we need stronger invariants on memory, such as the ones provided by separation-logic-based static analysis.

11.2.3 Separation Logic and Relational Separation-Logic-Based Analysis

Separation logic [35] is an extension of Floyd-Hoare logic [15] used to verify memory-manipulating programs. Its key feature is the separating conjunction connective "∗" (pronounced "and, separately"), that allows local reasoning by decomposition of a formula describing a large memory object into smaller formulas describing smaller objects.

Separation logic is widely used as the underlying logic for deductive verification of programs manipulating memory (e.g. [7, 20, 30]). But it has also been used for automated verification by abstract interpretation in tools such as Smallfoot [1], Infer [2], Xisa [5] or MemCad [6]: like our RMA plug-in, analyses based on abstract interpretation can infer the shape of memory without requiring manually written loop invariants.

In shape analyses based on separation logic, a separation logic formula is used to describe a set of memory objects, and standard static analysis operations (such as join or abstract transformer for assignment) perform updates on these formulas. In the following section, we present the meaning of these formulas on the `list_append` example.

11.2.3.1 Static Analysis Based on Separation Logic

Figure 11.4 presents two separation logic formulas in a graphical representation called *separating shape graphs*, one (on top) describing possible pre-states for `list_append`, given by the user, and one (in the bottom) describing possible post-states, which is computed by the RMA plug-in.

In this graphical representation, nodes correspond to values (which includes addresses), and edges correspond to memory regions. When a program variable (like

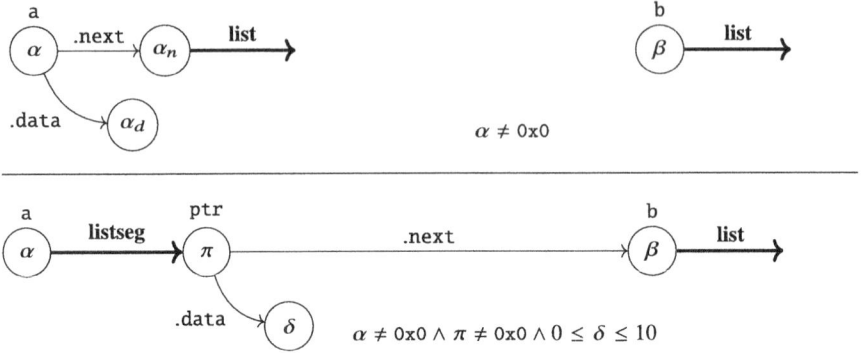

Fig. 11.4 State analysis of list_append based on separation logic : given pre-state (top) and computed post-state (bottom)

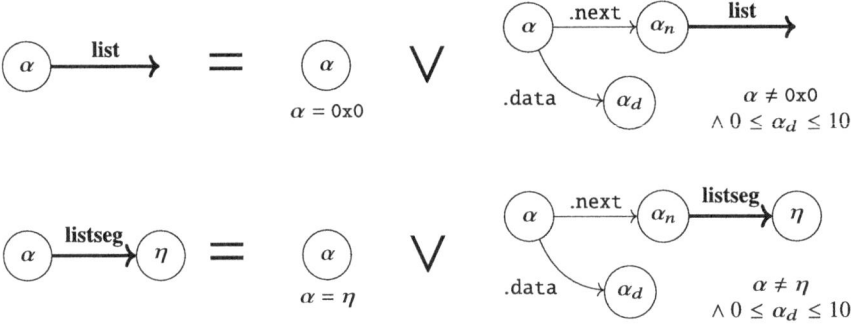

Fig. 11.5 Inductive definitions for **list** and **listseg**

a) is known to hold some value, its name is written next to the nodes that represent this value.

A thin arrow represents a single memory cell holding a word (called *points-to predicate* in separation logic). For instance, the arrow between α and α_n represents a memory cell at address α + .next holding value α_n, where .next is the offset of the field next (i.e. 0). The arrow between α and α_n represents a memory cell at address α + .data containing value α_d.

Thick arrows represent compound memory regions (called *inductive predicates* in separation logic), i.e. a *summary* representing possible sets of memory cells with some structure. For instance, the **list** edge starting from β can be either the empty list (in which case we must have $\beta = 0x0$), a list of length 1 (in which case **list** encodes two memory cells for the .data and .next fields of the list_node of that list), or a list of any length.

The precise meaning of inductive predicates is given by their inductive definition. The definition for **list** and **listseg** and is given on the top of Fig. 11.5. The definition for **list** says that a list starting at address α is either an empty list (in which case,

α must be null), or is a `list_node` followed by a **list** (in which case, α must be non-null, and the value in the data field must be between 0 and 10). The definition of a list segment (i.e., a part of a list) is similar.

A crucial point in the meaning of separating shape graphs is that different arrows represent different (or *separated*) memory regions—this contrasts with the representation for retained points-to predicates that we saw in Fig. 11.3, where points-to relations may alias each other. On the other hand, different nodes can represent the same value, e.g., α_d and α_n may be equal.

Thus, the lists in the pre-state of Fig. 11.4 are separated, and would thus exclude the wrong executions (7), (8) and (9) on our examples executions (in which there is an overlap between different nodes of the list). It also excludes execution (2), (3), and (4), as the inductive predicate only describes acyclic linked lists. Finally, observe that the list pointed by a is non-empty, and thus the pre-state would also exclude the faulty execution (5). Thus, the pre-state Fig. 11.4 represent only the executions (1) and (6), corresponding to the concatenation of acyclic linked lists; for these, it correctly deduces that the final state is that at the bottom of Fig. 11.4 (a points to a non-empty list segment pointing to the list pointed by b).

Note that it is possible, using an analysis based on separation logic, to have inductive predicates corresponding to list with cycles, to analyze states corresponding to executions (3) or (4) for instance; or to encode the situation corresponding to execution (2). However, this requires performing separate analyses (or using a disjunction of abstract domains) for the different cases.

11.2.3.2 Relational Separation Logic

The static analysis presented above can compute an over-approximation of the possible shape of the heap at different program states: it is what we call a *state analysis*. Our RMA plug-in can go beyond state analyses, and can also perform a *transformation analysis* (sometimes also called a relational analysis), resulting in an over-approximation of the transformation done by the program. The difference is that, instead of inferring a set of post-states from a set of given pre-states, we infer a relation saying how post-states can be obtained from a pre-state. In our case, this relation specifies which part of the memory have not changed, and how the parts of the memory that are modified change.

Consider Fig. 11.6, which is the relation that our analysis computes on `list_append` if we tell it that initially a and b are two separated acyclic linked lists. This relation is composed of three independent transformations, saying that 1. the list segment leading to the last node, and the `data` field of this node, are not modified; 2. the `next` field of this node, which was null, is modified to point to β, the address of the list contained in b; 3. the list pointed by b is not modified. In the figure, Id stands for absence of modification while the dashed arrow represents a modification; the formal semantics of these connectors, with an extension presenting more relations between the pre and post-state, is given in Illous et al. [19].

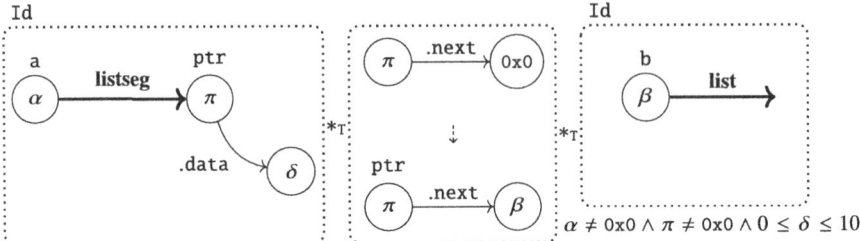

Fig. 11.6 Result of the transformational analysis of list_append

From this relation, it is easy to see that we can extract the pre-state and post-state given in Fig. 11.4—but in doing so we would lose some valuable information. For instance, the pre- and post-states both say that b points to a list starting at address β—but contrary to the relation of Fig. 11.6, they do not say that the list has not been modified: the same pair of pre- and post-state could represent transformations that reorder, add or remove elements of this list, or modify the data field. Similarly, the pair of pre and post-state do not say that the beginning of the list pointed by a is not modified.

Moreover, the transformation analysis is able to refine the possible pre-state to only those that will allow execution of the function. The relational summary of Fig. 11.6 can be obtained by the analysis giving it as a precondition that a and b are two separated, linked lists: if we do not provide the assumption that a is non-empty, the analysis will infer that this precondition is needed for the function to terminate. Thus our tool is able to refine a precondition, in addition to computing a postcondition.

One benefit of this relational information is that it can prove, as a special case, that if b is an empty linked list, then the function does not modify the heap, i.e., the property of the execution scenario (6) in Fig. 11.2.

Another benefit is that it allows compositional verification [18]: the relation of Fig. 11.6 can be used as a *function summary* to analyze some code calling the list_append function—without having to re-analyze list_append on each call site. This allows the analysis of large code bases to be dramatically more efficient in terms of computing resources.

11.2.3.3 Strengths and Weaknesses of the Separation-Logic-Based Analysis

The main strength of this analysis is the ability to compute strong invariants on the memory. The plug-in can distinguish lists that have cycles from those which have not, and verify that concatenating separated cycle-free linked lists results in creation of a cycle-free linked list (the analysis raises an alarm if a cyclic linked list may be created, e.g. if we concatenate **listseg**(α, β) to **listseg**(β, α)). The separation logic abstraction can verify advanced structural invariants, in addition to memory safety (both spatial and temporal).

Furthermore, the relational analysis computes valuable information about the transformations performed by the program (such as which part of the heap is modified), and can be used to perform efficient and compositional analysis.

The drawback of using this analysis is that the stronger invariant requires an analysis that can be heavier on resource usage, especially when there are lots of case splits to consider (but see Li et al. [24] for ways to alleviate this problem). Also, the analysis requires having more knowledge about the data structures manipulated by the function: e.g., we need to know whether the function may only be applied to separated acyclic linked lists, or if other usage scenarios are possible.

Finally, our RMA plug-in cannot currently handle complex, unstructured sharing patterns, nor low-level patterns such as unions or interior pointers (this is a limitation of our current plug-in, but note that some analyses based on separation logic can handle these features [23, 25]).

11.2.4 Discussion

The existence of two plug-ins for the same job (memory analysis, and proving the absence of memory-related undefined behaviors) can be disconcerting; but as we saw, each plug-in has its advantages and disadvantages, so the choice of which plug-in to use depends on the situation. Essentially, the analysis based on separation logic infers and uses stronger memory invariants than the type-based analysis—meaning that when the analysis based on separation logic succeeds, we know tighter constraints on the shape of the memory; but on the other hand, sometimes these invariants are too strong to describe the possible shapes of the memory (especially when there are a lot of weakly-structured aliases between the pointers), in which case the weaker but more general invariants of the type-based analysis work better.

For instance, it would be more difficult to use the separation-logic state analysis in the case where the `list_append` function may take both cyclic and acyclic linked lists as an argument, or if the arguments may not be separated, or if we simply do not know in what situations the function can be used. Thus, *if the sharing pattern is unclear, it is probably better to use the type-based analysis*.

On the other hand, the analysis based on separation logic allows inferring a much stronger invariant, such as the preservation of acyclicity of the linked lists. In particular, the type-based analysis currently cannot handle temporal memory safety (i.e., absence of use-after-free), which makes it a full solution for verifying memory safety only in the presence of a garbage collector, or on embedded systems where all the memory is statically allocated. Thus, *if strong properties of the data structure need to be verified, including temporal memory safety, it is better to use the analysis based on separation-logic*.

Finally, the RMA plug-in has the possibility to infer properties about the memory transformations done by a program, but also to infer necessary preconditions for a function. This inference of relational properties does not yet exist for the type-based static analysis. *Use the RMA plug-in if the verification of memory transformations properties is needed.*

The analyses are complementary: in a large program, there are parts where the sharing pattern is unstructured, not well known, or not very important, and only memory safety has to be verified, in which case the type-based static analysis will work well. There are other parts where the preservation of complex structural invariants is essential or where the goal is to verify relational properties, in which case the RMA plug-in will be better suited. In the future, we would like both kinds of analyses (type-based and separation-logic-based) to be blended in a single theoretical framework and Frama-C plug-in, which will make its usage simpler.

11.3 Abstract Interpretation Framework

11.3.1 Program States and Concrete Execution

The analyses that we perform reason about sets of possible program states or relations between these states. A *state* represents the value in every variable and allocated memory of the program. To simplify reasoning, we provide a simple low-level memory model corresponding more to the state of C programs after they have been compiled, but the presentation could be adapted to higher-level memory models of C. In particular, variables are not addressable (i.e., we cannot have pointers to local variables, as in C; this is fixed in the actual analysis by considering addressable variables like heap-allocated data).

In this model, all *values* $v \in \mathbb{V}$ are bitvectors; we denote by $v_1::v_2$ the concatenation of two bitvectors. We denote by $\mathbb{V}_n$ a bitvector of length n bits.

Addresses $a \in \mathbb{A}$ are a subset of all possible words, i.e. $\mathbb{A} \subseteq \mathbb{V}$. A *heap* $h \in \mathbb{H} \triangleq \mathbb{A} \to \mathbb{V}$ represents the contents of the program memory, and is a mapping from *addresses* to *values* of size 1 byte. For instance, on Fig. 11.2, $h_0, h_1, \ldots, h_4$ are all heaps.

To simplify notations, we allow accessing and storing values from ranges of addresses, i.e.

$$h[a..a+k] \triangleq h[a]::h[a+1]::\ldots::h[a+k]$$

Where :: represents the concatenation of bitvector values.

We have a set of *program variables* $x \in \mathbb{X}$. Variables are used by the program to refer to some values, and each state contains a *store* $\sigma \in \Sigma$ that maps program variables to the value that they store, i.e.

$$\Sigma \triangleq \mathbb{X} \to \mathbb{V}$$

Finally, a program state $s \in \mathbb{S}$ is a pair composed of a store and a heap, or is the error state named Ω:

$$\mathbb{S} \triangleq (\Sigma \times \mathbb{H}) \cup \{\Omega\}$$

11 Advanced Memory and Shape Analyses

Example state

On Fig. 11.2, $([a \mapsto \texttt{0x1000}, b \mapsto \texttt{0x1014}], h_0)$ is an example of a state.

We now turn to executions of the program. In this presentation, we omit the details of the program syntax and semantics, and simply assume that, given a C function f, the existence of a deterministic relation $\xrightarrow{f} \in S \times S$ representing the execution of f. We write $s_1 \xrightarrow{f} s_2$ if the program f, when given s_1 as its pre-state, ends the execution of f with the state s_2 as its post-state. Note that s_2 may be the error state. We write $s_1 \xrightarrow{f} \bot$ if the execution of f never finishes.

Example executions

On Fig. 11.2, executions (4) never finishes, executions (5) and (7) end in an error state, and all the other executions return a normal state.

11.3.2 Sound Static Analyses by Abstract Interpretation

Abstract interpretation [9, 36] is a general method for building static analyzers that automatically *infers program invariants* and verify program properties. Abstract interpreters are *sound*, in the sense that the invariants that they infer are correct by construction.

11.3.2.1 State Analysis

A *state analysis* infers *abstract states* that represent a superset of all the possible states that the program may reach. Abstract states are elements of a set $S^\sharp$, and the meaning of an abstract state is given using a *concretization function* $\gamma_{S^\sharp} \in S^\sharp \to \mathcal{P}(S)$.

Box abstract domain

An example abstract domain is the box abstract domain $\mathbb{B}^\sharp = \mathbb{X} \rightharpoonup \mathbb{I}$, that maps each program variable x to an interval $i \in \mathbb{I}$ containing all the values that x may take at runtime. An abstract state such as $b_1^\sharp = [\texttt{x} \mapsto [10; 20], \texttt{y} \mapsto [5; 25]]$ would be an element of this abstract domain. The box abstract domain constrains the states that it represents to have the value of their variable in the corresponding interval. This fact is captured by the concretization function of this domain, $\gamma_\mathbb{B}^\sharp : \mathbb{B}^\sharp \to \mathcal{P}(S)$, defined as:

$$\gamma_{\mathbb{B}}^{\sharp}(b^{\sharp}) = \{(\sigma, h) \in \mathbb{S} : \forall (x \mapsto [a; b]) \in b^{\sharp}, a \leq \sigma[x] \leq b\}$$

For instance, we have:

$$\gamma_{\mathbb{B}}^{\sharp}(b_1^{\sharp}) = \{(\sigma, h) \in \mathbb{S} : 10 \leq \sigma[x] \leq 20 \text{ and } 5 \leq \sigma[y] \leq 25\}$$

In addition to a concretization function, an abstract domain must provide *abstract semantics* that simulates, in the abstract, the transformations done by the program on the concrete states. More precisely, given a function f, its abstract semantics consists in a computable function:

$$[\![f]\!]^{\sharp} : \mathcal{P}(\mathbb{S}) \to \mathcal{P}(\mathbb{S})$$

such that the following property, called the soundness theorem, is fulfilled:

Property 11.1 (Soundness theorem for state analyses) *Given any abstract pre-state $s_i^{\sharp} \in \mathbb{S}^{\sharp}$. Let $s_o^{\sharp} \triangleq [\![f]\!]^{\sharp}(s_i^{\sharp})$ be the computed abstract post-state. We have:*

$$\{s_o \in \mathbb{S} : \exists s_i \in \gamma_{\mathbb{S}^{\sharp}}(s_i^{\sharp}) : s_i \xRightarrow{f} s_o\} \subseteq \gamma_{\mathbb{S}^{\sharp}}(s_o^{\sharp})$$

i.e. $s_o^{\sharp}$ is an over-approximation of all the post-states resulting from the execution of f for pre-states corresponding to $s_i^{\sharp}$, i.e. $[\![f]\!]^{\sharp}$ is a sound over-approximation of the behavior of f.

A sound state analysis allows in particular to prove the absence of runtime errors in the function: given an abstract state $s_i^{\sharp}$ describing all the states meeting the function precondition, it suffices to compute $s_o^{\sharp} = [\![f]\!]^{\sharp}(s_i^{\sharp})$ and check that the error state Ω does not belong to $\gamma_{\mathbb{S}^{\sharp}}(s_o^{\sharp})$.

11.3.2.2 Transformation Analysis

A *transformation analysis* infers relations between the program pre-state and post-state. In mathematics, a relation $\mathcal{R}$ is a set of pairs, the set of all pairs of elements that are related one with the other.

The result of a transformation analysis is thus an *abstract transformation relation* $t^{\sharp} \in \mathbb{T}^{\sharp}$, that represents a relation:

$$\gamma_{\mathbb{T}^{\sharp}} : \mathbb{T}^{\sharp} \to \mathcal{P}(\mathbb{S} \times \mathbb{S})$$

The modified variable abstract domain

We consider the example of a simple abstract domain representing an abstract transformation, the modified variable abstract domain $MV^\sharp = \mathcal{P}(\mathbb{X})$. An element of this domain is simply a set of variables, representing a superset of the variables that may have been modified between the pre-state and post-state, i.e

$$\gamma_{MV^\sharp} : MV^\sharp \to \mathcal{P}(\mathbb{S} \times \mathbb{S})$$
$$\gamma_{MV^\sharp}(mv) = \{((\sigma_i, h_i), (\sigma_o, h_o)) : \forall x \in \mathbb{X} : x \notin mv \implies \sigma_i[x] = \sigma_o[x]\}$$

For instance, the set $\{x, y\}$ represents a relation where except for x and y, the value of the other variables has not changed.

Note that this domain represents information that can not always be represented by state properties of the pre and post states. For instance, if z is mapped to the interval $[3; 3]$ in both the pre and post-state, we already know that its value has not changed; however if z is mapped to the interval $[0; 3]$ in both states, we cannot conclude that its values has not changed.

The abstract semantics for transformation is a computable function with signature:

$$[\![f]\!]_\mathcal{T}^\sharp : \mathbb{S}^\sharp \to \mathbb{T}^\sharp$$

The idea is that this function takes as input an abstract state (describing the function precondition), and outputs an abstract transformation, that over-approximates the relation between pre-states that match this precondition, and the "normal" post-states (excluding executions that do not terminate or end with an error).

This is expressed formally by proving, for a given transformation analysis, the following soundness theorem:

Property 11.2 (Soundness theorem for transformation analyses) *Given any abstract pre-state $s_i^\sharp \in \mathbb{S}^\sharp$. Let $t^\sharp \triangleq [\![f]\!]_\mathcal{T}^\sharp(s_i^\sharp)$ be the computed abstract transformation. We have:*

$$\{(s_i, s_o) \in \mathbb{S} \times \mathbb{S} : s_i \in \gamma_{\mathbb{S}^\sharp}(s_i^\sharp) \land s_i \xrightarrow{f} s_o \land s_o \neq \Omega\} \subseteq \gamma_{\mathbb{T}^\sharp}(t^\sharp)$$

i.e. $t^\sharp$ is an over-approximation of all the (pre-state, post-state) pairs of f such that the pre-state is described by $s_i^\sharp$ (and the post-state is not an error).

Note that the relations that we compute are given for a specific set of pre-states, given using a precondition. Computing relations true for every possible pre-states is possible but less precise: usually a function is called only in a specific context, and our transformation can take advantage of this.

Transformation analysis using the modified variable abstract domain

Let f be the following C function:

```
1    int u,v,x,y;
2    void f(int i){
3      switch(i){
4        case 0: u = 3; while(1);
5        case 1: v = 4; v = 1/0; return;
6        case 2: x = 5; return;
7        case 3: y = 6; return;
8      }
9    }
```

Sound results for a transformation analysis of f with an initial abstract state given as $[i \mapsto [0;2]]$ (using the box abstract domain), are $[\![f]\!]_\mathcal{T}^\sharp = \{x, v\}$, $[\![f]\!]_\mathcal{T}^\sharp = \{x, y\}$, or $[\![f]\!]_\mathcal{T}^\sharp = \{x\}$. The last one is the most precise result : y may not be modified because of the precondition that $i \in [0;2]$, and the modifications of the variables u and v respectively lead to an infinite loop or the error state, and thus do not need to be included.

Note that in practice, our plug-in computes the transformation between the prestate and every statement in the function, so it is also possible to retrieve the transformation for non-terminating or faulty executions if needed.

11.3.3 Base Abstractions

All of our memory analyses use some base abstract domains for representing numerical constraints and the contents of the store that we explain in this section. Representing constraints directly on program variables, as we did for the box abstract domain presented above, does not easily allow placing numerical constraints on the contents of the heap. To solve this problem, we instead introduce an intermediate notion of a *symbolic value*. A symbolic value $\alpha \in \mathcal{U}$ is a variable, noted using a greek letter, representing, at analysis time, some run-time values. Note that these symbolic values do not represent a memory location like a program variable, as done in variable-based analyses such as Eva (detailed in Chap. 3) instead it represents an (immutable) value that memory locations may contain (e.g. the same variable can appear twice in memory after a copy).

The numerical domain $\mathbb{N}^\sharp$ represents numerical constraints between symbolic values. Its concretization function $\gamma_{\mathbb{N}^\sharp}:\mathbb{N}^\sharp \to (\mathcal{U} \to \mathbb{V})$ concretizes into a set of *valuations* $v \in \mathcal{U} \to \mathbb{V}$, i.e. mapping of symbolic values to concrete values. In both of our plug-ins, the structure of the analyzers are completely independent of the numerical domain being used, so we represent the numerical domain simply as a conjunction of numerical equations.

The purpose of introducing these symbolic values (rather than attaching directly numerical information to program variables) is that it allows to easily attach numerical constraints not only to the program variables, but also for memory cells in the heap (as we did with δ in Figs. 11.3, 11.4 or Fig. 11.6).

To attach a numerical value to program variables, we will simply make use of the *abstract store domain* $\Sigma^\sharp = \mathbb{X} \to \mathbb{Q}$, that maps each program variable with a symbolic value, whose concretization is given below:

$$\gamma_{\Sigma^\sharp} : \Sigma^\sharp \to \mathcal{P}(\Sigma \times (\mathbb{Q} \to \mathbb{V}))$$
$$\gamma_{\Sigma^\sharp}(\sigma^\sharp) \triangleq \{(\sigma, \nu) \mid \forall x \in \mathbb{X} : \sigma[x] = \nu[\sigma^\sharp[x]]\}$$

Interval analysis with symbolic values

We reproduce the example given by the box abstract domain presented earlier, but this time introducing symbolic values. The domain consists in the numerical domain $\mathbb{N}^\sharp$ (used to hold constraints between symbolic value), used in conjunction [10] with an abstract store $\sigma^\sharp \in \Sigma^\sharp$.

An example of abstract state is $s^\sharp = (n^\sharp, \sigma^\sharp)$ where

$$n^\sharp = (10 \le \alpha \le 20 \land 5 \le \beta \le 25)$$
$$\sigma^\sharp = [x \mapsto \alpha, y \mapsto \beta]$$

The concretization for this symbolic box domain $\mathbb{SB}^\sharp = \mathbb{N}^\sharp \times \Sigma^\sharp$ is given as:

$$\gamma_{\mathbb{SB}^\sharp} : \mathbb{SB}^\sharp \to \mathcal{P}(\mathbb{S})$$
$$\gamma_{\mathbb{SB}^\sharp}(n^\sharp, \sigma^\sharp) \triangleq \{(\sigma, h) \in \Sigma \times \mathbb{H} \mid \exists \nu \in \gamma_{\mathbb{N}^\sharp}(n^\sharp) : (\sigma, \nu) \in \gamma_{\Sigma^\sharp}(\sigma^\sharp)\}$$

For instance, we have:

$$\gamma_{\mathbb{SB}}^\sharp(s^\sharp) = \{(\sigma, h) \in \mathbb{S} : 10 \le \sigma[x] \le 20 \text{ and } 5 \le \sigma[y] \le 25\}$$

11.4 Type-Based Static Analysis with the Codex Plug-in

In this section, we formalize what the type-based analysis for the Codex plug-in computes. For the sake of simplicity, we have removed the aspects of the analysis that are not needed to understand the example of Fig. 11.3, which we detail at the end of the section. We also do not provide the abstract semantics of our analysis, i.e., how the result is computed; details can be found in Nicole et al. [34].

The analysis is built on a core notion of well-typed states that we expose below. Our type abstraction then builds upon this notion (Sect. 11.4.2). Finally, we expose extensions needed when the type abstraction alone is insufficiently precise or when the program is not always well-typed (Sect. 11.4.3).

11.4.1 Well-Typed States

The analysis is parameterized by a set of *type definitions* $\mathcal{D} \in \mathcal{N} \to \mathbb{T}$, associating *type names* to a *type*. The grammar of types is given in Fig. 11.7, while an example of type definitions is given in Fig. 11.3. This example consists in three definitions:

1. The definition of type int, which is a bitvector of size 4 bytes (i.e., a word);
2. The definition of type node_data, which is a numerically constrained int;
3. The definition of type list_node, which is a record with a pointer and a node_data.

Note that pointers do not point to types, but to *types with offset*, which are pairs made of a type and a numerical offset within that type (i.e., the offset k in a type with offset $t.(k)$ should be such that $0 \leq k < \text{size}(t)$). The function size : $\mathbb{T} \to \mathbb{N}$, returns the size of the type, and is defined inductively by:

$$
\begin{array}{rll}
expr ::= & \texttt{self} & \text{(constrained type variable)} \\
 | & c & \text{(constant } c \in \mathbb{V}\text{)} \\
 | & expr \diamond expr & \text{(binary op., } \diamond \in \{+, -, \times, \cdots\}\text{)} \\
pred ::= & expr \bowtie expr & \text{(comparison, } \bowtie \in \{\leq, <, =, \neq\}\text{)} \\
 | & \neg pred \ | \ pred \wedge pred & \\
\mathbb{T} \ni t ::= & \texttt{word}_n & \text{(base type of size } n \text{ bytes)} \\
 | & \texttt{n} & \text{(named type with type name } \texttt{n} \in \mathcal{N}\text{)} \\
 | & t_o* & \text{(possibly null pointer)} \\
 | & \texttt{struct}\{t;t\} & \text{(record type)} \\
 | & t \ \texttt{with} \ pred & \text{(type with a refinement predicate)} \\
\mathbb{T}_O \ni t_o ::= & t.(k) & \text{(type with offset } k \in \mathbb{N}\text{)}
\end{array}
$$

Fig. 11.7 Definition of types and types with offset

11 Advanced Memory and Shape Analyses

$$\text{size}(\text{word}_n) = n \qquad \text{size}(t_o*) = 4 \qquad \text{size}(\text{n}) = \text{size}(\mathcal{D}(\text{n}))$$

$$\text{size}(\text{struct}\{t_1; t_2\}) = \text{size}(t_1) + \text{size}(t_2) \qquad \text{size}(t \text{ with } p) = \text{size}(t)$$

We exclude types for which *size* is not defined (e.g., **struct** that include each other).

Allocation maps are mappings $\mathcal{A} \in \mathbb{A} \to \mathbb{T}_O$ from addresses to types with offset, with the additional condition that the mapping is whole and contiguous: i.e. if $\mathcal{A}[a + k] = t.(k)$ for $0 \leq k < \text{size}(t)$, then $\mathcal{A}[a] = t.(0), \mathcal{A}[a + 1] = t.(1), \ldots,$ and $\mathcal{A}[a + \text{size}(t) - 1] = t.(\text{size}(t) - 1)$. To simplify the notations, in that case we will simply write that $\mathcal{A}[a..a + \text{size}(t) - 1] = t$.

For instance, the allocation map needed to describe Fig. 11.2 is defined as:

$\mathcal{A}[\texttt{0x1000..0x1007}] = \texttt{list\_node} \qquad \mathcal{A}[\texttt{0x1008..0x100f}] = \texttt{list\_node}$
$\mathcal{A}[\texttt{0x1010..0x1013}] = \texttt{word}_4$
$\mathcal{A}[\texttt{0x1014..0x101b}] = \texttt{list\_node} \qquad \mathcal{A}[\texttt{0x101c..0x1023}] = \texttt{list\_node}$
$\mathcal{A}[\texttt{0x1024..0x1027}] = \texttt{word}_4$
$\mathcal{A}[\texttt{0x1028..0x102f}] = \texttt{list\_node}$

The idea behind $\mathcal{A}[\texttt{0x1000..0x1007}] = \texttt{list\_node}$ is to say that these addresses should hold the content of a `list_node`. But note that the last four bytes of a `list_node` is always a `node_data`, so we would like to also say that $\mathcal{A}[\texttt{0x1004..0x1007}]$ corresponds to a `node_data`, and similarly that $\mathcal{A}[\texttt{0x1000..0x1003}]$ corresponds to `list_node.(0)*`. This is achieved using a *containment* relation $\preceq$ between types with offset, where $t.(k) \preceq u.(l)$ intuitively says that addresses tagged as $t.(k)$ in the allocation map can also be viewed as being $u.(l)$.

Definition 11.1 (*Containment relation between types with offset*) The relation $\preceq \in \mathbb{T}_O \times \mathbb{T}_O$ is defined inductively according to the rules below:

$$\frac{}{t.(k) \preceq t.(k)} \qquad \frac{t.(k) \preceq u.(l) \quad u.(l) \preceq v.(m)}{t.(k) \preceq v.(m)} \qquad \frac{t = \mathcal{D}(\text{n}) \quad 0 \leq k < \text{size}(t)}{\text{n}.(k) \preceq t.(k)}$$

$$\frac{0 \leq k < \text{size}(t_1)}{(\text{struct}\{t_1; t_2\}).(k) \preceq t_1.(k)} \qquad \frac{0 \leq k < \text{size}(t_2)}{(\text{struct}\{t_1; t_2\}).(\text{size}(t_1) + k) \preceq t_2.(k)}$$

What these rules say is that containment is reflexive and transitive; that a type name contains the type of its corresponding definition; that a record type contains its first and second fields at the matching offsets.

Intuitively, types have another role than tagging memory: it is also used to denote a set of values. For instance, values of type `node_data` are integers between 0 and 10, while values of type `list_node.(0)*` are addresses containing a `list_node`. This is formalized by the notion of an *interpretation of a type*.

Definition 11.2 (*Interpretation of types*) Given labeling $\mathcal{A}$, the *interpretation* function $(\!|\cdot|\!)_{\mathcal{A}} : \mathbb{T} \to \mathcal{P}(\mathbb{V})$ is defined by:

$$(\!|\texttt{word}_n|\!)_{\mathcal{A}} = \mathbb{V}_n$$
$$(\!|\texttt{n}|\!)_{\mathcal{A}} = (\!|\mathcal{D}(\texttt{n})|\!)_{\mathcal{A}}$$
$$(\!|t_o*|\!)_{\mathcal{A}} = \{a \in \mathbb{A} \mid \mathcal{A}(a) \preceq t_o\} \cup \{0\}$$
$$(\!|\texttt{struct}\{t_1; t_2\}|\!)_{\mathcal{A}} = \{v_1 :: v_2 \mid v_1 \in (\!|t_1|\!)_{\mathcal{A}} \text{ and } v_2 \in (\!|t_2|\!)_{\mathcal{A}}\}$$
$$(\!|t \texttt{ with } p|\!)_{\mathcal{A}} = \{v \in (\!|t|\!)_{\mathcal{A}} \mid p(v) = \textbf{true}\}$$

What this definition says is that a $\texttt{word}_n$ can hold any bitvector of size n; the values of a type name n are those of its definition; the values for a pointer to a t_o is all the addresses that can be viewed as a t_o in the allocation map, or 0; the values for a structure is the concatenation of the values of its two fields; and the values for a refinement type are the values of the initial types for which the predicate holds.

We now have everything we need to define what is a well-typed heap and a well-type state:

Definition 11.3 (*Well-typed heap*) A well-typed heap is a pair $(h, \mathcal{A})$ of a heap $h \in \mathbb{H}$ and an allocation map $\mathcal{A} \in \mathbb{A} \to \mathbb{T}_O$ such that for all address $a \in \mathbb{A}$ and type $t \in \mathbb{T}$

$$\mathcal{A}[a..a + \text{size}(t) - 1] = t \quad \Rightarrow \quad h[a..a + \text{size}(t) - 1] \in (\!|t|\!)_{\mathcal{A}}$$

This definition means that the contents of the heap should match the type assigned to the address by the allocation map.

Definition 11.4 (*Well-typed state*) A well-typed state is a 4-tuple $(\sigma, \Gamma, h, \mathcal{A})$ where $\sigma \in \Sigma$ is a store, $h \in \mathbb{H}$ is a heap, $\mathcal{A}$ is an allocation map, and $\Gamma : \mathbb{X} \to \mathbb{T}$ is an environment mapping variables to their type, such that:

1. $(h, \mathcal{A})$ is a well-typed heap;
2. The value of each variable in the store matches its type in the environment:

$$\forall x \in \mathbb{X} : \sigma[x] \in (\!|\Gamma[x]|\!)_{\mathcal{A}}.$$

11.4.2 The Type Abstraction

We can now describe the memory abstraction of our type-based analysis, namely, the $\Gamma^\sharp$ abstract domain. Each element of this domain is simply a Γ mapping:

$$\Gamma^\sharp \triangleq \mathbb{X} \to \mathbb{T}$$

The meaning of this domain is given by the following concretization:

$$\gamma_{\Gamma^\sharp}:\Gamma^\sharp \to \mathcal{P}(\mathbb{S})$$
$$\gamma_{\Gamma^\sharp}(\Gamma) = \{(\sigma,h) \in \mathbb{S} \mid \exists \mathcal{A} : (\sigma,\Gamma,h,\mathcal{A}) \text{ is a well-typed state}\}$$

Note that neither the heap h nor the allocation map $\mathcal{A}$ are explicitly represented in the domain: we just assume that they exist and are a well-typed heap. The fact that the invariants on the heap is represented only globally by the type definition, and thus does not change between program points, is a key to the high efficiency of this domain.

The complete analysis is obtained by using this memory domain in conjunction [10] with the base domains (abstract store $\Sigma^\sharp$ and numerical domain $\mathbb{N}^\sharp$) provided in Sect. 11.3.3.

$$\mathbb{TB}^\sharp = \mathbb{N}^\sharp \times \Sigma^\sharp \times \Gamma^\sharp$$
$$\gamma_{\mathbb{TB}^\sharp} : \mathbb{TB}^\sharp \to \mathcal{P}(\mathbb{S})$$
$$\gamma_{\mathbb{TB}^\sharp}(n^\sharp,\sigma^\sharp,\Gamma) = \{(h,\sigma) \mid \exists v \in \gamma_{\mathbb{N}^\sharp}(n^\sharp) : \quad (\sigma,v) \in \gamma_{\Sigma^\sharp}(\sigma^\sharp)$$
$$\wedge (h,\sigma) \in \gamma_{\Gamma^\sharp}(\Gamma)\}$$

Example illustrating an element of the type-based abstract domain $\mathbb{TB}^\sharp$

The abstract state used as a precondition in Fig. 11.3 gives an example of an element of the $\mathbb{TB}^\sharp$ domain. This abstract state contains a $\Gamma = [\mathtt{a} \mapsto \mathtt{list\_node}.(0)*; \mathtt{b} \mapsto \mathtt{list\_node}.(0)*]$, a $\sigma^\sharp = [\mathtt{a} \mapsto \alpha; \mathtt{b} \mapsto \beta]$, and a numerical domain $n^\sharp$ containing the constraint $\alpha \neq 0$. This abstract state describes, through the concretization $\gamma_{\mathbb{TB}^\sharp}$, a set of well-typed states $(\sigma,\Gamma,h,\mathcal{A})$ for which a contains a value in $(\!|\mathtt{list\_node}.(0)*|\!)_\mathcal{A} \setminus \{0\}$, b contains a value in $(\!|\mathtt{list\_node}.(0)*|\!)_\mathcal{A}$; then forgets about the Γ and $\mathcal{A}$ in these states (i.e. it performs some kind of type erasure).

To simplify this presentation, we omitted some existing extensions to the type domain that we briefly list here. A first extension is the use of *abstract types* as values in the $\Gamma^\sharp$ abstract domain, instead of concrete types. The main difference is that in abstract types, the offset in a pointer type can be a symbolic value. This allows a precision gain, as it gives an accurate representation for the cases where a pointer may point to several locations in a type.

A second extension is the presence of *array types*. This extension justifies the presence of the abstract types, as it is very common, when manipulating pointer-to-array types, that the pointer may point to different offsets in the array (corresponding to the different array cells).

A last extension is the possibility to parameterize the analysis by symbolic values. For instance, the length of arrays, or the predicates in refinement types, can refer to a shared symbolic value. This extension is useful, for instance, to verify code operating over arrays of unknown length.

11.4.3 Points-to Predicates

Here we present independent domains that extend our type-based analysis to improve precision in two important cases: first, when the invariant brought by the fact that heaps are well-typed is insufficient; second, when the program performs memory stores that temporarily produce ill-typed states.

For instance, when our plug-in must analyze line 12 on Fig. 11.1 on the second iteration, the state before line 12 will be $(n^\sharp, \sigma^\sharp, \Gamma)$ with

$n^\sharp = \alpha \neq 0$

$\sigma^\sharp = [\mathtt{a} \mapsto \alpha; \mathtt{b} \mapsto \beta; \mathtt{ptr} \mapsto \pi]$

$\Gamma = [\mathtt{a} \mapsto \mathtt{list\_node}.(0)*; \mathtt{b} \mapsto \mathtt{list\_node}.(0)*; \mathtt{ptr} \mapsto \mathtt{list\_node}.(0)*]$

This state does not exclude the fact that `ptr` will be non-null before dereferencing it, which will be caught by our system as possibly-null pointer dereferences cannot be typed. The problem is that the test at line 11 is not recorded, as it is a fact about the contents of memory, and our analysis is entirely storeless (i.e., stores nothing about the heap except the fact that it is well-typed).

The solution is to strengthen our invariant with properties about heap cells using another abstract domain. We call *retained predicates* these properties, as these are facts that are remembered about the heap. Formally, the retained points-to predicates abstract domain $\mathcal{P}^\sharp = \mathbb{Q} \rightharpoonup \mathbb{Q}$ consists in a partial function, i.e., a list of bindings of the form $\alpha \mapsto \beta$, where $\alpha \mapsto \beta$ means that in the heap, the word at address α contains the value β:

$$\gamma_{\mathbb{P}^\sharp} : \mathbb{P}^\sharp \to \mathcal{P}(\mathbb{H} \times (\mathbb{Q} \to \mathbb{V}))$$

$$\gamma_{\mathbb{P}^\sharp}(p^\sharp) = \{(h, v) \mid \forall (\alpha \mapsto \beta) \in p^\sharp : h[v(\alpha)..v(\alpha)+3] = v(\beta)\}$$

Furthermore, the type-based analysis with points-to predicates, $\mathbb{TBP}^\sharp$, becomes:

$$\mathbb{TBP}^\sharp = \mathbb{N}^\sharp \times \Sigma^\sharp \times \Gamma^\sharp \times \mathbb{P}^\sharp$$

$$\gamma_{\mathbb{TBP}^\sharp} : \mathbb{TBP}^\sharp \to \mathcal{P}(\mathbb{S})$$

$$\gamma_{\mathbb{TBP}^\sharp}(n^\sharp, \sigma^\sharp, \Gamma, p^\sharp) = \{(h, \sigma) \mid \exists v \in \gamma_{\mathbb{N}^\sharp}(n^\sharp) : \begin{array}{l} (\sigma, v) \in \gamma_{\Sigma^\sharp}(\sigma^\sharp) \\ \wedge (h, \sigma) \in \gamma_{\Gamma^\sharp}(\Gamma) \\ \wedge (h, v) \in \gamma_{\mathbb{P}^\sharp}(p^\sharp) \} \end{array}$$

Thanks to these retained points-to predicates, the abstract state at line 11 above can now record the result of the test at line 11 by adding the constraints $\pi \mapsto \delta \wedge \delta \neq 0$ to the abstract domain, i.e., adding a points-to predicate allowing to attach the symbolic value δ to the heap cell represented by `ptr->next`, which can be constrained numerically. Thanks to this, `ptr` will be recorded as null after line 12, and the fact that `ptr` is null becomes a loop invariant, making the error disappear.

Similarly to the retained points-to predicate, another kind of points-to predicate exists, the staged points-to predicates, allowing to delay some stores when they may produce ill-typed states; this happens in particular in calls to malloc, or when performing byte-per-byte copies of pointers or values with a refinement type.

11.5 Separation-Logic-Based Analysis with the RMA Plug-In

In this section, we present the shape analyses based on separation logic performed by our RMA plug-in: we begin by presenting our state analysis, before describing the transformation analysis that builds on the former.

11.5.1 Abstract Heaps and State Analysis

The heart of the state analysis is the use of separation logic formulas, used to describe properties of heaps $h \in \mathbb{H}$, in a strict format that is given below.

Definition 11.5 (*Syntax of abstract heaps*) Abstract heaps are separation logic formulas in the following syntax:

$$
\begin{aligned}
h^\sharp \; (\in \mathbb{H}^\sharp) ::= \; & \mathbf{emp} & & \text{empty heap} \\
| \; & \alpha.\mathtt{f} \mapsto \beta & (\alpha, \beta \in \mathbb{Q};\, .\mathtt{f} \in \mathbb{F}) & \text{points-to predicate} \\
| \; & h_1^\sharp *_S h_2^\sharp & (h_1^\sharp, h_2^\sharp \in \mathbb{H}^\sharp) & \text{separated conjunction} \\
| \; & \mathbf{ind} & (\mathbf{ind} \in \mathbb{I}^\sharp) & \text{inductive predicate} \\
\mathbf{ind} \; (\in \mathbb{I}^\sharp) ::= \; & \mathbf{list}(\alpha) & (\alpha \in \mathbb{Q}) & \text{linked list} \\
| \; & \mathbf{listseg}(\alpha, \beta) & (\alpha, \beta \in \mathbb{Q}) & \text{list segment}
\end{aligned}
$$

Abstract heaps are described using relations between symbolic values $\alpha \in \mathbb{Q}$. The basic relations are **emp**, representing an empty heap, and *points-to predicates* $\alpha.\mathtt{f} \mapsto \beta$, which represents a memory cell at address α (incremented with the offset of the field .f), and containing value β. These relations are composed using $*_S$, the *separated conjunction* operator of separation logic, which allows describing a heap by disjoint conjoining of the description of smaller heaps. Finally, *inductive predicates* **ind** $\in \mathbb{I}^\sharp$ allows representing memory regions of variable length.

The description of abstract heaps is actually parameterized by the set of inductive predicates $\mathbb{I}^\sharp$, which is provided by the user (any inductive predicate following the format of the definition operator Δ, explained later, can be used). In this presentation, we assume that inductive predicates describe either lists (**listseg**) or list segments (**listseg**), but it is easy to change the analysis to add more predicates (using the __add_inductive annotation).

Abstract heap

The abstract heap describing the heaps in the pre-state of list_append and represented graphically in Fig. 11.4 is given using the following formula:

$$h_5^\sharp \triangleq \alpha.\text{data} \mapsto \alpha_d *_S \alpha.\text{next} \mapsto \alpha_n *_S \mathbf{list}(\alpha_n) *_S \mathbf{list}(\beta)$$

The inductive predicates represent a *memory summary*, i.e., a finite description of an arbitrary number of memory cells. The meaning of these inductive predicates is defined coinductively using a definition operator $\Delta : \mathbb{I}^\sharp \to \mathcal{P}(\mathbb{H}^\sharp \times \mathbb{N}^\sharp)$ that unfolds the definition of a predicate into different cases, each case being described by a separation logic formula and a regular numerical formula. For the two inductive predicates **list** and **listseg**, their definition (which is graphically represented in Fig. 11.5) is the following:

Inductive predicates

$$\Delta : \mathbb{I}^\sharp \to \mathcal{P}(\mathbb{H}^\sharp \times \mathbb{N}^\sharp)$$

$$\Delta(\mathbf{list}(\alpha)) \triangleq \{(\mathbf{emp}, \alpha = 0), \qquad (\alpha_n \text{ and } \alpha_d \text{ fresh})$$
$$(\alpha.\text{data} \mapsto \alpha_d *_S \alpha.\text{next} \mapsto \alpha_n *_S \mathbf{list}(\alpha_d),$$
$$\alpha \neq 0 \wedge 0 \leq \alpha_d \leq 10)\}$$

$$\Delta(\mathbf{listseg}(\alpha, \beta)) \triangleq \{(\mathbf{emp}, \alpha = \beta), \qquad (\alpha_n \text{ and } \alpha_d \text{ fresh})$$
$$(\alpha.\text{data} \mapsto \alpha_d *_S \alpha.\text{next} \mapsto \alpha_n *_S \mathbf{listseg}(\eta, \beta),$$
$$\alpha \neq \beta \wedge 0 \leq \alpha_d \leq 10)\}$$

In the definition of the **list**(α) predicate, the first case corresponds to the empty list, i.e., it represents no memory cell and the pointer α must be null. The second case describes a list with one element which points to another separated list (and α cannot be the null pointer).

The definition of **listseg**(α, β) is similar, the difference being that the end of a list segment may be an arbitrary value β, instead of being only zero.

Now that we have fully described the syntax of abstract heaps, we can define their meaning using a concretization function $\gamma_{\mathbb{H}^\sharp}$. Here, we will define concrete heaps $\mathbb{H} \triangleq \mathbb{A} \rightharpoonup \mathbb{V}$ as partial functions from addresses to values, i.e., as a set of bindings $a \mapsto v$. Given a heap h, we represent its domain (i.e., the set of addresses that h contains) as $\mathbf{dom}(h) \triangleq \{a : (a \mapsto v) \in h\}$.

11 Advanced Memory and Shape Analyses

We use [] to represent the empty heap, and $[a \mapsto_s v]$ to represent a "unitary heap" mapping a word v of size s to address a, i.e., such that the first byte of v is mapped to address a, the second byte of v to address $a + 1$, etc.

We use $\uplus$ to represent the disjoint union:

$$h_1 \uplus h_2 = \{(a \mapsto v) : (a \mapsto v) \in h_1 \text{ or } (a \mapsto v) \in h_2\}$$

which is defined only when the two heaps describe disjoints set of addresses, i.e., $\mathbf{dom}(h_1) \cap \mathbf{dom}(h_2) = \emptyset$.

Finally, we assume that the size and offsets of different fields are described using functions offsetof $\in \mathbb{F} \to \mathbb{N}$ and size $\in \mathbb{F} \to \mathbb{N}$.

Definition 11.6 (*Concretization of abstract heaps*) The concretization function $\gamma_{\mathbb{H}^\sharp} \in \mathbb{H}^\sharp \to \mathcal{P}(\mathbb{H} \times (\mathbb{Q} \to \mathbb{V}))$ maps an abstract heap into a set of pairs made of a concrete heap and a valuation. It is defined by induction on the structure of abstract heaps as follows:

$$\gamma_{\mathbb{H}^\sharp}(\mathbf{emp}) = \{([], \nu) \mid \nu \in \mathbb{Q} \to \mathbb{V}\}$$
$$\gamma_{\mathbb{H}^\sharp}(\alpha.\mathtt{f} \mapsto \beta) = \{([\nu(\alpha) + \text{offsetof}(.\mathtt{f}) \mapsto \nu(\beta)], \nu) \mid \nu \in \mathbb{Q} \to \mathbb{V}\}$$
$$\gamma_{\mathbb{H}^\sharp}(h_1^\sharp *_S h_2^\sharp) = \{(h_1 \uplus h_2, \nu) \mid \quad (h_1, \nu) \in \gamma_{\mathbb{H}^\sharp}(h_1^\sharp) \wedge (h_2, \nu) \in \gamma_{\mathbb{H}^\sharp}(h_2^\sharp)$$
$$\wedge \mathbf{dom}(h_1) \cap \mathbf{dom}(h_2) = \emptyset \}$$
$$\gamma_{\mathbb{H}^\sharp}(\mathbf{ind}) = \bigcup_{(h^\sharp, n^\sharp) \in \Delta(\mathbf{ind})} \{(h, \nu) \in \gamma_{\mathbb{H}^\sharp}(h^\sharp) \mid \nu \in \gamma_{\mathbb{N}^\sharp}(n^\sharp)\}$$

Basically, **emp** represents the empty heap, a points-to predicate a unitary heap, separating conjunction represents the disjoint union between two heaps, and inductive predicates represents the union of all the cases to which it can unfold. The valuation is used to add numeric constraints on the values of the symbolic values in the heap, for instance, when unfolding the definition of inductive predicates.

The complete state analysis is obtained by using this memory domain in conjunction [10] with the base domains (abstract store $\Sigma^\sharp$ and numerical domain $\mathbb{N}^\sharp$) provided in Sect. 11.3.3.

$$\mathbb{SL}^\sharp = \mathbb{N}^\sharp \times \Sigma^\sharp \times \mathbb{H}^\sharp$$
$$\gamma_{\mathbb{SL}^\sharp} : \mathbb{SL}^\sharp \to \mathcal{P}(\mathbb{S})$$
$$\gamma_{\mathbb{SL}^\sharp}(n^\sharp, \sigma^\sharp, h^\sharp) = \{(\sigma, h) \mid \exists \nu \in \gamma_{\mathbb{N}^\sharp}(n^\sharp) : \quad (\sigma, \nu) \in \gamma_{\Sigma^\sharp}(\sigma^\sharp)$$
$$\wedge (h, \nu) \in \gamma_{\mathbb{H}^\sharp}(h^\sharp) \}$$

Abstract state in a separation-logic-based analysis

The abstract state describing the admissible pre-states in Fig. 11.4 is defined as:

$$sl_5^\sharp \triangleq (\alpha \neq 0), [\mathsf{a} \mapsto \alpha, \mathsf{b} \mapsto \beta], h_5^\sharp$$

where $h_5^\sharp$ is the separation logic formula given above. It represents all the states where a and b point to two separated acyclic linked lists, and the list pointed by a is not empty.

11.5.2 Abstract Heap Transformations and Transformation Analysis

Our transformation analysis relies on abstract heap transformations, that represent sets of pairs of heaps, where an abstract heap represented a set of heaps. Abstract heap transformations are defined on top of our definition of abstract heaps, as a new kind of formulas:

Definition 11.7 (*Syntax of abstract heap transformations*)

$$
\begin{aligned}
\mathsf{r}^\sharp \, (\in \mathbb{R}^\sharp) ::= \; & \mathtt{Id}(h^\sharp) & (h^\sharp \in \mathbb{H}^\sharp) & \quad \text{(identity)} \\
\mid \; & [h_i^\sharp \dashrightarrow h_o^\sharp] & (h_i^\sharp, h_o^\sharp \in \mathbb{H}^\sharp) & \quad \text{(transform-into relation)} \\
\mid \; & \mathsf{r}_1^\sharp *_\mathrm{T} \mathsf{r}_2^\sharp & (\mathsf{r}_1^\sharp, \mathsf{r}_2^\sharp \in \mathbb{R}^\sharp) & \quad \text{(separate transformations)}
\end{aligned}
$$

Abstract heap transformations describe either an unmodified memory region, a transformation from one region to another, or independent transformations. The meaning of abstract heap transformations is provided using a concretization function, which produces a set of triples representing an input heap, an output heap, and a valuation: each pair of input/output heap represents a possible transformation from the input to the output, and the valuation is used to further constrain the possible transformations (when abstract heap transformations are used in conjunction with numerical constraints on the symbolic values). Abstract heap transformations are also called abstract heap relations, as they produce a relation between the input and output heaps of a function.

11 Advanced Memory and Shape Analyses 515

Definition 11.8 (*Concretization of abstract heap transformations*) The concretization function $\gamma_{\mathbb{R}^\sharp} \in \mathbb{R}^\sharp \to \mathcal{P}(\mathbb{H} \times \mathbb{H} \times (\mathbb{Q} \to \mathbb{V}))$ is defined by induction on the structure of $r^\sharp$:

$$\gamma_{\mathbb{R}^\sharp}(\mathtt{Id}(h^\sharp)) = \{(h, h, \nu) \mid (h, \nu) \in \gamma_{\mathbb{H}^\sharp}(h^\sharp)\}$$
$$\gamma_{\mathbb{R}^\sharp}([h_i^\sharp \dashrightarrow h_o^\sharp]) = \{(h_i, h_o, \nu) \mid (h_i, \nu) \in \gamma_{\mathbb{H}^\sharp}(h_i^\sharp) \wedge (h_o, \nu) \in \gamma_{\mathbb{H}^\sharp}(h_o^\sharp)\}$$
$$\gamma_{\mathbb{R}^\sharp}(r_1^\sharp *_T r_2^\sharp) = \{(h_{i,1} \uplus h_{i,2}, h_{o,1} \uplus h_{o,2}, \nu) \mid$$
$$\mathbf{dom}(h_{i,1}) \cap \mathbf{dom}(h_{i,2}) = \emptyset \wedge \mathbf{dom}(h_{o,1}) \cap \mathbf{dom}(h_{o,2}) = \emptyset$$
$$\wedge (h_{i,1}, h_{o,1}, \nu) \in \gamma_{\mathbb{R}^\sharp}(r_1^\sharp) \wedge \mathbf{dom}(h_{i,1}) \cap \mathbf{dom}(h_{o,2}) = \emptyset$$
$$\wedge (h_{i,2}, h_{o,2}, \nu) \in \gamma_{\mathbb{R}^\sharp}(r_2^\sharp) \wedge \mathbf{dom}(h_{i,2}) \cap \mathbf{dom}(h_{o,1}) = \emptyset\}$$

This definition simply says that identity relation keeps the heap unchanged; that transform-into relations replace an input heap by an output heap, where both heaps are separately described, each using an abstract heap; and separated transformations allow merging transformations together. It is important to note that in a separate transformation $r_1^\sharp *_T r_2^\sharp$, $r_1^\sharp$ and $r_2^\sharp$ are really independent as, for instance, the output heap described by $r_1^\sharp$ cannot come from the input heap described by $r_2^\sharp$.

Abstract heap transformation

The abstract heap transformation describing the modifications to the heap in list_append for pre-states matching the precondition given by $sl_5^\sharp$ above, and represented graphically in Fig. 11.6 is given using the following formula:

$$r_5^\sharp \triangleq \quad \mathtt{Id}(\mathbf{listseg}(\alpha, \pi)) *_S \pi.\mathtt{data} \mapsto \delta)$$
$$*_T [\pi.\mathtt{next} \mapsto \mathtt{0x0} \dashrightarrow \pi.\mathtt{next} \mapsto \beta]$$
$$*_T \mathtt{Id}(\mathbf{list}(\beta))$$

This transformation indicates that the list starting at α and β is unmodified, except for the last element in the α list.

Finally, we can define the behavior of the complete analysis, which, in addition to the heap transformations, also records numerical constraints and changes in the store. Note that to record changes in the store, it suffices to use a pair of abstract stores: this comes from the fact that, unlike the heap, the contents of the store can be enumerated. The result of our analysis is a relation between the possible pre-states and post-states of the function.

$$\mathbb{RSL}^\sharp = \mathbb{N}^\sharp \times \Sigma^\sharp \times \Sigma^\sharp \times \mathbb{R}^\sharp$$
$$\gamma_{\mathbb{RSL}^\sharp} : \mathbb{RSL}^\sharp \to \mathcal{P}(\mathbb{S} \times \mathbb{S})$$
$$\gamma_{\mathbb{RSL}^\sharp}(n^\sharp, \sigma_i^\sharp, \sigma_o^\sharp, r^\sharp) =$$
$$\{((\sigma_i, h_i), (\sigma_o, h_o)) \mid \exists \nu \in \gamma_{\mathbb{N}^\sharp}(n^\sharp) : \quad (\sigma_i, \nu) \in \gamma_{\Sigma^\sharp}(\sigma_i^\sharp)$$
$$\land (\sigma_o, \nu) \in \gamma_{\Sigma^\sharp}(\sigma_o^\sharp)$$
$$\land (h_i, h_o, \nu) \in \gamma_{\mathbb{R}^\sharp}(r^\sharp) \}$$

Abstract transformation in a separation-logic-based transformation analysis

The abstract heap transformation describing the transformation of the states in list_append for pre-states matching the precondition given by $sl_5^\sharp$ above, and represented graphically in Fig. 11.6 is given using the following formula:

$$rsl_5^\sharp \triangleq n_5^\sharp, [\mathtt{a} \mapsto \alpha, \mathtt{b} \mapsto \beta], [\mathtt{a} \mapsto \alpha, \mathtt{b} \mapsto \beta, \mathtt{ptr} \mapsto \pi], r_5^\sharp$$

where $n_5^\sharp \triangleq \alpha \neq \mathtt{0x0} \land \pi \neq \mathtt{0x0} \land 0 \leq \delta \leq 10$

and $r_5^\sharp$ is defined above.

This transformation clearly shows that neither a nor b have been modified by the function.

For simplicity, we have omitted some extensions provided by the RMA plug-in. One of the most important is *abstract heap transformation predicates* [16, 19]. In the presentation above, almost no relational information is retained when the heap is modified, as we just encode this using a transform-into predicate $[h_i^\sharp \dashrightarrow h_o^\sharp]$ above that says that the input has been replaced by the output. Transformation predicates enrich this with additional information about the transformation. The plug-in currently encodes two such pieces of information: one says whether the transformation has allocated or deallocated memory; another says which of the fields in a data structure have been modified by a transformation. Together, these pieces of information allow proving, for instance, that a sorting algorithm performed an in-place permutation (i.e., it did not allocate or deallocate any memory, and it did not change the .data field).

A very important application of abstract transformations is the ability to perform modular analysis of functions using abstract transformations as a way to summarize the behavior of a function, and reusing this summary to avoid re-analyzing a function when called in different contexts. Our benchmarks showed that this allows up to exponential gains in analysis performance, with no loss of precision in practice [18].

11.6 Further Reading

About separation-logic-based static analyses The state analysis based on a combination of separation logic with parameterized inductive predicates and numeric invariants, as performed by our RMA plug-in, is based on the work of Chang and Rival [5]. The relational separation logic extensions that allow transformation analysis was initially described in Illous et al. [17]. A more comprehensive description, including the description of the abstract heap transformation predicates, is provided in Illous et al. [19]. The use of abstract transformations to perform an efficient modular analysis was described in Illous et al. [18]. The result of all these works is available in our RMA plug-in.

The work of Chang and Rival [5] has also been extended in several other interesting directions allowing usage of the plug-in on more code patterns, such as combination with array abstraction [26], handling of C union, [23], efficient handling of disjunctions [24], or handling of unstructured sharing [25]. These extensions are largely orthogonal to the relational separation logic extensions of the RMA plug-in, and could thus be integrated into the plug-in.

A survey of shape analysis techniques (including analyses based on separation logic) can be found in Chang et al. [4].

About type-based static analyses The use of types in a static analysis based on abstract interpretation is uncommon. Most type-based static analyses assume type safety instead of establishing type safety: this is the case, e.g., for type-based alias analyses [12] or for the structural analysis of Marron [27].

Most automated type-based verification methods use unification-based type inference instead: this method is less precise than methods based on abstract interpretation and is thus usually unable to verify type safety. For instance, the physical subtyping method of Chandra and Reps [3] does not guarantee type safety, and the memory-safe dialects of C (e.g., Cyclone [21], CCured [31], or CheckedC [13]) use runtime verification to check the code that the static analysis cannot verify. Liquid types [37] enhance their type inference with abstract interpretation and SMT solving, and are thus able to verify type safety. They discuss several limitations that are solved by our analysis technique, such as lack of support for interior pointers and conservative decisions of when to fold and unfold variables.

Type-based verification methods have also been used in the context of deductive verification, e.g. in VCC [8] or RefinedC [38].

11.7 Conclusion

Effective methods to reason about memory are paramount in C programs, as one of the main features of C is its explicit, low-level memory accesses. This high degree of control comes at a cost, which is a high chance of introducing a critical memory vulnerability.

Our work on both the **RMA** and **Codex** aims to provide cost-effective methods to verify memory-related properties of **C** programs, including memory safety. Both of these plug-ins are based on abstract interpretation [9], which is a sound and automated verification method, so as to reduce the amount of manual effort and level of expertise needed to perform the verification. Both plug-ins share a common philosophy which is to gain the precision needed to verify memory safety and preservation of structural invariants by requiring a short description of the data structure invariants used by the program that we must verify. The aim for these annotations is to be simple enough so that every **C** programmer would be able to add them directly. We hope to integrate these annotations in the **ACSL** specification language in the future, to improve collaboration with other plug-ins of the **Frama-C** platform.

The two plug-ins differ mostly by the main memory abstraction on which they rely (separation logic for the **RMA** plug-in, and type-based for the **Codex** plug-in), the separation logic allowing to describe more precise (but also more restrictive) invariants. As both share the same design principles, we hope to be able to integrate them in a single plug-in in the near future; by describing them in a single formal framework, this chapter is a first step towards that future.

References

1. Berdine J, Calcagno C, O'Hearn PW (2005) Smallfoot: Modular automatic assertion checking with separation logic. In: de Boer FS, Bonsangue MM, Graf S, de Roever WP (eds) Formal methods for components and objects, 4th international symposium, FMCO 2005, Amsterdam, The Netherlands, November 1–4, 2005, Revised lectures, vol 4111. Lecture notes in computer science. Springer, pp 115–137
2. Calcagno C, Distefano D, O'Hearn PW, Yang H (2011) Compositional shape analysis by means of bi-abduction. J ACM 58(6):26:1–26:66
3. Chandra S, Reps T (1999) Physical type checking for C. ACM SIGSOFT Softw Engin Notes 24(5):66–75
4. Chang BYE, Dragoi C, Manevich R, Rinetzky N, Rival X (2020) Shape analysis. Found Trends Program Lang 6(1–2):1–158
5. Chang BYE, Rival X (2008) Relational inductive shape analysis. In: Proceedings of the 35th annual ACM SIGPLAN-SIGACT symposium on principles of programming languages, POPL '08, pp 247–260. ACM, New York (2008)
6. Chang BYE, Rival X (2013) Modular construction of shape-numeric analyzers. In: Banerjee A, Danvy O, Doh KG, Hatcliff J (eds) Festschrift for dave schmidt, Festschrift for dave schmidt, vol 129. EPTCS, Manhattan, Kansas, United States
7. Chin WN, David C, Gherghina C (2011) A hip and sleek verification system. In: Proceedings of the ACM international conference companion on object oriented programming systems languages and applications companion, pp 9–10
8. Cohen E, Moskal M, Tobies S, Schulte W (2009) A precise yet efficient memory model for C. Electron Notes Theor Comput Sci 254:85–103. Proceedings of the 4th international workshop on systems software verification (SSV 2009)
9. Cousot P, Cousot R (1977) Abstract interpretation: a unified lattice model for static analysis of programs by construction or approximation of fixpoints. Conference record of the fourth annual ACM SIGPLAN-SIGACT symposium on principles of programming languages. ACM Press, New York, pp 238–252

10. Cousot P, Cousot R (1979) Systematic design of program analysis frameworks. Conference record of the sixth annual ACM SIGPLAN-SIGACT symposium on principles of programming languages. ACM Press, New York, pp 269–282
11. Cousot P, Halbwachs N (1978) Automatic discovery of linear restraints among variables of a program. Conference record of the fifth annual ACM SIGPLAN-SIGACT symposium on principles of programming languages. ACM Press, New York, pp 84–97
12. Diwan A, McKinley KS, Moss JEB (1998) Type-based alias analysis. In: Proceedings of the ACM SIGPLAN 1998 conference on programming language design and implementation, PLDI '98. ACM, New York, pp 106–117
13. Elliott AS, Ruef A, Hicks M, Tarditi D (2018) Checked C: making C safe by extension. In: 2018 IEEE cybersecurity development, SecDev 2018, Cambridge, MA, USA, September 30–October 2, 2018. IEEE Computer Society, pp 53–60
14. Google Project Zero: A year in review of 0-days used in-the- wild in 2019. https://googleprojectzero.blogspot.com/2020/07/detection-deficit-year-in-review-of-0.html
15. Hoare CAR (1969) An axiomatic basis for computer programming. Commun ACM 12(10)
16. Illous H (2019) Abstract heap relations for a compositional shape analysis. PhD thesis, Université Paris Sciences et Lettres. https://inria.hal.science/tel-02399767
17. Illous H, Lemerre M, Rival X (2017) A relational shape abstract domain. In: NASA formal methods symposium. Springer, pp 212–229
18. Illous H, Lemerre M, Rival X (2020) Interprocedural shape analysis using separation logic-based transformer summaries. In: International static analysis symposium. Springer, pp 248–273
19. Illous H, Lemerre M, Rival X (2021) A relational shape abstract domain. In: Formal methods in system design, pp 1–58
20. Jacobs B, Smans J, Philippaerts P, Vogels F, Penninckx W, Piessens F (2011) Verifast: a powerful, sound, predictable, fast verifier for c and java. In: NASA formal methods symposium. Springer, pp 41–55
21. Jim T, Morrisett JG, Grossman D, Hicks MW, Cheney J, Wang Y (2002) Cyclone: a safe dialect of C. In: USENIX annual technical conference, General Track, pp 275–288
22. Klein G, Andronick J, Elphinstone K, Murray T, Sewell T, Kolanski R, Heiser G (2004) Comprehensive formal verification of an OS microkernel. ACM Trans Comput Syst 32(1):2:1–2:70 (2014)
23. Laviron V, Chang BYE, Rival X (2010) Separating shape graphs. In: Proceedings of the 19th European conference on programming languages and systems, ESOP'10. Springer, Berlin, pp 387–406
24. Li H, Berenger F, Chang BE, Rival X (2017) Semantic-directed clumping of disjunctive abstract states. In: Castagna G, Gordon AD (eds) Proceedings of the 44th ACM SIGPLAN symposium on principles of programming languages, POPL 2017, Paris, France, January 18–20, 2017, pp 32–45. ACM
25. Li H, Rival X, Chang, BE (2015) Shape analysis for unstructured sharing. In: Blazy S, Jensen TP (eds) Static analysis - 22nd international symposium, SAS 2015, Saint-Malo, France, September 9–11, 2015, Proceedings. Lecture notes in computer science, vol 9291. Springer, pp 90–108
26. Liu J, Rival X (2015) Abstraction of arrays based on non contiguous partitions. In: International workshop on verification, model checking, and abstract interpretation. Springer, pp 282–299
27. Marron M (2012) Structural analysis: Shape information via points-to computation. arXiv e-prints p. arXiv:1201.1277
28. Miller M (2019) Trends, challenge, and shifts in software vulnerability mitigation (2019). https://www.youtube.com/watch?v=PjbGojjnBZQ
29. Miné A (2006) The octagon abstract domain. Higher-Order Symb Comput 19(1):31–100
30. Müller P, Schwerhoff M, Summers AJ (2016) Viper: a verification infrastructure for permission-based reasoning. In: International conference on verification, model checking, and abstract interpretation. Springer, pp 41–62
31. Necula GC, Condit J, Harren M, McPeak S, Weimer W (2005) CCured: type-safe retrofitting of legacy software. ACM Trans Program Lang Syst (TOPLAS) 27(3):477–526

32. Nicole O (2022) Automated verification of systems code using type-based memory abstractions. PhD thesis, Université Paris Sciences et Lettres (2022). https://cea.hal.science/tel-03962643
33. Nicole O, Lemerre M, Bardin S, Rival X (2021) No crash, no exploit: automated verification of embedded kernels. In: 2021 IEEE 27th real-time and embedded technology and applications symposium (RTAS), pp 27–39. https://doi.org/10.1109/RTAS52030.2021.00011
34. Nicole O, Lemerre M, Rival X (2022) Lightweight shape analysis based on physical types. In: International conference on verification, model checking, and abstract interpretation. Springer, pp 219–241
35. Reynolds JC (2002) Separation logic: a logic for shared mutable data structures. In: 17th annual IEEE symposium on logic in computer science, 2002. Proceedings. IEEE, pp 55–74
36. Rival X, Yi K (2020) Introduction to static analysis: an abstract interpretation perspective. MIT Press (2020)
37. Rondon PM, Kawaguchi M, Jhala R (2010) Low-level liquid types. In: Hermenegildo MV, Palsberg J (eds) Proceedings of the 37th ACM SIGPLAN-SIGACT symposium on principles of programming languages, POPL 2010, Madrid, Spain, January 17–23, 2010. ACM, pp 131–144
38. Sammler M, Lepigre R, Krebbers R, Memarian K, Dreyer D, Garg D (2021) RefinedC: automating the foundational verification of C code with refined ownership types. In: Freund SN, Yahav E (eds) PLDI '21: 42nd ACM SIGPLAN international conference on programming language design and implementation, virtual event, Canada, June 20–25, 2021. ACM, pp 158–174

Chapter 12
Analysis of Embedded Numerical Programs in the Presence of Numerical Filters

Franck Védrine, Pierre-Yves Piriou, and Vincent David

Abstract This chapter presents how Frama-C verifies some complex loop invariant for numerical embedded code and how to produce such invariants. Numerical embedded code usually defines an endless loop that takes inputs from sensors and that emits outputs for actuators. Moreover, such a code commonly uses some floating-point global memories to keep track of the input or output values from the previous loop cycles. These memories store a summary of previous values to filter the input or the output over time. Hence, recursive linear or Infinite Impulse Response (IIR) filters are very common in such code. Among such filters, low-pass filters are challenging for Frama-C since finding a loop invariant with complex relationships between the variables is never an evident task for the engineer. Hence, this chapter presents different approaches to find and organize inductive invariants for such numerical reactive systems and shows how Frama-C can prove them with its Eva and Wp plug-ins. It also exposes optimal theoretical results for first-order and higher-order filters to compare the quality of results of the different solutions. Several examples and several concrete solutions illustrate the generation/writing of such inductive invariants and their formal verification.

Keywords Numerical analysis · Linear digital filters · Static analysis · Inductive invariants · Numerical reactive systems

F. Védrine (✉)
Université Paris-Saclay, CEA, List, Palaiseau, France
e-mail: franck.vedrine@cea.fr

P.-Y. Piriou
EDF R&D PRISME Performance, Risque Industriel et Surveillance pour la Maintenance et l'Exploitation, Chatou, France
e-mail: pierre-yves.piriou@edf.fr

V. David
IRSN, Fontenay-aux-Roses, France
e-mail: vincent.david@irsn.fr

© The Author(s), under exclusive license to Springer Nature Switzerland AG 2024
N. Kosmatov et al. (eds.), *Guide to Software Verification with Frama-C*, Computer Science Foundations and Applied Logic, https://doi.org/10.1007/978-3-031-55608-1_12

12.1 Introduction

Internal memories are very common in embedded numerical code to implement industrial control systems (ICS). Digital components like delay blocks, linear digital filters, PID (Proportional-Integral-Derivative) controllers use an internal memory to work. It is a memory containing floating-point variables that record a summary of the history of the inputs, the outputs, and the computations.

For code with endless loops, Frama-C usually needs a loop invariant to check it. Annotations are mandatory to use the weakest precondition plug-in Wp of Frama-C that implements deductive verification paradigms. Wp generates from the given code and annotations a set of proof obligations that are checked both internally and in external tools like Alt-Ergo [4], CVC4 [1], Z3 [12]. The plug-in Eva of Frama-C works differently. It analyzes the source code by abstract interpretation [3] and computes an over-approximation of the least fixpoint of the collecting semantics of the program, that is the execution of the program with sets of values instead of concrete values. This least fixpoint for loops corresponds to the most precise inductive loop invariant. Since it is usually not precisely computable, Eva produces an over-approximation in finite time thanks to its embedded widening operator. With floating-point memories, Frama-C/Eva may generate $[-DBL\_MAX, DBL\_MAX]$ as invariant for every state variable, which is true but not acceptable for the user. When the inductive invariant takes the form of floating-point intervals without any relationships between the variables, Frama-C/Eva usually finds the invariant without the need of having additional annotations. For an inductive invariant involving relationships between variables, Frama-C/Eva often needs some help or some specific annotations from the engineer to infer precise bounds for the output values.

Linear digital filters offer the opportunity to study different structures for loop invariants. They have been intensively studied in the literature and the existing work proposes different kinds of loop invariants: ellipsoids [6, 7, 15, 16], zonotopes [5], sets of boxes [11], sets of zonotopes [9]. Integrating the associated numerical domains in Frama-C has many drawbacks: first, they should only be active on some parts of the code, and second, they pollute the Frama-C/Eva iterations since they dramatically increase the number of analysis cycles needed by the analysis to reach a fixpoint. A more efficient solution is a modular cooperation between the invariant generation and its formal verification. Hence, the objective and motivation of this chapter is to show that the analysis of convergent linear digital filters with Frama-C has currently different solutions depending on the type of the filter: for filters of order 1, Frama-C/Eva automatically generates and proves loop invariants for the floating-point computations (see Sect. 12.3); for filters of order 2 or more with real eigenvalues, this chapter proposes a manual way for writing invariants targeting Frama-C/Wp and verification with real numbers (see Sect. 12.4); for filters of order 2 or more with complex eigenvalues, this chapter proposes to find an invariant outside Frama-C and, then, to import the resulting property as an ACSL (ANSI/ISO C Specification Language [2]) annotation for an automatic proof in Frama-C (see Sect. 12.5). This chapter also exposes optimal theoretical results for such filters to compare the quality of results of the different solutions. The full examples and the complete experiments are available in the companion artifact [18].

In the next section, we propose to explain the theory of static analysis of linear digital filters. This helps the engineers to better understand the problem they have to solve in order to choose an appropriate way of solving it. We then identify three different kinds of filters that require different annotations and ways to prove the convergence of the filter with Frama-C. The last section proposes solutions to cooperate with external tools to exploit within Frama-C the loop invariants they may find.

12.2 Problematic Aspects of Linear Digital Filters

The *linear digital filter* components usually propose an implementation like in the code snippet of Fig. 12.1. The key instruction (transfer function) is

```
out = -ko[2]*out2 - ko[1]*out1 + ki[2]*in2 + ki[1]*in1 + ki[0]*in
```

It computes the value of the variable out at loop cycle n from inputs coming from sensors and from the values of out at previous loop cycles. The variable out1 (resp. in1) at cycle n contains the value of the variable out (resp. in) at cycle $n - 1$ and the variable out2 (resp. in2) at cycle n contains the value of the variable out (resp. in) at cycle $n - 2$. In the mathematical formula below, ko_i, ki_i, in^i_j, out^i_j are notations for the values of the parameters and variables stored in the variables ko[i], ki[i], in, ...in2, out, ...out2. Hence, out^i_j is the value of the variable out$j$ (out0 = out) at cycle i, and we have $out^{i+1}_{j+1} = out^i_j$ for $j < order$, where $order = 2$ is the order of the filter in the above example.

From Frama-C's point of view, when the inputs in are in the interval $[-10.0, 10.0]$, the value of ki[2]*in2 + ki[1]*in1 + ki[0]*in belongs

```
1  double ..., out2 = 0.0, out1 = 0.0; /* memories keeping history of outputs */
2  double ..., in2 = 0.0, in1 = 0.0;   /* memories keeping history of inputs */
3  const double ki[] = { ... };
4  const double ko[] = { ... };
5  int main() {
6    double in, out;
7    while (true) {
8      ...
9      ...; in2 = in1; in1 = in;
10     ...; out2 = out1; out1 = out;
11     in = Frama_C_double_interval(-10.0, +10.0); /* sensor inputs */
12     out = ... - ko[2]*out2 - ko[1]*out1 + ki[2]*in2 + ki[1]*in1 + ki[0]*in;
13     ...
14   }
15 }
```

Fig. 12.1 Generic implementation of a digital filter inside a wider C code

Fig. 12.2 Implementation of a filter of order 1

```
1  double out1;
2  const double ki[] = { 0.5 };
3  const double ko[] = { 1.0, -0.9 };
4  int main() {
5    double in, out;
6    while (true) {
7      ...
8      out1 = out;
9      in = Frama_C_double_interval(-10.0, +10.0);
10     out = -ko[1]*out1 + ki[0]*in;
11     ...
12   }
13 }
```

to the interval $[-10.0 \times (|ki_2| + |ki_1| + |ki_0|), 10.0 \times (|ki_2| + |ki_1| + |ki_0|)]$, which provides correct bounds.

Finding correct bounds for -ko[2]*out2 - ko[1]*out1 is much more complex. Nevertheless, such bounds exist if the filter is convergent, a notion (defined below in Sect. 12.2) expressed with the eigenvalues of the transformation matrix of the filter. Their existence is ensured by two mathematical theorems: the Jordan-Chevalley decomposition and the diagonalizability of the Frobenius companion matrix with the help of the Vandermonde matrix.

To introduce the transformation matrix, we arrange the values of out2, out1, out in a vector. This vector is linearly modified after every loop cycle with a linear forward transfer function represented by the following matrix M.

$$M = \begin{pmatrix} 0 & 1 \\ -ko_2 & -ko_1 \end{pmatrix} \text{ and } \begin{pmatrix} out_1^{n+1} \\ out_0^{n+1} \end{pmatrix} = M \times \begin{pmatrix} out_1^n \\ out_0^n \end{pmatrix} + \begin{pmatrix} 0 \\ ki_2 \times in_2^n + ki_1 \times in_1^n + ki_0 \times in_0^n \end{pmatrix}$$

Let us start with the simplest filter code depicted in Fig. 12.2. The parameters of interest are the constants $ki_0 = 0.5$ and $ko_1 = -0.9$.

At every loop cycle n, the value out_0^n of the variable out for this cycle is computed as $out_0^n = 0.9 \times_{float} out_0^{n-1} +_{float} 0.5 \times_{float} in_0^n$ in floating-point numbers, with $in_0^n \in [-10.0, 10.0]$. If the computations of the instruction were performed using ideal (real) numbers instead of floating-point numbers, an inductive reasoning over n would demonstrate that the output value out_0^n validates the following property without the *rounding errors* part:

$$out_0^n \in 0.5 \times [-10.0, 10.0] \times \left(\sum_{k=0}^{n-1} 0.9^k\right) + \text{rounding errors}$$

The accumulation of *rounding errors* occurs due to small differences between the floating-point operations ($+_{float}$, $\times_{float}$) and their ideal counterparts ($+$, $\times$) in $\mathbb{R}$. The IEEE-754 norm bounds the error introduced by one operation by half of the value of the Unit in the Last Place (ULP for short) – for instance, for a result between 1.0 and 2.0 excluded, the introduced error is bounded by 1.11023e-16.

The properties simplify into

$$\text{out}_0^n \in 0.5 \times [-10.0, 10.0] \times \frac{1 - 0.9^n}{1 - 0.9} + \textit{rounding errors}$$

$$\in [-50.0, 50.0] \times (1 - 0.9^n) + \textit{rounding errors}$$

$$\in [-50.0, 50.0] + \textit{rounding errors}$$

This suggests that out_0^n is bounded over time. We will look at the rounding errors in Sect. 12.4. Now, let us express the vector of output values in the generic case:

$$\begin{pmatrix} \text{out}_2^n \\ \text{out}_1^n \\ \text{out}_0^n \end{pmatrix} \in M \times \begin{pmatrix} \text{out}_2^{n-1} \\ \text{out}_1^{n-1} \\ \text{out}_0^{n-1} \end{pmatrix} + \begin{pmatrix} \ldots \\ 0 \\ \sum_{o=0}^{order} ki_o \times [-10.0, 10.0] \end{pmatrix} + \textit{rounding errors}$$

$$\in [-10.0, 10.0] \times \left(\sum_{k=0}^{n-1} M^k\right) \begin{pmatrix} \ldots \\ 0 \\ \sum_{o=0}^{order} |ki_o| \end{pmatrix} + \textit{rounding errors}$$

Hence, for sound static analysis tools like **Frama-C**, proving that the variable out is bounded over time requires to prove that all the coefficients of the matrix $\sum_{k=0}^{n-1} M^k$ are bounded.

In theory, this statement is true for any convergent linear filter, defined with a matrix M whose all complex eigenvalues have a norm strictly less than 1.0. Two theorems prove this: the Jordan-Chevalley decomposition and the specific properties of the Frobenius companion matrix. The reader mainly interested in the **Frama-C** results can skip these theorems and the remark on the rounding errors and go to the next section. Section 12.4 will exploit the coefficients of the Frobenius companion matrix to provide provable **ACSL** annotations. Section 12.5 will only refer to the theoretical results of the Jordan-Chevalley decomposition to entrust the generation of a loop invariant to third-party tooling before the integration of the found invariant in the **Frama-C** world.

Theorem 12.1 (Jordan-Chevalley decomposition) *Let M be a matrix of size order $\times$ order with entries in $\mathbb{R}$ (or in $\mathbb{C}$). There exists unique matrices S and N such that:*

$$M = S + N$$

- S is diagonalizable over $\mathbb{C}$,
- N is nilpotent,
- $S \times N = N \times S$.

S is diagonalizable means that there exists P invertible complex matrix and D diagonal complex matrix such that $S = P^{-1} \times D \times P$.

The linear digital filter is *convergent* if and only if all the coefficients of the diagonal matrix D have a norm strictly less than 1.0. These coefficients of the diagonal matrix may be complex since they are the roots of the characteristic polynomial in λ, that is, the determinant of $M - \lambda \times I$.

With such a decomposition, we have

$$\sum_{k=0}^{n-1} M^k = \sum_{k=0}^{n-1} (P^{-1} \times D \times P + N)^k = \sum_{k=0}^{n-1} \left(\sum_{j=0}^{k} C_k^j P^{-1} \times D^{k-j} \times P \times N^j \right)$$

where C_k^j are the binomial coefficients.

Since N is nilpotent, which means $N^k = 0$ for $k \geq order$, it is interesting to swap both sums, so

$$\sum_{k=0}^{n-1} M^k = \sum_{j=0}^{\min(n,order)-1} \left(\sum_{k=j}^{n-1} C_k^j P^{-1} \times D^{k-j} \times P \right) \times N^j$$

$$= \sum_{j=0}^{\min(n,order)-1} \left(\sum_{k=0}^{n-1-j} C_{k+j}^k P^{-1} \times D^k \times P \right) \times N^j = \sum_{j=0}^{\min(n,order)-1} P^{-1} \times \left(\sum_{k=0}^{n-1-j} C_{k+j}^k D^k \right) \times P \times N^j$$

For $0 \leq j \leq \min(n, order) - 1$, the coefficients of $\displaystyle\sum_{k=0}^{n-1-j} C_{k+j}^k D^k$ are bounded because

$$C_{k+j}^k = C_{k+j}^j \leq \frac{(k+order)^{order}}{order!} \quad \text{and} \quad 0 < |\lambda| \leq \frac{2 \times |\lambda|}{1+|\lambda|} < 1.0 < \frac{2}{1+|\lambda|}$$

for any λ coefficient of the diagonal of D. Hence,

$$\lim_{k \to +\infty} \frac{(k+order)^{order}}{order!} \times \left(\frac{1+|\lambda|}{2} \right)^k = 0$$

(the ln of this property has a magnitude similar to $order \times \ln(k + order) + k \times$ *negative-number* and tends towards $-\infty$), and

$$\exists k_0 \geq 0. \forall k \geq k_0. \frac{(k+order)^{order}}{order!} \leq \left(\frac{2}{1+|\lambda|} \right)^k$$

which means that $\displaystyle\sum_{k \geq k_0} C_{k+j}^k |\lambda|^k \leq \frac{\mu^{k_0}}{1-\mu}$ with $\mu = \frac{2 \times |\lambda|}{1+|\lambda|} < 1$.

In the case of Fig. 12.1, the M matrix has the specific form of the transpose of the Frobenius companion matrix of its characteristic polynomial

$$p(\lambda) = \ldots + ko_2 \times \lambda^{order-2} + ko_1 \times \lambda^{order-1} + 1.0 \times \lambda^{order}$$

Theorem 12.2 (Diagonalizability of the companion matrix) *If $p(\lambda)$ has distinct roots $\lambda_1, \ldots, \lambda_{order}$ (the eigenvalues of M), then M and ${}^t M$ are diagonalizable as follows:*

$$V^{-1} \times {}^t M \times V = {}^t V \times M \times {}^t V^{-1} = diag(\lambda_1, \ldots, \lambda_{order})$$

where V is the Vandermonde matrix corresponding to the λs:

$$V = \begin{pmatrix} 1 & 1 & \ldots & 1 \\ \lambda_1 & \lambda_2 & \ldots & \lambda_{order} \\ \lambda_1^2 & \lambda_2^2 & \ldots & \lambda_{order}^2 \\ \vdots & \vdots & \ddots & \vdots \\ \lambda_1^{order-1} & \lambda_2^{order-1} & \ldots & \lambda_{order}^{order-1} \end{pmatrix}$$

If $p(\lambda)$ has a non-simple root, then M is not diagonalizable (its Jordan canonical form contains one block for each distinct root). The Jordan-Chevalley decomposition nevertheless applies.

Remark on the accumulation of rounding errors: For the examples of the chapter we do not specially care about the rounding errors, which are the differences between the ideal computation with real numbers and the actual implementation with floating-point numbers. The reason is that the propagation of rounding errors err^n in linear digital filters follows similar transformation rules as the propagation of the values. To simplify the demonstration, we suppose that the constant coefficients ko_i and ki_i are exactly represented. Then, the norm IEEE-754 guarantees that the forward transfer function of the rounding errors is approximated by the matrix transform below in interval arithmetic

$$M' = \begin{pmatrix} 0 & 1 & 0 \\ 0 & 0 & 1 \\ \ldots & -ko_2 \times (1.0 + order \times [-\varepsilon, +\varepsilon]) & -ko_1(1.0 + order \times [-\varepsilon, +\varepsilon]) \end{pmatrix}$$

and $\begin{pmatrix} err_2^{n+1} \\ err_1^{n+1} \\ err_0^{n+1} \end{pmatrix} = M' \times \begin{pmatrix} err_2^n \\ err_1^n \\ err_0^n \end{pmatrix} + \begin{pmatrix} err\_in_2^n \\ err\_in_1^n \\ err\_in_0^n \end{pmatrix}$

with $\begin{pmatrix} err\_in_2^{n+1} \\ err\_in_1^{n+1} \\ err\_in_0^{n+1} \end{pmatrix} = (\ldots) \begin{pmatrix} in_2^n \\ in_1^n \\ in_0^n \end{pmatrix} + B \times order \times \left(\sum_{o=0}^{order} |ko_o| \right) \times [-\varepsilon, +\varepsilon]$

where ε is the Machine epsilon and B the bounds for the out_i^n in real numbers (see previous theorems).

For a literal explanation of the coefficients of M', let us consider the floating-point computations in the key instruction

```
out = -ko[2]*out2 - ko[1]*out1 + ki[2]*in2 + ki[1]*in1 + ki[0]*in.
```

The error (floating-point value minus ideal value) of out2 that is err^{n-1} is multiplied by ko_2 and another error is introduced by the IEEE-754 multiplication. This error for the first term is bounded by $0.5 \times \varepsilon \times$ float_value_of(ko[2]*out2). Since the float value is the ideal value plus the error, the multiplication error is bounded by $0.5 \times \varepsilon \times (|ko_2| \times (B + err^{n-1}))$. There is an additional (constant) term for the error if the result is a subnormal number; this term is then absorbed by the last $B \times order \times \ldots$ of the last equation above. The same reasoning applies for the second computation -ko[1]*out1. The addition between the two computations propagates the rounding errors of each computation by adding them and it creates a new error bounded by $0.5 \times \varepsilon \times$ float_value_of(ko[2]*out2 + ko[1]*out1). Moreover, since the float value is the ideal value plus the error, the new addition error is bounded by $0.5 \times \varepsilon \times ((|ko_2| + ko_1|) \times (B + err^{n-1} + err^n))$. The whole instruction contains $2 \times order - 1$ additions that create the final perturbation $|ko_i| \times order \times [-\varepsilon, +\varepsilon]$ in the coefficients of the matrix M' with respect to the matrix M.

Let us now consider $M^k = P^{-1} \times D^k \times P$. Consequently, as k tends to infinity, all the coefficients of M^k tend to zero. Therefore, the matrix norm induced by vector ∞-norm also tends to zero. This matrix norm [21] defined as $\sup_{v \neq 0} \frac{\|M^k(v)\|_\infty}{\|v\|_\infty}$ can be computed as $\|M^k\|_\infty = \max_{0 \leq i \leq n-1} \sum_{j=0}^{n-1} |a_{ij}|$ where a_{ij} are the coefficients of M^k. Hence,

$$\exists k_0 > 0. \, \forall k \geq k_0. \, \|M^k\|_\infty \leq 0.5$$

Let us revisit M' with a relative perturbation of all its coefficients bounded by $1.0 + order \times [-\varepsilon, +\varepsilon]$. Next, we can compute the matrix norm $\|M'^{k_0}\|_\infty$ using interval arithmetic; since the computation of the matrix norm is continuous and $\|M^{k_0}\|_\infty \leq 0.5$, the resulting value ρ should be less than 1. If this condition holds, the accumulation of rounding errors would be lower than $\sum_{k \geq 0} M'^k(e_k)$ where (e_k) represents the vector of numerical errors introduced at each step. Therefore, the accumulation of rounding errors remains bounded by

$$\left\| \sum_{k \geq 0} M'^k(e_k) \right\|_\infty \leq \sum_{k \geq 0} \left\| M'^k \right\|_\infty \|e_k\|_\infty$$

$$\leq \frac{1}{1-\rho} \left(\sum_{k=0}^{k_0-1} \left\| M'^k \right\|_\infty \right) \times order \times 0.5 \times \varepsilon \times (|ko_2| + ko_1|) \times B$$

The work of [19, 20] provides a solution to tightly bound such an error with a 6-step analysis. The solution is perfectly valid for our framework. For any value μ close to zero, their algorithm provides a matrix W_N such that $\|W - W_N\| < \mu$ for the Frobenius norm and where W is the Worst-Case Peak Gain matrix of the linear system. This provides a guaranteed invariant for the output of the filter since it is bounded by W multiplied by the vector of inputs as interval. Hence, it is sufficient to add $order \times \mu$ to the resulting vector output to obtain an over-approximation. The proof of the solution nevertheless requires an eigenvalue decomposition to perform matrix powering and some multiple-precision operations, two theories requiring provers like **Coq** or **HOLLight**.

The first step of [20] (see [20, Eq. (26)]) shows that if the spectral radius $\rho(M') < 1$, then the accumulation of rounding errors is bounded. We forget then the nilpotent matrix N' for the explanations. Hence, by considering the system's state for the error before the application of P'^{-1} at each cycle, the vector of error is multiplied by $D' \times P' \times P'^{-1}$, with D' having only diagonal coefficients with small interval error. Let us define ρ_1 the max of the norm of the complex. The multiplication of $P' \times P'^{-1}$ generates the identity matrix plus an error for each coefficient that we bound by $\varepsilon' > 0$. The spectral radius is defined by $\max_{||u||_2=1} ||D' \times P' \times P'^{-1} \times u||_2$. Here, an upper bound exists as $\rho = \rho_1 + order \times \varepsilon'$. If $\rho < 1.0$, the accumulation of rounding errors has a bound. Moreover, the vector

$$P'^{-1} \times \left(\sum_{k=0}^{n} \rho^k (Id) \right) \times (err\_in)$$

computed in interval arithmetic, bounds this accumulation for each cycle n.

The handling of rounding errors is not explicitly incorporated in **Frama-C**, except for the **Numerors** plug-in [8], which operates similarly to **Eva**. While it is possible to express the computation of errors through **ACSL** annotations for **Wp**, doing so requires a substantial number of annotations. As an example, given a specification in real semantics written in the **PVS** (Prototype Verification System [13]) language, an extension of **PRECiSA** [17] generates C code and **ACSL** annotations for error propagation, which can be proven by **Frama-C/Wp** using **PVS** certificates also produced by **PRECiSA**. These **ACSL** annotations target the executed control flow, which is proven to be identical in real and floating-point semantics, provided no warnings occur during execution. Future efforts could introduce specific annotations to reason about numeric errors, potentially simplifying the annotation-writing process significantly.

12.3 Analysis of a Filter of Order 1 with Frama-C

To analyze Fig. 12.2 with **Eva**, the `widen_hints` annotation of Fig. 12.3 suggests some bounds *MIN*, *MAX* for the variable `out`. The intention behind this annotation is to assist **Eva** in proving that `out` $\in [MIN, MAX]$ as a loop invariant during its fixpoint analysis. The **Eva** analyzer interprets the code of the loop using abstract domains until it reaches a fixpoint, that is the same interval of values for the variables that have been modified by a loop iteration. When the intervals grow gradually, **Eva** applies its widening operator to abruptly increase the upper bound, decrease the lower bound of the interval, and then ensure the convergence of the fixpoint algorithm with few iterations. By providing a `widen_hints` annotation, the user suggests an appropriate upper bound and lower bound for the first application of the widening

Fig. 12.3 Analysis of a filter of order 1

```
1  #include "__fc_builtin.h"
2  #include <stdbool.h>
3
4  double out1;
5  const double ki[] = { 0.5 };
6  const double ko[] = { 1.0, -0.9 };
7  int main() {
8    double in=0, out=0;
9    //@ widen_hints out, -50.0, 50.0;
10   while (Frama_C_interval(0, 1)) {
11     out1 = out;
12     in = Frama_C_double_interval(-10.0, +10.0);
13     out = -ko[1]*out1 + ki[0]*in;
14   }
15 }
```

operator with the hope that Eva proves that the suggested interval is an invariant and infers out ∈ [-50. .. 50.] inside and at the end of the loop.

Without the ACSL annotation //@ widen_hints out, -50.0, 50.0;, the variable out remains bounded but the interval is far larger, and depends on the Eva implementation of the widening operator.

The Eva analysis of a filter of order 1 is robust. If the widen_hints is greater than [−50.0, 50.0], the result will be good.

For example if the domain of out is [−60.0, 60.0] at the loop iteration, the domain of out at the end of the loop becomes [−60 × 0.9 − 5.0, 60 × 0.9 + 5.0] = [−59.0, 59.0], which is more precise than [−60.0, 60.0].

Low pass filters are also well handled by Eva. If ko[1] = −0.9999, the annotation //@ widen_hints out, -50000.0, 50000.0; helps Eva to ensure out ∈ [-50000. .. 50000.].

No need to care about rounding errors since Eva exploits the monotonicity of floating-point operations to implement safe interval floating-point abstractions: $[a, b] +_{float} [c, d] = [a +_{float} c, b +_{float} d]$ when $+_{float}$ is the floating-point operation + for the rounding mode "to nearest even". In this context, safe interval means ∀ floating-point values $x, y \in [a, b] \times [c, d]$. $x +_{float} y \in [a +_{float} c, b +_{float} d]$.

12.4 Analysis of a Filter of Order 2 Having Real Eigenvalues

Let us consider now the analysis of a filter of order 2 having real eigenvalues. Figure 12.4 shows an implementation of such a filter. First, we can check that its eigenvalues are real and distinct: the characteristic polynomial is $p(\lambda) = ko[0] \times$

Fig. 12.4 Implementation of a filter of order 2 having real eigenvalues

```
1  // parameters
2  const double ki[] = { 0.1 };
3  const double ko[] = { 1.0, -1.0, 0.2 };
4
5  // input range
6  const double IN_MIN = -10.0;
7  const double IN_MAX = 10.0;
8
9  int main()
10 {
11   double out, out1, out2, in;
12   in = Frama_C_double_interval(IN_MIN, IN_MAX);
13   out2 = 0.0;
14   out1 = 0.0;
15   out = ki[0]*in - ko[1]*out1 - ko[2]*out2;
16   //@ loop unroll 1000;
17   while (Frama_C_interval(0, 1)) {
18     in = Frama_C_double_interval(IN_MIN, IN_MAX);
19     out2 = out1;
20     out1 = out;
21     out = ki[0]*in - ko[1]*out1 - ko[2]*out2;
22   }
23 }
```

$\lambda^2 + ko[1] \times \lambda + ko[2] = \lambda^2 - \lambda + 0.2$. Its discriminant is positive: $\Delta = 0.2$, hence its 2 eigenvalues are real:

$$\begin{cases} \lambda_1 = \dfrac{-ko[1] - \sqrt{\Delta}}{2} \approx 0.27639320225 \\ \lambda_2 = \dfrac{-ko[1] + \sqrt{\Delta}}{2} \approx 0.72360679775 \end{cases}$$

The direct analysis with **Eva**, despite a huge loop unrolling, produces an alarm of kind `is_nan_or_infinite` on line 21, where `out` is computed.

As explained in Sect. 12.2, to prove that the output is bounded, we can state the problem in the matrix form, introducing the companion matrix of the characteristic polynomial and applying Theorem 12.2 to diagonalize it (the last equality is introduced to simplify further formulas):

$$\begin{aligned}
\begin{pmatrix} out_1^{n+1} \\ out_0^{n+1} \end{pmatrix} &= \begin{pmatrix} 0 & 1 \\ -ko[2] & -ko[1] \end{pmatrix} \begin{pmatrix} out_1^n \\ out_0^n \end{pmatrix} + \begin{pmatrix} 0 \\ ki_0 \times in_0^{n+1} \end{pmatrix} \\
&= \left(V \times \begin{pmatrix} \lambda_1 & 0 \\ 0 & \lambda_2 \end{pmatrix} \times V^{-1} \right) \begin{pmatrix} out_1^n \\ out_0^n \end{pmatrix} + \begin{pmatrix} 0 \\ ki_0 \times in_0^{n+1} \end{pmatrix} \\
&= \left(\dfrac{1}{\lambda_1 - \lambda_2} V \times \begin{pmatrix} \lambda_1 & 0 \\ 0 & \lambda_2 \end{pmatrix} \times (\lambda_1 - \lambda_2) V^{-1} \right) \begin{pmatrix} out_1^n \\ out_0^n \end{pmatrix} + \begin{pmatrix} 0 \\ ki_0 \times in_0^{n+1} \end{pmatrix}
\end{aligned}$$

with $V = \begin{pmatrix} 1 & 1 \\ \lambda_1 & \lambda_2 \end{pmatrix}$ and $V^{-1} = \dfrac{1}{\lambda_1 - \lambda_2} \begin{pmatrix} -\lambda_2 & 1 \\ \lambda_1 & -1 \end{pmatrix}$

Thus we have

$$\left((\lambda_1 - \lambda_2)V^{-1}\right)\begin{pmatrix} out_1^{n+1} \\ out_0^{n+1} \end{pmatrix} = \begin{pmatrix} \lambda_1 & 0 \\ 0 & \lambda_2 \end{pmatrix}\left((\lambda_1 - \lambda_2)V^{-1}\right)\begin{pmatrix} out_1^n \\ out_0^n \end{pmatrix} + \left((\lambda_1 - \lambda_2)V^{-1}\right)\begin{pmatrix} 0 \\ ki_0 \times in_0^{n+1} \end{pmatrix}$$

or

$$\begin{pmatrix} e_1^{n+1} \\ e_2^{n+1} \end{pmatrix} = \begin{pmatrix} \lambda_1 & 0 \\ 0 & \lambda_2 \end{pmatrix}\begin{pmatrix} e_1^n \\ e_2^n \end{pmatrix} + \begin{pmatrix} ki_0 \times in_0^{n+1} \\ -ki_0 \times in_0^{n+1} \end{pmatrix} \text{ with } \begin{cases} e_1 = -\lambda_2 \times out_1 + out_0 \\ e_2 = \lambda_1 \times out_1 - out_0 \end{cases}$$

This way, the filter of order 2 is transformed into a linear combination of two filters of order 1:

$$out_0 = \frac{\lambda_1 \times e_1 + \lambda_2 \times e_2}{\lambda_1 - \lambda_2} \text{ with } \begin{cases} e_1^{n+1} = \lambda_1 \times e_1^n + ki_0 \times in_0^{n+1} \\ e_2^{n+1} = \lambda_2 \times e_2^n - ki_0 \times in_0^{n+1} \end{cases}$$

The theoretic bounds of these filters are:

$$\begin{cases} e_1 \in \left[\dfrac{ki_0}{1-\lambda_1} \times \underline{IN}, \dfrac{ki_0}{1-\lambda_1} \times \overline{IN}\right] \\ e_2 \in \left[-\dfrac{ki_0}{1-\lambda_2} \times \overline{IN}, -\dfrac{ki_0}{1-\lambda_2} \times \underline{IN}\right] \end{cases} \text{ where } \left[\underline{IN}, \overline{IN}\right] \text{ are the bounds of } in_0.$$

As illustrated by Fig. 12.5, we can introduce a ghost implementation of the filters e1 and e2 in the implementation of Fig. 12.4 in order to prove with Wp the loop invariants stating the relation between out_0, e_1, and e_2. Moreover, with enough loop unrolling, Eva can infer the right bounds for e_1 and e_2 (cf. Sect. 12.3). Finally, Eva can infer precise bounds for out: out ∈ [-6.7082039325 .. 6.7082039325].

The command line

```
frama-c -wp -wp-prover alt-ergo,z3 -wp-model real filter2-fc.c
-then -eva filter2-fc.c
```

produces the following output:

```
[wp] Proved goals:   17/18
  Qed:               13  (4ms-25ms-52ms)
  Alt-Ergo 2.3.3:     2  (164ms-176ms) (127) (interrupted: 1)
  Z3 4.8.6:           2  (50ms-410ms) (1076863) (interrupted: 1)


[eva] ====== VALUES COMPUTED ======
[eva:final-states] Values at end of function main:
  out \in [-6.7082039325 .. 6.7082039325]
  e1 \in [-1.38196601125 .. 1.38196601125]
  e2 \in [-3.61803398875 .. 3.61803398875]
```

```c
#include "__fc_builtin.h"

#define delta    (ko[1]*ko[1] - 4.0*ko[2])
#define MIN_e1   (ki[0]/(1.0-l1)*IN_MIN)
#define MAX_e1   (ki[0]/(1.0-l1)*IN_MAX)
#define MIN_e2   (-ki[0]/(1.0-l2)*IN_MAX)
#define MAX_e2   (-ki[0]/(1.0-l2)*IN_MIN)

// parameters
const double ki[] = { 0.1 };
const double ko[] = { 1.0, -1.0, 0.2 };

// input range
const double IN_MIN = -10.0;
const double IN_MAX = 10.0;

/*@ ghost /@ requires arg >= 0.0;
  @           requires \is_finite(arg);
  @           assigns \nothing ;
  @           ensures \is_finite(\result);
  @           ensures \result == sqrt(arg); @/
  @ double ghost_sqrt(double arg);
  */

int main()
{
  double out, out1, out2, in;
  in = Frama_C_double_interval(IN_MIN, IN_MAX);
  out2 = 0.0;
  out1 = 0.0;
  out = ki[0]*in - ko[1]*out1 - ko[2]*out2;
  /*@ ghost
    @ double e1, e2, e1_1, e2_1, l1, l2, min_out, max_out;
    @ l1 = (-ko[1] - ghost_sqrt(delta)) / 2.0;
    @ l2 = (-ko[1] + ghost_sqrt(delta)) / 2.0;
    @ e1_1 = 0.0;
    @ e2_1 = 0.0;
    @ e1 = l1*e1_1 + ki[0]*in;
    @ e2 = l2*e2_1 - ki[0]*in;
    @ */
  /*@
    @ loop assigns e1, e2, e1_1, e2_1, Frama_C_entropy_source;
    @ loop invariant expr1: e1 == -l2*out1 + out;
    @ loop invariant expr2: e2 == l1*out1 - out;
    @ loop unroll 1000;
    @ */
  while (Frama_C_interval(0, 1)) {
    //@ assert expr_out: out == (l1*e1 + l2*e2) / (l1 - l2);
    in = Frama_C_double_interval(IN_MIN, IN_MAX);
    out2 = out1;
    out1 = out;
    out = ki[0]*in - ko[1]*out1 - ko[2]*out2;
    /*@ ghost
      @ e1_1 = e1;
      @ e2_1 = e2;
      @ e1 = l1*e1_1 + ki[0]*in;
      @ e2 = l2*e2_1 - ki[0]*in;
      @ */
  }
  //@ assert expr_out: out == (l1*e1 + l2*e2) / (l1 - l2);
  //@ ghost min_out = (MAX_e1 + MAX_e2) / (l1 - l2);
  //@ ghost max_out = (MIN_e1 + MIN_e2) / (l1 - l2);
  //@ assert out_bounded: min_out <= out <= max_out ;
}
```

Fig. 12.5 Analysis of a filter of order 2 having real eigenvalues

Of course, Wp cannot prove the last assertion (out_bounded), that is proven by Eva, using the penultimate assertion (expr_out). The limitation of this proof is that it relies on the *real* model of Wp. As stated by the remark on numeric error in Sect. 12.2, mechanizing the proof in the float model would be far more complicated. On the float model, Wp mainly uses interval reasoning, since it cannot use the associativity properties – false for floating-point arithmetic – required by linear algebra. Moreover, the transformation matrix D' is no longer diagonal in this case, and it now contains perturbed coefficients, preventing Wp from proving the proof obligations. Finally, let us note that this method is applicable for a filter of any order while its eigenvalues are real and simple (that is for a filter of order N with N distinct real eigenvalues).

12.5 Analysis of a Filter of Order 2 Having Complex Eigenvalues

Filters of order 2 having complex eigenvalues like $\lambda_1 = \alpha + \beta \times i$ and $\lambda_2 = \alpha - \beta \times i$ where

- $\alpha \in \mathbb{R}, \beta \in \mathbb{R}, |\alpha| + |\beta| \geq 1.0$
- $\alpha^2 + \beta^2 < 1.0$

are problematic for numerous analyzers based on interval arithmetic. Such a filter has no inductive invariant that can be expressed as an interval. And, this is also the case for any linear combination of the variables out, out1, out2.

Let us consider the concrete example of Fig. 12.6.

The characteristic polynomial is $p(\lambda) = ko[0] \times \lambda^2 + ko[1] \times \lambda + ko[2] = \lambda^2 - 1.5\lambda + 0.75$. Its discriminant is negative: $\Delta = 2.25 - 4 \times 0.75 = -0.75$, and its two

eigenvalues are: $\begin{cases} \lambda_1 = \dfrac{1.5 - \sqrt{0.75} \times i}{2} \approx 0.75 - 0.4330127 \times i \\ \lambda_2 = \dfrac{1.5 + \sqrt{0.75} \times i}{2} \approx 0.75 + 0.4330127 \times i \end{cases}$

With $\alpha = 0.75$ and $\beta = \dfrac{\sqrt{0.75}}{2}$, $|\alpha| + |\beta| \approx 1.1830127 \geq 1.0$ and $\alpha^2 + \beta^2 = 0.75 < 1.0$.

Just to show that no inductive invariant exists under the interval form, let us consider the following inductive invariant: $out_0 \in [-A, A]$ and $out_1 \in [-A, A]$ with $A > 0$. Then the instruction out = 1.5*out - 0.75*out1 + 0.5*in of the loop requires that $2.25 \times A + 5.0 \leq A$, which is impossible for $A > 0$.

Finding automatically a complex inductive invariant requires making many decisions in a huge research space. If some theories are present in Why3 libraries, these decisions could rely on simpler heuristics. In the other cases, several techniques exist to find invariant: as an ellipsoid with [7], as a zonotope with [5], as a set of boxes with [11], as a set of zonotopes with [9].

Fig. 12.6 Implementation of a filter of order 2 having complex eigenvalues

```
// parameters
const double ki[] = { 0.5 };
const double ko[] = { 1.0, -1.5, 0.75 };

// input range
const double IN_MIN = -10.0;
const double IN_MAX = 10.0;

int main()
{
  double out, out1, out2, in;
  in = Frama_C_double_interval(IN_MIN, IN_MAX);
  out2 = 0.0;
  out1 = 0.0;
  out = ki[0]*in - ko[1]*out1 - ko[2]*out2;
  //@ loop unroll 1000;
  while (Frama_C_interval(0, 1)) {
    in = Frama_C_double_interval(IN_MIN, IN_MAX);
    out2 = out1;
    out1 = out;
    out = ki[0]*in - ko[1]*out1 - ko[2]*out2;
  }
}
```

Hence, we present four different methods attached to the previously mentioned techniques that are compliant with a formal proof.

12.5.1 Invariant as a Set of Intervals

The principle of this invariant generation comes from the following fact related to over-approximated analyses: the thinner are the input intervals, the more precise are the guaranteed output intervals. As an instance, let us look at the function $f(x) = x \times x - x$ on the interval $[0.0, 1.0]$. The default interval analysis of Eva shows that $f([0.0, 1.0]) \subseteq [-1.0, 1.0]$. Nevertheless, the same Eva analysis shows that $f([0.0, 0.5]) \subseteq [-0.5, 0.25]$ and that $f([0.5, 1.0]) \subseteq [-0.75, 0.5]$. Combining these results strongly improves the initial analysis since it proves that $f([0.0, 1.0]) \subseteq [-0.75, 0.5]$. If we could infinitely subdivide the input intervals up to input points, we could obtain precise and optimal results that are $f([0.0, 1.0]) = [-0.25, 0.0]$.

The application of this principle coming from the constraint programming research field applies to the search of invariant and it is described in [11]. Algorithm 1 starts with an initial tuple of intervals, that over-approximates the invariant defined as the least-fixpoint containing the initial state $(out_i) = 0.0$. Note that if the initial tuple of intervals contains the initial state but not the least-fixpoint invariant, Algorithm 1 will generate an empty result.

1: *in* ← {initial-cube over-approximating an invariant}
2: **do**
3:   *new-in* ← ∅
4:   *new-out* ← ∅
5:   *has-finished* ← **true**
6:   **for** *interval-in* ∈ *in* **do**
7:     *interval-out* ← *transfer-function*(*interval-in*)   ▷ simulate Eva analysis
8:     *new-out* ← *new-out* ∪ *interval-out*
9:     **if** *interval-out* ⊆ *in* **then**
10:      *new-local-in* ← *interval-in*
11:    **else if** *interval-out* ∩ *in* ≠ ∅ **then**
12:      *new-local-in* ← division of *interval-in* in several parts   ▷ heuristic considerations
13:      *has-finished* ← **false**
14:    **else**
15:      *new-local-in* ← ∅
16:      *has-finished* ← **false**
17:    *new-in* ← *new-in* ∪ *new-local-in*
18:    *has-finished* ← *has-finished* ∧ *new-out* ⊆ *new-in*
19:    *in* ← *new-in* ∩ *new-out*
20: **while not** *has-finished*
21: **return** *in*

Algorithm 1 Algorithm to find an invariant as a set of intervals.

Then Algorithm 1 splits this tuple into several parts and, for each part, it intersects the output result of the transfer function with the union of the parts. The input parts that are not reached by at least one output result are removed from the invariant candidate. The split dynamically goes on until the algorithm finds a post fixpoint such that $f(X) \subseteq X$, where f is the transfer function of the filter.

The principle of the solution is robust and it finds an invariant for the actual implementation with floating-point types. Nevertheless, the heuristics concerning the initial cube, the choice between splitting or discarding, how the split and remove actions should interleave, how to split, may be complex to tune. An implementation can be found in the companion artifact [18]. Note that if the least-fixpoint invariant for out is $out_0 \in [-49.19, 49.19]$, this implementation generates a substantial over-approximation with $out_0 \in [-65.0, 65.0]$.

The final set of cubes for the example of Fig. 12.6 is displayed in Fig. 12.7.

Frama-C/Eva can prove the correctness of this invariant (see Fig. 12.8) because the generation method is based on intervals and on floating-point computations.

```
frama-c -eva -eva-slevel 600000 -eva-plevel 600 filter3-intervals.c
```

In the command-line above, `filter3-intervals.c` is the name of the file of Fig. 12.8. The option `-eva-plevel 600` prevents the analysis from merging the cells of the array `out_invariant` that contains about 600 cubes. `-eva-slevel 600000` prevents the analysis from merging the control flow of the cases where `has_found = true` and `has_found = false` in the nested loops. This is fundamental to prove the invariant.

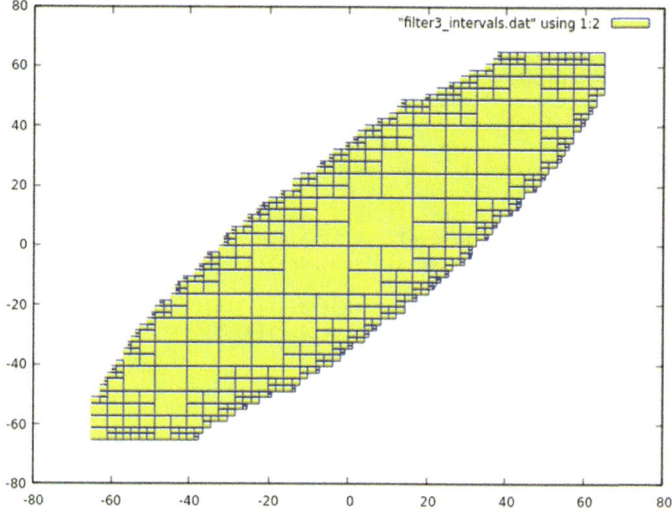

Fig. 12.7 Invariant as a set of cubes

Advantages of the method: This approach has many advantages; first, the proof is done in floating-point arithmetic, which corresponds to the implementation; second, Frama-C/Eva can check that the proof is correct with its arithmetic; finally, the user does not need to have a prior knowledge of the form of the invariant before applying the method that is relatively general for a wide range of floating-point invariant.

Drawbacks of the method: The main drawback lies in the generation of the invariant. Hence, if you need to modify the invariant, it is preferable not to edit it, but to provide a better generation algorithm. The invariant may also be strongly over-approximated, because its generation only evaluates the post-fixpoint properties. In particular, it does not consider the "least-fixpoint" character.

Note that the current proof is currently on a modified code intended to convince about the pertinence of the invariant for Eva. A better solution would use the `loop invariant` construction of ACSL. It is under experimentation.

12.5.2 Optimal Invariant with Linear Algebra

The diagonalization process presented in Sect. 12.2 can be used to find the "optimal" invariant as the least-fixpoint of the transfer function.

```
1  int main() {
2    double in, out;
3    /* defines out_inv */
4    #include "filter3_intervals.inch"
5    #define div_size sizeof(out_invariant)/(sizeof(double)*N*2)
6    //@ loop unroll div_size;
7    for (int div_index = 0; div_index < div_size; ++div_index) {
8      double (*out_div)[N][2] = &out_invariant[div_index];
9      in = Frama_C_double_interval(-10.0, +10.0);
10     in1 = Frama_C_double_interval(-10.0, +10.0);
11     in2 = Frama_C_double_interval(-10.0, +10.0);
12     outN = Frama_C_double_interval((*out_div)[N][0], (*out_div)[N][1]);
13     ...
14     out1 = Frama_C_double_interval((*out_div)[1][0], (*out_div)[1][1]);
15     out  = Frama_C_double_interval((*out_div)[0][0], (*out_div)[0][1]);
16     ...
17     ...; in2 = in1; in1 = in;
18     ...; out2 = out1; out1 = out;
19     in = Frama_C_double_interval(-10.0, +10.0);
20     out = ... - ko[2]*out2 - ko[1]*out1 + ki[2]*in2 + ki[1]*in1 + ki[0]*in;
21     ...
22     bool has_found = false;
23     //@ loop unroll div_size;
24     for (int rec_index = 0; !has_found && rec_index < div_size; ++rec_index) {
25       double (*rec_div)[N][2] = &out_invariant[rec_index];
26       has_found = (*rec_div)[N][0] <= outN && outN <= (*rec_div)[N][1] && ...
27          && (*rec_div)[1][0] <= out1 && out1 <= (*rec_div)[1][1]
28          && (*rec_div)[0][0] <= out && out <= (*rec_div)[0][1];
29     }
30     //@ assert has_found;
31   }
32 }
```

Fig. 12.8 Verification that the generation provides an invariant

We take the example of Fig. 12.6 to show the reasoning until obtaining the "optimal" invariant between the variable out and out1. The complete reasoning is available in [18].

Section 12.2 suggests writing the transfer function as

$$\begin{pmatrix} out_1^{n+1} \\ out_0^{n+1} \end{pmatrix} = \left(M = \begin{pmatrix} 0 & 1 \\ -0.75 & 1.5 \end{pmatrix} \right) \times \begin{pmatrix} out_1^n \\ out_0^n \end{pmatrix} + \begin{pmatrix} 0 \\ 0.5 \times in_0^n \end{pmatrix}$$

with $M = P^{-1} \times D \times P$ and P^{-1}, D, P defined by

$$P^{-1} = \begin{pmatrix} 1 + \frac{\sqrt{3}}{3}i & 1 - \frac{\sqrt{3}}{3}i \\ 1 & 1 \end{pmatrix}, D = \begin{pmatrix} 0.75 - \frac{\sqrt{3}}{4}i & 0 \\ 0 & 0.75 + \frac{\sqrt{3}}{4}i \end{pmatrix}, P = \begin{pmatrix} -\frac{\sqrt{3}}{2}i & 0.5 + \frac{\sqrt{3}}{2}i \\ \frac{\sqrt{3}}{2}i & 0.5 - \frac{\sqrt{3}}{2}i \end{pmatrix}$$

Since $M^k = P^{-1} \times D^k \times P$, we can prove by induction on n that

12 Analysis of Embedded Numerical Programs ...

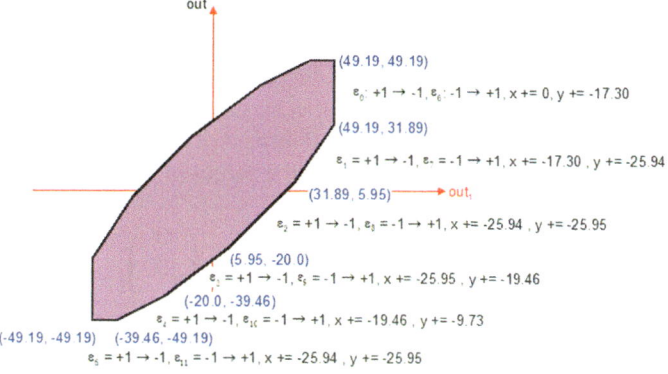

Fig. 12.9 Invariant as a optimal zonotope

$$\exists (\varepsilon_k \in [-1, 1])_{k \geq 0} \begin{cases} out_1^n \in [\underline{IN}, \overline{IN}] \times \sum_{k=0}^{n} \left(\frac{\sqrt{3}}{2}\right)^{k-1} \cos\left(\frac{\pi}{2} - k\frac{\pi}{6}\right) \varepsilon_k \\ out_0^n \in [\underline{IN}, \overline{IN}] \times \sum_{k=0}^{n} \left(\frac{\sqrt{3}}{2}\right)^{k} \cos\left(\frac{\pi}{3} - k\frac{\pi}{6}\right) \varepsilon_k \end{cases}$$

The inductive invariant is obtained with $\lim_{n \to +\infty}$. With some rewriting since 2π is a multiple of $\frac{\pi}{6}$, we obtain $\exists \varepsilon_k' \in ([-1.0, +1.0])_{0 \leq k < 12}$

$$\begin{cases} out_1 \in \dfrac{[\underline{IN}, \overline{IN}]}{1 - \frac{3^6}{2^{12}}} \times \sum_{k=0}^{11} \left(\left(\frac{\sqrt{3}}{2}\right)^{k-1} \cos\left(\frac{\pi}{2} - k\frac{\pi}{6}\right)\right) \times \varepsilon_k' \\ out_0 \in \dfrac{[\underline{IN}, \overline{IN}]}{1 - \frac{3^6}{2^{12}}} \times \sum_{k=0}^{11} \left(\left(\frac{\sqrt{3}}{2}\right)^{k} \cos\left(\frac{\pi}{3} - k\frac{\pi}{6}\right)\right) \times \varepsilon_k' \end{cases}$$

Contrary to a popular belief that sees the least-fixpoint (or optimal) invariant as an ellipsoid, this formula represents in fact a zonotope whose graphical representation is a zonohedron – it is a centrally symmetric convex polyhedron whose faces are also centrally symmetric convex polyhedra – of 12 faces (see Fig. 12.9) since there are only 6 different gradients between out_1 and out_0: $\left\{\frac{2}{\sqrt{3}} \times \frac{\cos(\frac{\pi}{3} - k\frac{\pi}{6})}{\cos(\frac{\pi}{2} - k\frac{\pi}{6})} / k \in [0, 5]\right\} = \{0, \frac{2}{3}, 1, \frac{4}{3}, 2, +\infty\}$.

The general form of this minimal invariant for filters with complex eigenvalues is usually not a zonotope, since the number of different gradients is infinite. The generic form is a "fractal" zonotope (see Fig. 12.11). The resulting "fractal" zonotope is the limit of the *out* vector defined with sequences of *in* having recent values in intervals and old values prior to *n* cycles at zero.

Formally, let us define $(In_k \in ([\underline{IN}, \overline{IN}])^{IN}$ and out_0^n, out_0^{n-1}, out_0^{n-2} the values of the variables out, out1, out2 at cycle n when the succession of input values is $in_0^k = In^{n-k}$ for $0 \leq k \leq n$. The domains of out_0^n are the same as the ones given by the semantics of the program, because the input values are in the same domain $[\underline{IN}, \overline{IN}]$ and we just rename (index) them from n to 0 instead of 0 to n.

out_n is a linear affine form with respect to the values of $(In_k)_{0 \leq k \leq n}$. Moreover,

$$\begin{pmatrix} out_0^{n-1} \\ out_0^n \\ out_0^{n+1} \end{pmatrix} - \begin{pmatrix} out_0^{n-2} \\ out_0^{n-1} \\ out_0^n \end{pmatrix} = M^n \begin{pmatrix} 0 \\ 0 \\ In_n \end{pmatrix}$$

Hence, the drawing in 3D of the zonotope expressing the relationships between out_0^{n+1}, out_0^n, out_0^{n-1} generates a zonohedron with the following steps from the previous zonohedron illustrating the relationships between out_0^n, out_0^{n-1}, out_0^{n-2} (see Fig. 12.10).

- Compute the 3D vector $M^n(0, 0, In^n)$. Its length tends to zero when n tends towards $+\infty$. It is a new generator to add to the previous zonohedron defined by out_0^n, out_0^{n-1}, out_0^{n-2}.
- Find the list of vertices in the previous zonohedron such that the line defined by the vertex and the 3D vector direction intersects the zonohedron in only one point, the vertex itself. This list of vertices defines a circumference of the convex hull of the previous zonotope.
- Split every vertex of the previous list in two and trace a segment corresponding to the 3D vector between the split vertices. The result is a new zonohedron.

Such a drawing shows that the initial faces of the zonotope never disappear and they remains flat. In the general case, the rotation angles θ of the complex eigenvalues $\rho e^{i\theta}$ of M has no rational relationship with π. Hence, any vertex of the zonohedron defined by out_0^n, out_0^{n-1}, out_0^{n-2} will be split in two by a further cycle $m > n$ in a new direction. This makes smoother the angles of the zonotopes, which creates the "fractal" zonotope when n tends toward $+\infty$ (see Fig. 12.11). A "fractal" zonotope has planar faces, but zooming on the vertices also shows a truncation of zonotope instead of a single point.

When an eigenvalue $\rho e^{i\theta}$ of M has a ρ close to 1, the "fractal" zonotope looks like an ellipsoid. But for convergent linear filters, ρ is fixed and is always strictly less than 1. Ellipsoids are another form of invariant but they over-approximate the least-fixpoint domain defined by the "fractal" zonotope – see the final ranges obtained in next subsections.

Advantages of the method: This approach provides an optimal invariant and therefore optimal bounds.

Drawbacks of the method: The proof is done in real arithmetic, which is an approximate model of the implementation; nevertheless, the transfer function of the accumulation of the rounding errors can also be written as a matrix transform, which provides

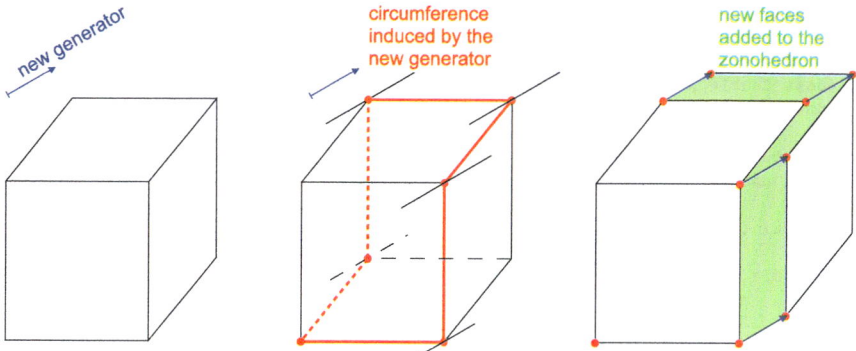

Fig. 12.10 Zonohedron construction with generators

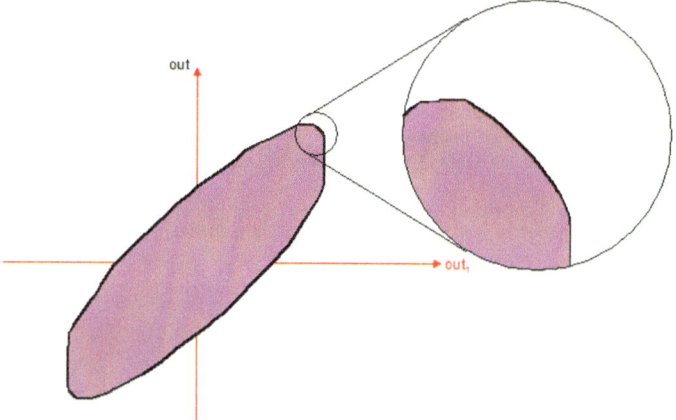

Fig. 12.11 Invariant as a fractal zonotope

an inductive invariant for the numeric error; an invariant for the floating-point implementation would be the sum of the inductive invariant for the ideal semantics and the inductive invariant for the numerical error. The second drawback is that the reasoning has been done by hand with the help of online diagonalization algorithms. We are currently looking at integrating it into **Frama-C/Wp** thanks to the **Why3** language. And, the final drawback is the form of the general invariant as fractal zonotope; it is difficult to manipulate this form in tools like **Frama-C** even if, from the filter output, the projection of the inductive invariant onto intervals provides a simple (non-inductive) invariant.

```
1  #include "__fc_builtin.h"
2
3  // parameters
4  const double ki[] = { 0.5 };
5  const double ko[] = { 1.0, -1.5, 0.75 };
6
7  // input range
8  const double IN_BOUND = 10.0;
9
10 int main()
11 {
12   double out, out1, out2, in;
13   in = Frama_C_double_interval(-IN_BOUND, IN_BOUND);
14   out2 = 0.0;
15   out1 = 0.0;
16   out = ki[0]*in - ko[1]*out1 - ko[2]*out2;
17   /*@ loop invariant ellipsis:
18     @    -49.5922*(IN_BOUND*0.5)*(IN_BOUND*0.5)
19     @    <= out*out + 0.75*out1*out1 - 1.5*out*out1
20     @    <= 49.5922*(IN_BOUND*0.5)*(IN_BOUND*0.5);
21     @ loop invariant add_ellipsis:
22     @    -7.5987*(IN_BOUND*0.5) <= out - out1 <= 7.5987*(IN_BOUND*0.5);
23     @ loop invariant add2_ellipsis:
24     @    -13.1974*(IN_BOUND*0.5) <= out - 1.5*out1 <= 13.1974*(IN_BOUND*0.5);
25     @ loop invariant out_bound:
26     @    -16.2632*(IN_BOUND*0.5) <= out <= 16.2632*(IN_BOUND*0.5);
27     @ */
28   while (Frama_C_interval(0, 1)) {
29     in = Frama_C_double_interval(-IN_BOUND, IN_BOUND);
30     out2 = out1;
31     out1 = out;
32     out = ki[0]*in - ko[1]*out1 - ko[2]*out2;
33   }
34 }
```

Fig. 12.12 Inductive loop invariant with ellipsoid

12.5.3 Invariant as an Ellipsoid

Finding an ellipsoid as an invariant like in [7] (see Fig. 12.12) addresses certain limitations of the previous method. The result is more comprehensible and can be verified manually. Moreover, it can be maintained even in case of modifications in the source code.

The principle of finding such an invariant consists first in expressing the invariant with a generic template form of an ellipsoid as below:

$$out_0^2 + a \times out_1^2 + b \times out_0 \times out_1 + c \times out_0 + d \times out_1 \leq e$$

Then a, b, c, d, e are symbolic values for which we look for a value with the method of variation of parameters. The transfer function of the digital filter propagates this property at the end and constraints on a, b, c, d, e are then generated to ensure that the previous property is an inductive invariant.

To isolate a common factor and to simplify the global system of constraints, it is necessary to introduce the following additional constraints:

$$\exists\, 0.0 < \lambda < 1.0. \begin{cases} a = \frac{0.5625}{\lambda} \\ b = \frac{-2.125}{\lambda + 0.75} \\ c = d = 0.0 \\ 2.25 - \lambda + \frac{0.5625}{\lambda} - \frac{3.375}{\lambda + 0.75} = 0 \end{cases}$$

Then with the solution $\lambda = 0.75$, the constraints simply becomes $3.25 \times 5.0^2 + 3 \times \sqrt{0.1875} \times 5.0 \times \sqrt{e} \leq \frac{e}{4}$ that has a minimal solution with $e = 1239.81$.

Hence, it is possible to verify by hand that

$$out_0^2 + 0.75 \times out_1^2 - 1.5 \times out_0 \times out_1 \leq 1239.81$$

is an inductive invariant (see the manual proof in [18]).

Its verification with Wp is not evident since the manual proof is complex. Why3 and Coq should provide a way to prove this invariant and then the proof could be imported in ACSL for Frama-C.

However, as mentioned in [10], a recent effort has achieved an automatic proof using Frama-C/Wp with a modified version of the solver Alt-Ergo [14]. This development is particularly significant within the Frama-C framework because the automatic proof succeeds by utilizing floating-point semantics, whereas the manual proof described earlier relies on ideal semantics.

Advantages of the method: The generated invariant is short, readable, understandable and verifiable by hand. It is possible to edit it as an ACSL annotation and to make it evolve.

Drawbacks of the method: The inductive invariant is only valid in real arithmetic. To generate the invariant, λ is found as a root of a polynomial of third degree – it may require the intervention of an external solver. If it is easy to write such an ACSL loop invariant, the Frama-C/Wp verification is hard (see the manual reasoning in [18]). The projection of this invariant in intervals is far from being optimal – it has an overapproximation of 65% in the example of Fig. 12.6 compared to the optimal invariant of previous section (see Fig. 12.13).

12.5.4 Invariant as a Zonotope

This part is a summary of some efforts spent by IRSN and CEA to synthesize an inductive loop invariant for the accuracy analysis of Fluctuat [5] for reactive numerical embedded programs of several tens of thousands of lines of code. Such programs are built over some numerical blocks. The blocks that prevent a quick and

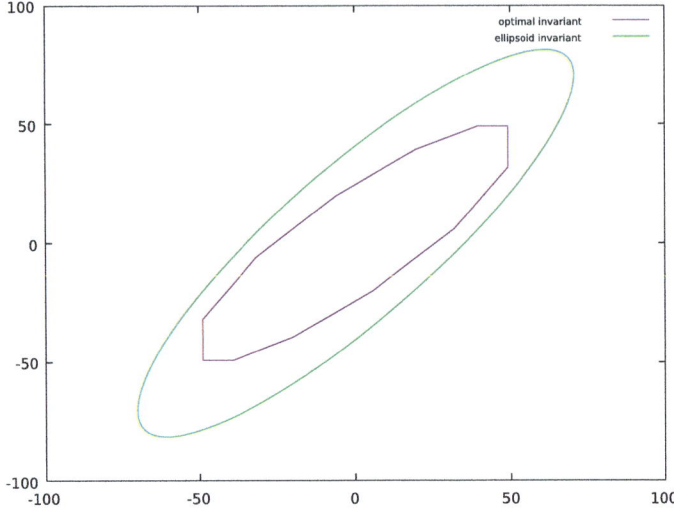

Fig. 12.13 Invariant as an ellipsoid

precise accuracy analysis are mainly the blocks with memories. It mainly concerns delay blocks, Proportional Integrative Derivative blocks, and linear digital filters with a cascade of filters. If the initial attempts use [5] to find an invariant as a zonotope, the obtained k-inductive invariant with $k \approx 500$ usually requires several thousands of analysis iterations with zonotopic abstract domains. The stabilization of these iterations considerably slows down the reaching of a fixpoint. This strategy is no longer suitable when the reactive loop contains several tens of thousands lines of code.

Recently, new attempts with the variation of parameters method have tried to synthesize an inductive loop invariant of the component alone as a zonotope that is understandable by the static analyzer. Then, the static analyzer widens its first iteration with this invariant. Since it is an inductive invariant, one cycle later after the widening, the static analysis just checks that a fixpoint is reached. The proof of invariance consists in an inclusion of zonotopes that requires some hints from the invariant synthesis. The same strategy can be applied with **Frama-C**.

The ellipsoid generation used a generic ellipsoid as template. The zonotope generation uses the first iterations of a static analysis by abstract interpretation to build a template. The affine forms chosen as abstract domains limit the over-approximations for linear digital filters components. Since the domains out_i for $0 \leq i < order$ (where $order$ is the order of the filter) are affine forms, the tuple of the $order$ domains forms a zonotopic domain.

Then, instead of applying a widening operator, which is common in the abstract interpretation framework, the static analysis is modified to introduce a symbolic perturbation over the affine forms. Then a symbolic propagation over one additional analysis cycle generates constraints over the symbolic perturbation, such that the perturbed affine form is a 1-inductive invariant. The symbolic perturbation is composed with a symbolic expansion factor $\frac{v_1}{64}, \frac{v_3}{64}, \frac{v_5}{64}$ and a symbolic stretch factor $\frac{v_2}{64}, \frac{v_4}{64}, \frac{v_6}{64}$ at every assignment of the variables `out2`, `out1`, and `out`. Since we impose constraints on the v_i to be in $[-1, +1]$, we are looking for an inductive invariant that is very close to the under-approximated iteration – at most $\frac{1}{64} + \frac{1}{64} = \frac{1}{32} \approx 3\%$. Hence, the searched inductive invariant is at most 3% greater than the optimal inductive invariant produced in Sect. 12.5.2.

The zonotope obtained after 33 loop iterations ($\alpha_i, \beta_i, \gamma_i$ are real constants in the formula below and ε_i, μ_i are free symbolic variables living in $[-1, 1]$) of the transfer function is perturbed as:

$$\begin{cases} out_2 = \alpha_0 + \sum \alpha_i \times \varepsilon_i \\ out_1 = \beta_0 + \sum \beta_i \times \varepsilon_i \\ out_0 = \gamma_0 + \sum \gamma_i \times \varepsilon_i \end{cases} \xrightarrow{\text{perturbation}} \begin{cases} out_2 = v_1 \left(\alpha_0 + \sum \alpha_i \times \varepsilon_i \right) + v_2 \times \mu_1 \\ out_1 = v_3 \left(\beta_0 + \sum \beta_i \times \varepsilon_i \right) + v_4 \times \mu_2 \\ out_0 = v_5 \left(\gamma_0 + \sum \gamma_i \times \varepsilon_i \right) + v_6 \times \mu_3 \end{cases}$$

The transfer function `out2 = out1; out1 = out; out = -0.75*out2 + 1.5*out1 + in;` infers a new perturbed zonotope at the end of the loop body:

$$\begin{cases} out_2 = \sum_i \left(\alpha'_i \sum_j \alpha'_j \times v_j \right) \times \varepsilon_i + \sum_i \left(\alpha''_i \sum_j \alpha''_j \times v_j \right) \times \mu_i \\ out_1 = \sum_i \left(\beta'_i \sum_j \beta'_j \times v_j \right) \times \varepsilon_i + \sum_i \left(\beta''_i \sum_j \beta''_j \times v_j \right) \times \mu_i \\ out_0 = \sum_i \left(\gamma'_i \sum_j \gamma'_j \times v_j \right) \times \varepsilon_i + \sum_i \left(\gamma''_i \sum_j \gamma''_j \times v_j \right) \times \mu_i \end{cases}$$

To obtain an inductive invariant, the new perturbed zonotope at the end of the loop has to be included in the perturbed zonotope at the beginning of the loop. For that, constraints over the v_i are generated in order to find (v_i) such that

$$\forall (\varepsilon'_i)_i \in [-1., +1]. \, \forall (\mu'_i)_i \in [-1., +1]. \, \exists (\varepsilon_i)_i \in [-1., +1]. \, \exists (\mu_i)_i \in [-1., +1].$$

$$\begin{cases} \sum_i \left(\alpha'_i \sum_j \alpha'_j \times v_j \right) \times \varepsilon_i + \sum_i \left(\alpha''_i \sum_j \alpha''_j \times v_j \right) \times \mu_i = v_1 \left(\alpha_0 + \sum \alpha_i \times \varepsilon'_i \right) + v_2 \times \mu'_1 \\ \sum_i \left(\beta'_i \sum_j \beta'_j \times v_j \right) \times \varepsilon_i + \sum_i \left(\beta''_i \sum_j \beta''_j \times v_j \right) \times \mu_i = v_3 \left(\beta_0 + \sum \beta_i \times \varepsilon'_i \right) + v_4 \times \mu'_2 \\ \sum_i \left(\gamma'_i \sum_j \gamma'_j \times v_j \right) \times \varepsilon_i + \sum_i \left(\gamma''_i \sum_j \gamma''_j \times v_j \right) \times \mu_i = v_5 \left(\gamma_0 + \sum \gamma_i \times \varepsilon'_i \right) + v_6 \times \mu'_3 \end{cases}$$

With heuristics like $\varepsilon'_i = \varepsilon_{i+1}$ and Fourier-Motzkin transformations, the prototype of CEA and IRSN sends the following constraints to a Simplex solver:

$$\begin{cases} -1.0 <= v_1, v_2, v_3, v_4, v_5, v_6 <= 1.0 \\ 3.562\text{e-}4 - 8.12\text{e-}2 \times v_3 - 1.1414\text{e-}1 \times v_4 + 9.6584\text{e-}2 \times v_5 - 7.625\text{e-}2 \times v_6 >= 0 \\ -1.318\text{e-}2 \times v_3 + 1.318\text{e-}2 \times v_5 <= 0 \\ 2.03584\text{e-}2 + 1.223\text{e-}2 \times v_3 - 1.522\text{e-}1 \times v_4 - 1.22\text{e-}1 \times v_5 + 1.525\text{e-}1 \times v_6 <= 0 \\ 1.525\text{e-}1 \times v_6 >= 0 \\ 2.0358\text{e-}2 + 1.319\text{e-}1 \times v_1 - 1.5187\text{e-}1 \times v_2 - 1.3159\text{e-}1 \times v_3 + 1.522\text{e-}1 \times v_4 <= 0 \\ 1.5218\text{e-}1 \times v_4 >= 0 \end{cases}$$

The solver then provides the following solution for the constraints

$$\begin{cases} v_1 = v_3 = v_5 = 1.0 \\ v_2 = 0.2722 \\ v_4 = 0.1358 \\ v_6 = 0.0 \end{cases}$$

This provides the inductive invariant below for the example of Fig. 12.6.

$$\begin{cases} out_2 = (-0.013781 \times \varepsilon_1 + 0.01837 \times \varepsilon_3 + 0.03675 \times \varepsilon_4 + 0.04900 \times \varepsilon_5 + 0.04900 \times \varepsilon_6 + 0.03267 \times \varepsilon_7 \\ \quad + -0.04355 \times \varepsilon_9 + -0.08711 \times \varepsilon_{10} + -0.11615 \times \varepsilon_{11} + -0.11615 \times \varepsilon_{12} + -0.07743 \times \varepsilon_{13} + 0.103241 \times \varepsilon_{15} \\ \quad +0.20648 \times \varepsilon_{16} + 0.27531 \times \varepsilon_{17} + 0.27531 \times \varepsilon_{18} + 0.18354 \times \varepsilon_{19} + -0.24472 \times \varepsilon_{21} + -0.48944 \times \varepsilon_{22} \\ \quad + -0.65259 \times \varepsilon_{23} + -0.65259 \times \varepsilon_{24} + -0.43506 \times \varepsilon_{25} + 0.58008 \times \varepsilon_{27} + 1.16016 \times \varepsilon_{28} + 1.546875 \times \varepsilon_{29} \\ \quad +1.546875 \times \varepsilon_{30} + 1.03125 \times \varepsilon_{31} + 0.041343 \times \mu_1) \times (IN\_BOUND \times 0.5) \\ out_1 = (-0.02067 \times \varepsilon_1 + -0.01378 \times \varepsilon_2 + 0.01837 \times \varepsilon_4 + 0.03675 \times \varepsilon_5 + 0.04900 \times \varepsilon_6 + 0.04900 \times \varepsilon_7 \\ \quad +0.03267 \times \varepsilon_8 + -0.04355 \times \varepsilon_{10} + -0.08711 \times \varepsilon_{11} + -0.11615 \times \varepsilon_{12} + -0.11615 \times \varepsilon_{13} + -0.07743 \times \varepsilon_{14} \\ \quad +0.10324 \times \varepsilon_{16} + 0.20648 \times \varepsilon_{17} + 0.27531 \times \varepsilon_{18} + 0.27531 \times \varepsilon_{19} + 0.18354 \times \varepsilon_{20} + -0.24472 \times \varepsilon_{22} \\ \quad + -0.48944 \times \varepsilon_{23} + -0.65259 \times \varepsilon_{24} + -0.65259 \times \varepsilon_{25} + -0.43506 \times \varepsilon_{26} + 0.58008 \times \varepsilon_{28} + 1.16016 \times \varepsilon_{29} \\ \quad +1.546875 \times \varepsilon_{30} + 1.546875 \times \varepsilon_{31} + 1.03125 \times \varepsilon_{32} + 0.02067 \times \mu_2) \times (IN\_BOUND \times 0.5) \\ out_0 = (-0.02067 \times \varepsilon_1 + -0.02067 \times \varepsilon_2 + -0.01378 \times \varepsilon_3 + 0.01837 \times \varepsilon_5 + 0.03675 \times \varepsilon_6 + 0.04900 \times \varepsilon_7 \\ \quad +0.04900 \times \varepsilon_8 + 0.03267 \times \varepsilon_9 + -0.04355 \times \varepsilon_{11} + -0.08711 \times \varepsilon_{12} + -0.11615 \times \varepsilon_{13} + -0.11615 \times \varepsilon_{14} \\ \quad + -0.077431 \times \varepsilon_{15} + -0.103241 \times \varepsilon_{17} + 0.206483 \times \varepsilon_{18} + 0.275311 \times \varepsilon_{19} + 0.275311 \times \varepsilon_{20} + 0.183540 \times \varepsilon_{21} \\ \quad + -0.244720 \times \varepsilon_{23} + -0.489441 \times \varepsilon_{24} + -0.652588 \times \varepsilon_{25} + -0.652588 \times \varepsilon_{26} + -0.435059 \times \varepsilon_{27} + 0.580078 \times \varepsilon_{29} \\ \quad +1.160156 \times \varepsilon_{30} + 1.546875 \times \varepsilon_{31} + 1.546875 \times \varepsilon_{32} + 1.03125 \times \varepsilon_{33}) \times (IN\_BOUND \times 0.5) \end{cases}$$

The verification of this invariant is available in [18]. It only involves hints to define the ε'_i, μ'_i as linear combinations of ε_i, μ_i. Then it is easy to check that the ε'_i, μ'_i are in $[-1, +1]$ with basic interval arithmetic. With these definitions, linear simplifications show that the property is an inductive invariant. The projection of this invariant onto intervals provides $out_0 \in [-50.32, 50.32]$, which is close to the optimal interval $[-49.19, 49.19]$ found in Sect. 12.5.2.

Advantages of the method: If this approach succeeds, it provides an abstraction very close to the optimal invariant – guaranteed to be at most 3% greater than the optimal invariant. The generated invariant, if not editable, is relatively small (100 coefficients instead of 2000 coefficients for the set of cubes approach). The reasoning used to prove the invariant is very basic – linear simplification and basic interval arithmetic. This approach should also work to provide bounds for the numeric error, which could give a bound for the floating-point implementation.

Drawbacks of the method: The inductive invariant is currently only valid in real arithmetic. The generation of the invariant involves many heuristics (because the constraints mix the abs function and linear algebra) and complex algorithms of linear algebra, like Fourier-Motzkin and the Simplex algorithm. Frama-C is currently not able to check this invariant, but, since the Eva analysis domain Numerors [8] is based on zonotopes, we hope that some modifications of this domain should be able to prove that the above property is an inductive invariant.

12.6 Conclusion

This chapter provides different strategies to analyze numerical reactive systems and to find bounds for variables that are valid at every cycle. It is focused on providing and proving inductive invariant and this is illustrated with emblematic components of systems such as linear digital filters.

For numerical reactive systems, we propose to provide invariants for the components before applying Frama-C/Eva on the system. The Frama-C team is currently working on integrating the theory of standard components like the linear digital filters in Wp with dedicated solvers like Colibri-2 (see Sects. 12.4 and 12.5.2). This should be a straightforward solution. This solution is built over the alternatives presented in this chapter by automating the reasoning of the engineer. Hence, knowing this reasoning is interesting to understand the different verification possibilities, how they provide solutions and their limitations.

- For filters of order 1, the natural form of inductive invariant is based on intervals; hence, Sect. 12.3 shows how Frama-C/Eva automatically finds and proves the bounds of the filter.
- For filters of order 2 or higher, we propose to extract the linear matrix that represents the transfer function and to find the eigenvalues of this matrix.
- If the eigenvalues are all real numbers (less than 1.0), then the solution proposed in Sect. 12.4 applies. Some ghost variables defined with the coefficients of the Vandermonde matrix indicate to Frama-C how to translate this problem into a problem containing only filters of order 1. This verification has some holes since the verification of the transformation is done with Wp in real numbers and the verification of the transformed invariant is done with Eva in floating-point numbers. The consideration on the accumulation of the rounding errors at the end of Sect. 12.2 gives confidence in the results. It is still possible (but it requires consequent Why3 modeling work) to express the equations of the rounding errors and to solve them with the introduction of new ghost variables defined with the coefficients of a new Vandermonde matrix. The new Vandermonde matrix is very close (but a bit different) to the previous one used for the real number domains.
- If some eigenvalues are complex and not real, then Sect. 12.5 applies. The engineer should choose between the following forms of invariant, like a set of intervals, an ellipsoid or a zonotope. Techniques based on constraint solving (set of intervals) or

on the variation of parameters (ellipsoid, zonotope) enable us to find an inductive invariant, that Frama-C could prove with more or less efforts.
- Invariants expressed as a set of cubes can be proven using Eva, and they provide bounds for the actual floating-point implementation. Such invariants are nevertheless generated by an external program, and they often rely on several hundreds of generated numbers.
- On the contrary, invariants expressed as an ellipsoid are easy to read as they depend only on 3 or 4 coefficients, which are generated in [10]. Their proof can be done manually for real numbers, but it requires Why3 and Coq imports in Frama-C/Wp. The work of [10] can perform an automatic proof for floating-point numbers of the generated ellipsoids using a modified version of Alt-Ergo [14].
- An intermediate possibility is to choose the invariant as a zonotope with about thirty coefficients to obtain a meaningful invariant. The proof then requires simple reasoning mixing interval and linear relationships. Like for the manual ellipsoid approach, it also relies on real numbers. The same invariant representation and generation technique should also apply on the numeric error, which provides a way to bound the implementation, defined as the addition of the ideal outputs with the rounding errors.

The abundance of possible verification techniques offered by Frama-C is a good starting point for the analysis of numerical software. Future work will offer a more intuitive support for such code, like the integration of a theory concerning the linear digital filters and the integration of equations for the rounding errors to link the results of Eva and Wp.

References

1. Barrett CW, Conway CL, Deters M, Hadarean L, Jovanovic D, King T, Reynolds A, Tinelli C (2011) CVC4. In: Gopalakrishnan G, Qadeer S (eds) Computer aided verification—23rd international conference, CAV 2011, Snowbird, UT, USA, July 14–20, 2011. proceedings, LNCS, vol 6806, pp 171–177. Springer. https://doi.org/10.1007/978-3-642-22110-1_14
2. Baudin P, Filliâtre JC, Marché C, Monate B, Moy Y, Prevosto V ACSL: ANSI/ISO C specification language. http://frama-c.com/acsl.html
3. Blazy S, Bühler D, Yakobowski B (2017) Structuring abstract interpreters through state and value abstractions. In: International conference on verification, model checking, and abstract interpretation (VMCAI). https://doi.org/10.1007/978-3-319-52234-0_7
4. Conchon S, Coquereau A, Iguernlala M, Mebsout AM (2018) Alt-Ergo 2.2. In: International workshop on satisfiability modulo theories, Oxford, United Kingdom
5. Delmas D, Goubault E, Putot S, Souyris J, Tekkal K, Védrine F (2009) Towards an industrial use of FLUCTUAT on safety-critical avionics software. In: Formal methods for industrial critical systems, FMICS. https://doi.org/10.1007/978-3-642-04570-7_6
6. Feret J (2004) Static analysis of digital filters. In: European symposium on programming (ESOP). https://doi.org/10.1007/978-3-540-24725-8_4
7. Feret J (2005) Numerical abstract domains for digital filters. In: International workshop on numerical & symbolic abstract domains (NSAD 2005)
8. Jacquemin M (2021) Arithmétiques relationnelles pour l'analyse par interprétation abstraite de propriétés de précision numérique. (Relational arithmetics for abstract interpretation based

analysis of numerical accuracy properties). PhD thesis, University of Paris-Saclay, France. https://theses.hal.science/tel-03566701. (In French)
9. Kabi B, Goubault E, Miné A, Putot S (2020) Combining zonotope abstraction and constraint programming for synthesizing inductive invariants. In: Software verification, VSTTE, and 13th international workshop, NSV 2020. https://doi.org/10.1007/978-3-030-63618-0_14
10. Khalife E, Garoche PL, Farhood M (2023) Code-level formal verification of ellipsoidal invariant sets for linear parameter-varying systems. In: NASA formal methods—15th international symposium, NFM 2023, Houston, TX, USA, May 16–18, 2023, proceedings, *LNCS*, vol 13903, pp 157–173. Springer. https://doi.org/10.1007/978-3-031-33170-1_10
11. Miné A, Breck J, Reps TW (2016) An algorithm inspired by constraint solvers to infer inductive invariants in numeric programs. In: Programming languages and systems, ESOP. https://doi.org/10.1007/978-3-662-49498-1_22
12. de Moura L, Bjørner N (2008) Z3: an efficient SMT solver. In: Ramakrishnan CR, Rehof J (eds) Tools and algorithms for the construction and analysis of systems, pp 337–340. Springer
13. Owre S, Rushby JM, Shankar N (1992) PVS: a prototype verification system. In: Automated deduction—CADE-11, 11th international conference on automated deduction, Saratoga Springs, NY, USA, June 15–18, 1992, Proceedings, *LNCS*, vol 607, pp 748–752. Springer. https://doi.org/10.1007/3-540-55602-8_217
14. Roux P, Iguernelala M, Conchon S (2018) A non-linear arithmetic procedure for control-command software verification. In: 24th international conference on tools and algorithms for the construction and analysis of systems (TACAS), April 2018, Thessaloniki, Greece, pp 132–151
15. Roux P, Iguernlala M, Conchon S (2018) A non-linear arithmetic procedure for control-command software verification. In: Beyer D, Huisman M (eds) Proceedings of the 24th international conference, TACAS 2018, Thessaloniki, Greece, April 14–20, 2018, Proceedings, pp 132–151. https://doi.org/10.1007/978-3-319-89963-3_8
16. Roux P, Jobredeaux R, Garoche PL, Féron E (2012) A generic ellipsoid abstract domain for linear time invariant systems. In: Proceedings of the 15th ACM international conference on hybrid systems: computation and control, HSCC 2012, Beijing, China, April 17–19, 2012. https://doi.org/10.1145/2185632.2185651
17. Titolo L, Moscato MM, Feliú MA, Muñoz CA (2020) Automatic generation of guard-stable floating-point code. In: Integrated formal methods—16th international conference, IFM 2020, Lugano, Switzerland, November 16–20, 2020, Proceedings, *LNCS*, vol. 12546, pp 141–159. Springer. https://doi.org/10.1007/978-3-030-63461-2_8
18. Vedrine F, Piriou PY, David V (2023) Examples, proofs and algorithms for the verification of loop invariant of linear filters with Frama-C. https://doi.org/10.5281/zenodo.7695668
19. Volkova A (2017) Towards reliable implementation of digital filters. PhD thesis, Université Pierre et Marie Curie–Paris VI. https://theses.hal.science/tel-01916214
20. Volkova A, Hilaire T, Lauter C (2015) Reliable evaluation of the worst-case peak gain matrix in multiple precision. In: 2015 IEEE 22nd symposium on computer arithmetic, pp 96–103
21. Wikipedia: Matrix norms induced by vector norms. https://en.wikipedia.org/wiki/Matrix_norm#Matrix_norms_induced_by_vector_norms

Part III
Case Studies and Industrial Applications

Chapter 13
An Exercise in Mind Reading: Automatic Contract Inference for Frama-C

Jesper Amilon, Zafer Esen, Dilian Gurov, Christian Lidström, and Philipp Rümmer

Abstract Using tools for deductive verification, such as Frama-C, typically imposes substantial work overhead in the form of manually writing annotations. In this chapter, we investigate techniques for alleviating this problem by means of automatic inference of ACSL specifications. To this end, we present the Frama-C plugin Saida, which uses the assertion-based model checker TriCera as a back-end tool for inference of function contracts. TriCera transforms the program, and specifications provided as **assume** and **assert** statements, into a set of constrained Horn clauses (CHC), and relies on CHC solvers for the verification of these clauses. Our approach assumes that a C program consists of one entry-point (main) function and a number of helper functions, which are called from the main function either directly or transitively. Saida takes as input such a C program, where the main function is annotated with an ACSL function contract, and translates the contract into a harness function, comprised mainly of **assume** and **assert** statements. The harness function, together with the original program, is used as input for TriCera and, from the output of the CHC solver, TriCera infers pre- and post-conditions for all the helper functions in the C program, and translates them into ACSL function contracts. We illustrate on several examples how Saida can be used in practice, and discuss ongoing work on extending and improving the plugin.

J. Amilon (✉) · D. Gurov · C. Lidström
KTH Royal Institute of Technology, Stockholm, Sweden
e-mail: jamilon@kth.se

D. Gurov
e-mail: gurov@kth.se

C. Lidström
e-mail: clid@kth.se

Z. Esen
Uppsala University, Uppsala, Sweden
e-mail: zafer.esen@it.uu.se

P. Rümmer
University of Regensburg, Regensburg, Germany
e-mail: philipp.ruemmer@ur.de

Keywords Deductive verification · Contract inference · Assertion-based model checking · Constrained Horn clauses

13.1 Introduction

Static program verification approaches, such as deductive verification [1, 13], abstract interpretation [10, 12], and model checking [11], offer a level of confidence in the *correctness* of software that testing alone cannot provide. Software development in certain industrial domains, in particular the ones concerned with systems of a safety-critical nature, increasingly relies on such *formal methods*, in part driven by safety standards such as the automotive ISO-26262 and avionics DO-178C standards, to provide the needed correctness guarantees. One of the most mature existing frameworks for the static analysis and verification of software written in the C programming language is Frama-C [28]. The framework is modular and supports, through its many plugins, a wealth of different static analyses.

The term *deductive verification* refers to the task of proving, by means of automated deduction, that a program is correct with respect to a specification that is provided by the programmer in the form of logical annotations. In the context of the C programming language, these annotations are typically pre- and post-conditions for C functions (called function *contracts*), loop invariants, or frame conditions describing the memory that is or is not allowed to be accessed by the C function. Frama-C supports deductive verification through its Wp plugin [5], and the language for writing annotations is called ACSL [6].

However, even when the correctness requirements for a C program are spelled out and provided to the programmer in some informal or semi-formal format, the task of producing ACSL annotations in the program code that capture these requirements is notoriously time-consuming, expertise-demanding, and error-prone, and often gives rise to an annotation overhead comparable with the size of the code itself. This has so far been one of the main obstacles to the wider-spread adoption of deductive verification in industry [33]. We see the *automation of the annotation process* as one of the primary means of addressing this obstacle. In this chapter, we describe one approach for tackling this problem.

In earlier work [3], we proposed a technique for automated inference of annotations based on *logical interpolation* applied to a representation of annotated programs as *constrained Horn clauses* (CHC). Model checkers based on these concepts include Eldarica [25], SeaHorn [8], and others [22]. In particular, we explored the TriCera [18] model checker to automatically infer function contracts as ACSL annotations for Frama-C. Our technique allows, from a given *top-level* specification for a C module (i.e., contract for its main function), to automatically infer contracts for all remaining (or *helper*) functions. Since software specifications for C programs are typically written in industry at the module-level rather than for individual functions, these specifications can be taken as (the starting point for) the contract for the

13 Automatic Contract Inference for Frama-C

```
1  int r, x;
2
3  void incr_x() {
4      x = x + 1;
5  }
6
7  /*@
8      requires \true;
9      ensures r == \old(x) + 2;
10  */
11 void main() {
12     incr_x();
13     incr_x();
14     r = x;
15 }
```

```
1  /*@
2      requires \true;
3      ensures x - \old(x) == 1 &&
4              r == \old(r);
5  */
6  void incr_x();
```

(a) Example code and function contract.

(b) Inferred contract for incr_x.

Fig. 13.1 Simple example C program with ACSL specification

module's entry-point, or main function, and contracts for all other functions called within the module can then be inferred.

We present in this chapter, building on our previous work, a Frama-C plugin called Saida. The plugin is still under development, but can already be used to infer useful contracts in a wide variety of cases.

Illustrating example. Consider the program in Fig. 13.1a, consisting of a main function making two calls to the helper function incr_x. The main function is annotated with a contract that specifies the desired properties of the program.[1] Verifying this program with procedure-modular approaches (such as the one used by the Wp plugin) requires that there is a contract for incr_x as well. To free the programmer from the need to manually provide contracts for helper functions such as incr_x, we propose to generate such contracts automatically by utilizing existing techniques for contract inference. In particular, we make use of the TriCera model checker. Our Saida plugin first transforms the program and the contract for main into what is called a *harness function* (see Sect. 13.3.1), which is then used as input for TriCera. When verifying harness functions, we instruct TriCera to infer contracts for all (helper) functions called from the main function. In this case, TriCera infers for incr_x the function contract shown in Fig. 13.1b. Saida then inserts the inferred contract into the original program and, thereafter, verification in Frama-C can proceed using, e.g., the Wp plugin.

Limitations. While the automation of the annotation process is crucial for reducing the manual annotation effort, it also comes with certain limitations. First, the inferred contracts are heavily dependent on the top-level specification: they are not necessarily the strongest (or most complete) ones, but are just sufficient to prove the contract of

---

[1] Note that the pre-condition true can be omitted since it is the default pre-condition of ACSL function contracts. We include it here for clarity.

the main function. Second, the resulting contracts are closely tied to the source code of the functions for which contracts are inferred and of the functions calling them. The contracts can therefore be seen as documenting the code rather than its *intended functionality*, which is what one usually expects from contracts. And third, since the resulting contracts are dependent on the context in which the function is called, they may not be suitable for reuse at arbitrary other call sites. Later in the chapter, we include a discussion on future work for addressing these limitations.

Related work. One area of research into inferring contracts focuses on the notion of *strongest post-conditions*, also called *function summaries*, and is the result of the pioneering work of Dijkstra [16]. Here, inference starts from a pre-condition and results in a post-condition that is the strongest possible contract for the given pre-condition. While this is useful for inferring a complete characterization of the possible final states, it has the downside of generating overly verbose assertions, and does not capture the *intention* of the program. Symbolic execution can be used to infer strongest post-conditions, as long as the program does not contain unbounded loops [21]. In [36], an algorithm is presented that converts the exponentially large post-conditions resulting from strongest post-condition computations into a more concise and usable form.

A similar approach is based on the dual notion of *weakest pre-conditions*, i.e., the most general condition that is sufficient to ensure a given post-condition. The original weakest pre-condition calculus, as pioneered by Dijkstra [15], and in which post-conditions are transformed into pre-conditions, is the basis of many deductive verification tools, such as the Wp plugin of Frama-C. Pre-conditions can also be computed by symbolic execution, or, alternatively, by a combination of abstract interpretation and quantifier elimination as proposed in [31]. While this method is able to infer pre-conditions in the presence of loops, the computed pre-conditions are not necessarily the weakest. In [34, 35], an algorithm based on Counter-Example Guided Abstraction Refinement (CEGAR) is presented. By starting from an over-approximation of weakest pre-conditions, the algorithm can iteratively cull invalid pre-states until sufficient (and necessary) pre-conditions remain. While the technique can handle loops, it is in general not guaranteed to terminate.

Most methods only take as input the code of the program for which specifications are to be inferred. Property-guided contract inference methods, on the other hand, aim to find contracts that are sufficient to verify a given property rather than the most general ones. The approach presented in this paper is one such method, and has the advantage that *partial contracts* can often be found, even when inferring complete contracts is infeasible. Weakest pre-condition methods, in particular *maximal specification inference*, as proposed in [2], can also start from existing specifications. Maximal specification synthesis is a generalization of weakest pre-condition computation, and considers the specifications of multiple functions. Related to this is the approach proposed in [14], where syntactic patterns are used to infer annotations for automatically generated code.

Structure. The chapter is organized as follows. Section 13.2 provides some theoretical background, defines the language fragment targeted, and describes the model checker TriCera. Section 13.3 explains how TriCera is used to infer contracts. Section 13.4 presents the Saida plugin, and describes how it can be used in practice, illustrated on examples. In Sect. 13.5, several current and potential future extensions of Saida are discussed. Finally, Sect. 13.6 concludes the chapter.

13.2 Preliminaries

We start by introducing the core concepts required in this chapter: deductive verification, the considered language fragment, and model checking based on constrained Horn clauses.

13.2.1 Hoare Logic and Contract-Based Verification

This section introduces Hoare logic [23], logical variables and how Hoare logic extends to contract-based verification in the form of Hoare logic contracts.

In Hoare logic, program verification is based on judgments in the form of Hoare triples. A Hoare triple is written $\{P\}S\{Q\}$, where P is called the pre-condition, Q the post-condition and S is the program to be verified. In this section, we consider P and Q as first-order predicates, and $\{P\}S\{Q\}$ is evaluated over program states as follows. $\{P\}S\{Q\}$ holds with respect to *partial correctness* if executing S from any state s where P holds, either diverges (never terminates) or terminates in a state s' where Q holds. $\{P\}S\{Q\}$ holds with respect to *total correctness* if executing S from any state s where P holds, terminates in a state s' where Q holds. For the remainder of this section, we shall consider partial correctness semantics only. The state s of such an execution is called the *pre-state* and s' the *post-state* of the execution. In this section, a state is considered a mapping from program variables to values.

As an example, consider the following Hoare triple over the program variables r, x and n, intended to specify that S calculates the value of x rem n, where rem is the remainder operator, and stores the result in r:

$$\{n > 0\} \ S \ \{r = x \text{ rem } n\}. \tag{13.1}$$

Figure 13.2a shows a C implementation of the modulo operator such that the Hoare triple $\{n > 0\}$ cmod() $\{r = x \text{ rem } n\}$ holds.

Logical variables. In a Hoare triple, the pre- and post-conditions typically aim to relate the values of certain variables before and after execution of the statement S. This requires the introduction of *logical variables*. Logical variables are not program variables and are hence never altered during program execution.

To understand why logical variables are needed, consider again the Hoare triple in (13.1). Since the values of x and n are evaluated in the post-state, the Hoare triple

```
  int x, n, r;

void cmod() {
    r = (x / n) * n;
    r = x - r;
}
```

(a) Real C implementation of the modulo operator.

```
int x, n, r;

void foo() {
    x = 1;
    n = 1;
    r = 0;
}
```

(b) Bogus C implementation of the modulo operator.

Fig. 13.2 Two implementations of the modulo operator

can be satisfied by letting S alter the values of x and n until Q holds. An example of such a program is shown in Fig. 13.2b, where function foo simply assigns 1 to x and n, and 0 to r and, since 1 rem 1 = 0, $\{n > 0\}$ foo() $\{r = x \text{ rem } n\}$ holds. However, foo does not fulfill the property *intended* to be specified with the Hoare triple in (13.1). Therefore, consider instead the following contract for the modulo operator, where x_0 and n_0 are logical variables.

$$\{n > 0 \wedge x = x_0 \wedge n = n_0\} \, S \, \{r = x_0 \text{ rem } n_0\} \tag{13.2}$$

In the post-condition, the value of r is now constrained by the values of x_0 and n_0, which, by the pre-condition, evaluate to the values of x and n in the pre-state. Therefore, with cmod given as in Fig. 13.2a,

$$\{n > 0 \wedge x = x_0 \wedge n = n_0\} \, \text{cmod}() \, \{r = x_0 \text{ rem } n_0\}$$

still holds, as expected. However for foo in Fig. 13.2b,

$$\{n > 0 \wedge x = x_0 \wedge n = n_0\} \, \text{foo}() \, \{r = x_0 \text{ rem } n_0\}$$

does not hold, as desired (below, we provide further intuition as to why this is true by considering Hoare triples as contracts).

Hoare logic contracts. Verification using Hoare triples can be lifted to contract-based verification, by separating the pre- and post-condition from the program implementation. That is, from a Hoare triple $\{P\}S\{Q\}$, we can extract the Hoare logic contract $C = (P, Q)$. Extracting the pre- and post-condition into a contract is often useful, e.g., for procedure-modular verification. As an example, consider the following contract extracted from the Hoare triple in (13.2):

$$C \stackrel{\text{def}}{=} (n > 0 \wedge x = x_0 \wedge n = n_0, \, r = x_0 \text{ rem } n_0) \tag{13.3}$$

To rely on contracts for verification, we must define formally when a program satisfies a contract. To this end, we introduce the notion of *interpretations* over logical variables. An interpretation $\mathcal{I}$ is simply a mapping from logical variables to values. For example, $\mathcal{I} = [x_0 \mapsto 5, n_0 \mapsto 2]$ is an interpretation of the logical variables x_0 and n_0 in (13.3).

Using the notion of interpretations over logical variables, we say that a program S satisfies the contract $C = (P, Q)$ with respect to $\mathcal{I}$, denoted $S \models_\mathcal{I} C$, if and only if $\{P\}S\{Q\}$ holds, where the values of logical variables in P and Q are given by $\mathcal{I}$. Furthermore, let $S \models C$ denote that S satisfies C, defined by quantifying over interpretations:

$$S \models C \quad \text{if and only if} \quad \forall \mathcal{I}.\ S \models_\mathcal{I} C$$

As an example, consider again the contract C for the modulo operator in (13.3) and let P and Q denote the pre- and post-condition of C, respectively. Also, let $\mathcal{I}$ be an interpretation over x_0 and n_0, and s be a program state such that P holds when evaluated over s and $\mathcal{I}$, i.e.,

$$\mathcal{I}(x_0) = s(x) \ \wedge\ \mathcal{I}(n_0) = s(n) \ \wedge\ s(n) > 0,$$

where $s(x)$ and $s(n)$ is the value of x and n in s, respectively, and $\mathcal{I}(x_0)$ is the value of x_0 in $\mathcal{I}$. Then, executing the cmod function in Fig. 13.2a will terminate in a state s' such that Q holds over s' and $\mathcal{I}$. Therefore, $\{P\}$ cmod() $\{Q\}$ holds with respect to $\mathcal{I}$, and thus cmod $\models_\mathcal{I} C$. Moreover, for any choices of s and $\mathcal{I}$ such that P does not hold, $\{P\}$cmod()$\{Q\}$ holds vacuously, wherefore cmod $\models C$. We leave it as an exercise for the reader to verify that the foo function in Fig. 13.2b does not satisfy C.

13.2.2 Target Programming Language and Semantics

In this chapter, we target C programs written in a subset of the C99 standard [27], extended with **assert** and **assume** statements. The subset of C99 includes integer variables, structs, loops and (recursive) function calls. We do not consider floating-point variables, pointers or arrays. However, in Sect. 13.5.2, we provide some experimental results and a discussion regarding pointers. The main reason for these restrictions are limitations in the TriCera model checker, which is currently under development. As specifications, we consider ACSL function contracts, which are restricted with analogous limitations as for the C programs. Furthermore, as detailed in Sect. 13.4.5, we do not, in general, allow quantified predicates or recursive predicates in ACSL contracts.

In Sect. 13.3, we provide some results concerning translations of programs and (contract) annotations between Frama-C and TriCera. Reasoning formally about these translations requires a formal semantics for the subset of C (and ACSL) considered. However, it is beyond the scope of this chapter to provide or describe a formal semantics for the C language, wherefore the reasoning is kept here at an informal level. Still, the intention is that our reasoning about C programs should be compatible with the informal semantics given in natural language in the C99 standard, as well as with standard formal semantics, such as the denotational semantics presented in [32]. Similarly, the semantics for ACSL contracts is assumed to follow traditional Hoare logic semantics, as described in Sect. 13.2.1, as well as being compatible with the semantics of ACSL, defined in natural language in the ACSL reference manual [6].

Semantics of assume and assert statements. For specification and verification purposes, one frequently uses in C programs the **assume** and **assert** statements, even though they are not officially part of the C language. The **assert** statement[2] is used to state conditions that have to hold at particular program points. The execution of the statement **assert(p)** will terminate normally if the Boolean expression **p** evaluates to *true* in the current program state, and will cause erroneous termination otherwise: the program will "go wrong" in the terminology of Flanagan and Saxe [19]. A program with **assume** and **assert** statements is considered *correct* when no execution of the program goes wrong.

The **assume** statement is useful to model partial execution: given the statement **assume(p)**, program execution will be blocked if **p** evaluates to *false*, and it will continue as normal otherwise. As a result, the only program states possible right after its execution are those satisfying **p**. Importantly, the **assume** statement never goes wrong, which distinguishes it from the **assert** statement.

When verifying programs, one can translate Hoare logic triples (explained in Sect. 13.2.1) into programs with **assert** and **assume** statements. To verify that a function satisfies its contract (given with a pre-condition and a post-condition), its pre-condition can be *assumed* to hold at entry, and its post-condition can be *asserted* to hold at exit.

13.2.3 Constrained Horn Clauses (CHCs)

Horn clauses are named after Alfred Horn, who first formulated their significance in 1951 [26]. A constrained Horn clause (CHC) is a logical formula of the following shape: H represents the *head* of the clause, which is either an application of a predicate over first-order terms or *false*; C is the constraint; and B_i is an application of a predicate over first-order terms. All terms appearing in a CHC are (implicitly) universally quantified.

$$\overbrace{H}^{\text{Head}} \leftarrow \overbrace{C \wedge B_1 \wedge \ldots \wedge B_n}^{\text{Body}}$$

In the context of program verification, CHCs can be used to easily encode programs, including programs with features such as function calls, recursion, or concurrency. In this sense, CHCs can be seen as an intermediate verification language that

---

[2] Note that we consider assert as a *statement*, so that it differs from the **assert** *macro* described in the C99 standard.

provides a unified framework for program verification and synthesis [8]. CHCs represent a fragment of first-order logic that can be processed efficiently by automated solvers, for instance Eldarica [24] or Z3/Spacer [29], which can thus reason about the absence of bugs in programs. These state-of-the-art solvers employ algorithms derived from methods in model checking to check whether a given set of CHCs is satisfiable.

The encoding of programs as CHCs differs fundamentally from the verification condition (VC) generation applied in deductive verification tools, for instance in the Wp plugin. VCs are commonly formulated in such a way that the solver result *unsat* corresponds to a program satisfying its specification (no violations exist). With CHCs, in contrast, the encoding is chosen such that satisfiability of the CHCs constitutes a proof that the considered program adheres to its specification, whereas unsatisfiability (and counterexamples to satisfiability) corresponds to incorrectness of the program. The intuition is that CHC solving represents a search for verification annotations (such as state invariants or contracts) that are sufficient to show program correctness. Demonstrating the satisfiability of a set of CHCs implies that suitable annotations have been found, whereas unsatisfiability demonstrates that no sufficient annotations can exist.

13.2.4 CHC-Based Model-Checking

We give a high-level overview of how imperative programs can be encoded as a set of CHCs. Consider the C program shown in Fig. 13.3a, which has a recursive function f that calculates the sum of values from 0 to its argument n and is called from main. At the end of main it is asserted that this program should only terminate in states in which x is greater than or equal to y.

A control-flow graph (Fig. 13.3b) for this program contains nodes for the program states and edges representing state transitions that are the result of executing statements. The edge labels state the conditions (or constraints) that must hold in order to take a transition. The unwanted state in which x is less than y can be represented with an error node, and this condition can be added to the edge leading to that node. In addition, the graph contains solid *call* and dashed *return* edges modeling the function calls.

Figure 13.3c shows the encoding of the same program in CHCs. Overall, the encoding of the program using CHCs is defined in such a way that the program is correct (the assertion in the program can never fail) if and only if the set of CHCs has a solution.

Program states of the main function are encoded using predicates $main_i$ that have the local and global variables as their arguments, and represent state invariants of the respective control states. The state transitions are modeled using logical implications (written from right to left), in the style of Floyd [20], and the constraints defining the effects of a transition are added as guards on the right-hand side. There are no

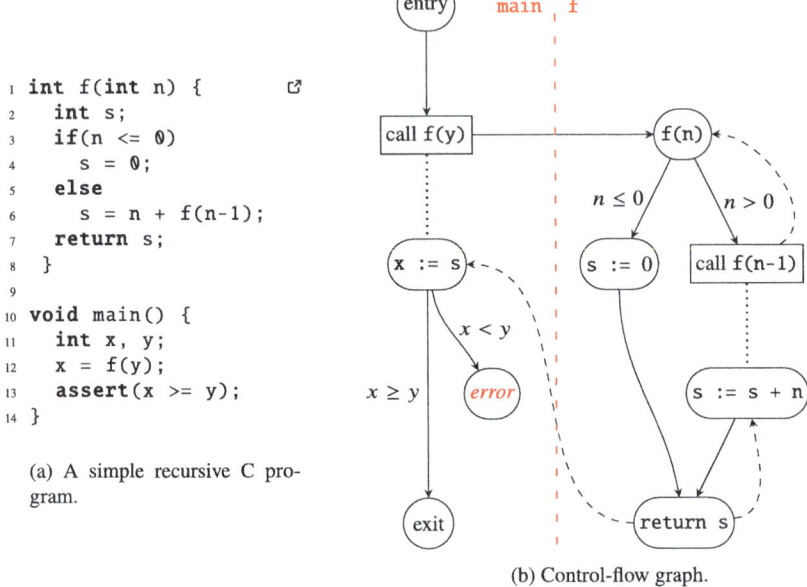

Fig. 13.3 Sample code with corresponding Control-Flow-Graph and Horn clauses. The clauses are satisfied by setting $f_{post}(n, s) \equiv s \geq n \land n \geq 0$, and interpreting the other predicates correspondingly

pre-conditions for main (the entry), which is encoded as *true* in the body of the first Horn clause encoding main.

The error state is represented using a clause with the head *false*. Recall that all free variables in a CHC are (implicitly) universally quantified. This means that the satisfiability of a set of CHCs implies the unreachability of the error state. In Fig. 13.3c, if the set of CHCs is satisfiable (which is the case), there does not exist any valuation of the encodings of the program variables x and y at $main_2$ such that $x < y$. Note that a set of CHCs that does not have any clause with *false* in its head (i.e., no **assert** statements in the encoded program) is always satisfiable by interpreting all uninterpreted predicates as *true*.

The uninterpreted predicates f_{pre} and f_{post} denote the pre- and post-conditions of the recursive function f, respectively, which we want to infer by finding an interpretation. f_{pre} is a predicate over the function argument n (and global program variables, if any). f_{post} represents a relation between the function argument n and the function result s (and between the pre- and post-values of global variables, if any). Each function invocation is encoded using two clauses: one clause that ensures that the pre-condition of f holds at the invocation point (through an assertion of f_{pre}), and one clause that describes the possible results of the function call and the effects in terms of global variables (through an assumption of f_{post}).

In the encoding of f, the recursive call is encoded in the same way: f_{pre} in the bodies of the clauses encoding f corresponds to assuming the pre-condition of f, and in the last clause where f_{post} is at the head corresponds to asserting its post-condition.

CHC solvers like Eldarica [24] and Z3/Spacer [29] can compute a solution for the given CHCs, and thus verify that the C program is correct.

It should be noted, though, that the representation given in Fig. 13.3 is an abstraction of the actual program semantics, as the variables are represented using mathematical integers. This is not a fundamental limitation of CHCs, as the variables could as well be represented using the SMT-LIB theories of bit-vectors, floating-point numbers, or real numbers.

13.2.5 The TriCera Model Checker

We employ the verification tool TriCera [18] for the translation from C programs to CHCs. TriCera is an automated program verification tool that accepts programs written in a subset of the C language and checks their safety. Safety is checked against both explicit assertions, expressed using the **assert** statement, and implicit assertions including memory and type safety of heap accesses. TriCera works by encoding the input program into a set of CHCs, as explained earlier in this section, and internally uses the CHC solver Eldarica in order to check if the generated clauses are solvable. Function calls can be handled either by inlining the function body, or using a translation with explicit pre- and post-conditions, as in Sect. 13.2.4. An overview of TriCera's architecture can be seen in Fig. 13.4.

If one of the assertions does not hold, a counterexample trace leading to the failing assertion is generated, which is useful in finding program bugs. A solution for each predicate appearing in the CHCs can automatically be generated by the CHC solver when the CHCs can be solved, and it is exactly this feature that makes TriCera useful in the context of this work: a function's pre- and post-conditions can

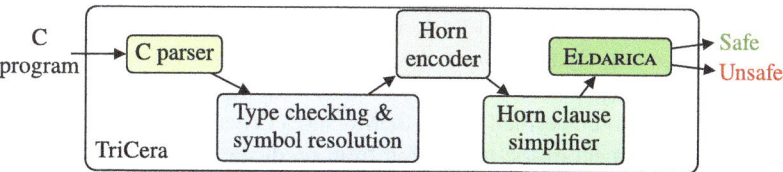

Fig. 13.4 TriCera architecture

be encoded as predicates, and finding a solution to these predicates corresponds to automatic contract inference. An example of this can be seen in Fig. 13.3, where f_{pre} and f_{post} are the pre- and post-conditions of the function f.

As of writing this chapter, TriCera supports most C programs over integers and structs. It also has some support for arrays, pointers and heap operations, and is being actively developed to support a larger subset of the C language.

13.3 Contract Inference Using Assertion-Based Model Checking

This section describes how Hoare logic contracts can be translated into **assume** and **assert** statements for verification using assertion-based verification methods. In particular, we show how a contract C and a C program S are translated into what we call a *harness function*, and then show how the harness function can be used for contract inference, by using the TriCera model checker (see 13.2.5). For simplicity, we further restrict the target subset of the C language in this section to programs without return values and parameters.

13.3.1 The Harness Function

The overall idea of the harness function is that a contract $C = (P, Q)$ and a C function f are translated into a C function that first assumes the pre-condition, then calls the function f and, thereafter, asserts the post-condition. Before defining the harness function explicitly, we first discuss how (and if) the predicates P and Q can be represented using C expressions. We also provide a discussion on how the harness function could be defined for quantified predicates.

13.3.1.1 Translating Predicates into C Expressions

We will now discuss how to translate the pre- and post-condition into C expressions. For a predicate P, we shall use P^C to denote its translation into a C expression. We do not give an explicit definition of P^C here, but provide a discussion of important aspects to consider and, in Sect. 13.4.1, we show how the Saida plugin translates ACSL predicates into C expressions. A similar translation is defined explicitly for Java programs in [7], where JML specifications are translated into **assume** and **assert** statements to be verified by the JBMC model checker.

One important aspect of the translation from predicates to C expressions is the treatment of logical variables. Since logical variables are not allowed in C expressions, our translation of a predicate P into P^C includes converting all logical variables into program variables. Note that this imposes a difference between the domains over which P and P^C are evaluated. In particular, P is evaluated over a program state and an interpretation of logical variables, while P^C is evaluated only over a program state. In order to unify the evaluation of P and P^C, let s be a program state and $\mathcal{I}$ an interpretation over logical variables $lv_1 \ldots lv_n$. Then, we denote with $s_\mathcal{I}$ the program state created by extending s with program variables $lv_1 \ldots lv_n$, such that $s_\mathcal{I}(lv_i) = \mathcal{I}(lv_i)$. For example, let $s = [x \mapsto 2, y \mapsto 2]$ and $\mathcal{I} = [x_0 \mapsto 1]$, then $s_\mathcal{I}$ becomes:

$$s_\mathcal{I} = [x \mapsto 2, y \mapsto 2, x_0 \mapsto 1]$$

The definition of $s_\mathcal{I}$ allows for reasoning about how P^C preserves the semantics of P. Recall that the C language does not have a Boolean data type, but that Boolean behavior is achieved by letting zero act as *false* and any non-zero value as *true*. Therefore, preserving the semantics means that P evaluates to *true* over state s and interpretation $\mathcal{I}$ if and only if P^C evaluates to non-zero over $s_\mathcal{I}$, i.e.:

$$s \models_\mathcal{I} P \iff s_\mathcal{I} \models P^C \neq 0 \qquad (13.4)$$

For example, let x_0 be a logical variable, x and y program variables and let P be the predicate defined as:

$$P \stackrel{\text{def}}{=} (x \geq x_0) \Rightarrow (y \geq x_0)$$

Then, P^C becomes the following C expression:

$$P^C \stackrel{\text{def}}{=} \;!(\texttt{x >= x}_0) \;||\; (\texttt{y >= x}_0)$$

Now, consider that we want to evaluate P over some state s and interpretation $\mathcal{I}$. Then, P evaluates to *true* whenever $s(x) \geq \mathcal{I}(x_0)$ and $s(y) \geq \mathcal{I}(x_0)$ or, vacuously, if $s(x) < \mathcal{I}(x_0)$, and to *false* otherwise. Similarly, by the semantics of C expressions, P^C evaluates to 1 if $s(x) < \mathcal{I}(x_0)$ or if $s(x) \geq \mathcal{I}(x_0)$ and $y \geq \mathcal{I}(x_0)$, and to 0 otherwise. Thereby, property (13.4) holds for this choice of P.

Unfortunately, property (13.4) is difficult to achieve for a full translation of first-order predicates into C expressions. For example, quantified predicates of first-order logic are, in general, not expressible as C expressions. To deal with such predicates, one can potentially use heuristic translations. In Sect. 13.5.4, we discuss such translations and how they can be used to infer what we call *partial contracts*, while in Sect. 13.3.1.3, we show how some special cases of quantified predicates can be directly translated into **assume** and **assert** statements.

13.3.1.2 The Translation Function Har(C, f)

The translation from a contract into a harness function can now be defined by relying on the translation P^C for a predicate P. Below follows a formal definition, after which we discuss and justify the definition.

Definition 13.1 (*Harness function*) Let $C = (P, Q)$ be a Hoare logic contract with logical variables $x_0, \ldots, x_n$ and let f be a C function. Then, the harness function Har(C, f) is defined as:

$$\text{Har}(C, \text{f}) \stackrel{\text{def}}{=} \begin{array}{l} \text{void harness() } \{ \\ \quad \text{int x0; } \ldots \text{ ; int xn;} \\ \quad \text{assume}(P^C); \\ \quad \text{f();} \\ \quad \text{assert}(Q^C); \\ \} \end{array}$$

The harness function starts by declaring logical variables as local variables. As discussed above, this is needed since C expressions do not allow logical variables. Our treatment of logical variables in the harness function relies on the following two important assumptions:

(i) Local variables are initialized non-deterministically.
(ii) The values of x0, ..., xn cannot be altered during a function call f().

Assumption (i) has support in the C99 standard [27], which states that the value of non-initialized local variables is indeterminate.[3] Assumption (ii) will always hold if f is a well-formed C program according to the C99 standard, where well-formed means absence of potential undefined behavior, such as buffer overflows.

After initializing the logical variables, the harness function continues by first assuming P^C, then calling the function f and, lastly, asserting Q^C. As an example, consider again the simple increment example in Fig. 13.1a. The ACSL contract in the example can be expressed as the following Hoare logic contract:

$$C \stackrel{\text{def}}{=} (x = x_0, \ r = x_0 + 2) \tag{13.5}$$

Then, for a given implementation main, the harness function Har(C, f) is shown in Fig. 13.5.

The central idea of the harness function is that, if there is any execution such that **assert**(Q^C) fails, then f does not satisfy C. Note that this can only hold if P^C and Q^C satisfy property (13.4). The following theorem formalizes this correctness property of the harness function.

---

[3] It should be noted here that reading from an indeterminate variable might be undefined behavior (if the value is a trap representation). However, since the harness function will only be used as input to TriCera, and not executed as a C program, we are not overly concerned about undefined behaviors here.

Fig. 13.5 Harness function for the increment example program

```
void harness() {
    int x0;
    assume(x == x0);
    main();
    assert(r == x0 + 2);
}
```

Theorem 13.1 (Correctness) *Let f be a C function and $C = (P, Q)$ a Hoare logic contract, and let the translated predicates P^C and Q^C satisfy property* (13.4). *Then, the harness function* Har(C, f) *has no failing executions if and only if* $f \models C$.

Proving this theorem formally requires defining and reasoning about the formal semantics of contracts and C statements, which is beyond the scope of this chapter. In [4], a proof is given for a subset of C and below we provide a semi-formal sketch of the reasoning required for the proof.

Proof (sketch) To show the if-direction, assume $f \models C$, which, by definition, means that: $\forall \mathcal{I}.\ f \models_{\mathcal{I}} C$. Now, let s be a state and $\mathcal{I}$ an interpretation of logical variables, and assume that we are executing f from s. Now, if P does not hold over s and $\mathcal{I}$, then, by property (13.4), P^C does not hold over $s_{\mathcal{I}}$, wherefore the **assume**(P^C) statement will not succeed; it will block program execution and thus, the execution will not fail. If, instead, P does hold over s and $\mathcal{I}$, then the **assume**(P^C) statement will succeed and the function call f() will be executed. Since $f \models C$, if executing f terminates, it will do so in a state s' such that Q holds over s' and $\mathcal{I}$. Therefore, again by (13.4), the **assert**(Q^C) statement is guaranteed to succeed. Thereby, whenever $f \models C$, Har(C, f) never fails.

For the only-if-direction, assume contra-positively that $f \not\models C$. Then, there must exist an interpretation $\mathcal{I}$ and states s and s' satisfying the following conditions:

1. P holds over s and $\mathcal{I}$.
2. There is an execution of f from s that terminates in s'.
3. Q does not hold over s' and $\mathcal{I}$.

Now, consider that the harness function is executed from s and the values of the logical variables are given by $\mathcal{I}$. Then, the **assume** statement will be executed from $s_{\mathcal{I}}$ and, by (13.4), it will succeed. Furthermore, there is an execution of f() terminating in $s'_{\mathcal{I}}$ so that, again by (13.4), the **assert** statement fails, showing that there is at least one failing execution of the harness function. □

Note that Theorem 13.1 embodies an equivalence result between verifying Hoare logic contracts and assertion-based model checking of harness functions. In particular, for a contract C and a function f, an assertion-based model checker (such as TriCera) will verify Har(C, f), if and only if $f \models C$.

13.3.1.3 Quantified Predicates as assume and assert Statements

As mentioned above, quantified predicates cannot, in general, be expressed as C expressions. However, as shown in [7], some cases of quantified predicates can be translated directly into **assume** or **assert** statements. Their idea is based on that, in a verification context, **assert** statements are implicitly universally quantified (over execution paths), while **assume** statements are implicitly existentially quantified. Therefore, the authors translate universally quantified predicates into **assert** statements and existentially quantified predicates into **assume** statements. In our harness function, we assume the pre-condition and assert the post-condition, which means that, following the approach of [7], we can allow existential quantification in the pre-condition and universal quantification in the post-condition. As an example, consider the following contract over program variable x:

$$C \stackrel{\text{def}}{=} (\exists x_0.\ x_0 < x,\ \forall x_1.\ x_1 \geq x)$$

Then, for a given function implementation foo, we can create the harness function by treating the bounded variables x_0 and x_1 as logical variables, i.e., introducing them non-deterministically as:

```
1  void harness() {
2      int x0, x1;
3      assume(x0 < x);
4      foo();
5      assert(x1 >= x);
6  }
```

Note that this translation method is not restricted to finite domains for the bounded variables.

Translations are also defined for any quantified predicate, given that the domain of the bounded variable is finite [7]. This is accomplished by using a loop with one iteration for each possible value in the domain of the bounded variable.

13.3.2 Inferring Contracts with TriCera

This section describes how the harness function can be used to verify the entire program, and (as a by-product) to infer missing function contracts. The verification is performed by relying on the assertion-based model checker TriCera, described in Sect. 13.2.5.

Recall from Sect. 13.2.5 that TriCera can be instructed to verify a program by inferring pre- and post-conditions for any function called in the program. By using the Harness function as input to TriCera, we can extract pre- and post-conditions inferred by TriCera during verification. For example, consider again the contract C in (13.5), specifying the simple increment example in Fig. 13.1a. Notice that, in

13 Automatic Contract Inference for Frama-C

Fig. 13.1a, the `main` calls the helper function `incr_x` twice. Now, consider also the harness function Har(C, `main`) for this program, shown in Fig. 13.5. When verifying Har(C, `main`), TriCera infers the pre- and post-condition for `incr_x` as part of the CHC solution, corresponding to the following Hoare logic contract:

$$C_{\text{incr}} \stackrel{\text{def}}{=} (x = x_0 \wedge r = r_0, \ x = x_0 + 1 \wedge r = r_0)$$

Using the contracts. The essence of our approach is that the inferred contracts can be reused in the context of deductive verification. Since TriCera verifies the harness function by relying on the inferred contracts, they should be strong enough also for the original top-level contract to rely on for procedure-modular verification. For the increment example in Fig. 13.1a, this means that the inferred contract C_{incr} should be strong enough to verify that `main` satisfies the contract C in (13.5). In the following Sect. 13.4, we show how inferred contracts can be used for deductive, procedure-modular verification with Frama-C.

Readability of the contracts. In general, one cannot expect contracts inferred by TriCera to be easily graspable by human readers. In particular, for larger functions the inferred contracts tend to become incomprehensible. This could be an issue for users interested in understanding or modifying the inferred contracts. However, inferred contracts often contain redundant clauses, wherefore existing techniques for clause elimination could be applied on the inferred contracts to improve readability.

13.4 The Saida Plugin

To evaluate the technique presented in Sect. 13.3 in practice, we present our experimental Frama-C plugin Saida. The plugin takes as input a C file, the entry point of which is already annotated with an ACSL function contract. Given the input file, Saida performs the following two tasks. First, Saida generates a harness function for the main function, as described in Sect. 13.3.1. Then, it attempts to infer contracts for all functions called directly or transitively from the main function, by verifying the harness function in TriCera, as described in Sect. 13.3.2. Below, we elaborate on how Saida performs these two tasks, followed by an illustration of how the inferred contracts can be used for verification. Finally, we state the limitations of Saida.

13.4.1 Generating the Harness Function

Overall, the harness function is created following the approach described in Sect. 13.3.1. That is, we first assume the pre-condition, then call the function to be verified and, lastly, assert the post-condition. Below, we first illustrate on an

```
1  int r, x;
2
3  /*@contract@*/
4  void incr_x() {
5    x = x + 1;
6  }
7
8  void main2() {
9    incr_x();
10   incr_x();
11   r = x;
12 }
13
14 extern int non_det_int();
15
16 void main()
17 {
18   //Non-det assignment of global variables
19   r = non_det_int();
20   x = non_det_int();
21
22   //Initialization of logical old-variables
23   int old_x;
24   assume(old_x == x);
25
26   //The requires-clauses translated into assumes
27   assume(1);
28
29   //Function call that the harness function verifies
30   main2();
31
32   //The ensures-clauses translated into asserts
33   assert((r == (old_x + 2)));
34 }
```

Fig. 13.6 Program code and harness function for the increment example program

example how Saida creates the harness function and, thereafter, describe how ACSL predicates are translated into C expressions.

To illustrate how Saida generates the harness function, we continue with the simple increment example from Fig. 13.1a. The program also contains an ACSL contract for the main function, which corresponds to the Hoare logic contract in (13.5). The following command runs Saida on the example program in Fig. 13.1a:

```
frama-c -saida -lib-entry increment_program.c
```

From the contract of the main function, Saida generates a harness function, and merges it with the source code, as shown in Fig. 13.6. Note that the harness function becomes the main function since it should be the entry point for TriCera, and the original main function is automatically renamed to main2.

13 Automatic Contract Inference for Frama-C

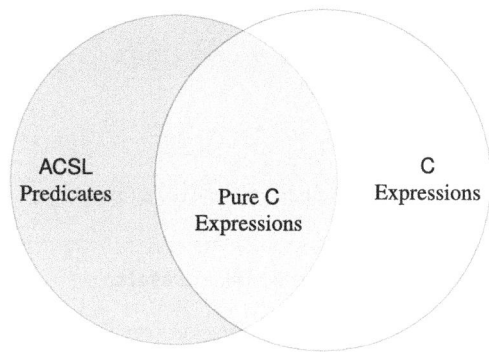

Fig. 13.7 Relationship between ACSL predicates and C expressions

Furthermore, observe that the harness function in Fig. 13.6 begins with assigning the global variables to the `non_det_int` function, which is declared above the harness function. This allows TriCera to treat the value of the global variables as non-deterministic. This behavior can be enabled or disabled with the Frama-C kernel option (-no)-lib-entry.

From ACSL predicates to C expressions. When creating the harness function, an important aspect is how ACSL predicates should be translated into C expressions. As stated in the ACSL documentation [6], the language of ACSL predicates extends the language of *pure C expressions*, which are expressions free from side-effects, i.e., expressions that do not assign any values or make any function calls. Figure 13.7 illustrates the relationship between the two languages; note that they intersect on the pure C expressions.

Currently, Saida supports ACSL predicates corresponding to pure C expressions as well as the \old, \result and \let constructs. Saida also supports logical operators and constants not native to C, such as <==>. In Sect. 13.4.5, we explicitly state the limitations of Saida. Since the languages of ACSL predicates and C expressions intersect on the pure C expression, we only have to translate the constructs of ACSL that extend the language of pure C expressions. Table 13.1 shows how logical ACSL operators and constants not native to the C language are translated. Below, we describe how Saida deals with \old, \result and \let.

Table 13.1 Translation from ACSL to C of operators and constants

Name	ACSL syntax	C Translation
Logical equivalence	p_1 <==> p_2	p_1 == p_2
Logical implication	p_1 ==> p_2	!p_1 || p_2
Logical xor	p_1 ^^ p_2	p_1 != p_2
true	\true	1
false	\false	0

```
1  /*@
2     requires \true;
3     ensures \result == \old(x)+1;
4  */
```

(a) ACSL contract using **\result**.

```
1  /*@
2     void main() {
3        int old_x;
4        assume(old_x == x);
5        assume(1);
6        int res = main2();
7        assert(res == old_x+1);
8     }
9  */
```

(b) Harness function using **\result**.

Fig. 13.8 Simple example using the **\result** construct

The **\old** construct is defined semantically by assuming that, for all program variables v, **\old**(v) = v holds in the pre-state [6], so that **\old** plays the part of logical variables in traditional Hoare logic. Therefore, our approach for dealing with **\old** follows the approach for dealing with logical variables given in Sect. 13.3.1. That is, for all program variables occurring in an **\old** context, Saida defines a new local variable in the harness function. For example, in Fig. 13.6, the harness function declares the local variable old_x and then assumes old_x == x, so that old_x captures the value of x in the pre-state.

The **\result** construct is handled by modifying the harness function so that the return value of the called function is saved to a local variable in the harness function, which can then be used in the **assert** statement for the post-condition. For example, consider the simple ACSL contract in Fig. 13.8a, specifying a function that returns the value of x+1. Let foo be any implementation of this contract; then the harness function is given in Fig. 13.8b, where the result of the call to foo is stored in the variable res. One may note here that, in Frama-C, there may be a semantic difference between the expression **\old**(x) + 1 in the post-condition in Fig. 13.8a and the expression old_x + 1. Namely that the former expression is over mathematical integers (and therefore no overflow can happen) while the second is over the C type int (which may overflow). TriCera supports arithmetic operations over both mathematical and machine integers; in the translation, we assume the former is used. We plan to extend TriCera so that mathematical integers can be specified even when the encoding is in machine arithmetic, but this is currently a limitation.

The **\let** construct is translated by simply replacing each occurrence of the defined variable with its given value. For example, consider this alternative contract for the function returning the value of x incremented with 1:

```
1  /*@
2     requires \true;
3     ensures \let x1 = \old(x) + 1; \result == x1;
4  */
```

This contract can be translated into a harness function by replacing all occurrences of x1 with its assigned value (i.e., \old(x) + 1). If we again let foo be the function to verify, then also this contract is translated into the harness function in Fig. 13.8b.

13.4.2 Inferring Contracts

After the generated harness function is merged with the source file, the merged result is used as input for TriCera. Before running TriCera, Saida also adds the annotation @contract@ above every non-main function in the source code, which instructs TriCera to rely on contract inference rather than on function inlining (see Sect. 13.2.5).

TriCera then attempts to verify the merged file and, in doing so, infers pre- and post-conditions for all non-main functions, as described in Sect. 13.3.2. Further, TriCera translates the inferred pre- and post-conditions into ACSL function contracts, which are then inserted by Saida into the source file in the appropriate locations. For the increment example program, TriCera outputs an ACSL contract for the incr_x function, which is shown in Fig. 13.9. The post-condition of the inferred contract states that x - \old(x) == 1, which is clearly strong enough for TriCera to verify that the last **assert** statement in the harness function never fails.

13.4.3 Using Inferred Contracts for Verification

This section describes how the contracts inferred by Saida can be used for verifying the original contract for the entry-point function. The output from Saida is a C file with the original program code, where the inferred contracts are added to their respective function. The intention is that the inferred contracts should be verifiable in Frama-C using, e.g., the Wp plugin. Furthermore, they should be strong enough for verification of the original contract for the main function. In the increment program example, the inferred contract in Fig. 13.9 is added as a function contract for the incr_x function. Running Wp on the output file for the program yields the following result:

```
1 /*@
2   requires \true;
3   ensures x - \old(x) == 1 && r == \old(r);
4 */
5 void incr_x() {
6   x = x + 1;
7 }
```

Fig. 13.9 The ACSL contract inferred by TriCera for the incr_x function

```
[wp] 3 goals scheduled
[wp] Proved goals:    3 / 3
Qed:                  2   (0.85ms-2ms-4ms)
Alt-Ergo 2.4.1:       1   (8ms) (12)
```

That is, the inferred contract can be verified by Wp and it is also strong enough for verification of the contract for the original main function, as expected.

13.4.4 Handling Recursive Functions

One advantage of our approach is that it naturally supports the inference of contracts for functions that are *recursive*. For example, consider again the simple recursive program in Fig. 13.3a. In Fig. 13.10, the main function has been annotated with a contract corresponding to the **assert** statement in Fig. 13.3a. Figure 13.10 also shows a contract inferred by Saida for the f function. The Wp plugin successfully verifies both contracts in Fig. 13.10. Note that, due to the recursive structure of f, this program would not be verifiable in Frama-C by inlining function f rather than using the inferred contract for f.

```
1  int x, y;
2
3  /* Inferred contract: */
4  /*@
5    requires \true;
6    ensures y == \old(y) && x == \old(x) &&
7            \result >= \old(n) && \result >= 0;
8  */
9  int f(int n) {
10   int s;
11   if(n <= 0)
12     s = 0;
13   else
14     s = n + f(n-1);
15   return s;
16 }
17
18 /*@
19   requires \true;
20   ensures x >= y;
21 */
22 void main() {
23   x = f(y);
24 }
```

Fig. 13.10 Simple program with inferred contract for the recursive function f

13.4.5 Limitations

As explained in Sect. 13.3.1.3, quantified predicates cannot, in general, be translated correctly into C expressions. Therefore, we currently only support existential quantification in the pre-condition and universal quantification in the post-condition, as detailed in Sect. 13.3.1.3. As of writing, Saida also lacks support for various other ACSL constructs, such as model variables, built-in logic functions and user-defined logical predicates and functions. See the Saida GitHub page[4] for the latest information on supported features.

The contract inference part of Saida also inherits any limitations of the TriCera model checker (see Sect. 13.2.5). For example, floating-point values are currently not supported.

13.5 Extensions and Ongoing Work

In this section, we discuss some additional techniques and ongoing research on inferring ACSL annotations using Saida.

13.5.1 Programs with Loops

Our approach supports naturally the inference of contracts for programs with *loops*, since TriCera can verify such programs automatically, i.e., without relying on a manually provided *loop invariant* for each loop. Figure 13.11 shows a program where a contract for foo is inferred by Saida, given the manually provided contract for main. The inferred contract for foo is strong enough for Wp to verify the original contract for main. However, since no loop invariant is added for the while loop, Wp cannot verify that foo satisfies the inferred contract (unless a loop invariant is provided manually).

To avoid having to manually add loop invariants in Frama-C, our plan is to extend Saida so that it also infers loop invariants from the CHC solution in TriCera. For the example program in Fig. 13.11, TriCera infers the loop invariant shown in Fig. 13.12. Note that the invariant has been manually translated into ACSL.

After annotating the while loop in Fig. 13.11 with this invariant, Wp verifies the entire program, including the loop invariant and the contracts for foo and main.

---

[4] https://github.com/rse-verification/saida.

```
1  int x, y;
2
3  /*@
4    requires y >= 1;
5    ensures y == \old(y) && x + -1 * \old(y) + -1 * \old(x) == 0 &&
6             \old(y) >= 1;
7  */
8  void foo() {
9    int i = 0;
10   while(i < y) {
11     x += 1;
12     i++;
13   }
14 }
15
16 /*@
17   requires y >= 0;
18   ensures x == \old(x + y) + 1;
19 */
20 void main() {
21   y++;
22   foo();
23 }
```

Fig. 13.11 Simple program with inferred contract for the function foo containing a while loop

Fig. 13.12 Inferred loop invariant for the while loop in Fig. 13.11, manually translated into ACSL

```
1  /*@
2    loop invariant
3         (y == \at(y, Pre)) &&
4         (x == \at(x, Pre) + i) &&
5         (y + \at(x, Pre) >= x) &&
6         (x >= \at(x, Pre)) &&
7         y >= 1;
8  */
```

13.5.2 Programs with Pointers and Heap Allocations

In this chapter we have mainly focused on programs over primitive data-types, in particular integers. Annotation inference becomes more challenging in the presence of non-primitive data-types: in C, in particular in programs with heap-based data-structures, and arrays. Such programs are currently only partly supported by Saida, and its extension to achieve full coverage of C is ongoing work.

On the side of TriCera, data-types are represented using a tailor-made theory of heaps [17]. This theory offers functions for reading and updating objects stored on the heap, similar to the theory of arrays [30], but also provides functions for allocating new objects and checking whether a pointer is valid. The theory also covers C-style arrays, in the sense that sequences of consecutive addresses can be allocated and updated. The theory of heaps is at this point supported both by Eldarica, which can solve CHCs with heap as the background theory, and by TriCera, which can translate

C programs to CHCs with heap. What is currently missing in Saida to achieve full heap and pointer support is a translator from formulas over the theory of heaps to ACSL annotations.

13.5.3 Inferring assigns Clauses

An obvious extension of our work on translating CHC solutions to ACSL contracts is the inference of **assigns** clauses. Opportunities for this can be seen, for instance, in Fig. 13.10, where sub-formulas like y == \old(y) point out non-assigned variables. The derivation of **assigns** clauses is relatively easy for global variables, but requires more research when heap data-structures and arrays are involved (see Sect. 13.5.2).

13.5.4 Inferring Partial Contracts

Our method is not always capable of inferring contracts. There are two principal reasons for this. The first one is when the original contract cannot be expressed as pure C expressions (e.g., when it contains quantified or recursive predicates). The second one is when TriCera is not able to verify the program together with the harness function. In both cases, there is still the option to *weaken* the contract for the main function, before translating it into a harness function. The weaker contract can then be used for inference of *partial contracts* for the helper functions, which can later be completed manually, so that they become strong enough for the verification of the original, stronger contract of the main function.

To illustrate the concept, consider the program in Fig. 13.13, where the main function is annotated with a function contract. It is currently not possible to use Saida on this program, since the program together with the harness function is too hard for TriCera to verify. However, if we relax the post-condition by removing the first conjunct $x == y * n_1 * n_2$, and only keeping the second one $x >= 0$, then TriCera does succeed in verifying the program with the harness function as produced by Saida. During verification, the following contract, with some redundant clauses removed for brevity, is inferred for the foo function:

```
1  /*@
2    requires  x == 0 && n2 >= 0 && n1 >= 0 && y >= 0;
3    ensures   x >= 0;
4  */
```

Wp can use this inferred contract to verify the easier contract of the main function. Furthermore, the contract can be manually completed, by adding the clause x == y * n1 * n2, after which Wp can verify also the original stronger contract of main. Note, however, that loop invariants are required for Wp to verify the inferred contracts. The loop invariants could of course also be partially inferred, by combining the approaches of this section and Sect. 13.5.1.

Fig. 13.13 Program with loops for which partial contract can be inferred

```
1  int x, y, n1, n2;
2
3  void foo() {
4    int i = 0; int j = 0;
5    while(i < n1) {
6      j = 0;
7      while(j < n2) {
8        x += y;
9        j++;
10     }
11     i++;
12   }
13 }
14
15 /*@
16   requires y >= 0 && n1 >= 0 && n2 >= 0;
17   ensures x == y*n1*n2 && x >= 0;
18 */
19 void main() {
20   x = 0;
21   foo();
22 }
```

13.5.5 Contracts with Uninterpreted Predicates

The handling of partial contracts described in the previous section can be generalized, by providing an interface through which the user can explicitly state which parts of a contract are to be inferred automatically. To this end, we envision the extension of ACSL with *uninterpreted predicates* that stand for unknown parts of a contract; instead of inferring complete contracts, as it is currently done by Saida, the goal of inference is then reduced to replacing uninterpreted predicates with concrete formulas.

As an example, consider again the program in Fig. 13.13. In this example, the inference of the post-condition x == y * n1 * n2 turned out to be too hard for the CHC solver, whereas other parts of the contract could be inferred successfully. To capture the inference of only parts of the contract, we propose to reformulate the annotated program to the version in Fig. 13.14, in which the post-condition x == y * n1 * n2 is now manually provided, but other parts of the contract of f are abbreviated using the uninterpreted predicates pre_ranges and post_ranges. The inference task now consists of finding formulas pre_ranges and post_ranges that make the program verify as a whole.

Such *contracts with holes* are already supported by the TriCera model checker, which can handle uninterpreted predicates occurring in **assume** and **assert** statements. It is planned that also future versions of Saida will be able to process programs like the one in Fig. 13.14.

It should be noted that the concept of uninterpreted predicates in annotations is general and rather powerful, though much research remains to be done. It could be

```
int x, y, n1, n2;

/*@
  u_predicate pre_ranges (int x, int y, int n1, int n2);
  u_predicate post_ranges(int x);
*/

/*@
  requires pre_ranges(x, y, n1, n2);
  ensures x == y*n1*n2 && post_range(x);
*/
void foo() {
  // ...
}

/*@
  requires y >= 0 && n1 >= 0 && n2 >= 0;
  ensures x == y*n1*n2 && x >= 0;
*/
void main() {
  // ...
}
```

Fig. 13.14 Function contracts with holes, represented by uninterpreted predicates

used, for instance, to guide the inference of complex loop invariants by imposing some intended structure of the formula to be inferred, like in the following example:

```
/*@
  loop invariant \forall integer j;
                 0 <= j < i ==> p(j, a[j]);
*/
for (int i = 0; i < n; ++i) {
  a[i] = i;
}
```

The inference task now consists of finding some binary predicate p that turns the provided loop invariant skeleton into a complete loop invariant. Methods to infer quantified invariants in this manner have been proposed in the context of CHC solving [9], and could be integrated into a general framework of contracts with uninterpreted predicates.

13.5.6 Inferring Contracts for Library/API Functions

The contracts inferred by Saida are heavily dependent on both the top-level contract, as well as the implementation of the function for which we infer a contract, and the implementations of its callers. This is because the resulting contracts are just *sufficient* to verify the top-level contract, in the specific context. This can be seen, for example, in Fig. 13.10, where the resulting contract essentially repeats the assertion in the top-level contract, whereas a stronger contract would specify the exact relation between the values of the variables in the pre- and post-state. Another example can be seen in Fig. 13.11, where the pre-condition of the inferred contract is directly related to the state just before it is called, and not a general pre-condition for calling the function. As such, many inferred contracts are not suitable to be reused when verifying other parts of the code base, as would be desirable, for example, when inferring contracts for library functions that are used in many parts of a system.

In general, then, contracts for all functions need to be inferred again whenever a function or contract is verified in a new context. However, after a certain amount of such weak contracts have been inferred, it may be possible to combine these into a stronger contract, which in turn can be reused in new places [3]. Say, for simplicity, we have inferred some contracts where the pre-condition is simply *true*. In such cases, we can combine the contracts by forming a new post-condition that is the conjunction of all inferred post-conditions. Now, say we have also inferred different pre-conditions in the contracts, and each contract C_i is a pair (P_i, Q_i), where P_i are pre-conditions and Q_i post-conditions. It is no longer possible to form a new, stronger contract by simple conjunction. Instead, we may then create, for each contract C_i, an implication $P_i \Rightarrow Q_i$ and form a new contract where the post-condition is the conjunction of all the implications: $\bigwedge_i (P_i \Rightarrow Q_i)$. As long as the pre-conditions are not required for the function to execute, but only serve as enabling conditions for their respective post-conditions, this will produce a new contract which is a valid specification for the function and is stronger than (or *refines*) the inferred contracts. Alternatively, one could make use of the ACSL behavior construct, for an equivalent effect. While this is not yet implemented as part of Saida, it is an obvious candidate for automation, being a simple syntactic transformation.

13.6 Conclusion

In this chapter, we have defined a translation from ACSL function contracts into *harness functions* containing **assume** and **assert** statements, to be used as input for the assertion-based model checker TriCera. Furthermore, we have presented the Frama-C plugin Saida, which, given a partially annotated C program, implements this translation from contracts to harness functions. Saida also calls TriCera for verification of the harness function, a process during which TriCera uses specification inference techniques to infer ACSL contracts for all helper functions in the original C program. Saida essentially implements the approach outlined in [3].

Future work will focus on the topics discussed in Sect. 13.5, as well as on extending the subsets of C and ACSL supported by Saida, e.g., floating-points, arrays, and logical ACSL functions. We also plan to investigate deeper the practical utility of Saida, by conducting case studies on C programs taken from the automotive industry.

References

1. Ahrendt W, Beckert B, Bubel R, Hähnle R, Schmitt PH, Ulbrich M (eds) (2016) Deductive software verification-the key book-from theory to practice. Lecture notes in computer science, vol 10001. Springer. https://doi.org/10.1007/978-3-319-49812-6
2. Albarghouthi A, Dillig I, Gurfinkel A (2016) Maximal specification synthesis. In: Annual symposium on principles of programming languages (POPL). https://doi.org/10.1145/2837614.2837628
3. Alshnakat A, Gurov D, Lidström C, Rümmer P (2020) Constraint-based contract inference for deductive verification. Springer International Publishing. https://doi.org/10.1007/978-3-030-64354-6_6
4. Amilon J (2021) Automated inference of ACSL function contracts using TriCera. Master's thesis, KTH, School of Electrical Engineering and Computer Science (EECS)
5. Baudin P, Bobot F, Correnson L, Dargaye Z, Blanchard A WP Plug-in Manual–Frama-C 23.1 (Vanadium). CEA LIST. https://frama-c.com/download/frama-c-wp-manual.pdf
6. Baudin P, Filliâtre JC, Marché C, Monate B, Moy Y, Prevosto V ACSL: ANSI/ISO C specification language. http://frama-c.com/acsl.html
7. Beckert B, Kirsten M, Klamroth J, Ulbrich M (2020) Modular verification of JML contracts using bounded model checking. In: Int. Symp. On Leveraging Applications of Formal Methods, Verification and Validation (ISoLA)
8. Bjørner N, Gurfinkel A, McMillan KL, Rybalchenko A (2015) Horn clause solvers for program verification. In: Beklemishev LD, Blass A, Dershowitz N, Finkbeiner B, Schulte W (eds) Fields of logic and computation II-essays dedicated to Yuri Gurevich on the occasion of His 75th birthday. Lecture notes in computer science. vol 9300. Springer. https://doi.org/10.1007/978-3-319-23534-9_2
9. Bjørner N, McMillan KL, Rybalchenko A (2013) On solving universally quantified horn clauses. In: Logozzo F, Fähndrich M (eds) International symposium on static analysis (SAS). Lecture notes in computer science, vol 7935. Springer. https://doi.org/10.1007/978-3-642-38856-9_8
10. Blazy S, Bühler D, Yakobowski B (2017) Structuring abstract interpreters through state and value abstractions. In: 18th international conference on verification model checking and abstract interpretation (VMCAI 2017). Proceedings of the international conference on verification model checking and abstract interpretation. LNCS, vol 10145. Paris, France, pp 112–130. https://doi.org/10.1007/978-3-319-52234-0_7. https://hal-cea.archives-ouvertes.fr/cea-01808886
11. Clarke EM, Henzinger TA, Veith H, Bloem R (eds) (2018) Handbook of model checking. Springer. https://doi.org/10.1007/978-3-319-10575-8
12. Cousot P (2012) Formal verification by abstract interpretation. In: Goodloe A, Person S (eds) 4th NASA formal methods symposium (NFM 2012). Lecture notes in computer science, vol 7226. Springer-Verlag, Heidelberg, pp 3–7. https://doi.org/10.1007/978-3-642-28891-3_3
13. Cuoq P, Kirchner F, Kosmatov N, Prevosto V, Signoles J, Yakobowski B (2012) Frama-C: a software analysis perspective. In: International conference on software engineering and formal methods (SEFM). https://doi.org/10.1007/s00165-014-0326-7
14. Denney E, Fischer B (2006) A generic annotation inference algorithm for the safety certification of automatically generated code. In: International conference on generative programming and component engineering (GPCE). https://doi.org/10.1145/1173706.1173725
15. Dijkstra EW (1975) Guarded commands, nondeterminacy and formal derivation of programs. Commun ACM 18(8):453–457. https://doi.org/10.1145/360933.360975
16. Dijkstra EW (1976) A discipline of programming. Prentice-Hall. http://www.worldcat.org/oclc/01958445
17. Esen Z, Rümmer P (2022) An SMT-LIB theory of heaps. In: Déharbe D, Hyvärinen AEJ (eds) International workshop on satisfiability modulo theories (SMT), vol 3185, pp 38–53. http://ceur-ws.org/Vol-3185/paper1180.pdf

18. Esen Z, Rümmer P (2022) TriCera: verifying C programs using the theory of heaps. In: Annual conference on formal methods in computer aided design (FMCAD)
19. Flanagan C, Saxe J (2001) Avoiding exponential explosion: generating compact verification conditions. In: Annual ACM symposium on principles of programming languages (POPL), vol 36. https://doi.org/10.1145/373243.360220
20. Floyd RW (1967) Assigning meanings to programs. In: Symposium on applied mathematics, vol 19. http://laser.cs.umass.edu/courses/cs521-621.Spr06/papers/Floyd.pdf
21. Gordon M, Collavizza H (2010) Forward with hoare. In: Reflections on the Work of CAR Hoare. Springer, pp 101–121. https://doi.org/10.1007/978-1-84882-912-1_5
22. Grebenshchikov S, Lopes NP, Popeea C, Rybalchenko A (2012) Synthesizing software verifiers from proof rules. In: International conference on programming language design and implementation (PLDI). https://doi.org/10.1145/2254064.2254112
23. Hoare CAR (1969) An axiomatic basis for computer programming. Commun ACM 12(10). https://doi.org/10.1145/363235.363259
24. Hojjat H, Rümmer P (2018) The ELDARICA horn solver. In: Bjørner N, Gurfinkel A (eds) Annual conference on formal methods in computer aided design (FMCAD). https://doi.org/10.23919/FMCAD.2018.8603013
25. Hojjat H, Rümmer P, Subotic P, Yi W (2014) Horn clauses for communicating timed systems. In: Workshop on horn clauses for verification and synthesis (HCVS). https://doi.org/10.4204/EPTCS.169.6
26. Horn A (1951) On sentences which are true of direct unions of algebras. J Symb Log 16(1). https://doi.org/10.2307/2268661
27. International organization for standardization (ISO) The ANSI C standard (C99). http://www.open-std.org/JTC1/SC22/WG14/www/docs/n1124.pdf
28. Kirchner F, Kosmatov N, Prevosto V, Signoles J, Yakobowski B (2015) Frama-C: a software analysis perspective. Form Asp Comput. https://doi.org/10.1007/s00165-014-0326-7
29. Komuravelli A, Gurfinkel A, Chaki S, Clarke EM (2013) Automatic abstraction in SMT-based unbounded software model checking. In: Sharygina N, Veith H (eds) International conference on computer aided verification (CAV). Lecture notes in computer science, vol 8044. Springer. https://doi.org/10.1007/978-3-642-39799-8_59
30. McCarthy J (1962) Towards a mathematical science of computation. In: Congress on information processing. North-Holland. https://dblp.org/rec/conf/ifip/McCarthy62.bib
31. Moy Y (2008) Sufficient preconditions for modular assertion checking. In: International conference on verification, model checking, and abstract interpretation (VMCAI). https://doi.org/10.1007/978-3-540-78163-9_18
32. Nielson HR, Nielson F (2007) Semantics with applications: an appetizer. Springer-Verlag. https://doi.org/10.1007/978-1-84628-692-6
33. Nyberg M, Gurov D, Lidström C, Rasmusson A, Westman J (2018) Formal verification in automotive industry: enablers and obstacles. In: Margaria T, Steffen B (eds) Leveraging applications of formal methods, verification and validation. Industrial practice. Springer International Publishing
34. Seghir MN, Kroening D (2013) Counterexample-guided precondition inference. In: European symposium on programming languages and systems (ESOP). https://doi.org/10.1007/978-3-642-37036-6_25
35. Seghir MN, Schrammel P (2014) Necessary and sufficient preconditions via eager abstraction. In: Asian symposium on programming languages and systems (APLAS). https://doi.org/10.1007/978-3-319-12736-1_13
36. Singleton JL, Leavens GT, Rajan H, Cok DR (2019) Inferring concise specifications of APIs. CoRR. http://arxiv.org/abs/1905.06847

Chapter 14
Exploring Frama-C Resources by Verifying Space Software

Rovedy Aparecida Busquim e Silva, Nanci Naomi Arai,
Luciana Akemi Burgareli, Jose Maria Parente de Oliveira,
and Jorge Sousa Pinto

Abstract The verification process is mandatory in the critical software realm. To improve this process, static analysis tools can make significant contributions. Static analysis meets a variety of goals, including error detection, security analysis, and program verification, which is why standards for critical software development recommend the use of static analysis to identify errors that are difficult to detect at run-time. Thus, this chapter presents a case study on the use of Frama-C as a static analyzer for formal verification of critical software and a lightweight semantic-extractor tool; the former uses an abstract interpretation technique, and the latter allows for the extraction of semantic information to provide a better understanding of source code. In practical terms, the chapter shows how Frama-C can support the development life cycle of an inertial system in aerospace applications, reporting a list of pros and cons. The final results indicate the benefits obtained in terms of software safety, software quality assurance and, consequently, the software verification process.

R. A. Busquim e Silva (✉) · N. N. Arai · L. A. Burgareli
Division of Aerodynamics, Control and Structures (ACE), Institute of Aeronautics and Space (IAE), São José dos Campos, SP 12228-904, Brazil
e-mail: rovedyrabs@fab.mil.br

N. N. Arai
e-mail: nancinna@fab.mil.br

L. A. Burgareli
e-mail: lucianalab1@fab.mil.br

J. M. Parente de Oliveira
Division of Computer Science, Aeronautics Institute of Technology (ITA), São José dos Campos, SP 12228-900, Brazil
e-mail: parente@ita.br

J. Sousa Pinto
High-Assurance Software Laboratory (HASLab), Institute for Systems and Computer Engineering, Technology and Science (INESC TEC) and University of Minho, 4710-57 Braga, Portugal
e-mail: jsp@di.uminho.pt

© The Authors(s), under exclusive license to Springer Nature Switzerland AG 2024
N. Kosmatov et al. (eds.), *Guide to Software Verification with Frama-C*, Computer Science Foundations and Applied Logic, https://doi.org/10.1007/978-3-031-55608-1_14

Keywords Embedded aerospace software · Formal verification · **Frama-C** · Software safety · Static analysis

14.1 Is Static Analysis Worthwhile?

A verification process is mandatory in the critical software development life cycle. Static analysis can be included in this process to make it more efficient. Furthermore, it is recommended by space standards and researchers in the scientific community.

The European Cooperation for Space Standardization standard ECSS-E-ST-40C on Space engineering–Software (2009) recognizes that static analysis is a useful resource in terms of code verification activity, e.g., [22]:

> f. The supplier shall verify source code robustness (e.g. resource sharing, division by zero, pointers, run-time errors).
> AIM: use static analysis for the errors that are difficult to detect at run-time.

The National Aeronautics and Space Administration Procedural Requirements NPR-7150.2D–NASA Software Engineering Requirements (2022) recommend the use of static analysis to define software management requirements and software engineering life cycle requirements, e.g., [43]:

> 3.11.5 The project manager shall test the software and record test results for the required software cybersecurity mitigation implementations identified from the security vulnerabilities and security weaknesses analysis. [SWE-159]
> Note: Include assessments for security vulnerabilities during Peer Review/Inspections of software requirements and design. Utilize automated security static analysis as well as coding standard static analyses of software code to find potential security vulnerabilities.

> 3.11.7 The project manager shall verify that the software code meets the project's secure coding standard by using the results from static analysis tool(s). [SWE-185]

> 4.4.4 The project manager shall use static analysis tools to analyze the code during the development and testing phases to, at a minimum, detect defects, software security, code coverage, and software complexity. [SWE-135]

According to the International Standard ISO/IEC/IEEE 24765 (2017), static analysis is the process of evaluating a system or component based on its form, structure, content, or documentation [37]. From an implementation point of view, static analysis of source code is the science of computing synthetic information about the source code without executing it [4].

Static analysis can be applied to meet a variety of objectives: type checking, style checking, program understanding, program verification and property checking, bug finding, and security review [15].

To meet the aforementioned recommendations related to program understanding, bug finding, and program verification, in this chapter, we exploit the application of the Framework for Modular Analysis of C (Frama-C).[1] Frama-C gathers several static analysis techniques into a single collaborative framework [4].

The objective of this chapter is to present useful insights into the benefits and challenges associated with the use of Frama-C and its practical feasibility as a tool for use in the software verification process. We aim to share our experience in applying Frama-C in a case study from a real project in the aerospace domain with both the scientific community and software engineering teams interested in verifying their software.

14.2 Literature Review of Frama-C Applications

In the industrial and scientific communities, the most common use of Frama-C is related to plug-ins based on formal methods. An important benefit of formal methods is that they produce sound analyses, i.e., every finding is correct; they never assert a property to be true when it is not true. Formal methods help to find bugs during the early phases of the software development life cycle, particularly through the use of software verification techniques. Formal software verification can be employed in a manner that is complementary to testing and simulation activities to help ensure critical system reliability.

Source code testing and inspection activities can become very expensive depending on the size of the application and the coverage criterion adopted [49]. In addition, techniques such as model checking, run-time verification, deductive verification, and abstract interpretation can be used to complement these activities by trying to maintain the cost-benefit relationship inherent to each of them. Model checking analyzes a program's model automatically; however, it presents the well-known problem of state space explosion [16]. Run-time verification also automatically analyzes a concrete execution of a program but not exhaustively. Deductive verification generates logical formulas (proof obligations) to be proven by other mechanisms in a powerful way, but it is not completely automatic. Abstract interpretation overapproximates the behavior of a program nearly automatically but generates false alarms.

Several works have employed Frama-C to analyze systems from different application areas. Approaches that improve the usability of formal methods in the software development life cycle with Frama-C were presented in [40]. In another related work [20], the main goal was to establish a scalable methodology for using static analysis through the development process by a development team. The authors presented modular analysis and bottom-up strategies involving the Eva and Wp plug-ins. In the safety context, they proposed a framework for safe software development in which the integrated code is analyzed statically to verify the system's critical safety properties using Eva and the formal specification language ACSL [33].

In addition, several published works have demonstrated the application of formal methods in the verification of critical software in specific domains. We select some

---

[1] https://frama-c.com.

works that apply Frama-C in security- and safety-critical systems in the Internet of Things (IoT), avionics, and nuclear domains.

In the IoT domain, an example of how formal verification can be applied using Frama-C is shown in [7]. The growing number of connected devices around the world, applied or not to critical systems, require security and formal methods as a way to ensure improved reliability. The work is based on a combination of three verification techniques: static analysis to guarantee the absence of run-time errors (Eva plug-in), deductive verification for functional correctness (Wp plug-in), and dynamic verification for parts of the source code that cannot be proven using deductive verification (E-ACSL plug-in).

In the avionics domain, a method of detecting software security vulnerabilities in real time was developed in such a way that it automatically combines static Eva and dynamic E-ACSL analyses in the Frama-C suite [14]. The results confirmed that Frama-C is capable of identifying families of cybersecurity weaknesses and analyzing applications more robustly while delivering the required performance for industrial scale-up. Additionally, formal methods have been applied to the new avionics software products developed at Airbus [9]. The Wp plug-in and the ACSL specification language of Frama-C are used in the verification processes. Finally, the formal verification of the Compact Position Reporting (CPR) algorithm, which is part of the Automatic Dependent Surveillance-Broadcast (ADS-B) system for aircrafts, was performed [21]. The Wp plug-in was used to generate verification conditions, which were discharged with the aid of the Gappa and Alt-Ergo automatic solvers.

In the nuclear domain, a comparative analysis based on the abstract interpretation method of different tools, including the Frama-C Value plug-in, was used to assess safety-critical software employed in nuclear power plants [44]. The article describes practical experimentation and presents an overview of the results and limitations of these tools.

In addition to the aforementioned areas, research projects have applied Frama-C to carry out analyses in industrial case studies. To facilitate the development of safe software systems, an engineering method that consists of four stages was proposed: system modeling and validation, code generation and integration, static code analysis, and dynamic code analysis. In the static analysis stage, the source code was annotated using ACSL, and it was analyzed through the Eva plug-in to verify the critical safety properties. The proposed method was applied in several industrial and research projects [32]. The Verification Engineering of Safety- and security-Critical Dynamic Industrial Applications (VESSEDIA) project proposed to enhance and scale up modern software analysis tools, namely, the mostly open source Frama-C analysis platform, to allow developers to rapidly benefit from them when developing connected applications and to provide safety and security to many new software applications and devices [55].

Some works have evaluated Frama-C by comparison with other static analysis tools. Frama-C was appointed as one of the best open-source static analysis tools available for C even though it is not specifically focused on security [29]. Other research has evaluated how much time tools, including Frama-C, need to detect run-time errors in programs with and without slicing. In the case of Frama-C, the Slicing plug-in improved the verification process time [39].

Frama-C has also been used as a basis for developing the commercial tool TrustIn-Soft Analyzer, which is a C and C++ source code analyzer and the industrial version of the open-source Frama-C [54].

Two important aspects of Frama-C that must be considered are (1) its capability to satisfy tool qualification requirements, which shows that it is a reliable tool and worth the effort to use, and (2) its ability to help applications achieve the requirements established in certification standards, including, for example, those established by organizations such as the National Institute of Standards and Technology (NIST) and Bureau Veritas.

The Frama-C Eva plug-in is one of the few tools in the world to have passed the NIST 6th Static Analysis Tool Exposition (SATE VI) Ockham Sound Analysis Criteria. According to the three evaluated criteria, Frama-C Eva: (1) is a sound tool; (2) produces at least 75% of the program findings; and (3) all findings are correct [6]. A partnership between Bureau Veritas and CEA-List produced a guideline for Software Development & Assessment using Frama-C as part of its proof of concept and concluded that Frama-C makes it possible to analyze and verify software to ensure it meets recommended standards at optimal cost [12]. Furthermore, Frama-C is able to correlate outputs with the entries of the Common Weakness Enumeration (CWE) and of the SEI CERT C Coding Standard [17].

The fact that the Frama-C Eva plug-in has also satisfied the SATE VI criterion is an indicator that it can assist in accomplishing the certification requirements of critical software. A source code analysis activity must achieve two certification objectives: (1) accuracy and consistency and (2) verifiability [47]. The former determines the correctness of the source code (e.g., floating-point arithmetic, use of uninitialized variables, and unused variables), which can be shown by the absence of some classes of run-time errors that were established through the technique of abstract interpretation. The latter ensures that the source code does not contain statements and structures that cannot be verified and that the code does not have to be altered to test it.

14.3 Overview of IAE Satellite Launcher Projects

Brazilian space projects aim at scientific technological research, innovation, and development to consolidate national air and space power, space launch operations, and services in aviation, space and defense systems [41].

The Institute of Aeronautics and Space (IAE) is a military organization of the Aeronautical Command (COMAER) within the Department of Aerospace Science and Technology (DCTA). As shown in Fig. 14.1, IAE and several public and private institutions are involved in the space program coordinated by the Brazilian Space Agency (AEB) [10]. IAE has developed two satellite launcher projects and a series of sounding vehicles (VS) [1].

Fig. 14.1 Institutions coordinated by AEB [10]

IAE has been developing satellite launcher projects, as illustrated in Fig. 14.2. The Satellite Launch Vehicle 1 (VLS-1) was the first Brazilian satellite launch vehicle. This project aimed to provide Brazil with the ability to design, manufacture, launch, control, stabilize, and deliver a payload in orbit autonomously. A real-time embedded software, named SOAB, was developed for VLS-1 to control flight based on a set of control, navigation, and guidance algorithms as well as the sequence of events that occur in the various launching phases. SOAB was also responsible for sending telemetry data to the ground station [36].

Due to technological challenges and limited financial and human resources, the need to realign the VLS-1 development strategy has arisen, taking advantage of its legacy and adapting it to the current scenario. The new proposal encompasses the development of a simpler controlled suborbital vehicle [36].

The Microsatellite Launch Vehicle (VLM-1) project aims to develop a three-stage solid propellant vehicle for the launch of microsatellites in equatorial or re-entry low Earth orbits (LEO). IAE and the German Aerospace Center (Deutsches Zentrum für Luft- und Raumfahrt–DLR) are jointly developing the VLM-1. The Brazilian part of the project first comprises the development of VS-50 for in-flight qualification of the S50 motor, the electrical and pyrotechnic networks, and the guidance, control and navigation system that will later be used in VLM-1, mitigating the technical risks of the project [1, 2]. The VS-50 project concerns a two-stage solid propellant suborbital vehicle. The first stage of both vehicles is almost identical, and the second stage of VS-50 has the same configuration as the third stage of VLM-1 [1].

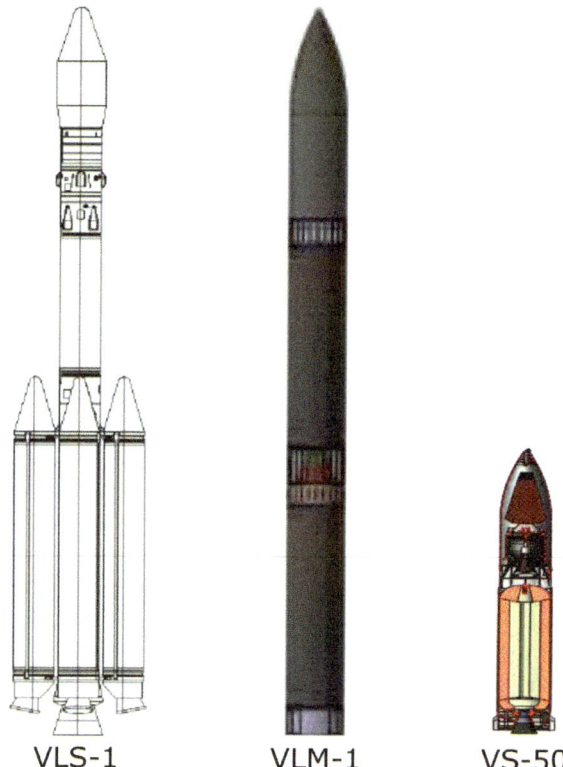

Fig. 14.2 IAE Launch vehicle projects. Adapted from [1]

14.3.1 Case Study: Embedded Aerospace INS Software

During the development of launch and suborbital vehicles at IAE, the need for reliable inertial navigation, guidance, and control systems has arisen [35]. Therefore, the SISNAV (*Sistema de Navegação Inercial*–inertial navigation system) is under development to provide a national inertial navigation system (INS), replacing the current imported system. To avoid a trade embargo, the objective is to update the involved technology. Initially, conceived to be used in VLS-1, SISNAV will be a subsystem in VS-50 and VLM-1 [35, 45].

An INS consists of an inertial measurement unit (IMU) and a computational unit. The SISNAV is a strapdown INS whose IMU has three accelerometers and three fiber optic gyrometers (FOGs) that measure acceleration and angular velocity, respectively. SISNAV's computational unit encompasses a floating-point digital signal processor (DSP) with 256K internal memory, a field-programmable gate array (FPGA), a voltage-to-frequency (V/F) card (accelerometer interface), serial channels (connecting to FOG), and additional A/D and I/O interfaces [45].

SISNAV's software (SSISNAV) is a critical real-time software embedded in a DSP that processes data from the IMU and performs calculations to determine position, velocity, and attitude. SSISNAV implements algorithms for testing, calibration, and compensation of inertial sensor signals, self-alignment, navigation, and data storage and transmission [52]. A Very High Speed Integrated Circuit (VHSIC) Hardware Description Language (VHDL) code embedded in FPGA is responsible for data acquisition, which is out of the scope of this work.

SSISNAV is a natural candidate for Frama-C analysis because it is a highly critical embedded software implemented in the C programming language. Embedded code is adequate to be analyzed by Frama-C due to its characteristics, such as the absence of recursion, dynamic memory allocation, and calls to external libraries. SSISNAV source code consists of ninety-nine functions and 43,699 physical source lines of code (SLOC). It contains main algorithms (accelerometer and gyrometer calibration, coarse and fine alignments, and navigation) with many numerical calculations based on floating-point arithmetic, whose computation depends on six C library mathematical functions. The implementation contains function-like macros and static memory allocations. The SSISNAV team has been testing and modifying the source code and extending the documentation. Its development follows the processes proposed by the ECSS standards [22, 24].

Components have been developed and tested individually in the Laboratory of Identification, Control and Simulation (LINCS) [35, 45]. Furthermore, integrated configurations have been tested in roller coaster rides, aircraft flights and compact electric utility vehicle rides. The plan is that SISNAV will be embedded in a Brazilian launcher in a future flight. A series of tests is planned in advance, including a flight as a payload onboard the vehicle HANBIT-TLV, which was developed by the South Korean company INNOSPACE [34]. The performance of SISNAV is assessed using a hybrid simulation environment. This is a hardware-in-the-loop simulation of the complete system, including the following components: the onboard computer with the control laws, the vehicle dynamics, the actuator dynamics, and the sensor dynamics [13]. Since the simulation data are recorded according to the state, preparation or navigation, the static code analysis is executed in two parts. We used the simulation output obtained by the control system team. Because the data of the selected case study are classified, this work presents only a limited subset of the results. For the same reason, some information pertaining to the algorithms (such as function names) is not disclosed.

14.4 Employing Frama-C in Space Software

This section presents our experiments in applying Frama-C to space software. First, we provide an overview of the results obtained from our former experiment, whose approach relied on abstract interpretation and deductive verification techniques. Second, we focus on presenting our current experience in using Eva as a static analyzer for formal verification and Callgraph and SpareCode as lightweight semantic-extractor tools. Our current approach to static analysis by abstract interpretation slightly improves on our previous work, and we present here in detail our experiments in the context of SSISNAV. Furthermore, this section describes the tools and equipment used, the experiments carried out, and the results obtained.

14.4.1 Former Experiment: Embedded Aerospace Control Software

Our first experience in applying Frama-C started with a research project whose objective was to develop an approach to formal verification activity in the context of spatial critical software. Our case study was the SOAB, which is the software onboard the VLS-1. The results suggested that this activity is relevant in the software verification process and can be used to complement validation activities, including testing and simulation [50].

The main contribution of this research project was to disseminate our experience in applying formal software verification to the scientific community and to software engineering teams interested in verifying their software. We used formal static analysis based on two techniques: abstract interpretation [18] and deductive verification [31], which were implemented using Frama-C through the Value (currently named Eva) and Jessie plug-ins, respectively. Moreover, InOut and Metrics were used as auxiliary plug-ins.

At the end of the research project, the Frama-C analysis did not emit any alarms. This result was expected considering that the case study concerned a product that had been in production for many years. Nevertheless, we detected anomalies related to two software products: documentation and implementation. According to Value, three anomalies in the documentation and four in the source code were detected, while Jessie allowed us to detect seven anomalies in the documentation and six in the source code. Although the results obtained with Frama-C are limited to alarms, they can indirectly lead to identifying anomalies in the software documentation and implementation. Both kinds of anomalies were detected in our approach using Value and Jessie.

Our experience with the code inspection process for space software made it relatively easy and intuitive to detect anomalies visually (specifically implementation anomalies). The anomalies detected in the source code were related to unnecessary and/or duplicate library inclusions, dead code, redundant code, global variables passed as function arguments, inconsistency in the argument type between the function definition and function body, implicit typecasts, and incorrect comments.

Anomalies in the documentation were discovered by using Value when information about the sensor input data was needed. In the data dictionary (DD) of the software requirements specification (SRS), an incorrect variation domain was discovered, and it was related to sensor input data that were examined to parameterize the analysis. In the software design document (SDD) and DD of the SDD, the descriptions of two sensor vector components were inverted, and anomalies in documentation were found when ACSL function contracts were required to use Jessie.

In the SDD, anomalies in the structure charts were identified: one missing output variable and two incorrect parameters. In the DD of the SDD, there were anomalies in the type definition and names of variable and function parameters.

Perhaps these anomalies could remain unnoticed by other approaches to inspection. Although they did not directly impact the final result of the application, they might create difficulties in understanding, documenting, and maintaining the source code, which could potentially lead to serious failures in the future. For instance, during the software maintenance process, developers might implement some incorrect modifications in the source code based on incorrect documentation. Additionally, finding and reporting anomalies can assist in the evaluation of quality attributes such as reliability and productivity and in process improvement.

The preliminary results were obtained by using the **Frama-C** 6 (Carbon) release. The final results were later published in [50] with results updated using the **Frama-C** 10 (Neon) release.

14.4.2 Current Experiment: Embedded Aerospace INS Software

The context of the current selected case study is different from the previous one; SOAB was a legacy project, while SSISNAV is currently under development and has not yet been embedded in a launcher and evaluated during flight. On the other hand, we were able to reduce the effort necessary to carry out the current verification because it is based on the approach that was previously developed for the formal verification activity of SOAB, which provided us with a background for the abstract interpretation technique and familiarity with the tool. In addition to **Eva** and **Metrics**, which were already used in the previous project, the use of **Callgraph** and **SpareCode** is also investigated in this new case study. **Frama-C** plug-ins can be categorized according to the intended analyses. Specifically, the plug-ins employed in our experiment can be divided into the following [5]:

- Verification plug-in–**Eva**;
- Understanding plug-ins–**Callgraph** and **Metrics**;
- Simplification plug-in–**SpareCode**.

Frama-C is a collaborative platform, and some plug-ins work on results that were already computed by other plug-ins in the framework [17]. This dependency relationship can help the user establish a sequence for running the plug-ins. **SpareCode** relies on **Eva** except for the `-sparecode-rm-unused-globals` option. **Metrics** relies on **Eva** for the `-metrics-eva-cover` and `-eva` options. **Callgraph** does not depend on **Eva**, but its precision may be improved by using the `-eva` option. Figure 14.3 shows which plug-ins (identified by their prefix) are running according to the command and options entered by the user in the terminal. The figure also shows the dependencies (or absence thereof) and running order among the employed plug-ins. Parsing of files is executed by the kernel independently of any plug-in.

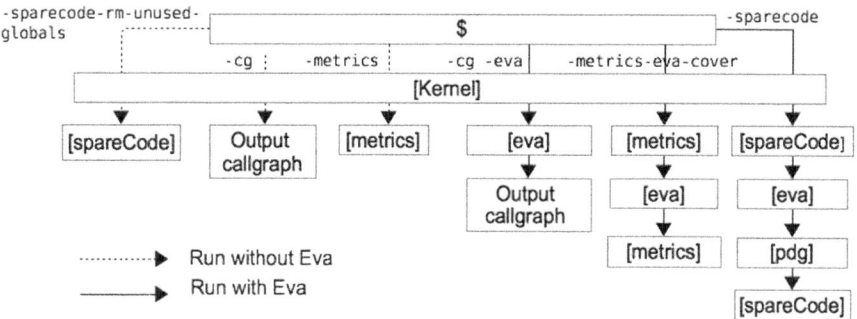

Fig. 14.3 Examples of dependencies among plug-ins running on the terminal

To conduct the SSISNAV experiments, we used the Frama-C 23 (Vanadium) release on a 2.9 GHz notebook with 8 GB of memory and a 1 TB hard disk running the MacOS X Yosemite 10.10.5 release. To plot the graphs, we used Plot2 version 2.6.17. To translate from DOT to PNG format, we used dot/graphviz version 2.40.1. To draw figures, we used the Papyrus release 2022–03 and Apache OpenOffice 4.1.2.

14.4.3 SSISNAV Analysis with the Callgraph Plug-In

The call graph is a directed graph that represents relationships between functions in the program: the nodes are functions, and the edges represent one or more invocations of a function by another function [48]. Call graphs can be used to aid software engineering teams in the following tasks: software documentation to improve program understanding [25], detection of program execution anomalies [28], and analysis of the impact of program changes [42].

The Callgraph plug-in automatically computes the program's call graph [3]. Its output is saved to a dot file [27] that requires an appropriate application to be visualized, or it can be translated into the PNG format.

We have generated call graphs for SSISNAV main functions with some auxiliary functions omitted for simplification purposes. For example, Fig. 14.4 shows the call graph of the main navigation function, which can be generated and translated into the PNG format with the following command lines:

```
$ frama-c -cg out.dot ./DIR1/f1.c ./DIR2/f2.c ./DIR2/f2.h
$ dot out.dot -Tpng -o f1.png
```

Figure 14.5 shows the call graph of the fine alignment function, and it can be observed that the *function5b4* is not called. Therefore, it is necessary to look at other call graphs to investigate if it is called in any other source code file; otherwise, it should be reported as an anomaly. A complete call graph generated from a main function could be used to detect functions that are never called. However, this call graph could be very large and difficult to visualize. Therefore, we have generated

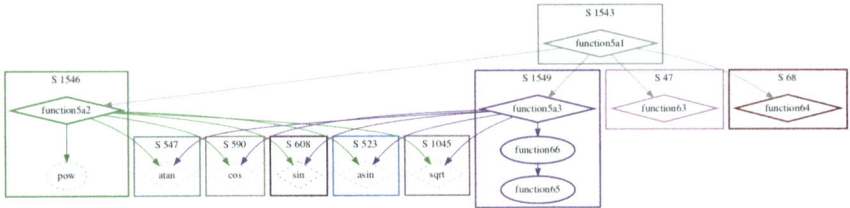

Fig. 14.4 Call graph of the main navigation function

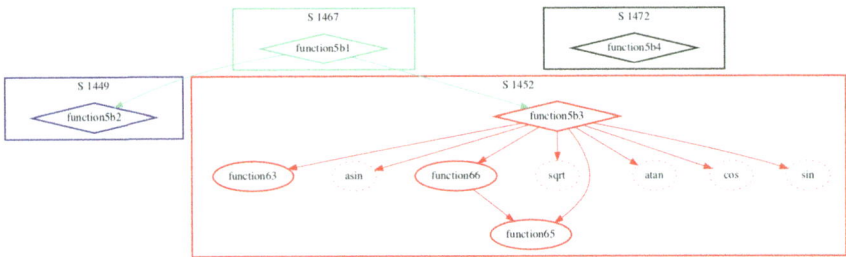

Fig. 14.5 Call graph of the fine alignment function

call graphs for the functions encoding the main algorithms, which are smaller and easier to examine.

After analyzing the applicability of call graphs in our context, we identified that call graphs can assist in writing the SDD section <5.4.6><*Subordinates*>, help detect functions that are never called, and contribute to program understanding because they depict the structure of the program with all subordinate functions.

The plug-in does not handle function-like macros, i.e., they were not shown in the call graphs. Additionally, we noticed that a missing function definition is shown as a dotted line, and any function called from it is not shown in the graph, generating an incomplete call hierarchy. This depends on how the source code files are organized. If all function definitions are implemented in a single file, the **Frama-C** command line will require only this file to generate a complete graph; otherwise, all of the files will be required. The #include directives of the analyzed function, e.g., for navigation or fine alignment, can be examined to select the necessary files.

The option -eva generates a call graph according to the analysis, so some branches may not be reached and functions called from them will not be shown. Comparing different call graphs generated with and without the -eva option for valid and invalid intervals can be useful in assessing the coverage of functions and visually identifying unused source code.

14.4.4 SSISNAV Analysis with the SpareCode Plug-In

The SpareCode plug-in removes *spare code*, i.e., code that does not contribute to the final results of the program, and it produces an output program that is guaranteed to be a compilable C program [4].

Attention

A piece of source code that does not contribute to the final state of the program is not necessarily dead code; we can only be sure that all its side effects are shadowed by further instructions [53].

The plug-in includes the option -sparecode-rmunused-globals, which removes unused global types and variables. Nevertheless, as it prevents the global variables from being listed explicitly, the generated code requires inspection (visually or through a file comparison tool) to detect unused variables. According to a discussion in the Frama-C forum, the debug key globs must be enabled, and the debugging level must be assigned the minimum value of 2 to generate the output, including unused global variables preceded by the expression remove var. Thus, the grep command can be employed to extract only the list of unused variables.

Detecting local static variables is slightly different. According to the same forum discussion, unused local static variables are removed while parsing the source code regardless of whether SpareCode is executed or not. Nevertheless, it is possible to extract the desired information using the option -kernel-msg-key parser:rmtmps.

There are two reasons to detect unused variables. First, this is one of several items that may be helpful as preliminary steps to obtain guarantees of code correctness [47]. Second, for embedded applications, a good programming practice is to keep the source code as lean as possible. Despite not improving safety assurance, it might impact the consumption/optimization of memory and impair code readability.

The command used to detect declared but unused global variables in the navigation algorithm in SSISNAV, and its output are the following:

```
$ frama-c navigation.c -sparecode-rmunused-globals -sparecode-msgkey globs -sparecode-debug 2 | grep "remove var"
[sparecode:globs] remove var senLc_2
[sparecode:globs] remove var senLc_4
```

For the navigation algorithm, there were two declared but unused global variables detected in the source code; for the sensor calibration and alignment algorithms, the plug-in did not detect any declared but unused global variables.

To detect declared but unused local static variables, the command and its output for the gyrometer calibration algorithm are given as follows:

```
$ frama-c -main function3a1 calibrationG.c -sparecode-
rmunused-globals -sparecode-msgkey globs -sparecode-
debug 2 -kernel-msg-key parser:rmtmps | grep "removing
local"
[kernel:parser:rmtmps] removing local: proc_giros_dado-
Giros
```

For the coarse alignment algorithm, the command and its output are the following:
```
$      frama-c -main function5c1 alignment1.c -sparecode-
rmunused-globals -sparecode-msgkey globs -sparecode-
debug 2 -kernel-msg-key parser:rmtmps | grep "removing
local"
[kernel:parser:rmtmps] removing local: alingr_ciclo_M2_
dVc
[kernel:parser:rmtmps] removing local: alingr_ciclo_M2_
dThetac
```

The number of declared but unused static local variables detected for the gyrometer calibration and coarse alignment algorithms was one and two, respectively; for the accelerometer calibration, navigation, and fine alignment algorithms, the command execution did not detect any unused local static variables.

14.4.5   SSISNAV Analysis with the **Metrics** Plug-In

Metrics are the only way to quantitatively assess the quality of development processes and products, and they are typically used to manage development [23, 24]. It is recommended to generate at least basic metrics such as size (code), complexity (code) and test coverage. An important metric is the cyclomatic complexity, which can directly affect the quality and maintenance of a source code. Standards recommend that safety-critical software components have a defined value of cyclomatic complexity [23, 43].

The **Metrics** plug-in implements the automatic computation of a set of measures on the source code, e.g., McCabe's cyclomatic complexity, Halstead complexity, SLOC, and the Eva coverage estimate [4]. The option -metrics-eva-cover provides the **Eva** coverage statistics, which include the semantic and syntactic reachability coverages, the coverage estimation, and the number of statements reached in each analyzed function. It can be used to compare the code effectively analyzed by **Eva** with what **Metrics** are considered reachable from the main function [8]. Therefore, the user is able to perform a coverage analysis to count how many unreached functions exist, accounting for the number of syntactically reachable functions. Additionally, **Metrics** provides the number of lines of code containing calls to non-analyzed functions.

The SSISNAV metrics shown in Table 14.1 are delivered to the software quality team because these metrics allow them to perform risk management, indicating whether risks have to be minimized and controlled and thereby reported to developers.

Table 14.1 Global metrics by phase in Scenario 2

	Accelerometer calibration	Gyrometer calibration	Coarse alignment	Fine alignment	Navigation
Cyclomatic complexity	4	6	30	28	25
SLOC	34	67	124	166	148

In our software development processes, there is no requirement related to complexity based on the cyclomatic complexity metric. Therefore, we adopted a maximum cyclomatic complexity level of 20 based on standards [23, 43]. It is necessary to assess the source code when this threshold is exceeded to determine whether it is acceptable or needs to be modified. For the latter, the suggested corrective actions, in addition to reviewing the algorithm logic to reduce complexity, include writing smaller functions, minimizing the number of decision structures, and removing unused source code.

14.4.6 SSISNAV Analysis with the Eva Plug-In

In the Frama-C 15 (Phosphorus) release, the support for the legacy Value plug-in was abandoned and replaced by Evolved Value Analysis (Eva), which provides more precise and extensible abstract domains [11]. The Eva plug-in is a forward dataflow analysis based on the principles of abstract interpretation, and it performs whole-program analyses [38]. Eva is both context-sensitive and path-sensitive [11], i.e., function calls are handled through a symbolic inlining of function bodies [38] and infeasible paths are excluded. Its main objective is to automatically compute sets of possible values for the variables in an analyzed program [11]. An additional purpose of the plug-in is to find alarms, i.e., errors that could occur at run-time and/or to demonstrate their absence. It might be employed to assist in the following tasks [11]: familiarization with foreign code, automatic document production, bug detection, and guaranteeing the absence of bugs.

With respect to our previous research project, the workflow of the software verification approach employing Eva was updated in relation to its inputs and outputs, which are highlighted in red in Fig. 14.6.

The approach begins with preparing the application context, i.e., identifying the information required to perform the verification. There are two scenarios in which the sensor measurements are treated differently. Scenario 1 considers the range of maximum and minimum values accepted by the sensors, which illustrates a typical application of Eva. Scenario 2 is an unusual application of Eva in which it is used as a C interpreter [19]. It becomes very useful when a specific input dataset is known in advance [26], e.g., the values of sensor measurements obtained from a simulation. These values are deterministic, i.e., a specific value obtained from the sensor is

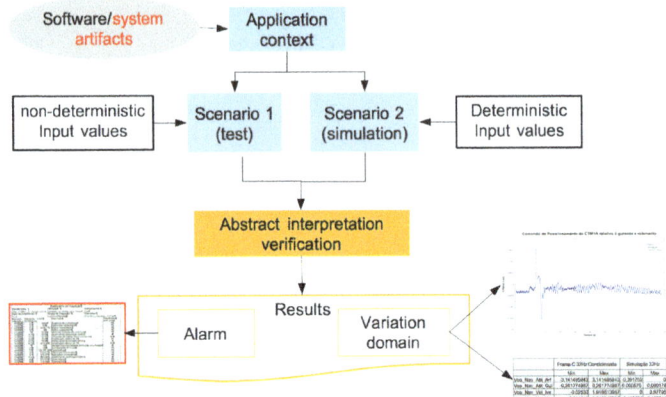

Fig. 14.6 Updated approach overview for verification by abstract interpretation with the Frama-C Eva plug-in. Adapted from [50]

considered for each instant, and thus, **Frama-C** acts as a simulation tool. Scenario 2 can adequately run a sufficiently deep analysis; it is completely deterministic and does not generate false alarms, while undefined behaviors (not detectable by testing) are still detected. The output results include alarms and the variation domains for the variables, which are presented in graphs and tables to facilitate data visualization and analysis.

The detailed approach is presented as an activity diagram with the updates highlighted in red in Fig. 14.7. The activities are grouped into two main phases: *context definition* and *implementation and refinement*. The former consists of executing the following activities: defining the entry point function, identifying the input data, and identifying the output data to be evaluated. For each inserted function, the latter consists of addressing library function inclusion, addressing missing functions, addressing alarm/nonalarm/message, and analyzing the variation domain. This process is iterative and interactive because it requires user intervention, and it is performed repeatedly until the intended result has been obtained. The user could adjust the accuracy of the analysis through the plug-in parameters to account for the trade-off between efficiency and accuracy.

In the context definition phase, an artifact set that may vary from project to project due to each particular need and availability is required. Some artifacts that assist in this phase are described below:

- Data flow diagram (DFD) from SRS, which contains the input data needed to execute the algorithms and their output data.
- DD from SRS, which contains detailed descriptions of each control and data flow.
- SDD, whose software component design section identifies the variables in the source code.
- DD from SDD, which contains the variable ranges (i.e., maximum and minimum values).

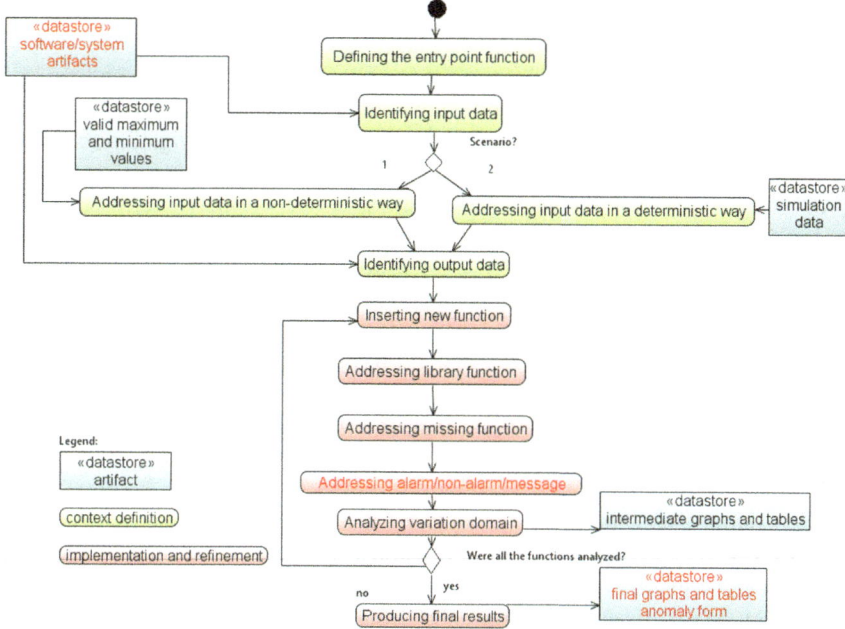

Fig. 14.7 Updated activity diagram for verification by abstract interpretation with the Frama-C Eva plug-in. Adapted from [50]

- System documentation, which can be useful for understanding source code that implements specific algorithms.
- Source code whose implementation itself and/or comments can assist in the identification of data, algorithms, etc.

Thus, with respect to our previous project, the artifact set was updated from *software documentation* to *software/system artifacts* to include system documentation and source code.

In the implementation and refinement phase, we expanded the activities from *Addressing alarms* to *Addressing alarm/non-alarm/message*, given that we are also handling nonalarms and messages that may correspond to loss of precision. An alarm is a guard against some undesirable behavior and corresponds to a particular warning category [3]. In practice, alarms are warning messages emitted by Eva that start with the prefix [eva:alarm]. Additionally, there are warning messages that are classified as nonalarms, starting with the word *Warning*. Finally, a message is only informative text that does not begin with the word *Warning* and is not prefixed with [eva:alarm] [11]. In relation to the output update, we have added an anomaly form that gathers all of the emitted alarms, which is described in Sect. 4.7.

14.4.6.1 SSISNAV Formal Verification

Eva analysis is executed in two parts corresponding to the preparation and navigation states. The sensor calibration and alignment algorithms are verified during the preparation state, and the navigation algorithm is verified in the navigation state. The experiments were run through the Eva plug-in batch mode.

Following the activity diagram shown in Fig. 14.7, we first describe the context definition phase. Figure 14.8 depicts the required input artifacts to each activity step to define the context for SSISNAV compared to SOAB. The software artifacts employed are SRS, SDD, and the source code, while the system artifact concerns the documentation related to the implementation of SSISNAV algorithms. The activities in this phase are succinctly described in the following paragraphs.

For the *defining the entry point function* activity, the main function in the source code was identified as the entry point function, which was corroborated by its call graph.

In the *identifying input data* activity, we used the context diagram (level 0 DFD)–SRS as an initial source together with the DD–SRS to understand the flows, as shown in Fig. 14.8. The input data required are mission, configuration, and sensor data. For both states and scenarios, it is necessary to set parameters related to specific mission requirement data, e.g., sensor calibration; reference data such as Earth's rotation; and launch site data such as local latitude. We used acceleration, angular velocity, and temperature (all raw data) as sensor input data in the preparation state and the calibrated sensor measurements in the navigation state. After that, we mapped the data/control flow into variables with the aid of the SDD, source code, and system documentation.

To *address the input data in a nondeterministic way*, we asked the system analyst to identify the numeric domain (range) of the input variables. We modified sensor reading functions, replacing hardware function calls with Frama-C built-in primitives to address intervals.

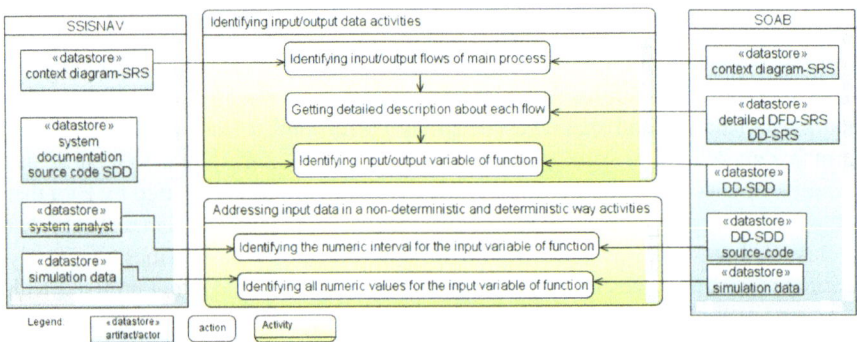

Fig. 14.8 Diagram for context definition of SSISNAV compared to SOAB

To *address the input data in a deterministic way*, we extracted deterministic values from two CSV simulation files, which contain 25,828 and 7,644 rows, for the preparation and navigation states, respectively. We wrote an additional program to extract data from these CSV files and created C functions that assign the extracted data to arrays, emulating the hardware functions. Then, the sensor reading functions were adapted to include these C functions in place of the hardware function calls.

When *identifying output data*, we used the context diagram–SRS and the DD–SRS. We used the following sensor output data: calibrated sensor measurements in the preparation state and inertial linear velocity, inertial position, and Euler angles in the navigation state. We modified the function that composes the telemetry data frames, replacing hardware function calls with Frama-C built-in primitives to display the telemetry data frames.

Now, we discuss the implementation and refinement phase and describe the related activities.

In the *inserting new function* activity, as proposed in our approach, we insert functions one by one to perform the analysis incrementally. In this activity, the *loop* global metric (i.e., number of loops) can be used as an input indicator of the number of loop unroll annotations required for each function.

During the *addressing library function* activity, we performed library inclusion treatment because Eva does not support certain external libraries (e.g., the hardware library). Therefore, because keeping these libraries in the code implies the introduction of a dependency chain that could not be provided, we decided to remove all included directives from the source code and later reinsert them incrementally as needed.

For the *addressing missing function* activity, we identified two missing function types. The input hardware functions were treated in one of two ways: either parameterizing the analysis by providing value intervals through Frama-C functions or using simulation data. The output hardware function was replaced by Frama-C functions to allow the results to be observed. We noticed the introduction of recent Frama-C improvements in its math library. Currently, Frama-C provides implementations for the six employed math functions that are different from our former experience when we had to write C code for functions such as sin, fabs, and modf.

In the *addressing alarm/non-alarm/message* activity, we evaluated and addressed the alarms as soon as they were detected, and we subsequently reran the analysis. Eva detected alarms related to accessing out-of-bounds indices in the navigation init function, uninitialized variables in the coarse alignment function, and NaN or infinite floating-point values in the fine alignment function. These results are presented in Table 14.4, identified by the expression Eva alarm. In addition to those alarms, others related to infinite floating-point values were detected in the alignment and navigation functions, which could not be eliminated. Nevertheless, Eva terminated the analysis, resulting in some intervals corresponding to the data type ranges that cannot be considered sufficiently accurate when evaluating the domain of an output variable.

For example, the emitted messages *starting to merge loop iterations* indicate the necessity of unrolling a loop. Instead of the option `-eva-slevel`, which was used in the previous experiment, we decided to annotate the loops because the source code is under development and it was easier for us to edit it and test other ways to unroll loops. Adding loop unroll annotations is more precise and stable than -eva-slevel; however, it has the drawback of requiring source code changes [11]. Even if modifications to the source code preserve the semantics, it is still recommended to keep the code used for testing and verification as close as possible to the original. For the preparation and navigation states, the numbers of loop unroll annotations inserted were 34 and 33, respectively. The nonalarms emitted by Eva are related to implicit function declarations, missing function specifications, and floating-point constant representations.

Comparing our previous and current Frama-C experiments, we noticed that the obtained nonalarms and messages are quite different for reasons related to differences in the case studies (e.g., implementation particularities) and to the evolution of Frama-C itself. This can make it difficult to use the *grep* command to obtain a specific message from a log file if the displayed messages are unknown. For this reason, it is necessary, after every analysis execution, to visually look for messages in the log file and catalog them. Thus, familiarity with Eva and its previous application in similar case studies make this activity easier.

For the *analysis of variation domain* activity, we applied the following treatment to make the analysis more precise: we used the option -eva-slevel to improve the analysis precision when evaluating *if* or *switch* conditional statements. In its default configuration, Eva produces results that are too approximate to handle loops. Loop unroll annotations ensure that Eva unrolls the loops and keeps the analysis precise; otherwise, Eva might generate alarms due to imprecisions [11]. After this treatment, the variation domain for the variables needs to be evaluated, and the analysis continues until valid values for the observed magnitudes have been obtained. To help in this evaluation, we observed intermediate graphs of some variables. We wrote an auxiliary program to extract data from the Eva output to be plotted in graph form.

In the *producing final results* activity, the final results of the verification are presented for each scenario. We generated six and twelve graphs for Scenario 1 and Scenario 2, respectively. We compared the Frama-C output to the simulation output to determine whether the computed variables were valid. In Scenario 1, we compared the intervals computed by Eva with the deterministic values obtained from simulation data. In Scenario 2, we compared the deterministic values computed by Eva with the deterministic simulation values.

The graphs generated for Scenario 1 plot the variation domain for the output variables of the gyrometer and accelerometer calibration algorithms. Figure 14.9 shows the maximum and minimum values at each time instant for the calibrated accelerometer data on the X-axis (AX). The graph shows that the AX intervals computed by Frama-C contain all simulation values, i.e., there exists no AX value from the simulation that is below the minimum limit or above the maximum limit of the interval computed by Frama-C. We remark that if the Frama-C bounds did not contain the simulation results, this would raise serious concerns regarding the correctness of the case study source code.

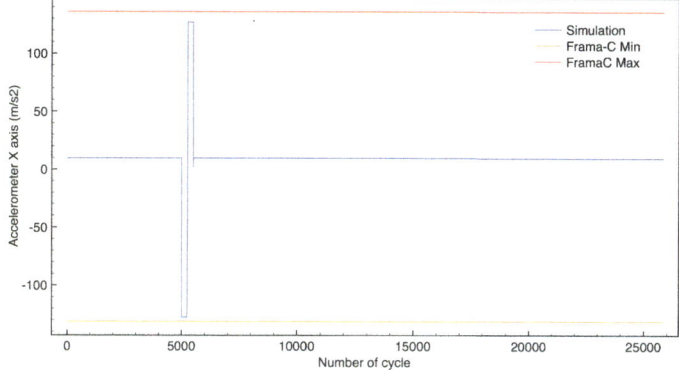

Fig. 14.9 Graph of calibrated accelerometer data on the X-axis in Scenario 1

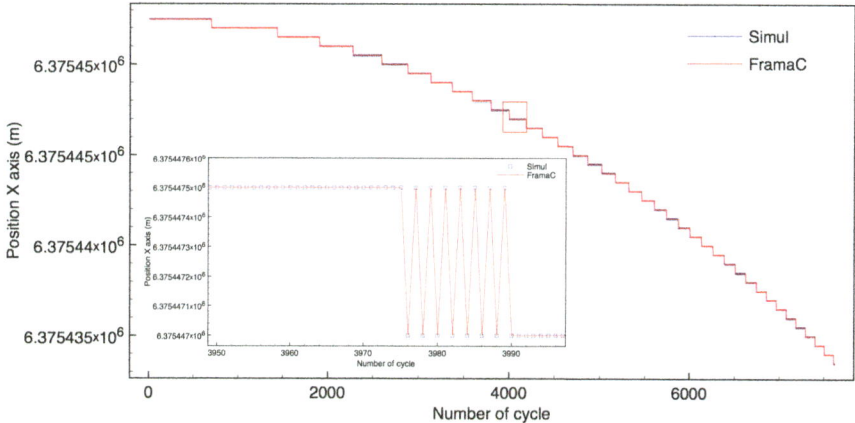

Fig. 14.10 Graph of position data on the X-axis in Scenario 2 with a zoomed-in section

Figures 14.10 and 14.11 show the graphs generated for Scenario 2 regarding the position data on the X-axis and the Euler angles, respectively. Eva produced profiles so similar to those obtained from the simulation that the curves are coincident.

During the implementation and refinement phase, the plug-in issues alarms that represent run-time errors and should thus be recorded as anomalies. In addition to these alarms, as was done in our former experiment as described in Sect. 14.4.1, we applied the same methodology to prepare the case study source code to be executed by Eva. We detected anomalies indirectly during this process in both the context definition and implementation and refinement phases. During implementation, when we found an imprecise or incorrect variable domain, we examined its dependencies in the source code, often finding multiple statements per line related to each dependency. This poor readability made the task of evaluating the domain more difficult. In the

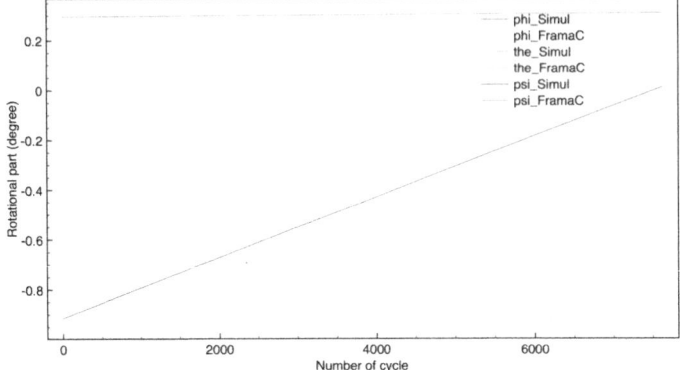

Fig. 14.11 Graph of Euler angles in Scenario 2

documentation, we detected one anomaly in the source code listing of the navigation algorithm, which is attached to a system document. This is presented in Sect. 4.7.

14.4.6.2 SSISNAV Metrics from Eva Coverage Statistics

The common method prescribed by standards to assure the dependability and safety of critical software is to apply measures such as source code statement coverage by test suites [22]. In the context of testing, statement coverage is a measure of the part of the program within which every executable source code statement has been invoked at least once [24].

Eva is currently able to produce semantic coverage metrics. Starting with the Frama-C 19 (Potassium) release, Eva prints an analysis summary outlining the analysis coverage through two measures: the ratio of functions that have been covered by the analysis and the percentage of statements that are reached within these functions. In our interactive approach, the percentage examination is useful for two reasons: it is possible to use these values to confirm if the insertion of a new function or an Eva resource improved the coverage during the analysis, and additionally, this ratio may help in reasoning about statements that are not reached at the end of the analysis.

Tip

The coverage percentages are useful in terms of observing the evolution of the interactive approach, e.g., when a missing function is provided, the coverage increases!

The metrics obtained from Eva and Metrics related to the number of functions analyzed and their reached statements were compared for the preparation phase

14 Exploring Frama-C Resources by Verifying Space Software

Table 14.2 Eva coverage statistics for Scenarios 1 and 2 in the preparation state

	Scenario 1	Scenario 2
Syntactically reachable functions	70 (out of 70)	76 (out of 76)
Semantically reached functions	70	76
Coverage estimation	100.0%	100.0%
Statements reached	1,563 (out of 1,904)	311,478 (out of 311,854)
% statements reached	82%	99%

in Scenario 2. Both plug-ins indicated 100% coverage related to the number of functions analyzed. For the statements reached in these functions, Eva indicated 99% coverage (311,429 statements reached), and Metrics indicated 99.9% coverage (311,478 statements reached). The source code analyzed by Eva was provided as an argument for the Metrics plug-in.

Table 14.2 presents the output of the analysis coverage using Metrics for Scenarios 1 and 2. Scenario 2 presented better results due to the limited dataset in contrast to Scenario 1, where intervals were used. Metrics depends on the results from Eva, but since Metrics itself runs much faster than Eva, the run-times of the Metrics and Eva plug-ins are similar. At the end of the Eva analysis, if any function has not been reached and detecting it by manual verification is difficult, there are two auxiliary options: the Metrics option `-metrics-cover` can identify the functions that are not syntactically reached from the main function, and Callgraph identifies functions that are not called.

The graph in Fig. 14.12 shows the number of functions grouped by the reached statement percentage for the preparation phase in Scenario 2. The unreached function statements depend on our context definition. The unreached statements are located inside branches related to the handling of invalid sensor input data and static memory allocation. This information is only shown by the Frama-C GUI in which it is highlighted.

The functions with the highest coverage mostly contain assignment statements, while the functions with lower coverage contain if—else statements to address error conditions (e.g., alignment functions). In our experiments, we have considered only valid input data obtained from a simulation; therefore, the condition for invalid inputs is never satisfied, which is one of the reasons why analysis does not move into some error-handling branches. Branches that are not analyzed must not be considered formally verified and may contain some errors. In poorly covered functions, it is not assured that the software meets the specified requirements correctly or reliably. In well-covered functions, this assurance is usually better, that is, the probability of guaranteeing that it is error-free is higher. One way to improve the coverage is to create scenarios for addressing non-analyzed functions. In addition, to prevent specific behaviors, the safety team develops new techniques and tools and puts in place new approaches to both system and software engineering. These are all intended to sup-

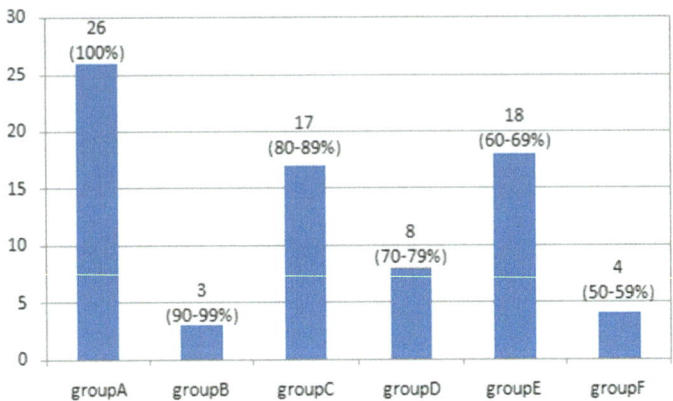

Fig. 14.12 Analyzed function number grouped by reachable statement percentage

port safety analysis, which belongs to the verification activity under the developer's responsibility.

14.4.6.3 SSISNAV Analysis Performance

Considering the high time consumption and cost of the human resources required for formal software verification activity, it is important to measure how time is employed by the analysis, which can be provided by **Eva** plug-in options.

Option -eva-show-perf computes and provides a summary of the time spent analyzing function calls [3] from two perspectives: the first shows the time of all function calls, and the second shows the time considering the sequential order of the function calls.

Option -eva-flamegraph, from the **Frama-C** 14 (Silicon) release, dumps a summary of the time spent to analyze function calls in a format suitable for the flame graph tool [3].

A flame graph is an adjacency diagram with an inverted icicle layout employed to view stack traces [30]. A stack trace is represented as a column of boxes, where each box represents a function. Flame graphs have been adopted by many languages, products, and companies and have become a standard tool for performance analysis [30]. They may help visualize the **Eva** analysis performance either during execution or later. Interestingly, the flame graph can be generated while the analysis is still running, which is useful when an analysis seems frozen at a certain point but does not display any messages. Running the "frama-c-script flamegraph" creates a flame graph of the probe's current state, while the trace file is still being updated by the probe in progress.

To generate a flame graph while running an analysis, it is sufficient to add the option -eva-flamegraph and a name for the output file, as shown in the first

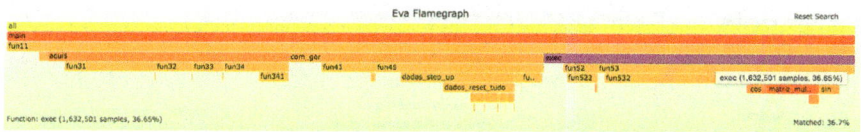

Fig. 14.13 Flame graph of the navigation algorithm for Scenario 2

command below. To visualize a flame graph, Frama-C offers a script to help the user, as shown in the second command below. The command generates the *flamegraph.svg* and *flamegraph.html* files in the default directory (or in the FRAMAC_SESSION directory) and opens the html file in a browser [3].

```
$ frama-c -eva file.c -evaflamegraph flamegraph.txt
$ frama-c-script flamegraph flamegraph.txt
```

Figure 14.13 shows the flame graph obtained for the navigation algorithm. The stacks start from the top with the main function (red color), and each called function is stacked underneath. The length of the bar corresponds to the amount of time spent in it. Hovering over a bar shows the function name and the percentage of time spent in it. In the example, clicking on a bar will zoom in on it, increasing the size of its children for better readability. Note that bar colors do not have any special meaning. In addition to hovering, it is possible to search for a function name. The result is a function bar shown in magenta displaying the consumed time percentage.

In addition to the Eva options, it is possible to obtain analysis processing times by using the Unix time command. In Scenario 2, the spent times were 108m 18.860 s and 38m 20.947 s in the preparation and navigation states, respectively, while in Scenario 1, they were 79m 56.044 s and 89m 49.306 s, respectively.

14.4.6.4 SSISNAV Analysis Limitations

The case study contains one volatile variable whose analysis results in a wide domain that makes it imprecise. Information concerning this restriction can be found in the Eva manual [11]. A second restriction we faced concerns the maximum number of assignments that are allowed in a function, but this can be easily circumvented by using additional functions. For Scenario 2, a function for storing sensor data in a two-dimensional array [25828][6] exceeded the number of statements that Eva is currently able to process in a single function. To solve this problem, we had to split the function into smaller functions.

14.4.7 Reporting of Anomalies Found in the Analyses

We have considered static analysis as a technique in our software verification process, in the software design & implementation engineering process, and as part of the code

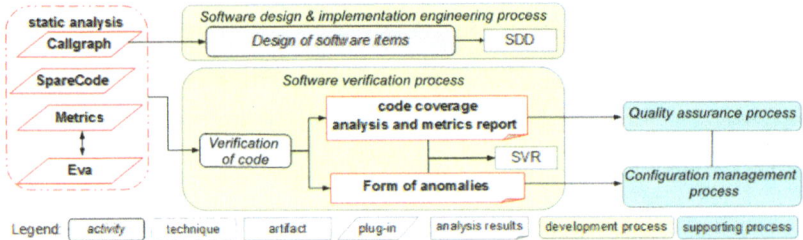

Fig. 14.14 Frama-C plug-ins applied to the software development activities

review activity [15]. After our experiments, we associated the **Frama-C** plug-ins with the processes and activities in the software development life cycle as proposed by ECSS-E-ST-40C and shown in Fig. 14.14.

In the software design & implementation engineering process, **Callgraph** can contribute to SDD elaboration because it helps developers by providing an accurate view of function calls. Additionally, the call graphs can be inserted as new figures and/or be compared to other control-flow graphs already present in the SDD.

During the software verification process, the static analysis results can be synthesized as follows: (1) the anomaly form is sent to the configuration management process, which involves discrepancies found in documents, code, and call graphs in the cases of functions that are never called; and (2) code coverage analysis and metrics reports, which are sent to the quality assurance process. Both reports can be useful in the elaboration of the software code verification report since the quality assurance process and the configuration management process, which are responsible for analyzing and approving the results, depend on each other to support the activities in the software development life cycle.

We proposed an anomaly form as shown in Table 14.3, which has been only partially filled in for illustrative purposes. The form fields are based on the IEEE Std 1044–2009 [51] and manual code inspection [46]. Each record of the form contains an anomaly with its attributes, where *asset* and *artifact* are the software asset (product, component, module, etc.) and the specific software work product that contains the anomaly, respectively; *lin/pg* specifies the anomaly location in the artifact, *description* defines the anomaly, *severity* is the degree of impact that the anomaly could cause (inconsequential, minor, major, critical), *suggested corrective action* is the alteration recommended to fix the anomaly, and *plug-in* is the **Frama-C** plug-in responsible for anomaly detection, either directly or indirectly. The *type* attribute encompasses the anomaly category. We propose the following abridged list of types:

- Function: an anomaly in a function definition, mapping, access, or use, e.g., unused return value of function calls;
- Variable and constant: an anomaly in data definition, initialization, mapping (data traceability), access, or use, e.g., declared but unused variable;
- Program flow control: an anomaly in control statements, e.g., loop variable not properly initialized, incremented and tested;

14 Exploring Frama-C Resources by Verifying Space Software

Table 14.3 Anomaly form with example records

Asset	Artifact	Lin/pg	Type	Description	Severity	Corrective action	Plug-in
Source code	function5a3	243	Comment	Misspelling	Inconsequential	Correct spelling	Eva
Source code	function5a2	137	Program flow control	Accessing index out of bounds	Critical	Correct code	Eva alarm
Source code	function5b4	98	Function	Function not called	Major	Check code	Callgraph
Source code	navigate.c	51	Variable and constant	Unused global variable	Minor	Variable removal	SpareCode
Source code	function5a3	305	Logic	Cyclomatic complexity = 25	Major	Check code	Metrics
Documentation	148-000/BOX	35	Function	Incorrect source code list	Major	Correct documentation	Eva

- Logic: an anomaly in sequencing, branching, or algorithm definition as found in natural language specifications or in implementation language, e.g., excessive branching;
- Defensive programming: an anomaly related to problematic issues that may arise due to unforeseen circumstances, e.g., divisors not tested for zero;
- Unused source code: unnecessary code, such as redundant code, unexecuted code due to a lack of flow control to reach it, code whose results are never used, or even commented code, e.g., code inserted for testing that should have been discarded later;
- Comment: an anomaly in the comments, e.g., incorrect comments that do not contribute to software understanding and maintenance;
- Readability and legibility: an anomaly that makes a program harder to read and understand and more difficult for its elements to be identified, e.g., the presence of multiple statements per line.

Table 14.4 presents a summary in which anomalies are categorized by type and associated with the number of occurrences and the plug-ins responsible for their detection. We remark that the items detected by Eva can occur both in the context definition phase and in the implementation and refinement phase. In the latter phase, the plug-in issues alarms representing errors that occur at run-time. In our case study, these alarms were reported as anomalies and classified as having critical severity. All of the emitted alarms must be corrected before running the analysis with Eva again.

In addition to anomalies, call graphs, code coverage analysis, and metrics reports, the context definition activities, which required a deep program understanding, have contributed to the elaboration of the DD-SRS, detailed DFDs, and DD-SDD.

14.5 Conclusion

The results we obtained indicate that it is a good practice to adopt static analysis techniques during the software verification process because they are capable of guaranteeing greater reliability in the final software product. Since it is not necessary to run the software, static analysis can be used in a step prior to the testing activity, making it possible to start detecting bugs and correcting them even before testing. In addition, it can be a way to validate the software's behavior and to make adjustments before actually starting the testing activity. This can save both money and time because it is easier to modify the software during an initial stage than at later stages of the development life cycle. Implementation efforts are minimized if bugs are found earlier since less reworking is needed.

The adoption of this technique can have a direct impact on other processes in the software development life cycle, including software product assurance, the software maintenance process, and the software design & implementation engineering process. We can say that the obtained benefits are valuable.

14 Exploring Frama-C Resources by Verifying Space Software

Table 14.4 Summary of anomalies related to the source code

Occurrence number	Type	Description	Plug-in
1	Program flow control	Loop variable not properly initialized, incremented and tested	Eva alarm
5	Variable and constant	Uninitialized variable	Eva alarm
2	Defensive programming	Divisor that is untested for zero or incorrect values	Eva alarm
36	Readability and legibility	More than one statement per line	Eva
2	Unused source code	Redundant code	Eva
3	Variables and constants	Equality comparison by the operator == between two floating-point numbers	Eva
17	Unused source code	Commented code, unused or in parts of the program that never runs	Eva
7	Comment	Comment that does not contribute to understanding and maintenance	Eva
12	Function	Return values from function calls that are unused	Eva
2	Variable and constant	Unused variable	SpareCode
3	Variable and constant	Unused variable	kernel
3	Logic	Excessive branching (cyclomatic complexity >20)	Metrics

Static analysis can aid in successfully achieving the very important software property of program correctness through the detection and removal of development errors to build safer, cheaper, and more reliable source code. Analysis code coverage is an important metric in terms of guaranteeing the quality of software products, which is essential for certified and critical software. The higher the percentage of code covered by analysis, the lesser the likelihood that it will contain errors. In general, the definition of acceptable code coverage shall be agreed upon between the customer and the supplier based on the software criticality degree.

Static analysis based on formal methods requires a deep level of understanding of the application to be verified, both in terms of execution and evaluation of the results. Before running an analysis, a study of the application context is required, e.g., which and how the external entities interact with software (input/output data). Furthermore, the interpretation of results depends on knowledge of the system algorithms and an existing database (e.g., simulation data) to compare the outputs. Software documentation is an important resource to support analysis. This obtained program understanding may help in performing and improving a software maintenance process and, in general, for software engineering processes for new or outsourced software systems of the same type.

The use of static analysis in the software design & implementation engineering process may well affect the coding and software item design activities. One of the most obvious applications of static analysis is in coding activity. It can identify weaknesses in source code that lead to potential vulnerabilities. Additionally, the elaboration of an SDD, an output of the software item design activity, can employ call graphs obtained from static analysis. On the one hand, users can face significant challenges when applying **Frama-C**. The interaction demanded by some plug-ins, such as **Eva** and **Wp**, requires considerable knowledge. Additionally, for experiments involving large volumes of input data, the analysis can be time-consuming.

Despite these challenges, the benefits outweigh the costs. **Frama-C** is an open-source platform organized into a modular architecture, and its collaborative extensible framework permits developers to implement their own plug-ins and make them available to other users. It is advantageous to have several analyzers in the same framework since the user can perform a variety of analyses using the same tool. **Frama-C** is regularly updated and effectively maintained by the **Frama-C** development team, which seeks to meet user requests, as exemplified by our need for a trigonometric function.

Finally, our work has shown that the described development process can be used in the industrial context as well. The integration of **Frama-C** in the industrial process can be a solution to improve quality by employing static analysis for run-time error detection.

Acknowledgements The authors would like to thank LINCS/IAE, ITA, HASLab, INESC TEC, the University of Minho, and all of the reviewers for their invaluable work on how to improve the chapter. This work is partially financed by the National Funds through the Portuguese funding agency, FCT—Fundação para a Ciência e a Tecnologia, within project LA/P/0063/2020.

References

1. Agência Espacial Brasileira (2022) Transporte Espacial. https://www.gov.br/aeb/pt-br/programa-espacial-brasileiro/transporte-espacial. Accessed 20 Sep 2022
2. Agência Força Aérea: IAE realiza Operação Santa Maria 1/2021 no Centro de Lançamento de Alcântara (CLA) (2021). https://www.fab.mil.br/noticias/mostra/37556. Accessed 24 Jan 2022
3. Alberti M, Antignac T, Barany G et al Help of frama-c tool
4. Alberti M, Antignac T, Barany G et al (2021) FRAMA-C. https://frama-c.com/. Accessed 15 Jan 2022

5. Baudin P, Bobot F, Bühler D, Correnson L, Kirchner F, Kosmatov N, Maroneze A, Perrelle V, Prevosto V, Signoles J, Williams N (2021) The dogged pursuit of bug-free C programs: the Frama-C software analysis platform. Commun ACM. https://doi.org/10.1145/3470569
6. Black PE, Walia KS (2020) SATE VI Ockham sound analysis criteria. Tech. rep., national institute of standards and technology—NIST. https://nvlpubs.nist.gov/nistpubs/ir/2020/NIST.IR.8304.pdf
7. Blanchard A, Kosmatov N, Loulergue F (2018) A lesson on verification of IoT software with Frama-C. In: Conference on high performance computing and simulation (HPCS). https://doi.org/10.1109/HPCS.2018.00018
8. Bonichon R, Yakobowski B Frama-C's metrics plug-in 24.0 (Chromium). https://www.frama-c.com/download/metrics-manual-24.0-Chromium.pdf
9. Brahmi A, Carolus MJ, Delmas D, Essoussi MH, Lacabanne P, Lamiel VM, Randimbivololona F, Souyris J (2020) Industrial use of a safe and efficient formal method based software engineering process in avionics. In: Conference on European congress embedded real time systems (ERTS)
10. Brazilian Space Agency (2022) The Brazilian Space Agency–The bridge to the future. https://www.gov.br/aeb/pt-br/centrais-de-conteudo/publicacoes/LivretoBrazilianSpaceAgency.pdf. Accessed 18 Jan 2022
11. Bühler D, Cuoq P, Yakobowski B, Lemerre M, Maroneze A, Perrelle V, Prevosto V Eva–the evolved value analysis plug-in 24.0 (Chromium). https://www.frama-c.com/download/eva-manual-24.0-Chromium.pdf
12. Bureau Veritas group (2016) Bureau Veritas releases a guide, co-written with the CEA, for enhancing the reliability and performance of embedded software. https://group.bureauveritas.com/fr/node/387. Accessed 27 Sep 2022
13. Carrijo DS, Oliva AP, de Castro Leite Filho W (2002) Hardware-in-loop simulation development. Int J Model Simul. https://doi.org/10.1080/02286203.2002.11442238
14. CEA List (2017) Dassault Aviation innovates in cybersecurity with Frama-C. https://www.cea-tech.fr/cea-tech/english/Pages/ec_2017/dassault-aviation-innovates-in-cybersecurity-with-frama-c-smart-digital-systems.aspx. Accessed 15 Jan 2022
15. Chess B, West J (2007) Secure programming with static analysis. Addison-Wesley
16. Clarke EM, Grumberg O, Peled DA (2000) Model checking. The MIT Press
17. Correnson L, Cuoq P, Kirchner F, Maroneze A, Prevosto V, Puccetti A, Signoles J, Yakobowski B (2021) Frama-C user manual release 24.0 (Chromium). http://frama-c.com/download/user-manual-24.0-Chromium.pdf. Accessed 13 Apr 2023
18. Cousot P, Cousot R (1977) Abstract interpretation: a unified lattice model for static analysis of programs by construction or approximation of fixpoints. In: Conference on ACM SIGACT-SIGPLAN symposium on principles of programming languages (POPL). https://doi.org/10.1145/512950.512973
19. Cuoq P, Monate B, Pacalet A, Prevosto V, Regehr J, Yakobowski B, Yang X (2012) Testing static analyzers with randomly generated programs. In: International symposium on NASA formal methods (NFM). Springer. https://doi.org/10.1007/978-3-642-28891-3_12
20. Duprat S, Lamiel VM, Kirchner F, Correnson L, Delmas D (2016) Spreading static analysis with Frama-C in industrial contexts. In: Conference on European congress embedded real time software and systems (ERTS)
21. Dutle A, Moscato M, Titolo L, Muñoz C, Anderson G, Bobot F (2021) Formal analysis of the compact position reporting algorithm. Form Asp Comput 33(1), 65–86. https://doi.org/10.1007/s00165-019-00504-0
22. ECSS: E-ST-40C space engineering—software (2009)
23. ECSS: Q-HB-80-04A Space product assurance—software metrication programme definition and implementation (2011)
24. ECSS: Q-ST-80C space product assurance—software product assurance (2017)
25. Eisenbarth T, Koschke R, Simon D (2001) Aiding program comprehension by static and dynamic feature analysis. In: IEEE international conference on software maintenance (ICSM). https://doi.org/10.1109/ICSM.2001.972777

26. Frama-C (2012) Frama-C news and ideas homepage, 2015 [Online]. https://frama-c.com/2012/01/16/Csmith-testing.html. Accessed 27 Sep 2022
27. Gansner ER, Koutsofios E, North S (2015) Drawing graphs with *dot*. https://www.graphviz.org/pdf/dotguide.pdf
28. Gao D, Reiter MK, Song D (2004) Gray-box extraction of execution graphs for anomaly detection. In: ACM conference on computer and communications security (CCS). https://doi.org/10.1145/1030083.1030126
29. Gentsch C (2020) Evaluation of open source static analysis security testing (SAST) tools for c. Tech. rep., DLR-German Aerospace Center, Jena. https://elib.dlr.de/133945/
30. Gregg B (2016) The flame graph: this visualization of software execution is a new necessity for performance profiling and debugging. ACM Queue. https://doi.org/10.1145/2927299.2927301
31. Hoare CAR (1969) An axiomatic basis for computer programming. Commun ACM. https://doi.org/10.1145/363235.363259
32. Hussein M, Nouacer R, Radermacher A (2017) Towards a safe software development environment. In: Euromicro conference on digital system design (DSD). https://doi.org/10.1109/DSD.2017.13
33. Hussein M, Nouacer R, Radermacher A, Puccetti A, Gaston C, Rapin N (2018) An end-to-end framework for safe software development. Microprocess Microsyst. https://doi.org/10.1016/j.micpro.2018.07.004
34. INNOSPACE: INNOSPACE (2023). http://www.innospc.com/myboard/sub04_02. Accessed 13 Apr 2023
35. Instituto de Aeronáutica e Espaço: Projeto SIA (2018). https://www.iae.dcta.mil.br/index.php/todos-os-projetos/todos-os-projetos-desenvolvidos/projetos-sia. Accessed 13 Apr 2023
36. Instituto de Aeronáutica e Espaço: VLS-1 (2019). https://iae.dcta.mil.br/index.php/todos-os-projetos/todos-os-projetos-desenvolvidos/projetos-vls1. Accessed 20 Jan 2022
37. International Organization for Standardization (2017) ISO/IEC/IEEE 24765 Systems and software engineering—Vocabulary
38. Kirchner F, Kosmatov N, Prevosto V, Signoles J, Yakobowski B (2015) Frama-C: a software analysis perspective. Form Asp Comput. https://doi.org/10.1007/s00165-014-0326-7
39. Kumar N, Neema S, Das M, Mohan BR (2021) Program slicing analysis with KLEE, DIVINE and Frama-C. In: International conference on automation and computing (ICAC). https://doi.org/10.23919/ICAC50006.2021.9594142
40. Maroneze A, Perrelle V, Kirchner F (2019) Advances in usability of formal methods for code verification with Frama-C. Electron Commun EASST. https://doi.org/10.14279/tuj.eceasst.77.1108. Interactive workshop on the industrial application of verification and testing (ETAPS)
41. Ministry of Defense (2012) Defense white paper–Livro Branco de Defesa Nacional. https://www.gov.br/defesa/pt-br/arquivos/estado_e_defesa/livro_branco/lbdna_2013a_inga_net.pdf. Accessed 18 Jan 2022
42. Musco V, Monperrus M, Preux P (2017) A large-scale study of call graph-based impact prediction using mutation testing. Softw Qual J. https://doi.org/10.1007/s11219-016-9332-8
43. NASA (2022) NPR 7150.2D NASA software engineering requirements
44. Ourghanlian A (2015) Evaluation of static analysis tools used to assess software important to nuclear power plant safety. Nucl Eng Technol. https://doi.org/10.1016/j.net.2014.12.009
45. Ramos F (2015) History and current status of SISNAV: a brief report. In: Simpósio Brasileiro de Engenharia Inercial (SBEIN). https://doi.org/10.13140/RG.2.1.3529.0323
46. Romani M, Takahashi P, Lahoz C (2009) A process of code inspection for space software. In: Conf. on Int. astronautical congress
47. RTCA/EUROCAE (2011) RTCA DO-178C software considerations in airborne systems and equipment certification
48. Ryder BG (1979) Constructing the call graph of a program. IEEE Trans Softw Eng. https://doi.org/10.1109/TSE.1979.234183
49. Signoles J (2020) Abstract interpretation and properties of c programs. http://ejcp2019.icube.unistra.fr/slides/js.pdf. Accessed 10 Dec 2020

50. Silva RAB, Arai NN, Burgareli LA, Oliveira JMP, Pinto JS (2016) Formal verification with Frama-C: a case study in the space software domain. IEEE Trans Reliab. https://doi.org/10.1109/TR.2015.2508559
51. Software & Systems Engineering Standards Committee (2009) IEEE 1044 standard classification for software anomalies
52. de Souza J, Filho W (2012) Sistema de Navegação Inercial SISNAV - Mecânica e Eletrônica Embarcada. In: Simpósio Brasileiro de Engenharia Inercial (SBEIN)
53. Stack Overflow (2019) Sparecode analysis in Frama-C. https://stackoverflow.com/questions/59240081/sparecode-analysis-in-frama-c
54. TrustInSoft (2022) https://trust-in-soft.com. Accessed 15 Jan 2022
55. VESSEDIA (2022) https://cordis.europa.eu/project/id/731453. Accessed 15 Jan 2022

Chapter 15
Ten Years of Industrial Experiments with Frama-C at Mitsubishi Electric R&D Centre Europe

Éric Lavillonnière, David Mentré, and Benoît Boyer

Abstract Mitsubishi Electric R&D Centre Europe (MERCE), the advanced European research laboratory of Mitsubishi Electric group, has been carrying research activity on Formal Methods for more than 10 years now. MERCE applied various formal methods in its projects and, among them, MERCE conducted several extensive experiments with Frama-C, from safety to security applications, in different industrial domains. Through three of these experiments, namely industrial code verification, automated test generation and verification of an industrial firewall, this chapter gives some feedback on several Frama-C's capabilities based on the plug-ins Eva, PathCrawler and Wp. Demonstrating the tractability of the solutions in industrial context is one of the leading issues when MERCE evaluates tools and methods. Frama-C appears as a mature technology which can be operated in very different contexts from classical testing to full-fledged formal functional verification.

Keywords Safety · Security · Test generation · Formal verification · Experience report

15.1 Introduction

Mitsubishi Electric R&D Centre Europe (MERCE) is the advanced European research laboratory of Mitsubishi Electric group, located in Rennes, France and Livingston, United Kingdom. Among its three divisions, the Communication and Information Systems (CIS) division develops since 2009 a research activity on Formal Methods. While not focusing on a specific kind of Formal Methods, MERCE extensively uses Frama-C for all its research activities related to verification of pro-

---

É. Lavillonnière (✉) · D. Mentré · B. Boyer
Mitsubishi Electric R&D Centre Europe (MERCE), Rennes, France
e-mail: E.Lavillonniere@fr.merce.mee.com

D. Mentré
e-mail: D.Mentre@fr.merce.mee.com

B. Boyer
e-mail: B.Boyer@fr.merce.mee.com

grams written in the C language. This is part of the various experience that we want to share share in this chapter through three kinds of experiments showing more and more complex use of Frama-C capabilities.

Firstly, we present the use of the Eva plug-in to analyze safety-critical embedded software. Here, Eva is used "as is" by the end user. We present the practical method used, the challenges for such kind of formal verification, and some proposed solutions as well as some of the "tips and tricks" that helped us.

Secondly, we show the use of the Eva and PathCrawler plug-ins within an automatic structural unit test generation technology for C programs. Here Eva and PathCrawler are used as external command line tools but nonetheless fully integrated within the test generation tool, the end user not being aware of their use. We detail the overall architecture of the tool with a particular focus on above plug-ins on the search for test cases and identification of unfeasible test objectives. We give some results about the performance achieved using these techniques.

Lastly, the third part shows an even tighter integration within Frama-C through its application on an industrial firewall. In this experiment, we automatically generate the ACSL-annotated C code for the packet processing logic of the firewall, from its filtering rules. Using Wp, the annotations are then automatically proved to demonstrate the absence of run-time errors and functional correctness.

Through this chapter, we hope to convince the reader that Frama-C is a mature technology, applicable to industrial software in fully automatic or mostly automatic way. Frama-C framework is versatile and tractable to very different contexts, from safety to security, from classical testing to full-fledged formal verification of functional behavior.

15.2 Industrial Code Verification with Eva

In this section we consider the application by MERCE of Abstract Interpretation techniques as implemented in Eva on an embedded industrial code used in a safety-critical system. The target source code was relatively big, about 300,000 lines in total divided into two independent applications of about 150,000 lines each. Those applications are bare metal concurrent embedded code (i.e. without operating system or other environment) without dynamic memory allocation. MERCE checked absence of run-time errors (i.e., undefined behaviors in C code) within one of those applications using TIS Analyzer, a commercial variant of Eva proposed by TrustInSoft company,[1] a spin-off from CEA. The methodology and techniques used in this study are not specific to TIS Analyzer and we will show how to apply them with Eva.[2]

One of the challenges of formal verification in an industrial context is the large size of target applications, which forces us to deploy techniques to scale up the analy-

---

[1] https://trust-in-soft.com/.
[2] Differences between TIS Analyzer and Eva are mostly on the name of commands used. TIS Analyzer has some additional capabilities to handle industrial code that we will not discuss here.

sis. Moreover, we also show techniques used to address particularities of considered embedded code, like direct access to hardware. Last but not least, as the analysis results are meant to be used as elements of the safety case of a safety-critical application, we will also show practical methods used to organize the analysis and keep track of results.

15.2.1 Organizing People for Verification

As presented previously in Chap. 3, Eva identifies situations when run-time errors could occur and raises alarms for them. Then a human must review those alarms and determine if they are true or false alarms. This is a complex step, relying on a good knowledge of C language but also understanding of the source code and the corresponding program behavior to identify the potential cause of an alarm. This step relies on usual debugging ability of a developer and goes beyond the scope of this chapter. Nonetheless, we found in our experiment that using Eva to explore the code makes Eva a kind of advanced debugger bringing useful insights on the program. Once true or false alarms are determined, if they are true alarms, she/he should correct the code. If they are false alarms, she/he should improve the analysis precision to remove the false alarms.

As it was the case in our experiments, the person doing the analysis (the analyst) might not be the person having such understanding of the source code. Two approaches are then possible. The first one is a joint work in front of the same screen of a developer having good understanding of the source code under analysis and an analyst. Faced with an alarm, both persons can discuss in order to get a common understanding of the source code and remove the cause of the alarm. The second approach is firstly for the analyst to do the analysis and try to diagnose the alarms, and secondly through meetings or e-mail exchanges between the analyst and source code developers to make a review and get a common understanding of the alarms.

The first approach is clearly preferable from efficiency point of view, allowing fast understanding of potential program behaviors that might explain the alarms through discussions and immediate correction of potential issues. However the second approach provides two independent reviews that might be beneficial from safety point of view. In the second approach, when the code is changed to correct true alarms, we found useful to insert a "reminder" alarm on the exact same line, by inserting code[3] like:

if (Frama_C_nondet(0,1)) { 1/0; } /*Comment on alarm*/

In both cases, it is important to keep a trace of raised alarms (file, function, line number and kind of alarm) and their outcomes (program change, no change, analysis change) with a short justification. Those results will be reused for the safety case. One can use a spreadsheet which initial content can be generated from the CSV outputs generated by Eva using the Report plug-in.

---

[3] It would be useful if the user could raise by herself/himself an alarm of a specific kind.

15.2.2 Method for Safety-Critical Code Verification

For the verification of safety-critical applications, we proposed the following method.

Firstly, the analyst needs to prepare for the analysis with the following tasks:

1. Identify safety-critical modules. Restricting the analysis on safety-critical modules[4] reduces the amount of source code to be analyzed, thus helping on the analysis' scalability.
2. Identify entry points. Find all possible entry points in the analyzed program. main() is an obvious candidate but you might also have interrupt handlers, spawning entries of concurrent parts or different ways to call the program.
3. Setup versioning of the analysis. You need to keep detailed records of the modifications in the program and in your analysis scripts. Use a Source Version Control system (e.g., git, SVN, ...) and put in it all analysis scripts and source codes.

We recommend following directories organization under Source Version Control for the analysis:

- scripts/: All scripts of the analysis. At least one script for each entry point
- drivers/: C source code of analysis drivers used by each analysis script
- src/modif/: Source code under analysis. It will be modified with code corrections or annotations
- src/orig/: Unmodified source code as released before the analysis. Keeping this source code in a separate directory allows to quickly identify made changes using diff -ru src/orig/ src/modif/ Unix command[5]
- out/: Volatile outputs of the analysis that can be discarded.

Tips
If your original source code were developed under Windows, you will have issues with the case of included files that might not match file names on the file system, e.g., # include <header.h> while source file is named Header.H. You can use Linux loop block device to create and mount a file that will contain the source code under Windows' file name case insensitive FAT file system format with the following Linux commands (replacing < user> with your username):

```
# 1. create the FAT file system in a file named
    'sourcecode.fat32img'
```

---

[4] One should notice that in this case it is also needed to justify isolation between the safety-critical and non safety-critical modules. InOut plug-in can help for this task but we have never done it.

[5] Versioning can also be used to track changes made (through versioning tags or branches), but we found more convenient to organize it that way, especially when the analysis is done on several versions of the source code.

```
2 dd if=/dev/zero of=sourcecode.fat32img bs=100MiB count=1 #
      100MB, adjust size to need
3 sudo losetup --find --show sourcecode.fat32img
4 [ a device like '/dev/loop0' is shown ]
5 sudo mkfs -t fat /dev/loop0
6
7 # 2. prepare directory to access FAT file system
8 mkdir /mnt/sourcecode
9
10 # 3. mount file system
11 sudo mount -o uid=<user>,gid=<user>,rw,noatime
      sourcecode.fat32img /mnt/sourcecode
12
13 # 4. update source code content into file system
14 cd /mnt/sourcecode
15 # do whatever you want with the source code on Windows file
      system
16 # e.g., unzip [path-to-a-ZIP-file].zip
17
18 # 5. unmount file system
19 sudo umount /mnt/sourcecode
```

You can repeat steps 3 and 5 each time you need to work on the source code.

Then the overall method for the analysis is the usual one: top to bottom starting from the previously identified entry points. Each alarm should be fixed according to the order of appearance raised by Eva (with code correction or analysis precision improvement), as very commonly an issue revealed by an alarm triggers other later alarms.

A very important point is to regularly review the so-called *"dead code"*, i.e. code not covered by Eva, displayed using a red background in the Frama-C GUI. If there is too much such deadcode, it probably means that the analysis is wrongly configured and does not consider enough of possible program values to reach such uncovered code. It is up to the human reviewer with her/his program knowledge to decide what is legitimate deadcode or not.

15.2.3 Practical Configuration of Eva

We now present the practical configuration of Eva to analyze an entry point. This work will have to be repeated for each entry point. In the following, we assume that we want to analyze the main() entry point.

To analyze an entry point, we use two shell scripts for analysis and display of results, and one driver in C language. We separate analysis and display into two different scripts because analysis can take a lot of time, sometimes tens of minutes. The first shell script, called main-analyze.sh, launches the analysis and stores its results in files. The second shell script, called main-load.sh, loads the results in Frama-C GUI for human review. So once the analysis is done, we can rapidly reload the results several times (and not lose important result if GUI is erroneously closed).

Analysis script   The basic structure of main-analyze.sh is as follows:

```bash
#!/bin/bash

LOG="out/main.log"
STATE="out/main.state"
AUDIT="out/main-audit.json"
SRC_DIR="src/modif/"

SLEVEL_FUNC=(
    main:10
    function2:5000
)

JOIN_SLEVEL_FUNC=$(printf ",%s" "${SLEVEL_FUNC[@]}")
JOIN_SLEVEL_FUNC=${JOIN_SLEVEL_FUNC:1}



echo "Log in: $LOG"
echo "State in: $STATE"

rm -f $LOG
rm -f $STATE

echo -n "==START== " > $LOG
date --rfc-3339=seconds >> $LOG

echo -n "Frama-C version: " >> $LOG
frama-c --version >> $LOG
echo >> $LOG

/usr/bin/time frama-c \
               -audit-prepare $AUDIT \
               -machdep ppc_32 \
               -absolute-valid-range 0x00000AA8-0x000FFFFF \
               -eva \
               -eva-undefined-pointer-comparison-propagate-all \
               -cpp-extra-args="-I$SRC_DIR/inc1 -I$SRC_DIR/inc2" \
               -cpp-extra-args="-Idrivers" \
               -main main_analysis_start \
               -eva-slevel-function $JOIN_SLEVEL_FUNC \
               -save $STATE \
               -kernel-msg-key pp \
               drivers/main_driver.c \
               $SRC_DIR/file1.c \
               >> $LOG 2>&1

echo -n "==END== " >> $LOG
date --rfc-3339=seconds >> $LOG
```

```
50  tail -5 $LOG
```

This script has two outputs, the **Frama-C** state (including **Eva** results) saved in the STATE file and the terminal output of **Frama-C** command stored in the LOG file. The latter also contains start and end times of the analysis as well as **Frama-C** version, useful for traceability. We display the last five lines of this file on the terminal so we can check that the script ran without error.

We take care to erase both STATE and LOG files at the beginning of the analysis to be sure that their content is aligned with the latest analysis run.

We recommend that the use the following options of **Frama-C**:

- `-audit-prepare $AUDIT`: store the detailed analysis configuration (e.g., list of all used source files with checksum) in a JSON file for later verification with `-audit-check` option
- `-machdep ppc_32`: set target architecture of the computer on which the analyzed software will run (in this example Big-Endian PowerPC 32 bits architecture)
- `-absolute-valid-range 0x00000AA8-0x000FFFFF`: set the range of memory addresses that are considered as valid for pointer dereferencing. **Eva** only supports one of such range
- `-eva`: apply **Eva**
- `-eva-undefined-pointer-comparison-propagate-all`: in case of comparison of memory addresses, **Eva** will consider, according to C standard, that such comparison are erroneous and should not be used. They are nonetheless used in embedded programs, e.g., in functions handling memory. This option forces **Eva** to consider both outcomes (0 and 1) of the comparison for correctness
- `-cpp-extra-args`: set source code's specific directories containing headers. In particular indicate the `drivers/` directory containing the drivers
- `-main main_analysis_start`: define **Eva** analysis entry point in your driver calling the `main()` analyzed entry point
- `-eva-slevel-function $JOIN_SLEVEL_FUNC`: define specific slevel of some functions to improve analysis precision
- `-save $STATE`: at end of analysis, save analysis result in STATE file
- `-kernel-msg-key pp`: give more information in **Frama-C** error messages
- `drivers/main_driver.c $SRC_DIR/file1.c`: set of source files to analyze, including drivers

Loading script Our second shell script, `main-load.sh`, used to load the analysis results is much simpler:

```
1  #!/bin/bash
2
3  STATE="out/main.state"
4
5  frama-c-gui -load $STATE
```

This script simply starts the **Frama-C GUI** loading its state from STATE file, including **Eva** results.

Analysis driver Our last file is the driver `drivers/main_driver.c`. It is used to configure the memory before calling `main()` function to start the analysis:

```
void main_analysis_start(void)
{
  // configure memory through appropriate C code

  main();
}
```

We will not detail how to configure the memory as it is specific to the analyzed program. The main idea is to use `Frama_C_make_unknown()` primitive to randomize relevant parts of the memory.

Once those three basic elements are ready, you can iterate with `main-analyze.sh` and `main-load.sh` scripts to launch the analysis and inspect its results.

You will notice that the above two shell scripts define the same STATE variable. Similarly, the definition of slevels will be the same for several analyses. We made this duplication for simplicity but for real analysis it is recommended to put all such definitions in a single file loaded into the various analysis scripts using `source` shell command. Sharing common definitions is important for avoiding errors, as in programming.

Tips
When you launch **Frama-C** on a remote server, the analysis could take some time. You might want to disconnect and reconnect frequently to the server during analysis without stopping launched commands from the shell. For that, we recommend to use tmux[6] terminal multiplexer program.

15.2.4 Challenges of Analyzing Big Source Code

One challenge of applying **Eva** on industrial code is to scale the analysis to its size. The default top to bottom analysis methodology starting from program's entry points does not scale up when the analyzed source code becomes too big.

The only available approach to analyze such code is to split the global analysis into multiple analyses applied on smaller parts of the code. A cutting point is a well-chosen function `f()` of the source code which becomes a new analysis entry point. On the one side, a dedicated analysis is built for `f()`, with associated analysis scripts and driver.

On the other side, inside all code calling `f()`, the source code is replaced by an ACSL contract or a simple stub in C code that describes its potential side effects on memory. Again, making such description can be a challenge. However, one does not need an exact contract but the minimum correct contract for analyzing callers of `f()`,

---

[6] https://github.com/tmux/tmux.

for example an over-approximation of the possible state of the variables known to be impacted. Usually, one can use her/his knowledge of the function f() to specify its potential effects. Another preferable but more complex option is to use the InOut plug-in on f() to identify all written variables and describe effects on them in the contract.

One potential issue of splitting an analysis in several parts is the correctness of the f() analysis driver with respect to the contract used when analyzing callers of f(). All the behaviors triggered when calling f() should be included in the driver used to analyze f(). This can be challenging but a safe assumption is to assume memory state before calling f() contains some unknown random content, thus addressing the widest range of potential memory states at the expense of analysis precision.

Finding a good function decomposition is for now more an art than a science. Nonetheless, from our own experience, the following hints can be considered:

- Error handling functions. Especially in safety critical programs, you will have a function that is called systematically each time an error occurs to put the program, and possibly the controlled system, in a safe state. Usually such function does not return. As it is called frequently, it is a good cutting point to accelerate analysis.
- Functions manipulating bits in memory, like CRC[7] computation functions. Such functions tend to mix bits of memory and Eva can hardly determine the resulting memory state they produce. Thus, they are good cutting points, replacing such functions by stubs telling that the returned computed CRC is just random and no memory is modified.
- Entry points in sub-modules. If your program is divided into well-identified sub-modules with their entry points, such sub-modules entry points could be interesting cutting points.

Handling of concurrency Another challenge for analysis scalability is the presence of concurrent behavior in the analyzed program. If several parts of the program run concurrently, it means making a separate analysis for each one of them, and then an analysis gathering side effects of concurrent parts on each other's, typically using Mthread plug-in. As the latter analysis needs to be applied on all the concurrent parts simultaneously, it is not very scalable. As far as we know, this approach has never been applied on industrial size programs up to now.

15.2.5 *Tips and Tricks to Improve Eva Analysis Precision*

When doing our analysis, we have faced various issues that most Eva users would also face. We now present typical cases of such issues and how to solve them.[8] Of course, the exact use of those techniques depends strongly on the program to be

---

[7] Cyclic Redundancy Check.

[8] We deeply thank TrustInSoft support that helped us when we ourselves faced such issues with TIS Analyzer.

analyzed and you might need several attempts before reaching the desired level of precision.

15.2.5.1 Case Splitting

Eva analysis is fundamentally non-relational, i.e. the domain of each variable is analyzed without considering potential relationships with other variables. In most situations, it works well. However in some situations, not considering those relations leads to an inaccurate analysis, even though those relations would be easily captured by the programmer.

As an example, we consider the following program where two functions unsplitted() and splitted() are called with two parameters a and b in range [0..9]. We have inserted calls to Frama_C_show_each() built-in function to show considered values by Eva for a, b and a+b just after the if (a + b >= 3) test.

```c
#include <__fc_builtin.h>

int g_array[20];

/*@ requires 0 <= a < 10 && 0 <= b < 10; */
void main(int a, int b)
{
  unsplitted(a, b);
  splitted(a, b);
}

void unsplitted(int a, int b)
{
  if (a + b >= 3) {
    Frama_C_show_each("a=", a, "b=", b, "a+b-3=", a+b-3);
    g_array[a + b - 3] = 42;
  }
}

void splitted(int a, int b)
{
  Frama_C_builtin_split(a, 10); // do case splitting on "a"
  if (a + b >= 3) {
    Frama_C_show_each("a=", a, "b=", b, "a+b-3=", a+b-3);
    g_array[a + b - 3] = 42;
  }
}
```

We apply Eva with frama-c -eva -eva-slevel 100 path_split.c command.

For function unsplitted(), we get an alarm:

```
[eva] path_split.c:15:
  Frama_C_show_each:
  {{ "a=" }}, [0..9], {{ "b=" }}, [0..9], {{ "a+b-3=" }},
    [-3..15]
```

```
4 [eva:alarm] path_split.c:16: Warning:
5   accessing out of bounds index. assert 0 <= (int)((int)(a + b)
      - 3);
```

Despite the test **if** (a + b > = 3) at line 14, Eva does not understand that the value of a + b is bigger than 3. Indeed, line 3 shows that the domains of a and b are considered independently in [0..9] range so a+b-3 is in [−3..15] range.

To improve Eva analysis precision, we insert a call to Frama_C_builtin_split() in function splitted() at line 22. This Eva built-in will consider several possible paths from the split call, one for each possible value of a. In that case, the output of the analysis is as follows:

```
1  [eva] path_split.c:24:
2    Frama_C_show_each:
3    {{ "a=" }}, {9}, {{ "b=" }}, [0..9], {{ "a+b-3=" }}, [6..15]
4  [eva] path_split.c:24:
5    Frama_C_show_each:
6    {{ "a=" }}, {8}, {{ "b=" }}, [0..9], {{ "a+b-3=" }}, [5..14]
7  [eva] path_split.c:24:
8    Frama_C_show_each:
9    {{ "a=" }}, {7}, {{ "b=" }}, [0..9], {{ "a+b-3=" }}, [4..13]
10 [eva] path_split.c:24:
11   Frama_C_show_each:
12   {{ "a=" }}, {6}, {{ "b=" }}, [0..9], {{ "a+b-3=" }}, [3..12]
13 [eva] path_split.c:24:
14   Frama_C_show_each:
15   {{ "a=" }}, {5}, {{ "b=" }}, [0..9], {{ "a+b-3=" }}, [2..11]
16 [eva] path_split.c:24:
17   Frama_C_show_each:
18   {{ "a=" }}, {4}, {{ "b=" }}, [0..9], {{ "a+b-3=" }}, [1..10]
19 [eva] path_split.c:24:
20   Frama_C_show_each: {{ "a=" }}, {3}, {{ "b=" }}, [0..9], {{
       "a+b-3=" }}, [0..9]
21 [eva] path_split.c:24:
22   Frama_C_show_each: {{ "a=" }}, {2}, {{ "b=" }}, [1..9], {{
       "a+b-3=" }}, [0..8]
23 [eva] path_split.c:24:
24   Frama_C_show_each: {{ "a=" }}, {1}, {{ "b=" }}, [2..9], {{
       "a+b-3=" }}, [0..7]
25 [eva] path_split.c:24:
26   Frama_C_show_each: {{ "a=" }}, {0}, {{ "b=" }}, [3..9], {{
       "a+b-3=" }}, [0..6]
```

You will notice that range of a+b-3 is always above 0, thus no alarm is triggered. We removed a false alarm!

This approach can be used in more complicate cases, improving precision by judiciously doing case splitting on relevant variables like array indexes. The only drawback is that this increase of potential analysis paths means an according increase of slevel to match the number of potential paths (e.g., 10 paths in our above simple example). Sometimes, in case of big loops or nested loops, it might not be possible.

Case splitting at function return A similar case splitting can be done when a function returns. By default, Eva merges all paths at the end of a function. Sometimes, you want to keep the various paths when the function returns to continue analysis along those paths, for example to consider a normal case and an erroneous case. You can use -eva-split-return-function option to do that.

Case splitting at loop end Similarly, you can use -eva-slevel-merge-after-loop Eva option to control the keeping of path splitting at end of loop. By default, Eva merges the paths but you might want to keep them distinct to increase precision.

15.2.5.2 Handling of Loops

Loops are the main hindrance to scalability with Eva. The basic approach to address this issue is to increase the slevel to match the number of loop iterations. For example, if you have a loop **for** (i=0; i< 100; i++) you increase slevel to at least 100 for this function. If you have **if** statements or other conditionals within the loop body, you can increase slevel to 500 or 1,000 to take into account potential additional paths within the loop body.

But when you have nested loop, you have to increase the slevel by the product of all nested loop bounds. For example, if you have following code:

```
for (i=0; i<100; i++) {
  for (j=0; j<300; j++) {
    ...
  }
}
```

Then the needed slevel is at least 30,000. The needed slevel rapidly increases with higher loop bounds and several nested loops, thus leading to non-acceptable analysis times.

Another issue with loops analysis is that Eva states are merged at the end of each loop iteration, therefore leading to a loss of precision.

Several approaches are possible to increase precision on loops. The first one is to increase the slevel, as presented above. Its main benefit is to increase the number of considered paths and precise effects on the state of program in each one of those paths, independently of the other paths, thus increasing precision. Its main drawback is its scalability.

Another approach is to unroll a specific loop using //@ **loop unroll** N; loop unrolling annotation where N is a constant or a variable of which content is known at analysis time. This approach allows to keep higher precision from one loop iteration to the next.

A third approach is to do syntactic loop unrolling using //@ **loop PRAGMA unroll** N; annotation where N is an integer. This directive will copy/paste the loop body N times. It has a very similar effect to the previous **loop unroll** annotation with higher overhead in Frama-C and thus not recommended. But it can be nonetheless useful when exploring the code because each loop unrolling is visible in the GUI and can be explored.

A fourth approach is to add an invariant to the loop through ACSL annotation. This loop invariant can constraint the range of a pointer within the loop and thus remove false alarms. Consider the following example where we scan a big array for the integer 42.

```
#include <__fc_builtin.h>

unsigned int big_array[10000];

int main(void)
{
  int i = 0;
  unsigned int *p = &big_array;

  big_array[8888] = Frama_C_nondet(42, 43);

  while (i < 10000) {
    if (42 == *p) {
      return i;
    }
    p++; i++;
  }

  return -1;
}
```

Without loop invariants, we would have one spurious warning on correct p pointer access:

```
[eva:alarm] loop_without_invariant.c:13: Warning:
  out of bounds read. assert \valid_read(p);
```

Now, we annotate this program with ACSL annotations:

```
#include <__fc_builtin.h>

unsigned int big_array[10000];

int main(void)
{
  int i = 0;
  unsigned int *p = &big_array;

  big_array[8888] = Frama_C_nondet(42, 43);

  /*@ loop invariant &big_array[0] <= p <= &big_array[10000];
      loop invariant p == &big_array[i];
      loop assigns p, i;
   */
  while (i < 10000) {
    //@ assert &big_array[0] <= p <= &big_array[9999];
    if (42 == *p) {
```

```
19      return i;
20    }
21    p++; i++;
22  }
23
24  return -1;
25 }
```

With the above loop invariants and assertion at line 17 we have no false alarm. However the correctness of the annotations themselves need to be proven or at least reviewed. Wp can be used to prove invariants and function safety as it is the case for this example by using -wp -wp-rte command line options.[9] However, it becomes quite complicated if there are function calls within the considered function (Wp needs a contract on each called function), if the logic of the function is complicated, or even impossible if the code is using **volatile** variables that Wp does not handle.

A last approach is to reduce the loop bounds, for example by reducing constants. You will not verify the original code, but at least a code with similar behavior and complexity and it might be sufficient if supported with other arguments (reduced loop bounds still cover complex behaviors, effect on the program of bigger bounds can be explained...).

15.2.5.3 Direct Access to Hardware Memory

Embedded programs usually do direct accesses to memory, for example to control hardware devices. Usually a memory map is defined for the program, with specific memory areas that the program can access.

Let us consider as an example the following program which initializes two memory areas A and B using two successive loops.

```
1  #define A_AREA_START   0x0A00
2  #define A_AREA_LENGTH  0x100
3
4  #define B_AREA_START   0x1B00
5  #define B_AREA_LENGTH  0x100
6
7  void main(void)
8  {
9    unsigned int *p = 0;
10   unsigned int i = 0;
11
12   for (p = A_AREA_START, i = 0; i <
         A_AREA_LENGTH/sizeof(*p)+3;//UNCAUGHT BUG!
13       p++, i++) {
14     *p = i;
15   }
16
```

---

[9] This example is proved without annotations by TIS Analyzer with slevel of 20,000 in about 10 s.

```
17    for (p = B_AREA_START, i = 0; i < B_AREA_LENGTH/sizeof(*p);
          p++, i++) {
18      *p = i;
19    }
20  }
```

This program can be analyzed with the following command line where we use -absolute-valid-range option to specify the valid memory range that includes both areas:

```
frama-c -eva -absolute-valid-range 0xA00-0x1BFF -eva-slevel 256
    hw_access.c
```

This analysis does not raise any error. However, you have probably noticed an error on line 12 were the initialization loop goes past the end of area A by three! This is because Eva considers the complete range 0xA00 to 0x1BFF as valid for memory accesses.

One approach to improve precision of the analysis is to define one C variable for each area and point to them. We can do this by adding following definitions at lines 9 to 18 in our program:

```
1  #define A_AREA_START   0x0A00
2  #define A_AREA_LENGTH  0x100
3
4  #define B_AREA_START   0x1B00
5  #define B_AREA_LENGTH  0x100
6  // unmodified program before
7
8  /**** points HW memory areas to C variables ****/
9  // define variables for HW memory areas
10 unsigned char a_area[A_AREA_LENGTH];
11 unsigned char b_area[B_AREA_LENGTH];
12
13 // let our program points to them
14 #undef  A_AREA_START
15 #define A_AREA_START &a_area
16
17 #undef  B_AREA_START
18 #define B_AREA_START &b_area
19 /**** end of redefinitions ****/
20
21 // unmodified program afterwards
22 void main(void)
23 {
24   unsigned int *p = 0;
25   unsigned int i = 0;
26
27   for (p = A_AREA_START, i = 0; i <
          A_AREA_LENGTH/sizeof(*p)+12; // CAUGHT BUG!
28        p++, i++) {
29     *p = i;
30   }
31
```

```
32  for (p = B_AREA_START, i = 0; i < B_AREA_LENGTH/sizeof(*p);
        p++, i++) {
33    *p = i;
34  }
35 }
```

Those lines define two C variables a_area and b_area and redefine A_AREA_START and B_AREA_START pre-processor definitions to point to them. No other modifications are needed. In a real program those modifications would be in some headers.

We can now analyze our modified program with following command where -absolute-valid-range option is no longer necessary:

frama-c -eva -eva-slevel 256 hw_access_var.c

This time, the memory access past the end of area A is caught by Eva! This technique allows a fine-grained check of memory accesses.[10]

15.2.5.4 Pre-loading of Binary Data Structure

The embedded program under analysis might use a data structure available in binary format at program start, for example in flash memory. To make an analysis where such binary data structure is available to Eva, we can use following approach.

Firstly, the binary data structure contained for example in memory_context.bin file is converted into a C array declaration with following Unix **xxd** command:

xxd -i memory_context.bin memory_context_driver.h

This command produces a C file of this form:

```
1 unsigned char memory_context_bin[] = {
2   0x41, 0x20, 0x62, 0x69, 0x6e, 0x61, 0x72, 0x79, 0x20, 0x73,
        0x74, 0x72,
3   0x69, 0x6e, 0x67, 0x00, 0x0a
4 };
5 unsigned int memory_context_bin_len = 17;
```

Then, within your analysis, you can load the data structure with the following typical C code where the data structure would be loaded at 0x1000 memory address:

```
1 #include "memory_context_driver.h"
2 #include <string.h>
3
4 #define MEM_BASE ((void*)0x1000)
5
6 void fill_memory() {
7   memcpy(MEM_BASE, memory_context_bin, memory_context_bin_len);
8 }
```

---

[10] TIS Analyzer now provides dedicated feature to take into account potential memory location of variables without resorting to such trick.

15.2.5.5 Other Techniques to Improve Precision

Handling of deeply nested data structures You might have deeply nested data structures in your program. By default, access to such data structure is precise up to an offset of 200 to index data structure fields. In case of deeply nested data structures this is far from enough. You can keep a precise analysis by increasing plevel parameter with following command: -eva-plevel 2000000.

Handling of compiler specific extensions of C language Sometimes, the compiler of a target embedded platform allows some specific keywords used within the analyzed source code and Frama-C cannot parse them as it mostly handles C99 language. You can use the C compiler preprocessor to remove those keywords. For example, you if you have __packed__ keyword in your source code, you can use -cpp-extra-args="-D__packed__=''" to remove them. Of course, removing such keywords might change the semantics of the program and a review of such code changes should be done.

15.2.6 Conclusion

We have shown in this section practical settings to setup and run Eva on a safety-critical embedded program, as well as tips and tricks to improve precision of an analysis. Armed with those techniques, you should be able to demonstrate the absence of run-time errors of programs found in real-world industrial contexts. Such verification needs some expertise in the use of Eva but, provided some time and experiments, satisfying results can be obtained.

If you plan to regularly apply Eva on programs produced by your organization, it might be worth defining coding rules that would help the application of Eva. For example such rules could mandate the use of intermediate variables for computations, constant loop bounds or clear identification of safety critical from non safety critical modules. We have developed such coding rules at MERCE.

Of course, no tool is perfect and we would appreciate improvement in Frama-C for such analyses. The first one would be on scalability, Eva being somewhat limited as code size grows.[11] Another useful enhancement would be a tool to help review legitimacy of "dead code", i.e. code not covered by Eva. One overall general improvement of Frama-C would be its tight integration with modern IDEs (Integrated Development Environments) like VSCode through LSP (Language Server Protocol) to display analysis results directly within the editor, facilitating immediate verification and correction of the source code.

---

[11] TIS Analyzer seems more scalable, even if real program size always becomes an issue at some points.

15.3 Automated Test Generation Using Eva and PathCrawler

Quality checking of software and more specifically safety critical software commonly relies on review and testing. Software testing is a time consuming and error-prone activity, mostly manual. On a typical safety critical industrial project, total testing amounts to 65% of total development time. Beyond cost, such long testing time impacts time-to-market and project agility.

One approach to increase productivity is to automatically generate tests. To achieve this in Mitsubishi Electric we are investigating automatic generation of structural unit tests of safety-critical embedded software in the C language with integers and real data types. Furthermore, we want to integrate seamlessly inside the existing test process, improving it without radically changing it. The tool takes as input unmodified source code and produces as output ready-to-use test sheets, i.e. description of test input values for a given function, in both human and machine-readable formats. To this end, Mitsubishi Electric R&D Centre Europe (MERCE) developed a tool combining several formal methods tools and a non-formal approach, namely Genetic Algorithms.

The formal approach was devised in collaboration with CEA, using Frama-C and PathCrawler/LTest [3, 6, 24] concolic test generation tool. The Genetic Algorithms approach was developed inside MERCE using a publicly available Genetic Algorithm library.

This tool developed in C++ language takes as input unmodified C language source code. It automatically adds to source code *labels* stating test objectives which are derived from proprietary coverage criterion defined by some Mitsubishi Electric's business unit. Then it generates stubs suitable for unit testing. Given those labels and stubs, it finds test cases satisfying test objectives. It finally produces test sheets and stubs for human and machine use in the remaining part of the test process.

Besides simple criterion (targeting a characteristic value of one condition, e.g. a specific boundary value for a variable), the tool targets MC/DC (Modified Condition/Decision Coverage) [17], which combines several conditions to achieve good test coverage in a reasonable time frame. Thanks to the smart combination of formal methods and genetic algorithms, we obtained a fully automatic, almost complete (i.e. beyond 95% of test objectives on real projects) and very fast structural unit test generation tool. For example, in a real-world industrial project, we demonstrated that our approach can reliably generate test cases when feasible or demonstrate they are unfeasible for 99% of the MC/DC test objectives in about half an hour for 82,000 lines of C code.

15.3.1 Technological Context

Industrial process We focus on unit testing, i.e., testing each function individually and independently of the others. This implies the necessity to develop *stubs*: replacement functions for each function that is called in the body of the function under test.

We consider structural tests, i.e. test objectives defined by the control structure of the code: on a given control statement (e.g., `if`, `while`, etc.), the structural coverage criterion defines, from the associated control condition, one or more test objectives, i.e., specific values that variables should have at this location at execution. The goal of our tool is to find test cases that fulfill those test objectives, i.e. to find the initial values of relevant input variables which lead, during execution, to expected values defined by the test objectives.

Our goal is to build a tool that could integrate seamlessly into existing industrial test processes. So, we target a fully automatic tool that processes unmodified C source code and produces ready-to-use test sheets, filled with input values for each test case. According to the test process, expected output values (*oracles*) should then be filled by the test engineer, and later checked by executing the code in an external tool.

MC/DC coverage criterion In addition to the specific business unit criterion we target MC/DC coverage criterion. MC/DC criterion exercises elementary Boolean expression called *condition* of a control statement in turn, to check that each of those conditions has influence on the statement's entire expression value (called *decision*) hence on the program control flow. More precisely, we target Masking MC/DC [17] criterion that takes into account the logical dependencies between conditions of a control statement decision.

Genetic algorithms Genetic algorithms have been widely used for generating test cases [2, 20]. The idea of genetic algorithms is to model Natural Selection. First, they *select*, from individuals (defined by their *genes*), the ones that fit best for solving some problem, according to their scores given by a *fitness* function. Then, they create new individuals by combining genes of candidate solutions during the *cross-over* phase. Lastly they also *mutate* some individuals' genes in order to avoid being stuck in local optimums. Finally they iterate this whole process, improving individuals' scores generation after generation, hopefully reaching a solution (which might not be optimal) of the initial problem at the end of the process.

Formal methods tools Genetic algorithms are known to have very good performances for finding test cases [2, 20], but they also have limits. In particular, they cannot detect unfeasible test objectives (typically a variable value target that is impossible at some location of the code, due to the control flow of the program), hence they can waste a lot of time trying to find solutions to problems that cannot be solved. Another limitation concerns the fact that they do not exhaustively consider the candidates test cases, hence sometimes do not find a solution for problems that can be solved. In our approach, we used several formal methods tools to alleviate those drawbacks.

- **Eva** [7] is a plug-in of the Frama-C framework [18] presented in Chap. 3. Eva implements Abstract Interpretation [10] technique to determine an over-approximation of the set of values that each variable can have at each point of the program and for all possible executions. It applies directly on C source code and is quite fast.
- **PathCrawler/LTest** [3, 6, 24], presented in Chap. 6, is also a plug-in of Frama-C framework [18]. It is a test generation tool implementing Concolic technique. PathCrawler mixes concrete execution of instrumented code to determine covered path and symbolic execution of code to determine constraints needed to satisfy a given path and thus needed input values to cover new paths. PathCrawler is sound (each test-case covers the test objective for which it was generated), and it is of medium speed (several seconds to minutes to find a test case) due to several calls to an external constraint solver.
- **CBMC** [8] is a Bounded Model Checker for C and C++. CBMC transforms C code into Static Single Assignment form and then into equations that can be given to an SMT solver to check a logical formula.

15.3.2 Tool Architecture

Our test generation tool architecture is organized into a six-step pipeline toolchain (Fig. 15.1).

Step 1 pre-processes source files, generates distinct stubs for the following steps, and computes labels for expressing test objectives for MC/DC criterion. The distinct stubs are created by specializing generic stubs for each of the context where it is used. Each of the steps 2–5 tries to identify unfeasible test objectives and/or find new test cases satisfying test objectives. At each step, the remaining unsolved test

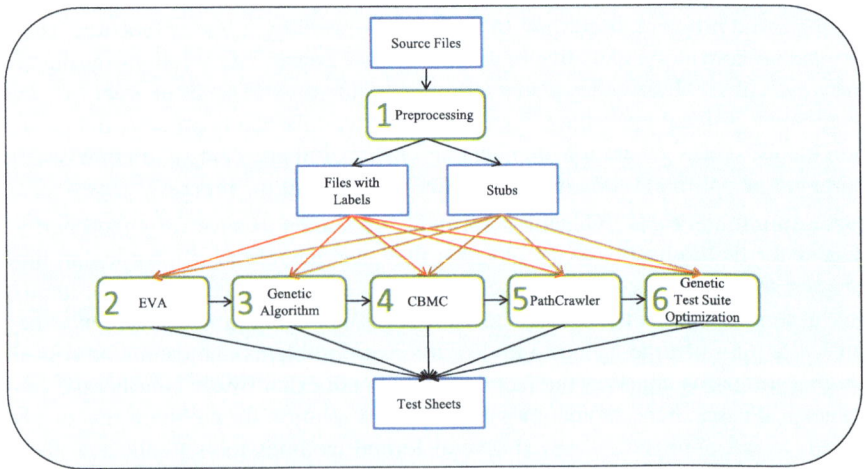

Fig. 15.1 Architecture of MERCE's toolchain

objectives are forwarded to the next step. Finally, step 6 optimizes the produced test suite, by combining the test cases in order to reduce their total number.

Step 1: Pre-processing This step first substitutes definitions and macros in C source code. Then, for each function under test, it identifies called functions (hence stubs to be defined), and the parameters (global variables, function inputs, etc.) that have an impact on at least one of the test objectives decisions of the function under test (in order to define the genes of individuals in step 3).

Stubs Since we target unit testing, stubs' implementations do not consider the original computational content of called functions. They are designed to have the greatest impact on the program data flow after they are called (i.e. they can modify any of their *outputs* with any value), in order to ease the classification of tests objectives as solved or unfeasible. Stubs are implemented in a way that allows them to change the value of any of their *outputs*: global variables that are read by the function under test, memory locations pointed by a pointer taken as input, and, of course, the returned value. Note that in the case we only search for unfeasible test objectives, we can suppose that all of these outputs can take a random value (in a domain defined by their types) at each call of the stub, while when searching for test cases, it is necessary to keep information of those output values for each different call of the stub, in order to be able to replay the test case afterwards. In practice, in the latter case, stubs contain arrays of values, for each output and each call of the stub. If the number of such values is less than the number of calls at execution (e.g. if the function is called in the body of a loop and we could not anticipate the number of calls), then the last values contained in the respective arrays are used as many times as necessary.

MC/DC labels Finally, we generate labels that express MC/DC test objectives. For example, if we have a control statement with a decision of the form **A && B**, The goal is to show that both conditions A and B have influence on the decision **A && B**. Since we consider here a conjunction between A and B, we shall generate labels (A true, B true) and (A false, B true) for condition A (and symmetrically for condition B). We take into account Masking MC/DC which allows us to discard inconsistent labels when they appear. Note that before generating labels, we normalize control statement decisions for maximizing the number of conditions of the form of an equality test between a variable and a constant, in order to provide the most efficient fitness function in step 3.

Step 2: Eva In step 2, Eva plug-in is used to identify unfeasible test objectives. We use a "reductio ad absurdum" reasoning. By using Eva mechanisms, we assert that the variable values necessary to fulfill the test objectives are not possible at their respective code locations. If Eva does not complain, that means that the over-approximations of the variables values domains given by Eva do not contain the values targeted by the test objectives. Eva considers all the possible behaviors before reaching a conclusion. Since the Eva plug-in is sound (a property is proven valid only if it is true for all possible executions of the program), it implies that the considered test objectives are unfeasible (as all possible executions have been considered by Eva).

Step 2 is crucial for the overall performance of the process, since it discards a lot of unfeasible test objectives that would substantially slow down the genetic algorithm at step 3.

Step 3: Genetic algorithm Step 3 uses a genetic algorithm to find test cases for test objectives that have not been determined as unfeasible during step 2.

Individuals' genes are defined as the values of the inputs of the function under test (global variables, actual input parameters of the function under test, and outputs of called functions (stubs)). We consider at each step populations composed of 100 individuals, and the initial population is randomly generated.

In practice, we first launch a global search considering all test objectives at the same time. After that, we launch a second search considering each of the remaining test objectives, one at a time. The **fitness** function evaluates the sum of the distances of the values of variable a to constant K for each condition (a==K) composing the considered test objective (the smaller the better). The **fitness** function also ranks better individuals that satisfy other conditions (not of the form (a==K)). In the former case, the **fitness** function additionally considers the number of test objectives satisfied by an individual and the number of test objectives code locations that are reached at execution (the bigger the better).

We implement **cross-over** as a complete random combination between two individuals' genes: which genes and how many genes are selected from each parent, are both random. **Mutation** is sometimes completely random: we select a random new value for a random gene of some individual. At other times, we also use a local hill climbing algorithm for defining mutations on randomly selected genes, in order to improve convergence of the genetic algorithm (i.e. we use mutations that maximize the **fitness** function locally).

Finally, for **selecting** the 100 individuals that will compose the next generation from individuals, their children obtained by cross-over and the mutated individuals, we use a roulette wheel selection: the selection's probability of an individual for next generation is proportional to the ranking of its score given by the fitness function.

Step 4: CBMC In step 4, we use CBMC to identify unfeasible test objectives and find new test cases, that were not identified/found by previous steps. CBMC is used with a timeout of 60 s for each test objective. As we will see in the results section, CBMC does classify some of the test objectives that could not be handled by the previous steps. CBMC is used before PathCrawler, since it is slightly faster to classify some of the test objectives.

Step 5: PathCrawler In step 5, we use PathCrawler/LTest to identify unfeasible test objectives and find new test cases. We use a global timeout of 120 s for each function under test. As we will see in the results section, PathCrawler is still able to classify some of the test objectives that were not handled by the previous steps.

Step 6: Genetic test suite optimization In step 6, Genetic Algorithms are applied again, not to classify remaining test objectives, but to optimize the already found test cases. In particular we optimize visibility of test cases at end of function under test, i.e. the side effects produced by a test case should be visible after the return statement of the function under test. This optimization allows easier handling of generated test cases by human testers who thus can distinguish between two test cases from their output during execution. At this step, we also reduce the total number of test cases by finding test cases satisfying several test objectives, in order to reduce burden on human testers that should fill expected test results (i.e. oracles) for each test case.

Implementation We have made a lot of effort to optimize performance of our tool. Our code is parallelized at file level, at function level and even sometimes at test objective level (e.g., with Genetic Algorithms to classify a specific test objective or with CBMC when several provers are run in parallel). In fact the main issue is not to generate parallelism but to bound parallelism so the running computer is not overloaded and remains efficient.

Overall, this architecture allows to build an automatic, fast and practically almost complete test generation tool. The tool is fully automatic and only needs unmodified source code. Speed is brought by steps 2 (Eva) and 3 (Genetic Algorithms), which are quick to respectively identify unfeasible test objectives and find test cases (see Sect. 15.3.5). Steps 4 (CBMC) and 5 (PathCrawler) are not that fast, but allow for classifying new test objectives that steps 2 and 3 could not handle (this concerns especially unfeasible test objectives in functions with a complex control flow). Despite increasing marginally the global coverage, steps 4 and 5 are still very important since human study of those complex test objectives would be particularly time consuming.

15.3.3 Generation of Test Objectives for Different Back-End Tools

Our intention was to drive several tools to find the values of test sheets while minimizing the software development to be done. To achieve this, we proceed in two steps. Starting from a C source file we produce an augmented C source file including the description of the values that key variables have to take at some given position of the C source code. It is coded with some C macro calls. Then in a second step these C macros are expanded depending on the tool that will do the search for the inputs of the test sheets:

- For the formal tools since we use a "reductio ad absurdum" reasoning, we assert the variable cannot get the expected values. Either the formal tool will prove that this is not true and potentially provide some input which will become input from the test sheet or will prove that this is true and so this test objective is not feasible.
- For the Genetic Algorithm the macros store the found values. The Genetic Algorithm analyzing the stored results after each run, drives the inputs to reach the expected values.

To avoid problems with formal tools which might not be able to treat complex conditions that a human could write in C language, we encode each check in a

Boolean C variable (an assignment of this variable is inserted in the treatment flow of the program). The formal tool will have only to check whether this C variable is true or false.

In practical terms starting from some tested function like this one:

```
void TestedFunction ( int a )
{
    if ( a > 3 ) {
        Something();
    } else {
        SomethingElse();
    }
}
```

We generate an augmented file along the following template. Two Boolean variables assert1 and assert2 are introduced to express test objectives on a > 3 condition trough calls to TASSERT macro. Then a TestedFunctionChecker() function is created to call the TestedFunction() and check results of the Boolean variables through calls to CASSERT macro.

```
char      assert1 = 0 ;
char      assert2 = 0 ;

void TestedFunction ( int a )
{
    TASSERT(assert1, a == 3);
    TASSERT(assert2, a == 4);
    if ( a > 3 ) {
        Something();
    } else {
        SomethingElse();
    }
}

void TestedFunctionChecker ( int a )
{
    TestedFunction(a);
    CASSERT(assert1 == false);
    CASSERT(assert2 == false);
}
```

The CASSERT and TASSERT macros use the corresponding assertion mechanism of the formal tool. For example, for Eva, CASSERT and TASSERT macros are expanded into:

```
char      assert1 = 0 ;
char      assert2 = 0 ;

void TestedFunction ( int a )
{
    if ( a == 3 )
        assert1 = 1 ;
    if ( a == 4 )
        assert2 = 1 ;
```

```
        if ( a > 3 ) {
            Something();
        } else {
            SomethingElse();
        }
    }

    void TestedFunctionChecker ( int a )
    {
        TestedFunction(a);
        //@ check assert1 == 0;
        //@ check assert2 == 0;
    }
```

15.3.4 Generation of Stubs

The function under test might call other functions. We are not interested in the behavior of the called functions. Our interest for them is only to what extent they could modify some input/output parameters or which value they could return. To be able to generate the test sheets we must also record the value they have generated at each call.

So for each called function we generate a stub taking values in some array. The following example, with simplified error handling, shows typical code generated to stub a function taking as input an integer parameter a and returning an integer.

```
#define TABSIZE 100

int outputValues [TABSIZE];
int stubFunctionPos = 0 ;

void Init ( void )
{
    for ( index = 0 ; index < TABSIZE ; index++ ) {
        outputValues [index] = RandomInt();
    }
}

int StubFunction ( int a )
{
    return outputValues [stubFunctionPos < TABSIZE ?
        stubFunctionPos++ : stubFunctionPos];
}
```

The Init() function initializes the inputValues array to random values. The RandomInt() function used for this purpose is provided by the formal tool in case of formal analysis, or by some C++ library in case of Genetic Algorithm analysis. For example, in the case of Eva, a suitable function for "RandomInt()" could be the following:

```
extern void Frama_C_make_unknown (char *p, size_t l) ;

long int RandomInt ( void )
{
   long int    result ;

   Frama_C_make_unknown((char *)&result, sizeof(long int));
   return result ;
}
```

The generated StubFunction() returns a different value taken inside the array for each call to it. Stubs also assign values to parameters if there are passed by reference. StubFunction() has no action on the value of a parameter since it cannot influence the calling code.

15.3.5 Results

Our tool has been implemented in about 70,000 lines of C++ code (the size of the externally used tools Eva, CBMC and PathCrawler is not considered) We tested it on a real industrial safety-critical project we will call project A, on a test machine running 64 bits Linux Fedora on an Intel I9-9980XE CPU with 18 cores (36 hardware threads), using about 32GB of RAM.

Project A contains about 82,000 lines of C code with integer data, spread in 230 files and 2,100 functions resulting in 23,550 test objectives. Among those 23,550 test objectives, 20,480 are in non-aborted functions, i.e. functions without errors (out-of-bound array access, division-by-zero, ...). Table 15.1 synthetizes the results of each step of the process on a typical run of the tool on project A. For each step, we detail the number of found test cases, test objectives determined as unfeasible, remaining test objectives, overall percentage of solved test objectives, and timing information. The vast majority of test objectives is found through Eva and Genetic Algorithms, CBMC and PathCrawler allowing to reach near complete coverage.

Those results were confirmed by running our tool on another project of same type and approximately same size: we obtained the same kind of results in terms of speed and coverage for each step of the process.

Table 15.1 Results on Project A

Step	Pre-proc	Eva	Gen. Alg.	CBMC	PathCrawler	Gen. Opt.	Total
Found	0	0	18,446	18	14	0	18,478
Unfeasible	0	1,709	0	115	56	0	1,880
Remaining	20,480	18,771	325	192	122	122	122
Solved	0%	8.34%	98.41%	99.06%	99.40%	99.40%	99.40%
Timing (min:sec)	3:02	2:50		0:41	4:34	17:30	28:37

Besides these very good performances, our tool has also good qualitative results. Firstly, as the test generation process is fully automated no error can be made, contrary to the human-based process in which we found several errors (e.g. consider a variable as 16 bits while it is 32 bits in the source code). Secondly, our tool allows quick identification of code lacking protections[12] against runtime errors like out-of-bound array access or division by zero. As our tool exercises thoroughly the function under test, it quickly finds missing protections which is of great help for the programmer. Typically on Project A, 10% of functions lack protection, and this can be detected by the lightweight configuration of our tool (configuration where the optimization step is skipped, thus greatly reducing the processing time).

15.3.6 Related Work

Each tool we use in the process has already been used for solving test objectives. PathCrawler is precisely designed for that purpose [3, 5, 6, 24]. CBMC has been used by Di Rosa et al. to generate test cases [12], but without identifying unfeasible ones. Genetic Algorithms are well known to be very efficient for generating structural tests [2, 20], in particular for MC/DC coverage criterion [21].

The combination of several formal methods has also been done in the past, for example Eva and Deductive Verification (Frama-C/Wp) with PathCrawler to extend the detection of unfeasible test objectives by Bardin et al. [4], but without addressing MC/DC coverage criteria.

However, to our knowledge there has been no previous work combining formal methods and genetic algorithms, targeting MC/DC criterion, with such good performance concerning both coverage completeness and generation speed.

15.3.7 Conclusion and Future Work

Our automatic structural unit test generation tool combining several formal methods (Abstract Interpretation, Bounded Model Checking and Concolic execution) and Genetic Algorithms to quickly generate ready-to-use test sheets extends the state of the art. We tested our tool on a real industrial safety critical project of 82,000 lines of code and demonstrated that it can solve test objectives of MC/DC coverage criterion for 99% of the 20,480 test objectives in about half an hour. We also showed that our tool could solve 98% of the test objectives, in less than 6 min, paving the way to its incorporation inside continuous integration process and development tools. While full unit testing process needs test oracles from the developer, our tool could still be used, without them, to quickly identify unprotected parts of the code, providing to the programmer useful test cases for understanding where problems come from.

---

[12] Code written to manage specific values in order to avoid runtime errors.

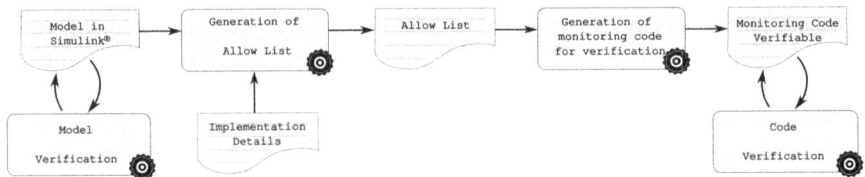

Fig. 15.2 Workflow of the proposed approach

15.4 Verifying That a Firewall Is Correctly Configured with Frama-C

Since the cyber-attacks by Stuxnet [23] and Dragonfly [22] worms, cyber security became a strong issue in industrial control systems (ICS) used in critical infrastructures such as electricity, gas and water. Experts commonly consider securing ICSs with "allow list"-based network monitor as one of the most suitable approach to upgrade security with a low impact on overall operability. Indeed, adding a network monitor is lightweight and easy to deploy on running infrastructures (no need to interrupt the systems). Moreover, upgrading security systems in case of creating new services or removing services simply relies on an update of the monitor configuration, i.e. the allow list. The allow list-based approach consists of determining the legitimate network communications. Many industrial processes implement Supervisory control and data acquisition (SCADA), a network-based architecture, which centralizes the supervision of devices and machinery through user access points. The communications of SCADA systems are cyclic and regular, which makes them really predictable. Thus, it is possible to build a model of the communications protocol which can be used as a reference to monitor all the traffic. The network monitor let all the packets complying with the communication model circulate, while all other packets are discarded and reported as alarms. Existing works on the specification-based method for ICS include [15, 16, 19]. In [15], the authors propose an intrusion detection approach for Modbus/TCP, a widely used communication protocol for ICS. Observing that these protocols are periodic, they propose to model the Modbus traffic as a deterministic finite automaton (DFA).

We developed a framework to design an allow list-based network monitor including formal verification integrated in Simulink. The toolchain generates the source code of the monitor kernel and uses Frama-C to verify soundness of the source code. The communication model is specified using the Simulink package StateFlow that allows to insert state machines in designs. Being fully automated, our approach aims to reduce the development cost and the need of security experts. Network engineers should be able to handle the entire process by themselves.

Once the communication model is specified, the allow list can be extracted from the model, and the corresponding C code is generated. The extraction and the generation are fully automated. Our approach also aims to ensure the reliability and security of the network monitor. To achieve this, our workflow includes two verification steps. The first step consists of verifying the communication model which is the starting point of the whole process: any specification error may lead to generate a non-compliant network monitor, which may induce some malfunctions or security threats. To mitigate this risk, we use the model-checker **Simulink Design Verifier** to check properties increasing the overall trust we can grant into the specification. The second step of verification, done by **Frama-C**, targets the soundness of generated monitor by checking that it is free of runtime errors and it accepts the network traffic from the specification only. After an overview of the modelling approach, next sections present the generated C code and its verification using **Frama-C/Wp**.

15.4.1 Modeling and Verification Using Simulink

The starting point of our approach is the specification of the communication model. From our experience, the specifications of industrial protocols focus on low-level details, e.g. kind of transactions, packet structure, valid data.... On the opposite, the communication model must be abstract and focus on application features like commands and answers. To do so, our approach uses the modeling and verification features of `StateFlow`. This commercial Model-Based Design solution is widely used to develop embedded software in various industrial domains, because **MathWorks**, commercializing **MATLAB**, provides a good support and a quite efficient Integrated Development Environment (IDE) from the design down to code generation. Moreover, **MathWorks** commercializes some certification kits addressing the safety and security requirements needed by standards (e.g., ISO-26262 for automotive, DO-178-B for aeronautics, EN-50128 for railways...). All these features make **MATLAB** and **Simulink** attractive for industrial developments.

By default, **Simulink** programming language implements the dataflow paradigm which is well adapted to the design of reactive systems. However, it is not the best option for network protocols for which the representation with state machines is more natural. The **MATLAB** package `StateFlow` extends **Simulink** with components defined as state machines. Figure 15.3 shows the minimal ICS architecture: a Human Machine Interface (HMI) overseeing and controlling some Programmable Logic Controller (PLC)which is a device (computer) responsible for the automation tasks. HMI and PLC are connected together over a local area network (LAN). The PLC has some field device(s)—like arm robot, conveyor, spinner...—which are not part of modeling, because only the PLC communicates with the HMI through the LAN. Regularly, the HMI sends requests to the PLC to get some status or to send orders. According to the requests received, the PLC either actuates its field devices or returns data coming from its internal state or from sensors. In the basic setup (Fig. 15.3), typical attacks relying on spoofing communications between the PLC and the HMI are not very difficult to set up, because ICS protocols commonly used do not feature any security mechanism.

As illustrated in the example, the use of communicating state machines is straightforward, and the periodic nature of the communications is easy to handle in this

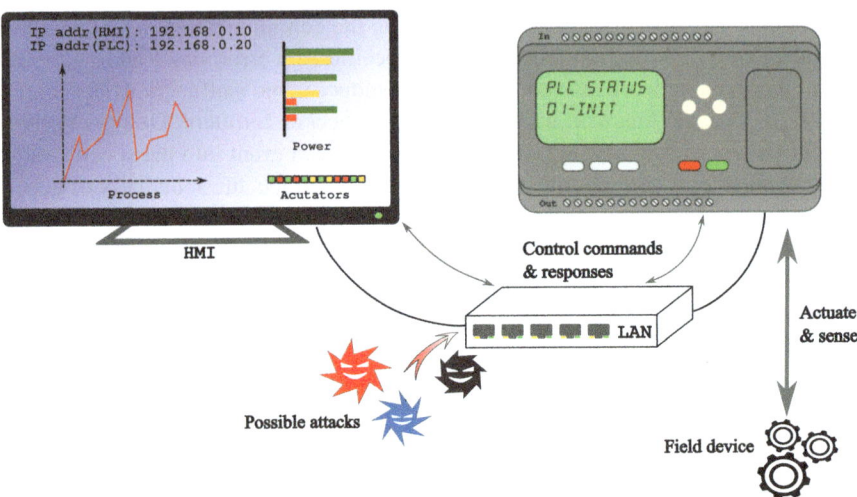

Fig. 15.3 Example of ICS

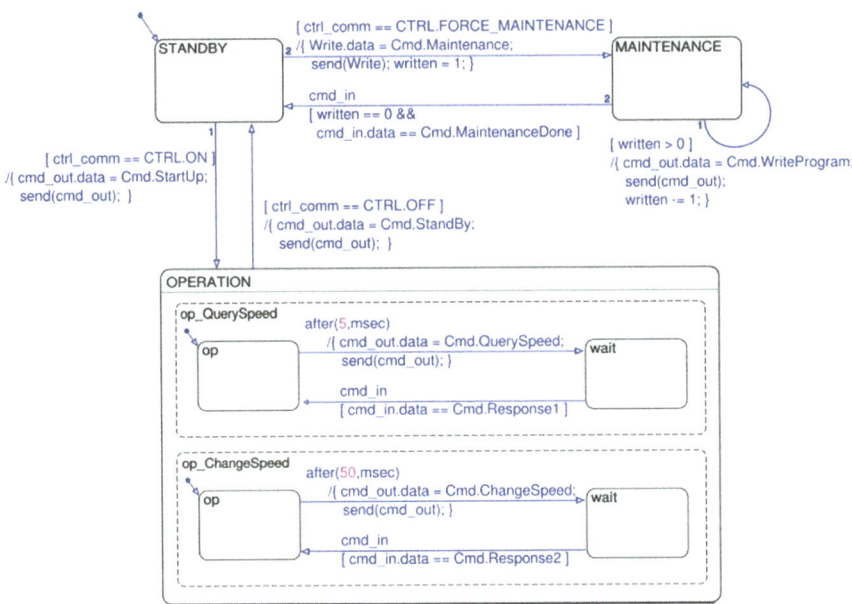

Fig. 15.4 Communication model for HMI

formalism. Figure 15.4 shows how we model protocol in Simulink. This example contains the typical characteristics of ICS, i.e., execution modes including cyclic and timed requests.

The protocol is divided in three modes STANDBY, MAINTENANCE and OPERATION. When the system is in the mode OPERATION, HMI queries PLC about its current speed every 5 ms and updates the speed target every 50 ms. PLC replies to each query by returning an acknowledgment containing requested data if any. The HMI's behavior is given in Fig. 15.4 as a hierarchical state machine in which the three main states match the modes. We observe that some transitions are timed according to the protocol description. The communications of the most complex mode (OPERATION) are described using an inner state machines which are executed in parallel. Besides, the HMI's behavior may depend on external events, like human interactions for instance, enabling the switching between the different modes. If need be, some additional control logic can model how and when the events may be triggered. This is particularly interesting to get a fully executable model which can be animated and analyzed during the modelling. In Fig. 15.4, we notice that some of the transitions between mode states are guarded by Boolean conditions (surrounded by brackets). In this way, the guard [`ctrl_comm` == CTRL.ON] ensures that the transition 1 is fired only when the button (denoted by the event [`CTRL.ON`]) for switching on the system is pushed. From an architectural point of view, the PLC is the slave of the HMI. Changing the system mode relies on navigating between the HMI's states STANDBY, MAINTENANCE or OPERATION. The protocol is in charge of notifying the PLC such that both HMI and PLC are always synchronized. Depending on the mode, the allowed messages differ: it is not possible to send messages for updating the PLC configuration while the system is in the mode OPERATION. Such messages can be exchanged in the MAINTENANCE mode exclusively. Message channels are unidirectional in Simulink. Two channels ensure the propagation from the state machine HMI to the state machine PLC, and conversely. In another way of modelling, separated channels could be used to isolate different kinds of messages and produce a more strengthened model. For instance, we could declare two enumeration types for operation and maintenance messages respectively. We could add two separate channels from HMI to PLC, each of which would be associated to a data type enumerating the allowed message to each channel. This approach structurally enforces the design by avoiding the mix of different kinds of messages in channels. This feature is particularly interesting for more complex systems. For sake of simplicity and readability, we limited our use of Stateflow features to the ones shown in this example: hierarchical and parallel state machines, timed transitions, transition junctions, signals, message-based communications and integer and Boolean arithmetic. This restriction may seem too strong, but it is important to recall that several modelling guidelines for Simulink have been defined to address safety, security and execution efficiency concerns in industry [9, 13, 14], because the language is very large and some construction of the language have a complex semantics, source of subtle modelling errors.

15.4.2 Simulating and Verifying the Model

In order to help the protocol modelling, the model validity is checked using the Simulink engine and verified using the model checker Simulink Design Verifier (SLDV). The simulation is useful to quickly detect and debug model issues. Animating the model is also interesting to understand its behavior. However, even long simulations are not sufficient to cover all reachable states and strongly assert the model correctness. This can be handled by model checking. The formal verification offers stronger guarantees about the modelling for some categories of properties. SLDV verifies some "Design properties" like the absence of runtime errors or for the dead logic in the model. These verifications are prerequisites to further verification steps. It is also able to check the range of integer values. This is very useful to check the arguments of messages in the protocol: for instance, it can be used to check that PLC always returned a valid speed value, i.e. a value between 0 and SPEED_MAX as defined in the functional requirements of the ICS. Dead logic denotes the unexecutable parts of the model; it is useful to identify transitions that cannot be fired and unreachable states, symptoms of an incorrect model. SLDV is also able to verify functional properties on the model. The targeted properties are expressed in Simulink as specific components tagged as "*verification components*". Each of these components implements a logical property as a Boolean expression. By monitoring some data of the model, the component evaluates the validity of each execution step against the logical property as a Boolean value. Then the model checker verifies that this resulting value is always true during any execution of the model. Whereas the approach is tractable for pure dataflow diagrams [11], the scalability becomes a strong issue when the model contains state machines. For instance, 7 min are required to check a property of the form "*whenever the message A sent, it can be followed by the message B only*" against the example model on a laptop running an Intel Core i7 with 16GB of RAM. To overcome the scalability issues of SLDV, it is necessary to split the model and isolate inner state machines to be able to do the verification of the specification.

15.4.3 Generating the Monitor

Once the protocol has been fully designed in Simulink, a C implementation of the monitor is automatically generated from the model. Since the model covers the whole protocol, valid packets are only the ones corresponding to the messages that can be emitted by transitions of the model. The C code implements the decision procedure in charge of accepting or discarding every network packet, in accordance with the functional specification. To be implemented, the protocol specification is converted then combined with some implementation details during the code generation process (see Fig. 15.2). This process is divided in two steps: (1) extracting the allow list from the model and (2) generating the C code from this intermediate representation. Once the generated code has been successfully verified with Frama-C (3), it is compiled

Table 15.2 Example of abstract allow list

InState	OutState	Sender	Receiver	Message	Conditions
Standby	Standby	HMI	Controller	StartUp	–
Standby	Operation	HMI	Controller	Maintenance	–
Operation	Operation	HMI	Controller	QuerySpeed	$t_0 = 5$ ms
Operation	Operation	HMI	Controller	ChangeSpeed	$t_1 = 50$ ms
...	...	...	...	...	...

and linked to a network packet parser in order to obtain a complete network monitor. The code generation takes the implementation details file describing the low-level encoding format as additional input. This file describes the abstract structure of network packets and the values for the basic commands (e.g., write, read...), memory addresses (e.g., registers) and application specific commands (e.g., speed, angle...). In ICS protocols, requesting some information consists in requesting a reading of some registers and sending an order relies on request for writing some value in some register. All request from the HMI are acknowledged by the PLC with a status code accompanied with data (in case of a reading request). The allow list is the list of transitions, named rules, that are labeled by a packet pattern. Matching one of the rules is enough to consider a packet legitimate. Otherwise, it becomes suspicious and it is blocked by the monitor. The C procedure decides whether each input packet matches one of the rules.

Extraction of the Allow List

The first step of the monitor generation consists of extracting the allow list from the model. This list is an intermediate representation which needs to be processed to add lower details. Whereas the Simulink model of the protocol is abstract and reflects only the functional features of the protocol, the concrete allow list is obtained by customizing it to a specific ICS platform. Thus, it is defined for the data structure of the packet (e.g., Modbus/TCP [1]). A rule essentially contains the addresses of the sender and the receivers, a command with its parameter constraints and some related time information for the timed transitions. A simple model parsing is sufficient to collect the list of abstract rules, as shown in Table 15.2.

Then, this information must be encoded in the low-level format. The implementation details (e.g., logical/hardware addresses, port numbers and message encoding that are used by the model and so on) are provided in an external configuration file: for instance, it maps every protocol entity defined in the abstract model to its network identity (IP address, TCP port), and each command string to its counterpart constant.

By applying this mapping on the abstract allow list, we get the concrete allow list as illustrated in Table 15.3. In this example, states are encoded using unsigned integers, and senders and receivers are characterized by a couple (IP address, TCP port). Integers are either decimal or hexadecimal constants. Time values are expressed by a decimal integer numbers of milliseconds. The special token '*' is used to denote any value. In Table 15.3, the token indicates that the TCP port used by HMI to

Table 15.3 Example of low-level allow list

In state	Out state	Sender		Receiver		Message	Conditions
		Addr.	Port	Addr.	Port		
0	1	0x0a80001	*	0xc0a80014	0x0f	0x2048	–
0	2	0x0a80001	*	0xc0a80014	0x0f	0x2080	–
1	1	0x0a80001	*	0xc0a80014	0x0c	0x2112	$t_0 = 5$
1	1	0x0a80001	*	0xc0a80014	0x0c	0x2144	$t_1 = 50$
2	2	0x0a80001	*	0xc0a80014	0x0f	0x2176	$t_2 = 5$
...	...	...	...	...	...	...	...

communicate can be any valid TCP port. Depending on the mode, i.e. the In State, two different ports are used by PLC to receive messages: the maintenance communication is done by the port 15 (0x0f) whereas the operation messages are received through the port 12 (0x0c).

Generating the C implementation

The allow list gives the necessary criteria to filter network packets. The C procedure in charge of analyzing packets must be configured with the allow list. Basically, the allow list is converted into a tree used by a decision procedure accepting (or rejecting) packets. Each layer of the tree denotes a criterion of the allow list. The role of the decision procedure is going through the decision tree until reaching a leaf or failing to switch to a child. When the procedure reaches a leaf, it means that the packet is matched by a rule. All paths from the root to some leaf corresponds to a rule of allow list, where each node matches one criterion of the rule. For example, children of the root are associated to the values *In State*. Considering the sub-tree of the node corresponding to the state `Operation` means that we consider all the rules having `Operation` as In State. The procedure explores the tree to evaluate each criterion until getting a matching rule. Since the decision tree is deterministic by construction, when a criterion is not satisfied in some sub-tree, the procedure can directly conclude a failure, backtracking is useless.

Our implementation of the decision tree traversal is generic. It handles any tree obtained from an allow list. Thereby, the C code generator generates the decision tree in C code only. The generic procedure does not require any update, unless the allow list format is modified which means that a new protocol is targeted.

The C generator parses the allow list and produces an abstract syntax tree following the allow list format. Then, it performs two tasks, generating the C code and the Frama-C specification. The task devoted to the construction of the decision tree is based on the following data type:

```
1  struct tree_st {
2      uint32_t    value;
3      uint32_t    info[2];
4      conditions_t    *conds;
5      uint32_t    size;
```

```
6      struct tree_st  *child;
7  };
8
9  typedef struct tree_st   tree_t;
```

The decision is statically declared in the header file rules.h, which is included into the source file monitor.c of the decision procedure. It embeds several fields corresponding to the possible data of the allow list columns.

Verifying the C Code

The second task concerns the generation of code annotations required for the verification of the C code. Verifying the source code is justified by the needs of having reliable and secure monitoring procedures as most as possible. Indeed, all the network security relies on the monitor, which is the main weakness of this architecture. Attacking and disabling the monitor is sufficient to render ICS vulnerable. Formal verification of the source code is an efficient way to improve the monitor robustness by reducing the possible vulnerabilities that could be exploited by attackers. To do so, we aim to obtain an implementation free of bugs and that fulfills its specification. One of the best candidates to achieve this goal is Frama-C, which provides a bettery of plug-ins for analyzing or/and verifying C programs. The verification of the monitor relies on the Wp plug-in 4. Wp is used to demonstrate (1) all the source code (decision tree and the procedure) is free of runtime errors, and (2) the procedure only accepts legitimate packets according to the allow list.

Securing C code starts by ensuring that the code is free of runtime errors. Typical attacks are built on top of invalid memory manipulations like buffer overflows, invalid cast…As such, removing causes of runtime errors avoids a lot of vulnerabilities. For this proof activity, Wp automatically detects the instructions which may produce a runtime error and annotates it with assertions corresponding to conditions ensuring that their execution will never crash. Our generic decision procedure is about 750 lines of code (loc). Wp adds 54 assertions relative to possible runtime errors.

```
1  //@assert signed_overflow: i+1<=2147483647;
2  i++;
3  //@assert mem_access: \valid_read(data);
4  if (*data == 1) {
5      ...
6  }
```

Many of these errors are relative to the validity of memory accesses into the tree data structure, while the decision procedure navigates into it. In particular, the procedure reads the fields conds and child which point to some memory allocations. The field conds is a reference to an abstract syntax tree conditions_t which is used to represent some basic pattern matching conditions for constraining the payload of network packets when necessary. The field child is an array referencing all the subtrees of the node. The number of children, i.e. the size of the array child, is given by the field size. While the fields value and info, which are statically allocated, do not require specific assumptions, it is necessary to have a predicate defining a correctly allocated tree. Since the type tree_t is defined inductively, we declare an inductive predicate valid_tree_t, which is an efficient way to express properties on such data structures in **ACSL** 5, in order to express the correct initialization of a decision tree, according to the data type.

```
/*@ inductive valid_tree_t{L}(tree_t t) {

  @ case valid_tree_leaf:
    \forall tree_t t; t.size == 0 ==>
      \valid(t.conds) ==> valid_t_conds(*t.conds) ==> valid_tree_t(t);

  @ case valid_tree_node:
    \forall tree_t t; t.size > 0 ==>
      \valid(t.child + (0..(t.size-1))) ==>
      (\forall integer i; 0 <= i < t.size ==> valid_tree_t(t.child[i])) ==>
      valid_tree_t(t);
  }
*/
```

We notice that the conditions are used in the leaves only. This is due to the design of the decision procedure that does pattern matching on the data at the last step. The correct initialization of data of type `conditions_t` is defined in the predicate `valid_t_conds` from which the case `valid_tree_leaf` depends on. The second case `valid_tree_node`, corresponding to the inductive case, states that a valid node with `size` children owns an array `child` containing `size` valid trees.

This definition is a key property because the well-formedness of the data structure is necessary to prove both runtime assertions and functional specification. Nonetheless, the whole soundness of the approach may be compromised if it is erroneous or too weak. As such, it seems relevant to make careful reviews of such definitions and predicates.

Proving the functional soundness

The monitoring activity plays a central role to ensure the security of the industrial network. An unreliable monitor could accept some illegitimate packets due to an implementation problem from the generated C code. Dually, a monitor rejecting valid packets interfer with the nominal behavior by raising false alarms. Despite all the efforts to get a strong specification, any bug in the final implementation may ruin the high reliability of the security mechanism. This problem could come from a bug in the generic decision procedure or from the tree generated by the C code generator. We notice that the bug could occur only for some particular types of decision trees. This reason motivates the need of verifying again the C code every time its configuration is changed.

ACSL contracts are used to give the specification of C functions. Here, we use contracts as target specification. We want to verify that our implementation of the function `monitor()` only accepts the packets matched by the rules. The resulting contract is quite simple:

```
/*@ requires \valid((tree_t*) rules)
         && valid_tree_t(*rules);
  @ requires parsed;
  @ ensures \result == 1 ==> matched;
  @ assigns outstate;            */
int monitor(void);
```

To ensure the sound execution of the function, the contract requires that the global variable `rules` points to a valid memory address containing a decision tree that satisfies the predicate `valid_tree_t`. It mainly requires that all the arrays `conds` and `child`, parts of the structure `tree_st`, are correctly initialized and all their cells contain `size` elements exactly. The predicate is inductively defined to cover all the nodes of the decision tree. In addition, the global environment defines a set of shared variables defining the interface between the network parser and the monitor. These variables are set with values of packet fields extracted by the parser. The predicate `parsed` requires that these variables are properly set by the parser. The clause `assigns outstate` specifies that `monitor` can set the global variable `outstate`, and there is no more side effect. The functional specification is given by the clause `@ensures`. It states that when `monitor` returns 1, the network packet is `matched` by one rule of the allow list. The definition of the predicate `matched` is also provided by the code generator. It is a very basic rewriting of each allow list rule as an ACSL predicate:

```
1  predicate rule_2 =
2    instate == 0                              &&
3    outstate == 1                       &&
4    send_info[IP] == 0x0a80001          &&
5    0 <= send_info[TCP] <= 65535        &&
6    recv_info[IP] ==    0xc0a80014      &&
7    recv_info[TCP] ==   0x0c            &&
8    command == 0x2112                   &&
9    timers[0] == 5;
```

The line `0 <= send_info[TCP] <= 65535` means any valid TCP port denoted by the wildcard symbol * in the allow list above. The identifiers `state`, `send_info`, `recv_info`, `command` and `timers` are the global variables fed with data collected by the network parser. The predicate `rule_2` is satisfied when each global variable satisfies its constraint in the predicate. Once every rule has been translated, the predicate `matched` is satisfied by any parsed packet satisfying one of these rules:

```
predicate matched = rule_0 || rule_1 || rule_2 || ...;
```

A second header file `spec.h` containing the definition of the predicates `rule_i` and `matched` is included into the file `monitor.c`. The application of the approach on an industrial use case of a small product line controlled by a PLC (80 in/out) supervised by a SCADA unit (for monitoring and basic administration), 28 allow list rules have been generated for a total of 650 lines of ACSL specification.

The traversal of the decision tree is done through several loops and recursive calls. This part of the code requires manual annotations to be proved by Wp; every loop needs an invariant that helps Frama-C to efficiently perform inductive reasoning on the loop body.

```
1  ...
2
3  tree_t *ni = NULL;
4
5  ni = match_value(rules, instate, NULL)
6  /*@ loop invariant ni ==> child_of(ni, rules);
7     ...
8     @ loop assigns ni; */
9     while (ni != NULL) {
```

```
10      ...
11
12      ni = match_value(rules, instate, ni);
13
14    }
15    ...
```

To process a node and iterate over its children, we designed several generic iterators like match_value in charge of returning the address of the next sub-tree having a value satisfying the variable considered by the decision procedure or the pointer NULL. Each loop is built following a similar pattern. At first, the procedure looks for a node which the value matches the variable instate. We can observe that the iterator is invoked on the root tree named rules. The third parameter of match_value, is the pointer returned at the last iteration or NULL to initialize the iteration. The loop ends as soon as the iteration ends, when the pointer NULL is returned.

Among the various guarantees written in the contract of the iterator function, the last two clauses requires are the most interesting. The first one claims that the function returns a child reference only if its value matches the one provided as input parameter. The second one states that the function is an iterator: it relies on the predicate child_of establishing that the returned pointer is one of the children of the parameter node. This information must be reported as a loop invariant, because it is necessary to ensure that the local iterator is valid at each iteration.

```
1  /*@ predicate child_of(tree_t *t1, tree_t *t2) =
2        \exists unsigned j; j < t2->size && t1 == t2->child+j;
3      @*/
4
5  /*@ requires \valid(node) && valid_tree_t(*node);
6      @ requires lm ==> child_of(lm, node);
7
8      @ ensures \result ==> \valid(\result) && valid_tree_t(*\result);
9      @ ensures \result ==> \result->value == value;
10     @ ensures \result ==> child_of(\result, node);
11     @ assigns \nothing;                                       */
12 tree_t *match_value(tree_t *node, uint32_t value, tree_t *lm);
```

In total, four different iterators have been specified and eight nested loops have been annotated to have the generic procedure monitor ready to be proven. Its generic architecture is a very powerful advantage from the verification point of view. Every time the code of a new allow list is generated, only the files rules.h and the Frama-C specification spec.h have to be re-created: the generic part of the monitor is not impacted. Neither important effort of writing manual annotations nor the verification the generic functions (e.g., iterators, constraints evaluators) need to be done again, because they have been stated for any allow list according to the format presented above. The proofs that must be replayed are related to the contract of monitor(), since the definition of matched and the global declaration of rules may have changed. For these proofs, Wp verification is still automated with no manual proof activity needed whatever allow list rules are provided to the code generator: automatic theorem provers are able to discharge the proof obligations requested by Frama-C. The verification of the industrial use case generates 296 proof obligations in case of a full replay of the verification. For the whole source code, spread over 4 main source files, the complete verification is done in less than 1 min on a regular desktop computer (Intel Core-i7 2.4GHz, 16GB of RAM).

```
Proof replay for 'conditions.c'...
[wp] 99 goals scheduled
[wp] Proved goals:    99 / 99

Proof replay for 'compare.c'...
[wp] 16 goals scheduled
[wp] Proved goals:    16 / 16

Proof replay for 'match.c'...
[wp] 74 goals scheduled
[wp] Proved goals:    74 / 74

Proof replay for 'monitor.c'...
[wp] 106 goals scheduled
[wp] Proved goals:    107 / 107
```

The internal prover is able to resolve 215 of the proof obligations quickly. However, the remaining ones are harder and require the combination of CVC4 with Z3 or Alt-Ergo.

15.4.4 Discussion

The novelty of this work is to propose a framework based on model-based engineering that provides a mechanism to generate an allow list-based network monitor with high confidence level for security purposes. Focusing on the functional features of the protocol, free of implementation details, makes models more readable, especially using a graphical representation for state machines. Combined with formal methods, this approach ensures a high-quality model from which a monitor free of bugs is built. Moreover, the different steps of formal verification are setup to be highly automated, such that they do not require strong expertise in formal methods to apply them. Finally, Simulink which is frequently used in industrial projects is the main user interface. As a proof of concept, all the process steps, i.e. code generation, verification with Frama-C and compilation, have been automated and integrated as a Simulink plug-in that directly generates a monitor kernel that can be loaded directly into the firewall equipment.

This should ease the adoption of such an approach: there is no need to learn specific tool (except Simulink Design Verifier) or language.

We may notice that use of a unique tool for building the allow list may introduce weaknesses in the overall process. There is no verification mechanism to improve the soundness of the transformation from the model to allow list and from the allow list to ACSL specification. We consider it acceptable because model extraction mechanism consists of a simple enumeration of the transitions followed by a rewriting into a table. Similarly, transforming the allow list into ACSL annotations is also a simple syntactic rewriting. Verifying these transformations is equivalent to reversing them and checking if the transformation combination leads to the identity. It does not bring so much more confidence since the algorithms are straightforward and very close from their reverse. In comparison, generating code that is sound and free of bugs is really more challenging. To achieve this last point, we also initially considered testing the generated source code. Testing requires developing more tools to preserve the full automation of the approach. Indeed, a test case generator is needed to generate a set network packets. For real industrial SCADA systems, the allow list size may be up to 700 rules, even if we consider additional test cases the approach is really tractable. However, an oracle is also needed to determine the output, i.e. acceptance or rejection. The oracle is not really different from the monitor itself, which raises question about the relevance of testing. Moreover, even with a good coverage criteria, testing is not fully efficient to ensure the absence of runtime errors. Test cases may run successfully while some memory remains improperly managed, because none of the test cases exploits the issuing area. These configurations lead us to consider formal verification as an approach more relevant than test.

The soundness of the monitor ensures that it detects all suspicious packets, but that is not sufficient in a practical manner. System operators need assistance to quickly distinguish actual attacks from false ones caused, for example, by network issues such as delay or loss of packets. Therefore, an oracle should be used to rank the alarms and measure the likelihood of attacks.

15.5 Conclusion

In this chapter, we have presented the use of Frama-C, from classical testing to formal proof, in various application domains, from safety to security. We have also shown an increasing degree of integration of Frama-C within MERCE methodology, from direct application by end users of Eva plug-in to fully integrated generation of ACSL annotated C code automatically demonstrated correct with Wp. These experiments show that Frama-C can now be applied in many different ways for an industrial user. When the right technology, i.e. well-chosen plug-in, is applied on a problem, one can solve it efficiently and at a reasonable cost. The resulting C programs are safer and more secure, without impact on development process efficiency or usability by regular engineers.

Of course, MERCE does not limit its research to the three presented use cases that have been chosen as illustrating samples. Going deeper in its understanding of Frama-C technology and its integration within its own technology, MERCE now develops its own plug-ins to increase the security of embedded programs or automate the verification and generation of concurrent programs. We now go towards a world were Formal Methods are used daily to satisfy increasing requirements on software, hardware and data in our digitalized society, and Frama-C is a key technology to engineer a world that we can trust.

References

1. Modbus Specifications and Implementation Guides. http://www.modbus.org/specs.php
2. Aggarwal R, Singh N (2017) Search based structural test data generations: a survey/ a current state of art. Int J Sci Eng Res 8:511–520
3. Bardin S, Chebaro O, Delahaye M, Kosmatov N (2014) An all-in-one toolkit for automated white-box testing. In: Proceedings of the 8th international conference on tests and proofs (TAP), LNCS, vol 8570. Springer, pp 53–60
4. Bardin S, Delahaye M, David R, Kosmatov N, Papadakis M, Le Traon Y, Marion JY (2015) Sound and quasi-complete detection of infeasible test requirements. In: International conference on software testing, verification and validation (ICST). IEEE, pp 1–10
5. Bardin S, Kosmatov N, Marre B, mentré d, williams n (2018) test case generation with pathcrawler/ltest: How to Automate an Industrial Testing Process. In: 8th international symposium ISoLA 2018, pp 104–120
6. Botella B, Delahaye M, Hong Tuan Ha S, Kosmatov N, Mouy P, Roger M, Williams N (2009) Automating structural testing of C programs: experience with PathCrawler. In: Proceedings of the 4th international workshop on the automation of software test. IEEE, pp 70–78
7. Canet G, Cuoq P, Monate B (2009) A value analysis for C programs. In: International working conference on source code analysis and manipulation
8. Clarke E, Kroening D, Lerda F (2004) A tool for checking ANSI-C programs. In: Tools and algorithms for the construction and analysis of systems (TACAS), LNCS, vol 2988. Springer, pp 168–176
9. Conrad M Model-based design for safety critical automotive applications. Technical report, samoconsult GmbH
10. Cousot P, Cousot R (1977) Abstract interpretation: a unified lattice model for static analysis of programs by construction or approximation of fixpoints. In: Symposium on principles of programming languages, pp 238–252 (1977)
11. Delebarre V, Etienne JF (2013) Proving global properties with the aid of the simulink desing verfier proof tool. Wiley. https://doi.org/10.1002/9781118561898.ch5
12. Di Rosa E, Giunchiglia E, Narizzano M, Palma G, Puddu A (2010) Automatic generation of high quality test sets via CBMC. In: VERIFY, pp 65–78 (2010)
13. Ferrari A, Fantechi A, Bacherini S, Zingoni N (2009) Modeling guidelines for code generation in the railway signaling context. In: NFM 2009, Moffett Field, California, USA
14. Fey I, Müller J (2008) Model-based design for safety-related applications
15. Goldenberg N, Wool A (2013) Accurate modeling of modbus/tcp for intrusion detection in scada systems. In: Int J Crit Infrastruct Prot
16. Hadeli R, Schierholz MB, Tuduce C (2009) Leveraging determinism in industrial control systems for advanced anomaly detection and reliable security configuration. In: Proceedings of the CETFA'2009
17. Kelly J, H, Dan S, V, John J, C, Leanna KR (2001) A practical tutorial on modified condition/decision coverage. Technical report, NASA Langley Research Center. https://ntrs.nasa.gov/api/citations/20010057789/downloads/20010057789.pdf
18. Kirchner F, Kosmatov N, Prevosto V, Signoles J, Yakobowski B (2015) Frama-C: a software analysis perspective. Formal aspects of computing, pp 573–609
19. Kleinmann A, Wool A (2014) Accurate modeling of the siemens s7 scada protocol for intrusion detection and digital forensics. In: JDFSL
20. McMinn P (2004) Search-based software test data generation: a survey. Softw Test Verif Reliab 14(2):105–156
21. Minj J (2013) Feasible test case generation using search based technique. Int J Comput Appl 70(28):51–54
22. Symantec: Dragonfly: Cyberespionage attacks against energy suppliers. Technical report, https://www.symantec.com/content/en/us/enterprise/media/security_response/whitepapers/Dragonfly_Threat_Against_Western_Energy_Suppliers.pdf

23. Symantec: W32.stuxnet dossier. Technical report, https://www.symantec.com/content/en/us/enterprise/media/security_response/whitepapers/w32_stuxnet_dossier.pdf
24. Williams N, Marre B, Mouy P, Roger M (2005) PathCrawler: automatic generation of path tests by combining static and dynamic analysis. In: Proceedings of the European dependable computing conference, pp 281–292

Chapter 16
Proof of Security Properties: Application to JavaCard Virtual Machine

Adel Djoudi, Martin Hána, and Nikolai Kosmatov

Abstract Security of modern software has become a major concern. One example of highly critical software is smart card software since smart cards often play a key role in user authentication and controlling access to sensitive services and data. To demonstrate the compliance of a smart card product to security requirements, its certification according to the Common Criteria is often recommended or even required. The highest, most rigorous levels of Common Criteria certification include formal verification. In this chapter, we show how an industrial smart card product—a JavaCard Virtual Machine—has been formally verified by Thales using Frama-C, a software verification platform for C code. We describe the main steps of this verification project, illustrate target security properties, present some lessons learned and discuss several directions for future work and necessary tool enhancements.

Keywords Deductive verification · JavaCard virtual machine · Confidentiality · Integrity · Certification · Common criteria

16.1 Introduction

Security of software has become a major concern today. Security issues can have a significant impact on public institutions, vital infrastructures, business and private life. Leaks of sensitive private data, unauthorized access to services or data, attacks

---

A. Djoudi
Thales Digital Identity & Security, Meudon, France
e-mail: adel.djoudi@thalesgroup.com

M. Hána
Thales Digital Identity & Security, Prague, Czech Republic
e-mail: martin.hana@thalesgroup.com

N. Kosmatov (✉)
Thales Research & Technology, Palaiseau, France
e-mail: nikolaikosmatov@gmail.com

© The Author(s), under exclusive license to Springer Nature Switzerland AG 2024
N. Kosmatov et al. (eds.), *Guide to Software Verification with Frama-C*, Computer Science Foundations and Applied Logic, https://doi.org/10.1007/978-3-031-55608-1_16

triggering malicious code execution are only a few examples of modern security threats.

Among various techniques used to prevent security flaws, formal verification of the target software code can be applied to ensure that the intended security protection mechanisms of the target software are correctly implemented and that the target software does not contain security vulnerabilities enabling the attacker to overcome those mechanisms.

To evaluate their compliance to expected security requirements, software products can undergo a security certification. An international standard for computer security certification is provided by the Common Criteria (CC) for Information Technology Security Evaluation [1]. Its highest Evaluation Assurance Levels EAL6–EAL7 require the developers to provide a formal Security Policy Model (SPM) and an associated mathematical proof of security properties using formal verification.

An example of highly critical software is smart card software since smart cards play an essential role in user authentication and controlling access to sensitive services and data. In the context of recent EAL6–EAL7 certification projects[1] of a smart card product, Thales conducted a successful formal verification of a JavaCard virtual machine [24] using the Frama-C verification platform [15]. This certification project was evaluated by CEA Leti (acting as IT Security Evaluation Facilities (ITSEF)) with the supervision of ANSSI (the French national cybersecurity agency, also being the French certification body), and the certificate was issued in 2021. The purpose of this chapter is to describe how formal verification with Frama-C was used in this industrial project.

Security of a smart card strongly relies on a set of isolation properties that must be ensured by the underlying JavaCard Virtual Machine (JCVM). According to the JavaCard protection profile published by the BSI [22], "an applet shall not read, write, compare a piece of data belonging to an applet that is not in the same context, or execute one of the methods of an applet in another context without its authorization". The corresponding access rules are usually implemented by a specific access control mechanism ensuring necessary isolation, called a *firewall*. Hence, target properties that must be verified include common security properties—such as integrity and confidentiality—that correspond to the required access control rules. Other examples of target properties are functional properties (partial for EAL6 and complete for EAL7 certification) and safety properties (such as the absence of invalid memory accesses and other runtime errors).

Another important benefit of formal verification is to ensure the correctness of the target software product and exclude the risk of bugs. The cost of a critical bug (that has to be fixed) is known to be high. In particular, for a smartcard product, it is very significant. It includes extra R&D effort to fix and qualify the bug, re-industrialization and re-certification of the product. It also requires a potential withdrawal of deployed products or a highly costly and complex patch deployment.

---

[1] EAL6 and EAL7 certificates are available, resp., at https://cyber.gouv.fr/sites/default/files/2021/11/certificat-anssi-cc-2021_42.pdf and https://cyber.gouv.fr/sites/default/files/document_type/Certificat-CC-2023_45fr_0.pdf.

This chapter presents the application of **Frama-C** to formal verification of a JavaCard virtual machine performed by Thales in an industrial certification project [9, 10]. The scope of verification includes the majority of functions of the JCVM with over 7,000 lines of C code. They were formally annotated in the **ACSL** specification language [5], and verified using the **Wp** and **MetAcsl** plugins of **Frama-C** [15]. Over 71,000[2] proof goals (verification conditions) were generated and successfully proved. We describe the main steps of this verification project, illustrate target security properties, present some lessons learned from this project, and outline future work directions and necessary tool enhancements. With respect to our earlier paper [9], this work presents a significantly extended version of the model ($\approx$71,000 proof goals compared to $\approx$52,000 proof goals in [9]), a detailed analysis of proof results, an extended presentation of lessons learned with new observations (e.g. on the usage of lemmas, proof scripts, other provers), and desired future improvements.

Related work. This work continues other formal verification efforts on the JavaCard platform. A classical verification approach used in several previous projects relies on a high-level formal model. Barthe at al. [3] propose an executable formal semantics of the JCVM and BCV (Bytecode Verifier, presented below in Sect. 16.2) realized in the **Coq** proof assistant. It contains 15,000 lines of **Coq** code. Nguyen and Chetali [19] introduce a refinement-based technique to demonstrate (again, using **Coq**) that a native JavaCard API function respects its specification. In general, in such case studies, the traceability between the expected security properties and the formally proven properties can be relatively hard to justify because of a larger distance between the formal model and the source code. An important difference of our approach with these previous efforts is that formal verification in our case is performed on the real-life code. This work is the first formal verification project for a successful certification up to the highest evaluation level (EAL7) where the real-life code was verified. Hence, our approach provides stronger guarantees of correctness for the product. We express all specified properties as **ACSL** annotations directly on the source code. Siveroni [27] and Eluard et al. [13] propose an operational semantics of a language modeling the JCVM behavior, including object ownership and the JavaCard firewall. This work is related to our formal specification. We also provide a full proof of target security properties on a real-life JCVM implementation. Previous efforts to ensure JavaCard platform security properties were also realized using other verification tools, such as **Krakatoa** [17], **Caduceus** [2] and **KeY** [18].

The present work is also broadly related to other verification case studies of real-life software [14]. The main goal remains to formally prove that the target program respects expected properties. Oortwijn and Huisman [20] present formal verification of an industrial traffic tunnel control system using the **VerCors** tool that mainly targets the verification of parallel programs. Among other recent case studies using **Frama-C**, we can also cite formal verification of the kLIBC library by Carvalho et al. [8]. Deductive verification of two real-life modules of the Contiki operating system was performed by Mangano et al. [16] and Blanchard et al. [6].

---

[2] All numbers are given for an extended version since the version described in [9, 10].

Dordowsky [11] describes an experience of using Frama-C on a real-world avionics example. Brahmi et al. [7] verify a large set of C functions of an avionics software [7]. Ebalard et al. [12] verify the X.509 parser and prove the absence of runtime errors in it. A novel approach for verification of global (in particular, security) properties with Frama-C using pervasive properties (expressed as *High-Level* ACSL *Requirements* (Hilare) or *metaproperties*) and the MetAcsl plugin was proposed by Robles et al. [25, 26]. In general, large functional properties become harder to prove as the code becomes larger and contains more low-level features. In our work, we focus on the challenging goal to prove global security properties on the real-life C code.

Outline. Previous chapters presented in detail the Frama-C plugins Wp, Rte, MetAcsl and the specification language ACSL (see Chaps. 1, 2, 4, and 10, respectively). This chapter is organized as follows. First, Sect. 16.2 provides necessary background on JCVM and its security properties. Section 16.3 gives an overview of the verified code with an illustrative example. Section 16.4 describes security property specification and verification with Wp and MetAcsl plugins of Frama-C. Section 16.5 presents our proof results. Section 16.6 outlines some lessons learned from this project. Finally, Sect. 16.7 discusses possible future work directions and improvements.

16.2 General Presentation of the JavaCard Virtual Machine

JavaCard Virtual Machine (JCVM) is a crucial component present on most current smartcards, where it plays an important role due to both functional and security aspects. The main functional goal of JCVM is to interpret *JavaCard bytecode*, that is, a sequence of opcodes, each one specifying one operation of the entire JavaCard program. Such operations include for instance arithmetic operations on the stack, writing a value from the stack into a heap object, an allocation of a new object of a given class, method invocations, etc. Thanks to this concept, interoperability is achieved, and the same JavaCard bytecode can be run without recompilation on JCVMs of different vendors. The only part that remains chip-dependent is the JCVM itself.

Several memory areas are managed by JCVM: *Java stack*, *data heap* and *code area*. As a result of Java compilation and a subsequent JavaCard conversion, we obtain a file in a special binary format optimized for a smartcard usage: a *CAP file*. When such a binary is loaded to the card, it is linked to already present code, which is imported by the loaded CAP file. After this process, the binary code remains immutable in the code area for the rest of its life cycle. This can be ensured by different means depending on the product (e.g. by a memory mapping).

Unlike the code area, the data heap can be modified dynamically when the bytecode is interpreted. It serves to store all static and instance data of the applet. JavaCard provides three types of heap memory, depending on the required life time of the data.

Transient deselect data is erased automatically when the owning applet is deselected, *transient reset* data is cleared during a card reset, while *persistent* data is preserved in all cases.[3]

Another important memory structure, Java stack, is used as a temporary location for data processing and argument passing. For example, the opcode `baload` is used to read a byte at a specific offset of an array of bytes. Before `baload` is called, other opcodes push the necessary arguments on the stack: the required offset and object reference. During `baload` interpretation, the object reference is used to find the corresponding object on the data heap. Security checks are then applied, for instance, to prevent an overflow of the array bounds. If all checks pass successfully, the value is read from the position in the given array determined by the given offset. All arguments are popped from the stack and the newly read value is pushed to the stack instead of them.

The main security role of JCVM is to ensure mutual applet isolation. Indeed, applets of different, mutually non-trusted vendors can be present on the card at the same moment. In particular, JCVM guarantees that data heap areas dedicated to an applet of one CAP file will not be read or modified by an applet of another CAP file. This isolation property is ensured by an oncard software component called a JavaCard firewall [23]. A firewall check is based mainly on the CAP file *context*, a unique value attributed to each CAP file during its loading to the card. Each allocated object is associated with such a context referred to as the object's *owner* (which is usually stored in the object's header). There are some well-defined exceptions to the firewall checks (static variables, global arrays, system context and, since JavaCard v.3.1, ArrayViews), but we will ignore them in this chapter for simplicity.

To make a first step towards a formalization of isolation properties, we can list the following examples of security properties, distinguishing confidentiality and integrity aspects:

$\mathcal{G}_{\text{integ}}^{\text{head}}$: Headers of already allocated objects cannot be modified during a VM run.

$\mathcal{G}_{\text{integ}}^{\text{data}}$: An element of a (persistent or transient reset) array can be modified only if the accessing context is the owner of the accessed object.

$\mathcal{G}_{\text{conf}}^{\text{data}}$: An elements of a (persistent or transient reset) array can be read only if the accessing context is the owner of the accessed object.

So, in the aforementioned example of the `baload` opcode, security checks must include a firewall check that the accessing context is the owner of the accessed object.

To understand the main principles of JavaCard security (and consequently the formal verification of JCVM), it is necessary to explain a mutual dependency between JCVM and another component of the JavaCard environment called *Bytecode Verifier* (BCV). BCV simulates the bytecode, but it focuses only on Java types, investigates all possible paths through the code and checks its type safety [21]. It depends on a concrete JCVM implementation whether it is defensive enough to provide isolation even if type safety was not checked (i.e. BCV is not applied), or if the BCV checks are assumed to be performed on the bytecode prior to its execution. In this chapter,

---

[3] The selection mechanism allows to deliver commands from the operating system to an applet [23].

we present a simplified example, in which we assume that type safety of the executed bytecode was checked by BCV. Therefore, to enable a successful proof, we introduce formal counterparts of BCV checks in the form of hypotheses in the ACSL specification of our example. An illustration of such hypotheses will be given below.

The dependencies between BCV and JCVM are bidirectional. As a hypothesis, BCV relies on the opcode specification [24], including the effect of opcodes on the memory managed by JCVM (e.g. number of slots popped and pushed on the Java stack). It is therefore essential to check that the functional specification of each opcode function assumed by BCV is indeed ensured by the implementation of each opcode, so that we can introduce properties guaranteed by BCV checks as hypotheses for the proof of JCVM.

16.3 Overview of the Secure Policy Model

In our certification approach (presented in detail in a recent paper [10]), the secure policy model is based on the real-life C code of the virtual machine. The code is annotated in the ACSL specification language [5] and then formally proved to respect the provided annotations using Frama-C/Wp. Let us illustrate our specification and verification approach on a toy example[4] initially presented in [9]. We split our example into Figs. 16.1, 16.2, 16.3, 16.4 and 16.5, where some less important fragments are omitted. The example gives a general idea of the approach but is not representative of the real-life code. It was strongly simplified to fit the chapter, and modified to prevent revealing real-life code features. It is too simple to provoke proof issues faced on real-life code, but it will allow us to illustrate when they occur and how we address them. We currently consider only one JCVM run,[5] and assume that allocated objects cannot be deleted, but new objects can be allocated.

16.3.1 Code and Heap Modeling

Figure 16.1 illustrates the modeling of the code and the heap. It is based on a certain number of program variables, ghost variables and validity predicates (linking program variables and ghost variables). Recall that *ghost variables* are variables added for the needs of the specification only: written in special comments and ignored by the compiler (and, hence, during program execution), they can be used inside ACSL annotations. Typically, they are used when extra variables are convenient to express some properties but adding them as program variables inside the C code is not desired. The set of such ghost variables is referred to as a *companion ghost*

---

[4] For convenience of the readers, the full example is available in the companion repository at https://git.frama-c.com/pub/frama-c-book-companion/-/tree/main/proof-of-security.

[5] This scope restriction implies that the selected application cannot be changed.

16 Proof of Security Properties: Application to JavaCard Virtual Machine

```
1  typedef unsigned char u1;
2  typedef unsigned short u2;
3  typedef unsigned int u4;
4  // === Code model and current Java context ===
5  #define CODE_SIZE 10000
6  u1   Code[CODE_SIZE], *JPC; // Code area and Java program counter
7  //@ ghost u4 gJPCOff;        // JPC offset in code area
8  u1   JCC;                    // Current Java context
9  // === Heap model ===
10 #define SEGM_SIZE 10000
11 #define MAX_OBJS 500
12 u1 ObjHeader[SEGM_SIZE];    // Object headers area: Header(8B),
13 //Header structure (8 Bytes), Bytes:Contents:
14 //   0:Owner, 1:Flags, 2-3:Class, 4-5:BodyOff, 6-7:BodySize
15 #define GET_OWN(addr)  ( *((u1*)addr + 0) )
16 #define GET_FLAG(addr) ( *((u1*)addr + 1) )
17 #define GET_OFF(addr) \
       ((u2)((*((u1*)addr+4))*256+*((u1*)addr+5)))
18 #define GET_SIZE(addr) \
       ((u2)((*((u1*)addr+6))*256+*((u1*)addr+7)))
19 u1 PersiData[SEGM_SIZE];    // Persistent objects data area
20 u1 TransData[SEGM_SIZE];    // Transient  objects data area
21
22 /*@ ghost // === Companion ghost memory model ===
23   u4 gNumOfObjs;             // Number of allocated objects
24   u1 gIsTrans   [MAX_OBJS];  // Nonzero for transient object
25   u4 gHeadStart [MAX_OBJS];  // Start offset of object header
26   u4 gDataStart [MAX_OBJS];  // Start offset of object data
27   u4 gDataEnd   [MAX_OBJS];  // End offset of object data
28   u4 gCurObj;    */          // Currently considered object number
```

Fig. 16.1 Illustrative example of JCVM: code and heap modeling

model. We start the names of ghost variables with a g. Validity predicates ensure the consistency of the corresponding model and will be maintained by all functions.[6]

Lines 1–3 in Fig. 16.1 define unsigned integer types containing 1, 2 and 4 bytes. A simple code model is defined by lines 5–8, where Java program counter JPC will be assumed to refer to an element inside the Code array as specified by the code model validity predicate (see lines 30–31 in Fig. 16.2). The offset gJPCOff is a ghost variable (line 7). It is used to avoid the need for an existentially quantified offset (e.g. in the validity predicate) and thus facilitates automatic proof. The current Java context is stored in JCC (line 8). The code validity predicate is maintained by all functions.

Lines 10–28 in Fig. 16.1 show a simplified model of the heap, in which we model only persistent and transient (reset) objects. We consider three separate memory segments: for objects headers (line 12), persistent object data (line 19) and transient object data (line 20). A header contains 8 bytes including the object's owner context (1 byte), flags (with a bit-level encoding inside 1 byte), class reference (2 bytes),

---

[6] Unless there is a critical exception and the JCVM execution is aborted.

```
29 /*@ // === Validity predicates ===
30 predicate valid_code_model = 0 <= gJPCOff < CODE_SIZE &&
31    JPC == &Code[gJPCOff];
32 predicate valid_heap_model =
33    0 <= gNumOfObjs <= MAX_OBJS &&
34 // headers of allocated objects are within ObjHeader segment
35    (\forall integer i; 0 <= i < gNumOfObjs ==>
36       0 <= gHeadStart[i] <= SEGM_SIZE - 8) &&
37 // no overlapping between headers (each header has 8 bytes)
38    (\forall integer i,j; 0 <= i < j < gNumOfObjs ==>
39       ( gHeadStart[i]>=gHeadStart[j]+8 ||
40         gHeadStart[j]>=gHeadStart[i]+8 )) &&
41 // IsTrans[i] encodes if i-th object's transient bit is set
42    (\forall integer i; 0 <= i < gNumOfObjs ==>
43       (gIsTrans[i]<==>(GET_FLAG(ObjHeader+gHeadStart[i])&0x08)))&&
44 // data of allocated objects is within a data segment
45    (\forall integer i; 0 <= i < gNumOfObjs ==>
46       gDataStart[i] == GET_OFF(ObjHeader+gHeadStart[i]) &&
47       gDataEnd[i] == gDataStart[i] +
48       GET_SIZE(ObjHeader+gHeadStart[i]) - 1 &&
49       0 <= gDataStart[i] < gDataEnd[i] < SEGM_SIZE) &&
50 // no overlapping between persistent object data
51    (\forall integer i,j; 0 <= i < j < gNumOfObjs && !gIsTrans[i]
                  &&
52       !gIsTrans[j] ==> ( gDataStart[i] > gDataEnd[j] ||
53       gDataStart[j] > gDataEnd[i] )) &&
54 // no overlapping between transient object data
55    (\forall integer i,j; 0<=i<j<gNumOfObjs && gIsTrans[i] &&
56       gIsTrans[j] ==> ( gDataStart[i] > gDataEnd[j] ||
57       gDataStart[j] > gDataEnd[i] )); */
58 // Lines 59-66 give declarations of functions updateJPC, etc.
```

Fig. 16.2 Illustrative example of JCVM: validity predicates

as well as the start offset of the object data (body) and its size, each over 2 bytes (cf. lines 13–14). Macros on lines 15–18 extract header fields that are used in our illustrative example. Ghost variable gNumOfObjs specifies the number of allocated objects. Allocated objects are supposed to be numbered starting from 0. For the object of index i, the offset of its header (in array ObjHeader) is modeled by a ghost array element gHeadStart[i], while gDataStart[i] and gDataSize[i] contain the offset and size of its body, located in one of the data segments (persistent or transient). The ghost array element gIsTrans[i] is nonzero iff the i-th object has transient data.

The consistency of the heap model is specified by the heap model validity predicate (lines 32–57). Line 33 specifies the interval of values for the number of allocated objects. Lines 34–40 state that object headers are within the bounds of the corresponding segment and do not overlap. Similarly, lines 44–57 state that object bodies are within the bounds of the corresponding data segments, the offset and size given in the header are correctly represented in the companion ghost model by arrays gDataStart and gDataSize, and two object bodies located in the same

```
67  /*@ // === A security property: object headers remain intact ===
68  predicate object_headers_intact{L1, L2} =
69    \forall integer i, off; 0 <= i < \at(gNumOfObjs,L1) &&
70      \at(gHeadStart[i],L1) <= off < \at(gHeadStart[i],L1) + 8 ==>
71      \at(ObjHeader[off],L1) == \at(ObjHeader[off],L2);
72
73  // === Memory footprint predicate and lemma example ===
74  predicate mem_model_footprint_intact{L1,L2} =
75    \at(gNumOfObjs,L1) <= \at(gNumOfObjs,L2) &&
76    ( \forall integer i; 0 <= i < \at(gNumOfObjs,L1) ==>
77      \at(gIsTrans[i],L1) == \at(gIsTrans[i],L2) &&
78      \at(gHeadStart[i],L1) == \at(gHeadStart[i],L2) &&
79      \at(gDataStart[i],L1) == \at(gDataStart[i],L2) &&
80      \at(gDataEnd[i],L1) ==\at(gDataEnd[i],L2) );
81
82  lemma vhm_preserved{L1,L2}: mem_model_footprint_intact{L1,L2} &&
83    object_headers_intact{L1,L2} && valid_heap_model{L1} &&
84    \at(gNumOfObjs,L1)==\at(gNumOfObjs,L2) ==>
          valid_heap_model{L2};
85  */
```

Fig. 16.3 Examples of a security property, a footprint-related predicate and a lemma

segment do not overlap. Lines 41–43 state that the truth value of the ghost array element gIsTrans[i] corresponds to the *transient bit* of the object flag. It is obtained from the flag byte with mask 0x08. If set, it indicates that the object data is located in the transient segment, otherwise in the persistent segment.

This predicate illustrates how we model the heap memory using a companion ghost model. Here, ghost variables are very convenient since some parts of the companion ghost model are not readily available in the C code. The heap validity predicate is maintained by all functions, including functions for new object allocation. In particular, in an allocation function (not shown in the chapter), the behavior when a new object is successfully allocated can be easily specified by stating that the number of objects is incremented: the validity predicate automatically ensures the necessary modeling assumptions for the new object (thanks to the universal quantifications) without the need to specify them separately for it.

Another benefit of the companion ghost model is to overcome proof scalability issues due to bit-level operations. Straightforward specification of the code with bit-related operations does not scale in our case study: automatic proof fails for many properties over the real-life code when numerous bits are involved. As illustrated for the transient bit, we duplicate the bit-level information by boolean ghost variables and maintain their equivalence (see lines 41–43). By expressing annotations using the ghost variable rather than the transient bit, we provide the provers with a parallel, companion view of bit-level information. It enhances their capacity of automatic proof in our project.

Another difficulty faced in our project is related to heterogeneous pointer casts, that is, casts between different pointer types. The definitions of macros of lines 17–18 in real-life code would use such casts:

```
86  /*@
87    requires vhm: valid_heap_model;
88    requires 0<=gCurObj<gNumOfObjs && ObjRef==gHeadStart[gCurObj];
89    assigns \nothing;
90    ensures \result <==> ( GET_OWN(ObjHeader+ObjRef) == JCC &&
91      gDataStart[gCurObj] + DestOff <= gDataEnd[gCurObj] );
92  */
93  u1 firewall(u4 ObjRef, u4 DestOff){
94    if(GET_OWN(ObjHeader+ObjRef) == JCC &&
95       DestOff < GET_SIZE(ObjHeader+ObjRef))
96      return 1;
97    return 0;
98  }
99  /*@
100   requires vhm: valid_heap_model;
101   requires vcm: valid_code_model;
102   admit requires 0 <= gCurObj < gNumOfObjs &&
103     ObjRef == gHeadStart[gCurObj];
104   assigns PersiData[0..(SEGM_SIZE-1)],
105     TransData[0..(SEGM_SIZE-1)], JPC, gJPCOff;
106   assigns JPC \from &Code[0]; // possible base address
107   ensures vhm: valid_heap_model;
108   ensures vcm: valid_code_model;
109   ensures oh: object_headers_intact{Pre,Post};
110   ensures mmf: mem_model_footprint_intact{Pre,Post};
111 */
112 void bastore(u4 ObjRef, u4 DestOff, u1 Val)
113 {
114   if( ! firewall(ObjRef,DestOff) )         // Check access and
115     return;                                 // exit if forbidden
116   if( GET_FLAG(ObjHeader+ObjRef) & 0x08 ) // If trans., write to
117     TransData[GET_OFF(ObjHeader+ObjRef)+DestOff]=Val; // tr.body
118   else                                     // Otherwise, write to
119     PersiData[GET_OFF(ObjHeader+ObjRef)+DestOff]=Val; //pers.body
120   updateJPC();
121 }
122 //Lines 123-196 contain other opcode functions and main_loop
```

Fig. 16.4 firewall and bastore functions with their ACSL contracts

```
197 /*@ // ===Metapr.: pers.object data written/read only by owner===
198 meta \prop,\name(meta_persi_objects_integrity),
199   \targets(\ALL),\context(\writing),
200   ( \forall integer i; 0 <= i < gNumOfObjs && !gIsTrans[i] &&
201   ObjHeader[gHeadStart[i] + 0] != JCC ==>
202   \separated(\written,PersiData+(gDataStart[i]..gDataEnd[i])) );
203 meta \prop,\name(meta_persi_objects_confident),
204   \targets(\ALL),\context(\reading),
205   ( \forall integer i; 0 <= i < gNumOfObjs && !gIsTrans[i] &&
206   ObjHeader[gHeadStart[i] + 0] != JCC ==>
207   \separated(\read,PersiData+(gDataStart[i]..gDataEnd[i])) ); */
```

Fig. 16.5 Metaproperties for persistent object data integrity/confidentiality

16 Proof of Security Properties: Application to JavaCard Virtual Machine   669

```
#define GET_OFF(addr)   ((u2)(*(u2*)(addr + 4)))  // real-life code
#define GET_SIZE(addr)  ((u2)(*(u2*)(addr + 6)))  // real-life code
```

In order to be able to use the Typed memory model of Wp, we rewrite such casts equivalently as shown on lines 17–18 of Fig. 16.1. The equivalence of rewriting can be checked separately, for instance, by an exhaustive enumeration. The need for this rewriting is explained by the fact that the Typed memory model of Wp [4] is both sound and efficient, but it is not able to support heterogeneous pointer casts. Lower-level models are either unsound or unable to reason efficiently on the code of our case study. Introducing ghost variables to store the resulting casted values (cf. lines 46–48) and using those ghost variables in annotations instead of cast-based operations was also beneficial for automatic proof.

Overall, the companion ghost model in our project has two main goals. First, memory-related properties are conveniently expressed using its ghost variables. Second, it facilitates automatic proof for bit-level operations and heterogeneous casts (rewritten with arithmetic operations).

16.3.2 Examples of Functions

The virtual machine executes opcodes in a main loop (also called a *dispatch loop*), where the next opcode is read from the code area (possibly with arguments) and the corresponding opcode function is called to execute the required actions. As we mentioned, the validity predicates are typically maintained by all opcode functions. They are also maintained as loop invariants of the dispatch loop.[7] We illustrate here simplified versions of an opcode function and the firewall function (see Fig. 16.4).

Function `bastore` (see lines 112–121 in Fig. 16.4) writes a given value into a given object at a given offset. The assigns clause (lines 104–105 in Fig. 16.4) specifies that the function can modify the contents of the persistent and transient memory and the Java program counter (with the associated offset ghost variable). For simplicity, we do not model the Java stack[8] in this simple example, so the arguments read from the stack in the real-life code are passed as function parameters in our example.

The `bastore` function calls the firewall to check the access, then tests the transient bit to choose the data segment where the object body is located, writes the target memory location and finally moves the program counter to a next opcode (using the function `updateJPC`, omitted here). The firewall function (see lines 93–98 in Fig. 16.4) allows the access if the current context is the object owner and the destination offset is within the bounds of the target object (as required by security properties $\mathcal{G}_{\text{integ}}^{\text{data}}$, $\mathcal{G}_{\text{conf}}^{\text{data}}$ in Sect. 16.2). The corresponding postcondition is stated by

---

[7] Except for a few specific cases such as critical errors that abort the VM execution, which are not detailed here.

[8] In the real-life code, this simplification is not made, therefore additional stack-related properties (such as the absence of overflows in method frames and system blocks) are verified to exclude the corresponding security vulnerabilities.

lines 90–91 in Fig. 16.4. Line 106 indicates the base address of memory locations pointer JPC can refer to after the function call. This information is used by an alias analysis in Wp recently introduced for pointers modified inside the function (see [4, Sect. 3.6]).

We mentioned in Sect. 16.2 that our simple example assumes that type safety of the bytecode has been verified by BCV, hence we introduce hypotheses based on the BCV checks. Such a hypothesis is illustrated by lines 102–103 of Fig. 16.4. They show a BCV assumption introduced by an **admit requires** clause, expressing a precondition assumed to be true without proof. It states that we assume the parameter ObjRef to refer to the beginning of an array header inside gHeadStart, which guarantees its correct alignment. Indeed, without this assumption, a correct alignment can be lost, and decisive fields inside the header (e.g. the object's owner) can be misinterpreted and access to non-authorized data can be granted.

Security properties and related clauses in the function contracts will be detailed in the next section.

16.4 Security Properties: Predicates and Metaproperties

In this section, we focus on specification and verification of security properties. They are specified in two ways, using predicates and metaproperties.

Figure 16.3 presents two predicates and a lemma. The first predicate (lines 68–71) states that the object headers of allocated objects remain unchanged between two given labels (that is, program points) L1, L2. It is used to specify the security property $\mathcal{G}_{integ}^{head}$ of Sect. 16.2 in function contracts, as illustrated by line 109 in Fig. 16.4.

The second predicate (lines 74–80 in Fig. 16.3) specifies that the variables of the companion model do not change between labels L1, L2 for objects that were already allocated at label L1, but new objects can have been allocated. As we mentioned, we do not consider object deletion. This second predicate is also used in the postconditions as illustrated by line 110 in Fig. 16.4. We say that this predicate expresses preservation of the *memory model footprint*. We emphasize that for a sound model, it is important to carefully specify the conditions under which all (program or ghost) variables used to specify target security properties can be modified.

Notice that for the firewall function, the indication of these additional properties is not needed since this function cannot modify any non-local variables, as specified by the **assigns \nothing** clause on line 89.

The lemma shown in Fig. 16.3 helps to propagate a validity predicate as long as companion ghost model variables and object headers are not modified. Such lemmas can be useful in particular in the dispatch loop, which reads the next opcode and chooses the opcode function to be called.

While some security properties can be stated as predicates, specification and verification of some other properties relies on metaproperties (see Chap. 10 for more detail on metaproperties and MetAcsl). Examples of metaproperties for integrity and confidentiality of persistent object data are shown in Fig. 16.5. They correspond to the persistent data case of properties $\mathcal{G}_{\text{integ}}^{\text{data}}$ and $\mathcal{G}_{\text{conf}}^{\text{data}}$ of Sect. 16.2. Instead of specifying directly that persistent object data is never written (resp., read), the main idea here is to specify that whenever some memory location is written (resp., read), that location is separated from any memory location inside persistent object data. The MetAcsl plugin automatically translates such properties into assertions in the C code at program points where some memory location is written (resp., read), and those assertions are then proved with Wp in the usual way.

16.5 Proof Results

In this section, we present proof results obtained in our project. These results were obtained by running Frama-C 25.0 (Manganese) on a virtual machine with Ubuntu 20.04, 12 cores and 64GB of RAM. Frama-C/Wp is run with option -wp-par 9, that is, up to 9 cores are given to Frama-C/Wp to run several solver instances in parallel. As an automatic prover, we use the SMT solver Alt-Ergo v.2.3.2 through Why3 v.1.5.0.

Given a C program with ACSL annotations, Frama-C/Wp automatically generates proof goals (also known as *verification conditions* or *proof obligations*). Each proof obligation is an elementary proof task. Proving that the C program fully complies with its ACSL annotations requires to discharge all proof obligations. We use the Rte plugin (via the option -wp-rte) to generate additional annotations that ensure the absence of undefined behaviors (also called runtime errors), such as invalid memory accesses, arithmetic overflows, division by 0. Those annotations are also proved by Wp.

Proof statistics are summarized in Table 16.1. The lines of the table correspond to various proof techniques and tools used to discharge proof goals (see Column 1): control-flow graph (CFG) based analysis, the simplification engine Qed of Wp, the SMT solver Alt-Ergo and by manually created proof scripts. The last line gives the total statistics for all of them.

ACSL annotations are inserted into C code either manually by the user or automatically by Frama-C plugins (in our case, by MetAcsl and Rte). The Wp plugin also offers the possibility to automatically introduce specific ACSL annotations (called *smoke-tests*) into the C code in order to check consistency of the specification. Logical inconsistencies may be introduced by an error in ACSL annotations. A logical inconsistency renders any property provable, including false properties. Fortunately, smoke-tests offer the possibility to automatically check if such false properties are provable (typically, by trying to prove **false** with a small timeout). Although smoke-tests do not guarantee to detect all logical inconsistencies, still they are very useful to check for them automatically, especially for large formal verification projects. Indeed, even after a careful review of thousands of manually written lines of ACSL specification, a risk of inconsistency cannot be fully excluded.

Table 16.1 Proof results with Frama-C/Wp

Proved by	User-provided annot.			MetAcsl	Rte	Smoke tests	Total	
	#Goals	#Asserts	#Lemmas	#Goals	#Goals	#Goals	#Goals	Proof time
CFG	7	7	0	356	14	7	384 (0.5%)	A few seconds (≈0s/goal)
Qed	19,152	131	0	33,946	572	4	53,674 (**75.3%**)	2h 05m 13s (0.14s/goal)
Alt-Ergo	6,544	194	2	7,412	980	1,882	16,818 (23.5%)	5h 36m 21s (1.2s/goal)
Script	465	**187**	1	68	20	0	553 (0.7%)	0h 22m 00s (2.4s/goal)
All	26,168 (**37%**)	519	3	41,782 (**58%**)	1,586 (2%)	1,893 (3%)	**71,429**	User **9h 09m 37s** Real **3h 14m 38s**

Columns 2–4 of the table show the proof goals coming from manually written annotations (with a total number of goals for all kinds of annotations and—more specifically—the subsets of goals coming from assertions and lemmas, typically provided to help the proof). For instance, 7 goals are proved by CFG, all of which come from user-provided assertions. Columns 5–7 show automatically generated proof goals, created, respectively, for metaproperties by MetAcsl, for runtime errors by Rte and for smoke tests by Wp. Column 8 gives the total numbers of goals for all those kinds and per each proof technique used to discharge them. Overall, 71,429 proof goals were generated. For selected cells, a percentage with respect to the total number of goals is indicated in addition between parentheses.

Column 9 gives the proof time. Proof durations for each technique are computed based on the statistics reported by Wp (which does not include the full time with all steps, as we discuss below, but gives a good indication for comparison). An average proof time per goal is given in parentheses. The last cell of Column 9 gives the total proof session duration with Frama-C/Wp, with CPU user time and real time, computed by the time utility. Real time is less than user time because several solvers are run in parallel. Thus the total proof time is about 3 hours 15 minutes.

The presentation is organized around three main questions (Q1–Q3, stated below) related to specification effort, proof statistics, automation and efficiency. The answers are provided according to the results of our particular industrial use case. Their generalization and comparison to other use cases deserve a dedicated investigation and are left as future work.

? Question Q1: Characterize the specification effort

In our project, the verified code contains over 7,000 lines of C code. The global amount of user-provided ACSL annotations is about 40,000 lines of code. Annotations include 54 metaproperties. The total amount of ACSL annotations automatically generated by MetAcsl from metaproperties is almost 600,000 lines of code, leading to 41,782 proof goals. This clearly shows the benefits of MetAcsl to facilitate the specification of security properties. The total amount of ACSL annotations generated by Rte is about 4,330 lines of code, leading to 1,586 proof goals.

? Question Q2: Characterize the number and complexity of generated proof obligations based on the proof results

Number of proof goals. Overall, 71,429 proof goals were generated in our verification project. The number of generated proof obligations may be spread over different kinds of ACSL annotations. In our project, as shown in Table 16.1, the majority of proof obligations (58%) are necessary to prove 54 metaproperties. The majority of them is proved automatically, and only 68 required proof scripts. This confirms the interesting level of abstraction offered by metaproperties in order to express high-level security properties over more than 7, 000 lines of C code. It also shows that a high level of automation is reached for the resulting assertions, instantiating metaproperties at different program locations. Respectively 2% and 3% of proof obligations are necessary to prove Rte assertions and to perform smoke-tests. The verification of Rte assertions is important because the absence of runtime errors (or *undefined behaviors*) is an assumption for soundness of Wp. Their relatively small number is reassuring because the analyzed critical C code is deemed to avoid undefined behaviors. 37% of proof obligations correspond to the core of the formal properties stated by the user. It is interesting to note that in addition to function and loop contracts, it is still necessary to add a few local assertions and lemmas to help the provers, but their number is relatively low.

Complexity. The proof context of each proof obligation encodes a (partial) control-flow of a C function enriched with ACSL annotations known to be true (and taken as hypotheses). Results in Table 16.1 show several categories of proof obligations with increasing complexity:

- Proof obligations discharged automatically by Frama-C/Wp. While only 0.5% of proof obligations have been discharged by control-flow analysis (being discovered as unreachable), the majority of proof obligations (75.3%) have been discharged by the simplification engine Qed of Wp. It is all the

more interesting that the average proof time per proof obligation is very short: 0.14s. Notice however that there is no timeout set for Qed simplifications, and some simplifications may take a long time (as we discuss below).
- Proof obligations discharged automatically and externally by an SMT solver. 23.5% of proof obligations are discharged fully automatically by the external SMT solver Alt-Ergo. The average proof time per goal is 1.2s which is quite efficient. Note that while the timeout fixed to prove each goal is 250s (and 2s per smoke-test), 90% of goals discharged by Alt-Ergo are discharged in less than 60s.
- Proof obligations discharged semi-automatically with Wp proof scripts. Automatic proof was not possible for 0.7% of proof obligations in our proof results with a 250s timeout for Alt-Ergo. This amounts to 533 proof obligations. Here the time cost for each proof obligation has to be considered differently. Actually, while the time depicted in Table 16.1 is quite efficient (2.4s/goal), it represents only the average time taken by Wp to apply tactics described in proof scripts. It does not include the manual effort required to create each proof script. A *proof script* indicates the first proof steps (or tactics) to be applied on the proof obligation. Applying such tactics results in one or several sub-goals, being typically simpler to discharge for automatic provers. In our project we estimate the average manual effort time to create such scripts at 30min/goal. This estimation is very rough. On the one hand, a tricky script may take one or several days to be designed. On the other hand, some common script patterns may be efficiently duplicated for several proof obligations. Proof scripts also have a bigger maintenance cost compared to automatic proof: they should be archived, checked and often updated during the migration to new versions of verification tools or modifications of the source code.

! Attention: time measurement and provers

All proof obligations are simplified by Qed, the simplification engine of Wp. In particular, proof goals discharged by the external SMT solver are first simplified by Qed. The time taken by such simplifications and some other internal steps of Frama-C is not explicitly reported by Frama-C and is not detailed in Table 16.1. Such steps include kernel preprocessing and normalization (not measured duration), generation of annotations with MetAcsl and Rte annotation ($\approx$2 min), proof goal computation by Wp ($\approx$20 min), Qed simplifications which do not succeed to discharge proof obligations (not measured duration) in addition to aborted intermediate calls to external provers when a script is more efficient (not measured duration). This explains why the sum of time durations reported in Column 9 for different proof goals is not equal to the overall user time of the proof session reported by the time utility (in the last cell). The reported user time shows the complete duration of the proof session, so it includes all those steps.

For some goals, we noticed that Qed simplifications were very long and lasted 45min. As one reason for that, the simplification feature of Qed that eliminates trivial branches of conditionals was diagnosed to be very time consuming and has been deactivated in our project (with option -wp-no-pruning of Frama-C/Wp). The C code has also been partly rewritten in order to reduce the number of control-flow branches in most complex functions and thus to reduce the Qed time.

Note that only one external prover (Alt-Ergo v.2.3.2) is used in this project. For comparison, the results obtained with Alt-Ergo v.2.4.1 are globally very similar, slightly less efficient in terms of proof time. There is no time spent trying other solvers.

In another proof experiment, we applied three external provers at once (Alt-Ergo, Z3 and CVC4). It increased the total user time to 16h instead of 9h with Alt-Ergo alone (and the total real time to 6h instead of 3h15min). Hence, adding extra provers leads to a significantly bigger duration of the proof session. Interestingly, using additional provers did not allow us to prove more goals automatically and to avoid any scripts in our project.

? Question Q3: Characterize the factors that influence proof automation

In an ideal verification project, all proof obligations would be automatically discharged in a sufficiently short time. In reality, and especially in a large industrial project, the C and ACSL inputs have a dramatic impact on the proof results, in particular, through the complexity of generated proof obligations. In such cases, scalability of the proof with Frama-C/Wp can be limited because of the need to create a large number of proof scripts required to achieve a full proof. For some proof goals, it can be very hard for industrial engineers to be able to prove the target goal even with a complicated proof script. As already mentioned, maintainability of a large number of proof scripts also leads to additional costs. Fortunately, a careful design of our ACSL specifications in addition to a local refactoring of a few big C functions allowed us to significantly alleviate this concern in our project.

Careful ACSL specification. Dealing with low level operations on bitfields is challenging for automatic provers. A careful usage of a companion ghost model allowed us to create a suitable abstraction of such operations using integer ghost variables and to maintain it readily available in the proof context (as we illustrated in Sect. 16.3). **Alt-Ergo** was able to reason more efficiently on the companion ghost variables instead of bit-fields directly.

A careful definition of predicates to describe global validity and security properties contributed to structure the proof context of proof obligations. This structure is intended to improve the readability of ACSL annotations and proof goals, as well as the efficiency of external provers to discharge the resulting proof obligations. The proof obligations resulting from metaproperties also benefit from this structure and are efficiently discharged by automatic provers.

A careful organization of predicates also facilitates the creation of proof scripts. Unfolding only a subset of relevant predicates is typically needed in proof scripts (where the concrete subset depends on the target property). For instance, unfolding predicates related to the code area would be necessary to prove properties about the code area, while unfolding predicates related to the data heap would be necessary to prove properties about object data.

Some operations using bit-masks were identified as hindering the automatic proof of many proof obligations. As it is generally done in such cases, a first solution was to define lemmas for some well-defined operation patterns.

Lemmas versus assertions. An interesting observation was that too many lemmas were also hindering the automatic proof. In theory, this can be expected since extra lemmas can create additional possibilities of reasoning during the proof (such as expression rewriting), which can put the prover on the wrong path in some cases. On the other hand, putting the same lemma as local assertions in several functions where its statement is needed makes it necessary to reprove the same statement several times and can also lead to loss of efficiency. In our project, we tried to measure this trade-off more precisely.

We created a second version of specification by replacing most lemmas by local assertions in functions where their statements are needed to help the prover. Table 16.1 presents the results for this second version. They show that 519 assertions with only 3 lemmas were defined to guide the automatic proof in the second version instead of 61 lemmas in the first version. Interestingly, the assertions required 187 scripts, several of which shared the same patterns and were easy to duplicate. Compared to the first version intensively relying on lemmas, the overall proof session time decreased by 1h20min thanks to replacing most lemmas by assertions. This shows that the usage of precisely located assertions leads to better performances in our project than adding lemmas, which are present in the proof context of all goals even when it is not necessary.

Proof scripts versus automatic proof. It may be tempting to intensively use proof scripts to discharge proof obligations whenever a proof script can do it faster than an automatic prover. We adopted this approach at the earlier stages of the project. In reality, it is not always a good practice. Proof scripts are difficult to maintain and are more fragile when the source code and the

ACSL specification evolve. Interestingly, we observed that at a large scale the average time cost is bigger for proof scripts than for external provers. It can be due to time computation in Wp which does not include all steps executed for proof scripts (such as more frequent Qed simplifications, for each subgoal, as we already mentioned). From our experience, the best trade-off is to minimize the manual effort as much as possible and enhance the tool efficiency for the automatic proof. In other words, manual proof scripts should be reserved only for properties which are not proved automatically within the desired timeout (or a little less, in order to avoid unstable proof results between several proof runs).

An interesting research question is to better understand whether certain patterns of code or kinds of proof obligations are more likely to require proof scripts. Our initial observations suggest that the usage of bit-level operations and integer casts are examples of code features that increase the need for proof scripts. A more precise analysis of this question is left as future work.

16.6 Summary of Lessons Learned and Challenges

Our application of deductive verification on a large industrial C program shows that formal verification of real-life industrial code has become feasible today. The way to write the source code and its formal specification is crucial for a successful proof of real-life code. In our project, it requires a careful combination of several ingredients: companion ghost code, preservation properties, proof-guiding assertions or lemmas, and proof scripts. This combination made it possible to efficiently reason about non-trivial code fragments involving bitwise operations without the use of external interactive tools (e.g. Coq) with a high level of automatic proof. The majority of proof goals (almost 99%) are proved automatically by the Qed simplification engine of Wp and an automatic SMT solver. Qed simplifications are quite efficient to facilitate automatic proof for the majority of proof obligations (75.3% with Qed alone, 23.5% with both Qed and Alt-Ergo). The remaining goals are successfully proved with manually created proof scripts. Proof scripts in Wp provide a powerful solution to finish the proof in the remaining cases, but require a more important effort for verification engineers. MetAcsl proved to offer a convenient and efficient technique for specification and verification of security-related properties.

An efficient support from the developers of Frama-C during the whole project was essential for its success. Some anomalies were reported and fixed, and several new features were requested and implemented. Examples of such features include the implementation of check-and-forget versions of all annotations (i.e. verified but not kept in the proof context), their usage for annotations generated by MetAcsl, new tactics helping to prove bit-level properties, as well as precise generation of memory model hypotheses necessary for a sound proof [4, Sect. 3.6].

With our current state of knowledge, there is no perfect solution how to make a maximal number of proof obligations provable automatically and efficiently. For large industrial software, proof obligations may be very complex and their number may be very large. Some proof obligations are not discharged automatically, and among those that are, some are not proved efficiently in a reasonable time. This remains a major challenge. We discussed above some observations on the complexity of proof goals, mainly related to low level bit-field operations and complex conditional branches. We also highlighted some solutions we adopted to tackle these issues using well-structured predicates, a companion ghost model and code refactoring.

For verification engineers, having to wait several hours to get a full proof status is a serious scalability problem. Having a very powerful machine is mandatory to work in bearable conditions but is still insufficient for solving current proof scalability issues. Using cached proof results provides a solution for one user running the proof on a unique machine, but is not practical enough for collaborative proof development on several machines or for using separate (virtual) machines for continuous integration and proof replay. Indeed, cached proof results in Wp depend on the host and cannot be shared between several machines. Parallel proof features of Wp and Why3 (activated by option -wp-propl) make it possible to run several prover instances for several goals in parallel. However, some steps are currently not parallelized (such as parsing, proof obligation computation, Qed simplifications). This limits the benefits of the parallelization: Frama-C/Wp is not capable to serve several processes, so that increasing their number does not sufficiently decrease the real proof time. We observe it in our proof results: while we use option `-wp-prop 9` allowing up to 9 provers in parallel, the factor between user time and real time (see Table 16.1) is in reality less than ×3.

Deductive verification of real-life industrial code requires a deep expertise from verification engineers. It goes without saying that a solid knowledge of C and ACSL semantics is a must. A deep understanding of proof tool characteristics and limitations is also necessary. Proof debugging remains difficult since a proof failure may be due to a bad (either incorrect or not optimal) ACSL specification, a bug in the C code or a proof tool limitation. Strictly speaking, even the presence of a bug in the proof tool itself cannot be excluded, so a careful analysis of the code, specification and proof results in case of doubt can be necessary.

Creating proof scripts requires a good knowledge of interactive proof tactics and their capacity to facilitate automatic proof by an SMT solver for a given proof goal. Wp may decide to rewrite some parts of the proof goal (during simplifications) or to split some complex proof obligations into several subgoals if the proof context contains many branches. Hence, a certain experience is needed to trace the proof context of a proof obligation back to the C code and ACSL annotations.

Regarding Common Criteria certification, our certification approach (presented in detail in a recent paper [10]) brings several benefits. The security policy model is based on the real-life C code, annotated in the ACSL specification language and then formally proved to respect the provided annotations using Frama-C/Wp. This approach was used for the first time for EAL6–EAL7 certifications in this work, and is now recognized as a formal method for such certifications by ANSSI. Compared to other formal verification approaches based on a high-level model of the actually implemented security features, our approach facilitates the traceability of proved properties since the SPM is (a subset of) the real-life code itself, with the same structure (same functions, variables, data structures, etc.). Therefore, its verification provides stronger guarantees of correctness for the product.

Another benefit of the Frama-C based approach is a better maintainability (in particular, in case of minor code updates or scope extensions). The specification effort is very significant, but it can be partly amortized during the following certifications. Typically, integration of new properties or functions for an EAL7 certification can significantly rely on the proof performed for the EAL6 level. The proposed approach strongly benefits from automation, which is particularly important for a large industrial product. The link between the SPM and the real-life code is explicit and can be automatically exploited by various tools. The call graph of the analyzed code provides a good view of fully verified functions (with a contract and body), functions that are included as stubs (with a contract and a declaration only) and excluded functions for the SPM scope. This approach thus allows for a good understanding of the proved code and properties by the evaluators. An interested reader will find a deeper analysis of the proposed methodology in the context of Common Criteria certification and its comparison with other approaches in [10].

16.7 Future Enhancements

In this section, we outline some directions of improvements that we identified during our case study.

Interactive proof features are essential to finalize the proof for goals that remain unproven by automatic provers. The study of the manually created proof scripts leads to an interesting conclusion: proof contexts tend to be overloaded with hypotheses that are irrelevant for the target proof goal. This is an intrinsic property of a large verification project with global validity and security properties expressed as ubiquitous global invariants in the analyzed program. An automatic prover is blurred with too much irrelevant information involving too many variables and memory locations. The filter tactic of Wp is useful but not always sufficient. A solution may be to enhance it with a better capacity to track memory updates and to identify irrelevant hypotheses.

Our initial experiments suggest that more powerful and flexible tactics would be useful to make it easier for verification engineers to create and maintain proof scripts. One idea is to create a tactic language that would allow the user to specify proof steps in a more convenient way. It can include a more practical manual application of some lemmas or hypotheses, instantiation of statements, removing some hypotheses that the user identifies as irrelevant or, on the contrary, unfolding some predicates that the user identifies as relevant, etc. The possibility to try a candidate tactic (and ignore it if it cannot be applied without failing), like the try command in Coq, can be convenient. The graphical user interface for simplified tactic application in Wp should be preserved and possibly enriched with additional user-defined (sequences of) tactics. Making it possible for the user to indicate some (sequences of) tactics to try automatically for some (subsets of) goals can be investigated. Copying some script patterns from one script to another more easily would be desirable. Finally, solutions to decrease the fragility of scripts to code and specification updates should be investigated. This would facilitate proof script creation and maintenance.

A larger support of the C language in Frama-C and Wp is required. Manual code transformations were used in our work as a workaround to some tool limitations. Such transformations may potentially hide a defect in the real-life code if they are used improperly. It is thus extremely important to reduce their usage to a strict minimum and to carefully review the transformations (or ideally to automate them and prove their soundness). Examples of features to be supported include setjmp/-longjmp instructions. This would avoid code transformations to mimic the effect of such instructions on the control flow of analyzed programs. Support for heterogeneous pointer casts in Wp is another future work direction. Solutions based on code rewriting to avoid such casts and to remain in the typed memory model are pragmatic and can be automated. An ongoing work targets the development of a new plugin of Frama-C, called Uncast, to perform necessary code transformations in order to avoid heterogeneous pointer casts so that the resulting code can be proved with a high-level (Typed) memory model. Unfortunately, low-level memory models do not scale up well enough to large programs today. As a more challenging perspective, the development of collaborative memory models of several levels would allow the tool to apply various models for various parts of code and hence move to low-level memory models only when necessary.

Having precise proof statistics is very useful for verification engineers. The fact that Qed simplifications are not clearly accounted (in the reported proof time or for the timeout) is very inconvenient. Precise durations of various proof steps should be recorded for all provers and scripts. The ongoing improvement of proof statistics reporting in Wp should be further continued and evaluated.

An interesting observation in our project was the duration of Qed simplifications, going up to 45 min per proof goal. A workaround to this problem can be code refactoring, but this limits the capacity of Frama-C/Wp to prove real-life code. A careful analysis and optimization of Qed simplifications, possibly with deactivation of the most costly ones, can be another future work direction.

A better parallelization of a proof session in Frama-C/Wp is another important future work. It can partly compensate for the complexity of proof for large industrial projects and decrease the global duration of a proof session. As we mentioned above, today some steps remain sequential (code parsing, proof goal computation, Qed simplifications, etc.). A deeper parallelization and an automatic distribution of proof tasks (for subsets of functions, or even for subsets of goals) between several Frama-C instances could allow the user to take better advantage of modern multi-core architectures. A step further will be to run a network of several proof workers on several machines. A better support of cached proof results (that can be shared between several machines) and partial proof replay (automatically identifying modified proof goals) would facilitate incremental proof development and continuous integration.

Metaproperties and their tool support by **MetAcsl** play a decisive role for a successful specification and verification of security properties. This is particularly beneficial in a certification context, since a small number of metaproperties are easy to review, and their automatic translation into assertions by **MetAcsl** allows for their efficient instantiation and verification. In this context, an interesting future work direction is to integrate informal requirements and hypotheses as global annotation artifacts that can be linked to formal **ACSL** annotations and metaproperies. That would improve the traceability of requirements and allow a stronger link between informal and formal properties and hypotheses, possibly with an automatic generation of some parts of documentation for certification. This can also facilitate the maintenance of the model for the developers and its review for the evaluators.

References

1. Common criteria for information technology security evaluation. Part 3: security assurance components (2017). Technical report, CCMB-2017-04-003. https://www.commoncriteriaportal.org/files/ccfiles/CCPART3V3.1R5.pdf
2. Andronick J, Chetali B, Paulin-Mohring C (2005) Formal verification of security properties of smart card embedded source code. In: International symposium of formal methods (FM 2005), LNCS, vol 3582, pp 302–317. Springer. https://doi.org/10.1007/11526841_21
3. Barthe G, Dufay G, Jakubiec L, Serpette BP, de Sousa SM (2001) A formal executable semantics of the JavaCard platform. In: 10th European symposium on programming on programming languages and systems, (ESOP 2001), held as part of the joint European conferences on theory and practice of software (ETAPS 2001), LNCS, vol 2028. Springer, pp 302–319. https://doi.org/10.1007/3-540-45309-1_20
4. Baudin P, Bobot F, Correnson L, Dargaye Z, Blanchard A (2020) WP plug-in manual. https://frama-c.com/download/frama-c-wp-manual.pdf
5. Baudin P, Filliâtre JC, Marché C, Monate B, Moy Y, Prevosto V, ACSL: ANSI/ISO C specification language. http://frama-c.com/acsl.html
6. Blanchard A, Kosmatov N, Loulergue F (2019) Logic against ghosts: comparison of two proof approaches for a list module. In: Proceedings of the 34th annual ACM/SIGAPP symposium on applied computing, software verification and testing track (SAC-SVT 2019). ACM, pp 2186–2195. ACM Best Software Development paper award. https://doi.org/10.1145/3297280.3297495
7. Brahmi A, Carolus MJ, Delmas D, Essoussi MH, Lacabanne P, Lamiel VM, Randimbivololona F, Souyris J (2020) Industrial use of a safe and efficient formal method based software engineering process in avionics. In: Embedded real time software and systems (ERTS 2020)
8. Carvalho N, da Silva Sousa C, Sousa Pinto J, Tomb A (2014) Formal verification of kLIBC with the WP frama-C plug-in. In: NASA formal methods (NFM 2014), LNCS, vol 8430. Springer, pp 343–358. https://doi.org/10.1007/978-3-319-06200-6_29
9. Djoudi A, Hána M, Kosmatov N (2021) Formal verification of a JavaCard virtual machine with Frama-C. In: Proceedings of the 24th international symposium on formal methods (FM 2021), LNCS, vol 13047. Springer, pp 427–444. https://doi.org/10.1007/978-3-030-90870-6_23. Long version available at https://nikolai-kosmatov.eu/publications/djoudi_hk_fm_2021.pdf

10. Djoudi A, Hána M, Kosmatov N, Kříženecký M, Ohayon F, Mouy P, Fontaine A, Féliot D (2022) A bottom-up formal verification approach for common criteria certification: application to JavaCard virtual machine. In: Proceedings of the 11th European congress on embedded real-time systems (ERTS 2022)
11. Dordowsky F (2015) An experimental study using ACSL and Frama-C to formulate and verify low-level requirements from a DO-178C compliant avionics project. Electron Proc Theor Comput Sci 187:28–41. https://doi.org/10.4204/EPTCS.187.3
12. Ebalard A, Mouy P, Benadjila R (2016) Journey to a RTE-free X.509 parser. In: Symposium sur la sécurité des technologies de l'information et des communications (SSTIC 2019). https://www.sstic.org/media/SSTIC2019/SSTIC-actes/journey-to-a-rte-free-x509-parser/SSTIC2019-Article-journey-to-a-rte-free-x509-parser-ebalard_mouy_benadjila_3cUxSCv.pdf
13. Éluard M, Jensen T, Denne E (2001) An operational semantics of the Java Card firewall. In: Smart card programming and security. Springer, pp 95–110. https://doi.org/10.1007/3-540-45418-7_9
14. Hähnle R, Huisman M (2019) Deductive software verification: from pen-and-paper proofs to industrial tools, LNCS, vol 10000. Springer, pp 345–373. https://doi.org/10.1007/978-3-319-91908-9_18
15. Kirchner F, Kosmatov N, Prevosto V, Signoles J, Yakobowski B (2015) Frama-C: a software analysis perspective. Formal Aspects Comput. https://doi.org/10.1007/s00165-014-0326-7
16. Mangano F, Duquennoy S, Kosmatov N (2016) Formal verification of a memory allocation module of Contiki with Frama-C: a case study. In: Proceedings of the 11th international conference on risks and security of internet and systems (CRiSIS 2016), LNCS, vol 10158. Springer, pp 114–120. https://doi.org/10.1007/978-3-319-54876-0_9
17. Marché C, Paulin-Mohring C, Urbain X (2004) The KRAKATOA tool for certification of Java/JavaCard programs annotated in JML. J Logic Algeb Program 58(1–2):89–106. https://doi.org/10.1016/j.jlap.2003.07.006
18. Mostowski W (2007) Fully verified Java Card API reference implementation. In: 4th international verification workshop in connection with CADE-21, CEUR Workshop Proceedings, vol 259. CEUR-WS.org. http://ceur-ws.org/Vol-259/paper12.pdf
19. Nguyen QH, Chetali B (2006) Certifying native java API by formal refinement. In: 7th IFIP WG 8.8/11.2 international conference on smart card research and advanced applications (CARDIS 2006), LNCS, vol 3928. Springer, pp 313–328. https://doi.org/10.1007/11733447_23
20. Oortwijn W, Huisman M (2019) Formal verification of an industrial safety-critical traffic tunnel control system. In: 15th international conference on integrated formal methods (IFM 2019), LNCS, vol 11918. Springer, pp 418–436. https://doi.org/10.1007/978-3-030-34968-4_23
21. Oracle (2002) Java card 2.2 off-card verifier, whitepaper. Technical report, Oracle. https://www.oracle.com/technetwork/java/embedded/javacard/documentation/offcardverifierwp-150021.pdf
22. Oracle (2020) Java Card system—open configuration protection profile, version 3.1. Technical report, Oracle. https://www.bsi.bund.de/SharedDocs/Downloads/DE/BSI/Zertifizierung/Reporte/ReportePP/pp0099V2b_pdf.pdf;jsessionid=6C3F5A7FB5FA0D928A1C310C1C0EF1CE.internet462?__blob=publicationFile&v=1
23. Oracle (2021) Java Card platform: runtime environment specification, classic edition, version 3.1. Technical report, Oracle, Oracle. https://docs.oracle.com/javacard/3.1/related-docs/JCCRE/JCCRE.pdf
24. Oracle (2021) Java Card platform: virtual machine specification, classic edition, version 3.1. Technical report, Oracle, Oracle. https://docs.oracle.com/javacard/3.1/related-docs/JCVMS/JCVMS.pdf
25. Robles V, Kosmatov N, Prevosto V, Rilling L, Le Gall P (2019) Tame your annotations with MetAcsl: specifying, testing and proving high-level properties. In: International conferences on tests and proofs (TAP), LNCS, vol 11823. Springer, pp 167–185. https://doi.org/10.1007/978-3-030-31157-5_11

26. Robles V, Kosmatov N, Prevosto V, Rilling L, Le Gall P (2021) Methodology for specification and verification of high-level properties with MetAcsl. In: Proceedings of the 9th IEEE/ACM international conference on formal methods in software engineering (FormaliSE 2021), pp 54–67. IEEE. https://doi.org/10.1109/FormaliSE52586.2021.00012
27. Siveroni IA (2004) Operational semantics of the Java Card Virtual Machine. J Logic Algeb Program 58(1–2):3–25. https://doi.org/10.1016/j.jlap.2003.07.003

Index

Symbols
!, 26
==, 26
==>, 9, **26**
_ ? _ : _, **27**
&&, **26**
<==>, **26**
⊔, *see* Join operator146
⊑, *see* Inclusion operator
∇, *see* Widening

A
`-absolute-valid-range`, 183, **109**, 622, 623, 631, 632
Abstract heap, 511
Abstract interpretation, 136, **141**, 217, 248, 342, 344, 431, 437, 438, 488, 489, 491, 495, 501, 517, 518, 522, 544, 554, 556, 618, 636, 643
Abstract state, **144**, 162
Abstract syntax tree, *see* AST, *see*
Abstract value, **162**
ACSL, **3**
ACSL++ *(language)*, 87
Ada *(language)*, 6, 264, 265
AddressSanitizer *(tool)*, 265, 283, 285, 286, 295, 296
`admit` clause, **34**
AEB, 587
Alarm, 95, 101, 103, 107, 108, 113, 114, 121, 408, 619, 621, 656
  conversion overflow, **179**
  dangling pointer, **179**
  division by zero, **179**
  false, **132**, 133, 437, 619, 629

integer overflow, 134, **179**
invalid memory access, 134, **178**
invalid pointer arithmetic, **178**
invalid shift operands, **179**
shift of negative integers, **179**
special floating-point values, **179**
true, 437, 619
uninitialized memory, 134, **179**
Alias *(Frama-C plug-in)*, **90**, 433
Aliasing, 90
Alt-Ergo *(tool)*, 189, 195, 229, 440, 522, 543, 548, 586, 655, 671, 672, 674–677
Analysis script, 620, 621
Anomaly form, 599, 608
ANSI-ISO C Specification Language, *see* ACSL
ANSSI *(agency)*, 660
Aoraï *(Frama-C plug-in)*, 88, 185, 277, 278, 444, 457, 458, 473, 475–484
`-aorai-automata`, 277
API, **344**, 345, 350
Apron *(library)*, 158, 161
Arithmetic promotion, 102
Array, 564, 576
ASan *(tool)*, 283, 285
Assembly code, 106
`assert` clause, 629
`assert` statement, 559, **560**, 563, 564, 568, 574
Assertion (ACSL), **34**
`assert` clause, **34**
`assigns` clause, **19**, 176, 288, 577, 652–654
`assumes` clause, 22
`assume` statement, 559, **560**, 564, 568

AST, **82**, 92, 94, 95, 99, 100, 103–105, 110, **342**, 344, 351, 358, 361, 371, 404, 408

\at, **35**, 267, 288, 300
Attribute, 96
Audit mode, **119**
-audit-check, **121**
-audit-prepare, **119**, 622, 623
AutoCorres *(tool)*, 10
Automatic Dependent Surveillance Broadcast, 586
Automaton, 473, 474, 476
  deterministic finite, 644
axiom, **65**
axiomatic, **65**, 268
Axiomatic definition, 227

B

B *(language)*, 256
Backward analysis, 389
\base_addr function, **49**
bash *(language)*, 412
BCV, 661, 663, 664, 670
behavior, **22**
Behavioral interface specification language, 5, 264, 295
BISL, *see* Behavioral interface specification language, *see*
Bit-vector, 563
BLAST *(tool)*, 257
Blast *(tool)*, 452
\block_length function, **49**
Boogie *(tool)*, 254
_Bool, 108
Boolean, 565
boolean type, **59**
Branch coverage, **306**
break statement, 41
Build EAR *(tool)*, 93, 418
Bytecode Verifier, 661

C

C++ *(language)*, 87, 93, 253, 587, 634, 636, 641
C# *(language)*, 265
Caduceus *(tool)*, 5, 661
CaFE *(Frama-C plug-in)*, **86**, 88
Call graph, 131
Callgraph *(Frama-C plug-in)*, **89**, 590, 592, 593, 605, 608, 609
CaRet *(language)*, 86
Cartesian Hoare logic, 484

Cast, 102
  invalid, 651
CAVEAT *(tool)*, 212
CBMC *(tool)*, 452, 636, 638, 639, 642, 643
CCured *(tool)*, 517
CEA *(agency)*, 543, 546
CEA *(company)*, 634
CEA Leti *(institute)*, 660
CEGAR, 424, 426, 427, 556
CegarMC *(Frama-C plug-in)*, 425, 427–429, 431–433, 451
Certification, 660, 681
CFG, **365**, 382, 388, 389
CH2O *(tool)*, 10
CHC, 554, **560**, 561, 563, 575–577
-check, 396
Check'n'Crash *(tool)*, 452
CheckedC *(tool)*, 517
Check**check** clause, **34**
CI, 83, 128, 300
Cil *(library)*, 82, 94, 428
Clang *(tool)*, 83, 93, 98, 273, 413
Cleanup, 95
cloc *(tool)*, 406
CMake *(tool)*, 93, 418
cmake *(tool)*, 418
Code coverage, 406
Code review activity, 607
Code specialization, 87
Codex *(tool)*, 487, **488**, 488, 489, 506, 518
Coding rule, 633
Coercion, 292
COLIBRI *(tool)*, 310
Colibri-2 *(tool)*, 547
Collaboration
  parallel, **84**, 99
  sequential, **84**, 99, 109
Command line, 82, 92
Common Criteria for Information Technology Security Evaluation, 660
Common sub-expression elimination, 301
Compact Position Reporting, 586
CompCert *(tool)*, 10, 256
complete behaviors, **23**
Complete execution, 12
Concolic execution, 634, 636, 643
Concretization function, 145, 501
Concurrency, 283, 618, 625, 656
Condition coverage, **306**
Condition-decision coverage, **306**
Conditional (ACSL), **27**
Confidentiality, 458, 459, 660, 663, 670
Conformity (ACSL), **16**

Subject Index 687

Conjunction (ACSL), **26**
Consistency, 117
Constant propagation, 301
`-constfold`, 95
Constfold *(Frama-C plug-in)*, **90**
Containment relation, **507**
Context-sensitive analysis, 132
continue statement, 41
Continuous integration, *see* CI, *see*
Contract, 554, 555, 561, 564, 568, 573, 575,
   624, 630
   inference, 555, 564, 573, 577
Control-flow graph, 90, *see* CFG
Convergent filter, 522–525, **525**, 540
Conversion
   implicit, 102
Coq *(tool)*, 10, 189, 252, 256, 473, 529, 543,
   548, 661, 677, 680
Correctness, 554, 561, 567
   partial, **16**, **557**
   total, **557**
Counterexample, 424, 426, 444, 447, 448,
   452, 563
Counterexample guided abstraction refinement, *see* CEGAR, *see*
Coverage criterion, 287, 305, **306**, 306, 635
CPAchecker *(tool)*, 257, 427–429, 451
`-cpp-command`, **93**
`-cpp-extra-args`, 110, 622, 623, 633
`-cpp-extra-args`, **93**, 359
`-cpp-extra-args-per-file`, **93**
`-cpp-extra-args-per-file`, 416
CRC, *see* Cyclic redundancy check
C-Reduce *(tool)*, 420, 421
Cross-over, 635, 638
CSV *(format)*, 89, 123, 124, 601, 619
CURSOR *(tool)*, 282
CUTE *(tool)*, 335
CVC4 *(tool)*, 189, 522, 655
CVC5 *(tool)*, 189
Cyclic redundancy check, 625
Cyclomatic complexity, 406, 596
Cyclone *(tool)*, 517

D

Dafny *(tool)*, 452
Daikon *(tool)*, 452
Dassault Aviation *(company)*, 287
Dataflow analysis, 289, 296, 342, 344, 347,
   365, 376, 388, **389**, 409
DD, 591, 598
Dead code, 90, 621, 633

Debugging, 403
Decision coverage, *see* Branch coverage
decreases clause, **47**
Deductive verification, 187, **189**, 189, 190,
   196, 198, 215, 217, 233, 237, 239, 240,
   245, 253–256, 258, 266, 336, 424, 432,
   434, 440, 443, 446, 449, 451, 452, 495,
   517, 522, 554, 561, 643, 677
Defensive programming, 264
Dependency graph, 89
Design by Contract, 264
DFA, *see* Automaton, deterministic finite
DFD, 598
disjoint behaviors, **23**
Disjunction (ACSL), **26**
Dive *(Frama-C plug-in)*, 89, 90, 181, 404,
   409–411
Division by zero, 292, 643
Domain
   Apron
      linear equalities, 161
      octagons, 161
      polyhedra, 161
   abstract, **142**, 493
   cvalue, **151**
   equality, **156**
   gauges, **159**
   non-relational, 626
   numerors, **161**
   octagon, **158**
   relational, 158
   symbolic locations, **157**
   taint, **162**
Dominator, 90
Dominators *(Frama-C plug-in)*, **90**, 365
Dr. Memory *(tool)*, 285, 286
Dragonfly, 644
Driver, 620, 621
Dsd Crasher *(tool)*, 452
DSE, *see* Dynamic symbolic execution
Dune *(tool)*, 348, 349, 393–399
dune-release *(tool)*, 399
Dynamic analysis, 452
Dynamic memory analyzer, *see* Memory
   debugger
Dynamic symbolic execution (DSE), **310**
Dynamic tool, 282
`dynamic_split` annotation, 172

E

`-e-acsl`, 272
E-ACSL *(Frama-C plug-in)*, 84, 86, 88, 112,
   117, 185, 263–269, **270**, 270–301, 303,

332, 333, 347, 367, 434–436, 438–444, 457, 458, 464, 468, 475, 478
E-ACSL *(language)*, 6, **267**, 267–270, 586
E-ACSL runtime library, 271, 272, 288, 296, 301
e-acsl-gcc.sh *(tool)*, 272, 273, 275, 276, 279

`-e-acsl-replace-libc-functions`, **275**
`-e-acsl-validate-format-strings`, **276**
`__e_acsl_assert`, **279**, 281, 282, 288
EAL, *see* Evaluation Assurance Level
ECLiPSe *(tool)*, 310
Efficiency, **266**, 291, 296, 300
Eiffel *(language)*, 5, 264
Eldarica *(tool)*, 554, 561, 563, 576
Electron *(library)*, 82, 90, 344, 345
Ellipsoid, 522, 534
ensures clause, **6**, 652, 654
Entry point, 620, 621, 623–625
Equality (ACSL), **26**
Equivalence (ACSL), **26**
Error, 124
ESC-Java *(tool)*, 5, 452
ESC-Modula-3 *(tool)*, 5
-eva, 83, 113, 115, 119, 122, 123, 125, 126, **133**, 622, 623, 626, 631, 632
`-eva-auto-loop-unroll`, **171**
`-eva-context-depth`, 176
`-eva-context-width`, 176
`-eva-domains`, 143, **150**
Eva *(Frama-C plug-in)*, 46, 84, 86, 88–91, 96, 101, 103, 109, 111–115, 117–119, 121, 122, 124, 125, **131**, 131–137, 139, 141–143, 145, 147, 149–153, 155–185, 188, 197, 235, 236, 248, 263, 269, 287, 300, 332–336, 344, 347, 367, 397, 404–406, 408–411, 415, 418–420, 429, 431–433, 435, 437–440, 444, 445, 450, 457, 458, 475, 478, 480, 483, 504, 521, 522, 529–532, 534–537, 547, 548, 585–587, 590–592, 596–607, 609–612, 617–619, 621, 623–628, 631–634, 636, 637, 639–643, 656
-eva-flamegraph, 177, 420
`-eva-interprocedural-history`, **174**
`-eva-interprocedural-splits`, 173
`-eva-no-alloc-returns-null`, 183

`-eva-octagon-through-calls`, 161
`-eva-partition-history`, **174**
-eva-plevel, 633
`-eva-precision`, **135**, 171
`-eva-show-perf`, 177
`-eva-slevel`, **174**, 626, 631, 632
-eva-slevel-function, 622, 623
-eva-slevel-merge-after-loop, 628
`-eva-split-limit`, 172
-eva-split-return-function, 628

`-eva-undefined-pointer-comparison-propagate-all`, 622, 623
`-eva-widening-delay`, 142
`-eva-widening-period`, 142
Evaluation Assurance Level, 660
Evaluation order, 99
`\exists`, **26**, 654
Expression, **99**
  pure, 103
Expressiveness, **266**, 296
extern, 94

F
F* *(tool)*, 256
False alarm, *see* Alarm, false
Firewall, 660, **663**, 663, 669, 670
Fitness function, 635, 637, 638
Flame graph, 420
Floating-point number, 563
Floating-point operation, 108
Floyd-Hoare logic, 495
Fluctuat *(tool)*, **543**
`\forall`, **26**, 652
Formal specification, **5**
Fortran *(language)*, 264
Forward analysis, 389
Fourier-Motzkin transformation, 545
Frama_C_builtin_split, 626, 627
Frama_C_interval, 177, 183
Frama-Clang *(Frama-C plug-in)*, **87**, 87, 93
Frama_C_make_unknown, 624
Frama_C_nondet, 619
frama-c-ptests *(tool)*, 393–396
frama-c-script, **411**, 418
  build, 418, 419
  configure, 418
  creduce, 420
  estimate-difficulty, 417

Subject Index

flamegraph, 420
make-wrapper, 419
Frama_C_show_each, **142**, 626
Frame condition, 554
Frobenius companion matrix, 524–526
From *(Frama-C plug-in)*, **89**, 89, 90, 371
Function
    library, 579
    recursive, 562, 574
Functional dependency, 89
Function summary, 498, 556
Fuzzing, 287

G

Gappa *(tool)*, 586
Gcc *(tool)*, 83, 93, 96, 98, 106, 270, 272–274, 309, 413, 414, 435
Gene, 635
Genetic algorithm, 634, 635, 638, 639, 641, 643
ghost code, **54**
ghost else, **56**
\ghost qualifier, **56**
ghost qualifier, 381
Gillian *(tool)*, 255
Github *(tool)*, 399
Git *(tool)*, 399
Global annotation (ACSL), **58**
Gmp *(library)*, 265, 271, 272, 290–294, 301

GNATprove *(tool)*, 255
GNU Autoconf *(tool)*, 418
GNU extended ASM, 106
Google *(company)*, 283–285
Graphical user interface, *see* gui
Gtk3 *(library)*, 82
GUI *(Frama-C plug-in)*, **82**, 82, 83, 89–91, 111–115, 117, 119, 178, 180–182, 190, 191, 193, 248, 250, 344, 345, 404, 405, 408, 419, 428, 605, 621–623, 629

H

Harness function, 555, **564**, **566**, 567–569, 572
Heap, 564
Heap allocation, 576
-help, 83
Here label, **35**
High-level ACSL requirement, *see* hilare
Hilare *(language)*, 88, 277, **460**, 461, 662
HMI, *see* Human machine interface
Hoare logic, 254, **557**, 560

Hoare memory model, **200**, **202**, 202
Hoare triple, **557**
HOLLight *(tool)*, 529
Horn clause, 560
    constrained, *see* CHC
HTML *(format)*, 331, 398, 406, 420
Human machine interface, 645

I

IAE, 587
ICS, 522, 644, 646
IDE, 633, 645
Impact *(Frama-C plug-in)*, **89**
Implementation-defined, 411
Implementation-defined behavior, **10**, 103, 107
Implication (ACSL), **26**
Inclusion operator, **147**
Inconsistency, 66, **239**
Indeterminate value, 152
inductive, **63**, 652
Inductive invariant (ACSL), **44**
Industrial control system, *see* ICS, *see*
Infeasible path, 313, 314, 316
Infer *(tool)*, 495
Inference
    maximal specification, 556
Infinity, 108
\initialized predicate, **52**, 295
InOut *(Frama-C plug-in)*, **89**, 89, 591, 620, 625
Instantiate *(Frama-C plug-in)*, **87**
Instruction, **99**
Instruction coverage, **306**
Instrumentation, 458, 473, 475, 478
integer type, **28**
Integrated development environment, *see* IDE, *see*
Integrity, 458, 660, 663, 670
Internet of things, 586
Interpretation, 558
Interpretation function, 508
Interpreted automata, 345, 365, **388**
Interval, 291
Invariant
    loop, 554, 575
    state, 561
IoT, *see* Internet of things
IRSN *(institute)*, 543, 546
Isabelle/HOL *(tool)*, 10, 256
ISO C11, 95
ISO C99, 95
IT Security Evaluation Facilities, 660

ITSEF, *see* IT Security Evaluation Facilities

Ivette *(Frama-C plug-in)*, 82, 89, 90, 344, 345, 404–410

J
Java *(language)*, 5, 128, 255, 265, 468, 564
JavaCard *(language)*, 660
JavaCard Virtual Machine, *see* JCVM
JavaScript *(language)*, 255, 344, 405
JBMC *(tool)*, 564
JCDB, 417, **418**
JCrasher *(tool)*, 452
JCVM, 660–662
Jessie *(tool)*, 254, 591
JML *(language)*, **5**, 5, 6, 20, 265, 288, 468, 484, 564
Join operator, **146**
Jordan-Chevalley decomposition, 524, **525**, 525
JSON *(format)*, 89, 119–121, 124–126, 344, 406, 417, 418, 623
JSON Compilation Database, *see* JCDB
-json-compilation-database, **93**

K
KCC *(tool)*, 10
-keep-unused-specified-functions, 95
-keep-unused-types, 95
Kernel (Frama-C), **344**
Kernel API, 288
  abort, 353
  AST, 376
  Annotations, 367, 368, 371, 378
  Ast.Untyped.get, 359
  Ast.compute, 359
  Ast.get, 359
  Cabs_debug, 352
  ChangeDoChildrenPost, 383
  ChangeTo, 382
  Cil, 360
  Cil_builder, 360, 368, 370
  Cil_datatype, 372
  Cil_state_builder, 375
  Cil_types_debug, 352
  Dataflow2, 388
  Dataflows, 388
  Datatype, **372**
  Db, 350
  Db.Postdominators, 365, 366
  debug, 353, 355, 371
  DoChildren, 382
  Dominators, 365
  Emitter, 368, 370
  enuminfo, 364
  enumitem, 364
  error, 353
  expr, 379
  Extlib, 368
  failure, 353
  False, 351
  fatal, 353, 371
  feedback, 353, 355
  File, 362, 368
  Filepath, 357
  get_filling_actions, 387
  GEnumTag, 364
  Globals, 360, 361, 363, 370, 371
  instr, 379
  Integer, 372
  Interpreted_automata, 390
  Interpreted_automatanew, 388
  JustCopy, 388
  Kernel_function, 360, 361, 368, 372
  Kernel_function_map, 357
  Logic_const, 360, 368
  Log, 353, 354
  not_yet_implemented, 354
  Plugin.Register, 349, 351, 353
  pp_list, 352
  Pretty_utils, 352
  Printer, 352, 371
  Project, 272, 277, **374**
  Property, 367
  Property_status, 367
  register_category, 354
  register_warn_category, 356
  result, 353
  SkipChildren, 379, 382, 388
  State_builder, 374
  stmt, 361, 372, 376
  term, 372
  varinfo, 376, 381
  Visitor, 378, 386
  Visitor_behavior, 387
  warning, 353
  Writes, 371
  WritesAny, 371
-kernel-help, 83, **107**
-kernel-log, 119
-kernel-msg-key, 622, 623
KeY *(tool)*, 255, 452, 661
KLEE *(tool)*, 314

Krakatoa *(tool)*, 5, 661

L
Label, 634, 637
Label (ACSL), **35**
Language server protocol, 633
LAnnotate *(Frama-C plug-in)*, 449
Larch *(language)*, 5
Lazyness, 269
Least fixpoint, 522
Legacy code, 95
Lemma, **60**, 227, 661, 670, 676, 677, 679
`\let`, **27**, 571, 572
Lexing, 92, 94
L4·project *(tool)*, 256
`-lib-entry`, 183, 570, 571
libc *(library)*, 87, 95–98, 179, 183, 275, 314, 418, 428, 433
Linear digital filter, **522**, 522, 523, 547
Linear temporal logic, *see* LTL
Linking, 92, 94, 415, 417
`-load`, 109, 110, **111**, 114, 177, 372, 375, 623
`-load-plugin`, 91, 92
Loading state, 82, **111**
`logic`, **58**
 first-order, 561
Logical interpolation, 554
Logical variable, 557
Logic theory
 equality, **28**
 integer arithmetic, **28**
 real arithmetic, **29**
Loop, 104, 575, 628
 invariant, 629
Loop *(Frama-C plug-in)*, **88**
`loop assigns` clause, **42**, 629, 653
Loop invariant, **39**, 140, 522
 polynomial, 88
`loop invariant` clause, **39**, 579, 629, 653
`loop PRAGMA unroll`, 628
`loop unroll` annotation, 170, 628
Loop unrolling, **165**, 170
`loop variant` clause, **46**
LReplay *(tool)*, 87, 449
LSP, *see* Language server protocol
LTest *(Frama-C plug-in)*, 87, 336, 449, 450, 634, 636, 638
LTL *(language)*, 86, **475**
LUncov *(Frama-C plug-in)*, 449

M
Machdep, 96, **98**, 114, 411
`-machdep`, **98**, 183, 359, 413, 622, 623
Machine-dependent behavior, 98
`-main`, 83, 113, 133, 622, 623
Makefile, 111, 415, 419
`malloc`, 83
MathWorks *(company)*, 645
MATLAB *(tool)*, 645
Maude *(tool)*, 10
MC/DC, **306**, 634–637, 643
 masking, 635
MDR *(Frama-C plug-in)*, 83, 89, 126
`-mdr-gen`, 126
`-mdr-out`, 126
MemCad *(tool)*, 495
MemCheck *(tool)*, 285, 286, 295
`memcpy`, 275
Memoization, 376, 388
Memory
 hardware, 630
Memory debugger, 265, **282**, **295**
Memory model, 188, **198**, 200–205, 208, 209, 239, 258
 E-ACSL, **296**
 runtime, **295**
Memory safety, 487, 488, 495, 499, 500, 518

Memory shadowing, **295**
MemorySanitizer *(tool)*, 283
MERCE, *see* Mitsubishi Electric R&D Centre Europe
`merge` annotation, 173
Message, 119
`-meta`, 110
Meta-segment, 296
MetAcsl *(Frama-C plug-in)*, 88, 110, 185, 197, 235, 236, 277, 346, 444, 457, 458, 460–464, 469, 483, 484, 661, 662, 670–674, 677, 681
Metaproperty, *see* hilare, 670, 673, 676, 681
Metrics *(Frama-C plug-in)*, 89, 404, **406**, 406, 407, 591, 592, 596, 604, 605, 609, 611
MiniTypeState *(Frama-C plug-in)*, 346, 350, 351, 354, 357, 359, 360, 380, 388, 390, 394–399
Mitigation, 300
Mitsubishi Electric *(company)*, 336, 617, 634, 636, 656
Mitsubishi Electric R&D Centre Europe *(company)*, 336, 617, 634

MLton *(tool)*, 87
Model checking, 424, 431, 452, 554, 561, 645
  bounded, 636, 643
  CHC-based, 561
Modified condition-decision coverage, *see* MC/DC
Monitoring
  inline, **266**
  online, **266**
  outline, **301**
Monomorphisation, 87
-<plugin>-msg-key, 355
MSVC *(tool)*, 96, 98
Mthread *(Frama-C plug-in)*, 183, 625
Multicore, 128
Multiple condition coverage, **306**

N
NaN, 108
NASA software engineering requirements, 584
Negation (ACSL), **26**
NIST *(agency)*, 283
Nix *(tool)*, 396
-no-autoload-plugins, **91**, 91, 92
-no-frama-c-stdlib, **97**
-no-keep-switch, 95
Non-compliance, 446–448, 452
NonTerm *(Frama-C plug-in)*, **90**
Non-terminating code, 90
Normalization, 95, **99**
\nothing, **20**, 654
NPM *(tool)*, 405
Numerors *(Frama-C plug-in)*, **529**, 547

O
Obfuscator *(Frama-C plug-in)*, **88**, 91
\object_pointer predicate, **52**
OCaml *(language)*, 94, 124, 128, 252, 265, 310, 342, 344, 345, 348–350, 352–354, 370, 372, 373, 376–378, 393, 396, 398, 399, 451
Occurrence *(Frama-C plug-in)*, **89**, 90
-ocode, 272
odoc *(tool)*, 350
Offensive programming, 264
\offset function, **49**
Offsetmap, **153**
\old, **21**, 288, 555, 571–573, 577
Old label, **35**
opam *(tool)*, 396, 399

OpenJML *(tool)*, 265, 484
OpenSSL *(library)*, 275, 440
Oracle, **308**, 308, 309, 312, 314, 315, 321, 322
Ortac *(tool)*, 265
Out-of-bound array access, 643
Over-approximation, **132**, 389, 408, 409
Overflow, 114
  arithmetic, 102, 290
  buffer, 132, 295, 651
  heap, 283, 295
  integer
    unsigned, 83
  stack, 283
Overhead
  memory, 283, 286, 290
  time, 283, 285, 290

P
Parallelism, 128, 639
Parsing, 82, **92**, 92, 94, 111
Partitioning directive, **168**
Partitioning key, **168**
Path conditions, *see* Path predicate
Path coverage, **306**
PathCrawler *(Frama-C plug-in)*, 87, 89, 305–311, 313–325, 327, 329, 331–337, 444, 446–449, 451, 617, 618, 634, 636, 638, 639, 642, 643
PathCrawler-online *(tool)*, 305, 306, 327, 330–332, 335
Path predicate, 309, 332, 334, 335
Path prefix, 312–314, 332
PDG, *see* Program dependence graph
Pdg *(Frama-C plug-in)*, 89, **90**
Perl *(language)*, 420
Pervasive properties, **458**, 458
PEX *(tool)*, 314
Physical type, 517
Pilat *(Frama-C plug-in)*, **88**
Pivot table, **407**
PLC, *see* Programmable logic controller
-plugins, **91**, 91, 92
Pointer, 564, 576
Points-to predicate, **496**
Polymorphism removal, 87
PolySpace *(tool)*, 438
Portability, 95, 96, 98
POSIX, 96, 97, 124, 183, 413, 414, 418
Post-condition
  strongest, 556
Post-condition (ACSL), **6**

Subject Index 693

Post-state, **557**
Postdominator, 90
Postdominators *(Frama-C plug-in)*, **90**, 365
Pre-condition
 weakest, 556
Pre-condition (ACSL), **6**
Pre-state, **557**
PRECiSA *(tool)*, 529
Predicate
 inductive, 268
 quantified, 568
 uninterpreted, 578
`predicate`, **59**, 653, 654
Predicate abstraction, 425, 426
Preprocessing, 92, 417, 418
Pre label, **35**
`-print`, **99**, 105, 272
`-print-as-is`, **99**, 104
`printf`, 276
`-print-plugin-path`, **91**, 91
Program dependence graph, 90
Programmable logic controller, 645
Program slicing, 445, 452
Project, **110**
Prolog *(language)*, 310, 319
Proof assistant, **189**, 242, 245, 252, 255, 256

Proof failure, 424, 446–448, 452
Proof obligation, *see* Verification condition
Proof script, 661, 671, 673–680
Propagation, 137
Property, 112, 124
 arithmetic, 282, 289, 290, 300
 block-level, **283**
 call traces, 458, **473**
 high-level, 277
 liveness, 266
 memory, 282, 289, 295, 300
 non-interference, 483
 relational, 458, **464**
 state, **266**, 266
 temporal, 266
Property status, 110, **112**
Proposition (ACSL), **26**
Prover, **189**, 639, 654
Pure expression, **571**
PVS *(language)*, 529
Python *(language)*, 412

Q

Qed *(library)*, 188, 195, 199, 202, 203, 212, 238, 239, **242**, 242–247, 254, 258, 671–675, 677, 678, 680

Quantification, 267
Quantifier elimination, 556

R

RAC, *see* Runtime annotation checking
Rational number, 268, 291
Reachability, 114, 131, **136**
Reactive system, 645
React *(library)*, 82, 90, 344, 405
`reads` clause, **66**
Real number, 268, 563, 634
`real` type, **29**
Recovery, 300
Reduction *(Frama-C plug-in)*, **89**
Reference memory model, **203**
RefinedC *(tool)*, 255, 517
Refinement type, 508, 509, 511
Relational analysis, 497
Relational Hoare logic, 484
Remote Procedure Call, *see* RPC
Remote server, 624
`-report`, **122**
Report *(Frama-C plug-in)*, 89, 122–124, 178, 180, 197, 619
`-report-classify`, 125
`-report-csv`, **123**, 177
`-report-output`, 125
`-report-rules`, 125
`requires` clause, **6**, 652, 654
`\result`, 7, **27**, 571, 572, 652, 654
Retained points-to predicates, **494**
Retained predicate, **510**
`return`, 105
Review, 634
Rice theorem, 84
RMA *(tool)*, 487, **488**, 488, 489, 495, 497, 499, 500, 511, 516–518
RPC, 345
RPP *(Frama-C plug-in)*, 88, 277, 346, 444, 457, 458, 464–466, 468–473, 483, 484

`-rte`, 110
Rte *(Frama-C plug-in)*, 68, **88**, 88, 91, 110, 112, 188, 190, 193, 194, 196, 197, 213, 235, 236, 273–275, 277, 282, 285, 287, 300, 301, 333, 440, 444, 662, 671–674

RTL, *see* E-ACSL runtime library
Runtime annotation checking, **264**, 295, 300, 434
Runtime assertion checking, *see* Runtime annotation checking
Runtime error, 132, 643

Runtime verification, **266**, 434, 437, 440, 443, 444, 451
Runtime Verification Inc. *(company)*, 283
Rust *(language)*, 128
RV, *see* Runtime verification
RV-Match *(tool)*, 283–285

S
Safety
  memory, 563
  type, 563
Safety of execution, **12**
SAGE *(tool)*, 314
Saida *(Frama-C plug-in)*, 553, 555, 557, 564, 569–580
SANTE *(Frama-C plug-in)*, 445, 446, 452
SARD-100 *(benchmark)*, 283, 284
SARIF *(format)*, 83, 126–128
SATABS *(tool)*, 427, 451
SATE, *see* Static analysis tool exposition
Satisfiability modulo theories, *see* SMT
-**save**, **111**, 113, 177, 372, 375, 622, 623
Saving state, 82, **111**
SCADA, **644**, 653, 656
scanf, 276
Scope *(Frama-C plug-in)*, **89**, 90
SDD, 591, 598
SeaHorn *(tool)*, 554
Security Policy Model, 660
SecuritySlicing *(Frama-C plug-in)*, **90**
Self-checking Program, **264**
Self-composition, 468
\separated predicate, **53**, 101
Separating conjunction, **495**
Separating shape graph, **495**
Separation logic, 255, 256, 488, 491, **495**, 495–499, 511, 512, 514, 517, 518
Server *(Frama-C plug-in)*, 90, 91, 345, 405
Set, 267
  comprehension, 267
Set of boxes, 522, 534
Set of zonotopes, 522, 534
-set-project-as-default, **110**, 110
Shadow
  primary, **298**
  secondary, **298**
Shadow encoding
  heap, 296
  stack, 298, 299
Shadow memory, **295**
Shape analysis, 488, **491**, 491, 495, 511, 517

Shell *(language)*, 280
Side effect, 99, 100
Simplex solver, 546
Simulink Design Verifier *(tool)*, 645, 648, 655
Simulink *(language)*, 644–646, 648, 649, 655
SISNAV, 589
SLDV *(tool)*, 648
-slice-assert, 110
Slicing *(Frama-C plug-in)*, **90**, 110, 374, 445, 587
-slicing-project-name, 110
-slicing-project-postfix, 110
Smallfoot *(tool)*, 495
SML *(language)*, 87
Smoke test, **239**, 240, 241, 258
SMT, **189**, 239, 255, 636
SMT-LIB *(library)*, 563
Software design & implementation engineering process, 607
Software development life cycle, 287, 585, 607
Software verification process, 607
Solver, 188, **189**, 189, 192, 197, 198, 201, 203–205, 208, 209, 211, 212, 228, 229, 240, 242, 244, 245, 247, 248, 250–253, 255, 561, 636
Soundness, 97, 132, **144**, 266, 282, 291, 300, 636, 637, 644, 656
Spacer *(tool)*, 561, 563
SpareCode *(Frama-C plug-in)*, **90**, 590, 592, 595, 609, 611
Spark2014 *(language)*, 6, 255, 265, 452
SPEC CPU *(benchmark)*, 285, 286
Spec# *(language)*, 265
split annotation, 172
SPM, *see* Security Policy Model
SQL *(language)*, 283
SRS, 591
SSA, *see* Static single assignment
SSINAV, 589
StaDy *(Frama-C plug-in)*, 89, 336, 447, 448, 468
Staged points-to predicates, **494**
Standard
  DO-178-B, 645
  DO-178C, 554
  EN-50128, 645
  IEEE-754, 108, 151
  ISO-26262, 554, 645
Standard C library, *see* libc
State abstraction, 493

State analysis, 497
State machine
 hierarchical, 647
State partitioning, **167**
Static analysis, 131, 266, 288, 332, 335, 336, 444, 449, 450, 452
Static analysis framework, **150**
Static analysis tool exposition, 587
Static single assignment, 636
Status, *see* Validity status
STL *(library)*, 253
Strategy, **252**, 252
Strengthening annotations, 45
String, 275
Stub, **96**, 309, 314–316, 334, 624, 625, 634, **635**, 636, 637, 641
Studia *(Frama-C plug-in)*, 89, 90, 404, 408–411
Stuxnet, 644
Subcontract weakness, 446–448
Subsumption, 292
Supervisory control and data acquisition, *see* SCADA
SVG *(format)*, 420
Symbolic execution, **310**, 310, 311, 314, 336, 556, 636
Symbolic output, 332
Synergy *(tool)*, 452

T
Tactic, **248**, 249–251
`taint` annotation, 162
`\tainted` predicate, 162
Term (ACSL), **26**
`terminates` clause, 267
Test, 300, 554, 634, 656
 concolic, **310**
 context, 309
 coverage, 449, 634
 criterion, 634
 generation, 306, 308–310, 314, 315, 318–320, 323, 330, 335, 424, 444–446, 449–452, 634, 636, 643
 mutation, 287, 635, 638
 objective, 449
  infeasible, 309, 336, 449, 450
 oracle, 635, 639, 656
 parameter, 330
 structural, 305, **306**, 306, 308, 332–335, 618, 634, **635**, 643
 unit, 618, 634, **635**, 637
`-then`, 83, **109**, 110, 122–124, 126, 351

`-then-last`, **110**, 110, 272, 277
`-then-on`, **110**, 110
`-then-replace`, **110**
ThreadSanitizer *(tool)*, 283
TIP *(tool)*, 188, **242**, 242, 246, 248–253, 258, 405
TIS Analyzer *(tool)*, 618, 625, 630, 632, 633

tmux *(tool)*, 624
Toyota ITC *(benchmark)*, 283, 284
Transfer function, **146**
Transformation
 syntactic, 95
Transformation analysis, 497
Transition function, 389
Transparency, **266**, 282
TriCera *(tool)*, 553–555, 557, 559, 563, 564, 566–573, 575–578, 580
Type safety, 491, 517
Typestates, **345**
Type system, 289, 291
Typing, 92, 94

U
UBSan *(tool)*, 283
`-ulevel`, 95
Uncast *(tool)*, 680
Undefined behavior, **11**, 86, 88, **95**, 99, 100, 102, 113, 131, 133, 273, 290, 300, 301, 618, 619, 651, 656
UndefinedBehaviorSanitizer *(tool)*, 283
Undefinedness, **269**
 problem, **269**
Under-approximation, 389
Understanding code, 619
`-unsafe-arrays`, **109**
Unspecified behavior, **10**, **95**, 99, 102, 107
`-unspecified-access`, **100**, **101**, 108
Unspecified sequence, **100**
Update
 strong, **154**
 weak, **154**
Users *(Frama-C plug-in)*, **89**, 89

V
V&V, *see* Validation and verification
Valgrind *(tool)*, 285, 295
`\valid` predicate, **51**, 265, 295, 296, 652, 654
`\valid_read` predicate, **52**, 651
Validation and verification, 300
Validity (ACSL), **15**

Validity status, **114**
Value analysis, 424, 429, 431, 432, 434, 437, 438, 445, 446, 449, 451, 452
Value *(Frama-C plug-in)*, 445, 586, 591, 597

Vandermonde matrix, 524
Variable-length array, 96
Variadic *(Frama-C plug-in)*, **87**, 94, 351
Variadic function, **87**, 94
VC, *see* Verification condition
VCC *(tool)*, 254, 255, 517
-<plugin>-verbose, 354
VerCors *(tool)*, 661
Verdict, **308**
VeriFast *(tool)*, 255
Verification condition, 188, **189**, 189, 191, 193, 205, 210, 214, 215, 217, 226, 228, 241, 242, 245, 250, 252–255, 258, 561

Verification cost, 287
Version control system, 620
Visitor mechanism, 342, 344, **378**, 407
VLA, *see* Variable-length array
VLM-1, 588
VLS-1, 588
Volatile, 630
VSCode, 633
VST *(tool)*, 256
VST-Floyd *(tool)*, 255, 256
Vulnerability, 651
   basic XSS, 283
   buffer overflow
      heap, *see* Overflow, heap
      stack, *see* Overflow, stack
   command injection, 283
   double free, 283
   format string, 276, 283
   hard-coded password, 283
   heap inspection, 283
   leftover debug code, 283
   memory leak, 283
   null dereference, 283
   pointer scaling, 283
   race condition, 283
   resource injection, 283
   SQL injection, 283
   string management, 283
   string termination, 283
   unchecked error, 283
   uninitialized variable, 283
   unrestricted lock, 283
   use-after-free, 283

W
-warn-invalid-bool, **108**
-warn-invalid-pointer, **108**
-<plugin>-warn-key, 355
-warn-left-negative-shift, **108**
-warn-pointer-downcast, **108**, 179
-warn-right-negative-shift, **108**

-warn-signed-downcast, **103**, 103, **108**, 110, 179
-warn-signed-overflow, **108**, 121, 179
-warn-special-float, **108**, 179
-warn-unsigned-downcast, **108**, 179
-warn-unsigned-overflow, **108**, 179
Warning, 124
Well-typed, 491–494, 506, 508–510
Well-typed heap, **508**
Well-typed state, **508**
Why *(tool)*, 5, 6, 254
Why3 *(tool)*, 5, 6, 189, 244, 252, 254–256, 534, 541, 543, 547, 548, 671, 678
WhyML *(language)*, 254
widen_hint, 150
Widening, **139**, 148, 149
Widening operator, 522
-wp, 110, 117, 630
Wp *(Frama-C plug-in)*, 4, 33, 34, 39, 44, 55, 57, 60, 67, 68, 78, 84, 86–89, 91, 96, 110, 112, 115–118, 178, 181, 185, **187**, 187–217, 219, 221, 224–230, 232, 233, 235–245, 247–255, 257, 258, 263, 266, 269, 287, 300, 332–336, 347, 367, 405, 432, 433, 440–445, 450, 457, 458, 463, 468–470, 472, 475, 478, 482, 483, 521, 522, 529, 532, 534, 541, 543, 547, 548, 554–556, 561, 573–575, 577, 585, 586, 612, 617, 618, 630, 643, 645, 651, 653, 654, 656, 661, 662, 664, 669–675, 677–680
-wp-rte, 117, 630

X
Xisa *(tool)*, 495
xxd *(tool)*, 632

Y
Ya *(language)*, 277, 475, **476**, 476, 478–481, 483, 484

Subject Index

Z
Zonotope, 522, 534

Z3 *(tool)*, 189, 254, 255, 522, 561, 563, 655

SPRINGER NATURE

GPSR Compliance

The European Union's (EU) General Product Safety Regulation (GPSR) is a set of rules that requires consumer products to be safe and our obligations to ensure this.

If you have any concerns about our products, you can contact us on ProductSafety@springernature.com

In case Publisher is established outside the EU, the EU authorized representative is:

Springer Nature Customer Service Center GmbH
Europaplatz 3
69115 Heidelberg, Germany

The manufacturer's authorised representative in the EU is Springer Nature Customer Service Centre GmbH, Europaplatz 3, 69115 Heidelberg, Germany. If you have any concerns regarding our products, please contact ProductSafety@springernature.com

Printed and bound by CPI Group (UK) Ltd, Croydon, CR0 4YY

25/03/2026

02078171-0015